T0214213

Lecture Notes in Computer Science 11829

Founding Editors

Gerhard Goos
Karlsruhe Institute of Technology, Karlsruhe, Germany
Juris Hartmanis
Cornell University, Ithaca, NY, USA

Editorial Board Members

Elisa Bertino
Purdue University, West Lafayette, IN, USA
Wen Gao
Peking University, Beijing, China
Bernhard Steffen
TU Dortmund University, Dortmund, Germany
Gerhard Woeginger
RWTH Aachen, Aachen, Germany
Moti Yung
Columbia University, New York, NY, USA

More information about this series at http://www.springer.com/series/7410

Yi Mu · Robert H. Deng · Xinyi Huang (Eds.)

Cryptology and Network Security

18th International Conference, CANS 2019
Fuzhou, China, October 25–27, 2019
Proceedings

 Springer

Editors
Yi Mu (iD)
Fujian Normal University
Fuzhou, China

Robert H. Deng (iD)
Singapore Management University
Singapore, Singapore

Xinyi Huang
Fujian Normal University
Fuzhou, China

ISSN 0302-9743   ISSN 1611-3349 (electronic)
Lecture Notes in Computer Science
ISBN 978-3-030-31577-1   ISBN 978-3-030-31578-8 (eBook)
https://doi.org/10.1007/978-3-030-31578-8

LNCS Sublibrary: SL4 – Security and Cryptology

© Springer Nature Switzerland AG 2019
This work is subject to copyright. All rights are reserved by the Publisher, whether the whole or part of the material is concerned, specifically the rights of translation, reprinting, reuse of illustrations, recitation, broadcasting, reproduction on microfilms or in any other physical way, and transmission or information storage and retrieval, electronic adaptation, computer software, or by similar or dissimilar methodology now known or hereafter developed.
The use of general descriptive names, registered names, trademarks, service marks, etc. in this publication does not imply, even in the absence of a specific statement, that such names are exempt from the relevant protective laws and regulations and therefore free for general use.
The publisher, the authors and the editors are safe to assume that the advice and information in this book are believed to be true and accurate at the date of publication. Neither the publisher nor the authors or the editors give a warranty, expressed or implied, with respect to the material contained herein or for any errors or omissions that may have been made. The publisher remains neutral with regard to jurisdictional claims in published maps and institutional affiliations.

This Springer imprint is published by the registered company Springer Nature Switzerland AG
The registered company address is: Gewerbestrasse 11, 6330 Cham, Switzerland

Preface

The 18th International Conference on Cryptology and Network Security (CANS 2019) was held in Fuzhou, China, during October 25–27, 2019, and was organized and hosted by Fujian Normal University, in corporation with the Fujian Provincial Key Laboratory of Network Security and Cryptology, International Association for Cryptologic Research, and Springer.

CANS is a recognized annual conference, focusing on cryptology, computer and network security, and data security and privacy, attracting cutting-edge research findings from world-renowned scientists in the area. Earlier editions were held in Taipei (2001), San Francisco (2002), Miami (2003), Xiamen (2005), Suzhou (2006), Singapore (2007), Hong Kong (2008), Kanazawa (2009), Kuala Lumpur (2010), Sanya (2011), Darmstadt (2012), Parary (2013), Crete (2014), Marrakesh (2015), Milan (2016), Hong Kong (2017), and Naples (2018).

This year the conference received 55 anonymous submissions. All the submissions were reviewed on the basis of their significance, novelty, technical quality, and practical impact. The Program Committee (PC) consisted of 49 members with diverse backgrounds and broad research interests. The review process was double-blind. After careful reviews by at least three experts in the relevant areas for each paper, and discussions by the PC members, 21 submissions were accepted as full papers (an acceptance rate 38%) and 8 submissions as short papers for presentation in the conference and inclusion in the conference proceedings. The accepted papers cover multiple topics, from algorithms to systems and applications.

CANS 2019 was made possible by the joint efforts of many people and institutions. There is a long list of people who volunteered their time and energy to put together the conference and who deserve special thanks.

We would like to thank all the PC members for their great effort in reading, commenting, and finally selecting the papers. We also thank all the external reviewers for assisting the PC in their particular areas of expertise.

We would like to emphasize our gratitude to the general chair Professor Changping Wang, for his generous support and leadership to ensure the success of the conference. We are deeply grateful for the tireless efforts of the local and Organizing Committee in the capable hands of Professors Xinyi Huang and Li Xu. Thanks also go to publicity chairs Wei Wu and Shangpeng Wang, and web chair Hong Zhao.

We sincerely thank the authors of all submitted papers and all the conference attendees. Thanks are also due to the staff at Springer for their help with producing the proceedings and to the developers and maintainers of the EasyChair software, which greatly helped simplify the submission and review process.

October 2019

Robert H. Deng
Yi Mu

Organization

Program Chairs

Robert Deng   Singapore Management University, Singapore
Yi Mu   Fujian Normal University, China

General Chair

Changping Wang   Fujian Normal University, China

Local Organization Chairs

Xinyi Huang   Fujian Normal University, China
Li Xu   Fujian Normal University, China

Publicity Chairs

Wei Wu   Fujian Normal University, China
Shangpeng Wang   Fujian Normal University, China

Web Chair

Hong Zhao   Fujian Normal University, China

Steering Committee

Yvo Desmedt (Chair)   University of Texas at Dallas, USA
Juan A. Garay   Yahoo! Labs, USA
Amir Herzberg   Bar-Ilan University, Israel
Yi Mu   Fujian Normal University, China
David Pointcheval   CNRS and ENS Paris, France
Huaxiong Wang   Nanyang Technological University, Singapore

Program Committee

Masayuki Abe   NTT, Japan
Man Ho Au   The Hong Kong Polytechnic University, SAR China
Reza Azarderakhsh   Florida Atlantic University, USA
Elisa Bertino   Purdue University, USA
Erik-Oliver Blass   Airbus Group Innovations, France
Zhenfu Cao   East China Normal University, China
Aniello Castiglione   University of Naples Parthenope, Italy

Rongmao Chen   National University of Defense Technology, China
Sherman S. M. Chow   The Chinese University of Hong Kong, SAR China
Robert Deng (PC Co-chair)   Singapore Management University, Singapore
Xuhua Ding   Singapore Management University, Singapore
Josep Domingo-Ferrer   Universitat Rovira i Virgili, Italy
Pooya Farshim   ENS, France
Chaya Ganesh   Aarhus University, Denmark
Peter Gaži   IOHK Research, Slovakia
Dieter Gollmann   Hamburg University of Technology, Germany
Dawu Gu   Shanghai Jiao Tong University, China
Fuchun Guo   University of Wollongong, Australia
Amir Herzberg   University of Connecticut, USA
Julia Hesse   IBM Research Zurich, Switzerland
Sotiris Ioannidis   FORTH-ICS, Greece
Frank Kargl   Ulm University, Germany
Kaoru Kurosawa   Ibaraki University, Japan
Yingjiu Li   Singapore Management University, Singapore
Dongdai Lin   Chinese Academy of Sciences, China
Joseph Liu   Monash University, Australia
Peng Liu   The Pennsylvania State University, USA
Zhe Liu   Nanjing University of Aeronautics and Astronautics,
                              China
Javier Lopez   University of Malaga, Spain
Di Ma   University of Michigan, USA
Mark Manulis   University of Surrey, UK
Evangelos Markatos   University of Crete, FORTH-ICS, Greece
Chris Mitchell   Royal Holloway, University of London, UK
Yi Mu (PC Co-chair)   Fujian Normal University, China
Chandrasekaran   Indian Institute of Technology, Madras, India
  Pandurangan
Stefano Paraboschi   Universita di Bergamo, Italy
Pierangela Samarati   University of Milan, Italy
Dominique Schroeder   Friedrich-Alexander-Universiät Erlangen-Nürnberg,
                              Germany
Luisa Siniscalchi   University of Salerno, Italy
Willy Susilo   University of Wollongong, Australia
George Theodorakopoulos   Cardiff University, UK
Vijay Varadharajan   The University of Newcastle, UK
Frederik Vercauteren   KU Leuven, ESAT/COSIC, Belgium
Cong Wang   City University of Hong Kong, SAR China
Ding Wang   Peking University, China
Huaxiong Wang   Nanyang Technological University, China
Lei Wang   Shanghai Jiao Tong University, China
Qian Wang   Wuhan University, China
Jian Weng   Jinan University, China

Min Yang   Fudan University, China
Yu Yu   Shanghai Jiao Tong University, China

Additional Reviewers

Anglés-Tafalla, Carles
Bamiloshin, Michael
Beullens, Ward
Chillotti, Ilaria
Ciampi, Michele
Diamantaris, Michalis
Gardham, Daniel
Gunasinghe, Hasini
Hamburg, Mike
Hou, Lin
Huang, Jianye
Huang, Xinyi
Karmakar, Angshuman
Kfir, Yehonatan
Lai, Jianchang

Li, Kang
Liu, Feng Hao
Liu, Jia-Nan
Ma, Jack P. K.
Markatou, Evangelia
    Anna
Mefenza, Thierry
Nieto, Ana
Ogata, Wakaha
Ohkubo, Miyako
Orsini, Emmanuela
Papadogiannaki, Eva
Sengupta, Binanda
Tian, Yangguang
Tomida, Junichi

Wan, Ming
Wang, Geng
Wang, Qian
Wang, Wenhao
Wang, Yi
Wong, Harry W. H.
Wu, Baofeng
Wu, Ge
Xie, Congge
Xue, Haiyang
Yang, Jingchun
Yang, Zhichao
Yu, Yang
Zhou, Junwei

Contents

Blockchains

Cloud Security

Secret Sharing and Interval Test

LWE

Encryption, Data Aggregation, and Revocation

Signature, ML, Payment, and Factorization

Homomorphic Encryption

Bi-homomorphic Lattice-Based PRFs and Unidirectional Updatable Encryption

Vipin Singh Sehrawat[1(✉)] and Yvo Desmedt[1,2]

[1] Department of Computer Science, The University of Texas at Dallas,
Richardson, USA
vipin.sehrawat.cs@gmail.com
[2] Department of Computer Science, University College London, London, UK

Abstract. We define a pseudorandom function (PRF) $F : \mathcal{K} \times \mathcal{X} \to \mathcal{Y}$
to be bi-homomorphic when it is fully Key homomorphic and partially
Input Homomorphic (KIH), i.e., given $F(k_1, x_1)$ and $F(k_2, x_2)$, there
is an efficient algorithm to compute $F(k_1 \oplus k_2, x_1 \ominus x_2)$, where $\oplus$ and
$\ominus$ are (binary) group operations. The homomorphism on the input is
restricted to a fixed subset of the input bits, i.e., $\ominus$ operates on some
pre-decided m-out-of-n bits, where $|x_1| = |x_2| = n, m < n$, and the
remaining $n - m$ bits are identical in both inputs. In addition, the output
length, ℓ, of the operator $\ominus$ is not fixed and is defined as $n \le \ell \le 2n$,
hence leading to Homomorphically induced Variable input Length (HVL)
as $n \le |x_1 \ominus x_2| \le 2n$. We present a learning with errors (LWE) based
construction for a HVL-KIH-PRF family. Our construction is inspired
by the key homomorphic PRF construction due to Banerjee and Peikert
(Crypto 2014). We use our novel PRF family to construct an updatable
encryption scheme, named QPC-UE-UU, which is quantum-safe, post-
compromise secure and supports unidirectional ciphertext updates, i.e.,
the *tokens* can be used to perform ciphertext updates, but they cannot
be used to undo completed updates. Our PRF family also leads to the
first left/right key homomorphic constrained-PRF family with HVL.

Keywords: Bi-homomorphic PRFs · LWE · Lattice-based · Updatable
encryption · Unidirectional updates · Post-compromise security

1 Introduction

In a PRF family [1], each function is specified by a short, random key, and can be
easily computed given the key. Yet the function behaves like a random one, in the
sense that if you are not given the key, and are computationally bounded, then
the input-output behavior of the function looks like that of a random function.
Since their introduction, PRFs have been one of the most fundamental building
blocks in cryptography. For a PRF F_s, the index s is called its key or seed.
Many variants of PRFs with additional properties have been introduced and
have found a plethora of applications in cryptography.

© Springer Nature Switzerland AG 2019
Y. Mu et al. (Eds.): CANS 2019, LNCS 11829, pp. 3–23, 2019.
https://doi.org/10.1007/978-3-030-31578-8_1

Key Homomorphic (KH) PRFs: A PRF family F is KH-PRF if the set of keys has a group structure and if there is an efficient algorithm that, given $F_s(x)$ and $F_t(x)$, outputs $F_{s+t}(x)$ [2]. Multiple KH-PRF constructions have been proposed via varying approaches [2–5]. These functions have many applications in cryptography, such as, symmetric-key proxy re-encryption, updatable encryption, and securing PRFs against related-key attacks [3]. But, lack of input homomorphism limits the privacy preserving applications of KH-PRFs. For instance, while designing solutions for searchable symmetric encryption [6], it is highly desirable to hide the search patterns, which can be achieved by issuing random queries for each search. But, this feature cannot be supported if the search index is built by using a KH-PRF family, since it would require identical query (i.e., function input) to perform the same search.

Constrained PRFs (CPRFs): Constrained PRFs (also called delegatable PRFs) are another extension of PRFs. They enable a proxy to evaluate a PRF on a strict subset of its domain using a trapdoor derived from the CPRF secret key. A trapdoor is constructed with respect to a certain policy predicate that determines the subset of the input values for which the proxy is allowed to evaluate the PRF. Introduced independently by Kiayias et al. [7], Boneh et al. [8] and Boyle et al. [9] (termed functional PRFs), CPRFs have multiple interesting applications, including broadcast encryption, identify-based key exchange, batch query supporting searchable symmetric encryption and RFID. Banerjee et al. [10], and Brakerski and Vaikuntanathan [11] independently introduced KH-CPRFs.

Variable Input Length (VIL) PRFs: VIL-PRFs [12] serve an important role in constructing variable length block ciphers [13] and authentication codes [12], and are employed in prevalent protocols like Internet Key Exchange (IKEv2). No known CPRF or KH-CPRF construction supports variable input length.

Updatable Encryption (UE): In data storage, key rotation refers to the process of (periodically) exchanging the cryptographic key material that is used to protect the data. Key rotation is a desirable feature for cloud storage providers as it can be used to revoke old keys, that might have been comprised, or to enforce data access revocation. All major cloud storage providers (e.g. Amazon's Key Management Service [14], Google Cloud Platform [15]) implement some variant of data-at-rest encryption and hybrid encryption techniques to perform key rotation [16], which although efficient, do not support full key rotation as the procedures are designed to only update the data encapsulation key but not the long-term key. Symmetric updatable encryption, introduced by Boneh et al. [3] (BLMR henceforth), supports full key rotation without performing decryption, i.e., the ciphertexts created under one key can be securely updated to ciphertexts created under another key with the help of a re-encryption/update token.

Everspaugh et al. [16] pointed out that the UE scheme from BLMR addresses relatively weak confidentiality goals and does not even consider integrity. They proposed a new security notion, named re-encryption indistinguishability, to better capture the idea of fully refreshing the keys upon rotation. They also presented an authenticated encryption based UE scheme, that satisfies the requirements of their new security model. Recently, Lehmann and Tackmann [17] observed that the previous security models/definitions (from BLMR and Everspaugh et al.) do not capture post-compromise security, i.e., the security guarantees after a key compromise. They also proved that neither of the two schemes is post-compromise secure under adaptive attacks. In the same paper, they presented the first UE scheme with post-compromise security. However, the security of that scheme is based on the Decisional Diffie Hellman (DDH) assumption, rendering it vulnerable to quantum computers [18]. It is important to note that all existing UE schemes only support bidirectional updates, i.e., the update token used to refresh the ciphertext for the current epoch can also be used to revert back to the previous epoch's ciphertext. Naturally, this is an undesirable feature. Hence, unidirectional updates [17], where the update tokens cannot be used to undo the ciphertext updates, is a highly desirable feature for UE schemes. But as mentioned earlier, no existing UE scheme achieves unidirectional updates.

1.1 Our Contributions

Our contributions can be classified into the following two broad classes.
1. Novel PRF Classes and Constructions. We introduce fully Key homomorphic and partially Input Homomorphic (KIH) PRFs with Homomorphically induced Variable input Length (HVL). A PRF, F, from such function family satisfies the condition that given $F(k_1, x_1)$ and $F(k_2, x_2)$, there exists an efficient algorithm to compute $F(k_1 \oplus k_2, x_1 \ominus x_2)$, where $|x_1 \ominus x_2| \geq |x_1| \ (= |x_2|)$ and the input homomorphism effects only some fixed m-out-of-n bits. We present a Learning with Errors (LWE) [19] based construction for such a PRF family. Our construction is inspired by the KH-PRF construction from Banerjee and Peikert [4]. A restricted case of our PRF family leads to another novel PRF class, namely left/right KH-CPRF with HVL.
2. Quantum-Safe Post-Compromise Secure UE Scheme with Unidirectional Updates. We use our HVL-KIH-PRF family to construct the first quantum-safe, post-compromise secure updatable encryption scheme with unidirectional updates. We know that the KH-PRF based UE scheme from BLMR is not post-compromise secure because it never updates the nonce [17]. Since our HVL-KIH-PRF family supports input homomorphism, in addition to key homomorphism, it allows updates to the nonce (i.e., the PRF input). Hence, we turn the BLMR UE scheme post-compromise secure by replacing their KH-PRF by our HVL-KIH-PRF, and introducing randomly sampled nonces for the ciphertext updates. The bi-homomorphism of our PRF family also allows us to enforce unidirectional updates. Since our PRF construction is based on the learning with errors (LWE) problem [19], our UE scheme is also quantum-safe.

1.2 Organization

Section 2 recalls the necessary background for the rest of the paper. Section 3 introduces and defines KIH-PRF, HVL-KIH-PRF and HVL-KH-CPRF. In

Sect. 4, we present a LWE-based construction for a HVL-KIH-PRF family, provide its proof of correctness and discuss the different types of input homomorphisms that it supports. Section 5 gives the security proof for our HVL-KIH-PRF construction, while Sect. 6 analyzes its time complexity. Section 7 presents the construction of left/right HVL-KH-CPRF, that follows from a restricted case of our HVL-KIH-PRF family. In Sect. 8, we use of our HVL-KIH-PRF family to construct the first quantum-safe, post-compromise secure updatable encryption scheme with unidirectional updates. Section 9 discusses an interesting open problem, solving which would lead to a novel searchable encryption scheme.

2 Background

This section recalls the necessary definitions required for the rest of the paper.

2.1 Learning with Errors (LWE)

The LWE problem requires to recover a secret s given a sequence of 'approximate' random linear equations on it. LWE is known to be hard based on certain assumptions regarding the worst-case hardness of standard lattice problems such as GapSVP (decision version of the Shortest Vector Problem) and SIVP (Shortest Independent Vectors Problem) [19,20]. Many cryptosystems have been constructed whose security can be proven under the LWE problem, including (identity-based, leakage-resilient, fully homomorphic, functional) encryption [19,21–26], PRFs [27], KH-PRFs [3,4], KH-CPRFs [10,11], etc.

Definition 1 [19]. (Decision-LWE) For positive integers n and q, such that $q = q(n) \geq 2$, and an error distribution $\chi = \chi(n)$ over $\mathbb{Z}_q$, the decision-$LWE_{n,q,\chi}$ problem is to distinguish between the following pairs of distributions:

$$(\mathbf{A}, \mathbf{A}^T \mathbf{s} + \mathbf{e}) \quad \text{and} \quad (\mathbf{A}, \mathbf{u}),$$

where $m = poly(n)$, $\mathbf{A} \xleftarrow{\$} \mathbb{Z}_q^{n \times m}$, $\mathbf{s} \xleftarrow{\$} \mathbb{Z}_q^n$, $\mathbf{e} \xleftarrow{\$} \chi^m$, and $\mathbf{u} \xleftarrow{\$} \mathbb{Z}_q^m$.

Definition 2 [19]. (Search-LWE) For positive integers n and q, such that $q = q(n) \geq 2$, and an error distribution $\chi = \chi(n)$ over $\mathbb{Z}_q$, the search-$LWE_{n,q,\chi}$ problem is to recover $\mathbf{s} \in \mathbb{Z}_q^n$, given $m(= poly(n))$ independent samples of $(\mathbf{A}, \mathbf{A}^T \mathbf{s} + \mathbf{e})$, where $\mathbf{A} \xleftarrow{\$} \mathbb{Z}_q^{n \times m}$, $\mathbf{s} \xleftarrow{\$} \mathbb{Z}_q^n$, and $\mathbf{e} \xleftarrow{\$} \chi^m$.

2.2 Learning with Rounding (LWR)

Banerjee et al. [27] introduced the LWR problem, in which instead of adding a random small error as done in LWE, a *deterministically* rounded version of the sample is released. In particular, for some $p < q$, the elements of $\mathbb{Z}_q$ are divided into p contiguous intervals of roughly q/p elements each. The rounding function is defined as: $\lfloor \cdot \rceil_p : \mathbb{Z}_q \to \mathbb{Z}_p$, that maps $x \in \mathbb{Z}_q$ into the index of the interval that

x belongs to. Note that the error is introduced only when $q > p$, with "absolute" error being roughly equal to q/p, resulting in the "error rate" (relative to q) to be on the order of $1/p$. The LWR problem was shown to be as hard as LWE for a setting of parameters where the modulus and modulus-to-error ratio are super-polynomial [27]. Following [27], Alwen et al. [28] gave a LWR to LWE reduction that works for a larger range of parameters, allowing for a polynomial modulus and modulus-to-error ratio. Bogdanov et al. [29] gave a more relaxed and general version of the theorem proved in [28].

2.3   LWE-Based KH-PRFs

Due to small error being involved, LWE based KH-PRF constructions [3,4] only achieve 'almost homomorphism', which is defined as:

Definition 3 [3]. Let $F : \mathcal{K} \times \mathcal{X} \rightarrow \mathbb{Z}_p^m$ be an efficiently computable function such that $(\mathcal{K}, \oplus)$ is a group. We say that the tuple $(F, \oplus)$ is a γ-almost key homomorphic PRF if the following two properties hold:

1. F is a secure pseudorandom function.
2. For every $k_1, k_2 \in \mathcal{K}$ and every $x \in \mathcal{X}$, there exists a vector $\mathbf{e} \in [0, \gamma]^m$ such that: $F_{k_1}(x) + F_{k_2}(x) = F_{k_1 \oplus k_2}(x) + \mathbf{e} \pmod{p}$.

Based on [27], BLMR gave the first standard-model constructions of KH-PRFs using lattices/LWE. They proved their PRF to be secure under the m-dimensional (over an m-dimensional lattice) LWE assumption, for error rates (as defined in Sect. 2.2) $\alpha = m^{-\Omega(l)}$. Following [3], Banerjee and Peikert [4] gave KH PRFs from substantially weaker LWE assumptions, e.g., error rates of only $\alpha = m^{-\Omega(\log l)}$, which yields better key sizes and runtimes.

3   Novel PRF Classes: Definitions

In this section, we formally define KIH-PRF, HVL-KIH-PRF and HVL-KH-CPRF. For convenience, our definitions assume seed and error matrices instead of vectors. Note that such interchange does not affect the hardness of LWE [30].

Notations. We begin by defining the important notations.

1. $x = x_\ell || x_r$, where $1 \leq |x_\ell| \leq \lfloor |x|/2 \rfloor$ and $|x_r| = |x| - |x_\ell|$.
2. $x_a = x_{a.\ell} || x_{a.r}$, where $1 \leq |x_{a.\ell}| \leq \lfloor |x_a|/2 \rfloor$ and $|x_{a.r}| = |x_a| - |x_{a.\ell}|$.

Assumption. Since the new PRF classes exhibit partial input homomorphism, without loss the generality, we assume x_r to be the homomorphic portion of the input x, with x_ℓ being the static/fixed portion. Obviously, the definitions remain valid if these are swapped, i.e., if x_ℓ is taken to be the homomorphic portion of the input with fixed/static x_r.

3.1   KIH-PRF

Definition 4. Let $F : \mathcal{K} \times \mathcal{X}' \times \mathcal{X} \to \mathbb{Z}_p^{m \times m}$ be a PRF family, such that $(\mathcal{K}, \oplus)$ and $(\mathcal{X}, \ominus)$ are groups. We say that the tuple $(F, \ominus, \oplus)$ is a γ-almost fully key and partially input homomorphic PRF if the following condition holds:

– For every $k_1, k_2 \in \mathcal{K}$, with $x_{1.r}, x_{2.r} \in \mathcal{X}$ and $x_{1.\ell}, x_{2.\ell} \in \mathcal{X}'$, such that $x_{1.\ell} = x_{2.\ell} = x_\ell$, there exists a vector $\mathbf{E} \in [0, \gamma]^{m \times m}$ such that:

$$F_{k_1}(x_{1.\ell}||x_{1.r}) + F_{k_2}(x_{2.\ell}||x_{2.r}) + \mathbf{E} = F_{k_1 \oplus k_2}(x_\ell||x_r) \pmod{p},$$

where $x_r = x_{1.r} \ominus x_{2.r}$.

3.2   HVL-KIH-PRF

Definition 5. Let $\mathcal{X} \subset \mathcal{Y}$, with $\ominus$ defining the surjective mapping: $\mathcal{X} \ominus \mathcal{X} \to \mathcal{Y}$. Let $F : \mathcal{K} \times \mathcal{X}' \times \mathcal{X} \to \mathbb{Z}_p^{m \times m}$ and $F' : \mathcal{K} \times \mathcal{X}' \times \mathcal{Y} \to \mathbb{Z}_p^{m \times m}$ be two PRF families, where $(\mathcal{K}, \oplus)$ is a group. We say that the tuple $(F, \ominus, \oplus)$ is a γ-almost fully key and partially input homomorphic PRF with homomorphically induced variable input length, if the following condition holds:

– For every $k_1, k_2 \in \mathcal{K}$, and with $x_{1.r}, x_{2.r} \in \mathcal{X}$ and $x_{1.\ell}, x_{2.\ell} \in \mathcal{X}'$, such that $x_{1.\ell} = x_{2.\ell} = x_\ell$, there exists a vector $\mathbf{E} \in [0, \gamma]^{m \times m}$ such that:

$$F_{k_1}(x_{1.\ell}||x_{1.r}) + F_{k_2}(x_{2.\ell}||x_{2.r}) + \mathbf{E} = F'_{k_1 \oplus k_2}(x_\ell, y) \pmod{p},$$

where $y = x_{1.r} \ominus x_{2.r}$.

3.3   Left/Right KH-CPRF with HVL

Definition 6. (Left KH-CPRF with HVL:) Let $\mathcal{X} \subset \mathcal{Y}$, with $\ominus$ defining the surjective mapping: $\mathcal{X} \ominus \mathcal{X} \to \mathcal{Y}$. Let $F : \mathcal{K} \times \mathcal{W} \times \mathcal{X} \to \mathbb{Z}_p^{m \times m}$ and $F' : \mathcal{K} \times \mathcal{W} \times \mathcal{Y} \to \mathbb{Z}_p^{m \times m}$ be two PRF families, where $(\mathcal{K}, \oplus)$ is a group. We say that the tuple $(F, \ominus, \oplus)$ is a left key homomorphic constrained-PRF with homomorphically induced variable input length, if for any $k_0 \in \mathcal{K}$ and a fixed $w \in \mathcal{W}$, given $F_{k_0}(w||x) \in \mathcal{F}$, where $x \in \mathcal{X}$, there exists an efficient algorithm to compute $F'_{k_0 \oplus k_1}(w, y) \in \mathcal{F}'$, for all $k_1 \in \mathcal{K}$ and $y \in \mathcal{Y}$.

(Right KH-CPRF with HVL:) Let $\mathcal{X} \subset \mathcal{Y}$, with $\ominus$ defining the surjective mapping: $\mathcal{X} \ominus \mathcal{X} \to \mathcal{Y}$. Let $F : \mathcal{K} \times \mathcal{X} \times \mathcal{W} \to \mathbb{Z}_p^{m \times m}$ and $F' : \mathcal{K} \times \mathcal{Y} \times \mathcal{W} \to \mathbb{Z}_p^{m \times m}$ be two PRF families, where $(\mathcal{K}, \oplus)$ is a group. We say that the tuple $(F, \ominus, \oplus)$ is a right key homomorphic constrained-PRF with homomorphically induced variable input length, if for any $k_0 \in \mathcal{K}$ and a fixed $w \in \mathcal{W}$, given $F_{k_0}(x||w) \in \mathcal{F}$, where $x \in \mathcal{X}$, there exists an efficient algorithm to compute $F'_{k_0 \oplus k_1}(y, w) \in \mathcal{F}'$, for all $k_1 \in \mathcal{K}$ and $y \in \mathcal{Y}$.

4   LWE-Based HVL-KIH-PRF Construction

In this section, we present the first construction for a HVL-KIH-PRF family. Our construction is based on the LWE problem, and is inspired from the KH-PRF construction by Banerjee and Peikert [4].

4.1   Rounding Function

Let λ be the security parameter. Define a rounding function, $\lfloor \cdot \rceil : \mathbb{Z}_q \rightarrow \mathbb{Z}_p$, where $q \geq p \geq 2$, as:

$$\lfloor x \rceil_p = \left\lfloor \frac{p}{q} \cdot x \right\rceil$$

That is, if $\lfloor x \rceil_p = i$, then $i \cdot \lfloor q/p \rceil$ is the integer multiple of $\lfloor q/p \rceil$ that is nearest to x. So, x is deterministically rounded to the nearest element of a sufficiently "coarse" public subset of $p \ll q$, well-separated values in $\mathbb{Z}_q$ (e.g., a subgroup). Thus, the "error term" comes solely from deterministically rounding x to a relatively nearby value in $\mathbb{Z}_p$. As described in Sect. 2.2, the problem of distinguishing such rounded products from uniform samples is called the decision-learning with rounding ($LWR_{n,q,p}$) problem. The rounding function is extended component wise to vectors and matrices over $\mathbb{Z}_q$.

4.2   Definitions

Let $l = \lceil \log q \rceil$ and $d = l + 1$. Define a gadget vector as:

$$\mathbf{g} = (0, 1, 2, 4, \ldots, 2^{l-1}) \in \mathbb{Z}_q^d.$$

Define a deterministic decomposition function $\mathbf{g}^{-1} : \mathbb{Z}_q \rightarrow \{0,1\}^d$, such that $\mathbf{g}^{-1}(a)$ is a "short" vector and $\forall a \in \mathbb{Z}_q$, it holds that: $\langle \mathbf{g}, \mathbf{g}^{-1}(a) \rangle = a$, where $\langle \cdot \rangle$ denotes the inner product. The function $\mathbf{g}^{-1}$ is defined as:

$$\mathbf{g}^{-1}(a) = (x', x_0, x_1, \ldots, x_{l-1}) \in \{0,1\}^d,$$

where $x' = 0$, and $a = \sum_{i=0}^{l-1} x_i 2^i$ is the binary representation of a. The gadget vector is used to define the gadget matrix $\mathbf{G}$ as:

$$\mathbf{G} = \mathbf{I}_n \otimes \mathbf{g} = diag(\mathbf{g}, \ldots, \mathbf{g}) \in \mathbb{Z}_q^{n \times nd},$$

where $\mathbf{I}_n$ is the $n \times n$ identity matrix and $\otimes$ denotes the Kronecker product. The binary decomposition function, $\mathbf{g}^{-1}$, is applied entry-wise to vectors and matrices over $\mathbb{Z}_q$. Thus, $\mathbf{g}^{-1}$ is extended to get another deterministic decomposition function $\mathbf{G}^{-1} : \mathbb{Z}_q^{n \times m} \rightarrow \{0,1\}^{nd \times m}$, such that, $\mathbf{G} \cdot \mathbf{G}^{-1}(\mathbf{A}) = \mathbf{A}$. The addition operations *inside* the binary decomposition functions $\mathbf{g}^{-1}$ and $\mathbf{G}^{-1}$ are performed as simple integer operations (over all integers $\mathbb{Z}$), and not done in $\mathbb{Z}_q$.

4.3   Main Construction

We begin by going over the frequently used notations for this section.

1. $x_{\ell h}$: left half of x, such that $|x_{\ell h}| = \lfloor |x/2| \rfloor$.
2. x_{rh}: right half of x, such that $|x_{rh}| = \lceil |x/2| \rceil$.
3. $x[i]$: the ith bit of x.

Let T be a full binary tree with at least one node, with $T.r$ and $T.\ell$ denoting its right and left subtree, respectively. For random matrices $\mathbf{A}_0, \mathbf{A}_1 \in \mathbb{Z}_q^{n \times nd}$, define function $\mathbf{A}_T : \{0,1\}^{|T|} \to \mathbb{Z}_q^{n \times nd}$ recursively as:

$$\mathbf{A}_T(x) = \begin{cases} \mathbf{A}_x & \text{if } |T| = 1 \\ \mathbf{A}_{T.\ell}(x_\ell) + \mathbf{A}_{x[0]}\mathbf{G}^{-1}(\mathbf{A}_{T.r}(x_r)) & \text{otherwise,} \end{cases}$$

where $x = x_\ell || x_r$, $x_\ell \in \{0,1\}^{|T.\ell|}$, $x_r \in \{0,1\}^{|T.r|}$, and $|T|$ denotes the number of leaves in T. Based on the random seed $\mathbf{S} \in \mathbb{Z}_q^{n \times nd}$, the KIH-PRF family, $\mathcal{F}_{(\mathbf{A}_0, \mathbf{A}_1, T, p)}$, is defined as:

$$\mathcal{F}_{(\mathbf{A}_0, \mathbf{A}_1, T, p)} = \left\{ F_{\mathbf{S}} : \{0,1\}^{2|T|} \longrightarrow \mathbb{Z}_p^{nd \times nd} \right\}.$$

Two *seed dependent* matrices, $\mathbf{B}_0, \mathbf{B}_1 \in \mathbb{Z}_q^{n \times nd}$, are defined as:

$$\mathbf{B}_0 = \mathbf{A}_0 + \mathbf{S} \qquad\qquad \mathbf{B}_1 = \mathbf{A}_1 + \mathbf{S},$$

Using the seed dependent matrices, a function $\mathbf{B}_T^{\mathbf{S}}(x)$ is defined recursively as:

$$\mathbf{B}_T^{\mathbf{S}}(x) = \begin{cases} \mathbf{B}_x & \text{if } |T| = 1 \\ \mathbf{B}_{T.\ell}^{\mathbf{S}}(x_\ell) + \mathbf{A}_{x[0]}\mathbf{G}^{-1}(\mathbf{B}_{T.r}^{\mathbf{S}}(x_r)) & \text{otherwise,} \end{cases}$$

Let $R : \{0,1\}^{|T|} \to \mathbb{Z}_q^{nd \times n}$ be a pseudorandom generator. Let $y = y_{\ell h} || y_{rh}$, where $y_{\ell h}, y_{rh} \in \{0,1\}^{|T|}$. In order to keep the length the equations in check, we represent the product $R(y_{\ell h}) \cdot \mathbf{A}_{y[0]}$ by the *notation: $R_0(y_{\ell h})$*. A member of the KIH-PRF family is indexed by the seed $\mathbf{S}$ as:

$$F_{\mathbf{S}}(y) := \lfloor \mathbf{S}^T \cdot \mathbf{A}_T(y_{\ell h}) + R_0(y_{\ell h}) \cdot \mathbf{G}^{-1}(\mathbf{B}_T^{\mathbf{S}}(y_{rh})) \rceil_p. \tag{1}$$

Let $\bar{0} = 00$, i.e., it represents two consecutive 0 bits. We define the following function family:

$$\mathcal{F}'_{(\mathbb{A}, T, p)} = \left\{ F_{\mathbf{S}}' : \{0,1\}^{|T|} \times \{0,1,\bar{0}\}^{|T|} \longrightarrow \mathbb{Z}_p^{nd \times nd} \right\},$$

where $\mathbb{A} = \{\mathbf{A}_0, \mathbf{A}_1, \mathbf{B}_0, \mathbf{B}_1, \mathbf{C}_0, \mathbf{C}_1, \overline{\mathbf{C}}_0\}$, and the matrices $\mathbf{C}_1, \mathbf{C}_0, \overline{\mathbf{C}}_0$ are defined by the seed $\mathbf{S} \in \mathbb{Z}_q^{n \times nd}$ as:

$$\mathbf{C}_1 = \mathbf{A}_0 + \mathbf{B}_1; \quad \overline{\mathbf{C}}_0 = \mathbf{A}_0 + \mathbf{B}_0; \quad \mathbf{C}_0 = \mathbf{A}_1 + \mathbf{B}_1.$$

Define a function $\mathbf{C}_T : \{0,1\}^{|T|} \times \{0,1,\bar{0}\}^{|T|} \to \mathbb{Z}_q^{n \times nd}$ recursively as:

$$\mathbf{C}_T^{\mathbf{S}}(x) = \begin{cases} \mathbf{C}_x & \text{if } |T| = 1 \\ \overline{\mathbf{C}}_0 & \text{if } |T| > 1 \wedge x[i] = x[i+1] = 0 \\ \mathbf{C}_{T.\ell}^{\mathbf{S}}(x_\ell) + \mathbf{A}_{x[0]}\mathbf{G}^{-1}(\mathbf{C}_{T.r}^{\mathbf{S}}(x_r)) & \text{otherwise,} \end{cases}$$

i.e., $\overline{\mathbf{C}}_0$ denotes two bits. Hence, during the evaluation of $\mathbf{C}_T^{\mathbf{S}}(x)$, a leaf in T may represent one bit or two bits. Let $z = z_0 || z_1$, where $z_0 \in \{0,1\}^{|T|}$ and $z_1 \in \{0,1,\bar{0}\}^{|T|}$. A member of the function family $\mathcal{F}'_{(\mathbb{A}, T, p)}$ is defined as:

$$F_{\mathbf{S}}'(z_0, z_1) := \lfloor \mathbf{S}^T \cdot \mathbf{A}_T(z_0) + R_0(z_0) \cdot \mathbf{G}^{-1}(\mathbf{C}_T^{\mathbf{S}}(z_1)) \rceil_p, \tag{2}$$

where $R_0(z_0) = R(z_0) \cdot \mathbf{A}_{z_0[0]}$. Similar to the KH-PRF construction from [4], bulk of the computation performed while evaluating the PRFs, $F_{\mathbf{S}}(x)$ and $F'_{\mathbf{S}}(x_0, x_1)$, is in computing the functions $\mathbf{A}_T(x), \mathbf{B}_T^{\mathbf{S}}(x), \mathbf{C}_T^{\mathbf{S}}(x)$. While computing these functions on an input x, if all the intermediate matrices are saved, then $\mathbf{A}_T(x')$, $\mathbf{B}_T^{\mathbf{S}}(x'), \mathbf{C}_T^{\mathbf{S}}(x')$ can be incrementally computed for a x' that differs from x in a single bit. Specifically, one only needs to recompute the matrices for those internal nodes of T which appear on the path from the leaf representing the changed bit to the root. Hence, saving the intermediate matrices, that are generated while evaluating the functions on an input x can significantly speed up successive evaluations on the related inputs x'.

4.4   Proof of Correctness

The homomorphically induced variable length (HVL) for our function family follows from the fact that $\{0,1\}^{|T|} \subset \{0,1,\bar{0}\}^{|T|}$. So, we move on to defining and proving the fully key and partially input homomorphic property for our function family. We begin by introducing a commutative binary operation, called 'almost XOR', which is denoted by $\bar{\oplus}$ and defined by the truth table given in Table 1.

Table 1. Truth table for 'almost XOR' operation, $\bar{\oplus}$

$1\bar{\oplus}1$	$= 0$
$0\bar{\oplus}0$	$= 00$
$0\bar{\oplus}1$	$= 1$

Theorem 1. *For any inputs $x, y \in \{0,1\}^{|T|}$ and a full binary tree $|T|$ such that: $x_{lh} = y_{lh} = z_0$ and $x_{rh}\bar{\oplus}y_{rh} = z_1$, where $z_0 \in \{0,1\}^{|T|}$, and $z_1 \in \{0,1,\bar{0}\}^{|T|}$, the following holds:*

$$F'_{(\mathbf{S}_1+\mathbf{S}_2)}(z_0, z_1) = F_{\mathbf{S}_1}(x) + F_{\mathbf{S}_2}(y) + \mathbf{E}, \tag{3}$$

where $||\mathbf{E}||_\infty \leq 1$.

Proof. We begin by making an important observation, and arguing about its correctness.

Observation 1. The HVL-KIH-PRF family, defined in Eq. 1, requires both addition and multiplication operations for each function evaluation. Hence, adding the outputs of two functions, $F_{\mathbf{S}_1}$ and $F_{\mathbf{S}_2}$, from the function family $\mathcal{F}$ translates into adding the outputs per node of the tree T. As a result, the decomposition function $\mathbf{G}^{-1}$ for each node in T takes one of the following three forms:

1. $\mathbf{G}^{-1}(\mathbf{A}_b + \mathbf{A}_{x[0]} \cdot \mathbf{G}^{-1}(\cdot))$,

2. $\mathbf{G}^{-1}(\mathbf{A}_b + \mathbf{A}_{x[0]} \cdot \mathbf{G}^{-1}(\cdot) + \cdots + \mathbf{A}_{x_0} \cdot \mathbf{G}^{-1}(\cdot))$,
3. $\mathbf{G}^{-1}(\mathbf{A}_b)$, $b \in \{0,1\}$,

where $\mathbf{G}^{-1}(\cdot)$ denotes possibly nested $\mathbf{G}^{-1}$. Note that each term $\mathbf{A}_{x[0]} \cdot \mathbf{G}^{-1}(\cdot)$ or a summation of such terms, i.e., $\mathbf{A}_{x[0]} \cdot \mathbf{G}^{-1}(\cdot) + \cdots + \mathbf{A}_{x[0]} \cdot \mathbf{G}^{-1}(\cdot)$, yields some matrix $\check{\mathbf{A}} \in \mathbb{Z}_q^{n \times nd}$. Hence, each decomposition function $\mathbf{G}^{-1}$ in the recursively unwound function $F'_{(\mathbf{s}_1+\mathbf{s}_2)}(z_0, z_1)$ has at most two "direct" inputs/arguments (i.e., $\mathbf{A}_b$ and $\check{\mathbf{A}}$). $\square$

Recall from Sect. 4.2 that for the "direct" arguments of $\mathbf{G}^{-1}$, addition operations are performed as simple integer operations (over all integers $\mathbb{Z}$) instead of being done in $\mathbb{Z}_q$. We know from Observation 1, that unwinding of the recursive function $F'_{\mathbf{S}}(z_0, z_1)$ yields each binary decomposition function $\mathbf{G}^{-1}$ with at most two "direct" inputs. We also know that binary decomposition functions ($\mathbf{g}^{-1}$ and $\mathbf{G}^{-1}$) are linear, provided there is no carry bit or there is an additional bit to accommodate the possible carry. Hence, by virtue of the extra bit, $d - l$ (where $l = \lceil \log q \rceil$, and $d = l + 1$), each $\mathbf{G}^{-1}$ behaves as a linear function during the evaluation of our function families, i.e., the following holds:

$$\mathbf{G}^{-1}(\mathbf{A}_i + \mathbf{A}_j) = \mathbf{G}^{-1}(\mathbf{A}_i) + \mathbf{G}^{-1}(\mathbf{A}_j), \tag{4}$$

where $\mathbf{G}^{-1}(\mathbf{A}_i) + \mathbf{G}^{-1}(\mathbf{A}_j)$ is component-wise vector addition of the n, d bits long bit vectors of the columns, $[\mathbf{v}_1, \ldots, \mathbf{v}_{nd}] \in \mathbb{Z}^{1 \times nd}$, of $\mathbf{G}^{-1}(\mathbf{A}_i)$ with the n, d bits long bit vectors of the columns, $[\mathbf{w}_1, \ldots, \mathbf{w}_{nd}] \in \mathbb{Z}_q^{1 \times nd}$, of $\mathbf{G}^{-1}(\mathbf{A}_j)$.

We are now ready to prove Eq. 3. Since $y_{lh} = x_{lh}$, we use x_{lh} to represent both, as that helps clarity. Let $\mathbf{S} = \mathbf{S}_1 + \mathbf{S}_2$, then by using Eq. 2, we can write the LHS of Eq. 3 as: $\lfloor \mathbf{S}^T \cdot \mathbf{A}_T(z_0) + R_0(z_0) \cdot \mathbf{G}^{-1}(\mathbf{C}_T^{\mathbf{S}}(z_1)) \rceil_p$.

Similarly, from Eq. 1, we get RHS of Eq. 3 equal to:

$$\lfloor \mathbf{S}_1^T \cdot \mathbf{A}_T(x_{lh}) + R_0(x_{lh}) \cdot \mathbf{G}^{-1}(\mathbf{B}_T^{\mathbf{S}_1}(x_{rh})) \rceil_p$$
$$+ \lfloor \mathbf{S}_2^T \cdot \mathbf{A}_T(x_{lh}) + R_0(x_{lh}) \cdot \mathbf{G}^{-1}(\mathbf{B}_T^{\mathbf{S}_2}(y_{rh})) \rceil_p + \mathbf{E}.$$

We know that $\lfloor a + b \rceil_p = \lfloor a \rceil_p + \lfloor b \rceil_p + e$. We further know that $x_{lh} = z_0$, and $R_0(x_{lh}) = R_0(z_0)$. Thus, from Eq. 4, the RHS can be written as:

$$\lfloor (\mathbf{S}_1 + \mathbf{S}_2)^T \cdot \mathbf{A}_T(z_0) + R_0(z_0) \cdot \mathbf{G}^{-1}(\mathbf{B}_T^{\mathbf{S}_1}(x_{rh}) + \mathbf{B}_T^{\mathbf{S}_2}(y_{rh})) \rceil_p$$
$$= \lfloor \mathbf{S}^T \cdot \mathbf{A}_T(z_0) + R_0(z_0) \cdot \mathbf{G}^{-1}(\mathbf{C}_T^{\mathbf{S}}(x_{rh} \bar{\oplus} y_{rh})) \rceil_p$$
$$= \lfloor \mathbf{S}^T \cdot \mathbf{A}_T(z_0) + R_0(z_0) \cdot \mathbf{G}^{-1}(\mathbf{C}_T^{\mathbf{S}}(z_1)) \rceil_p = LHS.$$

$\square$

5   Security Proof

The security proofs as well as the time complexity analysis of our construction depend on the tree T. Left and right depth of T are respectively defined as the maximum left and right depths over all leaves in T. The modulus q, the underlying LWE error rate, and the dimension n needed to obtain a desired level of provable security, are largely determined by two parameters of T. The first one, called *expansion e(T)* [4], is defined as:

$$e(T) = \begin{cases} 0 & \text{if } |T| = 1 \\ max\{e(T.\ell) + 1, e(T.r)\} & \text{otherwise}, \end{cases}$$

For our construction, $e(T)$ is the maximum number of terms of the form $\mathbf{G}^{-1}(\cdot)$ that get consecutively added together when we unwind the recursive definition of the function $\mathbf{A}_T$. The second parameter is called *sequentiality* [4], which gives the maximum number of nested $\mathbf{G}^{-1}$ functions, and is defined as:

$$s(T) = \begin{cases} 0 & \text{if } |T| = 1 \\ max\{s(T.\ell), s(T.r) + 1\} & \text{otherwise}, \end{cases}$$

For our function families, over the uniformly random and independent choice of $\mathbf{A}_0, \mathbf{A}_1, \mathbf{S} \in \mathbb{Z}_q^{n \times nd}$, and with the secret key chosen uniformly from $\mathbb{Z}_q^n$, the modulus-to-noise ratio for the underlying LWE problem is: $q/r \approx (n \log q)^{e(T)}$. Known reductions [19,20,31] (for $r \geq 3\sqrt{n}$) guarantee that such a LWE instantiation is at least as hard as approximating hard lattice problems like GapSVP and SIVP, in the worst case to within $\approx q/r$ factors on n-dimensional lattices. Known algorithms for achieving such factors take time exponential in $n/\log(q/r) = \tilde{\Omega}(n/e(T))$. Hence, in order to obtain provable 2^λ (where λ is the input length) security against the best known lattice algorithms, the best parameter values are the same as defined for the KH-PRF construction from [4], which are:

$$n = e(T) \cdot \tilde{\Theta}(\lambda) \quad \text{and} \quad \log q = e(T) \cdot \tilde{\Theta}(1). \tag{5}$$

5.1 Overview of KH-PRF from [4]

As mentioned earlier, our construction is inspired by the KH-PRF construction from [4]. Our security proofs rely on the security of that construction. Therefore, before moving to the security proofs, it is necessary that we briefly recall the KH-PRF construction from [4]. Although that scheme differs from our KIH PRF construction, certain parameters and their properties are identical. The rounding function $\lfloor \cdot \rceil_p$, binary tree T, gadget vector/matrix $\mathbf{g}/\mathbf{G}$, the binary decomposition functions $\mathbf{g}^{-1}/\mathbf{G}^{-1}$ and the base matrices $\mathbf{D}_0, \mathbf{D}_1$ in that scheme are defined similarly to our construction. There is a difference in the definitions of the decomposition functions, which for our construction are defined as (see Sect. 4.2): $\mathbf{g}^{-1} : \mathbb{Z}_q \to \{0,1\}^d$ and $\mathbf{G}^{-1} : \mathbb{Z}_q^{n \times m} \to \{0,1\}^{nd \times m}$, i.e., the dimensions of the output space for our decomposition functions has $d(= l+1)$ instead of $l = \lceil \log q \rceil$ as in [4]. Recall that the extra (carry) bit ensures that Eq. 4 holds.

KH-PRF Construction from [4]. Given two uniformly selected matrices, $\mathbf{D}_0, \mathbf{D}_1 \in \mathbb{Z}_q^{n \times nl}$, define function $\mathbf{D}_T(x) : \{0,1\}^{|T|} \to \mathbb{Z}_q^{n \times nl}$ as:

$$\mathbf{D}_T(x) = \begin{cases} \mathbf{D}_x & \text{if } |T| = 1 \\ \mathbf{D}_{T.\ell}(x_{lt}) \cdot \mathbf{G}^{-1}(\mathbf{D}_{T.r}(x_{rt})) & \text{otherwise}, \end{cases} \tag{6}$$

where $x = x_{lt}||x_{rt}$, for $x_{lt} \in \{0,1\}^{|T.\ell|}, x_{rt} \in \{0,1\}^{|T.r|}$. The KH-PRF function family is defined as:

$$\mathcal{H}_{\mathbf{D}_0, \mathbf{D}_1, T, p} = \left\{ H_{\mathbf{s}} : \{0,1\}^{|T|} \rightarrow \mathbb{Z}_p^{nl} \right\}.$$

where $p \leq q$ is the modulus. A member of the function family $\mathcal{H}$ is indexed by the seed $\mathbf{s} \in \mathbb{Z}_q^n$ as: $H_{\mathbf{s}}(x) = \lfloor \mathbf{s} \cdot \mathbf{D}_T(x) \rceil_p$.

Main Security Theorem [4]. For the sake of completeness, we recall the main security theorem from [4].

Theorem 2. *[4] Let T be any full binary tree, χ be some distribution over $\mathbb{Z}$ that is subgaussian with parameter $r > 0$ (e.g., a bounded or discrete Gaussian distribution with expectation zero), and*

$$q \geq p \cdot r \sqrt{|T|} \cdot (nl)^{e(T)} \cdot \lambda^{\omega(1)},$$

where λ is the input size. Then over the uniformly random and independent choice of $\mathbf{D}_0, \mathbf{D}_1 \in \mathbb{Z}_q^{n \times nl}$, the family $\mathcal{H}_{\mathbf{D}_0, \mathbf{D}_1, T, p}$ with secret key chosen uniformly from $\mathbb{Z}_q^n$ is a secure PRF family, under the decision-LWE$_{n,q,\chi}$ assumption.

5.2   Security Proof of Our Construction

The dimensions and bounds for the parameters r, q, p, n, m and χ in our construction are the same as in [4]. We begin by defining the necessary terminology.

1. Reverse-LWE: is an LWE instance $\mathbf{S}^T \mathbf{A} + \mathbf{E}$ with secret lattice-basis $\mathbf{A}$ and public seed matrix $\mathbf{S}$.
2. Reverse-LWR: is defined similarly, i.e., $\lfloor \mathbf{S}^T \mathbf{A} \rceil_p$ with secret $\mathbf{A}$ and public $\mathbf{S}$.
3. If H represents the binary entropy function, then we know that for uniformly random $\mathbf{A} \in \mathbb{Z}_q^{n \times nd}$ and a random seed $\mathbf{S} \in \mathbb{Z}_q^{n \times nd}$, it holds that: $H(\mathbf{A}) = H(\mathbf{S})$. Hence, it follows from elementary linear algebra that reverse-LWR$_{n,q,p}$ and reverse-LWE$_{n,q,\chi}$ are at least as hard as decision-LWR$_{n,q,p}$ and decision-LWE$_{n,q,\chi}$, respectively.

Observation 2. Consider the function family $\mathcal{F}_{(\mathbf{A}_0, \mathbf{A}_1, T, p)}$. We know that a member of the function family is defined by a random seed $\mathbf{S} \in \mathbb{Z}_q^{n \times nd}$ as:

$$F_{\mathbf{S}}(x) = \lfloor \mathbf{S}^T \cdot \mathbf{A}_T(x_{lh}) + R_0(x_{lh}) \cdot \mathbf{G}^{-1}(\mathbf{B}_T^{\mathbf{S}}(x_{rh})) \rceil_p,$$
$$= \underbrace{\lfloor \mathbf{S}^T \cdot \mathbf{A}_T(x_{lh}) \rceil_p}_{\mathbf{L}_T(x_{lh})} + \underbrace{\lfloor R_0(x_{lh}) \cdot \mathbf{G}^{-1}(\mathbf{B}_T^{\mathbf{S}}(x_{rh})) \rceil_p}_{\mathbf{R}_T(x_{rh})} + \mathbf{E}.$$

Observation 3. For $|T| \geq 1$ and $x \in \{0,1\}^{2|T|}$, each $\mathbf{L}_T(x_{lh})$ is the sum of the following three types of terms:

1. Exactly one term of the form: $\lfloor \mathbf{S}^T \cdot \mathbf{A}_{x[0]} \rceil_p$, corresponding to the leftmost child of the full binary tree T and the most significant bit, $x[0]$, of x.

2. At least one term of the following form:

$$\lfloor \mathbf{S}^T \cdot \mathbf{A}_{x[0]} \cdot \mathbf{G}^{-1}(\mathbf{A}_{x[i]}) \rceil_p; \qquad (1 \leq i \leq 2|T|),$$

corresponding to the right child at level 1 of the full binary tree T.

3. Zero or more terms with nested $\mathbf{G}^{-1}$ functions of the form:

$$\lfloor \mathbf{S}^T \cdot \mathbf{A}_{x[0]} \cdot \mathbf{G}^{-1}(\mathbf{A}_{x[i]} + \mathbf{A}_{x[0]} \cdot \mathbf{G}^{-1}(\cdot) + \ldots) \rceil_p; \qquad (1 \leq i \leq 2|T|).$$

We prove that for appropriate parameters and input length, each one of the aforementioned terms is a pseudorandom function on its own.

Lemma 1. *Let n be a positive integer, $q \geq 2$ be the modulus, χ be a probability distribution over $\mathbb{Z}$, and m be polynomially bounded (i.e. $m = poly(n)$). For a uniformly random and independent choice of $\mathbf{A}_0, \mathbf{A}_1 \in \mathbb{Z}_q^{n \times nd}$ and a random seed vector $\mathbf{S} \in \mathbb{Z}_q^{n \times nd}$, the function family $\lfloor \mathbf{S}^T \cdot \mathbf{A}_{x[i]} \rceil_p$ for the single bit input $x[i]$ ($1 \leq i \leq 2|T|$) is a secure PRF family under the decision-LWE$_{n,q,\chi}$ assumption.*

Proof. We know from [27] that $\lfloor \mathbf{S}^T \cdot \mathbf{A}_{x[i]} \rceil_p$ is a secure PRF under the decision-LWR$_{n,q,p}$ assumption, which is at least as hard as solving the decision-LWE$_{n,q,\chi}$ problem. □

Corollary 1 *(To Theorem 2). Let n be a positive integer, $q \geq 2$ be the modulus, χ be a probability distribution over $\mathbb{Z}$, and $m = poly(n)$. For uniformly random and independent matrices $\mathbf{A}_0, \mathbf{A}_1 \in \mathbb{Z}_q^{n \times nd}$ with a random seed $\mathbf{S} \in \mathbb{Z}_q^{n \times nd}$, the function $\lfloor \mathbf{S}^T \cdot \mathbf{A}_{x[0]} \cdot \mathbf{G}^{-1}(\mathbf{A}_{x[i]}) \rceil_p$ for the two bit input: $x[0]||x[i]$, is a secure PRF family, under the decision-LWE$_{n,q,\chi}$ assumption.*

Proof. For the two bit input $x[0]||x[i]$, the expression $\lfloor \mathbf{S}^T \cdot \mathbf{A}_{x[0]} \cdot \mathbf{G}^{-1}(\mathbf{A}_{x[i]}) \rceil_p$ is an instance of the function $\mathcal{H}_{\mathbf{A}_0, \mathbf{A}_1, T, p}$ (see Sect. 5.1). Hence, it follows from Theorem 2 that $\lfloor \mathbf{S}^T \cdot \mathbf{A}_{x[0]} \cdot \mathbf{G}^{-1}(\mathbf{A}_{x[i]}) \rceil_p$ is a secure PRF family under the decision-LWE$_{n,q,\chi}$ assumption. □

Corollary 2 *(To Theorem 2). Let n be a positive integer, $q \geq 2$ be the modulus, χ be a probability distribution over $\mathbb{Z}$, and $m = poly(n)$. Given uniformly random and independent $\mathbf{A}_0, \mathbf{A}_1 \in \mathbb{Z}_q^{n \times nd}$, and a random seed $\mathbf{S} \in \mathbb{Z}_q^{n \times nd}$, the function: $\lfloor \mathbf{S}^T \cdot \mathbf{A}_{x[0]} \cdot \mathbf{G}^{-1}(\mathbf{A}_{x[i]} + \mathbf{A}_{x[0]} \cdot \mathbf{G}^{-1}(\cdot) + \ldots) \rceil_p$ is a secure PRF family under the decision-LWE$_{n,q,\chi}$ assumption.*

Proof. Since, $\mathbf{A}_0, \mathbf{A}_1 \in \mathbb{Z}_q^{n \times nd}$ are random and independent, $\mathbf{A}_{x[0]} \cdot \mathbf{G}^{-1}(\cdot)$ is statistically indistinguishable from $\mathbf{A}_{x[i]} \cdot \mathbf{G}^{-1}(\cdot)$, as defined by the function $\mathbf{B}_T(x)$ (see Eq. 6), where $\mathbf{G}^{-1}(\cdot)$ represents possibly nested $\mathbf{G}^{-1}$. Hence, it follows from Theorem 2 that for the "right spine" (with leaves for all the left children) full binary tree T, $\lfloor \mathbf{S}^T \cdot \mathbf{A}_{x[0]} \cdot \mathbf{G}^{-1}(\mathbf{A}_{x[i]} + \mathbf{A}_{x[0]} \cdot \mathbf{G}^{-1}(\cdot) + \ldots) \rceil_p$ defines a secure PRF family under the decision-LWE$_{n,q,\chi}$ assumption. □

Corollary 3 *(To Theorem 2). Let n be a positive integer, $q \geq 2$ be the modulus, χ be a probability distribution over $\mathbb{Z}$, and $m = poly(n)$. For uniformly random and independent matrices $\boldsymbol{A}_0, \boldsymbol{A}_1 \in \mathbb{Z}_q^{n \times nd}$, and a random seed $\boldsymbol{S} \in \mathbb{Z}_q^{n \times nd}$, the function $\boldsymbol{R}_T(x_{rh}) = \lfloor R_0(x_{lh}) \cdot \boldsymbol{G}^{-1}(\boldsymbol{B}_T^S(x_{rh})) \rceil_p$ is a secure PRF family under the decision-$LWE_{n,q,\chi}$ assumption.*

Proof. We know that $\mathbf{A}_0, \mathbf{A}_1, \mathbf{S} \in \mathbb{Z}_q^{n \times nd}$ are generated uniformly and independently. Therefore, the secret matrices, $\mathbf{B}_0, \mathbf{B}_1$, defined as: $\mathbf{B}_0 = \mathbf{A}_0 + \mathbf{S}$ and $\mathbf{B}_1 = \mathbf{A}_1 + \mathbf{S}$, have the same distribution as $\mathbf{A}_0, \mathbf{A}_1$. As $R : \{0,1\}^{|T|} \to \mathbb{Z}_q^{nd \times n}$ is a PRG, $R(x_{lh})$ is a valid seed matrix for decision-LWE, making $\mathbf{R}_T(x_{rh})$ an instance of reverse-$LWR_{n,q,p}$, which we know is as hard as the decision-$LWR_{n,q,p}$ problem. Hence, it follows from Theorem 2 that $\mathbf{R}_T(x_{rh})$ defines a secure PRF family for secret $R(x_{lh})$. $\square$

Theorem 3. *Let T be any full binary tree, χ be some distribution over $\mathbb{Z}$ that is subgaussian with parameter $r > 0$ (e.g., a bounded or discrete Gaussian distribution with expectation zero), $R : \{0,1\}^{|T|} \to \mathbb{Z}_q^{nd \times n}$ be a PRG, and*

$$q \geq p \cdot r\sqrt{|T|} \cdot (nd)^{e(T)} \cdot \lambda^{\omega(1)}.$$

Then over the uniformly random and independent choice of $\boldsymbol{A}_0, \boldsymbol{A}_1 \in \mathbb{Z}_q^{n \times nd}$ and a random seed $\boldsymbol{S} \in \mathbb{Z}_q^{n \times nd}$, the family $\mathcal{F}_{(\boldsymbol{A}_0, \boldsymbol{A}_1, T, p)}$ is a secure PRF under the decision-$LWE_{n,q,\chi}$ assumption.

Proof. From Observations 2 and 3, we know that each member $F_\mathbf{S}$ of the function family $\mathcal{F}_{(\mathbf{A}_0, \mathbf{A}_1, T, p)}$ is defined by the random seed $\mathbf{S}$, and can be written as:

$$F_\mathbf{S}(x) = \lfloor \mathbf{S}^T \cdot \mathbf{A}_{x[0]} \rceil_p + d_1 \cdot \lfloor \mathbf{S}^T \cdot \mathbf{A}_{x[0]} \cdot \mathbf{G}^{-1}(\mathbf{A}_{x[i]}) \rceil_p +$$
$$d_2 \cdot \lfloor \mathbf{S}^T \cdot \mathbf{A}_{x[0]} \cdot \mathbf{G}^{-1}(\mathbf{A}_{x[i]} + \mathbf{A}_{x[0]} \cdot \mathbf{G}^{-1}(\cdot) + \ldots) \rceil_p +$$
$$d_3 \cdot \lfloor R(x_{lh}) \cdot \mathbf{A}_{x[0]} \cdot \mathbf{G}^{-1}(\mathbf{B}_T^\mathbf{S}(x_{rh})) \rceil_p + \mathbf{E},$$

where $d_1, d_2, d_3 \in \mathbb{Z}$, such that, $1 \leq d_1, d_3 \leq |T|$ and $0 \leq d_2 \leq |T|$. From Lemma 1, and Corollaries 1, 2 and 3, we know that the following are secure PRFs under the decision-$LWE_{n,q,\chi}$ assumption:

1. $\lfloor \mathbf{S}^T \cdot \mathbf{A}_{x[0]} \rceil_p$, $\lfloor \mathbf{S}^T \cdot \mathbf{A}_{x[0]} \cdot \mathbf{G}^{-1}(\mathbf{A}_{x[i]}) \rceil_p$,
2. $\lfloor \mathbf{S}^T \cdot \mathbf{A}_{x[0]} \cdot \mathbf{G}^{-1}(\mathbf{A}_{x[i]} + \mathbf{A}_{x[0]} \cdot \mathbf{G}^{-1}(\cdot) + \ldots) \rceil_p$,
3. $\lfloor R(x_{lh}) \cdot \mathbf{A}_{x[0]} \cdot \mathbf{G}^{-1}(\mathbf{B}_T^\mathbf{S}(x_{rh})) \rceil_p$.

Hence, it follows that the function family $\mathcal{F}_{(\mathbf{A}_0, \mathbf{A}_1, T, p)}$ is a secure PRF under the decision-$LWE_{n,q,\chi}$ assumption.

Corollary 4. *Let T be any full binary tree, χ be some distribution over $\mathbb{Z}$ that is subgaussian with parameter $r > 0$ (e.g., a bounded or discrete Gaussian distribution with expectation zero), $R : \{0,1\}^{|T|} \to \mathbb{Z}_q^{nd \times n}$ be a PRG, and*

$$q \geq p \cdot r\sqrt{|T|} \cdot (nd)^{e(T)} \cdot \lambda^{\omega(1)}.$$

Then over the uniformly random and independent choice of $\boldsymbol{A}_0, \boldsymbol{A}_1 \in \mathbb{Z}_q^{n \times nd}$ and a random seed $\boldsymbol{S} \in \mathbb{Z}_q^{n \times nd}$, the family $\mathcal{F}'_{(\mathbb{A}, T, p)}$ is a secure PRF under the decision-$LWE_{n,q,\chi}$ assumption.

6   Time Complexity Analysis

In this section, we analyze the time complexity of our HVL-KIH-PRF construction. The asymptotic time complexity of our construction is similar to that of the KH-PRF family from [4]. We know that the time complexity of the binary decomposition function $\mathbf{g}^{-1}$ is $O(\log q)$, and that $\mathbf{G}^{-1}$ is simply $\mathbf{g}^{-1}$ applied entry-wise. The size of the public matrices, $\mathbf{A}_0, \mathbf{A}_1 \in \mathbb{Z}_q^{n \times nd}$, is $\Theta(n^2 \log q)$, which by Eq. 5 is $e(T)^4 \cdot \tilde{\Theta}(\lambda^2)$ bits. The secret matrix, $\mathbf{S} \in \mathbb{Z}_q^{n \times nd}$, also has the same size, i.e., $e(T)^4 \cdot \tilde{\Theta}(\lambda^2)$. Computing $\mathbf{A}_T(x), \mathbf{B}_T(x)$ or $\mathbf{C}_T(x)$ requires one decomposition with $\mathbf{G}^{-1}$, one $(n \times nd)$-by-$(nd \times nd)$ matrix multiplication and one $(n \times nd)$-by-$(n \times nd)$ matrix addition over $\mathbb{Z}_q$, per internal node of T. Hence, the total time complexity comes out to be $\Omega(|T| \cdot n^\omega \log^2 q)$ operations in $\mathbb{Z}_q$, where $\omega \geq 2$ is the exponent of matrix multiplication.

7   Left/Right HVL-KH-CPRFs

In this section, we present the construction of another novel PRF class, namely left/right HVL-KH-CPRFs, as a special case of our HVL-KIH-PRF family. Let $F' : \mathcal{K} \times \mathcal{X} \times \mathcal{Y} \to \mathcal{Z}$ be the PRF defined by Eq. 2. The goal is to derive a *constrained* key PRF, $k_{x,left}$ or $k_{x,right}$ for every $x \in \mathcal{X}$ and $k_0 \in \mathcal{K}$, such that $k_{x,left} = F_{k_0}(x||\cdot)$ (where $F : \mathcal{K} \times \mathcal{X} \times \mathcal{X} \to \mathcal{Z}$ is the PRF family defined by Eq. 1) enables the evaluation of the PRF function $F_k'(x, y)$ for the key $k = k_0 \oplus k_1$, where $k_1 \in \mathcal{K}$, and the subset of points $\{(x, y) : y \in \mathcal{Y}\}$, i.e., all the points where the left portion of the input is x. Similarly, the constrained key $k_{x,right} = F_{k_0}(\cdot||x)$ enables the evaluation of the PRF function $F_k'(x, y)$ for the key $k = k_0 \oplus k_1$, where $k_1 \in \mathcal{K}$, and the subset of points $\{(y, x) : y \in \mathcal{Y}\}$, i.e., all the points where the right side of the input is x.

KH-CPRF Construction. We begin by giving a construction for left KH-CPRF, without HVL, and then turn it into a HVL-KH-CPRF construction. Our HVL-KIH-PRF function, defined in Eq. 1, is itself a left KH-CPRF when evaluated as: $F_{k_0}(x_0||\mathbf{1})$, i.e., the key is $k_0 \in \mathcal{K}$, the left side of the input is $x_0 \in \mathcal{X}$, and the right half is an all one vector, $\mathbf{1} = \{1\}^{\log |\mathcal{X}|}$. Now, to evaluate $F_k'(x_0, x_1)$ at a key $k = k_0 + k_1$, and any right input $x_1 \in \mathcal{Y}$, first evaluate $F_{k_1}(x_0||x_1')$ and add its output with that of the given constrained function, $F_{k_0}(x_0||\mathbf{1})$, i.e., compute: $F_k'(x_0, x_1) = F_{k_1}(x_0||x_1') + F_{k_0}(x_0||\mathbf{1})$, where $x_1' \in \mathcal{X}$, and $x_1 = x_1' \bar{\oplus} \mathbf{1}$, with $k = k_0 + k_1$. Recall from Table 1 that 'almost XOR', $\bar{\oplus}$, differs from XOR only for the case when both inputs are zero. Hence, having $\mathbf{1}$ as the right half effectively turns $\bar{\oplus}$ into $\oplus$, and ensures that all possible right halves $x_1 \in \mathcal{X}$ can be realized via $x_1' \in \mathcal{X}$.

Similarly, right KH-CPRF can be realized by provisioning the constrained function $F_{k_0}(\mathbf{1}||x_0)$, where $\mathbf{1} = \{1\}^{|\log \mathcal{X}|}$ is an all ones vector and $x_0 \in \mathcal{X}$. This interchange allows one to evaluate a different version of our HVL-KIH-PRF function, where the left portion of the input exhibits homomorphism (see Sect. 3). Hence, for all $k_1 \in \mathcal{K}$ and $x_1 \in \mathcal{Y}$, it supports evaluation of $F_{k_0+k_1}'(x_1||x_0)$.

Achieving HVL. In order to achieve HVL for the left/right KH-CPRF given above, simply replace the all ones vector, $\mathbf{1}$, with an all zeros vector, $\mathbf{0}$, of the same dimensions. Hence, the new constrained functions are $k_{right} = F_{k_0}(x_0||\mathbf{0})$ and $k_{left} = F_{k_0}(\mathbf{0}||x_0)$. This enables us to evaluate F' at any input $x_1 \in \mathcal{Y}$ via $x_1 = x_1' \bar{\oplus} \mathbf{0}$, where $x_1' \in \mathcal{X}$. The pseudorandomness and security follow from that of our HVL-KIH-PRF family.

8   QPC-UE-UU

In this section, we present the first quantum-safe (Q) post-compromise (PC) secure updatable encryption (UE) scheme with unidirectional updates (UU) as an example application of our KIH-HVL-PRF family. An updatable encryption scheme, UE, contains algorithms for a data owner and a host. We begin by recalling the definitions of these algorithms.

Definition 7 [17]. *An updatable encryption scheme* UE *for message space* $\mathcal{M}$ *consists of a set of polynomial-time algorithms* UE.setup, UE.next, UE.enc, UE.dec, *and* UE.upd *satisfying the following conditions:*
UE.setup: *The algorithm UE:setup is a probabilistic algorithm run by the owner. On input a security parameter* λ, *it returns a secret key* $k_0 \xleftarrow{\$} $ UE.setup(λ).
UE.next: *This probabilistic algorithm is also run by the owner. On input a secret key ke for epoch e, it outputs the secret key,* k_{e+1}, *and an update token,* Δ_{e+1}, *for epoch* $e + 1$. *That is,* $(k_{e+1}, \Delta_{e+1}) \xleftarrow{\$} $ UE.next(k_e).
UE.enc: *This probabilistic algorithm is run by the owner, on input a message* $m \in \mathcal{M}$ *and key* k_e *of some epoch e returns a ciphertext* $C_e \leftarrow \$ $ UE.enc(k_e, m).
UE.dec: *This deterministic algorithm is run by the owner, on input a ciphertext* C_e *and key* k_e *of some epoch e returns* $\{m', \bot\} \leftarrow $ UE.dec(k_e, C_e).
UE.upd: *This probabilistic algorithm is run by the host. Given ciphertext* C_e *from epoch e and the update token* Δ_{e+1}, *it returns the updated ciphertext* $C_{e+1} \leftarrow $ UE.upd(Δ_{e+1}, C_e). *After receiving* Δ_{e+1}, *the host first deletes* Δ_e. *Hence, during some epoch* $e + 1$, *the update token* Δ_{e+1} *is available at the host, but the tokens from earlier epochs have been deleted.*

8.1   Settings and Notations

Let $i \xleftarrow{\$} \mathcal{X}$ be the identifier for data block d_i. Let $F : \mathcal{K} \times \mathcal{X} \times \mathcal{X} \rightarrow \mathcal{Z}$ and $F' : \mathcal{K} \times \mathcal{X} \times \mathcal{Y} \rightarrow \mathcal{Z}$ be the functions defined in Eqs. 1 and 2, respectively. Let KeyGen(λ) be the key generation algorithm for F, where λ is the security parameter. As described in Fig. 1, QPC-UE-UU generates a random nonce per key rotation, hence ensuring that the encryption remains probabilistic, despite our PRF family being deterministic.

- QPC-UE-UU.setup(λ): Generate a random encryption key $k_0 \xleftarrow{\$} $ F.KeyGen(λ), and sample a random nonce $N_0 \xleftarrow{\$} \mathcal{X}$. Set $e \leftarrow 0$, and return the key for epoch $e = 0$ as: $ki_0 = (k_0, N_0)$.

- QPC-UE-UU.enc(ki_e, m): Let $i \xleftarrow{\$} \mathcal{X}$ be randomly sampled element that is shared by the host and the owner. Parse $ki_e = (k_e, N_e)$, and return the ciphertext for epoch e as: $C_e = (F'_{k_e}(i, N_e) + m)$. Note that the random nonce ensures that the encryption is not deterministic.

- QPC-UE-UU.dec(ki_e, C_e): Parse $ki_e = (k_e, N_e)$. Return $m \leftarrow C_e - F'_{k_e}(i, N_e)$.

- QPC-UE-UU.next(ki_e):

  1. Sample a random nonce $N_{e+1} \xleftarrow{\$} \mathcal{X}$. Parse ki_e as (k_e, N_e).

  2. For epoch $e + 1$, generate a random encryption key $k_{e+1} \xleftarrow{\$} $ F.KeyGen(λ), and return $\Delta_{e+1} = (\Delta_{e+1}^k, \Delta_{e+1}^N)$, where $\Delta_{e+1}^N = N_e \bar{\oplus} N_{e+1}$ is the nonce update token, and $\Delta_{e+1}^k = k_{e+1} - k_e$ is the encryption key update token. The key for epoch $e + 1$ is $ki_{e+1} = (2k_e - k_{e+1}, N_{e+1})$, where $2k_e - k_{e+1}$ is the encryption key.

- QPC-UE-UU.upd(Δ_{e+1}, C_e): Update the ciphertext as: $C_{e+1} = C_e - F'_{\Delta_{e+1}^k}(i, \Delta_{e+1}^N)$
$= F'_{k_e}(i, N_e) + m - F'_{k_{e+1}-k_e}(i, N_e \bar{\oplus} N_{e+1})$
$= F'_{-k_{e+1}+2k_e}(i, N_e - (N_e \bar{\oplus} N_{e+1})) + m = F'_{2k_e - k_{e+1}}(i, N_{e+1}) + m$.

Fig. 1. Quantum-safe, post-compromise secure updatable encryption scheme with unidirectional updates, (QPC-UE-UU).

8.2   Proof of Unidirectional Updates

As explained in Sect. 8.1, the random nonce ensures that the encryption in our scheme is probabilistic. Hence, the security of our QPC-UE-UU scheme follows from the pseudorandomness of our KIH-HVL-PRF family. We defer the detailed security proof to the full version of the paper. We move on to proving that the ciphertext updates performed by our scheme are indeed unidirectional. In schemes with unidirectional updates, an update token Δ_{e+1} can only move ciphertexts from epoch e into epoch $e+1$, but not vice versa. The notations used in the proof are the same as in Fig. 1.

Lemma 2. *For the HVL-KIH-PRF family, $\mathcal{F}' : \mathcal{X} \times \mathcal{Y} \rightarrow \mathcal{Z}$, as defined in Corollary 4 with $\mathcal{X} = \{0,1\}^{|T|}$, $\mathcal{Y} = \{0,1,\bar{0}\}^{|T|}$ and $\mathcal{Z} = \mathbb{Z}_p^{nd \times nd}$: given ciphertext, C_{e+1} and the update token $\Delta_{e+1} = (k_{e+1} - k_e, N_e \bar{\oplus} N_{e+1})$ for epoch $e + 1$, the following holds for a polynomial adversary $\mathcal{A}$, all randomly sampled $Q \xleftarrow{\$} \mathcal{Z}$ and security parameter λ:*

$$Pr[C_e] = Pr[Q] \pm \epsilon(\lambda),$$

where $\epsilon(\lambda)$ is a negligible function.

Proof. The main idea of the proof is that due to the bi-homomorphic property of our HVL-KIH-PRF family, $\mathcal{F}'$, the adversary, $\mathcal{A}$, can only revert back to either the key k_e or the nonce N_e, but not both. In other words, $\mathcal{A}$ cannot recover

$C_e = F'_{k_e}(i, N_e) + m$. We split the proof into two portions, one concentrating on reverting back to k_e as the function key, and the other one focusing on moving back to N_e as the function input. We prove that these two goals are mutually exclusive for our scheme, i.e., both of them cannot be achieved together.

Case 1: Reverting to k_e. Recall from Table 1 that the only well-defined operation for the operand $\bar{0}$ is $\bar{0} = 0+0$. We know that the ciphertext update for epoch $e + 1$ is performed as: $C_{e+1} = C_e - F'_{\Delta^k_{e+1}}(i, \Delta^N_{e+1})$. Since $\mathcal{F}'$ is a PRF family, the only way to revert back to $F'_{k_e} \in \mathcal{F}'$ via C_{e+1} and Δ_{e+1} is by computing:

$$C_{e+1} + F'_{\Delta^k_{e+1}}(i, \Delta^N_{e+1}) = F'_{2k_e - k_{e+1}}(i, N_{e+1}) + m + F'_{k_{e+1} - k_e}(i, N_e \bar{\oplus} N_{e+1})$$
$$= F'_{k_e}(i, N_{e+1} \bar{\oplus}(N_e \bar{\oplus} N_{e+1})).$$

Due to the key homomorphism exhibited by $\mathcal{F}'$, no other computations would lead to the target key k_e. We know that $\Delta^N_{e+1}(= N_e \bar{\oplus} N_{e+1}) \in \mathcal{Y}$, and that $N_{e+1}, N_e \in \mathcal{X}$. Therefore, the output of the above computation is not well-defined since it leads to $\Delta^N_e \bar{\oplus} N_{e+1} \notin \mathcal{Y}$ as being the input to $F'_{k_e}(i, \cdot) \in \mathcal{F}'$. Hence, when $\mathcal{A}$ successfully reverts back to the target key k_e, the nonce deviates from N_e (and the domain $\mathcal{Y}$ itself).

Case 2: Reverting to N_e. Given C_{e+1} and Δ_{e+1}, $\mathcal{A}$ can revert back to (i, N_e) as the function input by computing: $C_{e+1} - F'_{\Delta^k_{e+1}}(i, \Delta^N_{e+1}) = F'_{3k_e - 2k_{e+1}}(i, N_e) + m$. By virtue of the almost XOR operation and the operand $\bar{0}$, the only way to revert back to $N_e \in \mathcal{X}$ from $\Delta^N_{e+1}(= N_e \bar{\oplus} N_{e+1}) \in \mathcal{Y}$ is via subtraction. But, as shown above, subtraction leads to $F'_{3k_e - 2k_{e+1}}(i, N_e)$ instead of $F'_{k_e}(i, N_e)$. Hence, the computation that allows $\mathcal{A}$ to successfully revert back to N_e as the function input, also leads the function's key to deviate from k_e.

Considering the aforementioned arguments, it follows from Corollary 4 that $\forall Q \xleftarrow{\$} \mathcal{Z}$, it holds that: $Pr[C_e] = Pr[Q] \pm \epsilon(\lambda)$. $\square$

9   Open Problem: Novel Searchable Encryption Schemes

Searchable symmetric encryption (SSE) [6] allows one to store data at an untrusted server, and later search the data for records (or documents) matching a given keyword. A search pattern [6] is defined as any information that can be derived or inferred about the keywords being searched from the issued search queries. In a setting with multiple servers hosting unique shares of the data (generated via threshold secret sharing [32]), our HVL-KIH-PRF family may be useful in realizing SSE scheme that hides search patterns. For instance, if there are n servers $S_1, S_2, \ldots, S_n$, then n random keys $k_1, k_2, \ldots, k_n$ can be distributed among them in a manner such that any t-out-of-n servers can combine their respective keys to generate $k = \sum_{j=1}^{n} k_j$.

If the search index is generated via our HVL-KIH-PRF function $F'_k(i, \cdot) \in \mathcal{F}'$, where i is a fixed database identifier, then to search for a keyword x, the data owner can generate a unique, random query x_j for each server S_j ($1 \leq j \leq n$).

Similar to the key distribution, the data owner sends the queries to the servers such that any t of them can compute $x = \bigoplus_{j=1}^{n} x_j$. On receiving search query x_j, server S_j uses its key k_j to evaluate $F'_{k_j}(i, x_j)$. If at least t servers reply, the data owner can compute $\sum_{j=1}^{t} F'_{k_j}(i, x_j) = F'_k(i, x)$. Designing a compact search index that does not leak any more information than what is revealed by the PRF evaluation is an interesting open problem, solving which would complete this SSE solution.

References

1. Goldreich, O., Goldwasser, S., Micali, S.: How to construct random functions. J. ACM (JACM) **33**, 792–807 (1986)
2. Naor, M., Pinkas, B., Reingold, O.: Distributed pseudo-random functions and KDCs. In: Stern, J. (ed.) EUROCRYPT 1999. LNCS, vol. 1592, pp. 327–346. Springer, Heidelberg (1999). https://doi.org/10.1007/3-540-48910-X_23
3. Boneh, D., Lewi, K., Montgomery, H., Raghunathan, A.: Key homomorphic PRFs and their applications. In: Canetti, R., Garay, J.A. (eds.) CRYPTO 2013. LNCS, vol. 8042, pp. 410–428. Springer, Heidelberg (2013). https://doi.org/10.1007/978-3-642-40041-4_23
4. Banerjee, A., Peikert, C.: New and improved key-homomorphic pseudorandom functions. In: Garay, J.A., Gennaro, R. (eds.) CRYPTO 2014. LNCS, vol. 8616, pp. 353–370. Springer, Heidelberg (2014). https://doi.org/10.1007/978-3-662-44371-2_20
5. Parra, J.R., Chan, T., Ho, S.-W.: A Noiseless key-homomorphic PRF: application on distributed storage systems. In: Liu, J.K., Steinfeld, R. (eds.) ACISP 2016. LNCS, vol. 9723, pp. 505–513. Springer, Cham (2016). https://doi.org/10.1007/978-3-319-40367-0_34
6. Curtmola, R., Garay, J., Kamara, S., Ostrovsky, R.: Searchable symmetric encryption: improved definitions and efficient constructions. In: CCS 2006, pp. 79–88 (2006)
7. Kiayias, A., Papadopoulos, S., Triandopoulos, N., Zacharias, T.: Delegatable pseudorandom functions and applications. In: CCS 2013, pp. 669–684 (2013)
8. Boneh, D., Waters, B.: Constrained pseudorandom functions and their applications. In: Sako, K., Sarkar, P. (eds.) ASIACRYPT 2013. LNCS, vol. 8270, pp. 280–300. Springer, Heidelberg (2013). https://doi.org/10.1007/978-3-642-42045-0_15
9. Boyle, E., Goldwasser, S., Ivan, I.: Functional signatures and pseudorandom functions. In: Krawczyk, H. (ed.) PKC 2014. LNCS, vol. 8383, pp. 501–519. Springer, Heidelberg (2014). https://doi.org/10.1007/978-3-642-54631-0_29
10. Banerjee, A., Fuchsbauer, G., Peikert, C., Pietrzak, K., Stevens, S.: Key-homomorphic constrained pseudorandom functions. In: Dodis, Y., Nielsen, J.B. (eds.) TCC 2015. LNCS, vol. 9015, pp. 31–60. Springer, Heidelberg (2015). https://doi.org/10.1007/978-3-662-46497-7_2
11. Brakerski, Z., Vaikuntanathan, V.: Constrained key-homomorphic PRFs from standard lattice assumptions. In: Dodis, Y., Nielsen, J.B. (eds.) TCC 2015. LNCS, vol. 9015, pp. 1–30. Springer, Heidelberg (2015). https://doi.org/10.1007/978-3-662-46497-7_1

12. Bellare, M., Canetti, R., Krawczyk, H.: Pseudorandom functions revisited: the cascade construction and its concrete security. In: FOCS 1996, pp. 514–523 (1996)

13. Bellare, M., Rogaway, P.: On the construction of variable-input-length ciphers. In: Knudsen, L. (ed.) FSE 1999. LNCS, vol. 1636, pp. 231–244. Springer, Heidelberg (1999). https://doi.org/10.1007/3-540-48519-8_17

14. Protecting data using client-side encryption. http://docs.aws.amazon.com/AmazonS3/latest/dev/UsingClientSideEncryption.html

15. Managing data encryption. http://cloud.google.com/storage/docs/encryption/

16. Everspaugh, A., Paterson, K., Ristenpart, T., Scott, S.: Key rotation for authenticated encryption. In: Katz, J., Shacham, H. (eds.) CRYPTO 2017. LNCS, vol. 10403, pp. 98–129. Springer, Cham (2017). https://doi.org/10.1007/978-3-319-63697-9_4

17. Lehmann, A., Tackmann, B.: Updatable encryption with post-compromise security. In: Nielsen, J.B., Rijmen, V. (eds.) EUROCRYPT 2018. LNCS, vol. 10822, pp. 685–716. Springer, Cham (2018). https://doi.org/10.1007/978-3-319-78372-7_22

18. Shor, P.W.: Algorithms for quantum computation: discrete logarithms and factoring. In: FOCS 1994, pp. 124–134 (1994)

19. Regev, O.: On lattices, learning with errors, random linear codes, and cryptography. In: STOC 2005, pp. 84–93 (2005)

20. Peikert, C.: Public-key cryptosystems from the worst-case shortest vector problem. In: STOC 2009, pp. 333–342 (2009)

21. Gentry, C., Peikert, C., Vaikuntanathan, V.: Trapdoors for hard lattices and new cryptographic constructions. In: STOC 2008, pp. 197–206 (2008)

22. Akavia, A., Goldwasser, S., Vaikuntanathan, V.: Simultaneous hardcore bits and cryptography against memory attacks. In: Reingold, O. (ed.) TCC 2009. LNCS, vol. 5444, pp. 474–495. Springer, Heidelberg (2009). https://doi.org/10.1007/978-3-642-00457-5_28

23. Lyubashevsky, V., Peikert, C., Regev, O.: On ideal lattices and learning with errors over rings. In: Gilbert, H. (ed.) EUROCRYPT 2010. LNCS, vol. 6110, pp. 1–23. Springer, Heidelberg (2010). https://doi.org/10.1007/978-3-642-13190-5_1

24. Agrawal, S., Freeman, D.M., Vaikuntanathan, V.: Functional encryption for inner product predicates from learning with errors. In: Lee, D.H., Wang, X. (eds.) ASIACRYPT 2011. LNCS, vol. 7073, pp. 21–40. Springer, Heidelberg (2011). https://doi.org/10.1007/978-3-642-25385-0_2

25. Brakerski, Z., Vaikuntanathan, V.: Fully homomorphic encryption from ring-LWE and security for key dependent messages. In: Rogaway, P. (ed.) CRYPTO 2011. LNCS, vol. 6841, pp. 505–524. Springer, Heidelberg (2011). https://doi.org/10.1007/978-3-642-22792-9_29

26. Goldwasser, S., Kalai, Y., Popa, R.A., Vaikuntanathan, V., Zeldovich, N.: Reusable garbled circuits and succinct functional encryption. In: STOC 2013, pp. 555–564 (2013)

27. Banerjee, A., Peikert, C., Rosen, A.: Pseudorandom functions and lattices. In: Pointcheval, D., Johansson, T. (eds.) EUROCRYPT 2012. LNCS, vol. 7237, pp. 719–737. Springer, Heidelberg (2012). https://doi.org/10.1007/978-3-642-29011-4_42

28. Alwen, J., Krenn, S., Pietrzak, K., Wichs, D.: Learning with rounding, revisited. In: Canetti, R., Garay, J.A. (eds.) CRYPTO 2013. LNCS, vol. 8042, pp. 57–74. Springer, Heidelberg (2013). https://doi.org/10.1007/978-3-642-40041-4_4

29. Bogdanov, A., Guo, S., Masny, D., Richelson, S., Rosen, A.: On the hardness of learning with rounding over small modulus. In: Kushilevitz, E., Malkin, T. (eds.)

TCC 2016. LNCS, vol. 9562, pp. 209–224. Springer, Heidelberg (2016). https:// doi.org/10.1007/978-3-662-49096-9_9

30. Pietrzak, K.: Subspace LWE. In: Cramer, R. (ed.) TCC 2012. LNCS, vol. 7194, pp. 548–563. Springer, Heidelberg (2012). https://doi.org/10.1007/978-3-642-28914-9_31

31. Brakerski, Z., Langlois, A., Peikert, C., Regev, O., Stehle, D.: Classical hardness of learning with errors. In: STOC 2013, pp. 575–584 (2013)

32. Shamir, A.: How to share a secret. Commun. ACM **22**, 612–613 (1979)

Practical Fully Homomorphic Encryption for Fully Masked Neural Networks

Malika Izabachène, Renaud Sirdey, and Martin Zuber[✉]

CEA, LIST, 91191 Gif-sur-Yvette Cedex, France
{malika.izabachene,renaud.sirdey,martin.zuber}@cea.fr

Abstract. Machine learning applications are spreading in many fields and more often than not manipulate private data in order to derive classifications impacting the lives of many individuals. In this context, it becomes important to work on privacy preserving mechanisms associated to different privacy scenarios: protecting the training data, the classification data, the weights of a neural network. In this paper, we study the possibility of using FHE techniques to address the above issues. In particular, we are able to evaluate a neural network where both its topology and its weights as well as the user data it operates on remain sealed in the encrypted domain. We do so by relying on Hopfield neural networks which are much more "FHE friendly" than their feed-forward counterparts. In doing so, we thus also argue the case of considering different (yet existing) Neural Network models better adapted to FHE, in order to more efficiently address real-world applications.The paper is concluded by experimental results on a face recognition application demonstrating the ability of the approach to provide reasonable recognition timings ($\approx 0.6\,$s) on a single standard processor core.

Keywords: Fully Homomorphic Encryption · LWE · GSW · Hopfield neural networks · Face recognition · FHE performances

1 Introduction

A Fully Homomorphic Encryption (FHE) scheme is, ideally, an encryption scheme permeable to any kind of operation on its encrypted data. With any input x and function f, with E the encryption function, we can obtain $E\left(f(x)\right)$ non interactively from $E(x)$. This property is particularly valuable when privacy-preserving computations on remote servers are required. Unfortunately, all FHE known today still imply high computation overheads and tackling real applications euphemistically remain a challenge. Nevertheless, since Gentry's theoretical breakthrough around 2010, there has been steady progress towards more and more efficient (or less and less inefficient) FHE cryptosystems. In recent years, machine learning applications have been wider spread than ever in various domains. In this context, a growing concern is put on data privacy and confidentiality for both sensitive personal information and valuable intellectual

© Springer Nature Switzerland AG 2019
Y. Mu et al. (Eds.): CANS 2019, LNCS 11829, pp. 24–36, 2019.
https://doi.org/10.1007/978-3-030-31578-8_2

property (e.g. models). In this context, Fully Homomorphic Encryption has the potential to provide an interesting counter-measure but its state of the art does not (yet?) allow to directly apply existing FHE schemes to the highly complex real-world machine learning-based systems. While looking for faster and more efficient FHE schemes is of paramount importance, our paper fall into a line of work which addresses the needs of finding new FHE-friendly Machine Learning methods possibly by revisiting old or less conventional tools from that field and to tune existing FHE specifically to running these methods as efficiently as possible in the encrypted domain.

Prior Work. Most of the existing works on the subject have done so with artificial feed-forward neural networks. These kind of neural networks are the most popular in the Machine Learning field and successfully applying FHE on them in an efficient manner remains a challenging task. There have been different ways this has been tried and not always through the use of HE. For instance, in [2,17] and [13] the authors restrict themselves to a set-up in which the training data is shared between two parties and never fully revealed to any one of them, yet allowing the network to be trained with the whole data-set. Feed-forward networks require to have non-linear activation functions, such as a sigmoid, which are challenging to implement efficiently (and exactly) over FHE. Previous works have handled this by either using the sign function instead (as with [3]) or a polynomial function (as with [16]). Another issue with neural networks that previous work has grappled with is how to deal with its (multiplicative) depth. With that respect, [16] provides a leveled scheme that has its parameters grow with the depth of the network. It was improved most recently in [4]. Alternatively, using an efficient bootstrapping algorithm from [5] which was then refined in [3,6] proposes a scheme where the size of the parameters does not depend on the depth of the network hence demonstrating the practical relevance of this new breed of FHE schemes in the context of Machine Learning applications. To the best of our knowledge, what all the previous works on the subject have in common is the fact the network itself is always in the clear domain and that only the data that are aimed to be classified reside in the encrypted domain. What this paper aims for is to encrypt both the weights of a neural network as well as the data to be classified. Yet, doing this for feed forward networks appears very difficult as building on previous works - for instance the most recent work by [3] - to fit these constraints would lead to unreasonable timings. What we choose to do in the face of this difficulty is to solve the same problem by switching to a different kind of neural network: Hopfield networks.

Our Contribution. In this paper, we present a method to classify encrypted objects by means of a fully encrypted neural network with practically relevant timings and accuracy. We specifically present different approaches to homomorphic face recognition using neural networks. It is worth emphasizing that, because of the nature of the network used (it is discrete and uses a sign function as an activation function), no approximations are made in any of the homomorphic computations. Our main contributions are twofold. First, we provide results in the case where encrypted data are classified by a clear network. This method is both quite fast and lightweight as it takes less than ≈0.2 s to perform a face recognition operation with e.g. 110-bit security. Then, we provide the specifications for a second scheme in the case where encrypted data are classified by an encrypted network. We introduce a method that uses FHE to allow for an arbitrary number of rounds with parameters that are depth-invariant. It classifies in less than 0.6 s with 80-bit security. This paper is organized as follows. We start by reviewing the challenges that combining neural networks and fully homomorphic encryption presents. Some of those challenges have already been tackled in part in the recent literature and some have yet to be. For self-containedness, we then go into the basics of discrete Hopfield Networks at least to the extent that we use them in this paper. This is then followed by a brief presentation of the LWE encryption scheme as well as of TFHE, the LWE-based FHE scheme and library that we use. Then, we present the building blocks that make the bulk of our contribution: several encoding, encryption and computational methods designed for different types of computations involved in various flavors of Hopfield network homomorphic evaluation. Experimental results are finally provided with an emphasis on the very reasonable timings that we obtain for classification on a masked neural network and other variants, on a face-recognition application.

2   Homomorphic Neural Networks

2.1   Models for Evaluation of Encrypted Neural Networks

There are not many works yet on the subject of FHE applied to neural networks. To the best of our knowledge, all of the previous works on the question have focused on the classification of an encrypted input sent by a client but through the use of a public or clear-domain neural network. However, this is only one of the ways one could wish to apply FHE to neural networks. One could want a number of things to be masked:

Encrypted Training Data. Good-quality (labeled) training data is costly to obtain and some companies make a business out of providing such data. Therefore, one might very understandably want to protect the data one spent time or money acquiring when sending it to a remote server to be used for training.

Encrypted Weights. Training an efficient NN is generally harder than it seems and requires both computation time and know-how. Hence, the owner of a NN might want to avoid for its weights to be leaked or for some entity (e.g. the server homomorphically evaluating the network) to be able to use it without restriction.

Hidden Topology of the NN. Training a NN is as much about the training data and the training algorithm as it is about the topology of the NN (number of hidden layers and neurons per layer for instance) on which will depend a lot of the properties of the NN (too large and it has a tendency to generalize badly; too small and it might not be precise enough in its classification). So, in some cases, there may be an interest in masking the network topology and not only its weights, again w. r. t. the server homomorphically evaluating the network.

Encrypted Classification Data. This is arguably the first sought-after property as it pertains to user privacy and allows the network to run on encrypted data therefore providing a service without access to the actual user information.

On top of finding new, efficient ways to classify over encrypted data, this paper also aims to hide the topology and the weights of the network as well. The way we achieve this is by switching from a feed-forward network - one that was used by all previous works - to a different kind of NN: a Hopfield network (HN).

2.2 Hopfield Networks Basics

Hopfield networks are recursive neural networks with only one layer of neurons, all connected. They were first introduced by Hopfield in 1982 in [10,14]. See also [12] and [8] for further uses in the literature. Let us thus consider a network made up of M neurons with each having a value a_i and weights $w_{i,j}$ between neurons i and j.

During the evaluation of a network, the single layer updates the values it stores at every iteration according to a correlation property. The weight $w_{i,j}$ between two neurons i and j represents the correlation that exists between them. An update of neuron i translates to a weighted sum of all other neuron values with an activation function θ: $a_i = \theta\left(\sum_{j=1}^{n} a_j w_{i,j}\right)$ Therefore every neuron is "drawn" towards the values of the neurons most correlated to it. This correlation is symmetric ($w_{i,j} = w_{j,i}$) and a neuron is not correlated with itself ($w_{i,i} = 0$). In practice what the Hopfield network does naturally is storing a number of patterns by highly correlating the bits of the patterns. The network does not need to globally converge to be useful. If we define certain bits as "classification bits" (for example, in the face recognition experiment of Sect. 4.1 only 3 bits out of 256 are used to classify between up to 8 patterns), then only those bits need to be updated to obtain a first result. With a sign function as an activation function these networks are computationally lighter than their feed-forward counterparts.

The training of the network (determining the weights) requires us to have (as usual for NNs) three sets of data: the training set, the validation set and the testing set. We trained the network used in this work ourselves.

3 LWE Encryption Scheme

In this work, we use the TFHE encryption scheme by Chillotti et al. [5,6] for an homomorphic evaluation of Hopfield network with both encrypted weights and encrypted activation values. We refer to those papers for an in-depth presentation of TFHE and only refer here to what is necessary for the understanding of our work.

Notation. We use $\mathbb{B}$ to denote the set $\{0, 1\}$. We denote the reals $\mathbb{R}$ and use $\mathbb{T}$ to denote the real torus mod 1. The ring $\mathbb{Z}[X]/(X^N + 1)$ is denoted $\mathfrak{R}$ and $\mathbb{B}_N[X]$ is its subset composed of the polynomials with binary coefficients. We write $\mathbb{T}_N[X]$ the quotient $\mathbb{R}[X]/(X^N + 1)$ mod 1 where N is a fixed integer. Vectors are denoted with an arrow as such: $\vec{*}$. Ciphertexts are denoted with boldface letters. In the following, λ is the security parameter, M is the size of the network as presented in Sect. 2.2 and B is the maximum value taken by the weights of the network.

3.1   TFHE Basics

TRLWE is the ring version of LWE introduced in [11] over the torus. It encrypts messages in $\mathbb{T}_N[X]$. TLWE is its "scalar" version and encrypts messages in $\mathbb{T}$. TRGSW is the ring version of the GSW scheme introduced in [7]. A TRGSW ciphertext encrypts messages in $\mathfrak{R}$. We note here that in the rest of the paper, we use bold lower-case letters for TLWE ciphertexts and bold upper-case letters for TRLWE and TRGSW ciphertexts. A TLWE encryption c of μ with standard deviation σ and with key s will be noted as $c \in \text{TLWE}_{s,\sigma}(\mu)$ or $c \in \text{TLWE}(\mu, s, \sigma)$. The same notations are valid for TRLWE and TRGSW encryptions.

Parameters. We have the following parameters involved for TLWE, TRLWE and TRGSW encryption schemes respectively.

TLWE parameters: Given a minimal noise overhead α and a security parameter λ, we derive a minimal key size n.

TRLWE parameters: we call n the key size and we have $n = k \times N$. The fact we use n for both TLWE and TRLWE parameters is not a problem. We see in the next paragraph that through the TRLWE to TLWE extraction process one obtains a TLWE ciphertext with $n = k \times N$. In the rest of the paper we will have $k = 1$.

TRGSW parameters: We have decomposition parameters ℓ, B_g. To define homomorphic multiplication, we decompose a TRGSW ciphertext as a small linear combination of rows of a gadget matrix defined with respect to a basis B_g as a ℓ repeated super-decreasing sequence $(1/B_g, \ldots, 1/B_g^\ell)$. Since we use an approximated decomposition, we have an additional precision parameter ε. We will take $\beta = B_g/2$ and $\varepsilon = \frac{1}{2B_g^l}$.

Known Operations over TFHE Ciphertexts. We present linear algebra computations for homomorphic neural network evaluation. These are all operations presented in the TFHE papers [5,6].

Linear combination over TLWE: $\text{TLWE}^p \rightarrow \text{TLWE}$

For p TLWE ciphertexts c_i of $\mu_i \in \mathbb{T}$, and for p integers δ_i, we can obtain an encryption of $\sum_{i=1}^p \delta_i \mu_i$ through the operation $\sum_{i=1}^p \delta_i \cdot c_i$. We write the operation $\text{TLWEScalarProduct}(c, \delta)$. For $p = 2$ and $\delta_1 = \delta_2 = 1$, we write $c_1 + c_2 = \text{AddTLWE}(c_1, c_2)$. The TRLWE equivalent is AddTRLWE.

Extraction: $\text{TRLWE} \rightarrow \text{TLWE}$

From a TRLWE ciphertext C of $\mu = \sum_{i=0}^{N-1} \mu_i \in \mathbb{T}_N[X]$, it is possible to extract a TLWE ciphertext c of a single coefficient of μ_p at a position $p \in [0, N-1]$. We write this operation $c = \mathsf{SampleExtract}(C, p)$.

External Product: TRLWE × TRGSW → TRLWE
The external product between a TRGSW ciphertext C_1 of $\delta \in \mathbb{Z}_N[X]$ and a TRLWE ciphertext C_2 of $\mu \in \mathbb{T}_N[X]$ produces a TRLWE ciphertext C of the product $\delta \cdot \mu \in \mathbb{T}_N[X]$. We write $C = \mathsf{ExternalProduct}(C_2, C_1)$.

Public Rotation: TRLWE → TRLWE
Given a TRLWE ciphertext C of μ and an integer p we can obtain a TRLWE ciphertext C' of $\mu \times X^p$ at no cost to the variance of the ciphertext. We write $C' = \mathsf{RotateTRLWE}(C, p)$.

Key-switch: TLWE → TLWE
We give a TLWE ciphertext $c \in \mathsf{TLWE}(\mu, s)$. For a given TLWE key s and two integers base and t. Given $\mathsf{KS}_{i,j} \in \mathsf{TLWE}(s_i/\mathsf{base}^j, s', \gamma)$ for $j \in [1, t]$ and $i \in [1, n]$ and with γ a given standard deviation. The key-switching procedure outputs $c \in \mathsf{TLWE}_{s'}(\mu)$. We write $\mathsf{KS}_{s \to s'} = (\mathsf{KS}_{i,j})_{(i,j) \in [1,n] \times [1,t]}$ and we call it the key-switching key.

Public Functional key-switch: TLWE^p → TRLWE
We give p TLWE ciphertext $c_i \in \mathsf{TLWE}(\mu_i, s)$, and a public R-lipschitzian morphism $f : \mathbb{T}^p \mapsto \mathbb{T}_N[X]$ of $\mathbb{Z}$-modules. For a given TRLWE key s' and two integers base and t, we take $\mathsf{KS}_{i,j} \in \mathsf{TRLWE}(s_i/\mathsf{base}^j, s', \gamma)$ for $j \in [1, t]$ and $i \in [1, n]$ and with γ a given standard deviation. The functional keyswitching outputs $C \in \mathsf{TRLWE}_{s'}(f(\mu_1, \ldots, \mu_p))$. We write $\mathsf{KS}_{s \to s'} = (\mathsf{KS}_{i,j})_{(i,j) \in [1,n] \times [1,t]}$ and we call it the key-switching key. We exclusively use a very specific function in this paper. We will call it the identity function and refer to the key-switch operation as $\mathsf{Keyswitch}_{\mathsf{Id}}$ It is defined for any $t \in \mathbb{T}: t \mapsto t \cdot X^0$.

Bootstrapping: TLWE → TLWE
Given an integer S, a TLWE ciphertext $c \in \mathsf{TLWE}(\mu, s)$, n TRGSW ciphertexts $\mathsf{BK}_i \in \mathsf{TRGSW}(s_i, s', \alpha_b)$ where the s_i are coefficients of the equivalent TLWE key, we can obtain a ciphertext $c_o \in \mathsf{TLWE}(\mu_o, s', \alpha_{\mathsf{boot}})$ where $\mu_o = 1/S$ if $\mu \in [0, \frac{1}{2}]$ and $\mu_o = -1/S$ if $\mu \in [\frac{1}{2}, 1]$. Most importantly, the output standard deviation α_{boot} is fixed by the parameters of the bootstrapping key BK.

Homomorphic Operations and Variance Overhead. Table 1 summarizes the elementary operations we use in this paper, and the noise propagation they induce. We set several constants which depend on the parameters of the underlying encryption schemes. We set δ to be a plaintext integer encrypted as a TRGSW ciphertext in the external product operation.

Data Encoding. There are two types of data we want to encrypt: the activation values a_i which are equal to 1 or -1 and the weight values $w_{i,j} \in [-B, B]$ where $B \in \mathbb{Z}$. For TRLWE and TLWE encryption, we encode integers into torus values as follows: we divide the torus into S slices and encode any $x \in \mathbb{Z}/S\mathbb{Z}$

Table 1. Elementary operations and their variance overhead The ϑ_cs, the ϑ_Cs and the ϑ_Cs are the noise variances of their respective ciphertexts. ϑ_{KS} is the variance for the encryption of the key-switching key KS and ϑ_{BK} for the bootstrapping key BK. In the case of the external product, we write δ the message encrypted by the input TRGSW ciphertext C_2. Furthermore, $B = 2\ell N\beta^2$, $C = \epsilon^2(1+N) \cdot ||\delta||_2^2$, $D = ||\delta||_2^2$. As for the bootstrapping, the output variance overhead is given by: $\vartheta_{\mathrm{boot}} = 4Nn\ell\beta^2 \times \vartheta_{\mathsf{BK}} + n(1+N)\epsilon^2$.

Operation	Variance
$\mathsf{TLWEScalarProduct}(c, \delta)$	$||\delta||_2^2 \cdot \vartheta_c$
$\mathsf{AddTLWE}(c_1, c_2)$	$\vartheta_1 + \vartheta_2$
$\mathsf{AddTRLWE}(C_1, C_2)$	$\vartheta_1 + \vartheta_2$
$\mathsf{SampleExtract}(C, j)$	ϑ_C
$\mathsf{Keyswitch}(c)$	$\vartheta_c + nt\vartheta_{\mathsf{KS}} + n\mathsf{base}^{-2(t+1)}$
$\mathsf{Keyswitch}_{\mathrm{Id}}(c), p = 1$	$\vartheta_c + nt\vartheta_{\mathsf{KS}} + n\mathsf{base}^{-2(t+1)}$
$\mathsf{ExternalProduct}(C_1, C_2)$	$B\vartheta_2 + C + D\vartheta_1$
$\mathsf{Bootstrapping}$	$\vartheta_{\mathrm{boot}}$

as its corresponding slice in the torus representation. In other words, we have $\frac{x}{S}\mathrm{mod}1 \in \mathbb{T}$. In this paper, we only encode activation values in the torus and weights stay integers. The bigger the slice S is, the more error propagation can be allowed before the error makes the decryption overflow into the adjacent slice. Choosing to reduce the number of slices in order to relax the constraints on the parameters while still ensuring a correct output is a choice made also in [3]. Heuristically, we found that $S \geq M$ is appropriate for our case. And in effect we will choose $S = M$. Finally, we will call ϑ_{max} be the maximum ciphertext variance that still allows the message to be deciphered correctly. It depends only on S.

3.2   Encrypted Inputs Through a Clear Network

In this paper, we present several methods for classifying encrypted inputs via a Hopfield Network. In this section, we present the case where the network is not encrypted, but the input activation values are. We encrypt the activation values a_i as TLWE ciphertexts $c \in \mathsf{TLWE}(\frac{1+2a_i}{2S})$. Now, in order to update the activation value of one neuron, we need to compute a scalar product and take the sign of the result and then insert it back into the network for further computations. For a given neuron p, this corresponds to the following computation: $a_p = \mathsf{sign}\left(\sum_{i=1}^{M} a_i w_{p,i}\right)$ The TLWE ciphertexts $c_i \in \mathsf{TLWE}(a_i)$ are grouped in a vector $\vec{c}$ and the weights $w_{i,p}$ are in clear. Assuming we want to update the pth neuron, we have an algorithm depicted in Fig. 1 below. Then our scheme consists of applying this algorithm on a given number of activation values to update them. Once the number has been reached the resulting encrypted activation values can be sent back to the owner of their key.

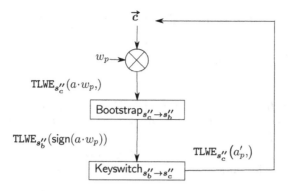

Fig. 1. This figure illustrates the "clear-case" update algorithm. The output of the Keyswitch operation is reincorporated into the vector of ciphertexts as its pth component.

If we choose the parameters of the scheme and ϑ_c such that

$$MB^2 \left(\vartheta_{\text{boot}} + nt\vartheta_{\text{KS}} + n\text{base}^{-2(t+1)} \right) \leq \vartheta_{\max} \quad \text{and} \quad \vartheta_c = \frac{\vartheta_{\max}}{MB^2}$$

then our computation is fully homomorphic.

Scheme. We now present a scheme based on this algorithm. For this we define three actors:

- "the network owner" (or "owner" for short). She trained a Hopfield network on a given type of formatted data and owns the resulting network.
- "the cloud". It has the computational ability to perform tasks with the owner's network at a low cost, making it interesting for other actors to use its services.
- "the client". She owns formatted data compatible with the owner's network and which she would like to run through the Hopfield network.

We present here the case where **the owner** does not encrypt the network. She therefore shares it with the cloud which has access to the weights in clear. She also shares instructions for the updates to perform for a classification. Furthermore, she has to determine the parameters to use for encryption both for activation values and the weights. She shares these parameters with the client.

The client does want to protect her data. She chooses a TLWE key, a TRLWE key and creates a bootstrapping key BK and a key-switching key KS. Finally she encrypts her data and the ciphertexts are sent, along with BK and KS, to the cloud.

The cloud then performs the necessary updates, outputs an updated ciphertext and sends it back to the client.

The client decrypts and has the result to the classification problem.

3.3 Encrypted Inputs Through an Encrypted Network

We now consider the case where both the weights and the activation values are encrypted. We have $\vec{a} = (a_0, \ldots, a_{M-1})$ and $\vec{w_p} = (w_{0,p}, \ldots, w_{M-1,p})$ for every p.

Encryption Methodology. We use the TRLWE and TRGSW encryption schemes to encrypt the activation values and the weights respectively. We are going to use the same key for both: s_c.

We constrain $N \geq M$. We first encode the data as two $(N-1)$-degree polynomials (with 0 coefficients to pad if need be):

$$\forall p \in [0, M-1], \quad W_p = \sum_{i=0}^{N-1} w_{i,p} \cdot X^{N-1-i}, \quad A = \sum_{i=0}^{N-1} a_i \cdot X^i \in \mathbb{T}_N[X]$$

We then encrypt both the weight polynomials and the activation polynomial respectively as TRGSW and TRLWE ciphertexts:

$$C_w^{(p)} \in \text{TRGSW}(W_p, s_c, \alpha_c) \quad \text{and} \quad C_a \in \text{TRLWE}(A, s_c, \alpha_c)$$

These are both encrypted using the same standard deviation α_c. We also have M TRLWE ciphertext for each individual activation value:

$$\forall p \in [0, M-1], \quad C_p \in \text{TRLWE}(a_p X^p, s_c, \alpha_c')$$

We group them in a vector of ciphertexts $\vec{C} = (C_0, \ldots, C_{M-1})$.

Update Algorithm. Figure 2 presents an update of the pth activation value for a given p. Note that the Keyswitch operation hides a rotation operation that transforms $\text{TRLWE}(a_p', s_c)$ into $\text{TRLWE}(a_p' X^p, s_c)$ at no cost to the error propagation and virtually no time cost. Keeping TRLWE ciphertexts of individual activation values (the C_i ciphertexts) allows us to rebuild a new and updated C_a ciphertext by summing them all up at the end.

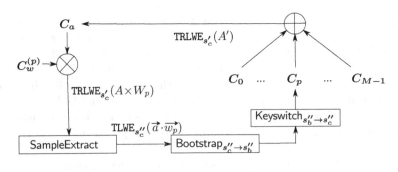

Fig. 2. This figure illustrates the update of the pth neuron in the "masked case".

Correctness. There are two sets of parameters defined both for the initial ciphertexts and for the bootstrapping key BK. We will therefore use the notation $*_c$ for the initial ciphertext parameters and $*_b$ for the bootstrapping key parameters. As for the key-switching key KS, we will use the notation $*_s$. We have $N_s = N_c$. This can work because all of the input ciphertexts have the same parameters with only one exception: the variance of the C_p ciphertext which is not equal to ϑ_c; we will refer to it by ϑ'_c. For this next theorem, we set some constants in order to simplify notations. We set $\mathbf{A} = 4N_cN_bl_b\beta_b^2$; $\mathbf{B} = N_c(1+N_b)\epsilon_b^2$; $\mathbf{C} = N_ct$; $\mathbf{D} = N_c\mathsf{base}^{-2(t+1)}$; $\mathbf{E} = MB^2$; $\mathbf{F} = \epsilon_c^2(1+N_c)$ and $\mathbf{G} = 2\ell_cN_c\beta_i^2$.

Then, if we choose the parameters of the scheme and ϑ_c, ϑ'_c such that

$$M(\mathbf{G} + \mathbf{E}) \times (\mathbf{A}\vartheta_{\mathsf{BK}} + \mathbf{B} + \mathbf{C}\vartheta_{\mathsf{KS}} + \mathbf{D}) + \mathbf{F} \times \mathbf{E} \leq \vartheta_{\max} \qquad (1)$$

$$\vartheta_c \leq \frac{\vartheta_{\max} - \mathbf{F} \times \mathbf{E}}{\mathbf{G} + \mathbf{E}} \quad \text{and} \quad \vartheta'_c = \mathbf{A}\vartheta_{\mathsf{BK}} + \mathbf{B} + \mathbf{C}\vartheta_{\mathsf{KS}} + \mathbf{D} \qquad (2)$$

then our computation is fully homomorphic.

Scheme. This scheme is identical to the one presented in Sect. 3.2 with the only difference that the owner and the client share the same key s_c. This means collusion between the client and the cloud gives them access to the owner's network. The same goes for collusion between the owner and the cloud. There are ways around this but none is perfect. One way is to introduce a Trusted Third Party (TTP).

4   Experimental Results

4.1   Face Recognition via Hopfield Networks

Although their applicability is less universal than CNN, Hopfield networks are known to perform well for certain tasks including image analysis and recognition. In this section, we provide experimental results in the case of a face recognition function. We used a free database available for research purposes. It consists of pictures of 153 individuals - male and female alike - with 20 pictures per individual with different facial expressions on each person. The database can be downloaded from here: https://cswww.essex.ac.uk/mv/allfaces/faces94.html. The feature extraction for the images was done using the LTP (Local Ternary Patterns) introduced in [15] and from the Java library Openimaj (see [9]). From a face image, the feature extraction gives us 128 80 × 80 matrices of raw binary data that we use for the classification. For this test we selected 8 random faces. By training the network on those patterns we wish to be able to classify a picture of a person as one of those 8 faces. For this we apply one iteration of classification on 3 neurons of a 256-neuron network. Empirically, the network stabilizes after the first iteration, hence we do not find better classification results with more iterations. We find that reducing the number of neurons to 256 maximizes the time/performance ratio, giving us an 86.2% classification performance (however we are confident that better results could be obtained with deeper machine learning know-how).

4.2 Parameter Choices

Choosing the parameters, we have to take into account the security constraints from the underlying LWE problem on top of the correctness constraints for the two schemes we present. The overall security of the scheme relies entirely on the security of the underlying TFHE scheme and therefore the LWE problem on which it is built. For a given size of the key and a given security parameter, we can obtain a minimum variance using the `lwe-estimator`[1] script. The estimator is based on the work presented in [1]. The structural parameters of the network are $M = 256$ and $B = 8$. We have $S = M = 256$.

Parameters for the Clear Hopfield Network. For both an 80-bit security and a 110-bit security, the parameters in Table 2 are appropriate. However we need to choose different values for n. We set $n = 470$ for an 80-bit security; And $n = 650$ for a 110-bit security.

Table 2. Parameter values for both security settings in the case of a clear network. We have $\alpha_s = \alpha_c$ and k is set to 1.

N	α_b	α_c	Bg_b	ℓ_b	t	base
1024	4.26e-13	2.41e-06	64	4	3	32

Parameters for the Masked Hopfield Network. Table 3 presents the parameters used for the initial ciphertext encryption, the bootstrapping key and the key-switching key. These parameters are valid for an 80-bit security. We were not able to go beyond this security level: the parameter constraints did not hold anymore.

Table 3. Parameter values for initial ciphertext encryption (top left), for the bootstrapping key BK (top right) and for the key-switching key KS (bottom) for an 80-bit security

λ	α_c	α'_c	N_c	Bg_c	ℓ_c	λ	α_b	N_b	Bg_b	ℓ_b
80	4.26e-13	4.26e-13	1024	64	4	80	8.88e-16	2048	256	5

λ	α_s	N_s	t	base
80	4.26e-13	1024	4	128

We provide the sizes of the ciphertexts associated with these parameters. The size of a single initial TRLWE ciphertext is 16.5 KBytes and the size of the initial TRGSW ciphertext 132 KBytes. The size of the Bootstrapping key is 337 MBytes

---

[1] https://bitbucket.org/malb/lwe-estimator/raw/HEAD/estimator.py.

and the size of the Key-switching key 135 MBytes. The total size of the encrypted network is 33.7 MBytes and the total size of the encrypted activation values 4.22 MBytes.

4.3 Network Performance

In this subsection, we present the timing results for our classification. All timings are measured on an Intel Core i7-6600U CPU. The performances are given in Table 4. Overall we can see that a 1-iteration classification does not take more than 0.21 s in the clear case and under 0.6 s for a fully masked, fully homomorphic, 1-iteration classification.

Table 4. Algorithm timings (in seconds) for the clear-case classification implementation (top) and for the masked case (bottom). As a reminder, a classification requires 3 updates.

TLWEScalarProduct	Bootstrap	Keyswitch
7.5×10^{-7}	0.06	0.009
Total time for 1 update		≤ 0.07

ExternalProduct	Bootstrap	Keyswitch	AddTRLWE
3×10^{-4}	1.6×10^{-1}	1.4×10^{-2}	5×10^{-4}
Total time for 1 update		≤ 0.2	

5 Conclusion

In this paper, we investigated a method to classify encrypted objects by means of a fully encrypted neural network with practically relevant timings and accuracy on a face recognition application, thus achieving both network and data privacy for the first time to the best of our knowledge. We have done so by considering Hopfield networks, a kind of neural network well-known in the Machine Learning community yet more amenable to practical FHE performances than their feed-forward counterparts. As such, this paper also provides insights as to how to intricate an FHE scheme with an algorithm in order to achieve decent FHE execution performances. As a matter of perspective, our next step is to achieve the training of a Hopfield network on an encrypted data set.

References

1. Albrecht, M.R., Player, R., Scott, S.: On the concrete hardness of learning with errors. Cryptology ePrint Archive, Report 2015/046 (2015)
2. Bansal, A., Chen, T., Zhong, S.: Privacy preserving back-propagation neural network learning over arbitrarily partitioned data. Neural Comput. Appl. **20**, 143–150 (2011)

3. Bourse, F., Minelli, M., Minihold, M., Paillier, P.: Fast homomorphic evaluation of deep discretized neural networks. In: Shacham, H., Boldyreva, A. (eds.) CRYPTO 2018. LNCS, vol. 10993, pp. 483–512. Springer, Cham (2018). https://doi.org/10.1007/978-3-319-96878-0_17

4. Chabanne, H., de Wargny, A., Milgram, J., Morel, C., Prouff, E.: Privacy-preserving classification on deep neural network. IACR Cryptology ePrint Archive 2017 (2017)

5. Chillotti, I., Gama, N., Georgieva, M., Izabachène, M.: Faster fully homomorphic encryption: bootstrapping in less than 0.1 seconds. In: Cheon, J.H., Takagi, T. (eds.) ASIACRYPT 2016. LNCS, vol. 10031, pp. 3–33. Springer, Heidelberg (2016). https://doi.org/10.1007/978-3-662-53887-6_1

6. Chillotti, I., Gama, N., Georgieva, M., Izabachène, M.: Improving TFHE: faster packed homomorphic operations and efficient circuit bootstrapping. Cryptology ePrint Archive, Report 2017/430 (2017). https://eprint.iacr.org/2017/430

7. Gentry, C., Sahai, A., Waters, B.: Homomorphic encryption from learning with errors: conceptually-simpler, asymptotically-faster, attribute-based. In: Canetti, R., Garay, J.A. (eds.) CRYPTO 2013. LNCS, vol. 8042, pp. 75–92. Springer, Heidelberg (2013). https://doi.org/10.1007/978-3-642-40041-4_5

8. Gurney, K.: An introduction to neural networks (1997)

9. Hare, J.S., Samangooei, S., Dupplaw, D.P.: OpenIMAJ and ImageTerrier: Java libraries and tools for scalable multimedia analysis and indexing of images. In: Proceedings of the 19th ACM International Conference on Multimedia. ACM (2011)

10. Hopfield, J.J.: Neural networks and physical systems with emergent collective computational abilities. Proc. Natl. Acad. Sci. **79**, 2554–2558 (1982)

11. Lyubashevsky, V., Peikert, C., Regev, O.: On ideal lattices and learning with errors over rings. In: Gilbert, H. (ed.) EUROCRYPT 2010. LNCS, vol. 6110, pp. 1–23. Springer, Heidelberg (2010). https://doi.org/10.1007/978-3-642-13190-5_1

12. MacKay, D.: Information theory, inference, and learning algorithms (2003)

13. Shokri, R., Shmatikov, V.: Privacy-preserving deep learning. In: 53rd Annual Allerton Conference on Communication, Control, and Computing (Allerton) (2015)

14. Sulehria, H.K., Zhang, Y.: Hopfield neural networks: a survey. In: Proceedings of the 6th WSEAS Conference on Artificial Intelligence, Knowledge Engineering and Data Bases. WSEAS (2007)

15. Tan, X., Triggs, B.: Enhanced local texture feature sets for face recognition under difficult lighting conditions. Trans. Image Process. **19**, 1635–1650 (2010)

16. Xie, P., Bilenko, M., Finley, T., Gilad-Bachrach, R., Lauter, K.E., Naehrig, M.: Crypto-Nets: neural networks over encrypted data. CoRR (2014)

17. Yuan, J., Yu, S.: Privacy preserving back-propagation neural network learning made practical with cloud computing. IEEE Trans. Parallel Distrib. Syst. **25**, 212–221 (2014)

SIKE and Hash

SIKE Round 2 Speed Record on ARM Cortex-M4

Hwajeong Seo[1]($\boxtimes$), Amir Jalali[2], and Reza Azarderakhsh[2,3]

[1] IT Department, Hansung University, Seoul, South Korea
hwajeong84@gmail.com
[2] Department of Computer and Electrical Engineering and Computer Science,
Florida Atlantic University, Boca Raton, FL, USA
{ajalali2016,razarderakhsh}@fau.edu
[3] PQSecure Technologies, LLC, Boca Raton, USA

Abstract. We present the first practical software implementation of Supersingular Isogeny Key Encapsulation (SIKE) round 2, targeting NIST's 1, 2, and 5 security levels on 32-bit ARM Cortex-M4 microcontrollers. The proposed library introduces a new speed record of SIKE protocol on the target platform. We achieved this record by adopting several state-of-the-art engineering techniques as well as highly-optimized hand-crafted assembly implementation of finite field arithmetic. In particular, we carefully redesign the previous optimized implementations of filed arithmetic on 32-bit ARM Cortex-M4 platform and propose a set of novel techniques which are explicitly suitable for SIKE/SIDH primes. Moreover, the proposed arithmetic implementations are fully scalable to larger bit-length integers and can be adopted over different security levels. The benchmark result on STM32F4 Discovery board equipped with 32-bit ARM Cortex-M4 microcontrollers shows that the entire key encapsulation over p434 takes about 326 million clock cycles (i.e. 1.94 s @168 MHz). In contrast to the previous optimized implementation of the isogeny-based key exchange on low-power 32-bit ARM Cortex-M4, our performance evaluation shows feasibility of using SIKE mechanism on the target platform. In comparison to the most of the post-quantum candidates, SIKE requires an excessive number of arithmetic operations, resulting in significantly slower timings. However, its small key size makes this scheme as a promising candidate on low-end microcontrollers in the quantum era by ensuring the lower energy consumption for key transmission than other schemes.

Keywords: Post-quantum cryptography · SIDH · SIKE · Montgomery multiplication · ARM Cortex-M4

1 Introduction

The hard problems of traditional PKC (e.g. RSA and ECC) can be easily solved by using Shor's algorithm [26] and its variant on a quantum computer. The traditional PKC approaches cannot be secure anymore against quantum attacks. A

© Springer Nature Switzerland AG 2019
Y. Mu et al. (Eds.): CANS 2019, LNCS 11829, pp. 39–60, 2019.
https://doi.org/10.1007/978-3-030-31578-8_3

number of post-quantum cryptography algorithms have been proposed in order to resolve this problem. Among them, Supersingular Isogeny Diffie-Hellman key exchange (SIDH) protocol proposed by Jao and De Feo is considered as a premier candidate for post-quantum cryptosystems [17]. Its security is believed to be secure even for quantum computers. SIDH is the basis of the Supersingular Isogeny Key Encapsulation (SIKE) protocol [2], which is currently under consideration by the National Institute of Standards and Technology (NIST) for inclusion in a future standard for post-quantum cryptography [27]. One of the attractive features of SIDH and SIKE is their relatively small public keys which are, to date, the most compact ones among well-established quantum-resistant algorithms. In spite of this prominent advantage, the "slow" speed of these protocols has been a sticking point which hinders them from acting like the post-quantum cryptography. Therefore, speeding up SIDH and SIKE has become a critical issue as it judges the practicality of these isogeny-based cryptographic schemes. In CANS'16, Koziel et al. presented first SIDH implementations on 32-bit ARM Cortex-A processors [22]. In 2017, Jalali et al. presented first SIDH implementations on 64-bit ARM Cortex-A processors [16]. In CHES'18, Seo et al. improved previous SIDH and SIKE implementations on high-end 32/64-bit ARM Cortex-A processors [25]. At the same time, the implementations of SIDH on Intel and FPGA are also successfully evaluated [3,10,19,21]. Afterward, in 2018, first implementation of SIDH on low-end 32-bit ARM Cortex-M4 microcontroller was suggested [20]. The paper shows that an ephemeral key exchange (i.e. SIDHp751) on a 32-bit ARM Cortex-M4@120 MHz requires 18.833 s to perform - too slow to use on low-end microcontrollers.

In this work, we challenge to the practicality of SIKE round 2 protocols for NIST PQC competition (i.e. SIKEp434, SIKEp503, and SIKEp751) on low-end microcontrollers. We present new optimized implementation of modular arithmetic for the case of low-end 32-bit ARM Cortex-M4 microcontroller. The proposed modular arithmetic, which is implemented on top of the SIKE round 2 reference implementation [1], demonstrates that the supersingular isogeny-based protocols are practical on 32-bit ARM Cortex-M4 microcontrollers.

2   Optimized SIKE/SIDH Arithmetic on ARM Cortex-M4

2.1   Multiprecision Multiplication

In this work, we describe the multi-precision multiplication method in multiplication structure and rhombus form.

Figures 1, 2, and 3 illustrate different strategies for implementing 256-bit multiplication on 32-bit ARM Cortex-M4 microcontroller. Let A and B be operands of length m bits each. Each operand is written as $A = (A[n-1], ..., A[1], A[0])$ and $B = (B[n-1], ..., B[1], B[0])$, where $n = \lceil m/w \rceil$ is the number of words to represent operands, and w is the computer word size (i.e. 32-bit). The result $C = A \cdot B$ is represented as $C = (C[2n-1], ..., C[1], C[0])$. In the rhombus form,

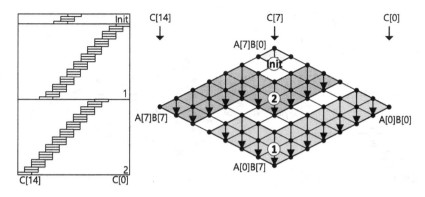

Fig. 1. 256-bit Operand Caching multiplication at the word-level where e is 3 on ARM Cortex-M4 [11], Init: initial block; ① → ②: order of rows.

the lowest indices ($i, j = 0$) of the product appear at the rightmost corner, whereas the highest indices ($i, j = n - 1$) appear at the leftmost corner. A black arrow over a point indicates the processing of a partial product. The lowermost points represent the results $C[i]$ from the rightmost corner ($i = 0$) to the leftmost corner ($i = 2n - 1$).

There are several works in the literature that studied the use of UMAAL instructions to implement multi-precision multiplication or modular multiplication on 32-bit ARM Cortex-M4 microcontrollers [8,9,11,13,20,23]. Among them, Fujii et al. [11], Haase et al. [13], and Koppermann et al. [20] provided the most relevant optimized implementations to this work, targeting Curve25519 and SIDHp751 by using optimal modular multiplication and squaring methods.

In [11], authors combine the UMAAL instruction with (Consecutive) Operand Caching (OC) method for Curve25519 (i.e. 256-bit multiplication). The UMAAL instruction handles the carry propagation without additional costs in Multiplication ACcumulation (MAC) routine. The detailed descriptions are given in Fig. 1. The size of operand caching is 3, which needs three rows ($3 = \lceil 8/3 \rceil$) for 256-bit multiplication on 32-bit ARM Cortex-M4. The multiplication starts from initial block and performs rows 1 and 2, sequentially. The inner loop follows column-wise (i.e. Product-Scanning) multiplication.

In [13], a highly-optimized usage of registers and the partial products are performed with the Operand Scanning (OS) method, targeting Curve25519 (i.e. 256-bit multiplication). The detailed descriptions are given in Fig. 2. In particular, the order of partial products has an irregular pattern which only works for the target operand length (i.e. 256-bit multiplication) due to the extremely compact utilization of available registers in each partial product. However, for a larger length integer multiplication, this greedy approach is not suitable since the number of register is not enough to cache sufficient operands and intermediate results to achieve the optimal performance.

In [20], authors proposed an implementation of 1-level additive Karatsuba multiplication with Comba method (i.e. Product Scanning) as the underlying

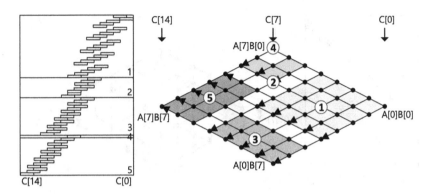

Fig. 2. 256-bit Operand Scanning multiplication at the word-level on ARM Cortex-M4 [13], ① → ② → ③ → ④ → ⑤: order of rows.

multiplication strategy, targeting 768-bit multiplication. They integrated their arithmetic library into SIDHp751 and reported the first optimized implementation of SIDH on ARM Cortex-M4 microcontrollers. However, the product scanning is inefficient with the UMAAL instruction, since all the intermediate results for long integer multiplication cannot be stored into the small number of available registers. In order to improve their results, we studied the performance evaluation of 448/512/768-bit multiplication by replacing the Comba method with OC method, using the 1-level additive/subtractive Karatsuba multiplication. However, we realized that the Karatsuba approach is slower than original OC method with UMAAL instruction for large integer multiplication on Cortex-M4, due to the excessive number of number of addition, subtraction, bit-wise exclusive-or, and loading/storing intermediate results inside Karatsuba method. Furthermore, 32-bit ARM Cortex-M4 microcontroller provides same latency (i.e. 1 clock cycle) for both 32-bit wise unsigned multiplication with double accumulation (i.e. UMAAL) and 32-bit wise unsigned addition (i.e. ADD).

We acknowledge that on low-end devices, such as 8-bit AVR microcontrollers, Karatsuba method is one of the most efficient approaches for multi-precision multiplication. In these platforms, the MAC routine requires at least 5 clock cycles [14]. This significant overhead is efficiently replaced with relatively cheaper 8-bit addition/subtraction operation (i.e. 1 clock cycle). However, UMAAL instruction in ARM Cortex-M4 microcontroller can perform the MAC routine within 1 clock cycle. For this reason, it is hard to find a reasonable trade-off between MAC (i.e. 1 clock cycle) and addition/subtraction (i.e. 1 clock cycle) on the ARM Cortex-M4 microcontroller. Following the above analysis, we adopted the OC method for implementing multiplication in our proposed implementation. Moreover, in order to achieve the most efficient implementation of SIKE protocol on ARM Cortex-M4, we proposed three distinguished improvements to the original method which result in significant performance improvement compared to previous works. We describe these techniques in the following.

Table 1. Comparison of multiplication methods, in terms of memory-access complexity. The parameter d defines the number of rows within a processed block.

Method	Load	Store
Operand Scanning	$2n^2 + n$	$n^2 + n$
Product Scanning [5]	$2n^2$	$2n$
Hybrid Scanning [12]	$2\lceil n^2/d \rceil$	$2n$
Operand Caching [15]	$2\lceil n^2/e \rceil$	$\lceil n^2/e \rceil + n$
Refined Operand Caching (This work)	$2\lceil n^2/(e+1) \rceil + 3(\lfloor n/(e+1) \rfloor)$	$\lceil n^2/(e+1) \rceil + n$

Table 2. Comparison of multiplication methods for different Integer sizes, in terms of the number of memory access on 32-bit ARM Cortex-M4 microcontroller. The parameters d and e are set to 2 and 3, respectively.

Method	448-bit			512-bit			768-bit		
	Load	Store	Total	Load	Store	Total	Load	Store	Total
OS	406	210	616	528	272	800	1,176	600	1,776
PS	392	28	420	512	32	544	1,152	48	1,200
HS	196	28	224	256	32	288	576	48	624
OC	132	80	212	172	102	274	384	216	600
R-OC	107	63	170	140	80	220	306	168	474

Efficient Register Utilization. The OC method follows the product-scanning approach for inner loop but it divides the calculation (i.e. outer loop) into several rows [15]. The number of rows directly affects the overall performance, since the OC method requires to load the operands and load/store the intermediate results by the number of rows[1]. Table 1 presents the comparison of memory access complexity depending on the multiplication techniques. Our optimized implementation (i.e. Refined Operand Caching) is based on the original OC method but we optimized the available registers and increased the operand caching size from e to $e+1$. In the equation, the number of memory load by $3(\lfloor n/(e+1) \rfloor)$ indicates the operand pointer access in each row.

Moreover, larger bit-length multiplication requires more memory access operations. Table 2 presents the number of memory access operations in OC method for different multi-precision multiplication size. In this table, our proposed R-OC method requires the least number memory access for different length multiplication. In particular, in comparison with original OC implementation, our proposed implementation reduces the total number of memory accesses by 19.8%, 19.7%, and 21% for 448-bit, 512-bit, and 768-bit, respectively[2].

In order to increase the size of operand caching (i.e. e) by 1, we need at least 3 more registers to retain two 32-bit operand limbs and one 32-bit

---

[1] The number of rows is $r = \lfloor n/e \rfloor$, where the number of needed words ($n = \lceil m/w \rceil$), the word size of the processor (w) (i.e. 32-bit), the bit-length of operand (m), and operand caching size (e) are given.

[2] Compared with original OC implementation, we reduce the number of row by 1 ($4 \rightarrow 3$), 2 ($5 \rightarrow 3$), and 2 ($7 \rightarrow 5$) for 448-bit, 512-bit, and 768-bit, respectively.

Table 3. Comparison of register utilization of the proposed method with previous works.

Registers	Fujii et al. [11]	Haase et al. [13]	This work
R0	Result pointer	Temporal pointer	Temporal pointer
R1	Operand A pointer	Operand A #1	Temporal register #1
R2	Operand B pointer	Operand B #1	Operand A #1
R3	Result #1	Operand B #2	Operand A #2
R4	Result #2	Operand B #3	Operand A #3
R5	Result #3	Operand B #4	Operand A #4
R6	Operand A #1	Operand B #5	Operand B #1
R7	Operand A #2	Result #1	Operand B #2
R8	Operand A #3	Result #2	Operand B #3
R9	Operand B #1	Result #3	Operand B #4
R10	Operand B #2	Result #4	Result #1
R11	Operand B #3	Result #5	Result #2
R12	Temporal register #1	Temporal register #1	Result #3
R13; SP	Stack pointer	Stack pointer	Stack pointer
R14; LR	Temporal register #2	Temporal register #2	Result #4
R15; PC	Program counter	Program counter	Program counter

intermediate result value. To this end, we redefine the register assignments inside our implementation. We saved one register for the result pointer by storing the intermediate results into stack. Moreover, we observed that in the OC method, both operand pointers are not used at the same time in the row. Therefore, we don't need to maintain both operand pointers in the registers during the computations. Instead, we store them to the stack and load one by one on demand.

Using the above techniques, we saved three available registers and utilized them to increase the size of operand caching by 1. In particular, three registers are used for operand A, operand B, and intermediate result, respectively. We state that our utilization technique imposes an overhead in memory access for operand pointers. However, since in each row, only three memory accesses are required, the overall overhead is negligible to the obtained performance benefit. We provide a detailed comparison of register assignments of this work with previous implementations in Table 3.

Optimized Front Parts. As it is illustrated in Fig. 3, our R-OC method starts from an initialization block (Init section). In the Init section, both operands are loaded from memory to registers and the partial products are computed. From the row 1, only one operand pointer is required in each column. The front part (i.e. I-F and 1-F) requires partial products by increasing the length of column to 4.

Fujii et al. [11] implemented the front parts using carry-less MAC routines. In their approach, they initialized up to two registers to store the intermediate results in each column. Figure 4 illustrates their approach. Since the UMLAL

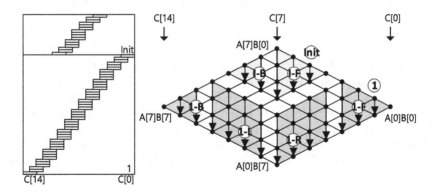

Fig. 3. Proposed 256-bit Refined Operand Caching multiplication at the word-level where e is 4 on ARM Cortex-M4, Ⓘnit: initial block; ①: order of rows; Ⓕ: front part; Ⓡ: middle right part; Ⓛ: middle left part; Ⓑ: back part.

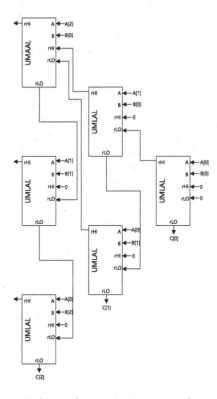

Fig. 4. 3-word integers with the product scanning approach using the UMLAL and UMAAL instructions for front part of OC method [11].

and UMAAL instructions need to update current values inside the registers, the initialized registers are required.

In order to optimize the explicit register initialization, we redesign the front part with product scanning. In contrast to Fujii's approach, we used UMULL and UMAAL instructions. As a result, the register initialization is performed together with unsigned multiplication (i.e. UMULL). This technique improves the overall clock cycles since each instruction directly assigns the results to the target registers. In particular, we are able to remove all the register initialization routines, which is 9 clock cycles for each front part compared to [11]. Moreover, the intermediate results are efficiently handled with carry-less MAC routines by using the UMAAL instructions. Figure 5 presents our 4-word strategy in further details.

Efficient Instruction Ordering. The ARM Cortex-M4 microcontrollers are equipped with 3-stage pipeline in which the instruction fetch, decode, and execution are performed in order. As a result, any data dependency between consecutive instructions imposes pipeline stalls and degrades the overall performance considerably. In addition to the previous optimizations, we reordered the MAC routine instructions in a way which removes data dependency between instructions, resulting in minimum pipeline stalls. The proposed approach is presented in Fig. 5 (1-R section). In this Figure, the operand and intermediate result are loaded from memory and partial products are performed column-wise as follows:

```
    ⋮
LDR    R6, [R0, #4 * 4] //Loading operand B[4] from memory
LDR    R1, [SP, #4 * 4] //Loading intermediate result C[4] from memory
UMAAL R14, R10, R5, R7 //Partial product (B[1]*A[3])
UMAAL R14, R11, R4, R8 //Partial product (B[2]*A[2])
UMAAL R14, R12, R3, R9 //Partial product (B[3]*A[1])
UMAAL R1, R14, R2, R6  //Partial product (B[4]*A[0])
    ⋮
```

The intermediate result ($C[4]$) is loaded to the R1 register. At this point, updating R1 register in the next instruction results in pipeline stall. To avoid this situation, first, we updated the intermediate results into other registers (R10, R11, R12, R14), while R1 register was updated during the last step of MAC. We followed a similar approach in 1-L section, where operand (A) pointer is loaded to a temporary register, and then the column-wise multiplications are performed with the operands ($A[4]$, $A[5]$, $A[6]$, and $A[7]$). In the back part (i.e. 1-B), the remaining partial products are performed without operand loading. This is efficiently performed without carry propagation by using the UMAAL instructions.

To compare the efficiency of our proposed techniques with previous works, we evaluated the performance of our 256-bit multiplication with the most relevant works on Cortex-M4 platform. To obtain a fair and uniform comparison, we

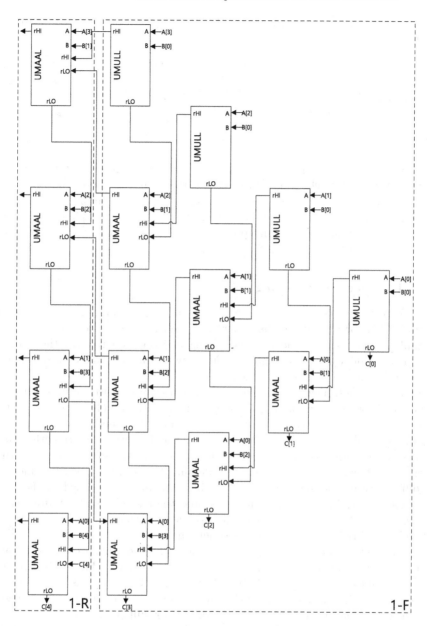

Fig. 5. 4-word integers with the product scanning approach using the UMULL and UMAAL instructions for front part of OC method.

benchmarked the proposed implementations in [11,13][3,4] with our implementation on our development environment.

---

[3] Fujii et al. https://github.com/hayatofujii/curve25519-cortex-m4.
[4] Haase et al. https://github.com/BjoernMHaase/fe25519.

Table 4. Comparison results of 256-bit multiplication on ARM Cortex-M4 microcontrollers.

Methods	Timings [cc]	Scalability	Bit length
Fujii et al. [11]	239	✓	256
Haase et al. [13]	212	✗	256
This work	196	✓	256

Table 4 presents the performance comparison of our library with previous works in terms of clock cycles. We observe that our proposed multiplication implementation method is faster than previous optimized implementation on the same platform. Furthermore, in contrast to the compact implementation of 256-bit multiplication in [13], our approach provides scalability to larger integer multiplication without any significant overhead.

2.2 Multiprecision Squaring

Most of the optimized implementations of cryptography libraries use optimized multiplication for computing the square of an element. However, squaring can be implemented more efficiently since using one operand reduces the overall number of memory accesses by half, while many redundant partial products can be removed (i.e. $A[i] \times A[j] + A[j] \times A[i] = 2 \times A[i] \times A[j]$).

Similar to multiplication, squaring implementation consists of partial products of the input operand limbs. These products can be divided into two parts: the products which have two operands with the same value and the ones in which two different values are multiplied. Computing the first group is straightforward and it is only computed once for each limb of operand. However, computing the latter products with different values and doubling the result can be performed in two different ways: doubled-result and doubled-operand. In doubled-result technique, partial products are computed first and the result is doubled afterwards $(A[i] \times A[j] \rightarrow 2 \times A[i] \times A[j])$, while in doubled-operand, one of the operands is doubled and then multiplied to the other value $(2 \times A[i] \rightarrow 2 \times A[i] \times A[j])$.

In the previous works [11,13], authors adopted the doubled-result technique inside squaring implementation. Figures 6 and 7 show their techniques for implementing optimized squaring on Cortex-M4 platform. The red parts in the figures present the partial products where the input values are the same and the black dots with gray background represent the doubled-result products.

Figure 6 demonstrates Sliding Block Doubling (SBD) based squaring method in [11]. This method is based on the product scanning approach. The squaring consists of two routines: initialization and row 1 computation. The intermediate results are doubled column-wise as the row 1 computations are performed.

Figure 7 presents the Operand Scanning (OS) based squaring method in [13]. In contrast to previous method, computations are performed row-wise. However, the intermediate results are doubled in each column. Note that in this method, the order of computation is designed explicitly for 256-bit operand to maximize

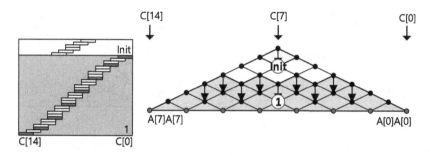

Fig. 6. 256-bit Sliding Block Doubling squaring at the word-level on ARM Cortex-M4, Init: initial block; ①: order of rows [11].

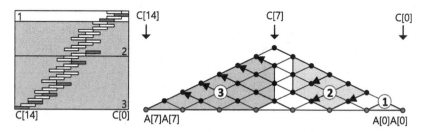

Fig. 7. 256-bit Operand Scanning squaring at the word-level on ARM Cortex-M4, ① → ② → ③: order of rows [13].

the operand caching. Similar to their multiplication implementation, the proposed method does not provide scalability to larger bit-length multiplications.

In this work, we proposed a hybrid approach for implementing a highly-optimized squaring operation which is explicitly suitable for SIKE/SIDH application. In general, doubling operation may result in one bit overflow which requires an extra word to retain. However, in the SIDH/SIKE settings, moduli are smaller than multiple of 32-bit word (434-bit, 503-bit, and 751-bit) which provide an advantage for optimized arithmetic design. Taking advantage of this fact, we designed our squaring implementation based on doubled-operand approach. We divided our implementation into three parts: one sub-multiplication and two sub-squaring operations. We used R-OC for sub-multiplication and SBD for sub-squaring operations. Figure 8 illustrates our hybrid method in detail. First, the input operand is doubled and stored into the stack memory. Taking advantage of doubled-operand technique, we perform the initialization part by using R-OC method.

Second, the remaining rows 1 and 2 are computed based on SBD methods. In contrast to previous SBD method, all the doubling operations on intermediate results are removed during MAC routines. This saves several registers to double the intermediate results since doubled-results have been already computed. Furthermore, our proposed method is fully scalable and can be simply adopted to larger integer squaring.

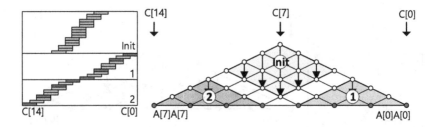

Fig. 8. 255-bit proposed squaring at the word-level on ARM Cortex-M4, Init: initial block; ① → ②: order of rows.

Table 5. Comparison results of 255/256-bit squaring on ARM Cortex-M4 microcontrollers.

Methods	Timings [cc]	Scalability	Bit length
Fujii et al. [11]	218	✓	256
Haase et al. [13]	141	✗	256
This work	136	✓	255

In order to verify the performance improvement of our proposed approach, we benchmarked our 255-bit squaring implementation with the most optimized available implementations in the literature. Table 5 presents the performance comparison of our method with previous implementations on our target platform.

Our hybrid method outperforms previous implementations of 256-bit squaring, while in contrast to [13], it is scalable to larger parameter sets. In particular, it enabled us to implement the same strategy for computing SIKE/SIDH arithmetic over larger finite fields.

2.3 Modular Reduction

Modular multiplication is a performance-critical building block in SIDH and SIKE protocols. One of the most well-known techniques used for its implementation is Montgomery reduction [24]. We adapt the implementation techniques described in Sects. 2.1 and 2.2 to implement modular multiplication and squaring operations. Specifically, we target the parameter sets based on the primes p434, p503, and p751 for SIKE round 2 protocol [1,6]. Montgomery multiplication can be efficiently exploited and further simplified by taking advantage of so-called "Montgomery-friendly" modulus, which admits efficient computations, such as *all-zero* words for lower part of the modulus.

The efficient optimizations for the modulus were first pointed out by Costello et al. [6] in the setting of SIDH when using modulus of the form $2^x \cdot 3^y - 1$ (referred to as "SIDH-friendly" primes) are exploited by the SIDH library [7].

In CHES'18, Seo et al. suggested the variant of Hybrid-Scanning (HS) for "SIDH-friendly" Montgomery reduction on ARM Cortex-A15 [25]. Similar to OC method, the HS method also changes the operand pointer when the row

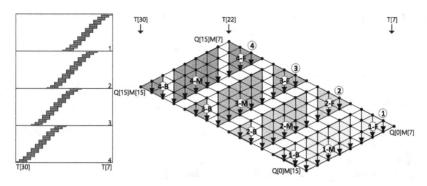

Fig. 9. 503-bit "SIDH-friendly" Montgomery reduction at the word-level, where d is 4 on ARM Cortex-M4, ① → ② → ③ → ④: order of rows; Ⓕ: front part; Ⓜ: middle part; Ⓑ: back part; where M, R, T, and Q are modulus, Montgomery radix, intermediate results, and quotient ($Q \leftarrow T \cdot M'$ mod R).

is changed. By using the register utilization described in Sect. 2.1, we increase the parameter d by 1 (3 → 4. Moreover, the initial block is also optimized to avoid explicit register initialization and the MAC routine is implemented in the pipeline-friendly approach. Compared with integer multiplication, the Montgomery reduction requires fewer number of registers to be reserved. Since the intermediate result pointer and operand Q pointer are identical value (i.e. stack), we only need to maintain one address pointer to access both values. Furthermore, the modulus for SIKE (i.e. operand M; SIKEp434, SIKEp503, and SIKEp751) is a static value. As a result, instead of obtaining values from memory, we assign the direct values to the registers. This step can be performed with the two instructions, such as MOVW and MOVT. The detailed 32-bit value assignment (e.g. 0x87654321) to register R1 is given as follows:

> ⋮
> MOVW R1, #0x4321 //R1 = #0x4321
> MOVT R1, #0x8765 //R1 = #0x8765 ≪ 16 | R1
> ⋮

In Fig. 9, the 503-bit "SIDH-friendly" Montgomery reduction on ARM Cortex-M4 microcontroller is described. The Montgomery reduction starts from row 1, 2, 3, to 4.

In the front of row 1 (i.e. 1-F), the operand Q is loaded from memory and the operand M is directly assigned using constant value. The multiplication accumulates the intermediate results from memory using the operand Q pointer and stored them into the same memory address. In the middle of row 1 (i.e. 1-M), the operand Q is loaded and the intermediate results are also loaded and stored, sequentially. In the back of row 1 (i.e. 1-B), the remaining partial products are computed. Furthermore, the intermediate carry values are stored into stack and used in the following rows.

Table 6. Comparison results of modular multiplication and squaring for SIDH on 32-bit ARM Cortex-M4 microcontrollers.

Methods	Timings [cc]			Modulus	Processor
	$\mathbb{F}_p$ mul	$\mathbb{F}_p$ sqr	reduction		
This work	1,110	981	544	$2^{216} \cdot 3^{137} - 1$	ARM Cortex-M4
SIDH v3.0 [7]	25,399	–	10,917	$2^{250} \cdot 3^{159} - 1$	ARM Cortex-M4
This work	1,333	1,139	654		
Bos et al. [4]	–	–	3,738	$2^{372} \cdot 3^{239} - 1$	ARM Cortex-A8
SIDH v3.0 [7]	55,178	–	23,484		
Koppermann et al. [20]	7,573	–	3,254		ARM Cortex-M4
This work	2,744	2,242	1,188		

Using the above techniques, we are able to reduce the number of row by 1 ($5 \rightarrow 4$), 2 ($6 \rightarrow 4$), and 2 ($8 \rightarrow 6$) for 448-bit, 512-bit, and 768-bit, respectively, compared to original implementation of HS based Montgomery reduction.

Recently, Bos et al. [4] and Koppermann et al. [20] proposed highly optimized techniques for implementation of modular multiplication. They utilized the product-scanning methods for modular reduction. However, our proposed method outperfoms both implementations in terms of clock cycles. In particular, our proposed method provides more than 2 times faster result compared to Bos et al. [4], while the benchmark results in [4] were obtained on the high-end ARMv7 Cortex-A8 processors which is equipped with 15 pipeline stages and is dual-issue super-scalar. Table 6 shows the detailed performance comparison of multiplication, squaring, and reduction over SIDH/SIKE primes in terms of clock cycles. We state that, the benchmark results for [7] are based on optimized C implementation and they are presented solely as a comparison reference between portable and target-specific implementations.

2.4 Modular Addition and Subtraction

Modular addition operation is performed as a long integer addition operation followed by a subtraction from the prime. To have a fully constant-time arithmetic implementation, the final reduction is performed using a masked bit. In this case, even if the addition result is inside the field, a redundant subtraction is performed, so the secret values cannot be retrieved using power and timing attacks. The detailed operations are presented in the following:

- Modular addition: (A+B) mod P
  ① C←A+B   ② {M,C}←C−P   ③ C←C+(P&M).
- Modular subtraction: (A−B) mod P
  ① {M,C}←A−B   ② C←C+(P&M).

Previous optimized implementations of modular addition on Cortex-M4 [20,25], provided the simple masked technique using hand-crafted assembly. However, In this work, we optimized this approach further by introducing three techniques:

- Proposed modular addition: (A+B) mod P
① {M,C}←A+B-P   ② C←C+(P&M).

First, we take advantage of the special shape of SIDH-friendly primes which have multiple words equal to 0xFFFFFFFF. Since this value is the same for multiple limbs, we load it once inside a register and use it for multiple single-precision subtraction. This operand re-using technique reduces the number of memory access by n and $\frac{n}{2}$ for modular addition and modular subtraction, where the number of needed words ($n = \lceil m/w \rceil$), the word size of the processor (w) (i.e. 32-bit), and the bit-length of operand (m) are given, respectively.

Second, we combine Step ① (addition) and ② (subtraction) into one operation ({M,C}←A+B-P). In order to combine both steps, we catch both intermediate carry and borrow, while we perform the combined addition and subtraction operation.

Figure 10 illustrates the proposed technique in details. In this Figure, first, 4-word addition operations ($A[0 \sim 3] + B[0 \sim 3]$) compute the addition result. Subsequently, a single register is set to constant (i.e. 0xFFFFFFFF), which is used for the carry catching step. In Fig. 10, this step is shown in the last row of fourth column. When the carry overflow happens from fourth word addition (i.e. $A[3] + B[3] + CARRY$), the carry catcher register is set to $2^{32} - 1$ (i.e. 0xFFFFFFFF ← 0xFFFFFFFF + 0xFFFFFFFF + 0x00000001) by using the constant (i.e. 0xFFFFFFFF) in last row of fourth column (Constant + Constant + Carry). Otherwise, the carry catcher register is set to $2^{32} - 2$ (i.e. 0xFFFFFFFE ← 0xFFFFFFFF + 0xFFFFFFFF + 0x00000000).

This addition operation stores the carry bit to the first bit of carry catcher register. The carry value in carry catcher register is used for the following addition steps (second column in the Fig. 10).

The stored carry in the first bit is shifted to the 32nd bit by using the barrel-shifter module. Afterward, the value is added to the constant (i.e. 0xFFFFFFFF). If the first bit of carry catcher is set, the carry happens (i.e. 0x00000001≪31 + 0xFFFFFFFF). Otherwise, no carry happens (i.e. 0x00000000≪31 + 0xFFFFFFFF).

Similarly, we obtained the borrow bit. The results of 4-word addition operations ($A[0 \sim 3] + B[0 \sim 3]$) are subtracted by modulus ($P[0 \sim 3]$) in the third column. When the borrow happens from fourth word subtraction (i.e. $A[3]+B[3]-P[3]-BORROW$), the borrow catcher register is set to $2^{32}-1$ (i.e. 0xFFFFFFFF ← 0x00000000 - 0x00000001) in last row of third column (Zero - Borrow). Otherwise, the borrow catcher register is set to 0 (i.e. 0x00000000 ← 0x00000000 - 0x00000000). The borrow bit in borrow catcher register is used for the following subtraction steps. To obtain the borrow bit, the zero constant is subtracted by the borrow catcher register. For one constant register optimization, we used the address pointer instead of zero constant.

Since the address pointer of 32-bit ARM Cortex-M4 microcontroller is aligned by 4-byte (i.e. 32-bit), the address is always ranging from 0 (i.e. 0x00000000) to $2^{32} - 4$ (0xFFFFFFFC). When the borrow catcher register is set, we can get the borrow bit through subtraction (e.g. Pointer - 0xFFFFFFFF where pointer

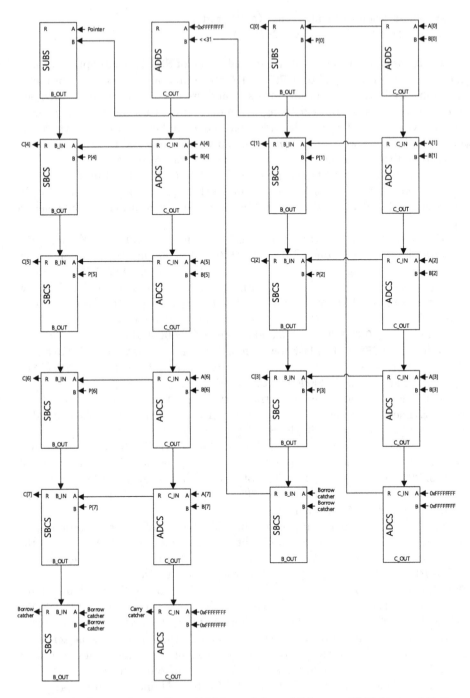

Fig. 10. Initial part of step ① in 512-bit modular addition on ARM Cortex-M4 (i.e. A[0~7]+B[0~7]-P[0~7]).

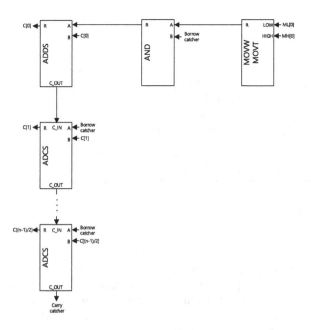

Fig. 11. Initial part of of step ② in 512-bit modular addition/subtraction on ARM Cortex-M4 (i.e. `C[0~(n-1)/2]+(P[0~(n-1)/2]&M)`).

is ranging from 0 to $2^{32} - 4$). Otherwise, no borrow happens. The combined modular addition routine reduces the number of memory access by $2n$ since we can avoi4d both loading and storing the intermediate results.

In addition to the above techniques, the masked addition routine is also optimized. This is shown as Step ② of modular addition and subtraction. When the mask value is set to `0xFFFFFFFF`, the lower part of SIDH modulus is also `0xFFFFFFFF`. Otherwise, both values are set to zero. We optimized the modulus setting (`MOVW/MOVT`) and masking operation (`AND`) for lower part of SIDH modulus. The detailed descriptions for initial part of step ② in 512-bit modular addition/subtraction are given in Fig. 11.

Using the above optimization techniques, we are able to reduce the number of memory access for modular addition and subtraction by $3n$ ($9n \rightarrow 6n$) and $n/2$ ($6n \rightarrow 11n/2$), respectively.

We benchmarked the proposed optimized addition and subtraction implementations on our target platform. We provide the performance evaluation of this work and previous works over different security levels in Table 7. Compared to previous works, the proposed method improved the performance by 16.7% and 14.7% for modular addition and subtraction, respectively.

3   Performance Evaluation

In this section, we present the performance evaluation of our proposed SIDH/SIKE implementations on 32-bit ARM Cortex-M4 microcontrollers. We

Table 7. Comparison results of modular addition and subtraction for SIDH/SIKE on ARM Cortex-M4 microcontrollers.

Methods	Timings [cc]		Modulus	Processor
	$\mathbb{F}_p$ add	$\mathbb{F}_p$ sub		
This work	254	208	$2^{216} \cdot 3^{137} - 1$	ARM Cortex-M4
SIDH v3.0 [7]	1,078	740	$2^{250} \cdot 3^{159} - 1$	ARM Cortex-M4
Seo et al. [25]	326	236		
This work	275	223		
SIDH v3.0 [7]	1,579	1,092	$2^{372} \cdot 3^{239} - 1$	ARM Cortex-M4
Koppermann et al. [20]	559	419		
Seo et al. [25]	466	333		
This work	388	284		

implemented highly-optimized arithmetic, targeting SIKE round 2 primes adapting our optimized techniques for multiplication, squaring, reduction, and addition/subtraction. We integrate our arithmetic libraries to the SIKE round 2 reference implementation [1] to evaluate the feasibility of adopting this scheme on low-end Cortex-M4 microcontrollers.

All the arithmetic is implemented in ARM assembly and the libraries are compiled with GCC with optimization flag set to -O3.[5]

Tables 8 and 9 present the comparison of our proposed library with highly optimized implementations in the literature over different security levels. The optimized C implementation timings by Costello et al. [7] and the reference C implementation of SIKE [1] illustrate the importance of target-specific implementations of SIDH/SIKE low-end microcontrollers such as 32-bit ARM Cortex-M4. In particular, compared to optimized C Comba based implementation in SIDH v3.0, the proposed modular multiplication for 503-bit and 751-bit provide 19.05x and 20.10x improvement, respectively.

The significant achieved performance improvement in this work is the result of our highly-optimized arithmetic library. Specifically, our tailored multiplication minimizes pipeline stalls on ARM Cortex-M4 3-stage pipeline, resulting in remarkable timing improvement compared to previous works.

Moreover, the proposed implementation achieved 362 and 977 million clock cycles for total computation of SIDHp503 and SIDHp751, respectively. The results are improved by 10.51x and 12.97x for SIDHp503 and SIDHp751, respectively. In comparison with the most relevant work, our proposed modular multiplication and SIDHp751 outperforms the optimized implementation in [20] by 2.75x and 4.35x, respectively.

Compared with other NIST PQC round 2 schemes, the SIKE protocol shows slower execution time but the SIKE protocols show the most competitive memory

---

[5] Our library will be publicly available in the near future.

Table 8. Comparison of SIDHp434, SIDHp503, and SIDHp751 protocols on the ARM Cortex-M4 microcontrollers. Timings are reported in terms of clock cycles.

Implementation	Language	Timings [cc]				Timings [cc × 10^6]				
		F_p add	F_p sub	F_p mul	F_p sqr	Alice R1	Bob R1	Alice R2	Bob R2	Total
SIDHp434										
This work	ASM	254	208	1,110	981	65	74	54	62	255
SIDHp503										
SIDH v3.0 [7]	C	1,078	740	25,399	–	986	1,086	812	924	3,808
This work	ASM	275	223	1,333	1,139	95	104	76	87	362
SIDHp751										
SIDH v3.0 [7]	C	1,579	1,092	55,178	–	3,246	3,651	2,669	3,112	12,678
Koppermann et al. [20]	ASM	559	419	7,573	–	1,025	1,148	967	1,112	4,252
This work	ASM	388	284	2,744	2,242	252	284	205	236	977

Table 9. Comparison of NIST PQC round 2 protocols on the ARM Cortex-M4 microcontrollers. Timings are reported in terms of clock cycles. Koppermann et al. [20] does not provide results on SIKE implementations.

Implementation	Language	Timings [cc]				Timings [cc × 10^6]				Memory [bytes]		
		F_p add	F_p sub	F_p mul	F_p sqr	KeyGen	Encaps	Decaps	Total	KeyGen	Encaps	Decaps
SIKEp434												
This work	ASM	254	208	1,110	981	74	122	130	326	6,580	6,916	7,260
SIKEp503												
SIDH v3.0 [7]	C	1,078	740	25,399	–	1,086	1,799	1,912	4,797	–	–	–
This work	ASM	275	223	1,333	1,139	104	172	183	459	6,204	6,588	6,974
SIKEp751												
SIDH v3.0 [7]	C	1,579	1,092	55,178	–	3,651	5,918	6,359	15,928	–	–	–
This work	ASM	388	284	2,744	2,242	282	455	491	1,228	11,116	11,260	11,852
NIST PQC Round 2 [18]												
Frodo640-AES	ASM	–	–	–	–	42	46	47	135	31,116	51,444	61,820
Frodo640-CSHAKE	ASM	–	–	–	–	81	86	87	254	26,272	41,472	51,848
Kyber512	ASM	–	–	–	–	0.7	0.9	1.0	2.6	6,456	9,120	9,928
Kyber768	ASM	–	–	–	–	1.2	1.4	1.4	4.0	10,544	13,720	14,880
Kyber1024	ASM	–	–	–	–	1.7	2.1	2.1	5.9	15,664	19,352	20,864
Newhope1024CCA	ASM	–	–	–	–	1.2	1.9	1.9	5	11,152	17,448	19,648
Saber	ASM	–	–	–	–	0.9	1.2	1.2	3.3	12.616	14,896	15,992

utilization for encapsulation and decapsulation[6]. Furthermore, small key size of SIKE ensures the lower energy consumption for key transmission than other schemes. The low-energy consumption is the most critical requirement for low-end (battery-powered) microcontrollers.

In Table 10, we evaluated the practicality of SIDH protocols on both high-end ARM Cortex-A family of processors and low-end ARM Corex-M4 microcontrollers by measuring the timing in seconds.

The fastest implementations of SIDHp503 on 64-bit ARMv8 Cortex-A53 and Cortex-A72 only require 0.041 s and 0.021 s, respectively. For the case of 32-bit ARMv7 Cortex-A15, the SIDHp751 protocol is performed in 0.157 s. This results emphasize that SIDH protocol is already a practical solution for those "high-end" processors.

---

[6] SIKEp434 requires more memory than SIKEp503 since SIKEp434 allocates more temporal storage than SIKEp503 in Fermat based inversion.

Finally, prior to this work, supersingular isogeny-based cryptography was assumed to be unsuitable to use on low-end devices due to the nonviable performance evaluations [20][7]. However, in contrast to benchmark results in [20], our SIKE and SIDH implementation for NIST's 1, 2, and 5 security levels are practical and can be used in real settings. The proposed implementation of SIDHp434 only requires 0.813 s, which shows that the quantum-resistant key exchange from isogeny of supersingular elliptic curve is a practical solution on low-power microcontrollers.

Table 10. Comparison of SIDH based key exchange protocols on high-end (ARM Cortex-A series) processors and low-end (ARM Cortex-M4) microcontrollers. Timings are reported in terms of seconds.

Protocol	Implementation	Platform	Freq [MHz]	Latency [sec.]		Comm. [bytes]	
				Alice	Bob	A→B	B→A
High-end ARM Processors							
SIDHp503	[22]	32-bit ARMv7 Cortex-A8	1,000	0.216	0.229	378	378
	[22]	32-bit ARMv7 Cortex-A15	2,300	0.064	0.067	378	378
	[25]		2,000	0.042	0.046	378	378
	[2]	64-bit ARMv8 Cortex-A53	1,512	0.061	0.050	378	378
	[25]		1,512	0.050	0.041	378	378
	[2]	64-bit ARMv8 Cortex-A72	1.992	0.030	0.025	378	378
	[25]		1.992	0.025	0.021	378	378
SIDHp751	[22]	32-bit ARMv7 Cortex-A8	1,000	1.406	1.525	564	564
	[22]	32-bit ARMv7 Cortex-A15	2,300	0.340	0.368	564	564
	[25]		2,000	0.135	0.157	564	564
Low-end ARM Microcontrollers							
SIDHp434	This work	32-bit ARMv7 Cortex-M4	168	0.715	0.813	326	326
SIDHp503	This work		168	1.028	1.143	378	378
SIDHp751	[20]		120	16.590	18.833	564	564
	This work		168	2.727	3.099	564	564

Acknowledgement. The authors would like to thank the reviewers for their comments.

This work of Hwajeong Seo was partly supported by the National Research Foundation of Korea (NRF) grant funded by the Korea government (MSIT) (No. NRF-2017R1C1B5075742) and partly supported by the MSIT (Ministry of Science and ICT), Korea, under the ITRC (Information Technology Research Center) support program (IITP-2019-2014-1-00743) supervised by the IITP (Institute for Information & communications Technology Planning & Evaluation).

References

1. Azarderakhsh, R., et al.: Supersingular Isogeny Key Encapsulation - Submission to the NIST's post-quantum cryptography standardization process, round 2 (2019). https://csrc.nist.gov/projects/post-quantum-cryptography/round-2-submissions/SIKE.zip

---

[7] Authors reported 18 s to key exchange on the ARM Cortex-M4 @120 MHz processor.

2. Azarderakhsh, R., et al.: Supersingular Isogeny Key Encapsulation - Submission to the NIST's post-quantum cryptography standardization process (2017). https://csrc.nist.gov/CSRC/media/Projects/Post-Quantum-Cryptography/documents/round-1/submissions/SIKE.zip

3. Bos, J., Friedberger, S.: Arithmetic considerations for isogeny based cryptography. IEEE Trans. Comput. (2018)

4. Bos, J.W., Friedberger, S.: Faster modular arithmetic for isogeny based crypto on embedded devices. IACR Cryptology ePrint Archive 2018:792 (2018)

5. Comba, P.G.: Exponentiation cryptosystems on the IBM PC. IBM Syst. J. **29**(4), 526–538 (1990)

6. Costello, C., Longa, P., Naehrig, M.: Efficient algorithms for supersingular isogeny Diffie-Hellman. In: Robshaw, M., Katz, J. (eds.) CRYPTO 2016. LNCS, vol. 9814, pp. 572–601. Springer, Heidelberg (2016). https://doi.org/10.1007/978-3-662-53018-4_21

7. Costello, C., Longa, P., Naehrig, M.: SIDH Library (2016–2018). https://github.com/Microsoft/PQCrypto-SIDH

8. de Groot, W.: A Performance Study of X25519 on Cortex-M3 and M4. Ph.D. thesis, Eindhoven University of Technology, September 2015

9. De Santis, F., Sigl, G.: Towards side-channel protected X25519 on ARM Cortex-M4 processors. In: Proceedings of Software Performance Enhancement for Encryption and Decryption, and Benchmarking, Utrecht, The Netherlands, pp. 19–21 (2016)

10. Faz-Hernández, A., López, J., Ochoa-Jiménez, E., Rodríguez-Henríquez, F.: A faster software implementation of the supersingular isogeny Diffie-Hellman key exchange protocol. IEEE Trans. Comput. **67**(11), 1622–1636 (2018)

11. Fujii, H., Aranha, D.F.: Curve25519 for the cortex-M4 and beyond. Prog. Cryptol.-LATINCRYPT **35**, 36–37 (2017)

12. Gura, N., Patel, A., Wander, A., Eberle, H., Shantz, S.C.: Comparing elliptic curve cryptography and RSA on 8-bit CPUs. In: Joye, M., Quisquater, J.-J. (eds.) CHES 2004. LNCS, vol. 3156, pp. 119–132. Springer, Heidelberg (2004). https://doi.org/10.1007/978-3-540-28632-5_9

13. Haase, B., Labrique, B.: AuCPace: efficient verifier-based PAKE protocol tailored for the IIoT. IACR Trans. Cryptogr. Hardw. Embed. Syst. 1–48 (2019)

14. Hutter, M., Schwabe, P.: Multiprecision multiplication on AVR revisited. J. Cryptogr. Eng. **5**(3), 201–214 (2015)

15. Hutter, M., Wenger, E.: Fast multi-precision multiplication for public-key cryptography on embedded microprocessors. In: Preneel, B., Takagi, T. (eds.) CHES 2011. LNCS, vol. 6917, pp. 459–474. Springer, Heidelberg (2011). https://doi.org/10.1007/978-3-642-23951-9_30

16. Jalali, A., Azarderakhsh, R., Kermani, M.M., Jao, D.: Supersingular isogeny Diffie-Hellman key exchange on 64-bit ARM. IEEE Trans. Dependable Secure Comput. (2017)

17. Jao, D., De Feo, L.: Towards quantum-resistant cryptosystems from supersingular elliptic curve isogenies. In: Yang, B.-Y. (ed.) PQCrypto 2011. LNCS, vol. 7071, pp. 19–34. Springer, Heidelberg (2011). https://doi.org/10.1007/978-3-642-25405-5_2

18. Kannwischer, M.J., Rijneveld, J., Schwabe, P., Stoffelen, K.: PQM4: Post-quantum crypto library for the ARM Cortex-M4. https://github.com/mupq/pqm4

19. Kim, S., Yoon, K., Kwon, J., Hong, S., Park, Y.-H.: Efficient isogeny computations on twisted Edwards curves. Secur. Commun. Netw. (2018)

20. Koppermann, P., Pop, E., Heyszl, J., Sigl, G.: 18 seconds to key exchange: limitations of supersingular isogeny Diffie-Hellman on embedded devices. Cryptology ePrint Archive, Report 2018/932 (2018). https://eprint.iacr.org/2018/932

21. Koziel, B., Azarderakhsh, R., Mozaffari-Kermani, M.: Fast hardware architectures for supersingular isogeny Diffie-Hellman key exchange on FPGA. In: Dunkelman, O., Sanadhya, S.K. (eds.) INDOCRYPT 2016. LNCS, vol. 10095, pp. 191–206. Springer, Cham (2016). https://doi.org/10.1007/978-3-319-49890-4_11

22. Koziel, B., Jalali, A., Azarderakhsh, R., Jao, D., Mozaffari-Kermani, M.: NEON-SIDH: efficient implementation of supersingular isogeny Diffie-Hellman key exchange protocol on ARM. In: Foresti, S., Persiano, G. (eds.) CANS 2016. LNCS, vol. 10052, pp. 88–103. Springer, Cham (2016). https://doi.org/10.1007/978-3-319-48965-0_6

23. Liu, Z., Longa, P., Pereira, G., Reparaz, O., Seo, H.: FourQ on embedded devices with strong countermeasures against side-channel attacks. In: International Conference on Cryptographic Hardware and Embedded Systems, CHES 2017, pp. 665–686 (2017)

24. Montgomery, P.L.: Modular multiplication without trial division. Math. Comput. **44**(170), 519–521 (1985)

25. Seo, H., Liu, Z., Longa, P., Hu, Z.: SIDH on ARM: faster modular multiplications for faster post-quantum supersingular isogeny key exchange. IACR Trans. Cryptogr. Hardw. Embed. Syst. 1–20 (2018)

26. Shor, P.W.: Algorithms for quantum computation: discrete logarithms and factoring. In: Proceedings of the 35th Annual Symposium on Foundations of Computer Science, pp. 124–134. IEEE (1994)

27. The National Institute of Standards and Technology (NIST). Post-quantum cryptography standardization (2017–2018). https://csrc.nist.gov/projects/post-quantum-cryptography/post-quantum-cryptography-standardization

Improved Cryptanalysis on SipHash

Wenqian Xin, Yunwen Liu, Bing Sun, and Chao Li[✉]

College of Liberal Arts and Sciences, National University of Defense Technology,
Changsha 410073, China
Wenqian_Xin@163.com, univerlyw@hotmail.com, hppy_come@163.com,
Lichao_nudt@sina.com

Abstract. SipHash is an ARX-based pseudorandom function designed by Aumasson and Bernstein for short message inputs. Recently, Ashur et al. proposed an efficient analysis method against ARX algorithm— "Rotational-XOR cryptanalysis". Inspired by their work, we mount differential and Rotational-XOR cryptanalysis on two instances of SipHash-1-x and SipHash-2-x in this paper, where SipHash-1-x (or SipHash-2-x) represents the Siphash instance with one (or two) compression round and x finalization rounds.

Firstly, we construct the search model for colliding characteristic and RX-colliding characteristic on SipHash. Based on the model, we find the colliding characteristics and RX-colliding characteristics of SipHash by the SMT-based automatic search tool. Moreover, we give a formula for the selection of initial constants to improve the resistance of Siphash against Rotational-XOR cryptanalysis to make the algorithm safer. In addition, we find an RX-colliding characteristic with probability $2^{-93.6}$ for a revised version of SipHash-1-x with one message block, and an RX-colliding characteristic with probability 2^{-160} for a revised version of SipHash-1-x with two message blocks. With the SMT-based technique, which outputs one message pair of the RX-collision if the given characteristic has a nonzero probability. Finally, with the RX-colliding characteristic we found earlier, we give the RX-collision with message pair and key of a revised version of SipHash-1-x with one message block.

Keywords: Differential cryptanalysis · Rotational-XOR cryptanalysis · SipHash · Initial constant · SMT software

1 Introduction

SipHash is a family of pseudorandom functions optimized for short inputs proposed by Aumasson and Bernstein at Indocrypt 2012 [3]. Message string of any length can be processed by SipHash with a 128-bit key to obtain a 64-bit or 128-bit output. The parameters c' and d' in SipHash-c'-d' represent the number of compression rounds and finalization rounds, respectively.

This work was supported by the Natural Science Foundation of China (NSFC) under Grant 61772545, Grant 61672530, and Grant 61902414.

© Springer Nature Switzerland AG 2019
Y. Mu et al. (Eds.): CANS 2019, LNCS 11829, pp. 61–79, 2019.
https://doi.org/10.1007/978-3-030-31578-8_4

One main design goal of SipHash is to ensure its security for short inputs. As far as we know, the earliest study on SipHash is given by Siddappa et al. [15], they proposed a SAT-based technique for a key recovery attack on SipHash-1-0, by converting the SipHash primitive to a Conjunctive Normal Form (CNF) and fed to a SAT solver. One of the best cryptanalytic results on SipHash is given by Dobraunig et al. [7] using differential cryptanalysis. They find the differential characteristic for SipHash-1-x with three message blocks and SipHash-2-x with one message block, and show that SipHash-1-x is resistant against differential cryptanalysis. It remains an interesting research question to check the security of SipHash against other cryptanalytical methods, such as Rotational-XOR cryptanalysis. Rotational-XOR cryptanalysis [4] is a recently proposed technique by Ashur et al. in 2016 for analysing ARX ciphers. It generalises rotational cryptanalysis and differential cryptanalysis to deal with the constants. Notice that the propagation rule of Rotational-XOR differences allows more possible paths than differential transitions by containing the normal XOR-difference propagation rule. It means that one may obtain more possible characteristics by Rotational-XOR cryptanalysis comparing with differential cryptanalysis. As implied by an example in SPECK32 [4], Rotational-XOR cryptanalysis is an effective technique for the ARX structure, and it may generate characteristics covering more rounds than differential ones. Moreover, the constants have a vital influence on the details of the RX-characteristics, unlike differential cryptanalysis in both single-key and related-key settings. Therefore, with Rotational-XOR cryptanalysis, it helps to detect "weak" realizations of the constants which may leads to a vulnerability.

Our Contributions. In this paper, we focus on the application of differential cryptanalysis and Rotational-XOR cryptanalysis to pseudorandom functions by searching for colliding characteristics and RX-colliding characteristics, by studying a special internal collision (as we call it "internal RX-collision") derived from RX-characteristics of certain properties. Especially, our target ciphers are the instances SipHash-1-x and SipHash-2-x. With differential cryptanalysis and Rotational-XOR cryptanalysis, our results are as follows.

- Combining the search model and the automatic search tool, we give a colliding characteristic with probability 2^{-278} of SipHash-1-x with four message blocks and a colliding characteristic with probability 2^{-241} of SipHash-2-x with one message block. Besides, we find an RX-colliding characteristic with probability 2^{-280} of SipHash-1-x with two message blocks.
- We give the theoretical proof for the colliding characteristics search results of SipHash-1-x with one and two message blocks, and the RX-colliding characteristic search result of SipHash-1-x with one message block.
- With Rotational-XOR cryptanalysis, we give an initial constants selection formula to make the algorithm safer. In addition, we find an RX-colliding characteristic with probability $2^{-93.6}$ for a revised version of SipHash-1-x introducing one block message, and an RX-colliding characteristic with probability 2^{-160} for a revised version of SipHash-1-x introducing two blocks message (Table 1).

- Moreover, we give an SMT-based technique to verity the exact probability of an RX-characteristic (or a differential-characteristic) is zero or nonzero, with the RX-colliding characteristic of a revised version of SipHash-1-x with one message block which we found earlier, we get an RX-collision with message pair and key by the SMT-based technique.

Organization of the Paper. The rest of this paper is organized as follows. A brief introduction of SipHash and an overview of differential cryptanalysis and Rotational-XOR cryptanalysis are given in Sect. 2. In Sect. 3, we construct the search model with differential cryptanalysis and Rotational-XOR cryptanalysis. The results of colliding characteristics and RX-colliding characteristics for SipHash are given in Sect. 4. In Sect. 5, we study the influence of constants selection on SipHash, and show an SMT-based technique for verifying the validity of characteristics, and we apply it in finding an internal RX-collision under different constants. We conclude in Sect. 6.

Table 1. Best found characteristics

Version	Type	Blocks	Probability	Reference
SipHash-1-x	DC	1	0	Sect. 4.1
SipHash-1-x	DC	2	0	Sect. 4.1
SipHash-1-x	DC	3	$2^{-169(*)}$	[7]
SipHash-1-x	DC	4	2^{-278}	Sect. 4.1
SipHash-2-x	DC	1	$2^{-242(*)}$	[7]
SipHash-2-x	DC	1	2^{-241}	Sect. 4.1
SipHash-1-x	RX	1	0	Sect. 4.2
SipHash-1-x	RX	2	2^{-280}	Sect. 4.2
Revised SipHash-1-x	RX	1	$2^{-93.6}$	Sect. 5.3
Revised SipHash-1-x	RX	2	2^{-160}	Sect. 5.3

*Supposed that there are two independent random inputs for every modular addition in this paper, the colliding characteristics probability of SipHash-1-x with three message blocks and SipHash-2-x with one message block in [7] are 2^{-169} and 2^{-242}, respectively.
**"DC" denotes differential cryptanalysis, and "RX" denotes Rotational-XOR cryptanalysis.

2 Preliminaries

The notations we use in this paper are summarised in Table 2.

Table 2. Notations

Symbol	Meaning
$(b_{1,i}, a_{1,i}, c_{1,i}, d_{1,i})$	Inputs of the i-th round
$(b_{2,i}, a_{2,i}, c_{2,i}, d_{2,i})$	Outputs of the i-th round
$u_{j,i}, v_{j,i}, w_{j,i}, z_{j,i}$	The intermediate values of i-th round
$(\Delta b_{1,i}, \Delta a_{1,i}, \Delta c_{1,i}, \Delta d_{1,i})$	Difference introduced in $(b_{1,i}, a_{1,i}, c_{1,i}, d_{1,i})$
$(\Delta b_{2,i}, \Delta a_{2,i}, \Delta c_{2,i}, \Delta d_{2,i})$	Difference introduced in $(b_{2,i}, a_{2,i}, c_{2,i}, d_{2,i})$
$(\overleftarrow{\Delta} b_{1,i}, \overleftarrow{\Delta} a_{1,i}, \overleftarrow{\Delta} c_{1,i}, \overleftarrow{\Delta} d_{1,i})$	RX-difference introduced in $(b_{1,i}, a_{1,i}, c_{1,i}, d_{1,i})$
$(\overleftarrow{\Delta} b_{2,i}, \overleftarrow{\Delta} a_{2,i}, \overleftarrow{\Delta} c_{2,i}, \overleftarrow{\Delta} d_{2,i})$	RX-difference introduced in $(b_{1,i}, a_{1,i}, c_{1,i}, d_{1,i})$
$\overleftarrow{x}$	$x \lll 1$

2.1   Description of SipHash

SipHash is a family of pseudorandom functions based on ARX. The integer parameters c' and d' in SipHash-c'-d' represent the number of compression rounds and the number of finalization rounds, respectively. Next, take SipHash-1-x as an example to introduce SipHash. The processes of authentication splits into three stages: Initialization, Compression and Finalization. The details of SipHash-1-x with one message block can be found in Figs. 1 and 2 as shown below. Here $(a_1, b_1, c_1, d_1) \rightarrow (a_2, b_2, c_2, d_2)$ are the intermediate values of the single round encryption process of SipHash.

1. Initialization. The initial constants V_0, V_1, V_2 and V_3 are four 64-bit constants. The only requirement for the constants is that (V_0, V_1) differ from (V_2, V_3) for avoiding some symmetry in the states. The following set is given by the original design document [3] as a possible choice of the constants.

$$
\begin{aligned}
V_0 &= \text{736f6d6570736575}, \\
V_1 &= \text{646f72616e646f6d}, \\
V_2 &= \text{6c7967656e657261}, \\
V_3 &= \text{7465646279746573}.
\end{aligned}
\tag{1}
$$

Four 64-bit words of internal state a_0, b_0, c_0, d_0 are initialized as

$$
\begin{aligned}
a_0 &= V_0 \oplus k_0, \quad b_0 = V_1 \oplus k_1, \\
c_0 &= V_2 \oplus k_0, \quad d_0 = V_3 \oplus k_1.
\end{aligned}
$$

where k_0, k_1 are 64-bit words derived from the key.

2. Compression. SipHash-1-x processes a b-byte string m by parsing it as $w = \lceil (b+1)/8 \rceil$ 64-bit words $m_0, \cdots, m_{w-1}$. m_{w-1} includes the last 8 bytes of m. Taking SipHash-1-x with one message block as an example. The message m_0 is initially XORed to the state.

$$
\begin{aligned}
a_{1,1} &= a_0, \quad b_{1,1} = b_0, \\
c_{1,1} &= c_0, \quad d_{1,1} = d_0 \oplus m_0.
\end{aligned}
\tag{2}
$$

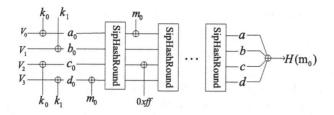

Fig. 1. SipHahs-1-x with one message block

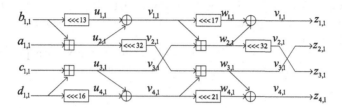

Fig. 2. SipHashRound

Then the intermediate values are obtained,

$$
\begin{aligned}
u_{1,1} &= b_{1,1} \lll 13, & v_{1,1} &= u_{1,1} \oplus u_{2,1}, \\
u_{2,1} &= a_{1,1} \boxplus b_{1,1}, & v_{2,1} &= u_{2,1} \lll 32, \\
u_{3,1} &= c_{1,1} \boxplus d_{1,1}, & v_{3,1} &= u_{3,1}, \\
u_{4,1} &= d_{1,1} \lll 16, & v_{4,1} &= u_{3,1} \oplus u_{4,1}.
\end{aligned}
\tag{3}
$$

$$
\begin{aligned}
w_{1,1} &= v_{1,1} \lll 17, & z_{1,1} &= w_{1,1} \oplus w_{2,1}, \\
w_{2,1} &= v_{1,1} \boxplus v_{3,1}, & z_{2,1} &= w_{2,1} \lll 32, \\
w_{3,1} &= v_{2,1} \boxplus v_{4,1}, & z_{3,1} &= w_{3,1} \\
w_{4,1} &= v_{4,1} \lll 21, & z_{4,1} &= w_{3,1} \oplus w_{4,1}.
\end{aligned}
\tag{4}
$$

$$
\begin{aligned}
b_{2,1} &= z_{1,1}, & a_{2,1} &= z_{3,1}, \\
c_{2,1} &= z_{2,1}, & d_{2,1} &= z_{4,1}.
\end{aligned}
\tag{5}
$$

where $u_{j,1}, v_{j,1}, w_{j,1}, z_{j,1}, j \in \{1, 2, 3, 4\}$ are four 64-bit words of intermediate values in SipHashRound.

Finally, m_0 is XORed to $b_{2,1}$: $b_{1,2} = b_{2,1} \oplus m_0$.

3. Finalization. The constant $\mathtt{0xff}$ is XORed to the state $c_{2,1}$,

$$
c_{1,2} = c_{2,1} \oplus \mathtt{0xff}.
$$

After x iterations of SipHashRound, the finalization phase returns a 64-bit value

$$
H(m_0) = a \oplus b \oplus c \oplus d.
$$

where a, b, c, d are 64-bit outputs of the SipHashRound.

2.2   Differential Cryptanalysis on ARX Structure

As a generally applicable cryptanalytic technique, differential cryptanalysis has been extensively studied on ARX ciphers. Nyberg et al. [13] have studied the differential propagations in modular addition by an automaton description, and an explicit formula was derived by Schulte-Geers [14] in 2013 for calculating the differential probabilities of independent inputs in modular addition. The studies on differential properties of modular addition have greatly promoted the analysis processes of ARX algorithms. A number of ARX ciphers are analyzed, such as a number of results obtained for differential cryptanalysis on SPECK [1, 6], differential cryptanalysis on SPARX by Ankele et al. [2], and the results of differential cryptanalysis in ARX structure [9,10,12]. In addition, there are many articles on the automatic search of differential characteristics in ARX ciphers [5,8,16,17].

Differences pass through the linear operations in ARX with probability 1, whereas the differential probability of modular addition can be characterised by the following equation

$$(x \oplus \Delta x) \boxplus (y \oplus \Delta y) = ((x \boxplus y) \oplus \Delta z).$$

When the inputs x and y are independent random variables, the following lemma depicts the propagation rule of the differences through modular addition.

Lemma 1 ([14]). *Suppose that $x, y \in \mathbb{F}_{2^n}$ are independent random variables, and let $\Delta x = x \oplus x'$, $\Delta y = y \oplus y'$, and $\Delta z = z \oplus z'$ be differences in $\mathbb{F}_{2^n}$, where $z = x \boxplus y$. Then, we have*

$$Pr[(x \oplus \Delta x) \boxplus (y \oplus \Delta y) = (z \oplus \Delta z)]$$
$$= 1_{(I \oplus SHL)(\Delta x \oplus \Delta y \oplus \Delta z) \preceq SHL((\Delta x \oplus \Delta z)|(\Delta y \oplus \Delta z))} \cdot 2^{-|SHL((\Delta x \oplus \Delta z)|(\Delta y \oplus \Delta z))|}.$$

$$(6)$$

where $SHL(x) = x \ll 1$ and $(I \oplus SHL)(x) = x \oplus SHL(x)$. $1_{x \preceq y} = 1$ is the characteristic function which evaluates to 1 when for all $i : x_i \le y_i, 0 \le i \le n$, otherwise to 0.

2.3   Rotational-XOR Cryptanalysis on ARX Structure

Rotational-XOR cryptanalysis is a recent cryptanalysis method for ARX algorithms proposed by Ashur et al. [4]. Rotational-XOR cryptanalysis was applied to the block cipher SPECK [11] by Liu et al., where they obtained characteristics covering more rounds than previous results. Rotational-XOR cryptanalysis works with a new notion of difference called RX-difference.

Definition 1 ([4]). *Rotational-XOR difference (or RX-difference in short) with rotational offset 1 of two bit-strings x and x' is defined as*

$$\overleftarrow{\Delta} x = x \oplus (x' \lll 1).$$

Here, we give a quick overview of the basic propagation rule of the RX-difference in ARX algorithms. Similar to differential cryptanalysis, the propagations of RX-difference through XOR and rotation is with probability 1. Whereas the RX-differential probability of modular addition can be characterised by the following equation.

$$((x \oplus \overleftarrow{\Delta} x) \ggg 1) \boxplus ((y \oplus \overleftarrow{\Delta} y) \ggg 1) = ((z \oplus \overleftarrow{\Delta} z) \ggg 1),$$

Where $\overleftarrow{\Delta} x, \overleftarrow{\Delta} y, \overleftarrow{\Delta} z$ denote the RX-difference introduced at x, y, z, and $z = x \boxplus y$. For the nonlinear operation modular addition, the probability is evaluated by the following proposition.

Proposition 1 ([4]). *Suppose that $x, y \in \mathbb{F}_{2^n}$ are independently uniform random variables, $z = x \boxplus y$. Let $\overleftarrow{\Delta} x, \overleftarrow{\Delta} y$ and $\overleftarrow{\Delta} z$ be constants in $\mathbb{F}_{2^n}$, which are the RX-differences. Then,*

$$Pr[((x \oplus \overleftarrow{\Delta} x) \ggg 1) \boxplus ((y \oplus \overleftarrow{\Delta} y) \ggg 1) = (z \oplus \overleftarrow{\Delta} z) \ggg 1]$$
$$= 1_{(I \oplus SHL)(\delta_x \oplus \delta_y \oplus \delta_z) \oplus 1 \preceq SHL((\delta_x \oplus \delta_z)|(\delta_y \oplus \delta_z))}$$
$$\cdot 2^{-|SHL((\delta_x \oplus \delta_z)|(\delta_y \oplus \delta_z))|} \cdot 2^{-3} \tag{7}$$
$$+ 1_{(I \oplus SHL)(\delta_x \oplus \delta_y \oplus \delta_z) \preceq SHL((\delta_x \oplus \delta_z)|(\delta_y \oplus \delta_z))}$$
$$\cdot 2^{-|SHL((\delta_x \oplus \delta_z)|(\delta_y \oplus \delta_z))|} \cdot 2^{-1.415}.$$

where $\delta_x = \overleftarrow{\Delta} x \gg 1$, $\delta_y = \overleftarrow{\Delta} y \gg 1$, and $\delta_z = \overleftarrow{\Delta} z \gg 1$.

In words: Proposition 1 gives the propagation law of the input RX-differences $\overleftarrow{\Delta} x, \overleftarrow{\Delta} y$ through modular addition to output RX-difference $\overleftarrow{\Delta} z$. More information please refer to [4] and [11].

3   SMT-Based Search Model of SipHash

To search the characteristics of SipHash which could lead to an internal collision (or internal RX-collision), we use an automatic search tool—Z3 solver based on SMT. To make use of such a tool to get the characteristics, several key processes need to be considered.

- Combine SipHash encryption processes with the propagation law of difference (or RX-difference) through modular addition, designing the search model of collision (or RX-collision).
- Convert the search model into SMT language with SMTLIB format.
- Run the solver to search for characteristics with a certain probability.

SMT-Based Colliding Characteristics Search Model. In this subsection, by combining the propagation law of difference proposed by Schulte-Geers [14] with an SMT solver, we construct a search model of differential characteristics in SipHash-1-x and SipHash-2-x. For convenience, a differential characteristic that leads to an internal collision will be referred to as a colliding characteristic. Taking the case of SipHash-1-x with one message block as an example, we describe the model as follow.

The notations of the input differences and the output differences in $(b_{1,i}, a_{1,i}, c_{1,i}, d_{1,i}), (b_{2,i}, a_{2,i}, c_{2,i}, d_{2,i})$, and the differences introduced in $u_{j,i}, v_{j,i}, w_{j,i}, z_{j,i}$ are showed in Table 2. To construct an internal collision, the difference Δm_0 introduced by the message propagates through the round function of SipHash, and needs to be cancelled by the same difference on a different branch. Hence, it is necessary to find a possible differential characteristic

$$(\Delta b_{1,1}, \Delta a_{1,1}, \Delta c_{1,1}, \Delta d_{1,1}) \rightarrow (\Delta b_{2,1}, \Delta a_{2,1}, \Delta c_{2,1}, \Delta d_{2,1}).$$

in SipHash round which satisfies the active pattern

$$(0, 0, 0, \Delta m_0) \rightarrow (0, \Delta m_0, 0, 0).$$

Similar to Eqs. (2)–(5), constraints of the difference propagations on SipHash-1-x with one message block are described in the following.

$$\Delta u_{1,1} = \Delta b_{1,1} \lll \alpha_1, \quad \Delta v_{1,1} = \Delta u_{1,1} \oplus \Delta u_{2,1},$$
$$(\Delta a_{1,1}, \Delta b_{1,1}) \overset{\boxplus}{\rightarrow} \Delta u_{2,1}, \Delta v_{2,1} = \Delta u_{2,1} \lll 32,$$
$$(\Delta c_{1,1}, \Delta d_{1,1}) \overset{\boxplus}{\rightarrow} \Delta u_{3,1}, \Delta v_{3,1} = \Delta u_{3,1},$$
$$\Delta u_{4,1} = \Delta d_{1,1} \lll \beta_2, \quad \Delta v_{4,1} = \Delta u_{3,1} \oplus \Delta u_{4,1},$$

(8)

where $\alpha_1 = 13$, $\alpha_2 = 17$, $\beta_2 = 16$, and $\beta_2 = 21$. We can get the values $\Delta w_{j,1}, \Delta z_{j,1}, j \in \{1, 2, 3, 4\}$ similar to $\Delta u_{j,1}, \Delta v_{j,1}, j \in \{1, 2, 3, 4\}$, by replacing α_1, β_1 with α_2, β_2.
The propagations of

$$(\Delta a_{1,1}, \Delta b_{1,1}) \overset{\boxplus}{\rightarrow} \Delta u_{2,1}, \quad (\Delta c_{1,1}, \Delta d_{1,1}) \overset{\boxplus}{\rightarrow} \Delta u_{3,1},$$
$$(\Delta v_{1,1}, \Delta v_{3,1}) \overset{\boxplus}{\rightarrow} \Delta w_{2,1}, \quad (\Delta v_{2,1}, \Delta v_{4,1}) \overset{\boxplus}{\rightarrow} \Delta w_{3,1}$$

are expected to satisfy the constraints in Lemma 1, and $(\Delta z_{1,1}, \Delta z_{3,1}, \Delta z_{2,1}, \Delta z_{4,1})$ equals to $(\Delta b_{2,1}, \Delta a_{2,1}, \Delta c_{2,1}, \Delta d_{2,1}) = (0, \Delta m_0, 0, 0)$.

SMT-Based RX-Colliding Characteristics Search Model. Similar to the internal collision based on differential distinguishers, if the RX-differences injected by the messages and the initial constants are cancelled by the RX-differences after the application of message injection phase, the input RX-differences to the finalization phase are all zero which leads to a rotational relation at the output of the pseudorandom function with a rotational probability. The RX-characteristic that leads to an internal RX-collision will be referred to as an RX-colliding characteristic.

Comparing with differential propagations, RX-difference has one extra condition to pass through modular addition, which implies that there may exist possible RX-characteristics even when the difference propagations are impossible. Therefore, it motivates us to study SipHash with Rotational-XOR cryptanalysis. Taking SipHash-1-x with one message block as an example, the RX-colliding characteristics need to satisfy the active pattern in Lemma 2.

Lemma 2. *Suppose that there exists an RX-characteristic $(\overleftarrow{\Delta} b_{1,1}, \overleftarrow{\Delta} a_{1,1}, \overleftarrow{\Delta} c_{1,1}, \overleftarrow{\Delta} d_{1,1}) \to (\overleftarrow{\Delta} b_{2,1}, \overleftarrow{\Delta} a_{2,1}, \overleftarrow{\Delta} c_{2,1}, \overleftarrow{\Delta} d_{2,1})$ which produces an internal RX-collision for SipHash-1-x with one message block. Then, it has the following active pattern,*

$$(C_1, C_0, C_2, C_3 \oplus \Delta) \to (0, \Delta, C_4, 0),$$

where $C_0 = V_0 \oplus \overleftarrow{V_0}, C_1 = V_1 \oplus \overleftarrow{V_1}, C_2 = V_2 \oplus \overleftarrow{V_2}, C_3 = V_3 \oplus \overleftarrow{V_3}, C_4 = 0xff \oplus \overleftarrow{0xff}$, Δ represents the RX-difference of message block is introduced in this case.

Proof. We give a theoretical explanation of this theorem according to the SipHash-1-x encryption processes and the RX-difference definition. The full proof is given in Appendix A.   □

Similar to the colliding characteristics search model of SipHash-1-x with one message block, the propagations of

$$(\overleftarrow{\Delta} a_{1,1}, \overleftarrow{\Delta} b_{1,1}) \overset{\boxplus}{\to} \overleftarrow{\Delta} u_{2,1}, \quad (\overleftarrow{\Delta} c_{1,1}, \overleftarrow{\Delta} d_{1,1}) \overset{\boxplus}{\to} \overleftarrow{\Delta} u_{3,1}$$

$$(\overleftarrow{\Delta} v_{1,1}, \overleftarrow{\Delta} v_{3,1}) \overset{\boxplus}{\to} \overleftarrow{\Delta} w_{2,1}, \quad (\overleftarrow{\Delta} v_{2,1}, \overleftarrow{\Delta} v_{4,1}) \overset{\boxplus}{\to} \overleftarrow{\Delta} w_{3,1}$$

are constrained by Proposition (1).

If the output RX-differences $(0, \Delta, C_4, 0)$ after the SipHash round are all zero, the right message pairs and right key pairs satisfying the characteristic can be found. Then after the finalization phase, the 64-bit output values $H(m_0), H(m_0')$ have a rotational relation $H(m_0) = \overleftarrow{H(m_0')}$.

Converting the Search Model into SMT Language. For the second process, all we need to do is converting the search model into SMT language with SMTLIB format. Take "z=((a≪16)⊕ b)⊞ (d ≪32)" for example, it can be converted by "(assert(= z (bvadd (bvxor ((_ rotate_left 16) a) b) ((_ rotate_left 32) d))))". We will do a characteristic search in the next section by the SMT software.

4   Colliding Characteristics and RX-Colliding Characteristics Search of SipHash

In this section, we will use the automatic tool to search the colliding characteristics and the RX-colliding characteristics for SipHash-1-x with multiple message blocks and SipHash-2-x with one message block, respectively.

4.1   Colliding Characteristics Search of SipHash

With the colliding characteristics search model and SMT software, we can obtain the colliding characteristics of SipHash-1-x with four message blocks and SipHash-2-x with one message block which are shown in Tables 3 and 4. The experimental search results show that no colliding characteristic is found for SipHash-1-x with one or two message blocks.

Table 3. Colliding characteristic of SipHash-1-x with four message blocks with probability 2^{-278}.

Δm_0	0000000600000000	Δk_0	00000000000000000
Δm_1	3fa0fe5a000000d0	Δk_1	00000000000000000
Δm_2	d0ce9cdda0101a69		
Δm_3	00000040000180000		
$\Delta a_{1,1}$	0000000000000000	$\Delta b_{1,1}$	0000000000000000
$\Delta c_{1,1}$	0000000000000000	$\Delta d_{1,1}$	0000000600000000
$\Delta a_{2,1}$	0000e000200000000	$\Delta b_{2,1}$	00000000200000000
$\Delta c_{2,1}$	0000000000000002	$\Delta d_{2,1}$	0004e0002000000c0
$\Delta a_{1,2}$	0000e000400000000	$\Delta b_{1,2}$	00000000200000000
$\Delta c_{1,2}$	0000000000000002	$\Delta d_{1,2}$	03feefe5800000010
$\Delta a_{2,2}$	087ce8ee80029d092	$\Delta b_{2,2}$	0800cc0d200000046
$\Delta c_{2,2}$	0000000720018c0d2	$\Delta d_{2,2}$	070ce8cefffb61762
$\Delta a_{1,3}$	0b86e70b20029d042	$\Delta b_{1,3}$	0800cc0d200000046
$\Delta c_{1,3}$	0000000720018c0d2	$\Delta d_{1,3}$	0a00010325fa60d0b
$\Delta a_{2,3}$	0f0c69c9dffee0669	$\Delta b_{2,3}$	02000004000020000
$\Delta c_{2,3}$	00001400000000020	$\Delta d_{2,3}$	00000c40000180020
$\Delta a_{1,4}$	0200800405ffe1c00	$\Delta b_{1,4}$	02000004000020000
$\Delta c_{1,4}$	00001400000000020	$\Delta d_{1,4}$	00000c00000000020
$\Delta a_{2,4}$	00000040000180000	$\Delta b_{2,4}$	0000000000000000
$\Delta c_{2,4}$	0000000000000000	$\Delta d_{2,4}$	0000000000000000
Time	1094 m59.207 s		

Table 4. Colliding characteristic of SipHash-2-x with one message block with probability 2^{-241}.

Δm_0	d08ca994f84216ba	Δk_0	0000000000000000
		Δk_1	0000000000000000
$\Delta a_{1,1}$	0000000000000000	$\Delta b_{1,1}$	0000000000000000
$\Delta c_{1,1}$	0000000000000000	$\Delta d_{1,1}$	d08ca994f84216ba
$\Delta a_{1,2}$	9911cfad8e574746	$\Delta b_{1,2}$	31845fb478420fca
$\Delta c_{1,2}$	78420fca31845fb4	$\Delta d_{1,2}$	a7c210b5d684654c
$\Delta a_{2,2}$	d08ca994f84216ba	$\Delta b_{2,2}$	0000000000000000
$\Delta c_{2,2}$	0000000000000000	$\Delta d_{2,2}$	0000000000000000
Time	112 m52.949 s		

From the colliding characteristics search model, the colliding characteristics need to satisfy the active pattern $(0,0,0,\Delta m_0) \rightarrow (0,\Delta m_0,0,0)$. As shown by a careful analysis of the difference transitions, the following theorem shows that the differential characteristics of the above active pattern are all impossible.

Theorem 1. *Given any non-zero value $\Delta \in \mathbb{F}_2^{64}$, $(0,0,0,\Delta) \rightarrow (0,\Delta,0,0)$ is an impossible differential characteristic of SipHash-1-x with one message block.*

Proof. The full proof is given in Appendix A. □

It is shown by the above theorem that under the single-key model, one cannot find the effective colliding characteristic of SipHash-1-x with one message block by differential cryptanalysis. A similar result can be derived for SipHash-1-x with two message blocks, and we omit the details here.

4.2 RX-Colliding Characteristics Search of SipHash

In this subsection, we use the automatic search tool to search for the RX-colliding characteristics of SipHash-1-x, with the initial constants used in the design document of Siphash [3]. We find an RX-colliding characteristic of SipHash-1-x with two message blocks which is shown in Table 5.

The results of Rotational-XOR cryptanalysis experiments show that there exist no RX-colliding characteristic of SipHash-1-x with one message block. Combining Lemma 2 and Proposition 1, we give a theorem to characterize the Rotational-XOR cryptanalysis experimental result for this case.

Theorem 2. *For any non-zero $\Delta \in \mathbb{F}_2^{64}$, $(C_1, C_0, C_2, C_3 \oplus \Delta) \rightarrow (0, \Delta, C_4, 0)$ is an impossible RX-characteristic of SipHash-1-x with one message block.*

Proof. The full proof is given in Appendix A. □

Table 5. RX-colliding characteristic of SipHash-1-x with two message blocks with probability 2^{-280}.

$\overleftarrow{\Delta}m_0$	46c8f685a03378df	$\overleftarrow{\Delta}k_0$	0000000000000000
$\overleftarrow{\Delta}m_1$	00030008000001e0	$\overleftarrow{\Delta}k_1$	0000000000000000
$\overleftarrow{\Delta}a_{1,1}$	95b1b7af9095af9f	$\overleftarrow{\Delta}b_{1,1}$	acb196a3b2acb1b7
$\overleftarrow{\Delta}c_{1,1}$	b48ba9afb2af96a3	$\overleftarrow{\Delta}d_{1,1}$	da675a232bafd74a
$\overleftarrow{\Delta}a_{2,1}$	4e88f6a3a12d7c0b	$\overleftarrow{\Delta}b_{2,1}$	00404002010004d4
$\overleftarrow{\Delta}c_{2,1}$	c00000fcfefcf781	$\overleftarrow{\Delta}d_{2,1}$	c0030007010f1a60
$\overleftarrow{\Delta}a_{1,2}$	08400026011e04d4	$\overleftarrow{\Delta}b_{1,2}$	00404002010004d4
$\overleftarrow{\Delta}c_{1,2}$	c00000fcfefcf781	$\overleftarrow{\Delta}d_{1,2}$	c000000f010f1b80
$\overleftarrow{\Delta}a_{2,2}$	00030008000001e0	$\overleftarrow{\Delta}b_{2,2}$	0000000000000000
$\overleftarrow{\Delta}c_{2,2}$	0000000000000101	$\overleftarrow{\Delta}d_{2,2}$	0000000000000000
Time	263 m55.258 s		

4.3 Discussion on Differential Cryptanalysis and Rotational-XOR Cryptanalysis

Based on the analysis of SipHash-1-x and SipHash-2-x, our work can cover SipHash-1-x or SipHash-2-x with more message blocks. For example, let the differences in the remaining $n - 4$ blocks are 0, the result of differential cryptanalysis of SipHash-1-x with four message blocks can be extended to SipHash-1-x with n message blocks, where $n > 4$. In addition, taking SipHash-1-x with two message blocks as an example, and letting the RX-differences of the last $m - 2$ message blocks are 0, whose differences are not 0 with probability of almost 1, the result of Rotational-XOR cryptanalysis can also be extended to SipHash-1-x with $m > 2$ message blocks.

Compared with the RX-characteristics found in block ciphers [11], one possible reason may be attributed to the input constraints between the RX-characteristic search processes of SPECK and RX-colliding characteristic search processes of SipHash-1-x. The input form of the RX-colliding characteristic on SipHash needs be controlled, which is not the same case as that of a block cipher. Therefore, better cryptanalysis results may be obtained by changing the input form of RX-colliding characteristics search model.

5 The Influence of Initial Constants on SipHash

Based on the analysis of SipHash, the results show that one cannot get the RX-colliding characteristic of SipHash-1-x with one message block. Next, we will utilise the SMT solver to produce an RX-characteristic under a different set of constants, where the right pairs of messages and keys are found.

5.1   Theoretical Characterization of Initial Constants of SipHash-1-x with One Message Block

In this subsection, we will analyze the influence of initial constants on SipHash-1-x with one message block theoretically. The dependency between the initial constants and the internal RX-collision are characterized in details by the follow theorem.

Theorem 3. *Suppose that $C_0 = V_0 \oplus \overleftarrow{V_0}$, $C_1 = V_1 \oplus \overleftarrow{V_1}$, $C_4 = 0xff \oplus \overleftarrow{0xff}$, and $\alpha = (C_1 \lll 13) \oplus (C_4 \ggg 49)$. If V_0 and V_1 are values satisfying the following condition:*

$$((I \oplus SHL)(C_0' \oplus C_1' \oplus \alpha' \oplus t)) \& (SHL((C_0' \oplus \alpha')|(C_1' \oplus \alpha'))) \\ \neq ((I \oplus SHL)(C_0' \oplus C_1' \oplus \alpha' \oplus t)). \tag{9}$$

where $C_0' = C_0 \gg 1, C_1' = C_1 \gg 1, \alpha' = \alpha \gg 1$ and $t = 0x0$ or $0x1$. Then, SipHash-1-x with one message block instantiated by the constants has no internal RX-collision.

Experimental Analysis of Initial Constants. We verified whether the 128-bit V_0, V_1 of random selection satisfy the Eq. (9) by experiments. After 2^{28} experiments, there are 2^{19} sub-experiments making it doesn't satisfy the Eq. (9). That means there exits 2^{-9} probability that the initial constants selection may produce an internal RX-collision. The total degree of freedom of V_0, V_1, V_2, V_3 is 256, with the Eq. (9), we can avoid $2^{256} \times 2^{-9} = 2^{247}$ selections which may produce an internal RX-collision.

Table 6. RX-colliding characteristic of a revised version of SipHash-1-x with one message blocks with probability $2^{-93.6}$.

V_0	556f883b0003e5f0	V_2	1affff070080a170
V_1	3061a82efffffc23	V_3	1f03f5ebe0ff1f2f
$\overleftarrow{\Delta} a_{1,1}$	ffb0984d00042e10	$\overleftarrow{\Delta} b_{1,1}$	50a2f87300000465
$\overleftarrow{\Delta} c_{1,1}$	2f0001090181e390	$\overleftarrow{\Delta} d_{1,1}$	2100000801012071
$\overleftarrow{\Delta} a_{2,2}$	00041e3420000100	$\overleftarrow{\Delta} b_{2,2}$	0000000000000000
$\overleftarrow{\Delta} c_{2,2}$	0000000000000101	$\overleftarrow{\Delta} d_{2,2}$	0000000000000000
Time	0 m37.782 s		

5.2   RX-Colliding Characteristics of Revised SipHash-1-x

With the RX-colliding characteristic search model, we assign new variables for the possible choices of the constants V_0, V_1, V_2, V_3 to analyze SipHash-1-x with Rotational-XOR cryptanalysis. (9) as a necessary condition for revised SipHash

Table 7. RX-colliding characteristic of a revised version of SipHash-1-x with two messsage blocks with probability 2^{-160}.

V_0	3ff8f919d6faaa57	$\overleftarrow{\Delta}k_0$	0000000000000000
V_1	aa84e6ace90555a8	$\overleftarrow{\Delta}k_1$	0000000000000000
V_2	23e2c1dfdc5b0300	$\overleftarrow{\Delta}m_0$	05d1bda4e5f90381
V_3	2153d54348f33a81	$\overleftarrow{\Delta}m_1$	0000000000400000
$\overleftarrow{\Delta}a_{1,1}$	40090b2a7b0ffef9	$\overleftarrow{\Delta}b_{1\_1}$	ff8d2bf53b0ffef9
$\overleftarrow{\Delta}c_{1,1}$	6427426064ed0500	$\overleftarrow{\Delta}d_{1\_1}$	6625c2613cec4c02
$\overleftarrow{\Delta}a_{2,1}$	0591bfa4e5f90f84	$\overleftarrow{\Delta}b_{2\_1}$	0000020000000404
$\overleftarrow{\Delta}c_{2,1}$	8000010101818181	$\overleftarrow{\Delta}d_{2\_1}$	8000000001410383
$\overleftarrow{\Delta}a_{1,2}$	0040020000000c05	$\overleftarrow{\Delta}b_{1\_2}$	0000020000000404
$\overleftarrow{\Delta}c_{1\_2}$	8000010101818181	$\overleftarrow{\Delta}d_{1\_2}$	8000000001010383
$\overleftarrow{\Delta}a_{2\_2}$	0000000000400000	$\overleftarrow{\Delta}b_{2\_2}$	0000000000000000
$\overleftarrow{\Delta}c_{2\_2}$	0000000000000101	$\overleftarrow{\Delta}d_{2\_2}$	0000000000000000
Time	279 m15.155 s		

with RX-colliding characteristics, according to condition (9), we search an RX-colliding characteristic with probability $2^{-93.6} > 2^{-128}$ of a revised version of SipHash-1-x with one message block. And we get a RX-collision with it, which is shown in Table 6. In addition, the RX-colliding characteristic with probability 2^{-160} of a revised version of SipHash-1-x with two message blocks which is shown in Table 7.

5.3   SMT-Based Experimental Verification on the Possibility of Distinguishers

Next, we develop an experimental method for verifying the possibility of RX-characteristics and generating right pairs with an SMT solver.

To experimentally verify whether a differential distinguisher $(\Delta_i \rightarrow \Delta_o)$ encompasses any right pair, a general solution is to run through the input space and check the differences of the outputs. However, it is often infeasible due to the size of the input space and the probability of the differential distinguisher.

In this subsection, we encode an ARX cipher into an SMT model for describing the differential propagations, and execute the solver to find right pairs of inputs following the characteristics. More specifically, a pairs of messages with the input difference are defined by two variables. The operations in the ARX algorithms can be described by the functions and instructions in SMT language, since they are basically binary and arithmetic operations. After transforming the round function of an ARX cipher into an SMT description, one obtains the intermediate values and the final outputs as an SMT model. By setting a constraint on the output differences (and intermediate differences), we run the SMT solver

Algorithm 1. SMT-based experimental verification of an RX-distinguisher in SipHash-1-x with one message block.

1: **Input:** Initial constants state V_0, V_1, V_2, V_3 and output difference pattern $(\overleftarrow{\Delta} a_{2,1}, \overleftarrow{\Delta} b_{2,1}, \overleftarrow{\Delta} c_{2,1}, \overleftarrow{\Delta} d_{2,1})$.
2: **Output:** The possibility of the distinguisher
3: Define m_0, m'_0, k_0, k_1 as 64-bit initial variables
4: Derive the intermediate variables $a_{1,1}, b_{1,1}, c_{1,1}, d_{1,1}$ and $a'_{1,1}, b'_{1,1}, c'_{1,1}, d'_{1,1}$ as expressions on the initial variables
5: Derive the expressions for the intermediate variables $a_{2,1}, b_{2,1}, c_{2,1}, d_{2,1}$ and $a'_{2,1}, b'_{2,1}, c'_{2,1}, d'_{2,1}$
6: Assign the output differences to the output variables
7: Run the solver for a realisation of the initial variables
8: If *satisfiable*
9: **return** k_0, k_1 and m_0, m'_0
10: Otherwise
11: **return** The distinguisher is impossible.

and ask for a solution. If the solver returns unsatisfiable, then the differential distinguisher is impossible; otherwise, the solver returns the right pairs.

As an application to SipHash-1-x with one message block, Algorithm 1 gives a detailed description of our experimental ideas on the verification of a RX-distinguisher, the model is analogous for differential distinguishers hence it is omitted here. With the RX-colliding characteristic in Table 6, there exits an internal RX-collision of a revised version of SipHash-1-x with one message block is shown in Table 8.

Table 8. RX-collision of a revised version of SipHash-1-x with one message block.

k_0	097ff58b00007470	V_0	556f883b0003e5f0
k_1	9b8e2a7effff7e0b	V_1	3061a82efffffc23
m_0	680ddf951f0079a4	V_2	1affff070080a170
m'_0	d01fa11e1e00f248	V_3	1f03f5ebe0ff1f2f

The assignment of the variables takes several seconds by the solver. In addition, comparing with the automatic-search-based technique of finding RX-characteristics, our method finds the right pair of messages in addition to generating a characteristic, which allows us to have a more precise evaluation of Rotational-XOR cryptanalysis on SipHash-1-x with short message inputs.

6   Conclusions

In this paper, we apply the Schulte-Geers's differential cryptanalysis framework on ARX cipher to find the colliding characteristics of SipHash-1-x and

SipHash-2-x. Besides, we study the security of SipHash-1-x with Rotational-XOR cryptanalysis, a recently proposed technique against ARX ciphers. Analogously to internal collision, we convert a special form of RX-characteristics into a new type of collisions, which we call internal RX-collisions. As shown by theoretical analysis, there is no internal collision for SipHash-1-x with one and two message blocks, nor internal RX-collision for SipHash-1-x with one message block. In addition, we look into the underlying cause of the impossibility and gave a criteria on the initial constants where no internal RX-collisions can be found. Notice that internal RX-collisions are closely related to the initial constants, we develop an SMT-based technique for verifying the possibility of characteristics by checking the existence of the right pairs. More importantly, with a different choice of the constants, an internal RX-collision is found with the pair of input messages recovered, which implies that the security of SipHash is closely related to the choice of the constants. Our future work is to deal with application of SMT-based technique in searching for right pairs of other ARX ciphers.

A   Proofs

Lemma 2. Suppose that there exists an RX-characteristic $(\overleftarrow{\Delta} a_1, \overleftarrow{\Delta} b_1, \overleftarrow{\Delta} c_1, \overleftarrow{\Delta} d_1) \rightarrow (\overleftarrow{\Delta} a_2, \overleftarrow{\Delta} b_2, \overleftarrow{\Delta} c_2, \overleftarrow{\Delta} d_2)$ which produces an internal RX-collision for SipHash-1-x with one message block. Then, it has the following active pattern,

$$(C_1, C_0, C_2, C_3 \oplus \Delta) \rightarrow (0, \Delta, C_4, 0),$$

where $C_0 = V_0 \oplus \overleftarrow{V_0}, C_1 = V_1 \oplus \overleftarrow{V_1}, C_2 = V_2 \oplus \overleftarrow{V_2}, C_3 = V_3 \oplus \overleftarrow{V_3}, C_4 = \texttt{0xff} \oplus \overleftarrow{\texttt{0xff}}$.

Proof. For a pair of messages m_0, m'_0, RX-difference $\Delta = m_0 \oplus \overleftarrow{m'_0}$. a_1, b_1, c_1, d_1 and a'_1, b'_1, c'_1, d'_1 are the inputs of the first round to the compression phase

$$a_1 = V_0 \oplus k_0, \qquad\qquad a'_1 = V_0 \oplus \overleftarrow{k_0},$$
$$b_1 = V_1 \oplus k_1, \qquad\qquad b'_1 = V_1 \oplus \overleftarrow{k_1},$$
$$c_1 = V_2 \oplus k_0, \qquad\qquad c'_1 = V_2 \oplus \overleftarrow{k_0},$$
$$d_1 = V_3 \oplus k_1 \oplus m_0. \quad d'_1 = V_3 \oplus \overleftarrow{k_1} \oplus m'_0.$$

where V_0, V_1, V_2, V_3 and k_0, k_1 denote the initial constants states and two 64-bit keys, respectively. The message m'_0 is processed by the keys $\overleftarrow{k_0}, \overleftarrow{k_1}$.

Then, the RX-differences $\overleftarrow{\Delta} a_1, \overleftarrow{\Delta} b_1, \overleftarrow{\Delta} c_1, \overleftarrow{\Delta} d_1$ between a_1, b_1, c_1, d_1 and a'_1, b'_1, c'_1, d'_1 equal to $(C_1, C_0, C_2, C_3 \oplus \Delta)$. Similarly, we have (b_2, a_2, c_2, d_2) equal to $(0, \Delta, C_4, 0)$ □

Theorem 1. Given any non-zero value $\Delta \in \mathbb{F}_2^{64}$, $(0, 0, 0, \Delta) \rightarrow (0, \Delta, 0, 0)$ is an impossible differential characteristic of SipHash-1-x with one message block.

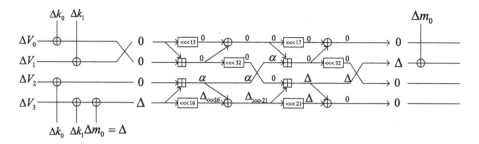

Fig. 3. The propagation of the intermediate differences in SipHahs-1-x processing one message block.

Proof. Suppose that there exists a non-zero $\Delta \in \mathbb{F}_2^{64}$, such that $(0, 0, 0, \Delta) \rightarrow (0, \Delta, 0, 0)$ is a possible differential characteristic. Then the differential propagation at the four modular addition in Fig. 3 should satisfy the differential propagation rules given by Lemma 1. We have

$$(0, 0) \xrightarrow{\boxplus} 0, \qquad (0, \Delta) \xrightarrow{\boxplus} \alpha,$$
$$(0, \alpha) \xrightarrow{\boxplus} 0, \qquad (0, \Delta \ggg 21) \xrightarrow{\boxplus} \Delta. \tag{10}$$

In each equation, the characteristic function defined in Lemma 1 is derived from the input and output differences, and evaluates to 1. Particular, when $(0, \alpha) \xrightarrow{\boxplus} 0$ is a possible differential characteristic, we have

$$((0 \oplus \alpha \oplus 0) \oplus (0 \oplus \alpha \oplus 0)_{<<1}) \preceq ((0 \oplus 0) | (\alpha \oplus 0)).$$

which is equivalent to $(\alpha \oplus \alpha_{\ll 1}) \preceq \alpha_{\ll 1}$. Therefore, we have

$$((\alpha_{63}, \cdots, \alpha_0) \oplus (\alpha_{62}, \cdots \alpha_0, 0)) \preceq (\alpha_{62}, \cdots, \alpha_0, 0). \tag{11}$$

Then $\alpha_0 = 0, \cdots, \alpha_{63} = 0$, namely, $\alpha = 0$. Analogously, we have $\Delta = 0$.

Hence, the characteristic $(0, 0, 0, \Delta) \rightarrow (0, \Delta, 0, 0)$ is trivial with $\Delta = 0$. So, we can't get right message pair that could lead to an internal collision when only one a message block is injected into SipHash-1-x.   □

Theorem 2. For any non-zero $\Delta \in \mathbb{F}_2^{64}$, $(C_1, C_0, C_2, C_3 \oplus \Delta) \rightarrow (0, \Delta, C_4, 0)$ is an impossible RX-characteristic of SipHash-1-x with one message block.

Proof. Figure 4 shows the notations for RX-differences in SipHash-1-x with one message block, $\overleftarrow{\Delta} m_0 = \Delta$ is injected before and after one round of SipHash. For the characteristic $(C_1, C_0, C_2, C_3 \oplus \Delta) \rightarrow (0, \Delta, C_4, 0)$, the RX-difference propagation at the modular additions in Fig. 4 should satisfy the propagation rule given by Proposition 1.

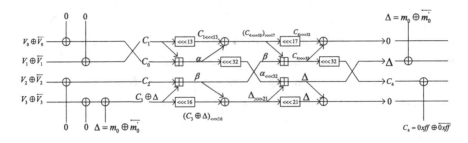

Fig. 4. Notations on RX-differences of SipHash-1-x with one message block

$$(C_1, C_0) \xrightarrow{\boxplus} \alpha,$$

$$(C_2, C_3 \oplus \Delta) \xrightarrow{\boxplus} \beta,$$

$$((C_4 \ggg 49), \beta) \xrightarrow{\boxplus} (C_4 \ggg 32),$$

$$((\alpha \lll 32), (\Delta \ggg 21)) \xrightarrow{\boxplus} \Delta. \tag{12}$$

where $C_1 = V_1 \oplus \overleftarrow{V_1} = $ 0xacb196a3b2acb1b7, $C_2 = V_2 \oplus \overleftarrow{V_2} = $ 0x95b1b7af9095af9f, $C_4 = $ 0xff $\oplus \overleftarrow{\text{0xff}}$, V_1, V_2 are the initial constants state given by SipHash design document.

By the relation between C_1, C_4, α from Fig. 4, one gets $\alpha = (C_1 \lll 13) \oplus ((C_4 \ggg 32) \ggg 17) = $ 0x32d4765596b67596. However, a necessary condition for the transition $(C_1, C_0) \xrightarrow{\boxplus} \alpha$ is $1_{(I \oplus SHL)(C_0 \oplus C_1 \oplus \alpha) \preceq SHL((C_0 \oplus \alpha)|(C_1 \oplus \alpha))}$ or $1_{(I \oplus SHL)(C_0 \oplus C_1 \oplus \alpha) \oplus 1 \preceq SHL((C_0 \oplus \alpha)|(C_1 \oplus \alpha))}$, which leads to a contradiction. Therefore, the characteristic $(C_1, C_0, C_2, C_3 \oplus \Delta) \to (0, \Delta, C_4, 0)$ is impossible. □

References

1. Abed, F., List, E., Lucks, S., Wenzel, J.: Differential cryptanalysis of round-reduced SIMON and SPECK. In: Cid, C., Rechberger, C. (eds.) FSE 2014. LNCS, vol. 8540, pp. 525–545. Springer, Heidelberg (2015). https://doi.org/10.1007/978-3-662-46706-0_27

2. Ankele, R., List, E.: Differential cryptanalysis of round-reduced Sparx-64/128. In: Preneel, B., Vercauteren, F. (eds.) ACNS 2018. LNCS, vol. 10892, pp. 459–475. Springer, Cham (2018). https://doi.org/10.1007/978-3-319-93387-0_24

3. Aumasson, J.-P., Bernstein, D.J.: SipHash: a fast short-input PRF. In: Galbraith, S., Nandi, M. (eds.) INDOCRYPT 2012. LNCS, vol. 7668, pp. 489–508. Springer, Heidelberg (2012). https://doi.org/10.1007/978-3-642-34931-7_28

4. Ashur T., Liu Y.: Rotational cryptanalysis in the presence of constants. In: IACR Transactions on Symmetric Cryptology, pp. 57–70 (2016)

5. Biryukov, A., Velichkov, V., Le Corre, Y.: Automatic search for the best trails in ARX: application to block cipher SPECK. In: Peyrin, T. (ed.) FSE 2016. LNCS, vol. 9783, pp. 289–310. Springer, Heidelberg (2016). https://doi.org/10.1007/978-3-662-52993-5_15

6. Dinur, I.: Improved differential cryptanalysis of round-reduced speck. In: Joux, A., Youssef, A. (eds.) SAC 2014. LNCS, vol. 8781, pp. 147–164. Springer, Cham (2014). https://doi.org/10.1007/978-3-319-13051-4_9

7. Dobraunig, C., Mendel, F., Schläffer, M.: Differential cryptanalysis of SipHash. In: Joux, A., Youssef, A. (eds.) SAC 2014. LNCS, vol. 8781, pp. 165–182. Springer, Cham (2014). https://doi.org/10.1007/978-3-319-13051-4_10

8. Fu, K., Wang, M., Guo, Y., Sun, S., Hu, L.: MILP-based automatic search algorithms for differential and linear trails for speck. In: Peyrin, T. (ed.) FSE 2016. LNCS, vol. 9783, pp. 268–288. Springer, Heidelberg (2016). https://doi.org/10.1007/978-3-662-52993-5_14

9. Leurent, G.: Analysis of differential attacks in ARX constructions. In: Wang, X., Sako, K. (eds.) ASIACRYPT 2012. LNCS, vol. 7658, pp. 226–243. Springer, Heidelberg (2012). https://doi.org/10.1007/978-3-642-34961-4_15

10. Leurent, G.: Construction of differential characteristics in ARX designs application to Skein. In: Canetti, R., Garay, J.A. (eds.) CRYPTO 2013. LNCS, vol. 8042, pp. 241–258. Springer, Heidelberg (2013). https://doi.org/10.1007/978-3-642-40041-4_14

11. Liu, Y., De Witte, G., Ranea, A., Ashur, T.: Rotational-XOR cryptanalysis of reduced-round SPECK. IACR Transactions on Symmetric Cryptology, **2017**(1), 24–36 (2017)

12. Mouha, N., Velichkov, V., De Cannière, C., Preneel, B.: The differential analysis of S-functions. In: Biryukov, A., Gong, G., Stinson, D.R. (eds.) SAC 2010. LNCS, vol. 6544, pp. 36–56. Springer, Heidelberg (2011). https://doi.org/10.1007/978-3-642-19574-7_3

13. Nyberg, K., Wallén, J.: Improved linear distinguishers for SNOW 2.0. In: Robshaw, M. (ed.) FSE 2006. LNCS, vol. 4047, pp. 144–162. Springer, Heidelberg (2006). https://doi.org/10.1007/11799313_10

14. Schulte-Geers, E.: On CCA-equivalence of addition mod 2^n. Des. Codes Crypt. **66**, 111–127 (2013)

15. Siddappa, S.K., Kaminsky, A.: SAT based attacks on SipHash. Rochester Institute of Technology (2014)

16. Song, L., Huang, Z., Yang, Q.: Automatic differential analysis of ARX block ciphers with application to SPECK and LEA. In: Liu, J.K., Steinfeld, R. (eds.) ACISP 2016. LNCS, vol. 9723, pp. 379–394. Springer, Cham (2016). https://doi.org/10.1007/978-3-319-40367-0_24

17. Sun, S., Hu, L., Wang, P., Qiao, K., Ma, X., Song, L.: Automatic security evaluation and (related-key) differential characteristic search: application to SIMON, PRESENT, LBlock, DES(L) and other bit-oriented block ciphers. In: Sarkar, P., Iwata, T. (eds.) ASIACRYPT 2014. LNCS, vol. 8873, pp. 158–178. Springer, Heidelberg (2014). https://doi.org/10.1007/978-3-662-45611-8_9

Lattice and Post-quantum Cryptography

Optimized Algorithms and Architectures for Montgomery Multiplication for Post-quantum Cryptography

Rami El Khatib[1]([✉]), Reza Azarderakhsh[1]([✉]),
and Mehran Mozaffari-Kermani[2]([✉])

[1] Florida Atlantic University, Boca Raton, FL, USA
{relkhatib2015,razarderakhsh}@fau.edu
[2] University of South Florida, Tampa, FL, USA
mehran2@usf.edu

Abstract. Finite field multiplication plays the main role determining the efficiency of public key cryptography systems based on RSA and elliptic curve cryptography (ECC). Most recently, quantum-safe cryptographic systems are proposed based on supersingular isogenies on elliptic curves which require large integer multiplications over extended prime fields. In this work, we present two Montgomery multiplication architectures for special primes used in a post-quantum cryptography system known as supersingular isogeny key encapsulation (SIKE). We optimize two existing Montgomery multiplication algorithms and develop area-efficient and time-efficient Montgomery multiplication architectures for hardware implementations of post-quantum cryptography. Our proposed time-efficient architecture is 32% to 42% faster than the leading one (depending on the prime size) available in the literature which has been used in original SIKE submission to the NIST standardization process. The area-efficient architecture is 42% to 50% smaller than the counterparts and is about 3% to 11% faster depending on the NIST security level.

Keywords: Hardware architectures · Isogeny-based cryptosystems · Montgomery multiplication · Post-quantum cryptography

1 Introduction

Post-quantum cryptography (PQC) refers to the research of cryptographic primitives (usually public-key cryptosystems) that are not efficiently breakable using quantum computers. Most notably, Shor's algorithm [1] can be efficiently implemented on a quantum computer to break standard Elliptic Curve Cryptography (ECC) and RSA cryptosystems. There exist some alternatives secure against quantum computing threats [2], such as lattice-based cryptosystems, hash-based signatures, code-based cryptosystems, multivariate public key cryptography, and isogeny-based cryptography [3].

© Springer Nature Switzerland AG 2019
Y. Mu et al. (Eds.): CANS 2019, LNCS 11829, pp. 83–98, 2019.
https://doi.org/10.1007/978-3-030-31578-8_5

Isogeny-based cryptography or more specifically supersingular isogeny Diffie-Hellman (SIDH) key exchange has been proposed by Jao and De Feo [4] as an alternative to Elliptic Curve Diffie-Hellman (ECDH) resistant to Shor's quantum attack. A more secure model of SIDH a.k.a. SIKE (supersingular isogeny key encapsulation) has been submitted to NIST standardization process [3]. SIKE computations constitute an algebraic map between supersingular elliptic curves, which appear to be resistant to quantum attacks. Existing results on the hardware implementations of SIKE have appeared in [5–8]. SIKE's lower level computations are mainly over $\mathbb{F}_{p^2}$ or extended prime fields. The prime size which decides the security of SIKE determines the size of arithmetic unit for lower level multiplication, addition, squaring, and inversion. Among these operations, multiplication plays the main role determining the performance of SIKE cryptosystem. Therefore, efficient and high-performance implementations of the multiplier is crucial. In comparison to the other post-quantum candidates, SIKE offers smallest key size for the same security level which is more attractive for bandwidth-constrained applications. However, SIKE is not the fastest quantum-safe candidate, and its performance still needs to be improved as stated in NIST submission [3].

In this paper, we focus on the optimization of arithmetic operations employed in SIKE and propose two new hardware architectures for modular multiplication algorithm targeting SIKE primes based on the well-known Montgomery modular multiplication algorithm [9]. Previous work on hardware implementation of Montgomery multiplication has been proposed in [10,11] for arbitrary primes and in [5,8] for SIKE primes. Since the primes employed in SIKE (SIKEp434, SIKEp503, SIKEp610, and SIKEp751 for NIST level-1, -2, -3, -5, respectively) have special forms, we developed a time-efficient implementation and an area-efficient implementation of Montgomery multiplication to be used in future work of SIKE. The time-efficient implementation reduces the latency and the area usage compared to previous work, maintaining high frequency while the area-efficient implementation significantly reduces the area.

Our Contributions

- We optimize two existing Montgomery multiplication algorithms for special primes used in post-quantum cryptography, SIKE.
- We provide efficient hardware architecture for the proposed Montgomery multiplication algorithms.
- We evaluate time and area performance of the proposed hardware architecture benchmarked on FPGA and compare with counterparts.

The organization of the paper is as follows. In Sect. 2, we discuss the Montgomery modular multiplication algorithm and two algorithms: Coarsely Integrated Operand Scanning (CIOS) and Finely Integrated Operand Scanning (FIOS) algorithms that perform Montgomery multiplication word-by-word. In Sect. 3, we provide optimization techniques for the CIOS and FIOS Montgomery multiplication algorithms. In Sect. 4, we propose efficient hardware architectures of the proposed Montgomery multiplier algorithms. In Sect. 5, we implement the

---

Algorithm 1. Montgomery Multiplication [9]

    Input : $p < 2^K$, $R = 2^K$, $p' = -p^{-1} \bmod R$, $a, b < p$
    Output: $a \cdot b \cdot R^{-1} \bmod p$
1 $T \leftarrow a \cdot b$
2 $m \leftarrow T \cdot p' \bmod R$
3 $T \leftarrow (T + m \cdot p)/R$
4 **if** $T > m$ **then return** $T - p$
5 **return** T

---

proposed hardware architectures on FPGA, provide area and timing results, and compare with counterparts available in the literature. Finally, in Sect. 6, we give our final thoughts and discuss future work.

2   Preliminaries: Montgomery Multiplication

Modular multiplication (i.e. $a \times b \bmod p$) of large integers (especially the ones used in SIKE) can be efficiently implemented using Montgomery multiplication. Montgomery multiplication [9] avoids the division operation which is difficult to implement in an efficient way in hardware. Montgomery multiplication has been used in recent hardware implementations of isogeny-based cryptography including SIDH [5–7] and SIKE [12].

2.1   Montgomery Multiplication Algorithm

Montgomery multiplication performs modular multiplication by transforming the division by p into division by a power of 2, which is a simple shift. The overhead cost of Montgomery multiplication is the need to convert the inputs into the Montgomery domain, then perform all arithmetic operations in the Montgomery domain, and finally convert back to the ordinary domain. For simple applications with few modular multiplications, this conversion would be expensive. However, in SIKE, this is extremely useful because of its high dependence on a large number of modular multiplications.

Montgomery multiplication algorithm (MontMult) takes two inputs a and b with the remaining inputs constants and produces a single output MontMult$(a, b) = a \cdot b \cdot R^{-1} \bmod p$ where R and p are co-prime. By taking R as a power of 2, the division becomes a simple shifting. Algorithm 1 shows the original Montgomery multiplication algorithm. The first part (line 1) performs the multiplication step while the second part (lines 2–4) performs the reduction step. Note that the same algorithm can be used to convert between the ordinary and Montgomery domain [9]. The conversion into the Montgomery domain can be done using MontMult$(a, R^2 \bmod p)$ where a is the input in the ordinary domain and the conversion into the ordinary domain can be done using MontMult$(a, 1)$ where a is the input in the Montgomery domain.

The final subtraction step (line 4) can be removed from the algorithm and performed once at the very end after converting to the ordinary domain. However, in this case, the output from MontMult, which is going to be the input for subsequent MontMult, is $<2p-1$ instead of $<p$. Therefore, the algorithm needs to work for any input $<2p-1$. The condition $(T + m \cdot p)/R < 2p - 1$ needs to be satisfied so that the output from line 3 is $<2p-1$. If R is taken such that it is 2 bits larger than the size of p, then this condition is satisfied.

Hardware implementations of MontMult for public key cryptography have been studied in [5,10,11,13–21].

2.2 SIKE Primes

In SIKE submission for post-quantum cryptography [12], the primes employed have a special form where Montgomery multiplication can be optimized. These primes are given in Table 1. The main advantage of the primes used in SIKE is that the least significant bits of the prime are all 1s. This form can be utilized in Montgomery multiplication algorithm variants that perform word by word computation such as the Separated Operand Scanning (SOS), Coarsely Integrated Operand Scanning (CIOS) or Finely Integrated Operand Scanning (FIOS) algorithms [22]. In these word-by-word variants, $p'_0 = p' \mod 2^w = -p^{-1} \mod 2^w$, where w is the number of bits in a word, can be used instead of p' [23]. When more than 2 words are used for SIKE primes, $p'_0 = 1$ and $m = T \cdot p'_0 \mod R = T \mod R$ in line 2 becomes a simple copy register.

Table 1. SIKE primes for post-quantum cryptography based on NIST standardization process [3]

Prime form	Classical/quantum security	Public key size (bytes)
$p_{434} = 2^{216}3^{137} - 1$	NIST level 1	330
$p_{503} = 2^{250}3^{159} - 1$	NIST level 2	378
$p_{610} = 2^{305}3^{192} - 1$	NIST level 3	462
$p_{751} = 2^{372}3^{239} - 1$	NIST level 5	564

2.3 Coarsely Integrated Operand Scanning (CIOS) Montgomery Multiplication

The CIOS Montgomery multiplication algorithm [22] is a method that performs word-by-word Montgomery multiplication by alternating between the multiplication and reduction steps. The inputs a and b and prime p are split into s words of w-bit wide each. CIOS is shown in Algorithm 2. As seen, lines 4–9 perform the multiplication step while lines 11–18 perform the reduction step. Hardware implementations of CIOS have been carried out in [10] by Mrabet *et al.* and in [13] by McIvor *et al.*

Algorithm 2. CIOS Montgomery Multiplication Algorithm [22]

  Input : $p < 2^K$, $R = 2^K$, w, s, $K = w \cdot s$, $p' = -p^{-1} \bmod 2^w$, $a, b < p$
  Output: MontMult(a, b)
1 $T \leftarrow 0$
2 **for** $i \leftarrow 0$ **to** $s - 1$ **do**
3     $C \leftarrow 0$
4   **for** $j \leftarrow 0$ **to** $s - 1$ **do**
5         $(C, S) \leftarrow T[j] + a[i] \cdot b[j] + C$
6         $T[j] \leftarrow S$
7     $(C, S) \leftarrow T[s] + C$
8     $T[s] \leftarrow S$
9     $T[s + 1] \leftarrow C$
10
11     $m \leftarrow T[0] \cdot p' \bmod 2^w$
12     $(C, S) \leftarrow T[0] + m \cdot p[0]$
13   **for** $j \leftarrow 1$ **to** $s - 1$ **do**
14         $(C, S) \leftarrow T[j] + m \cdot p[j] + C$
15         $T[j - 1] \leftarrow S$
16     $(C, S) \leftarrow T[s] + C$
17     $T[s - 1] \leftarrow S$
18     $T[s] \leftarrow T[s + 1] + C$
19 **return** T

2.4 Finely Integrated Operand Scanning (FIOS) Montgomery Multiplication

The FIOS Montgomery multiplication algorithm [22] is a method that performs word-by-word Montgomery multiplication by performing the multiplication and reduction steps in the same loop. Similar to CIOS, the inputs a,b and prime p are split into s words of w-bit each. FIOS is shown in Algorithm 3. In FIOS, the first multiplication and m must be computed (3–6) before performing the remaining multiplication and reduction steps (lines 7–15). The main difference between FIOS and CIOS is that in FIOS, the multiplication and reduction can be parallelized while in CIOS, the reduction has to wait for the multiplication step.

    Hardware implementation of FIOS has been conducted by McIvor et al. in [13]. It has been shown that FIOS performed slower than CIOS. The main reason for this is the need for carry propagation units (lines 4 and 9). To address, we will show in this paper that the carry propagation can be eliminated for 1-bit larger registers which adds minimal cost in hardware implementations and improves the critical path delay.

Algorithm 3. FIOS Montgomery Multiplication Algorithm [22]

Input : $p < 2^K$, $R = 2^K$, w, s, $K = w \cdot s$, $p' = -p^{-1} \mathrm{mod}\ 2^w$, $a, b < p$
Output: MontMult(a, b)

1   $T \leftarrow 0$
2   **for** $i \leftarrow 0$ **to** $s - 1$ **do**
3      $(C, S) \leftarrow T[0] + a[i] \cdot b[0]$
4      $ADD(t(1), C)$
5      $m \leftarrow S \cdot p' \mod 2^w$
6      $(C, S) \leftarrow S + m \cdot p[0]$
7   **for** $j \leftarrow 1$ **to** $s - 1$ **do**
8         $(C, S) \leftarrow T[j] + a[i] \cdot b[j] + C$
9         $ADD(T[j + 1], C)$
10         $(C, S) \leftarrow S + m \cdot p[j]$
11         $T[j - 1] \leftarrow S$
12      $(C, S) \leftarrow T[s] + C$
13      $T[s - 1] \leftarrow S$
14      $T[s] \leftarrow T[s + 1] + C$
15      $T[s + 1] \leftarrow 0$
16 **return** T

---

3 Proposed Finite Field Multiplier Algorithms

In this section, based on the information provided in the previous section, we propose an optimized CIOS and FIOS Montgomery multiplication algorithms for primes employed in SIKE.

3.1 Optimized CIOS (O-CIOS) Montgomery Multiplication Algorithm

We propose a new optimized coarsely integrated operand scanning multiplier (O-CIOS) for SIKE which requires less hardware units as shown in Algorithm 4. We show that by taking R 3 bits larger than the size of the prime p, we only require $s + 1$ registers. First, lines 11–12 in the original algorithm (Algorithm 2) are replaced by lines 9–10 for $s > 2$ since $p' = 1$ (as shown in Subsect. 2.2) and $p[0] = 2^w - 1$. Proposition 1 shows how lines 7–9 and 16–18 can be replaced by lines 7 and 14, respectively, for $w > 2$.

Proposition 1. *In Algorithm 2, lines 7–9 and 16–18 can be can be replaced by lines 7 and 14 in Algorithm 4, respectively, for $w > 2$.*

Proof. For $i = 0$, iteration $j = s - 1$ in line 5 has output $\leq (2^w - 1)(2^{w-2} - 1) + (2^w - 1) = 2^{w-2}(2^w - 1) \leq 2^{2w-2} - 1$ which implies $C \leq 2^{w-2} - 1 \leq 2^{w-1} - 1$. Therefore, $T[s] \leq 2^{w-1} - 1$ and $T[s + 1] = 0$ in lines 8 and 9. Iteration $j = s - 1$ in line 5 has output $\leq (2^w - 1) + (2^w - 1)(2^{w-3} - 1) + (2^w - 1) = (2^{w-3} + 1)(2^w - 1) \leq 2^{w-2}(2^w - 1) \leq 2^{2w-1} - 1$ for $w > 2$ which implies

---

Algorithm 4. Optimized CIOS Montgomery Multiplication Algorithm for SIKE primes

---

Input : $p < 2^{K-3}$, $R = 2^K$, $w > 2$, $s > 2$, $K = w \cdot s$, $a, b < 2p - 1$
Output: MontMult(a, b)

1 $T \leftarrow 0$
2 **for** $i \leftarrow 0$ **to** $s - 1$ **do**
3     $C \leftarrow 0$
4   **for** $j \leftarrow 0$ **to** $s - 1$ **do**
5        $(C, S) \leftarrow T[j] + a[i] \cdot b[j] + C$
6        $T[j] \leftarrow S$
7     $T[s] \leftarrow C$
8
9     $m \leftarrow T[0]$
10    $C \leftarrow T[0]$
11   **for** $j \leftarrow 1$ **to** $s - 1$ **do**
12       $(C, S) \leftarrow T[j] + m \cdot p[j] + C$
13       $T[j - 1] \leftarrow S$
14    $T[s - 1] \leftarrow T[s] + C$
15 **return** T

$C \le 2^{w-1} - 1$. Therefore, $T[s - 1] \le 2^{w-1} - 1$ and $T[s] = 0$ in lines 17 and 18. Now, proving for iteration $i = k$ where $k > 0$, iteration $j = s - 1$ in line 5 has output $\le (2^w - 1) + (2^w - 1)(2^{w-2} - 1) + (2^w - 1) = (2^{w-2} + 1)(2^w - 1) \le 2^{w-1}(2^w - 1) \le 2^{2w-1} - 1$ for $w > 1$ which implies $C \le 2^{w-1} - 1$. Therefore, $T[s] \le 2^{w-1} - 1$ and $T[s + 1] = 0$ in lines 8 and 9. Iteration $j = s - 1$ in line 5 has output $\le (2^w - 1) + (2^w - 1)(2^{w-3} - 1) + (2^w - 1) = (2^{w-3} + 1)(2^w - 1) \le 2^{w-2}(2^w - 1) \le 2^{2w-1} - 1$ for $w > 2$ which implies $C \le 2^{w-1} - 1$. Therefore, $T[s - 1] \le 2^{w-1} - 1$ and $T[s] = 0$ in lines 17 and 18. This complete the proof.

3.2 Optimized FIOS (O-FIOS) Montgomery Multiplication Algorithm

We also propose an optimized FIOS algorithm for Montgomery multiplication as shown in Algorithm 5. Similar to CIOS, lines 5 and 6 can be modified since $p' = 1$ and $p[0] = 2^w - 1$ for $s > 2$. As for the carry propagation in lines 4 and 9, they can be directly integrated inside the other carry C in lines 6 and 10. However, the carry C must use a register of size $w + 1$ bits to accommodate the extra accumulated bits. The changes are shown in lines 4–5 and line 7 in the optimized algorithm. We notice that the two multiplications in line 7 can be performed in the same cycle in parallel and without the need to propagate the result of the first multiplication which will decrease architecture complexity and routing delays. Proposition 2 shows how lines 12–15 can be replaced by 9 for $w > 2$.

Algorithm 5. Optimized FIOS Montgomery Multiplication Algorithm for SIKE primes

---

Input : $p < 2^{K-2}$, $R = 2^K$, $w > 2$, $s > 2$, $K = w \cdot s$, $a, b < 2p - 1$
Output: MontMult(a, b)

1   $T \leftarrow 0$
2   **for** $i \leftarrow 0$ **to** $s - 1$ **do**
3      $(C, S) \leftarrow T[0] + a[i] \cdot b[0]$
4      $m \leftarrow S$
5      $C \leftarrow C + S$
6   **for** $j \leftarrow 1$ **to** $s - 1$ **do**
7         $(C, S) \leftarrow T[j] + a[i] \cdot b[j] + m \cdot p[j] + C$
8         $T[j - 1] \leftarrow S$
9      $T[s - 1] \leftarrow \mathrm{LSW}(C)$
10 **return** T

Proposition 2. *In Algorithm 3, lines 12–15 can be replaced by line 9 in Algorithm 5 for $w > 2$.*

Proof. For $i = 0$, iteration $j = s - 1$ in line 7 in new algorithm has output $(\leq (2^w - 1)(2^{w-1} - 1) + (2^w - 1)(2^{w-2} - 1) + (2^w - 1) = (2^{w-1} + 2^{w-2} + 1)(2^w - 1) \leq 2^w(2^w - 1) = 2^{2w} - 1$ for $w > 1$ which implies $C \leq 2^w - 1$. Therefore, $T[s - 1] \leq 2^w - 1$ and $T[s] = 0$. Now, proving for iteration $i = k$ where $k > 0$, iteration $j = s - 1$ in line 7 in new algorithm has output $(\leq (2^w - 1) + (2^w - 1)(2^{w-1} - 1) + (2^w - 1)(2^{w-2} - 1) + (2^w - 1) = (2^{w-1} + 2^{w-2} + 2)(2^w - 1) \leq 2^w(2^w - 1) = 2^{2w} - 1$ for $w > 2$ which implies $C \leq 2^w - 1$. Therefore, $T[s - 1] \leq 2^w - 1$ and $T[s] = 0$. This complete the proof.

4   Proposed Efficient Architecture for O-CIOS and O-FIOS Montgomery Multiplication Algorithms

In this section, we propose a hardware architecture design for each of the new optimized algorithms O-CIOS and O-FIOS discussed in the previous section. The O-CIOS design focuses on minimizing area usage while the design for O-FIOS focuses on maximizing the frequency and minimizing the total multiplication time.

4.1   Proposed O-CIOS Architecture

The proposed O-CIOS architecture is illustrated in Fig. 1 which mainly improves the area usage in comparison to the ones adopted before for hardware implementations. The architecture is composed of several processing elements (PEs) cascaded to perform the multiplication and reduction steps as can be seen in

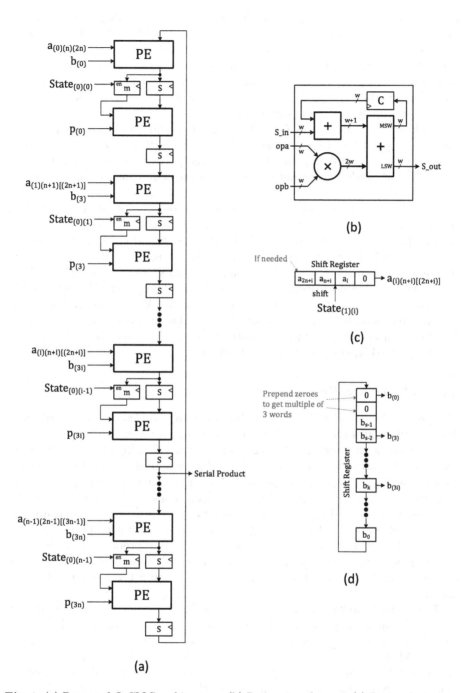

Fig. 1. (a) Proposed O-CIOS architecture. (b) Processing element. (c) Design for input b and p. (d) Design for input a.

Fig. 1(a). Each PE performs a multiplication and an addition in parallel followed by an addition corresponding to lines 5 and 12 in Algorithm 4 as shown in Fig. 1(b). Notice that unlike Mrabet's design [10], no final PEs for multiplication and reduction are required since lines 7 and 14 are respectively similar to lines 5 and 12 with some registers set to 0. In addition, line 10 can be integrated in the reduction PE (line 12) since $T[0] + m \cdot p[0] = T[0] \cdot 2^w$ which implies $C = T[0]$ before performing the first iteration.

Each *odd* PE performs the entire first inner loop (lines 4–6) and line 7 (multiplication step) of one outer iteration while each *even* PE performs entire second inner loop (lines 11–13) and line 14 (reduction step) of one outer iteration. There is one fan-out needed in the output of the *multiplication* PE to store m for the *reduction* PE.

Once the PE is finished with processing one iteration, it processes another iteration, which can be seen by the loop-back after the last PE. There is a delay of 3 cycles between two consecutive multiplication or reduction PEs corresponding to one multiplication, one reduction, and computing m cycles. Therefore, to minimize the number of cycles, the number of PEs used is $\lfloor (s+1)/3 \rfloor$. This means that each PE performs 2 or 3 iterations. Thus, each PE uses 2 or 3 different words of input a as can be seen in Fig. 1(c). Words from inputs b and p rotate across each PE as shown in Fig. 1(d). The number of cycles required for this architecture to compute Montgomery multiplication is $4s$ cycles. For instance, for p434, O-CIOS requires 112 clock cycles.

4.2   O-FIOS Architecture

The proposed O-FIOS architecture is illustrated in Figs. 2 and 3. As one can see, Fig. 2(a) illustrates the proposed systolic architecture based on PEs. This architecture focuses on improving timing results by parallelizing the multiplication and reduction steps. The initial PE (Fig. 2(b)) computes m and the first carry for each iteration (lines 3–5 in Algorithm 5). The remaining PEs (Fig. 3(c)) perform two consecutive iterations of the inner loop (lines 7–8). Therefore, each PE processes two words of input b and prime p following Fig. 3(d) and outputs the two words of the output following Fig. 3(e). Words for input a are pushed serially using a shift register (Fig. 3(f)) into the initial PE and propagated to the next PE after two cycles corresponding to the two consecutive iterations mentioned earlier. Similarly, m is propagated through each PE after being evaluated in the initial PE.

For the last line 9 of the algorithm, we tried feeding back the carry output of the last PE into its sum input. However, this has caused a routing delay in the FPGA we are using outweighing the cycle saved in the process. Therefore, we have decided that for *even* s, the last PE can process the last line by grounding the second b_{in}, second p_{in}, and S_{in} while in *odd* s, a simple register is used to store the result before fed-back into S_{in} of the last PE. The number of cycles required for this architecture to compute Montgomery multiplication is $3s$ cycles. For example, in p434, O-FIOS requires 84 cycles.

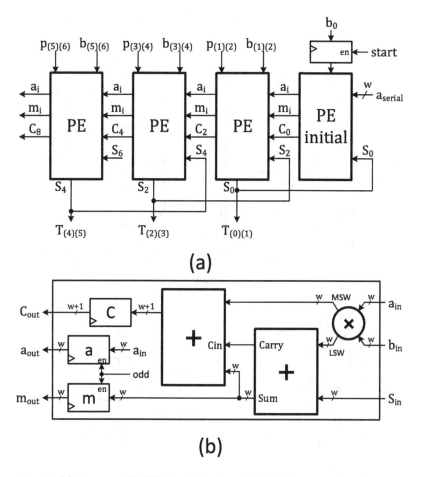

Fig. 2. (a) Proposed O-FIOS architecture. (b) Initial processing element.

4.3   Time Complexity Analysis

Table 2 provides a time complexity comparison between our O-CIOS and O-FIOS implementations and different Montgomery multiplication implementations. For a fair comparison, we have optimized Mrabet *et al.*'s implementation [13] for SIKE primes by changing the β-cell into a simple register since $p^{-1} = 1$. As for Koziel *et al.*'s implementation [5,11], we used a non-interleaved version of the multiplier. Our proposed O-CIOS uses less number of clock cycles while maintaining the same critical path delay of Mrabet's CIOS. Our proposed O-FIOS uses the least number of clock cycles of any design while maintaining the same critical path delay of Koziel's implementation. However, in the next section, our results show that O-FIOS perform at the same frequency of Mrabet's CIOS mainly because our design has minimal routing delays. Furthermore, our designs require less area which we will show in the next section after implementing in hardware.

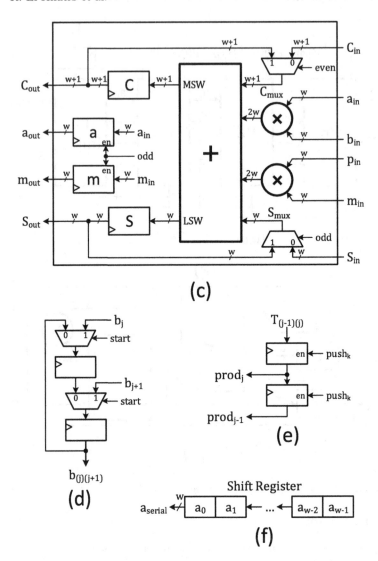

Fig. 3. Remaining components of O-FIOS architecture. (c) Single processing element. (d) Design for input b and p. (e) Design for input a. (f) Design for output.

5 Implementation Results

In this section, we are going to provide implementation results for the proposed Montgomery multiplication architectures, O-CIOS and O-FIOS, discussed in the previous sections. The implementations are performed in Xilinx Vivado 2018.2 for Xilinx Virtex-7 FPGA xc7vx690tffg1157-3. Table 2 reports area and timing results for O-CIOS and O-FIOS. As one can see, for NIST level-1, our proposed O-CIOS architecture operates in 233.5 MHz and occupies 770 Flip-flops, 1869

Table 2. Time complexity comparison of 448-bit Montgomery multipliers for SIKEp434. $T_{16\times}$ indicates the critical path of a 16-bit×16-bit multiplication. T_{32+} . indicates the critical path of a 32-bit addition.

Work	Critical path delay	Latency (cc)
Mrabet et al. [10] ($w = 16$)	$T_{16\times} + T_{32+}$	114
McIvor/Koziel et al. [5,11] ($w = 16$)	$T_{16\times} + 2T_{32+}$	87
This work O-CIOS ($w = 16$)	$T_{16\times} + T_{32+}$	112
This work O-FIOS ($w = 16$)	$T_{16\times} + 2T_{32+}$	84

Table 3. Implementation results and comparison of proposed O-FIOS and O-CIOS Montgomery multiplication architectures on a Xilinx Virtex-7 FPGA device, xc7vx690tffg1157-3

Prime	NIST level	Area				Time			Area × Time
		# FFs	# LUTs	# DSPs	# Slices	Freq. (MHz)	Latency (cc)	Total time (ns)	(×1000)
This work O-CIOS ($w = 16$)									
p_{434}	1	770	1869	40	447	233.481	112	479.696	214.424
p_{503}	2	851	2094	44	519	216.685	128	590.720	306.584
p_{610}	3	1075	2609	56	646	217.297	156	717.912	463.771
p_{751}	5	1309	3188	68	761	219.635	192	874.176	665.248
This work O-FIOS ($w = 16$)									
p_{434}	1	1119	1905	43	607	271.370	**84**	**309.540**	**187.891**
p_{503}	2	1290	2219	49	554	**267.380**	96	**359.040**	**198.908**
p_{610}	3	1568	2704	61	835	**267.380**	117	**437.580**	**365.379**
p_{751}	5	1794	3308	73	967	232.829	144	**618.480**	**598.070**
Mrabet et al. [10]* ($w = 16$)									
p_{434}	1	3492	3737	40	2959	**273.973**	114	416.100	1,231.240
p_{503}	2	3884	4240	44	3334	251.130	129	513.678	1,486.911
p_{610}	3	4741	5148	**54**	4041	247.158	159	643.314	1,831.187
p_{751}	5	5814	6288	**66**	4970	**245.459**	195	794.430	2,267.751
McIvor/Koziel et al. [5,11]* ($w = 16$)									
p_{434}	1	687	3177	84	895	160.694	87	541.401	484.554
p_{503}	2	784	3641	96	1044	162.470	99	609.345	636.156
p_{610}	3	952	4040	117	1315	162.101	120	740.280	973.468
p_{751}	5	1168	4193	144	1310	159.974	147	918.897	1,203.755

*Note that the original paper does not have the results for these primes. The numbers are based on our implementations using this work.

LUTs, 40 DSPs and 447 slices. The total time to perform one Montgomery multiplication is 480 ns in O-CIOS. On the other hand, O-FIOS operates at 271.4 MHz and occupies 1119 FFs, 1905 LUTs, 43 DSPs and 607 slices for NIST level-1. The total time to perform one Montgomery multiplication is 310 ns in O-FIOS (Table 3).

5.1   Comparison and Discussion

Table 2 compares our results with two different implementations; Mrabet *et al.* [10] and non-interleaved Koziel *et al.* [5,11]. Our O-FIOS is 22% to 31% faster than Mrabet's implementation and 32% to 42% faster than Koziel's implementation (a non-interleaved version of the one used in SIKE). In addition, the O-FIOS architecture uses less area compared to other implementations (other than O-CIOS). Our O-CIOS focuses on minimizing the area usage while maintaining high frequency and low total time. For example, in p434, the number of slices used in O-CIOS is 447 which is 2× smaller than Koziel's implementation at 895 slices and 6.5× smaller than Mrabet's implementation at 2959 slices. This implementation of O-CIOS is slightly slower (9% to 13%) than Mrabet's CIOS and slightly faster (3% to 11%) than Koziel's implementation.

6   Conclusion

In this paper, we discussed two optimized Montgomery multiplication algorithms O-CIOS and O-FIOS for SIKE primes. We then developed an architecture for each algorithm. The O-CIOS architecture focuses on minimizing the area usage while the O-FIOS architecture focuses on minimizing the total time. For performance evaluation and comparison, we implemented our proposed architectures in FPGA and showed area and timing results.

The Montgomery multiplication architectures developed in this paper show great potential in increasing the performance of SIKE. Our future work is to develop an optimized version of SIKE and employ the two Montgomery multiplier architectures developed in this paper.

Acknowledgement. The authors would like to thank the reviewers for their comments. This work is supported in parts by NSF CNS-1801341, NIST-60NANB16D246, and ARO W911NF-17-1-0311.

References

1. Shor, P.W.: Algorithms for quantum computation: discrete logarithms and factoring. In: 35th Annual Symposium on Foundations of Computer Science, FOCS 1994, pp. 124–134 (1994)
2. Chen, L., et al.: Report on Post-Quantum Cryptography (2016). NIST IR 8105
3. Jao, D., et al.: Supersingular isogeny key encapsulation. Submission to the NIST Post-Quantum Standardization Project (2019)
4. Jao, D., De Feo, L.: Towards quantum-resistant cryptosystems from supersingular elliptic curve isogenies. In: Yang, B.-Y. (ed.) PQCrypto 2011. LNCS, vol. 7071, pp. 19–34. Springer, Heidelberg (2011). https://doi.org/10.1007/978-3-642-25405-5_2
5. Koziel, B., Azarderakhsh, R., Mozaffari-Kermani, M., Jao, D.: Post-quantum cryptography on FPGA based on isogenies on elliptic curves. IEEE Trans. Circuit. Syst. I: Regul. Pap. **64**(1), 86–99 (2017)

6. Koziel, B., Azarderakhsh, R., Mozaffari-Kermani, M.: Fast hardware architectures for supersingular isogeny Diffie-Hellman key exchange on FPGA. In: Dunkelman, O., Sanadhya, S.K. (eds.) INDOCRYPT 2016. LNCS, vol. 10095, pp. 191–206. Springer, Cham (2016). https://doi.org/10.1007/978-3-319-49890-4_11
7. Koziel, B., Azarderakhsh, R., Mozaffari-Kermani, M.: A high-performance and scalable hardware architecture for isogeny-based cryptography. IEEE Trans. Comput. **67**(11), 1594–1609 (2018)
8. Roy, D.B., Mukhopadhyay, D.: Post quantum ecc on fpga platform. Cryptology ePrint Archive, Report 2019/568 (2019). https://eprint.iacr.org/2019/568
9. Montgomery, P.L.: Modular multiplication without trial division. Math. Comput. **44**(170), 519–521 (1985)
10. Mrabet, A., et al.: High-performance elliptic curve cryptography by using the CIOS method for modular multiplication. In: Cuppens, Frédéric, Cuppens, Nora, Lanet, Jean-Louis, Legay, Axel (eds.) CRiSIS 2016. LNCS, vol. 10158, pp. 185–198. Springer, Cham (2017). https://doi.org/10.1007/978-3-319-54876-0_15
11. McIvor, C., McLoone, M., McCanny, J.V.: High-radix systolic modular multiplication on reconfigurable hardware. In: IEEE International Conference on Field-Programmable Technology, pp. 13–18 (2005)
12. Jao, D., et al.: Supersingular isogeny key encapsulation. Submission to the NIST Post-Quantum Standardization Project (2017)
13. McIvor, C., McLoone, M., McCanny, J.V.: FPGA Montgomery multiplier architectures - a comparison. In: 12th Annual IEEE Symposium on Field-Programmable Custom Computing Machines, pp. 279–282, April 2004
14. Alrimeih, H., Rakhmatov, D.: Fast and flexible hardware support for ECC over multiple standard prime fields. IEEE Trans. Very Large Scale Integr. (VLSI) Syst. **22**(12), 2661–2674 (2014)
15. Blum, T., Paar, C.: High-radix montgomery modular exponentiation on reconfigurable hardware. IEEE Trans. Comput. **50**(7), 759–764 (2001)
16. Chen, G., Bai, G., Chen, H.: A high-performance elliptic curve cryptographic processor for general curves over GF(p) based on a systolic arithmetic unit. IEEE Trans. Circ. Syst. II: Express Briefs **54**(5), 412–416 (2007)
17. Eberle, H., Gura, N., Shantz, S.C., Gupta, V., Rarick, L., Sundaram, S.: A public-key cryptographic processor for RSA and ECC. In: Proceedings of 15th IEEE International Conference on Application-Specific Systems, Architectures and Processors, pp. 98–110, September 2004
18. Ghosh, S., Alam, M., Roy Chowdhury, D., Sen Gupta, I.: Parallel crypto-devices for GF(p) elliptic curve multiplication resistant against side channel attacks. Comput. Electr. Eng. **35**, 329–338 (2009)
19. Sakiyama, K., Mentens, N., Batina, L., Preneel, B., Verbauwhede, I.: Reconfigurable modular arithmetic logic unit for high-performance public-key cryptosystems. In: Bertels, K., Cardoso, J.M.P., Vassiliadis, S. (eds.) ARC 2006. LNCS, vol. 3985, pp. 347–357. Springer, Heidelberg (2006). https://doi.org/10.1007/11802839_43
20. Tenca, A.F., Koç, Ç.K.: A scalable architecture for montgomery nultiplication. In: Koç, Ç.K., Paar, C. (eds.) CHES 1999. LNCS, vol. 1717, pp. 94–108. Springer, Heidelberg (1999). https://doi.org/10.1007/3-540-48059-5_10
21. Vliegen, J., et al.: A compact FPGA-based architecture for elliptic curve cryptography over prime fields. In: ASAP 2010–21st IEEE International Conference on Application-specific Systems, Architectures and Processors, pp. 313–316, July 2010

22. Kaya Koc, C., Acar, T., Kaliski, B.S.: Analyzing and comparing montgomery multiplication algorithms. IEEE Micro **16**(3), 26–33 (1996)
23. Dussé, S.R., Kaliski, B.S.: A cryptographic library for the motorola DSP56000. In: Damgård, I.B. (ed.) EUROCRYPT 1990. LNCS, vol. 473, pp. 230–244. Springer, Heidelberg (1991). https://doi.org/10.1007/3-540-46877-3_21

Simplified Revocable Hierarchical Identity-Based Encryption from Lattices

Shixiong Wang[1,4], Juanyang Zhang[2], Jingnan He[3,4(✉)], Huaxiong Wang[4], and Chao Li[1]

[1] College of Computer, National University of Defense Technology, Changsha, China
[2] School of Information Engineering, Ningxia University, Yinchuan, China
[3] State Key Laboratory of Information Security, Institute of Information Engineering of Chinese Academy of Sciences, Beijing, China
hejingnan@iie.ac.cn
[4] School of Physical and Mathematical Sciences, Nanyang Technological University, Singapore, Singapore

Abstract. As an extension of identity-based encryption (IBE), revocable hierarchical IBE (RHIBE) supports both key revocation and key delegation simultaneously, which are two important functionalities for cryptographic use in practice. Recently in PKC 2019, Katsumata et al. constructed the first lattice-based RHIBE scheme with decryption key exposure resistance (DKER). Such constructions are all based on bilinear or multilinear maps before their work. In this paper, we simplify the construction of RHIBE scheme with DKER provided by Katsumata et al. With our new treatment of the identity spaces and the time period space, there is only one short trapdoor base in the master secret key and in the secret key of each identity. In addition, we claim that some items in the keys can also be removed due to the DKER setting. Our first RHIBE scheme in the standard model is presented as a result of the above simplification. Furthermore, based on the technique for lattice basis delegation in fixed dimension, we construct our second RHIBE scheme in the random oracle model. It has much shorter items in keys and ciphertexts than before, and also achieves the adaptive-identity security under the learning with errors (LWE) assumption.

Keywords: Lattices · Identity-based encryption · Revocation · Delegation

1 Introduction

Background. Identity-based encryption (IBE), envisaged by Shamir [19] in 1984, is an advanced form of public-key encryption (PKE) where any string such as an email address can be used as a public key. Hierarchical IBE (HIBE), an extension of IBE introduced by Horwitz and Lynn [10] in 2002, further supports a key delegation functionality. Moreover, just as many multi-user cryptosystems,

© Springer Nature Switzerland AG 2019
Y. Mu et al. (Eds.): CANS 2019, LNCS 11829, pp. 99–119, 2019.
https://doi.org/10.1007/978-3-030-31578-8_6

an efficient revocation mechanism is usually necessary and imperative in the (H)IBE setting. The public/private key pair of a system user may need to be removed for various reasons, such as that the user is no longer a legitimate system user, or that the private key is lost or stolen. Designing the revocable IBE (RIBE) or revocable HIBE (RHIBE) turned out to be a challenging problem.

In 2001, Boneh and Franklin [6] proposed a naive solution for RIBE, which requires users to periodically renew their private keys. This solution is too impractical to be used in large-scale system, since for the key generation center (denoted by KGC), the workload grows linearly in the number of users N. Later in 2008, Boldyreva et al. [5] utilized the complete subtree (CS) method of Naor et al. [16] to construct the first scalable RIBE, where KGC's workload is only logarithmic in N. RIBE requires three types of keys: a secret key SK, a key update KU, and a decryption key DK. For each time period t, the KGC broadcasts a key update $KU_{KGC,t}$ through a public channel, and only non-revoked identity ID at this time period t can derive a decryption key $DK_{ID,t}$ by combining its secret key SK_{ID} with the key update $KU_{KGC,t}$. In the security model of [5], the adversary only has the access to a secret key reveal oracle and a key update reveal oracle. However, leakage of decryption keys may also happen in practice. In 2013, Seo and Emura [18] introduced a new security notion called decryption key exposure resistance (DKER), and thus refined the security model, where the adversary also has the access to a decryption key reveal oracle. The works in [5] and [18] attracted a lot of followup works, and their RIBE schemes were also extended to RHIBE schemes. Note that before Katsumata et al.'s work [11] in 2019, the constructions of R(H)IBE schemes with DKER are all based on bilinear or multilinear maps, and they rely heavily on the so-called key re-randomization property.

This paper focuses on the lattice-based cryptography, which has faster arithmetic operations and conjectured security against quantum attacks. In 2012, Chen et al. [8] employed Agrawal et al.'s IBE [1] and the CS method [16] to construct the first lattice-based RIBE scheme without DKER. Then in 2017, Takayasu and Watanabe [20] presented a new lattice-based RIBE scheme secure against exposure of a-priori bounded number of decryption keys for every identity. Namely, their scheme only achieves bounded DKER. Later in 2019, Katsumata et al. [11] proposed the first lattice-based R(H)IBE scheme with DKER under the learning with errors (LWE) assumption. Specifically, they provided a generic construction of RIBE with DKER from any RIBE without DKER and two-level HIBE. This result directly implies the first lattice-based RIBE scheme with DKER. Furthermore, they constructed the first lattice-based RHIBE scheme with DKER by further exploiting the algebraic structure of lattices. Since lattices are ill-fit with the so-called key re-randomization property, Katsumata et al. [11] introduced new tools such as leveled ciphertexts, leveled secret keys, leveled decryption keys, and level conversion keys. Therefore, their techniques highly depart from previous works which are based on bilinear or multilinear maps.

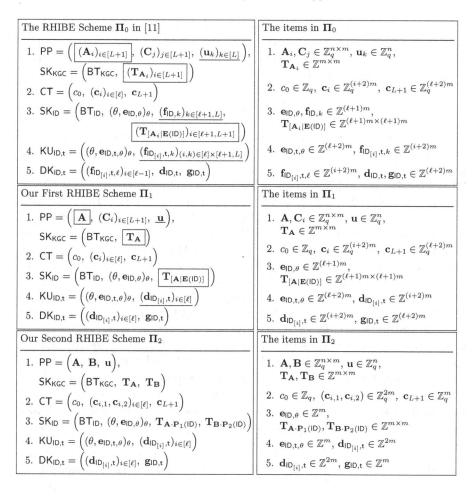

Fig. 1. Comparison of the RHIBE schemes Π_0, Π_1, Π_2

Our Contributions and Techniques. In this paper, we manage to simplify the construction of lattice-based RHIBE scheme with DKER in [11]. Specifically, we present two new RHIBE schemes Π_1 and Π_2, both of which are based on lattices and achieve DKER. Let Π_0 denote the RHIBE scheme with DKER in [11]. Then compared with Π_0, our first scheme Π_1 has fewer items in the public parameters, secret keys, and key updates. Furthermore, in our second scheme Π_2, the items in keys and ciphertexts are much shorter than Π_0, Π_1. The scheme Π_0 in [11] and our first scheme Π_1 are in the standard model, and they both satisfy the selective-identity security, assuming the hardness of the LWE problem. While our second scheme Π_2, which is in the random oracle model, achieves the adaptive-identity security under the LWE assumption.

In Fig. 1, we show the public parameters PP, the master secret key $\mathsf{SK_{KGC}}$ (the secret key of KGC), the ciphertext CT, the secret key $\mathsf{SK_{ID}}$, the key update $\mathsf{KU_{ID,t}}$, and the decryption key $\mathsf{DK_{ID,t}}$, together with the description of their items, for the schemes $\mathbf{\Pi}_0$, $\mathbf{\Pi}_1$ and $\mathbf{\Pi}_2$. In this figure, L is the maximum depth of the hierarchy, and we use $\ell := |\mathsf{ID}|$ to denote the depth of the corresponding ID explicitly in $\mathsf{SK_{ID}}$, $\mathsf{KU_{ID,t}}$, $\mathsf{DK_{ID,t}}$, or implicitly in CT, respectively. In addition, for $n_1, n_2 \in \mathbb{N}$, we set $[n_1, n_2] := \{n_1, n_1 + 1, \cdots, n_2\}$ if $n_1 \leqslant n_2$, or $[n_1, n_2] := \emptyset$ if $n_1 > n_2$, and then let $[n] := [1, n]$ for $n \in \mathbb{N}$. Figure 1 only provides a brief description of the RHIBE schemes $\mathbf{\Pi}_0, \mathbf{\Pi}_1, \mathbf{\Pi}_2$, and the notations in this figure will be clarified later in this paper when necessary. For example, the notation $\mathsf{BT_{KGC}}$ (or $\mathsf{BT_{ID}}$), which denotes a binary tree managed by KGC (or ID), is introduced in Sect. 2.3. The function $\mathbf{E}(\cdot)$ used in $\mathsf{SK_{ID}}$ for $\mathbf{\Pi}_0, \mathbf{\Pi}_1$ is described in Sect. 3, and the functions $\mathbf{P}_1(\cdot), \mathbf{P}_2(\cdot)$ used in $\mathsf{SK_{ID}}$ for $\mathbf{\Pi}_2$ are defined in Sect. 4. Actually, Fig. 1 is mainly for the comparison, from which we can see that our first scheme $\mathbf{\Pi}_1$ needs fewer items than $\mathbf{\Pi}_0$, and the sizes of items are much smaller in our second scheme $\mathbf{\Pi}_2$. Furthermore, with the help of Fig. 1, we can briefly introduce our techniques as follows.

In the RHIBE, each identity $\mathsf{ID} = (\mathsf{id}_1, \cdots, \mathsf{id}_\ell)$ at level $\ell \in [L]$ belongs to the hierarchical identity space $\mathcal{ID}_\mathsf{H} = (\mathcal{ID})^{\leqslant L} := \bigcup_{i \in [L]} (\mathcal{ID})^i$, where $\mathcal{ID}$ is the element identity space. The KGC, i.e., the key generation center, is the unique level-0 identity. For the construction of our scheme $\mathbf{\Pi}_1$, we introduce another space $\widetilde{\mathcal{ID}}$ such that $\mathcal{ID} \cap \widetilde{\mathcal{ID}} = \emptyset$, $|\mathcal{ID}| = |\widetilde{\mathcal{ID}}|$, and there is a one-to-one correspondence between $\mathsf{id} \in \mathcal{ID}$ and $\widetilde{\mathsf{id}} \in \widetilde{\mathcal{ID}}$. Suppose that in the encryption algorithm, a message M is encrypted under an identity $\mathsf{ID} = (\mathsf{id}_1, \cdots, \mathsf{id}_\ell) \in \mathcal{ID}_\mathsf{H}$ (and under a time period t). Then from Fig. 1, we know that both the schemes $\mathbf{\Pi}_0$ and $\mathbf{\Pi}_1$ will output the ciphertext $\mathsf{CT} = \left(c_0, (\mathbf{c}_i)_{i \in [\ell]}, \mathbf{c}_{L+1} \right) \in \mathbb{Z}_q \times (\mathbb{Z}_q^{3m} \times \mathbb{Z}_q^{4m} \times \cdots \times \mathbb{Z}_q^{(\ell+2)m}) \times \mathbb{Z}_q^{(\ell+2)m}$. However, for $\mathbf{\Pi}_0$ the item $\mathbf{c}_i$ in CT is generated from $\mathsf{ID}_{[i]} := (\mathsf{id}_1, \cdots, \mathsf{id}_{i-1}, \mathsf{id}_i)$, while the item $\mathbf{c}_i$ for our $\mathbf{\Pi}_1$ is created from $\widetilde{\mathsf{ID}_{[i]}} := (\mathsf{id}_1, \cdots, \mathsf{id}_{i-1}, \widetilde{\mathsf{id}}_i)$. As a consequence, our scheme $\mathbf{\Pi}_1$ only needs one short trapdoor base $\mathbf{T_A}$ (or $\mathbf{T}_{[\mathbf{A}|\mathbf{E}(\mathsf{ID})]}$) in the secret key $\mathsf{SK_{KGC}}$ (or $\mathsf{SK_{ID}}$), and accordingly the matrix $\mathbf{A}$ is used in PP instead of $(\mathbf{A}_i)_{i \in [L+1]}$, shown as in Fig. 1. In the security proof, the adversary $\mathcal{A}$ may issue a secret key reveal query on $\mathsf{ID}^*_{[i^*]}$ but not on any $\mathsf{ID}^*_{[j]}$ for $j \in [i^* - 1]$, where ID^* denotes the challenge identity and $i^* \leqslant |\mathsf{ID}^*|$. In this case, the LWE problem instance is used to construct $\mathbf{A}, \mathbf{u}$ in PP and $c_0, \mathbf{c}_{i^*}$ in CT for our scheme $\mathbf{\Pi}_1$. Though without the trapdoor $\mathbf{T_A}$, we are still able to construct $\mathbf{T}_{[\mathbf{A}|\mathbf{E}(\mathsf{ID}^*_{[i^*]})]}$ in $\mathsf{SK}_{\mathsf{ID}^*_{[i^*]}}$ for the adversary $\mathcal{A}$, since the simulated $\mathbf{c}_{i^*}$ is only related to $\widetilde{\mathsf{ID}^*_{[i^*]}}$, not $\mathsf{ID}^*_{[i^*]}$ itself. The construction of $\mathbf{T}_{[\mathbf{A}|\mathbf{E}(\mathsf{ID}^*_{[i^*]})]}$ will not succeed, if $\mathbf{c}_{i^*}$ is obtained in the way of the scheme $\mathbf{\Pi}_0$. This is also the reason why $\mathbf{\Pi}_0$ employs $L+1$ short trapdoor bases $(\mathbf{T}_{\mathbf{A}_i})_{i \in [L+1]}$ in $\mathsf{SK_{KGC}}$, and $L + 1 - \ell$ short trapdoor bases $(\mathbf{T}_{[\mathbf{A}_i|\mathbf{E}(\mathsf{ID})]})_{i \in [\ell+1, L+1]}$ in $\mathsf{SK_{ID}}$, just as Fig. 1 shows. Similarly, we also deal with the time period t differently in the encryption algorithm for our scheme $\mathbf{\Pi}_1$. As a result, we no longer need $\mathbf{T_A}$ to answer all the queries made by the adversary $\mathcal{A}$ in the security proof.

The items in $\boldsymbol{\Pi}_0, \boldsymbol{\Pi}_1$ related to the above changes are boxed in Fig. 1. Besides, we describe the underlined items in $\boldsymbol{\Pi}_0, \boldsymbol{\Pi}_1$ as follows (the items in $\mathsf{DK}_{\mathsf{ID},t}$ are not marked since there is no simplification). For the scheme $\boldsymbol{\Pi}_0$ in Fig. 1, the vector $\mathbf{f}_{\mathsf{ID},k}$ in $\mathsf{SK}_{\mathsf{ID}}$, the vector $\mathbf{f}_{\mathsf{ID},t,k}$ in $\mathsf{KU}_{\mathsf{ID},t}$ and the vector $\mathbf{d}_{\mathsf{ID},t}$ in $\mathsf{DK}_{\mathsf{ID},t}$, satisfy the condition $\mathbf{f}_{\mathsf{ID},t,k} = \mathbf{d}_{\mathsf{ID},t} + [\mathbf{f}_{\mathsf{ID},k} \| \mathbf{0}_{m \times 1}] \in \mathbb{Z}^{(\ell+2)m}$ for $k \in [\ell+1, L]$, where $[\cdot \| \cdot]$ denotes vertical concatenation of vectors, and $\ell = |\mathsf{ID}|$. Actually, as a preparation for achieving DKER, Katsumata et al. [11] also presented an RHIBE scheme without DKER, where the decryption key $\mathsf{DK}_{\mathsf{ID},t}$ does not contain the item $\mathbf{g}_{\mathsf{ID},t}$. Following this scheme without DKER, they introduced these vectors $\mathbf{f}_{\mathsf{ID},k}, \mathbf{f}_{\mathsf{ID}_{[i]},t,k}$ to avoid a trivial attack. For simplicity, one can imagine that if there is no $\mathbf{g}_{\mathsf{ID},t}$ in $\mathsf{DK}_{\mathsf{ID},t}$ for our scheme $\boldsymbol{\Pi}_1$ in Fig. 1, then the private $\mathsf{DK}_{\mathsf{ID},t}$ is totally contained in the public $\mathsf{KU}_{\mathsf{ID},t}$, which is obviously insecure. However, for the construction of RHIBE with the DKER setting, it can be proved that the item $\mathbf{g}_{\mathsf{ID},t}$ itself is sufficient to guarantee the security. Therefore, one no longer needs the items $(\mathbf{f}_{\mathsf{ID},k})_{k \in [\ell+1,L]}$ in $\mathsf{SK}_{\mathsf{ID}}$, or part of the items $(\mathbf{f}_{\mathsf{ID}_{[i]},t,k})_{(i,k) \in [\ell] \times [\ell+1,L]}$ in $\mathsf{KU}_{\mathsf{ID},t}$. Then in the public parameters PP we can also use only one vector $\mathbf{u}$, instead of $(\mathbf{u}_k)_{k \in [L]}$, and finally our scheme $\boldsymbol{\Pi}_1$ is obtained as a simplification of $\boldsymbol{\Pi}_0$, shown as in Fig. 1.

As for our second RHIBE scheme $\boldsymbol{\Pi}_2$, we follow the idea of our $\boldsymbol{\Pi}_1$, and adopt the technique for lattice basis delegation in fixed dimension introduced in [2]. Therefore, the sizes of items are much smaller than $\boldsymbol{\Pi}_0$ and $\boldsymbol{\Pi}_1$. For example, the ciphertext CT under an identity ID with $\ell = |\mathsf{ID}|$, is a vector in $\mathbb{Z}_q^{(2\ell+1)m+1}$ for our $\boldsymbol{\Pi}_2$. While in $\boldsymbol{\Pi}_0$ and $\boldsymbol{\Pi}_1$, CT is a vector in $\mathbb{Z}_q^{(\frac{1}{2}\ell^2 + \frac{7}{2}\ell + 2)m + 1}$. Moreover, as Fig. 1 shows, the items in $\mathsf{SK}_{\mathsf{ID}}, \mathsf{KU}_{\mathsf{ID},t}, \mathsf{DK}_{\mathsf{ID},t}$ for $\boldsymbol{\Pi}_2$ do not depend on $\ell = |\mathsf{ID}|$. They are either short matrices in $\mathbb{Z}^{m \times m}$, or short vectors in $\mathbb{Z}^m$ or $\mathbb{Z}^{2m}$. Unlike only one matrix $\mathbf{T_A}$ in $\mathsf{SK}_{\mathsf{KGC}}$ for $\boldsymbol{\Pi}_1$, we emphasize that the master secret key $\mathsf{SK}_{\mathsf{KGC}}$ in our scheme $\boldsymbol{\Pi}_2$ contains two trapdoor bases $\mathbf{T_A}, \mathbf{T_B}$. This comes from the different technique introduced in [2]. Following this, two trapdoor bases are necessary even for the construction of RIBE (not RHIBE) without DKER.

Organization. The rest of this paper is organized as follows. Section 2 reviews some background on lattices, the definitions for RHIBE re-formalized in [11], and the complete subtree method. Then in Sect. 3, we provide our first RHIBE scheme $\boldsymbol{\Pi}_1$, together with its analysis. The construction and the security proof for our second RHIBE scheme $\boldsymbol{\Pi}_2$, are presented in Sect. 4. Finally, the conclusion is given in Sect. 5.

2   Preliminaries

Notations. The acronym PPT stands for "probabilistic polynomial-time". We say that a function $\epsilon : \mathbb{N} \to \mathbb{R}$ is negligible, if for sufficient large $\lambda \in \mathbb{N}$, $|\epsilon(\lambda)|$ is smaller than the reciprocal of any polynomial in λ. The notation $\mathsf{negl}(\lambda)$ is used to denote a negligible function $\epsilon(\lambda)$. Besides, an event is said to happen with overwhelming probability if it happens with probability at least $1 - \mathsf{negl}(\lambda)$. The statistical distance of two random variables X and Y over a discrete domain Ω

is defined as $\Delta(X;Y) := \frac{1}{2}\sum_{s\in\Omega}|\Pr[X=s] - \Pr[Y=s]|$. If $X(\lambda)$ and $Y(\lambda)$ are ensembles of random variables, we say that X and Y are statistically close if $d(\lambda) := \Delta(X(\lambda);Y(\lambda))$ is equal to $\mathsf{negl}(\lambda)$. For a distribution χ, we often write $x \hookleftarrow \chi$ to indicate that we sample x from χ. For a finite set Ω, the notation $x \overset{\$}{\hookleftarrow} \Omega$ means that x is chosen uniformly at random from Ω. We treat vectors in their column form. For a vector $\mathbf{x} \in \mathbb{Z}^n$, denote $\|\mathbf{x}\|$ as the Euclidean norm of $\mathbf{x}$. For a matrix $\mathbf{A} \in \mathbb{Z}^{n\times m}$, denote $\|\mathbf{A}\|$ as the Euclidean norm of the longest column in $\mathbf{A}$, and denote $\|\mathbf{A}\|_{\mathrm{GS}}$ as $\|\mathbf{A}_{\mathrm{GS}}\|$, where $\mathbf{A}_{\mathrm{GS}}$ is the Gram-Schmidt orthogonalization of $\mathbf{A}$.

2.1 Background on Lattices

Integer Lattices. A (full-rank) integer lattice Λ of dimension m is defined as the set $\left\{\sum_{i\in[m]} x_i\mathbf{b}_i \mid x_i \in \mathbb{Z}\right\}$, where $\mathbf{B} := \{\mathbf{b}_1, \cdots, \mathbf{b}_m\}$ are m linearly independent vectors in $\mathbb{Z}^m$. Here $\mathbf{B}$ is called the basis of the lattice Λ. Let n, m and $q \geqslant 2$ be positive integers. For a matrix $\mathbf{A} \in \mathbb{Z}_q^{n\times m}$, define the m-dimensional lattice $\Lambda_q^\perp(\mathbf{A}) := \{\mathbf{x} \in \mathbb{Z}^m \mid \mathbf{A}\mathbf{x} = \mathbf{0} \bmod q\}$. For any $\mathbf{u}$ in the image of $\mathbf{A}$, define the coset $\Lambda_q^\mathbf{u}(\mathbf{A}) := \{\mathbf{x} \in \mathbb{Z}^m \mid \mathbf{A}\mathbf{x} = \mathbf{u} \bmod q\}$.

Discrete Gaussians over Lattices. Let Λ be a lattice in $\mathbb{Z}^m$. For any parameter $\sigma \in \mathbb{R}_{>0}$, define $\rho_\sigma(\mathbf{x}) := \exp(-\pi\|\mathbf{x}\|^2/\sigma^2)$ for $\mathbf{x} \in \mathbb{Z}^m$, and $\rho_\sigma(\Lambda) := \sum_{\mathbf{x}\in\Lambda}\rho_\sigma(\mathbf{x})$. The discrete Gaussian distribution over Λ with parameter σ is $\mathcal{D}_{\Lambda,\sigma}(\mathbf{y}) := \rho_\sigma(\mathbf{y})/\rho_\sigma(\Lambda)$, for $\mathbf{y} \in \Lambda$. Some properties are shown as follows.

Lemma 1 ([15]). For $\mathbf{A} \in \mathbb{Z}_q^{n\times m}, \mathbf{u} \in \mathbb{Z}_q^n$ with $q \geqslant 2, m > n$, let $\mathbf{T_A}$ be a basis for $\Lambda_q^\perp(\mathbf{A})$ and $\sigma \geqslant \|\mathbf{T_A}\|_{\mathrm{GS}} \cdot \omega(\sqrt{\log m})$, then $\Pr[\mathbf{x} \hookleftarrow \mathcal{D}_{\Lambda_q^\mathbf{u}(\mathbf{A}),\sigma} : \|\mathbf{x}\| > \sigma\sqrt{m}] \leqslant \mathsf{negl}(n)$.

Lemma 2 ([9]). Suppose that $n, m, q \in \mathbb{Z}_{>0}$, $\sigma \in \mathbb{R}_{>0}$, with q a prime, $m \geqslant 2n\log q$ and $\sigma \geqslant \omega(\sqrt{\log n})$. Then for $\mathbf{A} \overset{\$}{\hookleftarrow} \mathbb{Z}_q^{n\times m}, \mathbf{e} \hookleftarrow \mathcal{D}_{\mathbb{Z}^m,\sigma}$, the distribution of $\mathbf{u} := \mathbf{A}\mathbf{e} \pmod{q}$ is statistically close to uniform over $\mathbb{Z}_q^n$. Furthermore, for a fixed vector $\mathbf{u} \in \mathbb{Z}_q^n$ and a matrix $\mathbf{A} \overset{\$}{\hookleftarrow} \mathbb{Z}_q^{n\times m}$, the conditional distribution of $\mathbf{e} \hookleftarrow \mathcal{D}_{\mathbb{Z}^m,\sigma}$ given $\mathbf{A}\mathbf{e} = \mathbf{u} \pmod{q}$ is $\mathcal{D}_{\Lambda_q^\mathbf{u}(\mathbf{A}),\sigma}$ with overwhelming probability.

In addition, as in [2], we set $\sigma_\mathbf{R} := \sqrt{n\log q}\cdot\omega(\sqrt{\log m})$, and let $\mathcal{D}_{m\times m}$ denote the distribution on matrices in $\mathbb{Z}^{m\times m}$ defined as $(\mathcal{D}_{\mathbb{Z}^m,\sigma_\mathbf{R}})^m$ conditioned on the resulting matrix being $\mathbb{Z}_q$-invertible.

Algorithms about Lattices. Let us briefly review some algorithms which are useful for lattice-based cryptography. For these algorithms introduced below, we simply assume that $n, m, m_0, q \in \mathbb{Z}_{>0}$ with $q \geqslant 3$ a prime and $m = \Omega(n\log q)$. Besides, we note that according to [14], there exists a fixed full rank matrix $\mathbf{G} \in \mathbb{Z}_q^{n\times m}$, called the gadget matrix, such that the lattice $\Lambda_q^\perp(\mathbf{G})$ has a publicly known basis $\mathbf{T_G} \in \mathbb{Z}^{m\times m}$ with $\|\mathbf{T_G}\|_{\mathrm{GS}} \leqslant \sqrt{5}$.

$\mathsf{TrapGen}(1^n, 1^m, q) \to (\mathbf{A}, \mathbf{T_A})$ ([3,4,14]): On input n, m, q, output a matrix $\mathbf{A} \in \mathbb{Z}_q^{n\times m}$ and a basis $\mathbf{T_A}$ of $\Lambda_q^\perp(\mathbf{A})$, such that $\mathbf{A}$ is distributed statistically close to uniform over $\mathbb{Z}_q^{n \times m}$ and $\|\mathbf{T_A}\|_{GS} \leqslant O(\sqrt{n \log q})$ with overwhelming probability in n.

SamplePre$(\mathbf{A}, \mathbf{T_A}, \mathbf{u}, \sigma) \to \mathbf{e}$ ([9]): On input a full rank matrix $\mathbf{A} \in \mathbb{Z}_q^{n \times m}$, a basis $\mathbf{T_A}$ of $\Lambda_q^\perp(\mathbf{A})$, a vector $\mathbf{u} \in \mathbb{Z}_q^n$, and a Gaussian parameter $\sigma \geqslant \|\mathbf{T_A}\|_{GS} \cdot \omega(\sqrt{\log m})$, output a vector $\mathbf{e} \in \mathbb{Z}^m$ distributed statistically close to $\mathcal{D}_{\Lambda_q^\mathbf{u}(\mathbf{A}), \sigma}$.

SampleLeft$(\mathbf{A}, \mathbf{M}, \mathbf{T_A}, \mathbf{u}, \sigma) \to \mathbf{e}$ ([1,7]): On input a full rank matrix $\mathbf{A} \in \mathbb{Z}_q^{n \times m}$, a matrix $\mathbf{M} \in \mathbb{Z}_q^{n \times m_0}$, a basis $\mathbf{T_A}$ of $\Lambda_q^\perp(\mathbf{A})$, a vector $\mathbf{u} \in \mathbb{Z}_q^n$, and a Gaussian parameter $\sigma \geqslant \|\mathbf{T_A}\|_{GS} \cdot \omega(\sqrt{\log(m + m_0)})$, output a vector $\mathbf{e} \in \mathbb{Z}^{m+m_0}$ distributed statistically close to $\mathcal{D}_{\Lambda_q^\mathbf{u}([\mathbf{A}|\mathbf{M}]), \sigma}$.

SampleRight$(\mathbf{A}, \mathbf{H} \cdot \mathbf{G}, \mathbf{R}, \mathbf{T_G}, \mathbf{u}, \sigma) \to \mathbf{e}$ ([1,14]): On input a matrix $\mathbf{A} \in \mathbb{Z}_q^{n \times m}$, a matrix of the form $\mathbf{H} \cdot \mathbf{G} \in \mathbb{Z}_q^{n \times m}$ (where $\mathbf{H} \in \mathbb{Z}_q^{n \times n}$ is full rank and $\mathbf{G} \in \mathbb{Z}_q^{n \times m}$ is the gadget matrix [14]), a uniform random matrix $\mathbf{R} \xleftarrow{\$} \{-1, 1\}^{m \times m}$, a basis $\mathbf{T_G}$ of $\Lambda_q^\perp(\mathbf{G})$, a vector $\mathbf{u} \in \mathbb{Z}_q^n$, and a Gaussian parameter $\sigma \geqslant \|\mathbf{T_G}\|_{GS} \cdot \sqrt{m} \cdot \omega(\sqrt{\log m})$, output a vector $\mathbf{e} \in \mathbb{Z}^{2m}$ distributed statistically close to $\mathcal{D}_{\Lambda_q^\mathbf{u}([\mathbf{A}|\mathbf{AR}+\mathbf{HG}]), \sigma}$.

RandBasis$(\mathbf{T}, \sigma) \to \mathbf{T}'$ ([7]): On input a basis $\mathbf{T}$ of an m-dimensional lattice $\Lambda_q^\perp(\mathbf{A})$ and a Gaussian parameter $\sigma \geqslant \|\mathbf{T}\|_{GS} \cdot \omega(\sqrt{\log m})$, output a new basis $\mathbf{T}'$ of $\Lambda_q^\perp(\mathbf{A})$ such that $\mathbf{T}'$ is distributed statistically close to $\mathcal{D}_{Basis}(\Lambda_q^\perp(\mathbf{A}), \sigma)$ introduced below, and $\|\mathbf{T}'\|_{GS} \leqslant \sigma\sqrt{m}$ holds with overwhelming probability.

The distribution $\mathcal{D}_{Basis}(\Lambda_q^\perp(\mathbf{A}), \sigma)$ used above can be briefly described as follows. Let $\mathcal{O}(\Lambda_q^\perp(\mathbf{A}), \sigma)$ be an algorithm that generates samples from the distribution $\mathcal{D}_{\Lambda_q^\perp(\mathbf{A}), \sigma}$, and set m as the dimension of $\Lambda_q^\perp(\mathbf{A})$. For $i = 1, 2, \cdots, m$, run $\mathbf{v} \leftarrow \mathcal{O}(\Lambda_q^\perp(\mathbf{A}), \sigma)$ repeatedly until $\mathbf{v}$ is linearly independent of $\{\mathbf{v}_1, \cdots, \mathbf{v}_{i-1}\}$, and then set $\mathbf{v}_i \leftarrow \mathbf{v}$. After that, convert the set of vectors $\{\mathbf{v}_1, \cdots, \mathbf{v}_m\}$ to a basis $\mathbf{T_A}$ of $\Lambda_q^\perp(\mathbf{A})$ using Lemma 7.1 of [13] (and using some canonical basis of $\Lambda_q^\perp(\mathbf{A})$). The distribution of this $\mathbf{T_A}$ is then denoted as $\mathcal{D}_{Basis}(\Lambda_q^\perp(\mathbf{A}), \sigma)$. Actually, in the process of RandBasis$(\mathbf{T}, \sigma) \to \mathbf{T}'$, the input basis $\mathbf{T}$ is only used to run the algorithm SamplePre$(\mathbf{A}, \mathbf{T}, \mathbf{0}, \sigma)$, instead of the above algorithm $\mathcal{O}(\Lambda_q^\perp(\mathbf{A}), \sigma)$. Thus up to a negligible statistical distance, the distribution of the output basis $\mathbf{T}'$ does not depend on $\mathbf{T}$.

Using the distribution $\mathcal{D}_{Basis}(\Lambda_q^\perp(\mathbf{A}), \sigma)$ introduced above, we are able to describe the following algorithms for generating a random basis of some lattice.

SampleBasisLeft$(\mathbf{A}, \mathbf{M}, \mathbf{T_A}, \sigma) \to \mathbf{T}_{[\mathbf{A}|\mathbf{M}]}$ ([1,7]): On input a full rank matrix $\mathbf{A} \in \mathbb{Z}_q^{n \times m}$, a matrix $\mathbf{M} \in \mathbb{Z}_q^{n \times m_0}$, a basis $\mathbf{T_A}$ of $\Lambda_q^\perp(\mathbf{A})$, and a Gaussian parameter $\sigma \geqslant \|\mathbf{T_A}\|_{GS} \cdot \omega(\sqrt{\log(m + m_0)})$, output a basis $\mathbf{T}_{[\mathbf{A}|\mathbf{M}]} \in \mathbb{Z}^{(m+m_0) \times (m+m_0)}$ distributed statistically close to $\mathcal{D}_{Basis}(\Lambda_q^\perp([\mathbf{A} \mid \mathbf{M}]), \sigma)$.

SampleBasisRight$(\mathbf{A}, \mathbf{H} \cdot \mathbf{G}, \mathbf{R}, \mathbf{T_G}, \sigma) \to \mathbf{T}_{[\mathbf{A}|\mathbf{AR}+\mathbf{HG}]}$ ([1,14]): On input a matrix $\mathbf{A} \in \mathbb{Z}_q^{n \times m}$, a matrix of the form $\mathbf{H} \cdot \mathbf{G} \in \mathbb{Z}_q^{n \times m}$ (where $\mathbf{H} \in \mathbb{Z}_q^{n \times n}$ is full rank and $\mathbf{G} \in \mathbb{Z}_q^{n \times m}$ is the gadget matrix [14]), a uniform random matrix $\mathbf{R} \xleftarrow{\$} \{-1, 1\}^{m \times m}$, a basis $\mathbf{T_G}$ of $\Lambda_q^\perp(\mathbf{G})$, and a Gaussian parameter $\sigma \geqslant \|\mathbf{T_G}\|_{\mathrm{GS}} \cdot \sqrt{m} \cdot \omega(\sqrt{\log m})$, output a basis $\mathbf{T}_{[\mathbf{A}|\mathbf{AR}+\mathbf{HG}]} \in \mathbb{Z}^{2m \times 2m}$ distributed statistically close to $\mathcal{D}_{Basis}(\Lambda_q^\perp([\mathbf{A} \mid \mathbf{AR} + \mathbf{HG}]), \sigma)$.

BasisDel$(\mathbf{A}, \mathbf{R}, \mathbf{T_A}, \sigma) \rightarrow \mathbf{T}_{(\mathbf{AR}^{-1})}$ ([2]): On input a full rank matrix $\mathbf{A} \in \mathbb{Z}_q^{n \times m}$, a $\mathbb{Z}_q$-invertible matrix $\mathbf{R} \in \mathbb{Z}^{m \times m}$ sampled from $\mathcal{D}_{m \times m}$, a basis $\mathbf{T_A}$ of $\Lambda_q^\perp(\mathbf{A})$, and a Gaussian parameter $\sigma \geqslant \|\mathbf{T_A}\|_{\mathrm{GS}} \cdot \sqrt{nm \log q} \cdot \omega(\log^2 m)$, output a basis $\mathbf{T}_{(\mathbf{AR}^{-1})} \in \mathbb{Z}^{m \times m}$ distributed statistically close to $\mathcal{D}_{Basis}(\Lambda_q^\perp(\mathbf{AR}^{-1}), \sigma)$.

SampleRwithBasis$(\mathbf{A}, \sigma) \rightarrow (\mathbf{R}, \mathbf{T}_{(\mathbf{AR}^{-1})})$ ([2,7]): On input a full rank matrix $\mathbf{A} \in \mathbb{Z}_q^{n \times m}$, and a Gaussian parameter $\sigma \geqslant \sqrt{n \log q} \cdot \omega(\sqrt{\log m})$, output a $\mathbb{Z}_q$-invertible matrix $\mathbf{R} \in \mathbb{Z}^{m \times m}$ sampled from a distribution statistically close to $\mathcal{D}_{m \times m}$, and a basis $\mathbf{T}_{(\mathbf{AR}^{-1})} \in \mathbb{Z}^{m \times m}$ distributed statistically close to $\mathcal{D}_{Basis}(\Lambda_q^\perp(\mathbf{AR}^{-1}), \sigma)$.

Recall that the distribution $\mathcal{D}_{m \times m}$ used above has already been defined below Lemma 2. Besides, the algorithm SampleRwithBasis described above is actually a combination of the original algorithm SampleRwithBasis in [2] and the algorithm RandBasis in [7]. We directly describe this modified SampleRwithBasis just for convenience in the future proof of security.

Hardness Assumption. The learning with errors (LWE) problem, first introduced by Regev [17], plays a central role in lattice-based cryptography. The security of our schemes will rely on the following LWE assumption.

Assumption 1 (LWE). *Suppose that $n, m, q \in \mathbb{Z}_{>0}$, $\alpha \in (0, 1)$ with q a prime satisfy $\alpha q > 2\sqrt{n}$. For a PPT algorithm $\mathcal{A}$, the advantage for the learning with errors problem* $\mathsf{LWE}_{n,m,q,\mathcal{D}_{\mathbb{Z}^m},\alpha q}$ *of $\mathcal{A}$ is defined as* $|\Pr[\mathcal{A}(\mathbf{A}, \mathbf{A}^\top \mathbf{s} + \mathbf{x}) = 1] - \Pr[\mathcal{A}(\mathbf{A}, \mathbf{v}) = 1]|$, *where* $\mathbf{A} \overset{\$}{\leftarrow} \mathbb{Z}_q^{n \times m}, \mathbf{s} \overset{\$}{\leftarrow} \mathbb{Z}_q^n, \mathbf{x} \hookleftarrow \mathcal{D}_{\mathbb{Z}^m, \alpha q}, \mathbf{v} \overset{\$}{\leftarrow} \mathbb{Z}_q^m$. *We say that the LWE assumption holds if the above advantage is negligible for all PPT $\mathcal{A}$.*

2.2   Revocable Hierarchical Identity-Based Encryption

We briefly review the syntax, correctness and security definition for RHIBE, which are re-formalized in [11]. First of all, let us introduce some notations as follows.

Recall that the hierarchical identity space in RHIBE is denoted by $\mathcal{ID}_\mathsf{H} = (\mathcal{ID})^{\leqslant L} = \bigcup_{i \in [L]} (\mathcal{ID})^i$, where $\mathcal{ID}$ is the element identity space, and L is the maximum depth of the hierarchy. The KGC is the unique level-0 identity, and an identity $\mathsf{ID} \in \mathcal{ID}_\mathsf{H}$ at level $\ell \in [L]$ is expressed as a length-ℓ vector $\mathsf{ID} = (\mathrm{id}_1, \cdots, \mathrm{id}_\ell) \in (\mathcal{ID})^\ell$. For $k \in [\ell]$, we set $\mathsf{ID}_{[k]} := (\mathrm{id}_1, \cdots, \mathrm{id}_k)$ as the length-k prefix of ID, and define $\mathsf{prefix}(\mathsf{ID}) := \{\mathsf{ID}_{[1]}, \mathsf{ID}_{[2]}, \cdots, \mathsf{ID}_{[\ell]} = \mathsf{ID}\}$. Besides, we let $\mathsf{pa}(\mathsf{ID}) := \mathsf{ID}_{[\ell-1]}$ if $\ell \geqslant 2$, and $\mathsf{pa}(\mathsf{ID}) := \mathsf{KGC}$ if $\ell = 1$. Here $\mathsf{pa}(\mathsf{ID})$ is called the parent of ID. We use $\mathsf{ID}\|\mathcal{ID}$ to denote the subset of $(\mathcal{ID})^{\ell+1}$ which contains all the members that have $\mathsf{ID} \in (\mathcal{ID})^\ell$ as its parent. When $\mathsf{ID} = \mathsf{KGC}$ (i.e. $\ell = 0$), the notation $\mathsf{ID}\|\mathcal{ID}$ just denotes $\mathcal{ID}$.

Next, we introduce the notation $\mathsf{RL}_\mathsf{t} (\subseteq (\mathcal{ID})^{\leqslant L})$ to denote the revocation list on the time period t. If $\mathsf{ID} \in \mathsf{RL}_\mathsf{t}$, then implicitly we assume $\mathsf{ID}' \in \mathsf{RL}_\mathsf{t}$ also holds, where ID' is any descendant of ID. Besides, it is required that $\mathsf{RL}_{t_1} \subseteq \mathsf{RL}_{t_2}$ for $t_1 < t_2$. We set $\mathsf{RL}_{\mathsf{ID},t} := \mathsf{RL}_t \cap (\mathsf{ID}\|\mathcal{ID})$ as the revocation list managed by the identity ID on the time period t. Following these notations, when we write "$\mathsf{ID} \in \mathsf{RL}_t$", it means that user ID has been revoked on the time period t. For any $\mathsf{ID}' \in \mathsf{prefix}(\mathsf{ID})$ and any $t' \leqslant t$, we have $\mathsf{ID}' \in \mathsf{RL}_{\mathsf{pa}(\mathsf{ID}'),t'} \Rightarrow \mathsf{ID} \in \mathsf{RL}_t$. When we write "$\mathsf{ID} \notin \mathsf{RL}_t$", it means that user ID is not revoked on the time period t. We have $\mathsf{ID} \notin \mathsf{RL}_t \Leftrightarrow \mathsf{ID}' \notin \mathsf{RL}_{\mathsf{pa}(\mathsf{ID}'),t}, \forall \mathsf{ID}' \in \mathsf{prefix}(\mathsf{ID})$.

Syntax. As re-formalized in [11], an RHIBE scheme Π consists of the following six algorithms **Setup, Encrypt, GenSK, KeyUp, GenDK, Decrypt**. Here the "revoke" algorithm is not explicitly introduced, since it is a simple operation of appending revoked users into a revocation list.

Setup$(1^\lambda, 1^L) \to (\mathsf{PP}, \mathsf{SK}_{\mathsf{KGC}})$: This is the setup algorithm run by the KGC. On input a security parameter λ and the maximum depth of the hierarchy L, it outputs public parameters PP and the KGC's secret key $\mathsf{SK}_{\mathsf{KGC}}$.

Encrypt$(\mathsf{PP}, \mathsf{ID}, t, \mathsf{M}) \to \mathsf{CT}$: This is the encryption algorithm run by a sender. On input public parameters PP, an identity ID, a time period t, and a plaintext M, it outputs a ciphertext CT.

GenSK$(\mathsf{PP}, \mathsf{SK}_{\mathsf{pa}(\mathsf{ID})}, \mathsf{ID}) \to (\mathsf{SK}_{\mathsf{ID}}, \mathsf{SK}'_{\mathsf{pa}(\mathsf{ID})})$: This is the secret key generation algorithm run by $\mathsf{pa}(\mathsf{ID})$, the parent user of ID. On input public parameters PP, the parent user's secret key $\mathsf{SK}_{\mathsf{pa}(\mathsf{ID})}$, and the identity ID, it outputs a secret key $\mathsf{SK}_{\mathsf{ID}}$ for ID along with the parent user's "updated" secret key $\mathsf{SK}'_{\mathsf{pa}(\mathsf{ID})}$.

KeyUp$(\mathsf{PP}, t, \mathsf{SK}_{\mathsf{ID}}, \mathsf{RL}_{\mathsf{ID},t}, \mathsf{KU}_{\mathsf{pa}(\mathsf{ID}),t}) \to (\mathsf{KU}_{\mathsf{ID},t}, \mathsf{SK}'_{\mathsf{ID}})$: This is the key update generation algorithm run by the user ID. On input public parameters PP, a time period t, a secret key $\mathsf{SK}_{\mathsf{ID}}$, a revocation list $\mathsf{RL}_{\mathsf{ID},t}$, and the parent user's key update $\mathsf{KU}_{\mathsf{pa}(\mathsf{ID}),t}$, it outputs a key update $\mathsf{KU}_{\mathsf{ID},t}$ along with the "updated" secret key $\mathsf{SK}'_{\mathsf{ID}}$. (In the special case $\mathsf{ID} = \mathsf{KGC}$, since $\mathsf{KU}_{\mathsf{pa}(\mathsf{KGC}),t}$ is not needed, we just define $\mathsf{KU}_{\mathsf{pa}(\mathsf{KGC}),t} := \perp$ for all $t \in \mathcal{T}$.)

GenDK$(\mathsf{PP}, \mathsf{SK}_{\mathsf{ID}}, \mathsf{KU}_{\mathsf{pa}(\mathsf{ID}),t}) \to \mathsf{DK}_{\mathsf{ID},t}$ or $\perp$: This is the decryption key generation algorithm run by the user ID. On input public parameters PP, a secret key $\mathsf{SK}_{\mathsf{ID}}$, and the parent user's key update $\mathsf{KU}_{\mathsf{pa}(\mathsf{ID}),t}$, it outputs a decryption key $\mathsf{DK}_{\mathsf{ID},t}$, or the special "invalid" symbol $\perp$ which indicates that ID has been revoked.

Decrypt$(\mathsf{PP}, \mathsf{DK}_{\mathsf{ID},t}, \mathsf{CT}) \to \mathsf{M}$: This is the decryption algorithm run by the user ID. On input public parameters PP, a decryption key $\mathsf{DK}_{\mathsf{ID},t}$, and a ciphertext CT, it outputs the decrypted plaintext M.

Correctness. The correctness requirement for an RHIBE scheme Π states that, for all $\lambda, L \in \mathbb{Z}_{>0}$, $\ell \in [L]$, $\mathsf{ID} \in (\mathcal{ID})^\ell$, $t \in \mathcal{T}$, $\mathsf{M} \in \mathcal{M}$, $\mathsf{RL}_t \subseteq (\mathcal{ID})^{\leqslant L}$, if $\mathsf{ID} \notin \mathsf{RL}_t$, and all parties follow the above prescribed algorithms **Setup, GenSK, KeyUp, GenDK, Encrypt** to generate $\mathsf{PP}, \mathsf{DK}_{\mathsf{ID},t}, \mathsf{CT}$, then **Decrypt**$(\mathsf{PP}, \mathsf{DK}_{\mathsf{ID},t}, \mathsf{CT}) = \mathsf{M}$.

Security Definition. Let $\Pi = ($**Setup, Encrypt, GenSK, KeyUp, GenDK, Decrypt**$)$ be an RHIBE scheme. We first consider the selective-identity security, which is defined via the following game between an adversary $\mathcal{A}$ and a challenger $\mathcal{C}$.

At the beginning, $\mathcal{A}$ sends the challenge identity/time period pair $(\mathsf{ID}^*, \mathsf{t}^*) \in (\mathcal{ID})^{\leqslant L} \times \mathcal{T}$ to $\mathcal{C}$. After that, $\mathcal{C}$ runs $(\mathsf{PP}, \mathsf{SK}_{\mathsf{KGC}}) \leftarrow \mathbf{Setup}(1^\lambda, 1^L)$, and prepares a list SKList that initially contains $(\mathsf{KGC}, \mathsf{SK}_{\mathsf{KGC}})$. During the game, whenever a new secret key is generated or an existing secret key is updated for some identity $\mathsf{ID} \in \{\mathsf{KGC}\} \cup (\mathcal{ID})^{\leqslant L}$, the challenger $\mathcal{C}$ will store or update the identity/secret key pairs $(\mathsf{ID}, \mathsf{SK}_{\mathsf{ID}})$ in SKList, and we do not explicitly mention this addition/update. The global counter t_{cu}, which denotes the "current time period", is initialized with 1. Then $\mathcal{C}$ executes $(\mathsf{KU}_{\mathsf{KGC},1}, \mathsf{SK}'_{\mathsf{KGC}}) \leftarrow \mathbf{KeyUp}(\mathsf{PP}, \mathsf{t}_{\mathsf{cu}} = 1, \mathsf{SK}_{\mathsf{KGC}}, \mathsf{RL}_{\mathsf{KGC},1} = \emptyset, \perp)$ for $\mathsf{t}_{\mathsf{cu}} = 1$, and gives $\mathsf{PP}, \mathsf{KU}_{\mathsf{KGC},1}$ to $\mathcal{A}$.

From this point on, $\mathcal{A}$ may adaptively make the following five types of queries to $\mathcal{C}$.

Secret Key Generation Query: Upon a query $\mathsf{ID} \in (\mathcal{ID})^{\leqslant L}$ from $\mathcal{A}$, the challenger $\mathcal{C}$ checks whether the condition $(\mathsf{ID}, *) \notin \mathsf{SKList}$, $(\mathsf{pa}(\mathsf{ID}), \mathsf{SK}_{\mathsf{pa}(\mathsf{ID})}) \in \mathsf{SKList}$ is satisfied. If not, $\mathcal{C}$ just returns $\perp$. Otherwise, $\mathcal{C}$ executes $(\mathsf{SK}_{\mathsf{ID}}, \mathsf{SK}'_{\mathsf{pa}(\mathsf{ID})}) \leftarrow \mathbf{GenSK}(\mathsf{PP}, \mathsf{SK}_{\mathsf{pa}(\mathsf{ID})}, \mathsf{ID})$. Furthermore, if $\mathsf{ID} \in (\mathcal{ID})^{\leqslant L-1}$, then $\mathcal{C}$ executes $(\mathsf{KU}_{\mathsf{ID},\mathsf{t}_{\mathsf{cu}}}, \mathsf{SK}'_{\mathsf{ID}}) \leftarrow \mathbf{KeyUp}(\mathsf{PP}, \mathsf{t}_{\mathsf{cu}}, \mathsf{SK}_{\mathsf{ID}}, \mathsf{RL}_{\mathsf{ID},\mathsf{t}_{\mathsf{cu}}} = \emptyset, \mathsf{KU}_{\mathsf{pa}(\mathsf{ID}),\mathsf{t}_{\mathsf{cu}}})$, and returns $\mathsf{KU}_{\mathsf{ID},\mathsf{t}_{\mathsf{cu}}}$ to $\mathcal{A}$.

Secret Key Reveal Query: Upon a query $\mathsf{ID} \in (\mathcal{ID})^{\leqslant L}$ from $\mathcal{A}$, the challenger $\mathcal{C}$ checks whether the following condition is satisfied.
  - If $\mathsf{t}_{\mathsf{cu}} \geqslant \mathsf{t}^*$ and $\mathsf{ID} \in \mathsf{prefix}(\mathsf{ID}^*)$, then $\mathsf{ID} \in \mathsf{RL}_{\mathsf{t}^*}$.

If not, $\mathcal{C}$ just returns $\perp$. Otherwise, $\mathcal{C}$ finds $\mathsf{SK}_{\mathsf{ID}}$ from SKList, and returns it to $\mathcal{A}$.

Revoke & Key Update Query: Upon a query $\mathsf{RL} \subseteq (\mathcal{ID})^{\leqslant L}$ from $\mathcal{A}$, the challenger $\mathcal{C}$ checks whether the following conditions are satisfied simultaneously.
  - $\mathsf{RL}_{\mathsf{t}_{\mathsf{cu}}} \subseteq \mathsf{RL}$.
  - For $\mathsf{ID}, \mathsf{ID}' \in (\mathcal{ID})^{\leqslant L}$ with $\mathsf{ID}' \in \mathsf{prefix}(\mathsf{ID})$, if $\mathsf{ID}' \in \mathsf{RL}$, then $\mathsf{ID} \in \mathsf{RL}$.
  - If $\mathsf{t}_{\mathsf{cu}} = \mathsf{t}^* - 1$, and $\mathsf{SK}_{\mathsf{ID}}$ for some $\mathsf{ID} \in \mathsf{prefix}(\mathsf{ID}^*)$ has been revealed by the secret key reveal query, then $\mathsf{ID} \in \mathsf{RL}$.

If not, $\mathcal{C}$ just returns $\perp$. Otherwise, $\mathcal{C}$ increments the current time period by $\mathsf{t}_{\mathsf{cu}} \leftarrow \mathsf{t}_{\mathsf{cu}} + 1$, and then sets $\mathsf{RL}_{\mathsf{t}_{\mathsf{cu}}} \leftarrow \mathsf{RL}$. Next, for all $\mathsf{ID} \in \{\mathsf{KGC}\} \cup (\mathcal{ID})^{\leqslant L-1}$ with $(\mathsf{ID}, *) \in \mathsf{SKList}$, $\mathsf{ID} \notin \mathsf{RL}_{\mathsf{t}_{\mathsf{cu}}}$ in the breadth-first order in the identity hierarchy, $\mathcal{C}$ set $\mathsf{RL}_{\mathsf{ID},\mathsf{t}_{\mathsf{cu}}} \leftarrow \mathsf{RL}_{\mathsf{t}_{\mathsf{cu}}} \cap (\mathsf{ID}\|\mathcal{ID})$, and run $(\mathsf{KU}_{\mathsf{ID},\mathsf{t}_{\mathsf{cu}}}, \mathsf{SK}'_{\mathsf{ID}}) \leftarrow \mathbf{KeyUp}(\mathsf{PP}, \mathsf{t}_{\mathsf{cu}}, \mathsf{SK}_{\mathsf{ID}}, \mathsf{RL}_{\mathsf{ID},\mathsf{t}_{\mathsf{cu}}}, \mathsf{KU}_{\mathsf{pa}(\mathsf{ID}),\mathsf{t}_{\mathsf{cu}}})$. Finally, $\mathcal{C}$ returns all these generated key updates $\{\mathsf{KU}_{\mathsf{ID},\mathsf{t}_{\mathsf{cu}}}\}$ to $\mathcal{A}$.

Decryption Key Reveal Query: Upon a query $(\mathsf{ID},\mathsf{t}) \in (\mathcal{ID})^{\leqslant L} \times \mathcal{T}$ from $\mathcal{A}$, the challenger $\mathcal{C}$ checks whether the following condition is satisfied.
  - $\mathsf{t} \leqslant \mathsf{t}_{\mathsf{cu}}$, $\mathsf{ID} \notin \mathsf{RL}_\mathsf{t}$, $(\mathsf{ID},\mathsf{t}) \neq (\mathsf{ID}^*, \mathsf{t}^*)$.

If not, $\mathcal{C}$ just returns $\perp$. Otherwise, $\mathcal{C}$ finds $\mathsf{SK}_{\mathsf{ID}}$ from SKList, runs $\mathsf{DK}_{\mathsf{ID},\mathsf{t}} \leftarrow \mathbf{GenDK}(\mathsf{PP}, \mathsf{SK}_{\mathsf{ID}}, \mathsf{KU}_{\mathsf{pa}(\mathsf{ID}),\mathsf{t}})$, and returns $\mathsf{DK}_{\mathsf{ID},\mathsf{t}}$ to $\mathcal{A}$.

Challenge Query: $\mathcal{A}$ is allowed to make this query only once. Upon a query $(\mathsf{M}_0, \mathsf{M}_1)$ with $|\mathsf{M}_0| = |\mathsf{M}_1|$ from $\mathcal{A}$, the challenger $\mathcal{C}$ picks the challenge bit $b \xleftarrow{\$} \{0,1\}$, runs $\mathsf{CT}^* \leftarrow \mathbf{Encrypt}(\mathsf{PP}, \mathsf{ID}^*, \mathsf{t}^*, \mathsf{M}_b)$, and returns the challenge ciphertext CT^* to $\mathcal{A}$.

At some point, $\mathcal{A}$ outputs $b' \in \{0, 1\}$ as the guess for b and terminates.

The above completes the description of the game. In this game, $\mathcal{A}$'s selective-identity security advantage is defined by $\mathsf{Adv}_{\Pi,L,\mathcal{A}}^{\mathsf{RHIBE\text{-}sel}}(\lambda) := 2 \cdot |\Pr[b' = b] - 1/2|$, where λ is the security parameter. We say that an RHIBE scheme Π with depth L satisfies the selective-identity security, if the advantage $\mathsf{Adv}_{\Pi,L,\mathcal{A}}^{\mathsf{RHIBE\text{-}sel}}(\lambda)$ is negligible for any PPT adversary $\mathcal{A}$.

The game for the adaptive-identity security, is defined in the same way as the above game, except that the adversary $\mathcal{A}$ chooses the challenge identity/time period pair $(\mathsf{ID}^*, \mathsf{t}^*) \in (\mathcal{ID})^{\leqslant L} \times \mathcal{T}$ not at the beginning of the game, but at the time when $\mathcal{A}$ makes the challenge query. Formally, the challenge query is defined differently as follows.

Challenge Query: $\mathcal{A}$ is allowed to make this query only once. The query $(\mathsf{ID}^*, \mathsf{t}^*, \mathsf{M}_0, \mathsf{M}_1)$ from $\mathcal{A}$ must satisfy the following conditions simultaneously.
- $|\mathsf{M}_0| = |\mathsf{M}_1|$.
- If $\mathsf{t}_{cu} \geqslant \mathsf{t}^*$, and $\mathsf{SK}_{\mathsf{ID}}$ for some $\mathsf{ID} \in \mathsf{prefix}(\mathsf{ID}^*)$ has been revealed by the secret key reveal query, then $\mathsf{ID} \in \mathsf{RL}_{\mathsf{t}^*}$.
- If $\mathsf{t}_{cu} \geqslant \mathsf{t}^*$, then $\mathcal{A}$ has not submitted $(\mathsf{ID}^*, \mathsf{t}^*)$ as a decryption key reveal query.

After receiving this query $(\mathsf{ID}^*, \mathsf{t}^*, \mathsf{M}_0, \mathsf{M}_1)$, $\mathcal{C}$ picks the challenge bit $b \xleftarrow{\$} \{0, 1\}$, runs $\mathsf{CT}^* \leftarrow \mathbf{Encrypt}(\mathsf{PP}, \mathsf{ID}^*, \mathsf{t}^*, \mathsf{M}_b)$, and returns the challenge ciphertext CT^* to $\mathcal{A}$.

Besides, in the other queries, the conditions related to $\mathsf{ID}^*, \mathsf{t}^*$ are naturally omitted before $\mathcal{A}$ makes the above challenge query. Recall that at last $\mathcal{A}$ will output $b' \in \{0, 1\}$ as the guess for b. The adaptive-identity security advantage is then defined by $\mathsf{Adv}_{\Pi,L,\mathcal{A}}^{\mathsf{RHIBE\text{-}ad}}(\lambda) := 2 \cdot |\Pr[b' = b] - 1/2|$ for this modified game. Similarly, we say that an RHIBE scheme Π with depth L satisfies the adaptive-identity security, if the advantage $\mathsf{Adv}_{\Pi,L,\mathcal{A}}^{\mathsf{RHIBE\text{-}ad}}(\lambda)$ is negligible for any PPT adversary $\mathcal{A}$.

2.3 The Complete Subtree Method

Similar to the works in [5, 8], the RHIBE scheme Π_0 in [11], and our schemes Π_1, Π_2 constructed in this paper, all need the complete subtree (CS) method of Naor et al. [16] to achieve the revocation mechanism.

Shown as in Fig. 1, every identity ID, including the KGC, keeps a binary tree $\mathsf{BT}_{\mathsf{ID}}$ in its secret key $\mathsf{SK}_{\mathsf{ID}}$. Actually, each member that has ID as its parent, will be randomly assigned to a leaf node of $\mathsf{BT}_{\mathsf{ID}}$. For a leaf node η, we use $\mathsf{Path}(\mathsf{BT}_{\mathsf{ID}}, \eta)$ to denote the set of nodes on the path from η to the root in $\mathsf{BT}_{\mathsf{ID}}$ (both η and the root inclusive). For a non-leaf node θ, let θ_l, θ_r denote the left and right child of θ, respectively. Besides, recall that $\mathsf{RL}_{\mathsf{ID},\mathsf{t}}$ is the revocation list managed by the identity ID on the time period t. Then the algorithm KUNode, which takes $\mathsf{BT}_{\mathsf{ID}}$ and $\mathsf{RL}_{\mathsf{ID},\mathsf{t}}$ as input, can be described as follows: (1) $X, Y \leftarrow \emptyset$; (2) for each $\mathsf{ID}' \in \mathsf{RL}_{\mathsf{ID},\mathsf{t}}$, add $\mathsf{Path}(\mathsf{BT}_{\mathsf{ID}}, \eta_{\mathsf{ID}'})$ to X, where $\eta_{\mathsf{ID}'}$ denotes the leaf node to which ID' is assigned; (3) for each node $\theta \in X$, add θ_l to Y if $\theta_l \notin X$, and add θ_r to Y if $\theta_r \notin X$; (4) if $\mathsf{RL}_{\mathsf{ID},t} = \emptyset$, add the root node of $\mathsf{BT}_{\mathsf{ID}}$ to Y; (5) return Y as the output of $\mathsf{KUNode}(\mathsf{BT}_{\mathsf{ID}}, \mathsf{RL}_{\mathsf{ID},t})$.

Let us focus on the decryption key generation algorithm $\mathbf{GenDK}(\mathsf{PP}, \mathsf{SK}_{\mathsf{ID}'}, \mathsf{KU}_{\mathsf{ID},t})$ run by the user ID' with $\mathsf{pa}(\mathsf{ID}') = \mathsf{ID}$. Here the secret key $\mathsf{SK}_{\mathsf{ID}'}$ contains the set of nodes $\mathsf{P} := \mathsf{Path}(\mathsf{BT}_{\mathsf{ID}}, \eta_{\mathsf{ID}'})$. While the key update $\mathsf{KU}_{\mathsf{ID},t}$ contains the set of nodes $\mathsf{K} := \mathsf{KUNode}(\mathsf{BT}_{\mathsf{ID}}, \mathsf{RL}_{\mathsf{ID},t})$. If $\mathsf{ID}' \notin \mathsf{RL}_{\mathsf{ID},t}$, we have $\mathsf{P} \cap \mathsf{K} = \{\theta^*\}$, which contains exactly one node θ^*. Then ID' is able to generate its decryption key $\mathsf{DK}_{\mathsf{ID}',t}$, using some item related to θ^*. If $\mathsf{ID}' \in \mathsf{RL}_{\mathsf{ID},t}$, we have $\mathsf{P} \cap \mathsf{K} = \emptyset$, from which ID' can never obtain $\mathsf{DK}_{\mathsf{ID}',t}$. This is the general way to achieve the revocation mechanism from the CS method.

3   RHIBE Scheme in the Standard Model

In this section, we describe our first RHIBE scheme $\mathbf{\Pi}_1$ in Sect. 3.1, and then present its selective-identity security in Sect. 3.2. As a preparation, we need to explain our treatment of some spaces such as $\mathcal{T}, \mathcal{ID}, \mathcal{ID}_\mathsf{H} = (\mathcal{ID})^{\leqslant L}$, and introduce an encoding with full-rank differences used in the scheme $\mathbf{\Pi}_1$.

Treatment of Spaces. The element identity space $\mathcal{ID}$ is treated as a subset of $\mathbb{Z}_q^n \setminus \{\mathbf{0}_n\}$, namely, $\mathcal{ID} \subset \mathbb{Z}_q^n \setminus \{\mathbf{0}_n\}$. We need to define a function $f : \mathcal{ID} \to \widetilde{\mathcal{ID}}$ such that $f(\mathsf{id}_1) \neq f(\mathsf{id}_2)$ for $\mathsf{id}_1 \neq \mathsf{id}_2$. Here $\widetilde{\mathcal{ID}}$ is a new space satisfying $\widetilde{\mathcal{ID}} \subset \mathbb{Z}_q^n \setminus \{\mathbf{0}_n\}$ and $\mathcal{ID} \cap \widetilde{\mathcal{ID}} = \emptyset$. For simplicity, we just define

$$\mathcal{ID} := \{1\} \times \mathbb{Z}_q^{n-1}, \quad \widetilde{\mathcal{ID}} := \{2\} \times \mathbb{Z}_q^{n-1} \quad \text{and} \quad f(1\|\mathbf{v}) := 2\|\mathbf{v} \quad \text{for } \mathbf{v} \in \mathbb{Z}_q^{n-1}.$$

The time period space $\mathcal{T} = \{1, 2, \cdots, \mathsf{t}_{\max}\}$ is encoded into the set $\mathbb{Z}_q^{n-1}$. Here we note that one can also choose disjoint $\mathcal{ID}, \widetilde{\mathcal{ID}} \subset \mathbb{Z}_q^n \setminus \{\mathbf{0}_n\}$ such that $|\mathcal{ID}| = |\widetilde{\mathcal{ID}}| = \frac{1}{2}(q^n - 1)$, and set $\mathcal{T}$ as a subset of $\mathbb{Z}_q^n \setminus \{\mathbf{0}_n\}$ with $|\mathcal{T}| = \lfloor \frac{1}{L}(q^n - 1) \rfloor$. Besides, let us deal with the hierarchical identity space $\mathcal{ID}_\mathsf{H} = (\mathcal{ID})^{\leqslant L} = \bigcup_{i \in [L]} (\mathcal{ID})^i$. Define $\mathcal{F} : (\mathcal{ID})^{\leqslant L} \to \bigcup_{i \in [0, L-1]} (\mathcal{ID})^i \times \widetilde{\mathcal{ID}}$ as $\mathcal{F}(\mathsf{ID}) := (\mathsf{id}_1, \cdots, \mathsf{id}_{\ell-1}, f(\mathsf{id}_\ell))$ for $\mathsf{ID} = (\mathsf{id}_1, \cdots, \mathsf{id}_{\ell-1}, \mathsf{id}_\ell)$. Thus for $|\mathsf{ID}| = \ell \geqslant 2$, we have $\mathsf{ID} \neq \mathcal{F}(\mathsf{ID})$, $\mathsf{ID}_{[\ell-1]} = [\mathcal{F}(\mathsf{ID})]_{[\ell-1]}$. For simplicity, let us set $\widetilde{\mathsf{id}} := f(\mathsf{id})$, $\widetilde{\mathsf{ID}} := \mathcal{F}(\mathsf{ID})$, and use $\widetilde{\mathsf{ID}}_{[i]}$ to denote $\mathcal{F}(\mathsf{ID}_{[i]})$.

Encoding with Full-Rank Differences. We use the standard map H defined in [1] to encode vectors as matrices. The function $H : \mathbb{Z}_q^n \to \mathbb{Z}_q^{n \times n}$ is actually an encoding with full-rank differences for a prime q. Namely, the matrix $H(\mathsf{ch}_1) - H(\mathsf{ch}_2)$ is full rank for any two distinct $\mathsf{ch}_1, \mathsf{ch}_2 \in \mathbb{Z}_q^n$, and H is computable in polynomial time in $n \log q$. One can refer to [1] for the explicit construction of the map H. Finally, for $\mathsf{CH} = (\mathsf{ch}_1, \mathsf{ch}_2, \cdots, \mathsf{ch}_\ell) \in (\mathbb{Z}_q^n \setminus \{\mathbf{0}_n\})^{\leqslant L}$ and $i \in [L]$, $\mathsf{t} \in \mathbb{Z}_q^{n-1}$, we define the following functions:

- $\mathbf{E}(\mathsf{CH}) := [\mathbf{C}_1 + H(\mathsf{ch}_1)\mathbf{G} \mid \mathbf{C}_2 + H(\mathsf{ch}_2)\mathbf{G} \mid \cdots \mid \mathbf{C}_\ell + H(\mathsf{ch}_\ell)\mathbf{G}] \in \mathbb{Z}_q^{n \times \ell m}$,
- $\mathbf{F}(i, \mathsf{t}) := \mathbf{C}_{L+1} + H(i\|\mathsf{t})\mathbf{G} \in \mathbb{Z}_q^{n \times m}$.

Here $(\mathbf{C}_i)_{i \in [L+1]}$ are uniformly random matrices in $\mathbb{Z}_q^{n \times m}$ chosen in the setup algorithm of the scheme $\mathbf{\Pi}_1$ and $\mathbf{G}$ is the gadget matrix [14]. In addition, we can treat $i\|\mathbf{t}$ as a vector in $\mathbb{Z}_q^n$, since $L < q$ obviously holds due to the parameters selection given later.

3.1   Construction

Due to our new treatment of the identity spaces and the time period space, we can obtain a much simple RHIBE scheme $\mathbf{\Pi}_1$ in the standard model, which is described as follows. Here we let $\alpha, \alpha', (\sigma_\ell)_{\ell \in [0,L]}$ be positive reals denoting Gaussian parameters, and set N as the maximum number of children each parent manages. These parameters, together with positive integers n, m and a prime q, are all implicitly determined by the security parameter λ, and in particular we set $n(\lambda) := \lambda$.

Setup$(1^n, 1^L) \to (\mathsf{PP}, \mathsf{SK}_{\mathsf{KGC}})$:

Taking the security parameter n and the maximum depth of the hierarchy L as input, it performs the following steps.

1. Run $(\mathbf{A}, \mathbf{T_A}) \leftarrow \mathsf{TrapGen}(1^n, 1^m, q)$.
2. Select $\mathbf{C}_i \xleftarrow{\$} \mathbb{Z}_q^{n \times m}$ for $i \in [L+1]$, and $\mathbf{u} \xleftarrow{\$} \mathbb{Z}_q^n$.
3. Create a binary tree $\mathsf{BT}_{\mathsf{KGC}}$ with N leaf nodes, which denote N children users.
4. Output $\mathsf{PP} := \left(\mathbf{A}, (\mathbf{C}_i)_{i \in [L+1]}, \mathbf{u} \right)$, $\mathsf{SK}_{\mathsf{KGC}} := \left(\mathsf{BT}_{\mathsf{KGC}}, \mathbf{T_A} \right)$.

Here recall that $(\mathbf{C}_i)_{i \in [L+1]}$ define the functions $\mathbf{E}(\cdot)$ and $\mathbf{F}(\cdot)$ introduced before.

Encrypt$(\mathsf{PP}, \mathsf{ID}, \mathsf{t}, \mathsf{M}) \to \mathsf{CT}$:

For $\mathsf{M} \in \{0,1\}$, $|\mathsf{ID}| = \ell \in [L]$, it performs the following steps.

1. Select $\mathbf{s}_i \xleftarrow{\$} \mathbb{Z}_q^n$ for $i \in [\ell] \cup \{L+1\}$. Then sample $x \hookleftarrow \mathcal{D}_{\mathbb{Z}, \alpha q}$, $\mathbf{x}_i \hookleftarrow \mathcal{D}_{\mathbb{Z}^{(i+2)m}, \alpha' q}$ for $i \in [\ell]$, and $\mathbf{x}_{L+1} \hookleftarrow \mathcal{D}_{\mathbb{Z}^{(\ell+2)m}, \alpha' q}$.
2. Set
$$\begin{cases} c_0 := \mathbf{u}^\top(\mathbf{s}_1 + \mathbf{s}_2 + \cdots + \mathbf{s}_\ell) + \mathbf{u}^\top \mathbf{s}_{L+1} + x + \mathsf{M}\lfloor \frac{q}{2} \rfloor, \\ \mathbf{c}_i := [\mathbf{A} \mid \mathbf{E}(\widetilde{\mathsf{ID}_{[i]}}) \mid \mathbf{F}(i, \mathsf{t})]^\top \mathbf{s}_i + \mathbf{x}_i \quad \text{for} \quad i \in [\ell], \\ \mathbf{c}_{L+1} := [\mathbf{A} \mid \mathbf{E}(\mathsf{ID}) \mid \mathbf{F}(\ell, \mathsf{t})]^\top \mathbf{s}_{L+1} + \mathbf{x}_{L+1}. \end{cases}$$
3. Output $\mathsf{CT} := \left(c_0, (\mathbf{c}_i)_{i \in [\ell]}, \mathbf{c}_{L+1} \right) \in \mathbb{Z}_q \times (\mathbb{Z}_q^{3m} \times \mathbb{Z}_q^{4m} \times \cdots \times \mathbb{Z}_q^{(\ell+2)m}) \times \mathbb{Z}_q^{(\ell+2)m}$.

GenSK$(\mathsf{PP}, \mathsf{SK}_{\mathsf{pa}(\mathsf{ID})}, \mathsf{ID}) \to (\mathsf{SK}_{\mathsf{ID}}, \mathsf{SK}'_{\mathsf{pa}(\mathsf{ID})})$:

For $|\mathsf{ID}| = \ell \in [L]$, it performs the following steps.

1. Randomly pick an unassigned leaf node η_{ID} from $\mathsf{BT}_{\mathsf{pa}(\mathsf{ID})}$ and store ID in node η_{ID}. Then select $\mathbf{u}_{\mathsf{pa}(\mathsf{ID}), \theta} \xleftarrow{\$} \mathbb{Z}_q^n$ for node $\theta \in \mathsf{Path}(\mathsf{BT}_{\mathsf{pa}(\mathsf{ID})}, \eta_{\mathsf{ID}})$, if $\mathbf{u}_{\mathsf{pa}(\mathsf{ID}), \theta}$ is undefined. Here $\mathsf{pa}(\mathsf{ID})$ updates $\mathsf{SK}_{\mathsf{pa}(\mathsf{ID})}$ to $\mathsf{SK}'_{\mathsf{pa}(\mathsf{ID})}$ by storing new defined $\mathbf{u}_{\mathsf{pa}(\mathsf{ID}), \theta}$ in $\theta \in \mathsf{BT}_{\mathsf{pa}(\mathsf{ID})}$.

2. Run $e_{ID,\theta} \leftarrow$ SampleLeft($[A \mid E(pa(ID))], C_\ell + H(\widetilde{id_\ell})G, T_{[A\mid E(pa(ID))]}$, $u_{pa(ID),\theta}, \sigma_{\ell-1}$) for $\theta \in$ Path($BT_{pa(ID)}, \eta_{ID}$). Here $e_{ID,\theta} \in \mathbb{Z}^{(\ell+1)m}$ satisfies $[A \mid E(\widetilde{ID})]e_{ID,\theta} = u_{pa(ID),\theta}$.

3. Run $T_{[A\mid E(ID)]} \leftarrow$ SampleBasisLeft($[A \mid E(pa(ID))], C_\ell + H(id_\ell)G$, $T_{[A\mid E(pa(ID))]}, \sigma_{\ell-1}$).

4. Create a new binary tree BT_{ID} with N leaf nodes.

5. Output $SK_{ID} := \left(BT_{ID}, (\theta, e_{ID,\theta})_{\theta \in Path(BT_{pa(ID)}, \eta_{ID})}, T_{[A\mid E(ID)]}\right), SK'_{pa(ID)}$.

KeyUp($PP, t, SK_{ID}, RL_{ID,t}, KU_{pa(ID),t}$) $\rightarrow (KU_{ID,t}, SK'_{ID})$:
For $|ID| = \ell \in [0, L-1]$, it performs the following steps.

1. Select $u_{ID,\theta} \xleftarrow{\$} \mathbb{Z}_q^n$ for node $\theta \in$ KUNode($BT_{ID}, RL_{ID,t}$), if $u_{ID,\theta}$ is undefined. Here ID may update SK_{ID} to SK'_{ID} by storing new defined $u_{ID,\theta}$ in $\theta \in BT_{ID}$.

2. Run $e_{ID,t,\theta} \leftarrow$ SampleLeft($[A \mid E(ID)], F(\ell+1, t), T_{[A\mid E(ID)]}, u - u_{ID,\theta}, \sigma_\ell$) for $\theta \in$ KUNode($BT_{ID}, RL_{ID,t}$). Here $e_{ID,t,\theta} \in \mathbb{Z}^{(\ell+2)m}$ satisfies $[A \mid E(ID) \mid F(\ell+1, t)]e_{ID,t,\theta} = u - u_{ID,\theta}$.

3. If $\ell \geq 1$, run $DK_{ID,t} \leftarrow$ **GenDK**($PP, SK_{ID}, KU_{pa(ID),t}$), where **GenDK**($\cdot$) is defined below. Then extract $(d_{ID_{[i]},t})_{i\in[\ell]}$ from $DK_{ID,t}$.

4. Output $KU_{ID,t} := \left((\theta, e_{ID,t,\theta})_{\theta \in KUNode(BT_{ID}, RL_{ID,t})}, (d_{ID_{[i]},t})_{i\in[\ell]}\right), SK'_{ID}$.

GenDK($PP, SK_{ID}, KU_{pa(ID),t}$) $\rightarrow DK_{ID,t}$ or $\perp$:
For $|ID| = \ell \in [L]$, it performs the following steps.

1. Extract $P :=$ Path($BT_{pa(ID)}, \eta_{ID}$) in SK_{ID}, and $K :=$ KUNode($BT_{pa(ID)}$, $RL_{pa(ID),t}$) in $KU_{pa(ID),t}$. If $P \cap K = \emptyset$, output $\perp$. Otherwise, for the unique node $\theta^* \in P \cap K$, extract $e_{ID,\theta^*}, e_{pa(ID),t,\theta^*} \in \mathbb{Z}^{(\ell+1)m}$ in $SK_{ID}, KU_{pa(ID),t}$, respectively. Parse them as $e_{ID,\theta^*} = [e^L_{ID,\theta^*} \| e^R_{ID,\theta^*}]$, $e_{pa(ID),t,\theta^*} = [e^L_{pa(ID),t,\theta^*} \| e^R_{pa(ID),t,\theta^*}]$, where $e^L_{ID,\theta^*}, e^L_{pa(ID),t,\theta^*} \in \mathbb{Z}^{\ell m}$ and $e^R_{ID,\theta^*}, e^R_{pa(ID),t,\theta^*} \in \mathbb{Z}^m$. Then set $d_{ID,t} := [e^L_{ID,\theta^*} + e^L_{pa(ID),t,\theta^*} \| e^R_{ID,\theta^*} \| e^R_{pa(ID),t,\theta^*}] \in \mathbb{Z}^{(\ell+2)m}$.

2. If $\ell \geq 2$, extract $(d_{ID_{[i]},t})_{i\in[\ell-1]}$ from $KU_{pa(ID),t}$.

3. Run $g_{ID,t} \leftarrow$ SampleLeft($[A \mid E(ID)], F(\ell, t), T_{[A\mid E(ID)]}, u, \sigma_\ell$). Here $g_{ID,t} \in \mathbb{Z}^{(\ell+2)m}$ satisfies $[A \mid E(ID) \mid F(\ell, t)]g_{ID,t} = u$.

4. Output $DK_{ID,t} := \left((d_{ID_{[i]},t})_{i\in[\ell]}, g_{ID,t}\right)$.

Decrypt($PP, DK_{ID,t}, CT$) $\rightarrow M$:
For $|ID| = \ell \in [L]$, it performs the following steps.

1. Compute $c' := c_0 - \sum_{i=1}^{\ell} d^\top_{ID_{[i]},t}c_i - g^\top_{ID,t}c_{L+1} \in \mathbb{Z}_q$. Treat c' as an integer in $[q] \subset \mathbb{Z}$.

2. Output $M := 1$ if $|c' - \lfloor \frac{q}{2} \rfloor| < \lfloor \frac{q}{4} \rfloor$, and output $M := 0$ otherwise.

Correctness. Assume that ID has the depth $|ID| = \ell \in [L]$. If $ID \notin RL_t$, then one can obtain $DK_{ID,t} = \left((d_{ID_{[i]},t})_{i\in[\ell]}, g_{ID,t}\right)$. Recall that $d_{ID,t} = [e^L_{ID,\theta^*} + e^L_{pa(ID),t,\theta^*} \| e^R_{ID,\theta^*} \| e^R_{pa(ID),t,\theta^*}] \in \mathbb{Z}^{(\ell+2)m}$, where $\theta^* \in$ Path($BT_{pa(ID)}, \eta_{ID}$) $\cap$ KUNode($BT_{pa(ID)}, RL_{pa(ID),t}$). According to

$$[\mathbf{A} \mid \mathbf{E}(\widetilde{\mathsf{ID}})]\mathbf{e}_{\mathsf{ID},\theta^*} = \mathbf{u}_{\mathsf{pa}(\mathsf{ID}),\theta^*}, \quad [\mathbf{A} \mid \mathbf{E}(\mathsf{pa}(\mathsf{ID})) \mid \mathbf{F}(\ell,t)]\mathbf{e}_{\mathsf{pa}(\mathsf{ID}),t,\theta^*} = \mathbf{u} - \mathbf{u}_{\mathsf{pa}(\mathsf{ID}),\theta^*},$$
$$\mathbf{e}_{\mathsf{ID},\theta^*} = [\mathbf{e}^{\mathsf{L}}_{\mathsf{ID},\theta^*} \| \mathbf{e}^{\mathsf{R}}_{\mathsf{ID},\theta^*}], \qquad \mathbf{e}_{\mathsf{pa}(\mathsf{ID}),t,\theta^*} = [\mathbf{e}^{\mathsf{L}}_{\mathsf{pa}(\mathsf{ID}),t,\theta^*} \| \mathbf{e}^{\mathsf{R}}_{\mathsf{pa}(\mathsf{ID}),t,\theta^*}],$$

one can obtain $[\mathbf{A} \mid \mathbf{E}(\widetilde{\mathsf{ID}}) \mid \mathbf{F}(\ell,t)]\mathbf{d}_{\mathsf{ID},t} = \mathbf{u}$, $\mathbf{d}^{\top}_{\mathsf{ID},t}\mathbf{c}_\ell = \mathbf{u}^{\top}\mathbf{s}_\ell + \mathbf{d}^{\top}_{\mathsf{ID},t}\mathbf{x}_\ell$. Similarly, for $i \in [\ell - 1]$ we also have $[\mathbf{A} \mid \mathbf{E}(\widetilde{\mathsf{ID}}_{[i]}) \mid \mathbf{F}(i,t)]\mathbf{d}_{\mathsf{ID}_{[i]},t} = \mathbf{u}$, $\mathbf{d}^{\top}_{\mathsf{ID}_{[i]},t}\mathbf{c}_i = \mathbf{u}^{\top}\mathbf{s}_i + \mathbf{d}^{\top}_{\mathsf{ID}_{[i]},t}\mathbf{x}_i$. Besides, the vector $\mathbf{g}_{\mathsf{ID},t} \in \mathbb{Z}^{(\ell+2)m}$ satisfies $[\mathbf{A} \mid \mathbf{E}(\mathsf{ID}) \mid \mathbf{F}(\ell,t)]\mathbf{g}_{\mathsf{ID},t} = \mathbf{u}$, $\mathbf{g}^{\top}_{\mathsf{ID},t}\mathbf{c}_{L+1} = \mathbf{u}^{\top}\mathbf{s}_{L+1} + \mathbf{g}^{\top}_{\mathsf{ID},t}\mathbf{x}_{L+1}$. From the above, we can compute

$$c' = \mathbf{u}^{\top}(\textstyle\sum_{i=1}^{\ell} \mathbf{s}_i) + \mathbf{u}^{\top}\mathbf{s}_{L+1} + x + \mathsf{M}\lfloor \tfrac{q}{2} \rfloor - \textstyle\sum_{i=1}^{\ell} \mathbf{d}^{\top}_{\mathsf{ID}_{[i]},t}\mathbf{c}_i - \mathbf{g}^{\top}_{\mathsf{ID},t}\mathbf{c}_{L+1}$$
$$= \mathsf{M}\lfloor \tfrac{q}{2} \rfloor + (x - \textstyle\sum_{i=1}^{\ell} \mathbf{d}^{\top}_{\mathsf{ID}_{[i]},t}\mathbf{x}_i - \mathbf{g}^{\top}_{\mathsf{ID},t}\mathbf{x}_{L+1}).$$

Set $z := x - \sum_{i=1}^{\ell} \mathbf{d}^{\top}_{\mathsf{ID}_{[i]},t}\mathbf{x}_i - \mathbf{g}^{\top}_{\mathsf{ID},t}\mathbf{x}_{L+1}$ as the noise. Then according to the triangle inequality, the Cauchy-Schwarz inequality, and Lemma 1, the noise z can be bounded as follows with overwhelming probability:

$$|z| \leqslant |x| + \textstyle\sum_{i=1}^{\ell} \|\mathbf{d}_{\mathsf{ID}_{[i]},t}\| \cdot \|\mathbf{x}_i\| + \|\mathbf{g}_{\mathsf{ID},t}\| \cdot \|\mathbf{x}_{L+1}\|$$
$$\leqslant \alpha q + \textstyle\sum_{i=1}^{\ell} 2 \cdot \sigma_{i-1}\sqrt{(i+2)m} \cdot \alpha'q\sqrt{(i+2)m} + \sigma_\ell\sqrt{(\ell+2)m} \cdot \alpha'q\sqrt{(\ell+2)m}$$
$$= \alpha q + [\textstyle\sum_{i=1}^{\ell} 2(i+2)\sigma_{i-1} + (\ell+2)\sigma_\ell]m\alpha'q$$
$$\leqslant \alpha q + [2L(L+2) + (L+2)]\sigma_L m\alpha'q$$
$$= O(\alpha q + L^2\sigma_L m\alpha'q).$$

As a conclusion, if $O(\alpha q + L^2\sigma_L m\alpha'q) < q/5$, we know that $|z|$ is upper bounded by $q/5$ with overwhelming probability, and thus our RHIBE scheme $\mathbf{\Pi}_1$ only has negligible decryption error.

Parameters. The analysis for parameters selection is similar to that in [11]. We must consider the condition $O(\alpha q + L^2\sigma_L m\alpha'q) < q/5$ for the correctness requirement, and the condition $q > 2\sqrt{n}/\alpha$ for the hardness assumption of $\mathsf{LWE}_{n,m+1,q,\mathcal{D}_{\mathbb{Z}^{m+1},\alpha q}}$. Besides, we also need to make sure that algorithms such as SampleBasisLeft et al. can operate in the construction, and algorithms such as SampleBasisRight et al. can work in the security proof. Finally, we set the parameters used for our RHIBE scheme $\mathbf{\Pi}_1$ as follows:

$$m = 6n^{1+\delta} = O(Ln\log n), \qquad \alpha = [L^{\frac{5}{2}}m^{\frac{1}{2}L+2}\omega(\log^{\frac{1}{2}L+\frac{1}{2}} n)]^{-1}, \qquad \alpha' = O((Lm)^{\frac{1}{2}})\alpha,$$
$$q = L^{\frac{5}{2}}m^{\frac{1}{2}L+\frac{5}{2}}\omega(\log^{\frac{1}{2}L+\frac{1}{2}} n), \quad \sigma_\ell = m^{\frac{1}{2}\ell+\frac{1}{2}}\omega(\log^{\frac{1}{2}\ell+\frac{1}{2}} n) \text{ for } \ell \in [0,L],$$

and round up m to the nearest larger integer, and q to the nearest larger prime. Here we choose δ such that $n^\delta > \lceil \log q \rceil = O(L\log n)$.

3.2 Security

Theorem 1. *The RHIBE scheme* $\mathbf{\Pi}_1$ *satisfies the selective-identity security, assuming the hardness of the problem* $\mathsf{LWE}_{n,m+1,q,\chi}$ *where* $\chi = \mathcal{D}_{\mathbb{Z}^{m+1},\alpha q}$.

Let $\mathsf{ID}^* = (\mathsf{id}_1^*, \cdots, \mathsf{id}_{\ell^*}^*)$, t^* be the challenge identity and time period with $\ell^* := |\mathsf{ID}^*|$. Then the attack strategies taken by $\mathcal{A}$ can be divided into the following two types, which consist of $\ell^* + 1$ strategies in total.

- **Type-I**: $\mathcal{A}$ issues secret key reveal queries on at least one ID $\in$ prefix(ID*).
  - Further divided into **Type-I-i^*** ($i^* \in [\ell^*]$):
    $\mathcal{A}$ issues a secret key reveal query on $\mathsf{ID}^*_{[i^*]}$ but not on any ID $\in$ prefix($\mathsf{ID}^*_{[i^*-1]}$).
- **Type-II**: $\mathcal{A}$ does not issue secret key reveal queries on any ID $\in$ prefix(ID*).

Our security proof follows the framework in [12] (the full version of [11]). Firstly, we show that if a PPT adversary $\mathcal{A}$ follows the **Type-I-i^*** strategy for some $i^* \in [\ell^*]$, then its selective-identity security advantage is negligible, assuming the hardness of the problem $\mathsf{LWE}_{n,m+1,q,\chi}$. Secondly, the same also applies for a PPT adversary $\mathcal{A}$ following the **Type-II** strategy. Finally, we can complete the proof of Theorem 1 due to the "strategy-dividing lemma" introduced in [11,12]. Note that the proof makes heavy use of the algorithms SampleRight, SampleBasisRight, and Lemma 2. Due to space constraints, the details are given in the full version of this paper.

4   RHIBE Scheme in the Random Oracle Model

In this section, we describe our second RHIBE scheme $\mathbf{\Pi}_2$ in Sect. 4.1, and then provide its adaptive-identity security in Sect. 4.2. As a preparation, we need to explain our treatment of some spaces such as $\mathcal{T}, \mathcal{ID}, \mathcal{ID}_\mathsf{H} = (\mathcal{ID})^{\leqslant L}$, and introduce two random oracles used in the scheme $\mathbf{\Pi}_2$.

Treatment of Spaces. The time period space $\mathcal{T}$, the element identity space $\mathcal{ID}$, and the space $\widetilde{\mathcal{ID}}$ are all treated as subsets of $\{0,1,2\}^\omega$, such that $\mathcal{T} \cap \mathcal{ID} = \mathcal{T} \cap \widetilde{\mathcal{ID}} = \mathcal{ID} \cap \widetilde{\mathcal{ID}} = \emptyset$. Here ω is an integer determined by the security parameter. Similarly, we also need to define a function $f : \mathcal{ID} \to \widetilde{\mathcal{ID}}$ satisfying $f(\mathsf{id}_1) \neq f(\mathsf{id}_2)$ for $\mathsf{id}_1 \neq \mathsf{id}_2$. For simplicity, we just define

$$\mathcal{T} := \{0\} \times \{0,1,2\}^{\omega-1}, \ \mathcal{ID} := \{1\} \times \{0,1,2\}^{\omega-1}, \ \widetilde{\mathcal{ID}} := \{2\} \times \{0,1,2\}^{\omega-1},$$
$$\text{and} \ \ f(1\|\mathsf{ch}) := 2\|\mathsf{ch} \ \ \text{for} \ \ \mathsf{ch} \in \{0,1,2\}^{\omega-1}.$$

Note that here one can also choose $\mathcal{T}, \mathcal{ID}, \widetilde{\mathcal{ID}}$ as pairwise disjoint subsets of $\{0,1\}^*$. Next, let us deal with the hierarchical identity space $\mathcal{ID}_\mathsf{H} = (\mathcal{ID})^{\leqslant L} = \bigcup_{i \in [L]}(\mathcal{ID})^i$. We still define $\mathcal{F} : (\mathcal{ID})^{\leqslant L} \to \bigcup_{i \in [0,L-1]}(\mathcal{ID})^i \times \widetilde{\mathcal{ID}}$ as $\mathcal{F}(\mathsf{ID}) := (\mathsf{id}_1, \cdots, \mathsf{id}_{\ell-1}, f(\mathsf{id}_\ell))$ for $\mathsf{ID} = (\mathsf{id}_1, \cdots, \mathsf{id}_{\ell-1}, \mathsf{id}_\ell)$. Similarly, for $|\mathsf{ID}| = \ell \geqslant 2$, we have $\mathsf{ID} \neq \mathcal{F}(\mathsf{ID})$, $\mathsf{ID}_{[\ell-1]} = [\mathcal{F}(\mathsf{ID})]_{[\ell-1]}$. For simplicity, we still set $\widetilde{\mathsf{id}} := f(\mathsf{id})$, $\widetilde{\mathsf{ID}} := \mathcal{F}(\mathsf{ID})$, and use $\widetilde{\mathsf{ID}}_{[i]}$ to denote $\mathcal{F}(\mathsf{ID}_{[i]})$. In addition, for KGC and $\mathsf{ID} = (\mathsf{id}_1, \cdots, \mathsf{id}_\ell) \in (\{0,1,2\}^\omega)^\ell$, we define the notations $\mathsf{KGC}\|\mathsf{t} := \mathsf{t}$ and $\mathsf{ID}\|\mathsf{t} := (\mathsf{id}_1, \cdots, \mathsf{id}_\ell, \mathsf{t}) \in (\{0,1,2\}^\omega)^{\ell+1}$, and thus $(\mathsf{ID}\|\mathsf{t})_{[\ell]} = \mathsf{ID}$.

Random Oracles. We define two random oracles $\mathbf{H}_1, \mathbf{H}_2$ as follows:

- $\mathbf{H}_1 : (\{0,1,2\}^\omega)^{\leqslant L+1} \to \mathbb{Z}_q^{m \times m}$, $\mathsf{CH} \mapsto \mathbf{H}_1(\mathsf{CH}) \sim \mathcal{D}_{m \times m}$,
- $\mathbf{H}_2 : (\{0,1,2\}^\omega)^{\leqslant L} \to \mathbb{Z}_q^{m \times m}$, $\mathsf{CH}' \mapsto \mathbf{H}_2(\mathsf{CH}') \sim \mathcal{D}_{m \times m}$.

Here the outputs of $\mathbf{H}_1, \mathbf{H}_2$ are both distributed as $\mathcal{D}_{m \times m}$, which is defined below Lemma 2 in Sect. 2.1. Furthermore, for $\mathsf{CH} \in (\{0,1,2\}^{\omega})^{\leqslant L+1}$ with $\ell = |\mathsf{CH}|$, and $\mathsf{CH}' \in (\{0,1,2\}^{\omega})^{\leqslant L}$ with $\ell' = |\mathsf{CH}'|$, we define the following functions:

- $\mathbf{P}_1(\mathsf{CH}) := [\mathbf{H}_1(\mathsf{CH}_{[\ell]})\mathbf{H}_1(\mathsf{CH}_{[\ell-1]}) \cdots \mathbf{H}_1(\mathsf{CH}_{[1]})]^{-1} \in \mathbb{Z}_q^{m \times m}$,
- $\mathbf{P}_2(\mathsf{CH}') := [\mathbf{H}_2(\mathsf{CH}'_{[\ell']})\mathbf{H}_2(\mathsf{CH}'_{[\ell'-1]}) \cdots \mathbf{H}_2(\mathsf{CH}'_{[1]})]^{-1} \in \mathbb{Z}_q^{m \times m}$.

Therefore, after setting $\mathbf{P}_1(\mathsf{CH}_{[0]}), \mathbf{P}_2(\mathsf{CH}'_{[0]})$ as the identity matrix $\mathbf{I}_{m \times m}$, we have $\mathbf{P}_1(\mathsf{CH}_{[j]}) = \mathbf{P}_1(\mathsf{CH}_{[j-1]}) \cdot [\mathbf{H}_1(\mathsf{CH}_{[j]})]^{-1}$ for $j \in [\ell]$, and $\mathbf{P}_2(\mathsf{CH}'_{[j']}) = \mathbf{P}_2(\mathsf{CH}'_{[j'-1]}) \cdot [\mathbf{H}_2(\mathsf{CH}'_{[j']})]^{-1}$ for $j' \in [\ell']$.

4.1 Construction

In the following, we describe the construction of our RHIBE scheme $\mathbf{\Pi}_2$ in the random oracle model. Note that in this scheme the KGC's secret key $\mathsf{SK}_{\mathsf{KGC}}$ contains two trapdoor bases. Similar to Sect. 3.1, here we let $\alpha, (\sigma_\ell)_{\ell \in [0,L]}$ be positive reals denoting Gaussian parameters, and set N as the maximum number of children each parent manages. These parameters, together with positive integers n, m and a prime q, are all implicitly determined by the security parameter λ, and in particular we set $n(\lambda) := \lambda$. Besides, in the scheme $\mathbf{\Pi}_2$ we set $\tau_\ell := \sigma_\ell \sqrt{m} \cdot \omega(\sqrt{\log m})$ for $\ell \in [0, L]$ to make the algorithm SamplePre work.

Setup$(1^n, 1^L) \to (\mathsf{PP}, \mathsf{SK}_{\mathsf{KGC}})$:

    Taking the security parameter n and the maximum depth of the hierarchy L as input, it performs the following steps.

    1. Run $(\mathbf{A}, \mathbf{T_A}) \leftarrow \mathsf{TrapGen}(1^n, 1^m, q)$, and $(\mathbf{B}, \mathbf{T_B}) \leftarrow \mathsf{TrapGen}(1^n, 1^m, q)$.
    2. Select $\mathbf{u} \xleftarrow{\$} \mathbb{Z}_q^n$.
    3. Create a binary tree $\mathsf{BT}_{\mathsf{KGC}}$ with N leaf nodes, which denote N children users.
    4. Output $\mathsf{PP} := \left(\mathbf{A}, \mathbf{B}, \mathbf{u} \right)$, $\mathsf{SK}_{\mathsf{KGC}} := \left(\mathsf{BT}_{\mathsf{KGC}}, \mathbf{T_A}, \mathbf{T_B} \right)$.

Encrypt$(\mathsf{PP}, \mathsf{ID}, \mathsf{t}, \mathsf{M}) \to \mathsf{CT}$:

    For $\mathsf{M} \in \{0,1\}$, $|\mathsf{ID}| = \ell \in [L]$, it performs the following steps.

    1. Select $\mathbf{s}_i \xleftarrow{\$} \mathbb{Z}_q^n$ for $i \in [\ell] \cup \{L+1\}$. Then sample $x \hookleftarrow \mathcal{D}_{\mathbb{Z}, \alpha q}$, $\mathbf{x}_{i,j} \hookleftarrow \mathcal{D}_{\mathbb{Z}^m, \alpha q}$ for $(i, j) \in [\ell] \times [2]$, and $\mathbf{x}_{L+1} \hookleftarrow \mathcal{D}_{\mathbb{Z}^m, \alpha q}$.
    2. Define $\mathsf{ID}_{[0]} \| \mathsf{t} := \mathsf{t}$, and then set
$$\begin{cases} c_0 := \mathbf{u}^\top(\mathbf{s}_1 + \mathbf{s}_2 + \cdots + \mathbf{s}_\ell) + \mathbf{u}^\top \mathbf{s}_{L+1} + x + \mathsf{M} \lfloor \frac{q}{2} \rfloor, \\ \mathbf{c}_{i,1} := [\mathbf{A} \cdot \mathbf{P}_1(\widetilde{\mathsf{ID}_{[i]}})]^\top \mathbf{s}_i + \mathbf{x}_{i,1} \quad \text{for} \quad i \in [\ell], \\ \mathbf{c}_{i,2} := [\mathbf{B} \cdot \mathbf{P}_2(\mathsf{ID}_{[i-1]} \| \mathsf{t})]^\top \mathbf{s}_i + \mathbf{x}_{i,2} \quad \text{for} \quad i \in [\ell], \\ \mathbf{c}_{L+1} := [\mathbf{A} \cdot \mathbf{P}_1(\mathsf{ID} \| \mathsf{t})]^\top \mathbf{s}_{L+1} + \mathbf{x}_{L+1}. \end{cases}$$
    3. Output $\mathsf{CT} := \left(c_0, (\mathbf{c}_{i,1}, \mathbf{c}_{i,2})_{i \in [\ell]}, \mathbf{c}_{L+1} \right) \in \mathbb{Z}_q \times (\mathbb{Z}_q^m \times \mathbb{Z}_q^m)^\ell \times \mathbb{Z}_q^m$.

GenSK$(\mathsf{PP}, \mathsf{SK}_{\mathsf{pa}(\mathsf{ID})}, \mathsf{ID}) \to (\mathsf{SK}_{\mathsf{ID}}, \mathsf{SK}'_{\mathsf{pa}(\mathsf{ID})})$:

    For $|\mathsf{ID}| = \ell \in [L]$, it performs the following steps.

1. Randomly pick an unassigned leaf node η_{ID} from $\text{BT}_{\text{pa}(\text{ID})}$ and store ID in node η_{ID}. Then select $\mathbf{u}_{\text{pa}(\text{ID}),\theta} \xleftarrow{\$} \mathbb{Z}_q^n$ for node $\theta \in \text{Path}(\text{BT}_{\text{pa}(\text{ID})}, \eta_{\text{ID}})$, if $\mathbf{u}_{\text{pa}(\text{ID}),\theta}$ is undefined. Here $\text{pa}(\text{ID})$ updates $\text{SK}_{\text{pa}(\text{ID})}$ to $\text{SK}'_{\text{pa}(\text{ID})}$ by storing new defined $\mathbf{u}_{\text{pa}(\text{ID}),\theta}$ in $\theta \in \text{BT}_{\text{pa}(\text{ID})}$.

2. Define $\mathbf{P}_1(\text{KGC}), \mathbf{P}_2(\text{KGC})$ as the identity matrix $\mathbf{I}_{m \times m}$, and then run
$$\begin{cases} \mathbf{T}_{\mathbf{A} \cdot \mathbf{P}_1(\widetilde{\text{ID}})} \leftarrow \text{BasisDel}(\mathbf{A} \cdot \mathbf{P}_1(\text{pa}(\text{ID})), \mathbf{H}_1(\widetilde{\text{ID}}), \mathbf{T}_{\mathbf{A} \cdot \mathbf{P}_1(\text{pa}(\text{ID}))}, \sigma_{\ell-1}), \\ \mathbf{T}_{\mathbf{A} \cdot \mathbf{P}_1(\text{ID})} \leftarrow \text{BasisDel}(\mathbf{A} \cdot \mathbf{P}_1(\text{pa}(\text{ID})), \mathbf{H}_1(\text{ID}), \mathbf{T}_{\mathbf{A} \cdot \mathbf{P}_1(\text{pa}(\text{ID}))}, \sigma_{\ell-1}), \\ \mathbf{T}_{\mathbf{B} \cdot \mathbf{P}_2(\text{ID})} \leftarrow \text{BasisDel}(\mathbf{B} \cdot \mathbf{P}_2(\text{pa}(\text{ID})), \mathbf{H}_2(\text{ID}), \mathbf{T}_{\mathbf{B} \cdot \mathbf{P}_2(\text{pa}(\text{ID}))}, \sigma_{\ell-1}). \end{cases}$$

3. Run $\mathbf{e}_{\text{ID},\theta} \leftarrow \text{SamplePre}(\mathbf{A} \cdot \mathbf{P}_1(\widetilde{\text{ID}}), \mathbf{T}_{\mathbf{A} \cdot \mathbf{P}_1(\widetilde{\text{ID}})}, \mathbf{u}_{\text{pa}(\text{ID}),\theta}, \tau_{\ell-1})$ for $\theta \in \text{Path}(\text{BT}_{\text{pa}(\text{ID})}, \eta_{\text{ID}})$.

4. Create a new binary tree BT_{ID} with N leaf nodes.

5. Output $\text{SK}_{\text{ID}} := \left(\text{BT}_{\text{ID}}, \ (\theta, \mathbf{e}_{\text{ID},\theta})_{\theta \in \text{Path}(\text{BT}_{\text{pa}(\text{ID})}, \eta_{\text{ID}})}, \mathbf{T}_{\mathbf{A} \cdot \mathbf{P}_1(\text{ID})}, \mathbf{T}_{\mathbf{B} \cdot \mathbf{P}_2(\text{ID})} \right)$, $\text{SK}'_{\text{pa}(\text{ID})}$.

KeyUp$(\text{PP}, t, \text{SK}_{\text{ID}}, \text{RL}_{\text{ID},t}, \text{KU}_{\text{pa}(\text{ID}),t}) \rightarrow (\text{KU}_{\text{ID},t}, \text{SK}'_{\text{ID}})$:
For $|\text{ID}| = \ell \in [0, L-1]$, it performs the following steps.

1. Select $\mathbf{u}_{\text{ID},\theta} \xleftarrow{\$} \mathbb{Z}_q^n$ for node $\theta \in \text{KUNode}(\text{BT}_{\text{ID}}, \text{RL}_{\text{ID},t})$, if $\mathbf{u}_{\text{ID},\theta}$ is undefined. Here ID may update SK_{ID} to SK'_{ID} by storing new defined $\mathbf{u}_{\text{ID},\theta}$ in $\theta \in \text{BT}_{\text{ID}}$.

2. Run $\mathbf{T}_{\mathbf{B} \cdot \mathbf{P}_2(\text{ID}\|t)} \leftarrow \text{BasisDel}(\mathbf{B} \cdot \mathbf{P}_2(\text{ID}), \mathbf{H}_2(\text{ID}\|t), \mathbf{T}_{\mathbf{B} \cdot \mathbf{P}_2(\text{ID})}, \sigma_\ell)$, and then run $\mathbf{e}_{\text{ID},t,\theta} \leftarrow \text{SamplePre}(\mathbf{B} \cdot \mathbf{P}_2(\text{ID}\|t), \mathbf{T}_{\mathbf{B} \cdot \mathbf{P}_2(\text{ID}\|t)}, \mathbf{u} - \mathbf{u}_{\text{ID},\theta}, \tau_\ell)$ for $\theta \in \text{KUNode}(\text{BT}_{\text{ID}}, \text{RL}_{\text{ID},t})$.

3. If $\ell \geq 1$, run $\text{DK}_{\text{ID},t} \leftarrow \textbf{GenDK}(\text{PP}, \text{SK}_{\text{ID}}, \text{KU}_{\text{pa}(\text{ID}),t})$, where $\textbf{GenDK}(\cdot)$ is defined below. Then extract $(\mathbf{d}_{\text{ID}_{[i]},t})_{i \in [\ell]}$ from $\text{DK}_{\text{ID},t}$.

4. Output $\text{KU}_{\text{ID},t} := \left((\theta, \mathbf{e}_{\text{ID},t,\theta})_{\theta \in \text{KUNode}(\text{BT}_{\text{ID}}, \text{RL}_{\text{ID},t})}, (\mathbf{d}_{\text{ID}_{[i]},t})_{i \in [\ell]} \right)$, SK'_{ID}.

GenDK$(\text{PP}, \text{SK}_{\text{ID}}, \text{KU}_{\text{pa}(\text{ID}),t}) \rightarrow \text{DK}_{\text{ID},t}$ or $\perp$:
For $|\text{ID}| = \ell \in [L]$, it performs the following steps.

1. Extract $\text{P} := \text{Path}(\text{BT}_{\text{pa}(\text{ID})}, \eta_{\text{ID}})$ in SK_{ID}, and $\text{K} := \text{KUNode}(\text{BT}_{\text{pa}(\text{ID})}, \text{RL}_{\text{pa}(\text{ID}),t})$ in $\text{KU}_{\text{pa}(\text{ID}),t}$. If $\text{P} \cap \text{K} = \emptyset$, output $\perp$. Otherwise, for the unique node $\theta^* \in \text{P} \cap \text{K}$, extract $\mathbf{e}_{\text{ID},\theta^*}, \mathbf{e}_{\text{pa}(\text{ID}),t,\theta^*} \in \mathbb{Z}^m$ in $\text{SK}_{\text{ID}}, \text{KU}_{\text{pa}(\text{ID}),t}$, respectively. Then set $\mathbf{d}_{\text{ID},t} := [\mathbf{e}_{\text{ID},\theta^*} \| \mathbf{e}_{\text{pa}(\text{ID}),t,\theta^*}] \in \mathbb{Z}^{2m}$.

2. If $\ell \geq 2$, extract $(\mathbf{d}_{\text{ID}_{[i]},t})_{i \in [\ell-1]}$ from $\text{KU}_{\text{pa}(\text{ID}),t}$.

3. Run $\mathbf{T}_{\mathbf{A} \cdot \mathbf{P}_1(\text{ID}\|t)} \leftarrow \text{BasisDel}(\mathbf{A} \cdot \mathbf{P}_1(\text{ID}), \mathbf{H}_1(\text{ID}\|t), \mathbf{T}_{\mathbf{A} \cdot \mathbf{P}_1(\text{ID})}, \sigma_\ell)$, and then run $\mathbf{g}_{\text{ID},t} \leftarrow \text{SamplePre}(\mathbf{A} \cdot \mathbf{P}_1(\text{ID}\|t), \mathbf{T}_{\mathbf{A} \cdot \mathbf{P}_1(\text{ID}\|t)}, \mathbf{u}, \tau_\ell)$.

4. Output $\text{DK}_{\text{ID},t} := \left((\mathbf{d}_{\text{ID}_{[i]},t})_{i \in [\ell]}, \ \mathbf{g}_{\text{ID},t} \right)$.

Decrypt$(\text{PP}, \text{DK}_{\text{ID},t}, \text{CT}) \rightarrow \text{M}$:
For $|\text{ID}| = \ell \in [L]$, it performs the following steps.

1. Compute $c' := c_0 - \sum_{i=1}^{\ell} \mathbf{d}_{\text{ID}_{[i]},t}^\top [\mathbf{c}_{i,1} \| \mathbf{c}_{i,2}] - \mathbf{g}_{\text{ID},t}^\top \mathbf{c}_{L+1} \in \mathbb{Z}_q$. Treat c' as an integer in $[q] \subset \mathbb{Z}$.

2. Output $\text{M} := 1$ if $|c' - \lfloor \frac{q}{2} \rfloor| < \lfloor \frac{q}{4} \rfloor$, and output $\text{M} := 0$ otherwise.

Similar to Sect. 3.1, we can show that if $O(L\sigma_L m^{\frac{3}{2}}\omega(\sqrt{\log m})\alpha q) < q/5$, the above RHIBE scheme Π_2 only has negligible decryption error. Besides, we set the parameters used for Π_2 as follows:

$$m = 6n^{1+\delta} = O(Ln\log n), \qquad \alpha = [Lm^{\frac{3}{2}L+3}\omega(\log^{2L+\frac{5}{2}}n)]^{-1},$$
$$q = Lm^{\frac{3}{2}L+3}n^{\frac{1}{2}}\omega(\log^{2L+\frac{5}{2}}n), \qquad \sigma_\ell = m^{\frac{3}{2}\ell+\frac{3}{2}}\omega(\log^{2\ell+2}n) \text{ for } \ell \in [0, L],$$

and round up m to the nearest larger integer, and q to the nearest larger prime. Here we choose δ such that $n^\delta > \lceil\log q\rceil = O(L\log n)$. Due to space constraints, the detailed analysis for correctness and parameters selection is given in the full version of this paper.

4.2 Security

Theorem 2. *The RHIBE scheme Π_2 satisfies the adaptive-identity security, assuming the hardness of the problem $\mathsf{LWE}_{n,2m+1,q,\chi}$ where $\chi = \mathcal{D}_{\mathbb{Z}^{2m+1},\alpha q}$.*

It is shown in Sect. 3.2 that the attack strategies taken by $\mathcal{A}$ can be divided into the **Type-I** strategy (further divided into **Type-I-i^***) and the **Type-II** strategy. Suppose that a PPT adversary $\mathcal{A}$ follows the **Type-I** strategy, and its adaptive-identity security advantage is denoted by $\mathsf{Adv}^{\mathbf{Type\text{-}I}}_{\Pi_2,L,\mathcal{A}}(n)$. Then there exits a PPT algorithm $\mathcal{C}$, whose advantage for the $\mathsf{LWE}_{n,2m+1,q,\chi}$ problem is denoted by $\mathsf{Adv}^{\mathsf{LWE}}_{\mathcal{C}}(n)$, such that

$$\mathsf{Adv}^{\mathbf{Type\text{-}I}}_{\Pi_2,L,\mathcal{A}}(n) \leqslant (2L \cdot Q^L_{\mathbf{H}_1} \cdot Q_{\mathbf{H}_2}) \cdot \mathsf{Adv}^{\mathsf{LWE}}_{\mathcal{C}}(n) + \mathsf{negl}(n),$$

where $Q_{\mathbf{H}_1}, Q_{\mathbf{H}_2}$ denote the maximum numbers of queries made by $\mathcal{A}$ to the random oracles $\mathbf{H}_1, \mathbf{H}_2$, respectively. We note that the algorithm $\mathsf{SampleRwithBasis}$ and Lemma 2 will play an important role in the proof of the above inequality. Similarly, for a PPT adversary $\mathcal{A}$ following the **Type-II** strategy with adaptive-identity security advantage $\mathsf{Adv}^{\mathbf{Type\text{-}II}}_{\Pi_2,L,\mathcal{A}}(n)$, we have

$$\mathsf{Adv}^{\mathbf{Type\text{-}II}}_{\Pi_2,L,\mathcal{A}}(n) \leqslant (2L \cdot Q^{L+1}_{\mathbf{H}_1})\mathsf{Adv}^{\mathsf{LWE}}_{\mathcal{C}}(n) + \mathsf{negl}(n).$$

Finally, according to the "strategy-dividing lemma" introduced in [11,12], we get

$$
\begin{aligned}
\mathsf{Adv}^{\mathsf{RHIBE\text{-}ad}}_{\Pi_2,L,\mathcal{A}}(n) &\leqslant \mathsf{Adv}^{\mathbf{Type\text{-}I}}_{\Pi_2,L,\mathcal{A}}(n) + \mathsf{Adv}^{\mathbf{Type\text{-}II}}_{\Pi_2,L,\mathcal{A}}(n) \\
&\leqslant (2L \cdot Q^L_{\mathbf{H}_1} \cdot Q_{\mathbf{H}_2})\mathsf{Adv}^{\mathsf{LWE}}_{\mathcal{C}}(n) + \mathsf{negl}(n) + (2L \cdot Q^{L+1}_{\mathbf{H}_1})\mathsf{Adv}^{\mathsf{LWE}}_{\mathcal{C}}(n) + \mathsf{negl}(n) \\
&\leqslant 2L \cdot Q^L_{\mathbf{H}_1} \cdot (Q_{\mathbf{H}_2} + Q_{\mathbf{H}_1}) \cdot \mathsf{Adv}^{\mathsf{LWE}}_{\mathcal{C}}(n) + \mathsf{negl}(n).
\end{aligned}
$$

It is obtained that $\mathsf{Adv}^{\mathsf{LWE}}_{\mathcal{C}}(n) = \mathsf{negl}(n)$, assuming the hardness of the problem $\mathsf{LWE}_{n,2m+1,q,\chi}$. Since $2L \cdot Q^L_{\mathbf{H}_1} \cdot (Q_{\mathbf{H}_2} + Q_{\mathbf{H}_1})$ is polynomial in n, we know that $\mathsf{Adv}^{\mathsf{RHIBE\text{-}ad}}_{\Pi_2,L,\mathcal{A}}(n) \leqslant \mathsf{negl}(n)$, which completes the proof of Theorem 2. Due to space constraints, the detailed analysis for security proof is given in the full version of this paper.

5   Conclusion

In this paper, we present two new RHIBE schemes with DKER from lattices, and thus simplify the construction of RHIBE scheme provided by Katsumata et al. [11]. Our first scheme needs fewer items than that in [11], and the sizes of items are much smaller in our second scheme. The security of these two new schemes are both based on the hardness of the LWE problem, and our second scheme also achieves the adaptive-identity security.

Acknowledgments. The work in this paper is supported by the National Natural Science Foundation of China (Grant Nos. 11531002, 61572026 and 61722213), the Open Foundation of State Key Laboratory of Cryptology, and the program of China Scholarship Council (CSC) (No. 201703170302).

References

1. Agrawal, S., Boneh, D., Boyen, X.: Efficient lattice (H)IBE in the standard model. In: Gilbert, H. (ed.) EUROCRYPT 2010. LNCS, vol. 6110, pp. 553–572. Springer, Heidelberg (2010). https://doi.org/10.1007/978-3-642-13190-5_28
2. Agrawal, S., Boneh, D., Boyen, X.: Lattice basis delegation in fixed dimension and shorter-ciphertext hierarchical IBE. In: Rabin, T. (ed.) CRYPTO 2010. LNCS, vol. 6223, pp. 98–115. Springer, Heidelberg (2010). https://doi.org/10.1007/978-3-642-14623-7_6
3. Ajtai, M.: Generating hard instances of the short basis problem. In: Wiedermann, J., van Emde Boas, P., Nielsen, M. (eds.) ICALP 1999. LNCS, vol. 1644, pp. 1–9. Springer, Heidelberg (1999). https://doi.org/10.1007/3-540-48523-6_1
4. Alwen, J., Peikert, C.: Generating shorter bases for hard random lattices. Theory Comput. Syst. **48**(3), 535–553 (2011)
5. Boldyreva, A., Goyal, V., Kumar, V.: Identity-based encryption with efficient revocation. In: Proceedings of the 15th ACM Conference on Computer and Communications Security, pp. 417–426. ACM (2008)
6. Boneh, D., Franklin, M.: Identity-based encryption from the weil pairing. In: Kilian, J. (ed.) CRYPTO 2001. LNCS, vol. 2139, pp. 213–229. Springer, Heidelberg (2001). https://doi.org/10.1007/3-540-44647-8_13
7. Cash, D., Hofheinz, D., Kiltz, E., Peikert, C.: Bonsai trees, or how to delegate a lattice basis. In: Gilbert, H. (ed.) EUROCRYPT 2010. LNCS, vol. 6110, pp. 523–552. Springer, Heidelberg (2010). https://doi.org/10.1007/978-3-642-13190-5_27
8. Chen, J., Lim, H.W., Ling, S., Wang, H., Nguyen, K.: Revocable identity-based encryption from lattices. In: Susilo, W., Mu, Y., Seberry, J. (eds.) ACISP 2012. LNCS, vol. 7372, pp. 390–403. Springer, Heidelberg (2012). https://doi.org/10.1007/978-3-642-31448-3_29
9. Gentry, C., Peikert, C., Vaikuntanathan, V.: Trapdoors for hard lattices and new cryptographic constructions. In: Proceedings of the Fortieth Annual ACM Symposium on Theory of Computing, pp. 197–206. ACM (2008)
10. Horwitz, J., Lynn, B.: Toward hierarchical identity-based encryption. In: Knudsen, L.R. (ed.) EUROCRYPT 2002. LNCS, vol. 2332, pp. 466–481. Springer, Heidelberg (2002). https://doi.org/10.1007/3-540-46035-7_31

11. Katsumata, S., Matsuda, T., Takayasu, A.: Lattice-based revocable (Hierarchical) IBE with decryption key exposure resistance. In: Lin, D., Sako, K. (eds.) PKC 2019. LNCS, vol. 11443, pp. 441–471. Springer, Cham (2019). https://doi.org/10. 1007/978-3-030-17259-6_15

12. Katsumata, S., Matsuda, T., Takayasu, A.: Lattice-based Revocable (Hierarchical) IBE with Decryption Key Exposure Resistance. IACR Cryptology ePrint Archive, p 420 (2018). https://eprint.iacr.org/2018/420

13. Micciancio, D., Goldwasser, S.: Complexity of Lattice Problems: A Cryptographic Perspective, vol. 671. Springer, Heidelberg (2002)

14. Micciancio, D., Peikert, C.: Trapdoors for lattices: simpler, tighter, faster, smaller. In: Pointcheval, D., Johansson, T. (eds.) EUROCRYPT 2012. LNCS, vol. 7237, pp. 700–718. Springer, Heidelberg (2012). https://doi.org/10.1007/978-3-642-29011-4_41

15. Micciancio, D., Regev, O.: Worst-case to average-case reductions based on Gaussian measures. SIAM J. Comput. **37**(1), 267–302 (2007)

16. Naor, D., Naor, M., Lotspiech, J.: Revocation and tracing schemes for stateless receivers. In: Kilian, J. (ed.) CRYPTO 2001. LNCS, vol. 2139, pp. 41–62. Springer, Heidelberg (2001). https://doi.org/10.1007/3-540-44647-8_3

17. Regev O.: On lattices, learning with errors, random linear codes, and cryptography. In: Proceedings of the Thirty-Seventh Annual ACM Symposium on Theory of Computing, pp. 84–93. ACM (2005)

18. Seo, J.H., Emura, K.: Revocable identity-based encryption revisited: security model and construction. In: Kurosawa, K., Hanaoka, G. (eds.) PKC 2013. LNCS, vol. 7778, pp. 216–234. Springer, Heidelberg (2013). https://doi.org/10.1007/978-3-642-36362-7_14

19. Shamir, A.: Identity-based cryptosystems and signature schemes. In: Blakley, G.R., Chaum, D. (eds.) CRYPTO 1984. LNCS, vol. 196, pp. 47–53. Springer, Heidelberg (1985). https://doi.org/10.1007/3-540-39568-7_5

20. Takayasu, A., Watanabe, Y.: Lattice-based revocable identity-based encryption with bounded decryption key exposure resistance. In: Pieprzyk, J., Suriadi, S. (eds.) ACISP 2017. LNCS, vol. 10342, pp. 184–204. Springer, Cham (2017). https://doi. org/10.1007/978-3-319-60055-0_10

Lattice-Based Group Signatures with Verifier-Local Revocation: Achieving Shorter Key-Sizes and Explicit Traceability with Ease

Yanhua Zhang[1(✉)], Ximeng Liu[2], Yupu Hu[3], Qikun Zhang[1], and Huiwen Jia[4]

[1] Zhengzhou University of Light Industry, Zhengzhou 450002, China
{yhzhang,qkzhang}@zzuli.edu.cn
[2] Fuzhou University, Fuzhou 350108, China
snbnix@gmail.com
[3] Xidian University, Xi'an 710071, China
yphu@mail.xidian.edu.cn
[4] Guangzhou University, Guangzhou 510006, China
hwjia@gzhu.edu.cn

Abstract. For lattice-based group signatures (GS) with verifier-local revocation (VLR), it only requires the verifiers to possess up-to-date group information (i.e., a revocation list, RL, consists of a series of revocation tokens for revoked members), but not the signers. The first such scheme was introduced by Langlois et al. in 2014, and subsequently, a full and corrected version (to fix a flaw in the original revocation mechanism) was proposed by Ling et al. in 2018. However, both constructions are within the structure of a *Bonsai Tree*, and thus features bit-sizes of the group public-key and the member secret-key proportional to $\log N$, where N is the maximum number of group members. On the other hand, the tracing algorithm for both schemes runs in a linear time in N (i.e., one by one, until the real signer is traced). Therefore for a large group, the tracing algorithm of conventional GS-VLR is not convenient and both lattice-based constructions are not that efficient.

In this work, we propose a much more efficient lattice-based GS-VLR, which is efficient by saving the $\mathcal{O}(\log N)$ factor for both bit-sizes of the group public-key and the member secret-key. Moreover, we achieve this result in a relatively simple manner. Starting with Nguyen et al.'s efficient and compact *identity-encoding technique* in 2015 - which only needs a constant number of matrices to encode the member's identity, we develop an improved identity-encoding function, and introduce an efficient Stern-type statistical zero-knowledge argument of knowledge (ZKAoK) protocol corresponding to our improved identity-encoding function, which may be of independent cryptographic interest.

Furthermore, we demonstrate how to equip the obtained lattice-based GS-VLR with explicit traceability (ET) in some simple way. This attractive functionality, only satisfied in the non-VLR constructions, can enable the tracing authority in lattice-based GS-VLR to determine the signer's real identity in a constant time, independent of N. In the whole

© Springer Nature Switzerland AG 2019
Y. Mu et al. (Eds.): CANS 2019, LNCS 11829, pp. 120–140, 2019.
https://doi.org/10.1007/978-3-030-31578-8_7

process, we show that the proposed scheme is proven secure in the random oracle model (ROM) based on the hardness of the Short Integer Solution (SIS) problem, and the Learning With Errors (LWE) problem.

Keywords: Lattice-based group signatures · Verifier-local revocation · Stern-type zero-knowledge proofs · Identity-encoding technique · Explicit traceability

1   Introduction

Group signature (GS), put forward by Chaum and van Heyst [10], is a fundamental privacy-preserving primitive which allows any member to issue signatures on behalf of the whole group without compromising his/her identity information, and given a valid message-signature pair, the tracing authority (i.e., an opener) can find out the signer's real identity. These two properties, called *anonymity* and *traceability* respectively, allow GS to find several real-life applications. To construct such valid scheme is a interesting and challenging work for the research community, and over the last quarter-century, various GS constructions with different security requirements, different levels of efficiency, and based on different hardness assumptions have been proposed (e.g., [4–7, 13, 16] · · ·).

LATTICE-BASED GROUP SIGNATURES. Lattice-based cryptography, believed to be the most promising candidate for post-quantum cryptography (PQC), possesses several noticeable advantages over conventional number-theoretic cryptography: conjectured resistance against quantum computers, faster arithmetic operations and provable security under the *worst-case* hardness assumptions. Since the creative works of Ajtai [2], Regev [34], Micciancio and Regev [28], and Gentry et al. [12], lattice-based cryptography has attracted significant interest by the research community and become an exciting cryptographic research field. In recent ten years, lattice-based GS has been paid greet attention along with other primitives. The first construction was put forth by Gordon et al. [13], while their solution only obtains a low running efficiency, due to the linear-size of public-key and signature (i.e., linear in the security parameter n, and the maximum number of group members N). Camenisch et al. [8] introduced a variant of [13] to achieve the improvements with a shorter public-key and stronger anonymity while the signature size is still linear in N. The linear-size barrier problem is eventually overcome by Laguillaumie et al. [17], who provided the first logarithmic lattice-based GS scheme with relatively large parameters. Ling et al. [24] and Nguyen et al. [31] constructed more efficient schemes with $\mathcal{O}(\log N)$ signature size respectively. More recently, Libert et al. [20] developed a lattice-based accumulator from Merkle trees and based on which they designed the first lattice-based GS not requiring any GPV trapdoors. The first lattice-based GS realizations with message-dependent opening (MDO), forward-secure (FS), and without NIZK in the standard model (SM) were then proposed by Libert et al. [21], Ling et al. [26], and Katsumata and Yamada [14], respectively. For the lattice-based GS schemes mentioned above, all are designed for the static groups and analyzed in the security model of Bellare et al. [4], where no candidate member is allowed to join or leave after the whole group's preliminary setup.

For lattice-based GS schemes with dynamic features, member enrollment was firstly token into account by Libert et al. [19] and a dynamic construction in the model of Kiayias and Yong [16] and Bellare et al. [5] was introduced. Ling et al. [27] added some dynamic ingredients into a static accumulator constructed in [20] to construct the first lattice-based GS scheme with full dynamicity (i.e., candidate members can join and leave the group at will) in the model of Bootle et al. [7]. Recently, Ling et al. [25] introduced a constant-size lattice-based GS scheme (i.e., signature size is independent of N), meanwhile supporting dynamic member enrollments.

As an orthogonal problem of member enrollment, the support for membership revocation is also a desirable functionality of lattice-based GS. The verifier-local revocation (VLR) mechanism, which only requires the verifiers to possess some up-to-date group information (i.e., a revocation list, RL, consists of a series of revocation tokens for the revoked members), but not the signers, is more efficient than the accumulators, especially when considering a large group. The first such scheme was introduced by Langlois et al. [18] in 2014, and subsequently, a full and corrected version (to fix a flaw in original revocation mechanism) was proposed by Ling et al. [22], and two more schemes achieving different security notions were proposed by Perera and Koshiba [32,33] in 2018. However, all constructions are within the structure of a *Bonsai Tree* of hard random lattices [9], and thus features bit-sizes of the group public-key and the member secret-key proportional to $\log N$. The only two exceptions are [11,35] which adopt a identity-encoding function as introduced in [31] to encode the member's identity index and save a $\mathcal{O}(\log N)$ factor for both bit-sizes. However, the latter two constructions both involve a series of sophisticated encryption operations and zero-knowledge proof protocols in the signing phase, and on the other hand, the tracing algorithm for [11,18,22,35] runs in a linear time in N (i.e., one by one for all members, until the real signer is traced). For a large group, the tracing algorithm of conventional GS-VLR is not so convenient and almost of all lattice-based constructions are not that efficient. Thus these somewhat unsatisfactory state-of-affairs highlights the challenge of designing a simpler and more efficient lattice-based GS scheme with VLR, which can be more suitable for a large group.

OUR RESULTS AND MAIN TECHNIQUES. In this work, we reply positively to the problems discussed above. Specifically, we propose a new lattice-based GS-VLR achieving shorter key-sizes and explicit traceability. Here, by "shorter key-sizes", we mean saving a $\mathcal{O}(\log N)$ factor for both bit-sizes of the group public-key and the member secret-key; by "explicit traceability", we mean the tracing authority determining the signer's real identity in a constant time, independent of N. The proposed scheme is proven secure in the random oracle model (ROM) based on the hardness of the Short Integer Solution (SIS) problem, and the Learning With Errors (LWE) problem.

The comparisons between our scheme and previous works, in terms of asymptotic efficiency (i.e., key-sizes, explicit traceability), functionality (i.e., static or not) and anonymity, are shown in Table 1 (the security parameter is n, time period $T = 2^d$ and group size $N = 2^\ell = \mathsf{poly}(n)$).

Our construction operates in the model of Boneh and Shacham [6] for VLR, which enjoys the implicit traceability, and additionally, the explicit traceability is also obtained. Furthermore, we declare that the "shorter key-sizes" and "explicit traceability" can be obtained in a relatively simple manner, thanks to three main techniques discussed below.

Table 1. Comparisons of known lattice-based GS schemes.

Scheme	Group public-key size	Signer secret-key size	Explicit traceability	Functionality	Anonymity
GKV [13]	$N \cdot \widetilde{\mathcal{O}}(n^2)$	$\widetilde{\mathcal{O}}(n^2)$	yes	static	CPA
CNR [8]	$\widetilde{\mathcal{O}}(n^2)$	$\widetilde{\mathcal{O}}(n^2)$	yes	static	CCA
LLLS [17]	$\ell \cdot \widetilde{\mathcal{O}}(n^2)$	$\widetilde{\mathcal{O}}(n^2)$	yes	static	CPA
LLNW [18]	$\ell \cdot \widetilde{\mathcal{O}}(n^2)$	$\ell \cdot \widetilde{\mathcal{O}}(n)$	no	VLR	Selfless
LNW [24]	$\ell \cdot \widetilde{\mathcal{O}}(n^2)$	$\widetilde{\mathcal{O}}(n)$	yes	static	CCA
NZZ [31]	$\widetilde{\mathcal{O}}(n^2)$	$\widetilde{\mathcal{O}}(n^2)$	yes	static	CCA
LLNW [20]	$\widetilde{\mathcal{O}}(n^2 + n \cdot \ell)$	$\ell \cdot \widetilde{\mathcal{O}}(n)$	yes	static	CCA
LMN [21]	$\ell \cdot \widetilde{\mathcal{O}}(n^2)$	$\widetilde{\mathcal{O}}(n)$	yes	MDO	CCA
LLMNW [19]	$\ell \cdot \widetilde{\mathcal{O}}(n^2)$	$\widetilde{\mathcal{O}}(n)$	yes	enrollment	CCA
ZHGJ [35]	$\widetilde{\mathcal{O}}(n^2)$	$\widetilde{\mathcal{O}}(n)$	no	VLR	Selfless
LNWX [27]	$\widetilde{\mathcal{O}}(n^2 + n \cdot \ell)$	$\widetilde{\mathcal{O}}(n) + \ell$	yes	fully-dynamic	CCA
GHZW [11]	$\widetilde{\mathcal{O}}(n^2)$	$\widetilde{\mathcal{O}}(n)$	no	VLR	Selfless
LNWX [26]	$(\ell + d) \cdot \widetilde{\mathcal{O}}(n^2)$	$(\ell+d)^2 \cdot d \cdot \widetilde{\mathcal{O}}(n^2)$	yes	FS	CCA
LNLW [22]	$\ell \cdot \widetilde{\mathcal{O}}(n^2)$	$\ell \cdot \widetilde{\mathcal{O}}(n)$	no	VLR	Selfless
KP [33]	$\ell \cdot \widetilde{\mathcal{O}}(n^2)$	$\ell \cdot \widetilde{\mathcal{O}}(n)$	yes	VLR	almost-CCA
KP [32]	$\ell \cdot \widetilde{\mathcal{O}}(n^2)$	$\widetilde{\mathcal{O}}(n)$	yes	fully-dynamic	almost-CCA
LNWX [25]	$\widetilde{\mathcal{O}}(n)$	$\widetilde{\mathcal{O}}(n)$	yes	enrollment	CCA
KY [14]	$N \cdot \widetilde{\mathcal{O}}(n^2)$	$N \cdot \widetilde{\mathcal{O}}(n^2)$	yes	static	Selfless
Ours	$\widetilde{\mathcal{O}}(n^2)$	$\widetilde{\mathcal{O}}(n)$	yes	VLR	Selfless

Firstly, as we discussed earlier, adopting a *Bonsai Tree* structure to construct lattice-based GS-VLR results in a larger bit-sizes of the group public-key and the member secret-key. To realize a more efficient lattice-based GS-VLR with shorter key-sizes, we further need an efficient mechanism to encode the member's identity information, and a simpler zero-knowledge protocol to prove the signer's validity as a certified group member.

Towards the goal described as above, we utilize a compact *identity-encoding technique* introduced in [31] which only needs a constant number of matrices to encode the member's identity index. We consider the group of $N = 2^\ell$ members and each member is identified by a ℓ-bits string $\mathrm{id} = (d_1, d_2, \cdots, d_\ell) \in \{0,1\}^\ell$ which is a binary representation of his/her identity index $i \in \{1, \cdots, N\}$, that is, $\mathrm{id} = \mathrm{bin}(i) \in \{0,1\}^\ell$. Throughout this paper, let n be the security parameter, and other parameters N, m, q, β, s are the function of n and will be determined later (see Sect. 4). In our new VLR scheme, the group public-key only consists of a random vector $\mathbf{u} \in \mathbb{Z}_q^n$ and 4 random matrices $\mathbf{A}_0, \mathbf{A}_1^1, \mathbf{A}_2^2$ (used for identity-encoding) and $\mathbf{A}_3^3$ (only used for explicit traceability) over $\mathbb{Z}_q^{n \times m}$. For member i, instead of generating a trapdoor basis matrix for a hard random lattice as the signing secret-key for i as in [31], we sample some short $2m$-dimensional vector $\mathbf{e}_i = (\mathbf{e}_{i,1}, \mathbf{e}_{i,2}) \in \mathbb{Z}^{2m}$ satisfying $0 < \|\mathbf{e}_i\|_\infty \leq \beta$, and $\mathbf{A}_i \cdot \mathbf{e}_i = \mathbf{u} \bmod q$, where $\mathbf{A}_i = [\mathbf{A}_0 | \mathbf{A}_1^1 + i\mathbf{A}_2^2] \in \mathbb{Z}_q^{n \times 2m}$. Furthermore, for the VLR feature, the revocation token of i is constructed by $\mathbf{A}_0$ and $\mathbf{e}_{i,1} \in \mathbb{Z}^m$, that is, $\mathrm{grt}_i = \mathbf{A}_0 \cdot \mathbf{e}_{i,1} \bmod q$.

Secondly, the implicit tracing algorithm of conventional lattice-based GS-VLR runs in a linear time in N, and thus it is not so convenient, resulting in a low efficiency. To realize an efficient construction with explicit traceability, we further need an efficient mechanism to encrypt the identity index of member i (in our actual construction, it's to encrypt $\mathrm{bin}(i) \in \{0,1\}^\ell$) to obtain a ciphertext $\mathbf{c}$, and design a zero-knowledge argument to prove: $\mathbf{c}$ is a correct encryption of $\mathrm{bin}(i)$, namely, a lattice-based verifiable encryption protocol. Besides the public matrix $\mathbf{A}_0, \mathbf{A}_1^1,$ and $\mathbf{A}_2^2$ for identity-encoding, a fourth matrix $\mathbf{A}_3^3$ is required to encrypt $\mathrm{bin}(i)$ using the dual LWE cryptosystem [12]. This relation can be expressed as $\mathbf{c} = (\mathbf{c}_1 = \mathbf{A}_3^{3\top} \mathbf{s} + \mathbf{e}_1 \bmod q, \mathbf{c}_2 = \mathbf{G}^\top \mathbf{s} + \mathbf{e}_2 + \lfloor q/2 \rfloor \mathrm{bin}(i) \bmod q)$ where $\mathbf{G}$ is a random matrix, and $\mathbf{s}, \mathbf{e}_1, \mathbf{e}_2$ are random vectors having certain specific norm.

Thirdly, the major challenge for our construction lies in how to design a simpler and efficient zero-knowledge proof protocol to prove the following relations: (a) $[\mathbf{A}_0 | \mathbf{A}_1^1 + i\mathbf{A}_2^2] \cdot \mathbf{e}_i = \mathbf{u} \bmod q$; (b) $\mathrm{grt}_i = \mathbf{A}_0 \cdot \mathbf{e}_{i,1} \bmod q$; (c) $\mathbf{c} = (\mathbf{c}_1 = \mathbf{A}_3^{3\top} \mathbf{s} + \mathbf{e}_1 \bmod q, \mathbf{c}_2 = \mathbf{G}^\top \mathbf{s} + \mathbf{e}_2 + \lfloor q/2 \rfloor \mathrm{bin}(i) \bmod q)$. For relation (b), we utilize a creative idea introduced by Ling et al. [22] by drawing a matrix $\mathbf{B} \in \mathbb{Z}_q^{n \times m}$ from a random oracle and a vector $\mathbf{e}_0 \in \mathbb{Z}^m$ from the LWE error distribution, define $\mathbf{b} = \mathbf{B}^\top \mathrm{grt}_i + \mathbf{e}_0 = (\mathbf{B}^\top \mathbf{A}_0) \cdot \mathbf{e}_{i,1} + \mathbf{e}_0 \bmod q$, thus the member i's token grt_i is bound to a one-way and injective LWE function. For relation (c), we utilize a creative idea of Ling et al. [24] by constructing a matrix $\mathbf{P} \in \mathbb{Z}_q^{(m+\ell) \times (n+m+\ell)}$ (obtained from the public matrices $\mathbf{A}_3^3$ and $\mathbf{G}$, see Sect. 3 for details), and a vector $\mathbf{e} = (\mathbf{s}, \mathbf{e}_1, \mathbf{e}_2) \in \mathbb{Z}^{n+m+\ell}$, then let $\mathbf{c} = \mathbf{P}\mathbf{e} + (\mathbf{0}^m, \lfloor q/2 \rfloor \mathrm{bin}(i)) \bmod q$, thus the identity index i is now bound to this new form which is easy to construct a Stern-type statistical zero-knowledge proof protocol.

For relation (a), since $\mathbf{e}_i \in \mathbb{Z}^{2m}$ is a valid short solution to the Inhomogeneous Short Integer Solution (ISIS) instance $(\mathbf{A}_i, \mathbf{u})$ where $\mathbf{A}_i = [\mathbf{A}_0 | \mathbf{A}_1^1 + i\mathbf{A}_2^2]$, a direct way for member i to prove his/her validity as a certified group member without leaking $\mathbf{e}_i$ just by performing a Stern-type statistical zero-knowledge argument of knowledge (ZKAoK) as in [23]. However, in order to protect the anonymity of i, the structure of $\mathbf{A}_i$ should not be given explicitly. How to realize a Stern-type zero-knowledge proof without leaking $\mathbf{A}_i$ and $\mathbf{e}_i$ simultaneously? To solve this open problem, we transform matrix $\mathbf{A}_i$ to $\mathbf{A}'$ which enjoys a new form and is independent of the identity index i, i.e., $\mathbf{A}' = [\mathbf{A}_0|\mathbf{A}_1^1|\mathbf{g}_\ell \otimes \mathbf{A}_2^2] \in \mathbb{Z}_q^{n \times (\ell+2)m}$, where $\mathbf{g}_\ell = (1, 2, 2^2, \cdots, 2^{\ell-1})$ is a power-of-two vector, and the identity index i can be rewritten as $i = \mathbf{g}_\ell^\top \cdot \mathsf{bin}(i)$, the notation $\otimes$ denotes a concatenation with vectors or matrices, and the detailed definition will be given later (see Sect. 3). As a corresponding change to the member i's signing secret-key, $\mathbf{e}_i = (\mathbf{e}_{i,1}, \mathbf{e}_{i,2})$ is now transformed to $\mathbf{e}_i' = (\mathbf{e}_{i,1}, \mathbf{e}_{i,2}, \mathsf{bin}(i) \otimes \mathbf{e}_{i,2}) \in \mathbb{Z}^{(\ell+2)m}$. Thus, to argue the relation $\mathbf{A}_i \cdot \mathbf{e}_i = \mathbf{u} \bmod q$, we instead show that $\mathbf{A}' \cdot \mathbf{e}_i' = \mathbf{u} \bmod q$.

Putting the above transformations ideas and the versatility of the Stern-type argument system introduced by Ling et al. [23] together, we can construct an efficient Stern-type interactive protocol for the relations (a), (b) and (c).

To summarize, by incorporating the compact *identity-encoding technique* and the corresponding efficient Stern-type statistical ZKAoK into a lattice-based GS, we design a more efficient lattice-based GS-VLR. The proposed scheme obtains the shorter bit-sizes for the group public-key and the group member secret-key, furthermore, the explicit traceability, and thus, is more suitable for a large group. In addition, we believe that the innovative ideas and design approaches in our whole constructions may be of independent interest.

ORGANIZATION. In the forthcoming sections, we first recall some background on GS-VLR and lattice-based cryptography in Sect. 2. Section 3 turns to develop an improved identity-encoding technique, an explicit traceability mechanism and the corresponding new Stern-type statistical ZKAoK protocol that will be used in our construction. Our scheme is constructed and analyzed in Sect. 4.

2   Preliminaries

NOTATIONS. Assume that all vectors are in a column form. $\mathcal{S}_k$ denotes the set of all permutations of k elements, and $\xleftarrow{\$}$ denotes that sampling elements from a given distribution uniformly at random. Let $\|\cdot\|_\infty$ denote the infinity norm (ℓ_∞) of a vector. Given $e = (e_1, e_2, \cdots, e_n) \in \mathbb{R}^n$, $\mathsf{Parse}(e, k_1, k_2)$ denotes the vector $(e_{k_1}, e_{k_1+1}, \cdots, e_{k_2}) \in \mathbb{R}^{k_2-k_1+1}$ for $1 \le k_1 \le k_2 \le n$. $\log a$ denotes the logarithm of a with base 2. The acronym PPT stands for "probabilistic polynomial-time".

2.1   Group Signatures with VLR

A conventional GS-VLR scheme involves two entities: a group manager (also is a tracing authority) and a sets of group members. In order to support an explicit traceability we add an Open algorithm to conventional GS-VLR.

Syntax of GS-VLR with Explicit Traceability. A GS-VLR with the explicit traceability (GS-VLR-ET) consists of 4 polynomial-time algorithms: KeyGen, Sign, Verify, Open. Because of the page limitation, we omit the detailed definition, if any necessary, please contact the corresponding author for the full version.

Correctness and Security of GS-VLR-ET. As put forward by Boneh and Shacham [6], A conventional GS-VLR scheme should satisfy correctness selfless-anonymity, and traceability. Thus for GS-VLR-ET, these 3 requirements also should be satisfied. Due to the limited space, the details are presented in the full paper.

2.2 Background on Lattices

Ajtai [2] first introduced how to obtain a statistically close to uniform matrix $\mathbf{A}$ together with a low Gram-Schmidt norm basis for $\Lambda_q^{\perp}(\mathbf{A}) = \{\mathbf{e} \in \mathbb{Z}^m \mid \mathbf{A} \cdot \mathbf{e} = \mathbf{0} \bmod q\}$, then two improved algorithms were investigated by [3,30].

Lemma 1 ([2,3,30]). *Let integers $n \geq 1$, $q \geq 2$, and $m = 2n\lceil \log q \rceil$. There exists a PPT algorithm TrapGen(q, n, m) that outputs $\mathbf{A}$ and $\mathbf{R_A}$, such that $\mathbf{A}$ is statistically close to a uniform matrix in $\mathbb{Z}_q^{n \times m}$ and $\mathbf{R_A}$ is a trapdoor for $\Lambda_q^{\perp}(\mathbf{A})$.*

Lemma 2 ([12,30]). *Let integers $n \geq 1$, $q \geq 2$, and $m = 2n\lceil \log q \rceil$, given $\mathbf{A} \in \mathbb{Z}_q^{n \times m}$, a trapdoor $\mathbf{R_A}$ for $\Lambda_q^{\perp}(\mathbf{A})$, a parameter $s = \omega(\sqrt{n \log q \log n})$ and a vector $\mathbf{u} \in \mathbb{Z}_q^n$, there is a PPT algorithm SamplePre$(\mathbf{A}, \mathbf{R_A}, \mathbf{u}, s)$ that returns a short vector $\mathbf{e} \in \Lambda_q^{\mathbf{u}}(\mathbf{A})$ sampled from a distribution statistically close to $\mathcal{D}_{\Lambda_q^{\mathbf{u}}(\mathbf{A}),s}$.*

We recall 3 *average-case* lattices problems: ISIS, SIS (in the ℓ_∞ norm), LWE.

Definition 1. *The* (I)SIS$_{n,m,q,\beta}^\infty$ *problems are: Given a uniformly random matrix $\mathbf{A} \in \mathbb{Z}_q^{n \times m}$, a random syndrome vector $\mathbf{u} \in \mathbb{Z}_q^n$ and a real $\beta > 0$,*

- SIS$_{n,m,q,\beta}^\infty$: *to find a non-zero $\mathbf{e} \in \mathbb{Z}^m$ such that $\mathbf{A} \cdot \mathbf{e} = \mathbf{0} \bmod q$, $\|\mathbf{e}\|_\infty \leq \beta$.*
- ISIS$_{n,m,q,\beta}^\infty$: *to find a vector $\mathbf{e} \in \mathbb{Z}^m$ such that $\mathbf{A} \cdot \mathbf{e} = \mathbf{u} \bmod q$, $\|\mathbf{e}\|_\infty \leq \beta$.*

Lemma 3 ([12,29]). *For m, $\beta = \mathsf{poly}(n)$, and $q \geq \beta \cdot \widetilde{\mathcal{O}}(\sqrt{n})$, the average-case* (I)SIS$_{n,m,q,\beta}^\infty$ *problems are at least as hard as the SIVP$_\gamma$ problem in the worst-case to within $\gamma = \beta \cdot \widetilde{\mathcal{O}}(\sqrt{nm})$ factor. In particular, if $\beta = 1$, $q = \widetilde{\mathcal{O}}(n)$ and $m = 2n\lceil \log q \rceil$, then the* (I)SIS$_{n,m,q,1}^\infty$ *problems are at least as hard as SIVP$_{\widetilde{\mathcal{O}}(n)}$.*

Definition 2. *The* LWE$_{n,q,\chi}$ *problem is: Given a random vector $\mathbf{s} \in \mathbb{Z}_q^n$, a probability distribution χ over $\mathbb{Z}$, let $\mathcal{A}_{\mathbf{s},\chi}$ be the distribution obtained by sampling a matrix $\mathbf{A} \xleftarrow{\$} \mathbb{Z}_q^{n \times m}$, a vector $\mathbf{e} \xleftarrow{\$} \chi^m$, and outputting a tuple $(\mathbf{A}, \mathbf{A}^\top \mathbf{s} + \mathbf{e})$, to distinguish $\mathcal{A}_{\mathbf{s},\chi}$ and a uniform distribution $\mathcal{U}$ over $\mathbb{Z}_q^{n \times m} \times \mathbb{Z}_q^m$.*

Let $\beta \geq \sqrt{n} \cdot \omega(\log n)$, if q is a prime power, and χ is a β-bounded distribution (e,g., $\chi = \mathcal{D}_{\mathbb{Z}^m,s}$), then the LWE$_{n,q,\chi}$ *problem is as least as hard as SIVP$_{\widetilde{\mathcal{O}}(nq/\beta)}$.*

Lemma 4 ([1]). *Let $\mathbf{R}$ be an $m \times m$-matrix chosen at random from $\{-1,1\}^{m \times m}$, for vectors $\mathbf{e} \in \mathbb{R}^m$, $\Pr[\|\mathbf{R} \cdot \mathbf{e}\|_\infty > \|\mathbf{e}\|_\infty \cdot \omega(\sqrt{\log m})] < \mathsf{negl}(m)$.*

Lemma 5 ([1]). *Let $q \geq 3$, and $m > n$, $\mathbf{A}, \mathbf{B} \in \mathbb{Z}_q^{n \times m}$ and a real $s \geq \|\widetilde{\mathbf{R_B}}\| \cdot \sqrt{m} \cdot \omega(\log m)$. There is a PPT algorithm SampleRight$(\mathbf{A}, \mathbf{B}, \mathbf{R}, \mathbf{R_B}, \mathbf{u}, s)$ that given a trapdoor $\mathbf{R_B}$ for $\Lambda_q^{\perp}(\mathbf{B})$, a low-norm matrix $\mathbf{R} \in \{-1,1\}^{m \times m}$, and a vector $\mathbf{u} \in \mathbb{Z}_q^n$, outputs $\mathbf{e} \in \mathbb{Z}^{2m}$ distributed statistically close to $\mathcal{D}_{\Lambda_q^{\mathbf{u}}([\mathbf{A}|\mathbf{AR}+\mathbf{B}]),s}$.*

3 Preparations

3.1 The Improved of Identity-Encoding Technique

For an improved of identity-encoding technique, a public random vector $\mathbf{u} \in \mathbb{Z}_q^n$ is required, i.e., $\mathsf{Gpk} = (\mathbf{A}_0, \mathbf{A}_1^1, \mathbf{A}_2^2, \mathbf{A}_3^3, \mathbf{u})$, furthermore, the secret-key of i is not yet a trapdoor basis matrix for $\Lambda_q^\perp(\mathbf{A}_i)$, instead of a short $2m$-dimensional vector $\mathbf{e}_i = (\mathbf{e}_{i,1}, \mathbf{e}_{i,2})$ in the coset of $\Lambda_q^\perp(\mathbf{A}_i)$, i.e., $\Lambda_q^{\mathbf{u}}(\mathbf{A}_i) = \{\mathbf{e}_i \in \mathbb{Z}^{2m} \mid \mathbf{A}_i \cdot \mathbf{e}_i = \mathbf{u} \bmod q\}$, and thus, the revocation token of i is constructed by $\mathbf{A}_0$ and the first part of its secret-key, i.e., $\mathsf{grt}_i = \mathbf{A}_0 \cdot \mathbf{e}_{i,1} \bmod q$.

In order to design an efficient Stern-type ZKAoK protocol corresponding to the above new variant, we transform $\mathbf{A}_i = [\mathbf{A}_0 | \mathbf{A}_1^1 + i\mathbf{A}_2^2]$ corresponding to i to a new form. Before we do that, we first define 2 notations (we restate, in this paper, the group is of $N = 2^\ell$ members):

- $\mathbf{g}_\ell = (1, 2, \cdots, 2^{\ell-1})$: a power-of-2 vector, for $i \in \{1, 2, \cdots, N\}$, $i = \mathbf{g}_\ell^\top \cdot \mathsf{bin}(i)$ where $\mathsf{bin}(i) \in \{0, 1\}^\ell$ denotes a binary representation of i.
- $\otimes$: a concatenation with vectors or matrices, given $\mathbf{A} \in \mathbb{Z}_q^{n \times m}$, $\mathbf{e}' \in \mathbb{Z}_q^m$, and $\mathbf{e} = (e_1, e_2, \cdots, e_\ell) \in \mathbb{Z}_q^\ell$, define: $\mathbf{e} \otimes \mathbf{e}' = (e_1\mathbf{e}', e_2\mathbf{e}', \cdots, e_\ell\mathbf{e}') \in \mathbb{Z}_q^{m\ell}$, $\mathbf{e} \otimes \mathbf{A} = [e_1\mathbf{A} | e_2\mathbf{A} | \cdots | e_\ell\mathbf{A}] \in \mathbb{Z}_q^{n \times m\ell}$.

Next, we transform $\mathbf{A}_i$ to a public matrix $\mathbf{A}'$ that is independent of the index i, where $\mathbf{A}' = [\mathbf{A}_0 | \mathbf{A}_1^1 | \mathbf{A}_2^2 | 2\mathbf{A}_2^2 | \cdots | 2^{\ell-1}\mathbf{A}_2^2] = [\mathbf{A}_0 | \mathbf{A}_1^1 | \mathbf{g}_\ell \otimes \mathbf{A}_2^2] \in \mathbb{Z}_q^{n \times (\ell+2)m}$.

As a corresponding change to the group secret-key of member i, $\mathbf{e}_i = (\mathbf{e}_{i,1}, \mathbf{e}_{i,2})$ is now transformed to $\mathbf{e}_i'$, a vector with some special structure as for $\mathbf{e}_i$, that is, $\mathbf{e}_i' = (\mathbf{e}_{i,1}, \mathbf{e}_{i,2}, \mathsf{bin}(i) \otimes \mathbf{e}_{i,2}) \in \mathbb{Z}^{(\ell+2)m}$.

Thus, from the above transformations, the relation $\mathbf{A}_i \cdot \mathbf{e}_i = \mathbf{u} \bmod q$ is now transformed to a new form, (i) $\mathbf{A}_i \cdot \mathbf{e}_i = \mathbf{A}' \cdot \mathbf{e}_i' = \mathbf{u} \bmod q$.

For the revocation mechanism, as it was stated in [22], due to a flaw in the revocation mechanism of [18], a corrected technique which realizes revocation by binding signer's token grt_i to an LWE function was proposed, (ii) $\mathbf{b} = \mathbf{B}^\top \mathsf{grt}_i + \mathbf{e}_0 = (\mathbf{B}^\top \mathbf{A}_0) \cdot \mathbf{e}_{i,1} + \mathbf{e}_0 \bmod q$, where $\mathbf{B} \in \mathbb{Z}_q^{n \times m}$ is a uniformly random matrix from a random oracle, $\mathbf{e}_0 \in \mathbb{Z}^m$ is sampled from the LWE error χ^m.

For the explicit traceability mechanism, as it was shown in [24], the lattice-based dual LWE cryptosystem [12] can be used to encrypt the identity index of signer i. In our construction, the string $\mathsf{bin}(i) \in \{0, 1\}^\ell$ is treated as the plaintext and the ciphertext can be expressed as $\mathbf{c} = (\mathbf{c}_1, \mathbf{c}_2)$, where $\mathbf{c}_1 = \mathbf{A}_3^{3\top}\mathbf{s} + \mathbf{e}_1 \bmod q$, $\mathbf{c}_2 = \mathbf{G}^\top\mathbf{s} + \mathbf{e}_2 + \lfloor q/2 \rfloor \mathsf{bin}(i) \bmod q)$. Here, $\mathbf{G} \in \mathbb{Z}_q^{n \times \ell}$ is a random matrix, and $\mathbf{s}$, $\mathbf{e}_1$, $\mathbf{e}_2$ are random vectors sampled from the LWE error χ^n, χ^m, χ^ℓ, respectively. Thus, the above relation can be expressed as (iii) $\mathbf{c} = \mathbf{P} \cdot \mathbf{e} + (\mathbf{0}^m, \lfloor q/2 \rfloor \mathsf{bin}(i))$,

where $\mathbf{P} = \begin{pmatrix} \mathbf{A}_3^{3\top} \\ \cdots\cdots \\ \mathbf{G}^\top \end{pmatrix} \Big| \mathbf{I}_{m+\ell} \in \mathbb{Z}_q^{(m+\ell) \times (n+m+\ell)}$ and $\mathbf{e} = (\mathbf{s}, \mathbf{e}_1, \mathbf{e}_2) \in \mathbb{Z}^{n+m+\ell}$.

Putting all the above transformations ideas and the versatility of the Stern-extension argument system introduced by Ling et al. [23] together, we can construct an efficient Stern-type statistical ZKAoK protocol to prove the above new relations (i), (ii) and (iii).

3.2 A New Stern-Type Zero-Knowledge Proof Protocol

An efficient Stern-type ZKAoK protocol which allows $\mathcal{P}$ to convince any verifier $\mathcal{V}$ that $\mathcal{P}$ is a group member who signed M will be introduced, namely, $\mathcal{P}$ owns a valid secret-key, his/her token is correctly embedded into an LWE instance and the identity information is correctly hidden with the dual LWE cryptosystem.

Firstly, we recall some specific sets and techniques as in [17,18,22] that will be used in our VLR-ET construction. Due to the limited space, we give them in the full version and the readers can also refer to [17,18,22].

Secondly, we introduce the main building block, a new Stern-type interactive statistical zero-knowledge proof protocol, and we consider the group of $N = 2^\ell$ members and each member is identified by id $= (d_1, d_2, \cdots, d_\ell) \in \{0,1\}^\ell$ which is a binary representation of the index $i \in \{1, 2, \cdots, N\}$, namely, id $= \mathsf{bin}(i) \in \{0,1\}^\ell$. The underlying new Stern-type statistical ZKAoK protocol between $\mathcal{P}$ and $\mathcal{V}$ can be summarized as follows:

1. The public inputs include $\mathbf{A}' = [\mathbf{A}_0 | \mathbf{A}_1^1 | \mathbf{g}_\ell \otimes \mathbf{A}_2^2] \in \mathbb{Z}_q^{n \times (\ell+2)m}$, $\mathbf{B} \in \mathbb{Z}_q^{n \times m}$,

$$\mathbf{P} = \begin{pmatrix} \mathbf{A}_3^{3\top} \\ \cdots \\ \mathbf{G}^\top \end{pmatrix} \Big| \mathbf{I}_{m+\ell} \in \mathbb{Z}_q^{(m+\ell) \times (n+m+\ell)}, \mathbf{u} \in \mathbb{Z}_q^n, \mathbf{b} \in \mathbb{Z}_q^m, \mathbf{c} = (\mathbf{c}_1, \mathbf{c}_2).$$

2. $\mathcal{P}$'s witnesses include $\mathbf{e}' = (\mathbf{e}_1', \mathbf{e}_2', \mathsf{bin}(i) \otimes \mathbf{e}_2') \in \mathsf{Sec}_\beta(\mathsf{id})$ corresponding to a secret index $i \in \{1, \cdots, N\}$ and 4 short vectors $\mathbf{e}_0, \mathbf{s}, \mathbf{e}_1, \mathbf{e}_2$, the LWE errors.
3. $\mathcal{P}$'s goal is to convince $\mathcal{V}$ in zero-knowledge that:
   a. $\mathbf{A}' \cdot \mathbf{e}' = \mathbf{u} \bmod q$ where $\mathbf{e}' \in \mathsf{Sec}_\beta(\mathsf{id})$, while keeping id secret.
   b. $\mathbf{b} = (\mathbf{B}^\top \mathbf{A}_0) \cdot \mathbf{e}_1' + \mathbf{e}_0 \bmod q$ where $0 < \|\mathbf{e}_1'\|_\infty, \|\mathbf{e}_0\|_\infty \leq \beta$.
   c. $\mathbf{c} = \mathbf{Pe} + (\mathbf{0}^m, \lfloor q/2 \rfloor \mathsf{bin}(i)) \bmod q$, where $\mathbf{e} = (\mathbf{s}, \mathbf{e}_1, \mathbf{e}_2), 0 < \|\mathbf{e}\|_\infty \leq \beta$, while keeping $\mathsf{bin}(i) \in \{0,1\}^\ell$ secret.

Firstly, we sketch Group Membership Mechanism, i.e., $\mathcal{P}$ is a certified member and its goal is shown in a. $\mathcal{P}$ does as follows:

1. Parse $\mathbf{A}' = [\mathbf{A}_0 | \mathbf{A}_1^1 | \mathbf{A}_2^2 | \cdots | 2^{\ell-1} \mathbf{A}_2^2]$, use Matrix-Ext technique to extend it to $\mathbf{A}^* = [\mathbf{A}_0 | \mathbf{0}^{n \times 2m} | \mathbf{A}_1^1 | \mathbf{0}^{n \times 2m} | \cdots | 2^{\ell-1} \mathbf{A}_2^2 | \mathbf{0}^{n \times 2m} | \mathbf{0}^{n \times 3m\ell}]$.
2. Parse $\mathsf{id} = \mathsf{bin}(i) = (d_1, d_2, \cdots, d_\ell)$, extend it to $\mathsf{id}^* = (d_1, d_2, \cdots, d_{2\ell}) \in \mathsf{B}_{2\ell}$.
3. Parse $\mathbf{e}' = (\mathbf{e}_1', \mathbf{e}_2', \mathsf{bin}(i) \otimes \mathbf{e}_2') = (\mathbf{e}_1', \mathbf{e}_2', d_1 \mathbf{e}_2', d_2 \mathbf{e}_2', \cdots, d_\ell \mathbf{e}_2')$, use Dec, Ext techniques extending $\mathbf{e}_1'$ and $\mathbf{e}_2'$ to k vectors $\mathbf{e}_{1,1}', \mathbf{e}_{1,2}', \cdots, \mathbf{e}_{1,k}' \in \mathsf{B}_{3m}$, and k vectors $\mathbf{e}_{2,1}', \mathbf{e}_{2,2}', \cdots, \mathbf{e}_{2,k}' \in \mathsf{B}_{3m}$. For each $j \in \{1, 2, \cdots, k\}$, we define $\mathbf{e}_j' = (\mathbf{e}_{1,j}', \mathbf{e}_{2,j}', d_1 \mathbf{e}_{2,j}', d_2 \mathbf{e}_{2,j}', \cdots, d_{2\ell} \mathbf{e}_{2,j}')$, it can be checked that $\mathbf{e}_j' \in \mathsf{SecExt}(\mathsf{id}^*)$.

So $\mathcal{P}$'s goal in a is transformed to: $\mathbf{A}^* (\sum_{j=1}^k \beta_j \mathbf{e}_j') = \mathbf{u} \bmod q$, $\mathbf{e}_j' \in \mathsf{SecExt}(\mathsf{id}^*)$. To prove this new structure in zero-knowledge, we take 2 steps as follows:

1. Pick k random vectors $\mathbf{r}_1', \cdots, \mathbf{r}_k' \xleftarrow{\$} \mathbb{Z}_q^{(2\ell+2)3m}$ to mask $\mathbf{e}_1', \cdots, \mathbf{e}_k'$, then it can be checked that, $\mathbf{A}^* \cdot (\sum_{j=1}^k \beta_j (\mathbf{e}_j' + \mathbf{r}_j')) - \mathbf{u} = \mathbf{A}^* \cdot (\sum_{j=1}^k \beta_j \mathbf{r}_j') \bmod q$.
2. Pick two permutations $\pi, \varphi \in \mathcal{S}_{3m}$, one permutation $\tau \in \mathcal{S}_{2\ell}$, then it can be checked that, $\forall j \in \{1, 2, \cdots, k\}$, $\mathcal{T}_{\pi,\varphi,\tau}(\mathbf{e}_j') \in \mathsf{SecExt}(\tau(\mathsf{id}^*))$, where $\mathsf{id}^* \in \mathsf{B}_{2\ell}$ is an extension of $\mathsf{id} = \mathsf{bin}(i) \in \{0,1\}^\ell$.

Secondly, we sketch Revocation Mechanism, i.e., $\mathcal{P}$'s revocation token is correctly embedded in an LWE instance and its goal is shown in b. $\mathcal{P}$ does as follows:

1. Let $\mathbf{B}' = \mathbf{B}^{\top}\mathbf{A}_0 \bmod q \in \mathbb{Z}_q^{m \times m}$, and $\mathbf{e}'_{j,0} = \mathsf{Parse}(\mathbf{e}'_j, 1, m)$.
2. Parse $\mathbf{e}_0 = (e_1, e_2, \cdots, e_m) \in \mathbb{Z}^m$, use Dec, Ext techniques to extend $\mathbf{e}_0$ to k vectors $\mathbf{e}_1^0, \mathbf{e}_2^0, \cdots, \mathbf{e}_k^0 \in \mathsf{B}_{3m}$.
3. Let $\mathbf{B}^* = [\mathbf{B}'|\mathbf{I}^*]$ where $\mathbf{I}^* = [\mathbf{I}_m|\mathbf{0}^{n \times 2m}]$, $\mathbf{I}_m$ is identity matrix of order m.

So $\mathcal{P}$'s goal in b is transformed to: $\mathbf{b} = \mathbf{B}'(\sum_{j=1}^k \beta_j \mathbf{e}'_{j,0}) + \mathbf{I}^*(\sum_{j=1}^k \beta_j \mathbf{e}_j^0) = \mathbf{B}^* \cdot (\sum_{j=1}^k \beta_j(\mathbf{e}'_{j,0}, \mathbf{e}_j^0)) \bmod q$, $\mathbf{e}_j^0 \in \mathsf{B}_{3m}$. To prove this new structure in zero-knowledge, we take 2 steps as follows:

1. Let $\mathbf{r}'_{j,0} = \mathsf{Parse}(\mathbf{r}'_j, 1, m)$, pick k random vectors $\mathbf{r}_1, \cdots, \mathbf{r}_k \xleftarrow{\$} \mathbb{Z}_q^{3m}$ to mask $\mathbf{e}_1^0, \cdots, \mathbf{e}_k^0$, it can be checked that,

$$\mathbf{B}^* \cdot (\sum_{j=1}^k \beta_j(\mathbf{e}'_{j,0} + \mathbf{r}'_{j,0}, \mathbf{e}_j^0 + \mathbf{r}_j)) - \mathbf{b} = \mathbf{B}^* \cdot (\sum_{j=1}^k \beta_j(\mathbf{r}'_{j,0}, \mathbf{r}_j)) \bmod q$$

2. Pick $\phi \in \mathcal{S}_{3m}$, then it can be checked that, $\forall j \in \{1, 2, \cdots, k\}$, $\phi(\mathbf{e}_j^0) \in \mathsf{B}_{3m}$.

Thirdly, we sketch Explicit Traceability Mechanism, i.e., $\mathcal{P}$'s index is correctly embedded in a LWE cryptosystem and its goal is shown in c. $\mathcal{P}$ does as follows:

1. Let $\mathbf{P}^* = [\mathbf{P}|\mathbf{0}^{(m+\ell) \times 2(n+m+\ell)}]$ and $\mathbf{Q} = \begin{pmatrix} \mathbf{0}^{m \times \ell} & \mathbf{0}^{m \times \ell} \\ \cdots\cdots & \cdots\cdots \\ \lfloor q/2 \rfloor \mathbf{I}_\ell & \mathbf{0}^{\ell \times \ell} \end{pmatrix}$, where $\mathbf{I}_\ell$ is an identity matrix of order ℓ.
2. Parse $\mathbf{e} = (\mathbf{s}, \mathbf{e}_1, \mathbf{e}_2) \in \mathbb{Z}^{n+m+\ell}$, use Dec, Ext techniques to extend $\mathbf{e}$ to k vectors $\mathbf{e}^{(1)}, \mathbf{e}^{(2)}, \cdots, \mathbf{e}^{(k)} \in \mathsf{B}_{3(n+m+\ell)}$.
3. Let $\mathsf{id}^* = \mathsf{bin}(i)^* \in \mathsf{B}_{2\ell}$ be an extension of $\mathsf{id} = \mathsf{bin}(i) \in \{0,1\}^\ell$.

So $\mathcal{P}$'s goal in c is transformed to: $\mathbf{c} = \mathbf{P}^* \cdot (\sum_{j=1}^k \beta_j \mathbf{e}^{(j)}) + \mathbf{Q} \cdot \mathsf{id}^* \bmod q$, $\mathbf{e}^{(j)} \in \mathsf{B}_{3(n+m+\ell)}$, $\mathsf{bin}(i)^* \in \mathsf{B}_{2\ell}$. To prove this new structure in zero-knowledge, we take 2 steps as follows:

1. Pick a random vector $\mathbf{r}_{\mathsf{id}^*} \xleftarrow{\$} \mathbb{Z}_q^{2\ell}$ to mask $\mathsf{id}^* = \mathsf{bin}(i)^*$, k random vectors $\mathbf{r}''_1, \cdots, \mathbf{r}''_k \xleftarrow{\$} \mathbb{Z}_q^{3(n+m+\ell)}$ to mask $\mathbf{e}^{(1)}, \cdots, \mathbf{e}^{(k)}$, it can be checked that,

$$\mathbf{P}^* \cdot (\sum_{j=1}^k \beta_j(\mathbf{e}^{(j)} + \mathbf{r}''_j)) + \mathbf{Q} \cdot (\mathsf{id}^* + \mathbf{r}_{\mathsf{id}^*}) - \mathbf{c} = \mathbf{P}^* \cdot (\sum_{j=1}^k \beta_j \mathbf{r}''_j) + \mathbf{Q} \cdot \mathbf{r}_{\mathsf{id}^*} \bmod q$$

2. Pick $\rho \in \mathcal{S}_{3(n+m+\ell)}$, then it can be checked that, $\forall j \in \{1, 2, \cdots, k\}$, $\rho(\mathbf{e}^{(j)}) \in \mathsf{B}_{3(n+m+\ell)}$ and $\tau(\mathsf{id}^*) \in \mathsf{B}_{2\ell}$, where τ has been picked in the proof of group membership mechanism.

Putting the above techniques together, we can obtain a new Stern-type interactive statistical zero-knowledge proof protocol, the details will be given bellow.

In our VLR-ET construction, we utilize a statistically hiding, computationally blinding commitment scheme (COM) as proposed in [15]. For simplicity, we omit the randomness of COM. $\mathcal{P}$ and $\mathcal{V}$ interact as follows:

1. **Commitments:** $\mathcal{P}$ randomly samples the following random objects:

$$\begin{cases} \mathbf{r}'_1, \cdots, \mathbf{r}'_k \xleftarrow{\$} \mathbb{Z}_q^{(2\ell+2)3m}; \ \mathbf{r}_1, \cdots, \mathbf{r}_k \xleftarrow{\$} \mathbb{Z}_q^{3m}; \ \mathbf{r}''_1, \cdots, \mathbf{r}''_k \xleftarrow{\$} \mathbb{Z}_q^{3(n+m+\ell)}; \\ \pi_1, \cdots, \pi_k \xleftarrow{\$} \mathcal{S}_{3m}; \ \varphi_1, \cdots, \varphi_k \xleftarrow{\$} \mathcal{S}_{3m}; \ \phi_1, \cdots, \phi_k \xleftarrow{\$} \mathcal{S}_{3m}; \\ \rho_1, \cdots, \rho_k \xleftarrow{\$} \mathcal{S}_{3(n+m+\ell)}; \ \tau \xleftarrow{\$} \mathcal{S}_{2\ell}; \ \mathbf{r}_{\mathsf{id}^*} \xleftarrow{\$} \mathbb{Z}_q^{2\ell}. \end{cases}$$

Let $\mathbf{r}'_{j,0} = \mathsf{Parse}(\mathbf{r}'_j, 1, m)$, $j \in \{1, \cdots, k\}$, $\mathcal{P}$ sends $\mathsf{CMT} = (\mathbf{c}_1, \mathbf{c}_2, \mathbf{c}_3)$ to $\mathcal{V}$,

$$\begin{cases} \mathbf{c}_1 = \mathsf{COM}(\{\pi_j, \varphi_j, \phi_j, \rho_j\}_{j=1}^k, \tau, \mathbf{A}^* \cdot (\sum_{j=1}^k \beta_j \mathbf{r}'_j), \mathbf{B}^* \cdot (\sum_{j=1}^k \beta_j(\mathbf{r}'_{j,0}, \mathbf{r}_j)), \\ \qquad\qquad \mathbf{P}^* \cdot (\sum_{j=1}^k \beta_j \mathbf{r}''_j) + \mathbf{Q} \cdot \mathbf{r}_{\mathsf{id}^*}), \\ \mathbf{c}_2 = \mathsf{COM}(\{\mathcal{T}_{\pi_j, \varphi_j, \tau}(\mathbf{r}'_j), \phi_j(\mathbf{r}_j), \rho_j(\mathbf{r}''_j)\}_{j=1}^k, \tau(\mathbf{r}_{\mathsf{id}^*})), \\ \mathbf{c}_3 = \mathsf{COM}(\{\mathcal{T}_{\pi_j, \varphi_j, \tau}(\mathbf{e}'_j + \mathbf{r}'_j), \phi_j(\mathbf{e}^0_j + \mathbf{r}_j), \rho_j(\mathbf{e}^{(j)} + \mathbf{r}''_j)\}_{j=1}^k, \tau(\mathsf{id}^* + \mathbf{r}_{\mathsf{id}^*})). \end{cases}$$

2. **Challenge:** $\mathcal{V}$ chooses a challenge $\mathsf{CH} \xleftarrow{\$} \{1, 2, 3\}$ and sends it to $\mathcal{P}$.
3. **Response:** Depending on CH, $\mathcal{P}$ replies as follows:
   - $\mathsf{CH} = 1$. For $j \in \{1, 2, \cdots, k\}$, let $\mathbf{v}'_j = \mathcal{T}_{\pi_j, \varphi_j, \tau}(\mathbf{e}'_j)$, $\mathbf{w}'_j = \mathcal{T}_{\pi_j, \varphi_j, \tau}(\mathbf{r}'_j)$, $\mathbf{v}_j = \phi_j(\mathbf{e}^0_j)$, $\mathbf{w}_j = \phi_j(\mathbf{r}_j)$, $\mathbf{v}^{(j)} = \rho_j(\mathbf{e}^{(j)})$, $\mathbf{w}''_j = \rho_j(\mathbf{r}''_j)$, $\mathbf{t}_{\mathsf{id}} = \tau(\mathsf{id}^*)$ and $\mathbf{v}_{\mathsf{id}} = \tau(\mathbf{r}_{\mathsf{id}^*})$, define $\mathsf{RSP} = (\{\mathbf{v}'_j, \mathbf{w}'_j, \mathbf{v}_j, \mathbf{w}_j, \mathbf{v}^{(j)}, \mathbf{w}''_j\}_{j=1}^k, \mathbf{t}_{\mathsf{id}}, \mathbf{v}_{\mathsf{id}})$.
   - $\mathsf{CH} = 2$. For $j \in \{1, 2, \cdots, k\}$, let $\hat{\pi}_j = \pi_j$, $\hat{\varphi}_j = \varphi_j$, $\hat{\phi}_j = \phi_j$, $\hat{\rho}_j = \rho_j$, $\hat{\tau} = \tau$, $\mathbf{x}'_j = \mathbf{e}'_j + \mathbf{r}'_j$, $\mathbf{x}_j = \mathbf{e}^0_j + \mathbf{r}_j$, $\mathbf{x}''_j = \mathbf{e}^{(j)} + \mathbf{r}''_j$ and $\mathbf{x}_{\mathsf{id}} = \mathsf{id}^* + \mathbf{r}_{\mathsf{id}^*}$, define $\mathsf{RSP} = (\{\hat{\pi}_j, \hat{\varphi}_j, \hat{\phi}_j, \hat{\rho}_j, \mathbf{x}'_j, \mathbf{x}_j, \mathbf{x}''_j\}_{j=1}^k, \hat{\tau}, \mathbf{x}_{\mathsf{id}})$.
   - $\mathsf{CH} = 3$. For $j \in \{1, 2, \cdots, k\}$, let $\tilde{\pi}_j = \pi_j$, $\tilde{\varphi}_j = \varphi_j$, $\tilde{\phi}_j = \phi_j$, $\tilde{\rho}_j = \rho_j$, $\tilde{\tau} = \tau$, $\mathbf{h}'_j = \mathbf{r}'_j$, $\mathbf{h}_j = \mathbf{r}_j$, $\mathbf{h}''_j = \mathbf{r}''_j$ and $\mathbf{h}_{\mathsf{id}} = \mathbf{r}_{\mathsf{id}^*}$, define $\mathsf{RSP} = (\{\tilde{\pi}_j, \tilde{\varphi}_j, \tilde{\phi}_j, \tilde{\rho}_j, \mathbf{h}'_j, \mathbf{h}_j, \mathbf{h}''_j\}_{j=1}^k, \tilde{\tau}, \mathbf{h}_{\mathsf{id}})$.
4. **Verification:** Receiving RSP, $\mathcal{V}$ checks as follows:
   - $\mathsf{CH} = 1$. Check that $\mathbf{t}_{\mathsf{id}} \in \mathsf{B}_{2\ell}$, for each $j \in \{1, 2, \cdots, k\}$, $\mathbf{v}'_j \in \mathsf{SecExt}(\mathbf{t}_{\mathsf{id}})$, $\mathbf{v}_j \in \mathsf{B}_{3m}$, $\mathbf{v}^{(j)} \in \mathsf{B}_{3(n+m+\ell)}$, and that,

$$\begin{cases} \mathbf{c}_2 = \mathsf{COM}(\{\mathbf{w}'_j, \mathbf{w}_j, \mathbf{w}''_j\}_{j=1}^k, \mathbf{t}_{\mathsf{id}}), \\ \mathbf{c}_3 = \mathsf{COM}(\{\mathbf{v}'_j + \mathbf{w}'_j, \mathbf{v}_j + \mathbf{w}_j, \mathbf{v}^{(j)} + \mathbf{w}''_j\}_{j=1}^k, \mathbf{t}_{\mathsf{id}} + \mathbf{v}_{\mathsf{id}}). \end{cases}$$

   - $\mathsf{CH} = 2$. For $j \in \{1, 2, \cdots, k\}$, let $\mathbf{x}'_{j,0} = \mathsf{Parse}(\mathbf{x}'_j, 1, m)$, and check that,

$$\begin{cases} \mathbf{c}_1 = \mathsf{COM}(\{\hat{\pi}_j, \hat{\varphi}_j, \hat{\phi}_j, \hat{\rho}_j\}_{j=1}^k, \hat{\tau}, \mathbf{A}^* \cdot (\sum_{j=1}^k \beta_j \mathbf{x}'_j) - \mathbf{u}, \\ \qquad \mathbf{B}^* \cdot (\sum_{j=1}^k \beta_j(\mathbf{x}'_{j,0}, \mathbf{x}_j) - \mathbf{b}), \mathbf{P}^* \cdot (\sum_{j=1}^k \beta_j \mathbf{x}''_j) + \mathbf{Q}^* \cdot \mathbf{x}_{\mathsf{id}} - \mathbf{c}), \\ \mathbf{c}_3 = \mathsf{COM}(\{\mathcal{T}_{\hat{\pi}_j, \hat{\varphi}_j, \hat{\tau}}(\mathbf{x}'_j), \hat{\phi}_j(\mathbf{x}_j), \hat{\rho}_j(\mathbf{x}''_j)\}_{j=1}^k, \hat{\tau}(\mathbf{x}_{\mathsf{id}})). \end{cases}$$

   - $\mathsf{CH} = 3$. For $j \in \{1, 2, \cdots, k\}$, let $\mathbf{h}'_{j,0} = \mathsf{Parse}(\mathbf{h}'_j, 1, m)$, and check that,

$$\begin{cases} \mathbf{c}_1 = \mathsf{COM}(\{\tilde{\pi}_j, \tilde{\varphi}_j, \tilde{\phi}_j, \tilde{\rho}_j\}_{j=1}^k, \tilde{\tau}, \mathbf{A}^* \cdot (\sum_{j=1}^k \beta_j \mathbf{h}'_j), \\ \qquad \mathbf{B}^* \cdot (\sum_{j=1}^k \beta_j(\mathbf{h}'_{j,0}, \mathbf{h}_j)), \mathbf{P}^* \cdot (\sum_{j=1}^k \beta_j \mathbf{h}''_j) + \mathbf{Q}^* \cdot \mathbf{h}_{\mathsf{id}}), \\ \mathbf{c}_2 = \mathsf{COM}(\{\mathcal{T}_{\tilde{\pi}_j, \tilde{\varphi}_j, \tilde{\tau}}(\mathbf{h}'_j), \tilde{\phi}_j(\mathbf{h}_j), \tilde{\rho}_j(\mathbf{h}''_j)\}_{j=1}^k, \tilde{\tau}(\mathbf{h}_{\mathsf{id}})). \end{cases}$$

The verifier $\mathcal{V}$ outputs 1 iff all the above conditions hold, otherwise 0.

The associated relation $\mathcal{R}(n, k, \ell, q, m, \beta)$ in the above protocol is defined as:

$$
\mathcal{R} = \left\{
\begin{array}{l}
\mathbf{A}_0, \mathbf{A}_1^1, \mathbf{A}_2^2, \mathbf{B} \in \mathbb{Z}_q^{n \times m}, \mathbf{P} \in \mathbb{Z}_q^{(m+\ell) \times (n+m+\ell)}, \mathbf{u} \in \mathbb{Z}_q^n, \mathbf{b} \in \mathbb{Z}_q^m, \\
\mathbf{c} = (\mathbf{c}_1, \mathbf{c}_2) \in \mathbb{Z}_q^m \times \mathbb{Z}_q^\ell, \mathrm{id} = \mathrm{bin}(i) \in \{0, 1\}^\ell, \mathbf{e}_0 \in \mathbb{Z}^m, \\
\mathbf{e}' = (\mathbf{e}_1', \mathbf{e}_2', \mathrm{bin}(i) \otimes \mathbf{e}_2') \in \mathsf{Sec}_\beta(\mathrm{id}), \mathbf{e} \in \mathbb{Z}^{n+m+\ell};\ s.t. \\
0 < \|\mathbf{e}'\|_\infty, \|\mathbf{e}_0\|_\infty, \|\mathbf{e}\|_\infty \leq \beta, \mathbf{c} = \mathbf{P}\mathbf{e} + (\mathbf{0}^m, \lfloor q/2 \rfloor \mathrm{id}) \bmod q, \\
\mathbf{b} = (\mathbf{B}^\top \mathbf{A}_0) \cdot \mathbf{e}_1' + \mathbf{e}_0 \bmod q, [\mathbf{A}_0 | \mathbf{A}_1^1 | \mathbf{g}_\ell \otimes \mathbf{A}_2^2] \cdot \mathbf{e}' = \mathbf{u} \bmod q.
\end{array}
\right\}
$$

3.3 Analysis of the Protocol

The following theorem gives a detailed analysis of the above interactive protocol.

Theorem 1. *Let* COM *(as proposed in [15]) be a statistically hiding and computationally binding commitment scheme, for a given commitment* CMT, *3 valid responses* RSP_1, RSP_2 *and* RSP_3 *with respect to 3 different challenges* CH_1, CH_2 *and* CH_3, *the proposed protocol is a statistical zero-knowledge argument of knowledge for* $\mathcal{R}(n, k, \ell, q, m, \beta)$, *where each round has perfect completeness, soundness error* $2/3$, *argument of knowledge property and communication cost* $\widetilde{\mathcal{O}}(\ell n \log \beta)$.

Proof. The proof employs a list of standard techniques for Stern-type protocol as in [15, 18, 23]. Due to the limited space, the proof is presented in the full version.

4 The Lattice-Based GS-VLR-ET Scheme

4.1 Description of the Scheme

- KeyGen($1^n, N$): On input security parameter n, group size $N = 2^\ell = \mathsf{poly}(n)$. The prime modulus $q = \omega(n^2 \log n) > N$, dimension $m = 2n\lceil \log q \rceil$, Gaussian parameter $s = \omega(\sqrt{n \log q \log n})$, and the norm bound $\beta = \lceil s \cdot \log m \rceil$ such that $(4\beta + 1)^2 \leq q$. This algorithm specifies the following steps:

1. Run TrapGen(q, n, m) to generate $\mathbf{A}_0 \in \mathbb{Z}_q^{n \times m}$ and a trapdoor $\mathbf{R}_{\mathbf{A}_0}$.
2. Sample two matrices $\mathbf{A}_1^1, \mathbf{A}_2^2 \xleftarrow{\$} \mathbb{Z}_q^{n \times m}$ and a vector $\mathbf{u} \xleftarrow{\$} \mathbb{Z}_q^n$.
3. Run TrapGen(q, n, m) to generate $\mathbf{A}_3^3 \in \mathbb{Z}_q^{n \times m}$ and a trapdoor $\mathbf{R}_{\mathbf{A}_3^3}$.
4. As in [31], for group member with index $i \in \{1, 2, \cdots, N\}$, define a matrix $\mathbf{A}_i = [\mathbf{A}_0 | \mathbf{A}_1^1 + i\mathbf{A}_2^2] \in \mathbb{Z}_q^{n \times 2m}$, and do the followings:

   4.1. Sample $\mathbf{e}_{i,2} \xleftarrow{\$} \mathcal{D}_{\mathbb{Z}^m, s}$ and let $\mathbf{u}_i = (\mathbf{A}_1^1 + i\mathbf{A}_2^2) \cdot \mathbf{e}_{i,2} \bmod q$. Then run SamplePre($\mathbf{A}_0, \mathbf{R}_{\mathbf{A}_0}, \mathbf{u} - \mathbf{u}_i, s$) to obtain $\mathbf{e}_{i,1} \in \mathbb{Z}^m$.
   4.2. Let $\mathbf{e}_i = (\mathbf{e}_{i,1}, \mathbf{e}_{i,2}) \in \mathbb{Z}^{2m}$. Thus $\mathbf{A}_i \cdot \mathbf{e}_i = \mathbf{u} \bmod q$, $0 < \|\mathbf{e}_i\|_\infty \leq \beta$.
   4.3. Let the member i's group secret-key be $\mathsf{gsk}_i = \mathbf{e}_i$, and its revocation token be $\mathsf{grt}_i = \mathbf{A}_0 \cdot \mathbf{e}_{i,1} \bmod q$.

5. Output (Gpk, Gmsk, Gsk, Grt) where $\mathsf{Gpk} = (\mathbf{A}_0, \mathbf{A}_1^1, \mathbf{A}_2^2, \mathbf{A}_3^3, \mathbf{u})$, $\mathsf{Gmsk} = \mathbf{R}_{\mathbf{A}_3^3}$, $\mathsf{Gsk} = (\mathsf{gsk}_1, \mathsf{gsk}_2, \cdots, \mathsf{gsk}_N)$, $\mathsf{Grt} = (\mathsf{grt}_1, \mathsf{grt}_2, \cdots, \mathsf{grt}_N)$.

- Sign$(\mathsf{Gpk}, \mathsf{gsk}_i, M)$: Let $\mathcal{H} : \{0,1\}^* \to \{1,2,3\}^{\kappa=\omega(\log n)}$, $\mathcal{G} : \{0,1\}^* \to \mathbb{Z}_q^{n \times m}$ be two hash functions, modeled as random oracles. Let χ be a β-bounded distribution as in Definition 2. On input Gpk and a message $M \in \{0,1\}^*$, the member i with secret-key $\mathsf{gsk}_i = \mathbf{e}_i$ specifies the following steps:

1. Sample $\mathbf{v} \xleftarrow{\$} \{0,1\}^n$ and define $\mathbf{B} = \mathcal{G}(\mathbf{A}_0, \mathbf{A}_1^1, \mathbf{A}_2^2, \mathbf{u}, M, \mathbf{v}) \in \mathbb{Z}_q^{n \times m}$.
2. Sample $\mathbf{e}_0 \xleftarrow{\$} \chi^m$ and define $\mathbf{b} = \mathbf{B}^\top \mathsf{grt}_i + \mathbf{e}_0 = (\mathbf{B}^\top \mathbf{A}_0) \cdot \mathbf{e}_{i,1} + \mathbf{e}_0 \bmod q$.
3. Sample $\mathbf{G} \xleftarrow{\$} \mathbb{Z}_q^{n \times \ell}$, $\mathbf{s} \xleftarrow{\$} \chi^n$, $\mathbf{e}_1 \xleftarrow{\$} \chi^m$, $\mathbf{e}_2 \xleftarrow{\$} \chi^\ell$, define $\mathbf{c} = (\mathbf{c}_1, \mathbf{c}_2) \in \mathbb{Z}_q^m \times \mathbb{Z}_q^\ell$ where $\mathbf{c}_1 = \mathbf{A}_3^{3\top} \mathbf{s} + \mathbf{e}_1 \bmod q$, $\mathbf{c}_2 = \mathbf{G}^\top \mathbf{s} + \mathbf{e}_2 + \lfloor q/2 \rfloor \mathrm{bin}(i) \bmod q$,
4. Generate a zero-knowledge proof that the signer is indeed a group member who owns a valid secret-key, and has signed the message $M \in \{0,1\}^*$, and its revocation token is correctly embedded in $\mathbf{b}$, and its identity is correctly embedded in $\mathbf{c} = (\mathbf{c}_1, \mathbf{c}_2)$ constructed as above. This can be achieved by repeating $\kappa = \omega(\log n)$ times the Stern-type interactive protocol as in Sect. 3.3 with the public tuple $(\mathbf{A}_0, \mathbf{A}_1^1, \mathbf{A}_2^2, \mathbf{P}, \mathbf{u}, \mathbf{B}, \mathbf{b}, \mathbf{c} = (\mathbf{c}_1, \mathbf{c}_2))$ and a witness $(\mathsf{id}, \mathsf{gsk}_i, \mathbf{e}_0, \mathbf{e})$, then making it non-interactive via the Fiat-Shamir heuristic as a triple $\Pi = (\{\mathsf{CMT}_j\}_{j \in \{1, \cdots, \kappa\}}, \mathsf{CH}, \{\mathsf{RSP}_j\}_{j \in \{1, \cdots, \kappa\}})$ where $\mathsf{CH} = \{\mathsf{CH}_j\}_{j \in \{1, \cdots, \kappa\}} = \mathcal{H}(M, \mathbf{A}_0, \mathbf{A}_1^1, \mathbf{A}_2^2, \mathbf{P}, \mathbf{u}, \mathbf{B}, \mathbf{b}, \mathbf{c}, \{\mathsf{CMT}_j\}_{j \in \{1, \cdots, \kappa\}})$.
5. Output the signature $\Sigma = (M, \Pi, \mathbf{v}, \mathbf{b}, \mathbf{G}, \mathbf{c})$.

- Verify$(\mathsf{Gpk}, \mathsf{RL}, M, \Sigma)$: On input Gpk, a signature Σ on $M \in \{0,1\}^*$, a set of tokens $\mathsf{RL} = \{\mathsf{grt}_{i'}\}_{i' \leq N} \subseteq \mathsf{Grt}$, the verifier specifies the following steps:

1. Parse the signature $\Sigma = (M, \Pi, \mathbf{v}, \mathbf{b}, \mathbf{G}, \mathbf{c})$.
2. Let $\mathbf{P} = \begin{pmatrix} \mathbf{A}_3^{3\top} \\ \cdots \\ \mathbf{G}^\top \end{pmatrix} \mathbf{I}_{m+\ell}$, and check that if $\mathsf{CH} = \{\mathsf{CH}_1, \mathsf{CH}_2, \cdots, \mathsf{CH}_\kappa\} = \mathcal{H}(M, \mathbf{A}_0, \mathbf{A}_1^1, \mathbf{A}_2^2, \mathbf{P}, \mathbf{u}, \mathbf{B}, \mathbf{b}, \mathbf{c}, \{\mathsf{CMT}_j\}_{j \in \{1,2,\cdots,\kappa\}})$.
3. For $j \in \{1, 2, \cdots, \kappa\}$, run the verification steps of the protocol from Sect. 3.3 to check the validity of RSP_j with respect to CMT_j and CH_j.
4. Let $\mathbf{B} = \mathcal{G}(\mathbf{A}_0, \mathbf{A}_1^1, \mathbf{A}_2^2, \mathbf{u}, M, \mathbf{v}) \in \mathbb{Z}_q^{n \times m}$, and for each $\mathsf{grt}_{i'} \in \mathsf{RL}$, compute $\mathbf{e}_{i'} = \mathbf{b} - \mathbf{B}^\top \mathsf{grt}_{i'} \bmod q$, and check that if $\|\mathbf{e}_{i'}\|_\infty > \beta$.
5. If the above are all satisfied, output 1 and accept Σ, otherwise reject it.

- Open$(\mathsf{Gpk}, \mathsf{Gmsk}, M, \Sigma)$: On input Gpk, $\mathsf{Gmsk} = \mathbf{R}_{\mathbf{A}_3^3}$, a group signature Σ on $M \in \{0,1\}^*$, the tracing authority specifies the following steps:

1. Parse $\Sigma = (M, \Pi, \mathbf{v}, \mathbf{b}, \mathbf{G}, \mathbf{c})$, in particular, $\mathbf{G} = [\mathbf{g}_1, \mathbf{g}_2, \cdots, \mathbf{g}_\ell]$.
2. For $i \in \{1, 2, \cdots, \ell\}$, run SamplePre$(\mathbf{A}_3, \mathbf{R}_{\mathbf{A}_3^3}, \mathbf{g}_i, s)$ to obtain $\mathbf{f}_i \in \mathbb{Z}^m$, and define $\mathbf{F} = [\mathbf{f}_1, \mathbf{f}_2, \cdots, \mathbf{f}_\ell] \in \mathbb{Z}_q^{m \times \ell}$.
3. Compute $\mathsf{id}' = (d_1', d_2', \cdots, d_\ell') = \mathbf{c}_2 - \mathbf{F}^\top \mathbf{c}_1 \bmod q$. For $i \in \{1, 2, \cdots, \ell\}$, if d_i' is closer to 0 than to $\lfloor q/2 \rfloor$, define $d_i = 1$; otherwise, $d_i = 0$.
4. Let $\mathsf{id} = (d_1, d_2, \cdots, d_\ell)$ and output $i = \mathbf{g}_\ell^\top \cdot \mathsf{id}$.

4.2   Analysis of the Scheme

Efficiency and Correctness: For our lattice-based GS-VLR-ET, it only needs 3 public matrices for identity-encoding, and one more matrix for explicit traceability, thus the group public-key has bit-size $\tilde{\mathcal{O}}(n^2)$, the member secret-key has bit-size $\tilde{\mathcal{O}}(n)$ and the signature has bit-size $\ell \cdot \tilde{\mathcal{O}}(n) = \log N \cdot \tilde{\mathcal{O}}(n)$. Compared with the existing lattice-based GS-VLR constructions, our scheme saves a $\mathcal{O}(\log N)$ factor for both bit-sizes of the group public-key and the member secret-key, meanwhile, supporting the explicit traceability, thus is more suitable for a large group.

Theorem 2. *The proposed scheme is correct with overwhelming probability.*

Proof. To prove that for all Gpk, Gsk, Gmsk, Grt generated by KeyGen, all indexes $i \in \{1, 2, \cdots, N\}$, and all messages $M \in \{0,1\}^*$, the following holds true:

$$\text{Verify}(\text{Gpk}, \text{RL}, \text{Sign}(\text{Gpk}, \text{gsk}_i, M), M) = 1 \Leftrightarrow \text{grt}_i \notin \text{RL}.$$
$$\text{Open}(\text{Gpk}, \text{Gmsk}, \text{Sign}(\text{Gpk}, \text{gsk}_i, M), M) = i.$$

For the first 3 steps of Verify, a member i owning $(\mathbf{e}', \mathbf{e}_0) \in \text{Sec}_\beta(\text{id}) \times \chi^m$ can always generate a signature satisfying them. For step 4, $\mathbf{e}_{i'}$ can be expressed as $\mathbf{e}_{i'} = \mathbf{b} - \mathbf{B}^\top \text{grt}_{i'} = \mathbf{B}^\top \text{grt}_i + \mathbf{e}_0 - \mathbf{B}^\top \text{grt}_{i'} = \mathbf{B}^\top (\text{grt}_i - \text{grt}_{i'}) + \mathbf{e}_0 \mod q$.

1. To prove that, $\text{grt}_i \notin \text{RL} \Rightarrow \text{Verify}(\text{Gpk}, \text{RL}, \text{Sign}(\text{Gpk}, \text{gsk}_i, M), M) = 1$.
   Assume that $\text{grt}_i \notin \text{RL}$, we prove that, the step 4 is satisfied with overwhelming probability, namely, the infinity norm of vector $\mathbf{e}_{i'}$ is lager than β, and $\text{Verify}(\text{Gpk}, \text{RL}, \text{Sign}(\text{Gpk}, \text{gsk}_i, M), M) = 1$. For all $\text{grt}_{i'} \in \text{RL}$, we have that $\mathbf{B}^\top \cdot (\text{grt}_i - \text{grt}_{i'}) = \mathbf{e}_{i'} - \mathbf{e}_0 \mod q$.
   Let $\mathbf{s}_{i'} = \text{grt}_i - \text{grt}_{i'}$, we have that $\|\mathbf{B}^\top \mathbf{s}_{i'}\|_\infty \le \|\mathbf{e}_{i'}\|_\infty + \|\mathbf{e}_0\|_\infty \le \|\mathbf{e}_{i'}\|_\infty + \beta$. According to Lemma 4 of [22], $\|\mathbf{B}^\top \mathbf{s}_{i'}\|_\infty > 2\beta$ with overwhelming probability, thus $\|\mathbf{e}'_i\|_\infty > 2\beta - \beta = \beta$.
2. To prove that, $\text{Verify}(\text{Gpk}, \text{RL}, \text{Sign}(\text{Gpk}, \text{gsk}_i, M), M) = 1 \Rightarrow \text{grt}_i \notin \text{RL}$.
   Assume that $\text{Verify}(\text{Gpk}, \text{RL}, \text{Sign}(\text{Gpk}, \text{gsk}_i, M), M) = 1$. Thus for all $\text{grt}_{i'} \in \text{RL}$, we have $\|\mathbf{e}_{i'}\|_\infty > \beta$. Therefore, if there is an index i' satisfying $\text{grt}_i = \text{grt}_{i'}$, then we have $\mathbf{e}_{i'} = \mathbf{e}_0$, thus $\|\mathbf{e}_{i'}\|_\infty = \|\mathbf{e}_0\|_\infty \le \beta$, the signature cannot pass the verification of step 4, therefore, a contradiction appears.
3. To prove that, $\text{Open}(\text{Gpk}, \text{Gmsk}, \text{Sign}(\text{Gpk}, \text{gsk}_i, M), M) = i$ with overwhelming probability.
   We set the parameters so that the lattice-based dual LWE cyrptosystem is correct and a tracing authority owning the trapdoor for $\Lambda_q^\perp(\mathbf{A}_3^3)$ can compute an identity index belonging to the collection $\{1, 2, \cdots, N\}$ effectively, or a special symbol $\perp$ denoting the opening failure, which implies that our Open algorithm is also correct. This concludes the correctness proof.

Theorem 3. *If* COM *(as proposed in [15]) is a statistically hiding commitment scheme, then the proposed scheme is selfless-anonymous in* ROM.

Proof. To proof this theorem, we define a list of games as follows:
Game 0. It is the original **selfless-anonymity** game. $\mathcal{C}$ honestly does as follows:

1. Run KeyGen to obtain Gpk, Gmsk, Gsk, Grt. Set $RL = \varnothing$, $Corr = \varnothing$, and send Gpk to adversary $\mathcal{A}$.
2. If $\mathcal{A}$ queries the group secret-key of member i, $\mathcal{C}$ sets $Corr = Corr \cup \{i\}$ and returns gsk_i; if $\mathcal{A}$ queries the group signature on $M \in \{0,1\}^*$ of member i, $\mathcal{C}$ returns $\Sigma \leftarrow Sign(Gpk, gsk_i, M)$; if $\mathcal{A}$ queries the revocation token of member i, $\mathcal{C}$ sets $RL = RL \cup \{grt_i\}$ and returns it to $\mathcal{A}$.
3. $\mathcal{A}$ outputs a message $M^* \in \{0,1\}^*$, two members i_0 and i_1, and for each $b \in \{0,1\}$, $i_b \notin Corr$ and $grt_{i_b} \notin RL$.
4. $\mathcal{C}$ chooses $b \xleftarrow{\$} \{0,1\}$, and generates a valid VLR-ET group signature, $\Sigma^* = Sign(Gpk, gsk_{i_b}, M^*) = (M^*, \Pi, \mathbf{v}, \mathbf{b}, \mathbf{G}, \mathbf{c})$ and returns it to $\mathcal{A}$.
5. $\mathcal{A}$ can make queries as before, but it is not allowed to ask for gsk_{i_b} or grt_{i_b} for each $b \in \{0,1\}$.
6. Finally, $\mathcal{A}$ outputs a bit $b' \in \{0,1\}$.

Game 1: $\mathcal{C}$ does the same as that in Game 0, except that it simulates the signature generation in step 4 of Game 0 by programming the random oracle:

1. For the first 2 steps of algorithm Sign, work in the honest process, namely, sample $\mathbf{v} \xleftarrow{\$} \{0,1\}^n$, $\mathbf{e}_0, \mathbf{e}_1 \xleftarrow{\$} \chi^m$, $\mathbf{G} \xleftarrow{\$} \mathbb{Z}_q^{n \times \ell}$, $\mathbf{s} \xleftarrow{\$} \chi^n$, $\mathbf{e}_2 \xleftarrow{\$} \chi^\ell$. Let $\mathbf{B} = \mathcal{G}(\mathbf{A}_0, \mathbf{A}_1^1, \mathbf{A}_2^2, \mathbf{u}, M, \mathbf{v})$, $\mathbf{b} = \mathbf{B}^\top grt_{i_b} + \mathbf{e}_0 \bmod q$, and $\mathbf{c} = (\mathbf{c}_1, \mathbf{c}_2)$, where $\mathbf{c}_1 = \mathbf{A}_3^\top \mathbf{s} + \mathbf{e}_1 \bmod q$, $\mathbf{c}_2 = \mathbf{G}^\top \mathbf{s} + \mathbf{e}_2 + \lfloor q/2 \rfloor bin(i) \bmod q$.
2. The simulation algorithm does as in the proof of Theorem 1 and will be repeated $\kappa = \omega(\log n)$ times. $\mathcal{C}$ programs the random oracle $\mathcal{H}(M^*, \mathbf{A}_0, \mathbf{A}_1^1, \mathbf{A}_2^2, \mathbf{P}, \mathbf{u}, \mathbf{B}, \mathbf{b}, \mathbf{c}, CMT_1, \cdots, CMT_\kappa) = (CH_1, \cdots, CH_\kappa)$ and due to the statistically zero-knowledge of underlying argument of knowledge, the distribution of Π^* is statistically close to Π.
3. Finally, $\mathcal{C}$ outputs the simulated signature $\widehat{\Sigma}^* = (M^*, \Pi^*, \mathbf{v}, \mathbf{b}, \mathbf{G}, \mathbf{c})$.

Game 2: $\mathcal{C}$ does the same as that in Game 1, except that it computes the vector $\mathbf{b} = \mathbf{B}^\top \mathbf{r} + \mathbf{e}_0 \bmod q$. In Game 1, $\mathbf{b}$ is generated by the revocation token grt_{i_b}, which is unknown to $\mathcal{A}$ and statistically close to a uniform vector $\mathbf{r} \in \mathbb{Z}_q^n$. Thus the distribution of $\mathbf{b}$ is statistically close to that in Game 1, and Game 2 and 1 are statistically indistinguishable.

Game 3: $\mathcal{C}$ does the same as that in Game 2, except that it generates $(\mathbf{B}, \mathbf{b}) \xleftarrow{\$} \mathcal{U}$. In Game 2, $(\mathbf{B}, \mathbf{b})$ is generated by an $LWE_{n,q,\chi}$ instance, and according to Definition 2, this distribution is computationally close to a uniform distribution $\mathcal{U}$ over $\mathbb{Z}_q^{n \times m} \times \mathbb{Z}_q^m$. Thus Game 3 and 2 are computationally indistinguishable.

Game 4: $\mathcal{C}$ does the same as that in Game 3, except that it obtains $\mathbf{A}_3^3 \xleftarrow{\$} \mathbb{Z}_q^{n \times m}$. According to Lemma 1, $\mathbf{A}_3^3$ is statistically close to a uniform matrix in $\mathbb{Z}_q^{n \times m}$. Thus Game 4 and 3 are statistically indistinguishable.

Game 5: $\mathcal{C}$ does the same as that in Game 4, except that it generates $\mathbf{c} = (\mathbf{c}_1^*, \mathbf{c}_2^*)$, where $\mathbf{c}_1^* = \mathbf{z}_1$, $\mathbf{c}_2^* = \mathbf{z}_2 + \lfloor q/2 \rfloor bin(i)$, here $\mathbf{z}_1 \xleftarrow{\$} \mathbb{Z}_q^m$, $\mathbf{z}_2 \xleftarrow{\$} \mathbb{Z}_q^\ell$. According to Definition 2, the hardness of $LWE_{n,q,\chi}$ problem implies that Game 5 and 4 are computationally indistinguishable.

Game 6: $\mathcal{C}$ does the same as that in Game 5, except that it generates $\mathbf{c} = (\mathbf{c}_1^*, \mathbf{c}_2^*)$, where $\mathbf{c}_1^* = \mathbf{z}_1'$, $\mathbf{c}_2^* = \mathbf{z}_2'$, where $\mathbf{z}_1' \xleftarrow{\$} \mathbb{Z}_q^m$ and $\mathbf{z}_2' \xleftarrow{\$} \mathbb{Z}_q^\ell$. Thus it is easy to see Game 6 and 5 are statistically indistinguishable. Furthermore, Game 6 is independent of the bit b, thus the advantage $\mathsf{Adv}_{\mathcal{A}}^{\mathsf{self\text{-}anon}}$ of $\mathcal{A}$ in Game 6 is 0.

According to a series of Games 1 to 6 defined as above, the advantage $\mathsf{Adv}_{\mathcal{A}}^{\mathsf{self\text{-}anon}}$ in Game 1 is negligible, namely, the proposed scheme is selfless-anonymous.

Theorem 4. *If the $\mathsf{SIS}_{n,m,q,2\beta\cdot(1+\omega(\sqrt{\log m}))}^{\infty}$ problem is hard, then the proposed scheme is traceable in* ROM.

Proof. Without loss of generality (WLOG), we first assume that the commitment COM, mentioned in [15], is computationally binding.

Assume that there is a PPT forger $\mathcal{F}$ against our construction with advantage ϵ, we can use $\mathcal{F}$ to construct an algorithm $\mathcal{A}$ to solve the $\mathsf{SIS}_{n,m,q,2\beta\cdot(1+\omega(\sqrt{\log m}))}^{\infty}$ problem with non-negligible probability.

Given a SIS instance $\hat{\mathbf{A}} \in \mathbb{Z}_q^{n\times m}$, $\mathcal{A}$ is required to output a shorter non-zero vector $\hat{\mathbf{e}} \in \mathbb{Z}^m$ satisfying $\hat{\mathbf{A}} \cdot \hat{\mathbf{e}} = \mathbf{0} \bmod q$, and $0 < \|\hat{\mathbf{e}}\|_\infty \leq \mathsf{poly}(m)$.
Setup: $\mathcal{A}$ does as follows:

1. Sample $\mathbf{e}_1^{1*}, \mathbf{e}_2^{2*} \xleftarrow{\$} \mathcal{D}_{\mathbb{Z}^m,s}$, $\mathbf{R} \xleftarrow{\$} \{-1,1\}^{m\times m}$, an index $i^* \in \{1,2,\cdots,N\}$.
2. Run $\mathsf{TrapGen}(q,n,m)$ to generate $\mathbf{A}_2^2 \in \mathbb{Z}_q^{n\times m}$ and a trapdoor $\mathbf{R}_{\mathbf{A}_2^2}$.
3. Define $\mathbf{A}_0 = \hat{\mathbf{A}}$, $\mathbf{A}_1^1 = \mathbf{A}_0 \cdot \mathbf{R} - i^*\mathbf{A}_2^2 \bmod q$.
4. Run $\mathsf{TrapGen}(q,n,m)$ to generate $\mathbf{A}_3^3 \in \mathbb{Z}_q^{n\times m}$ and a trapdoor $\mathbf{R}_{\mathbf{A}_3^3}$.
5. Define $\mathbf{u} = \mathbf{A}_0 \cdot (\mathbf{e}_1^{1*} + \mathbf{R}_0 \cdot \mathbf{e}_2^{2*}) \bmod q$.
6. For $i = i^*$, let $\mathsf{gsk}_{i^*} = (\mathbf{e}_1^{1*}, \mathbf{e}_2^{2*})$, $\mathsf{grt}_{i^*} = \mathbf{A}_0 \cdot \mathbf{e}_1^{1*} \bmod q$.
7. For $i \in \{1,2,\cdots,N\} \setminus \{i^*\}$, define $\mathbf{A}_i = [\mathbf{A}_0 | \mathbf{A}_1^1 + i\mathbf{A}_2^2] \in \mathbb{Z}_q^{n\times 2m}$, and run $\mathsf{SampleRight}(\mathbf{A}_0, (i-i^*)\mathbf{A}_2^2, \mathbf{R}, \mathbf{R}_{\mathbf{A}_2^2}, \mathbf{u}, s)$ to obtain $\mathbf{e}_i = (\mathbf{e}_{i,1}, \mathbf{e}_{i,2}) \in \mathbb{Z}^{2m}$ and let $\mathsf{gsk}_i = \mathbf{e}_i$, $\mathsf{grt}_i = \mathbf{A}_0 \cdot \mathbf{e}_{i,1} \bmod q$.
8. Let $\mathsf{Gpk} = (\mathbf{A}_0, \mathbf{A}_1^1, \mathbf{A}_2^2, \mathbf{A}_3^3, \mathbf{u})$, $\mathsf{Gmsk} = \mathbf{R}_{\mathbf{A}_3^3}$, $\mathsf{Gsk} = (\mathsf{gsk}_1, \mathsf{gsk}_2, \cdots, \mathsf{gsk}_N)$, $\mathsf{Grt} = (\mathsf{grt}_1, \mathsf{grt}_2, \cdots, \mathsf{grt}_N)$, then send $(\mathsf{Gpk}, \mathsf{Gmsk}, \mathsf{Grt})$ to $\mathcal{F}$.

Queries: $\mathcal{F}$ can make a polynomially bounded number of queries as follows:

1. Corruption: Request for secret-key of i, $\mathcal{A}$ adds i to Corr, and returns gsk_i.
2. Signing: Request for a signature on $M \in \{0,1\}^*$ of member i. $\mathcal{A}$ returns $\Sigma \leftarrow \mathsf{Sign}(\mathsf{Gpk}, \mathsf{gsk}_i, M)$. For queries to oracle $\mathcal{H}$, uniformly random values in $\{1,2,3\}^{\kappa=\omega(\log n)}$ are returned. Assume that $q_{\mathcal{H}}$ is the number of queries to $\mathcal{H}$, for any $d \leq q_{\mathcal{H}}$, let r_d denote the answer to the d-th query.

Forgery: $\mathcal{F}$ outputs a message $M^* \in \{0,1\}^*$, a set of revocation tokens $\mathsf{RL}^* \subseteq \mathsf{Grt}$ and a non-trivial forged group signature $\Sigma^* = (M^*, \Pi^*, \mathbf{v}^*, \mathbf{b}^*, \mathbf{G}^*, \mathbf{c}^*)$, where $\Pi^* = (\{\mathsf{CMT}_j, \mathsf{CH}_j, \mathsf{RSP}_j\}_{j\in\{1,2,\cdots,\kappa\}})$, which satisfies the followings:

1. $\mathsf{Verify}(\mathsf{Gpk}, \mathsf{RL}^*, \Sigma^*, M^*) = 1$.
2. The tracing algorithm (no matter the implicit or explicit tracing) fails, or traces to a member outside of the coalition $\mathsf{Corr} \setminus \mathsf{RL}^*$.

$\mathcal{A}$ exploits the above forgery as follows:

1. Let $\mathbf{B}^* = \mathcal{G}(\mathbf{A}_0, \mathbf{A}_1^1, \mathbf{A}_2^2, \mathbf{u}, M^*, \mathbf{v}^*) \in \mathbb{Z}_q^{n \times m}$.
2. $\mathcal{A}$ must queried $\mathcal{H}$ on $(M^*, \mathbf{A}_0, \mathbf{A}_1^1, \mathbf{A}_2^2, \mathbf{P}, \mathbf{u}, \mathbf{B}^*, \mathbf{b}^*, \mathbf{c}^*, \{\mathsf{CMT}_j\}_{j \in \{1, \cdots, \kappa\}})$, since otherwise, the probability that $(\mathsf{CH}_1, \cdots, \mathsf{CH}_\kappa) = \mathcal{H}(M^*, \mathbf{A}_0, \mathbf{A}_1^1, \mathbf{A}_2^2, \mathbf{P}, \mathbf{u}, \mathbf{B}^*, \mathbf{b}^*, \mathbf{c}^*, \{\mathsf{CMT}_j\}_{j \in \{1, \cdots, \kappa\}})$ is at most $3^{-\kappa}$. Thus, there exists $d' \leq q_\mathcal{H}$ such that the d'-th hash query involves $(M^*, \mathbf{A}_0, \mathbf{A}_1^1, \mathbf{A}_2^2, \mathbf{P}, \mathbf{u}, \mathbf{B}^*, \mathbf{b}^*, \mathbf{c}^*, \{\mathsf{CMT}_j\}_{j \in \{1, \cdots, \kappa\}})$ with probability at least $\epsilon - 3^{-\kappa}$.
3. Let d' be the target point. $\mathcal{A}$ replays $\mathcal{F}$ many times with the same random tape and input as in the original execution. $\mathcal{F}$ is given the same answers to the first $d' - 1$ queries as in the original execution. From the d'-th query, $\mathcal{A}$ chooses fresh random values $r'_{d'}, \cdots, r'_{q_\mathcal{H}} \in \{1, 2, 3\}^\kappa$ as replies.

According to the Improved Forking Lemma of Pointcheval and Vaudenay, with a probability larger than $1/2$, algorithm $\mathcal{A}$ can obtain a 3-fork involving the tuple $(M^*, \mathbf{A}_0, \mathbf{A}_1^1, \mathbf{A}_2^2, \mathbf{P}, \mathbf{u}, \mathbf{B}^*, \mathbf{b}^*, \mathbf{c}^*, \{\mathsf{CMT}_j\}_{j \in \{1, 2, \cdots, \kappa\}})$ after at most $32 \cdot q_\mathcal{H} / (\epsilon - 3^{-\kappa})$ executions of $\mathcal{F}$. Let the answers of $\mathcal{A}$ corresponding to this 3-fork be $r_{d'}^1 = (\mathsf{CH}_1^1, \mathsf{CH}_2^1, \cdots, \mathsf{CH}_\kappa^1), r_{d'}^2 = (\mathsf{CH}_1^2, \mathsf{CH}_2^2, \cdots, \mathsf{CH}_\kappa^2), r_{d'}^3 = (\mathsf{CH}_1^3, \mathsf{CH}_2^3, \cdots, \mathsf{CH}_\kappa^3)$, then $\Pr[\exists i \in \{1, 2, \cdots, \kappa\} \ s.t. \ \{\mathsf{CH}_i^1, \mathsf{CH}_i^2, \mathsf{CH}_i^3\} = \{1, 2, 3\}] = 1 - (7/9)^\kappa$.

Thus, according to the existence of such i, one can parse these 3 forgeries corresponding to the fork to obtain $(\mathsf{RSP}_i^1, \mathsf{RSP}_i^2, \mathsf{RSP}_i^3)$ which are 3 valid responses corresponding to 3 different challenges for the same commitment CMT_i. Further, COM is computationally binding, using the knowledge extractor $\mathcal{K}$ as described in the proof of Theorem 1, one can extract a witness $(\mathrm{id} = \mathrm{bin}(i) \in \{0,1\}^\ell, \mathbf{e}_i = (\mathbf{e}_{i,1}, \mathbf{e}_{i,2}) \in \mathbb{Z}^{2m}, \mathbf{e}_0^*, \mathbf{e}_1^* \in \mathbb{Z}^m, \mathbf{s}^* \in \mathbb{Z}^n, \mathbf{e}_2^* \in \mathbb{Z}^\ell)$ such that,

1. $[\mathbf{A}_0 | \mathbf{A}_1^1 + i\mathbf{A}_2^2] \cdot \mathbf{e}_i = \mathbf{u} \bmod q$, and $\mathbf{e}_i \in \mathsf{Sec}_\beta(\mathrm{id})$.
2. $\mathbf{b}^* = (\mathbf{B}^{*\top} \mathbf{A}_0) \cdot \mathbf{e}_{i,1} + \mathbf{e}_0^* \bmod q$, and $0 < \|\mathbf{e}_0^*\|_\infty \leq \beta$.
3. $\mathbf{c}^* = (\mathbf{c}_1^*, \mathbf{c}_2^*) = (\mathbf{A}_3^{3\top} \mathbf{s}^* + \mathbf{e}_1^* \bmod q, \mathbf{G}^{*\top} \mathbf{s}^* + \mathbf{e}_2^* + \lfloor q/2 \rfloor \mathrm{bin}(i) \bmod q)$.

Now, we consider the following 2 cases:

1. If $i \neq i^*$, this event happens with a probability at most $1 - 1/N$, then $\mathcal{A}$ outputs $\bot$ and aborts.
2. If $i = i^*$, $\mathcal{A}$ returns $\hat{\mathbf{e}} = (\mathbf{e}_1^{1*} - \mathbf{e}_{i^*,1}) + \mathbf{R} \cdot (\mathbf{e}_2^{2*} - \mathbf{e}_{i^*,2})$ as a solution of the given SIS problem. By construction, we have

$$\hat{\mathbf{A}} \cdot \hat{\mathbf{e}} = \mathbf{A}_0 \cdot (\mathbf{e}_1^{1*} - \mathbf{e}_{i^*,1} + \mathbf{R} \cdot (\mathbf{e}_2^{2*} - \mathbf{e}_{i^*,2}))$$
$$= \mathbf{A}_0 \cdot (\mathbf{e}_1^{1*} + \mathbf{R} \cdot \mathbf{e}_2^{2*}) - \mathbf{A}_0 \cdot (\mathbf{e}_{i^*,1} + \mathbf{R} \cdot \mathbf{e}_{i^*,2}) = \mathbf{0} \bmod q.$$

Next, we show that $\hat{\mathbf{e}}$ is with high probability a short non-zero preimage of $\mathbf{0}$ under $\hat{\mathbf{A}}$.

1. $\|\hat{\mathbf{e}}\|_\infty \leq \mathsf{poly}(m)$. For $j \in \{1, 2\}$, $\|\mathbf{e}_j^{j*}\|_\infty, \|\mathbf{e}_{i^*,j}\|_\infty \leq \beta$, $\mathbf{R}$ is a low-norm matrix with coefficients ± 1, thus according to Lemma 4, with overwhelming probability, we have $\|\hat{\mathbf{e}}\|_\infty \leq (1 + \omega(\sqrt{\log m})) \cdot 2\beta = \mathsf{poly}(m)$.

2. $\hat{\mathbf{e}} \neq \mathbf{0}$. $\Sigma^* = (M^*, \Pi^*, \mathbf{v}^*, \mathbf{b}^*, \mathbf{G}^*, \mathbf{c}^*)$ is a valid forged signature, thus the tracing algorithm (no matter the implicit or explicit tracing) either fails, or traces to a member outside of the coalition $\mathsf{Corr} \backslash \mathsf{RL}^*$.

(2.1). If the tracing algorithm fails. $\mathsf{Verify}(\mathsf{Gpk}, \mathsf{grt}_{i^*}, \Sigma^*, M^*) = 1$ implies that $\mathbf{A}_0 \cdot \mathbf{e}_{i^*,1} \neq \mathsf{grt}_{i^*} = \mathbf{A}_0 \cdot \mathbf{e}_1^{1*} \bmod q$, thus $\mathbf{e}_{i^*,1} \neq \mathbf{e}_1^{1*}$.

(2.2). If the tracing algorithm traces to $j^* \notin \mathsf{Corr} \backslash \mathsf{RL}^*$. Clearly, we have 2 facts: $\mathsf{Verify}(\mathsf{Gpk}, \mathsf{grt}_{j^*}, \Sigma^*, M^*) = 0$, $\mathsf{Verify}(\mathsf{Gpk}, \mathsf{RL}^*, \Sigma^*, M^*) = 1$. Thus, we have the following conclusions:

a_1. $\mathsf{grt}_{j^*} \notin \mathsf{RL}^*$, thus $j^* \notin \mathsf{Corr}$.

a_2. Since $\|\mathbf{b}^* - \mathbf{B}^{*\top} \mathsf{grt}_{j^*}\|_\infty = \|\mathbf{B}^{*\top} \cdot (\mathbf{A}_0 \cdot \mathbf{e}_{i^*,1} - \mathsf{grt}_{j^*}) + \mathbf{e}_0^*\|_\infty \leq \beta$, $\|\mathbf{e}_0^*\|_\infty \leq \beta$, thus $\|\mathbf{B}^{*\top} \cdot (\mathbf{A}_0 \cdot \mathbf{e}_{i^*,1} - \mathsf{grt}_{j^*})\|_\infty \leq 2\beta$, furthermore, according to Lemma 4 of [22], we have that $\mathsf{grt}_{j^*} = \mathbf{A}_0 \cdot \mathbf{e}_{i^*,1} \bmod q$ with overwhelming probability.

Now, considering the following 2 cases:

b_1. If $\mathcal{F}$ has never requested gsk_{i^*}, then $(\mathbf{e}_1^{1*}, \mathbf{e}_2^{2*})$ cannot be known to $\mathcal{F}$, and thus $(\mathbf{e}_1^{1*}, \mathbf{e}_2^{2*}) \neq (\mathbf{e}_{i^*,1}, \mathbf{e}_{i^*,2})$ with overwhelming probability.

b_2. If $\mathcal{F}$ has requested gsk_{i^*}, then $i^* \in \mathsf{Corr}$, thus $i^* \neq j^*$, so $\mathsf{grt}_{i^*} \neq \mathsf{grt}_{j^*}$, which means $\mathbf{e}_{i^*,1} \neq \mathbf{e}_1^{1*}$.

Based on the above analysis, for an easy case, in (2.1) and b_2, suppose that $\mathbf{e}_2^{2*} = \mathbf{e}_{i^*,2}$, then we must have $\hat{\mathbf{e}} = \mathbf{e}_1^{1*} - \mathbf{e}_{i^*,1} \neq \mathbf{0}$. On the contrary, in (2.1), b_1 and b_2, $\mathbf{e}_2^{2*} \neq \mathbf{e}_{i^*,2}$, define $\hat{\mathbf{e}}_2 = \mathbf{e}_2^{2*} - \mathbf{e}_{i^*,2}$, in this case, we have $0 \neq \|\hat{\mathbf{e}}_2\|_\infty \leq 2\beta \ll q$, and thus there must be at least one coordinate of $\hat{\mathbf{e}}_2$ that is non-zero modulo q. WLOG, let this coordinate be the last one in $\hat{\mathbf{e}}_2$, and call it $\hat{e}$. Let $\mathbf{r}$ be the last column of $\mathbf{R}$, the expression of $\hat{\mathbf{e}}$ can be rewritten as $\hat{\mathbf{e}} = \mathbf{r} \cdot \hat{e} + \hat{\mathbf{e}}'$ where $\hat{\mathbf{e}}'$ does not depends on $\mathbf{r}$. The only information about $\mathbf{r}$ available to $\mathcal{F}$ is just contained in the last column of $\mathbf{A}_1 = \mathbf{A}_0 \cdot \mathbf{R}$. According to the leftover hash or pigeonhole principle, there are $\exp^{\mathcal{O}(m - n \log q) = \widetilde{\mathcal{O}}(n)}$ admissible and equally likely vectors $\mathbf{r}$ that are compatible with the view of $\mathcal{F}$, $\mathcal{F}$ cannot know the value of $\mathbf{r} \cdot \hat{e}$ with probability exceeding $\exp^{-\widetilde{\mathcal{O}}(n)}$, and at most one such value can result in a cancelation of $\hat{\mathbf{e}}$. Thus, $\hat{\mathbf{e}}$ is non-zero with a high probability $1 - \exp^{-\widetilde{\mathcal{O}}(n)}$.

Therefore, we deduce that $\hat{\mathbf{e}}$ is with a probability larger than $1/(2N) \cdot (1 - (7/9)^\kappa) \cdot (1 - \exp^{-\widetilde{\mathcal{O}}(n)}) \cdot \epsilon$ a short non-zero preimage of $\mathbf{0}$ under $\hat{\mathbf{A}}$, i.e., $\hat{\mathbf{A}} \cdot \hat{\mathbf{e}} = \mathbf{0} \bmod q$, $0 \neq \|\hat{\mathbf{e}}\|_\infty \leq 2\beta \cdot (1 + \omega(\sqrt{\log m})) = \mathsf{poly}(m)$. This concludes the proof.

Acknowledgments. The authors would like to thank the anonymous reviewers of CANS 2019 for their helpful comments and this research is supported by the National Natural Science Foundation of China under Grant 61772477.

References

1. Agrawal, S., Boneh, D., Boyen, X.: Efficient lattice (H)IBE in the standard model. In: Gilbert, H. (ed.) EUROCRYPT 2010. LNCS, vol. 6110, pp. 553–572. Springer, Heidelberg (2010). https://doi.org/10.1007/978-3-642-13190-5_28
2. Ajtai, M.: Generating hard instances of lattice problems (extended abstract). In: STOC, pp. 99–108. ACM (1996). https://doi.org/10.1145/237814.237838

3. Alwen, J., Peikert, C.: Generating shorter bases for hard random lattices. Theory Comput. Syst. **48**(3), 535–553 (2011). https://doi.org/10.1007/s00224-010-9278-3

4. Bellare, M., Micciancio, D., Warinschi, B.: Foundations of group signatures: formal definitions, simplified requirements, and a construction based on general assumptions. In: Biham, E. (ed.) EUROCRYPT 2003. LNCS, vol. 2656, pp. 614–629. Springer, Heidelberg (2003). https://doi.org/10.1007/3-540-39200-9_38

5. Bellare, M., Shi, H., Zhang, C.: Foundations of group signatures: the case of dynamic groups. In: Menezes, A. (ed.) CT-RSA 2005. LNCS, vol. 3376, pp. 136–153. Springer, Heidelberg (2005). https://doi.org/10.1007/978-3-540-30574-3_11

6. Boneh, D., Shacham, H.: Group signatures with verifier-local revocation. In: CCS, pp. 168–177. ACM (2004). https://doi.org/10.1145/1030083.1030106

7. Bootle, J., Cerulli, A., Chaidos, P., Ghadafi, E., Groth, J.: Foundations of fully dynamic group signatures. In: Manulis, M., Sadeghi, A.-R., Schneider, S. (eds.) ACNS 2016. LNCS, vol. 9696, pp. 117–136. Springer, Cham (2016). https://doi.org/10.1007/978-3-319-39555-5_7

8. Camenisch, J., Neven, G., Rückert, M.: Fully anonymous attribute tokens from lattices. In: Visconti, I., De Prisco, R. (eds.) SCN 2012. LNCS, vol. 7485, pp. 57–75. Springer, Heidelberg (2012). https://doi.org/10.1007/978-3-642-32928-9_4

9. Cash, D., Hofheinz, D., Kiltz, E., Peikert, C.: Bonsai trees, or how to delegate a lattice basis. In: Gilbert, H. (ed.) EUROCRYPT 2010. LNCS, vol. 6110, pp. 523–552. Springer, Heidelberg (2010). https://doi.org/10.1007/978-3-642-13190-5_27

10. Chaum, D., van Heyst, E.: Group signatures. In: Davies, D.W. (ed.) EUROCRYPT 1991. LNCS, vol. 547, pp. 257–265. Springer, Heidelberg (1991). https://doi.org/10.1007/3-540-46416-6_22

11. Gao, W., Hu, Y., Zhang, Y., Wang, B.: Lattice-Based Group Signature with Verifier-Local Revocation. J. Shanghai JiaoTong Univ. (Sci.) **22**(3), 313–321 (2017). https://doi.org/10.1007/s12204-017-1837-1

12. Gentry, C., Peikert, C., Vaikuntanathan, V.: Trapdoor for hard lattices and new cryptographic constructions. In: STOC, pp. 197–206. ACM (2008) https://doi.org/10.1145/1374376.1374407

13. Gordon, S.D., Katz, J., Vaikuntanathan, V.: A group signature scheme from lattice assumptions. In: Abe, M. (ed.) ASIACRYPT 2010. LNCS, vol. 6477, pp. 395–412. Springer, Heidelberg (2010). https://doi.org/10.1007/978-3-642-17373-8_23

14. Katsumata, S., Yamada, S.: Group signatures without NIZK: from lattices in the standard model. In: Ishai, Y., Rijmen, V. (eds.) EUROCRYPT 2019. LNCS, vol. 11478, pp. 312–344. Springer, Cham (2019). https://doi.org/10.1007/978-3-030-17659-4_11

15. Kawachi, A., Tanaka, K., Xagawa, K.: Concurrently secure identification schemes based on the worst-case hardness of lattice problems. In: Pieprzyk, J. (ed.) ASIACRYPT 2008. LNCS, vol. 5350, pp. 372–389. Springer, Heidelberg (2008). https://doi.org/10.1007/978-3-540-89255-7_23

16. Kiayias, A., Yung, M.: Secure scalable group signature with dynamic joins and separable authorities. Int. J. Secur. Netw. **1**(1/2), 24–45 (2006). https://doi.org/10.1504/ijsn.2006.010821

17. Laguillaumie, F., Langlois, A., Libert, B., Stehlé, D.: Lattice-based group signatures with logarithmic signature size. In: Sako, K., Sarkar, P. (eds.) ASIACRYPT 2013. LNCS, vol. 8270, pp. 41–61. Springer, Heidelberg (2013). https://doi.org/10.1007/978-3-642-42045-0_3

18. Langlois, A., Ling, S., Nguyen, K., Wang, H.: Lattice-based group signature scheme with verifier-local revocation. In: Krawczyk, H. (ed.) PKC 2014. LNCS, vol. 8383, pp. 345–361. Springer, Heidelberg (2014). https://doi.org/10.1007/978-3-642-54631-0_20

19. Libert, B., Ling, S., Mouhartem, F., Nguyen, K., Wang, H.: Signature schemes with efficient protocols and dynamic group signatures from lattice assumptions. In: Cheon, J.H., Takagi, T. (eds.) ASIACRYPT 2016. LNCS, vol. 10032, pp. 373–403. Springer, Heidelberg (2016). https://doi.org/10.1007/978-3-662-53890-6_13

20. Libert, B., Ling, S., Nguyen, K., Wang, H.: Zero-knowledge arguments for lattice-based accumulators: logarithmic-size ring signatures and group signatures without trapdoors. In: Fischlin, M., Coron, J.-S. (eds.) EUROCRYPT 2016. LNCS, vol. 9666, pp. 1–31. Springer, Heidelberg (2016). https://doi.org/10.1007/978-3-662-49896-5_1

21. Libert, B., Mouhartem, F., Nguyen, K.: A lattice-based group signature scheme with message-dependent opening. In: Manulis, M., Sadeghi, A.-R., Schneider, S. (eds.) ACNS 2016. LNCS, vol. 9696, pp. 137–155. Springer, Cham (2016). https://doi.org/10.1007/978-3-319-39555-5_8

22. Ling, S., Nguyen, K., Roux-Langlois, A., Wang, H.: A lattice-based group signature scheme with verifier-local revocation. Theor. Comput. Sci. **730**, 1–20 (2018)

23. Ling, S., Nguyen, K., Stehlé, D., Wang, H.: Improved zero-knowledge proofs of knowledge for the ISIS problem, and applications. In: Kurosawa, K., Hanaoka, G. (eds.) PKC 2013. LNCS, vol. 7778, pp. 107–124. Springer, Heidelberg (2013). https://doi.org/10.1007/978-3-642-36362-7_8

24. Ling, S., Nguyen, K., Wang, H.: Group signatures from lattices: simpler, tighter, shorter, ring-based. In: Katz, J. (ed.) PKC 2015. LNCS, vol. 9020, pp. 427–449. Springer, Heidelberg (2015). https://doi.org/10.1007/978-3-662-46447-2_19

25. Ling, S., Nguyen, K., Wang, H., Xu, Y.: Constant-size group signatures from lattices. In: Abdalla, M., Dahab, R. (eds.) PKC 2018. LNCS, vol. 10770, pp. 58–88. Springer, Cham (2018). https://doi.org/10.1007/978-3-319-76581-5_3

26. Ling, S., Nguyen, K., Wang, H., Xu, Y.: Forward-secure group signatures from lattices. In: Ding, J., Steinwandt, R. (eds.) PQCrypto 2019. LNCS, vol. 11505, pp. 44–64. Springer, Cham (2019). https://doi.org/10.1007/978-3-030-25510-7_3

27. Ling, S., Nguyen, K., Wang, H., Xu, Y.: Lattice-based group signatures: achieving full dynamicity with ease. In: Gollmann, D., Miyaji, A., Kikuchi, H. (eds.) ACNS 2017. LNCS, vol. 10355, pp. 293–312. Springer, Cham (2017). https://doi.org/10.1007/978-3-319-61204-1_15

28. Micciancio, D., Regev, O.: Worst-case to average-case reductions based on Gaussian measures. SIAM J. Comput. **37**(1), 267–302 (2007). https://doi.org/10.1137/s0097539705447360

29. Micciancio, D., Peikert, C.: Hardness of SIS and LWE with small parameters. In: Canetti, R., Garay, J.A. (eds.) CRYPTO 2013. LNCS, vol. 8042, pp. 21–39. Springer, Heidelberg (2013). https://doi.org/10.1007/978-3-642-40041-4_2

30. Micciancio, D., Peikert, C.: Trapdoors for lattices: simpler, tighter, faster, smaller. In: Pointcheval, D., Johansson, T. (eds.) EUROCRYPT 2012. LNCS, vol. 7237, pp. 700–718. Springer, Heidelberg (2012). https://doi.org/10.1007/978-3-642-29011-4_41

31. Nguyen, P.Q., Zhang, J., Zhang, Z.: Simpler efficient group signatures from lattices. In: Katz, J. (ed.) PKC 2015. LNCS, vol. 9020, pp. 401–426. Springer, Heidelberg (2015). https://doi.org/10.1007/978-3-662-46447-2_18

32. Perera, M.N.S., Koshiba, T.: Achieving strong security and verifier-local revocation for dynamic group signatures from lattice assumptions. In: Katsikas, S.K., Alcaraz, C. (eds.) STM 2018. LNCS, vol. 11091, pp. 3–19. Springer, Cham (2018). https://doi.org/10.1007/978-3-030-01141-3_1
33. Perera, M.N.S., Koshiba, T.: Zero-knowledge proof for lattice-based group signature schemes with verifier-local revocation. In: Barolli, L., Kryvinska, N., Enokido, T., Takizawa, M. (eds.) NBiS 2018. LNDECT, vol. 22, pp. 772–782. Springer, Cham (2019). https://doi.org/10.1007/978-3-319-98530-5_68
34. Regev, O.: On lattices, learning with errors, random linear codes, and cryptography. In: STOC, pp. 84–93. ACM (2005). https://doi.org/10.1145/1060590.1060603
35. Zhang, Y., Hu, Y., Gao, W., Jiang, M.: Simpler efficient group signature scheme with verifier-local revocation from lattices. KSII Trans. Internet Inf. Syst. $10(1)$, 414–430 (2016). https://doi.org/10.3837/tiis.2016.01.024

Tighter Security Proofs for Post-quantum Key Encapsulation Mechanism in the Multi-challenge Setting

Zhengyu Zhang[2], Puwen Wei[1,2(✉)], and Haiyang Xue[3]

[1] School of Cyber Science and Technology, Shandong University, Qingdao, China
`pwei@sdu.edu.cn`
[2] Key Laboratory of Cryptologic Technology and Information Security,
Shandong University, Ministry of Education, Jinan, China
`zhangzhengyu@mail.sdu.edu.cn`
[3] State Key Laboratory of Information Security,
Institute of Information Engineering, Chinese Academy of Sciences,
Beijing, China
`xuehaiyang@iie.ac.cn`

Abstract. Due to the threat posed by quantum computers, a series of works investigate the security of cryptographic schemes in the quantum-accessible random oracle model (QROM) where the adversary can query the random oracle in superposition. In this paper, we present tighter security proofs of a generic transformations for key encapsulation mechanism (KEM) in the QROM in the multi-challenge setting, where the reduction loss is independent of the number of challenge ciphertexts. In particular, we introduce the notion of multi-challenge OW-CPA (mOW-CPA) security, which captures the one-wayness of the underlying public key encryption (PKE) under chosen plaintext attack in the multi-challenge setting. We show that the multi-challenge IND-CCA (mIND-CCA) security of KEM can be reduced to the mOW-CPA security of the underlying PKE scheme (with δ-correctness) using $\text{FO}^{\perp}$ transformation. Then we prove that the mOW-CPA security can be tightly reduced to the underlying post-quantum assumptions by showing the tight mOW-CPA security of two concrete PKE schemes based on LWE, where one is the Regev's PKE scheme and the other is a variant of Frodo.

Keywords: KEM · QROM · CCA · Tight security

1 Introduction

Indistinguishability under adaptive chosen ciphertext attack (IND-CCA) security has become a standard notion for the security of public-key encryption (PKE) and key encapsulation mechanism (KEM). The most efficient constructions of those schemes are usually proposed in the random oracle model (ROM) [6]. Due to the threat to existing cryptographic systems posed by quantum

© Springer Nature Switzerland AG 2019
Y. Mu et al. (Eds.): CANS 2019, LNCS 11829, pp. 141–160, 2019.
https://doi.org/10.1007/978-3-030-31578-8_8

computers, Boneh et al. [9] argue that one needs to prove security in the quantum-accessible random oracle model (QROM) where the adversary can query the random oracle in superposition. Moreover they proved that a variant of Bellare-Rogaway encryption [6] is IND-CCA security in the QROM. Targhi and Unruh [26] proposed variants of the Fujisaki-Okamoto (FO) and OAEP, which are proven to be IND-CCA security in the QROM. Hofheinz, Hövelmanns and Kiltz [17] revisited the KEM version of FO transformation, and obtained several variants of FO transformation, which are $FO^{\perp}$, $FO^{\not\perp}$, $FO_m^{\perp}$, $FO_m^{\not\perp}$, $QFO_m^{\perp}$ and $QFO_m^{\not\perp}$.[1] Jiang et al. [19] proved the QROM security of two generic transformations, $FO^{\not\perp}$ and $FO_m^{\not\perp}$ without additional hash in the ciphertext, where the IND-CCA security of KEM is reduced to the standard OW-CPA security of the underlying PKE with quadratic security loss. Saito, Xagawa and Yamakawa [25] proposed a construction of an IND-CCA KEM based on a deterministic PKE (DPKE) scheme with a tight security reduction. However, the underlying DPKE in [25] should satisfy a non-standard security notion called disjoint simulatability and perfect correctness, and many practical lattice-based encryption schemes have a small correctness error, e.g., New Hope, Frodo and Kyber.

Standard security notions such as IND-CCA only consider one challenge ciphertext. The realistic setting for PKE and KEM, however, usually involves more ciphertexts, which is called multi-challenge setting. General result [5] shows that the single-challenge security implies security in the multi-challenge setting, and the reduction loss of the proof is t, where t denotes the number of challenge ciphertexts. Since the size of parameters is closely related to the loss of the security reduction, it is necessary to consider tighter security for the multi-challenge setting. A series works [1,7,8,11,13,16,18,21,28] explored cryptographic primitives with tight or almost tight security in the multi-challenge setting under the standard model. Considering the quantum adversary with multiple challenge ciphertexts, Katsumata et al. [20] presented almost tight security proofs for GPV-IBE (identity-based encryption) [14] in the QROM in the multi-challenge setting.

1.1   Our Contributions

In this paper, we investigate tighter security proofs of the generic transformations $FO^{\not\perp}$ [17] for KEM in the QROM in the multi-challenge setting. In order to reduce the security loss of $FO^{\not\perp}$ in the multi-challenge setting, we introduce the multi-challenge OW-CPA (mOW-CPA) security, which captures the one-wayness of the underlying PKE under chosen plaintext attack in the multi-challenge setting. Then we show that the multi-challenge IND-CCA (mIND-CCA) security of KEM can be reduced to the mOW-CPA security of the underlying PKE scheme (with δ-correctness) using $FO^{\not\perp}$ transformation, where the security loss can be independent of the number of the challenge ciphertexts.

---

[1] m (without m) means the ephemeral key $K = H(m)$ ($K = H(m,c)$), $\not\perp$ ($\perp$) means implicit (explicit) rejection in decryption algorithm, and Q means adding additional hash to the ciphertext in the QROM.

Note that although the standard hybrid argument proposed in [5] can still be applied in the QROM to achieve the multi-challenge security, it suffers a security loss t, which is the number of the challenge ciphertexts. Concretely, the single-challenge IND-CCA security of the KEM using $FO^{\perp}$ shown in [19] is about $q\sqrt{\delta} + q\sqrt{\varepsilon}$, where ε is the advantage of an adversary against the OW-CPA security of the PKE, and q is the number of adversary's queries to the oracles. By the standard hybrid argument in [5], the resulting multi-challenge security would be about $t(q\sqrt{\delta} + q\sqrt{\varepsilon})$. Using our method, however, the security bound can be reduced to about $q\sqrt{\delta} + q\sqrt{\varepsilon'}$, where ε' denotes the advantage of an adversary against the mOW-CPA security.

Moreover, we show that the mOW-CPA security can be tightly reduced to the underlying post-quantum assumptions. More precisely, we identify two PKE schemes based on LWE with tight mOW-CPA security. One is the Regev's PKE scheme [23] and the other is a slight variant of Frodo [10]. It is shown that the Regev's PKE has much tighter security proof than that of the latter, while the latter has much smaller public key size. By applying the generic transformations $FO^{\perp}$ to those schemes in the QROM, we can construct KEM with tighter multi-challenge IND-CCA security, where the security loss is independent on the number of the challenge ciphertexts.

1.2   Technical Overview

– **Generic Conversion.** For the classic security proof in random oracle model, a RO-query list is used to simulate the random oracle, where changing one or more points of the list usually does not provide much help for the PPT adversary to distinguish the random oracle and the simulated one. In the QROM, however, the adversary can query the random oracle in superposition, which means the adversary may find the difference even if only one point is changed. Fortunately, the one-way to hiding (OW2H) lemma introduced by [27] provided a useful tool to solve the above problem. But the OW2H lemma captures the indistinguishability between functions with only one point changed, and cannot be applied in the multi-challenge setting, where multiple points need to be reprogrammed. To overcome this problem, we take advantage of a more general OW2H lemma [3], which is extended to the multi-point setting while preserving the tightness. Applying the multi-point OW2H lemma, the mIND-CCA security of the generic transformations $FO^{\perp}$ [17] for KEM can be proven in the QROM in the multi-challenge setting.

– **Tight mOW-CPA Security.** Roughly speaking, the mOW-CPA security considers the following game. The adversary which can make the chosen plaintext attack is given multiple challenge ciphertexts and outputs a message. We say the adversary wins the game if the message is the plaintext of one of the challenge ciphertexts.

We show that the mOW-CPA security of Regev's PKE scheme [23] can be tightly reduced to the underlying LWE problem. Since Regev's PKE scheme is not very efficient due to its large public key size, we consider Frodo [10].

However, the proof techniques of the OW-CPA security for Frodo cannot be extended to the multi-challenge setting. Inspired by [20], we consider the mOW-CPA security of a slight variant of Frodo [10] using the "lossy mode" of the LWE problem, where a special distribution for a matrix A is computationally indistinguishable from the uniform distribution. That is, $(A, A^T \mathbf{s} + \mathbf{e}) = (A, \mathbf{b})$ leaks almost no information of the secret $\mathbf{s}$, if A samples from "lossy mode". The ciphertext of our modified Frodo is $c_1 = \mathbf{b}^T \mathbf{s}' + m\lceil q/2 \rceil$ and $c_2 = A^T \mathbf{s}' + \mathbf{e}'$. Note that, unlike Frodo, no additional error is added in c_1 due to the min-entropy of $\mathbf{s}'$. Here, the vector $\mathbf{b}^T$ can be viewed as a hash function, so that the leftover hash lemma is used to argue the entropy of the ciphertexts. To deal with multiple challenge ciphertexts, we need to extend the leftover hash lemma and the smoothed-variant of the leftover hash lemma [20] to the multi-challenge setting.

2   Preliminaries

2.1   Notation

We denote $[n]$ as the set $\{1, 2, \ldots, n\}$, where $n \in N$. For a distribution X, $x \leftarrow X$ denotes sampling x according to X, and $\mathbf{x} \leftarrow X$ denotes sampling independently t times according to X, where $\mathbf{x}$ is a t-dimensional vector. We write $s \leftarrow \mathcal{S}$ to denote uniformly choosing s from the set $\mathcal{S}$. $\mathbf{s} \leftarrow \mathcal{S}$ is defined in a similar way, where $\mathbf{s}$ is a vector. $|\mathcal{S}|$ denotes the number of elements in the set $\mathcal{S}$. The statistical distance $\Delta(X, Y)$ between two random variables X and Y, is defined as $\Delta(X, Y) = \frac{1}{2} \sum_s |\Pr[X = s] - \Pr[Y = s]|$. $\mathcal{H}$ denotes the set of all the functions H which maps $\mathcal{X}$ to $\mathcal{Y}$. Denote $||\mathbf{v}||_1$ and $||\mathbf{v}||_2$ or just $||\mathbf{v}||$ as the l_1-norm and l_2-norm for a vector $\mathbf{v} \in R^n$. A^T is the transposition of matrix A. We write $f(\mathbf{v})$ as the vector $(f(v_1), f(v_2), \ldots, f(v_t))^T$, where f is a function and $\mathbf{v} = (v_1, v_2, \ldots, v_t)^T$.

2.2   Quantum Random Oracle

We consider quantum adversaries which are given quantum access to the random oracles. For a oracle function $H : \{0,1\}^n \to \{0,1\}^m$, the quantum access to H is modeled by $O_H : |x\rangle|y\rangle \to |x\rangle|y \oplus H(x)\rangle$, where $x \in \{0,1\}^n$ and $y \in \{0,1\}^m$. The quantum adversary can access H in superposition by applying O_H. We denote $\mathcal{A}^{|H\rangle}$ that the oracle function H is quantum-accessible for $\mathcal{A}$. Next we recall some useful lemmas about quantum random oracle.

Lemma 1. *[29] For any quantum algorithm $\mathcal{A}$ making q times quantum queries to an oracle H, it cannot distinguish that H is drawn from truly random functions or $2q$-wise independent functions uniformly.*

Lemma 1 states that a quantum random function can be simulated by a $2q$-wise independent hash function, where q is the number of quantum oracle queries made by an adversary.

Another important lemma is the one-way to hiding lemma (OW2H) introduced by [27]. The OW2H lemma shows that if there exists an quantum adversary $\mathcal{A}$ making at most q queries to H that can distinguish $(x, H(x))$ from (x, y), where x and y are chosen uniformly, i.e., only one point of random oracle H is changed randomly, then we can find x by measuring one of the queries of $\mathcal{A}$. Then [3] extended the OW2H lemma to the multi-point setting, where more points are changed. That is, if $\mathcal{A}$ can distinguish $(\mathbf{x}, H(\mathbf{x}))$ and $(\mathbf{x}, \mathbf{y})$, where $\mathbf{x} = (x_1, x_2, \ldots, x_t)$ and $\mathbf{y} = (y_1, y_2, \ldots, y_t)$ are t-dimensional vectors chosen uniformly, we can find x_i for some i in $[t]$. In this paper, we adopt a modified description of the OW2H lemma in [3].

Lemma 2 Multiple One-way to Hiding (mOW2H). *[3] Let $\mathcal{H}$ be the function family of all the $H : \{0,1\}^n \to \{0,1\}^m$. Consider an oracle algorithm $\mathcal{A}$ that makes q queries to H, where $H \leftarrow \mathcal{H}$, and output a bit b'. Let $\mathcal{B}$ be an oracle algorithm that on input $\mathbf{x} \leftarrow \{0,1\}^n$ does the following: pick $i \leftarrow [q]$ and $\mathbf{y} \leftarrow \{0,1\}^m$, run $\mathcal{A}^{|H\rangle}(\mathbf{x}, \mathbf{y})$ until the i-th queries, measure the argument of the query in the computational basis , and output the measurement. (When $\mathcal{A}$ makes less than i queries, $\mathcal{B}$ outputs $\perp \notin \{0,1\}^n$.)*
Let $P_{\mathcal{A}}^1 = \Pr[b' = 1 : H \leftarrow \mathcal{H}, \mathbf{x} \leftarrow \{0,1\}^n, b' \leftarrow \mathcal{A}^{|H\rangle}(\mathbf{x}, H(\mathbf{x}))]$, $P_{\mathcal{A}}^2 = \Pr[b' = 1 : H \leftarrow \mathcal{H}, \mathbf{x} \leftarrow \{0,1\}^n, \mathbf{y} \leftarrow \{0,1\}^m, b' \leftarrow \mathcal{A}^{|H\rangle}(\mathbf{x}, \mathbf{y})]$, $P_{\mathcal{B}} = \Pr[x_1 \stackrel{?}{=} x' \lor x_2 \stackrel{?}{=} x' \lor, \ldots, \lor x_t \stackrel{?}{=} x' : H \leftarrow \mathcal{H}, \mathbf{x} \leftarrow \{0,1\}^n, x' \leftarrow \mathcal{B}^{|H\rangle}(\mathbf{x})]$, where $\mathbf{x} = (x_1, x_2, \ldots, x_t)$ and $\mathbf{y} = (y_1, y_2, \ldots, y_t)$ are t-dimensional vectors. Then $|P_{\mathcal{A}}^1 - P_{\mathcal{A}}^2| \leqslant 2q\sqrt{P_{\mathcal{B}}}$.

The proof of Lemma 2 is omitted since it is similar to that of the OW2H lemma in [3], with the exception that we do not need the assumption that all the xs are distinct. More precisely, in our proof, if x_i which is i-th choice of x equals to some x_j with $i < j$, we can set $y_i = y_j$.

Lemma 3. *[25] Let $H : \{0,1\}^n \times \mathcal{X} \to \mathcal{Y}$ and $H' : \mathcal{X} \to \mathcal{Y}$ be two independent random oracles. If an unbounded time quantum adversary $\mathcal{A}$ makes a query to H at most q_H times, then we have*

$$|\Pr[\mathcal{A}^{|H\rangle, |H(s,\cdot)\rangle}|s \leftarrow \{0,1\}^n] - \Pr[\mathcal{A}^{|H\rangle, |H'\rangle}]| \leqslant 2q_H \frac{1}{\sqrt{2^n}}.$$

Lemma 4 Generic search problem. *[4] Let $\lambda \in [0,1]$. The oracle $g : X \to \{0,1\}$, such that for each $x \in X$, $g(x)$ is distributed according to $\mathbf{B}_{\lambda_x}^2$, where $\lambda_x \leqslant \gamma$. For each x, $g'(x) = 0$. If an unbounded time quantum adversary $\mathcal{A}$ makes a query to g or g' at most q times, then we have*

$$|\Pr[b = 1|b \leftarrow \mathcal{A}^{|g\rangle}] - \Pr[b' = 1|b' \leftarrow \mathcal{A}^{|g'\rangle}]| \leqslant 2q\sqrt{\gamma}.$$

---

[2] $\mathbf{B}_\lambda$ is the Bernoulli distribution. I.e., $\Pr[g(x) = 1] = \lambda$ and $\Pr[g(x) = 0] = 1 - \lambda$.

2.3 Lattices

Lattices. An n-dimensional lattice is defined as the set of all integer combinations of n linearly independent vectors. The form of an n-dimensional lattice Λ is $\{\sum_{i=1}^{n} x_i b_i | x_i \in Z\}$ where $B = \{b_1, b_2, \ldots, b_n\}$ are n linearly independent vectors. B is called basis of the lattice Λ.

Gaussian. For a vector $\mathbf{x}$ and any $s > 0$, let $\rho_s(\mathbf{x}) = \exp(-\pi ||\mathbf{x}/s||^2)$ be a Gaussian function scaled by a factor of s. Define $\rho_s(\Lambda) = \sum_{\mathbf{x} \in \Lambda} \rho_s(\mathbf{x})$ for a lattice Λ. The discrete Gaussian probability distribution $D_{\Lambda,s}$ is

$$\forall \mathbf{x} \in \Lambda, D_{\Lambda,s} = \frac{\rho_s(\mathbf{x})}{\rho_s(\Lambda)}.$$

Lemma 5. *[25] For $\sigma \geqslant \omega(\sqrt{\log(n)})$, $\Pr_{e \leftarrow D_{Z^n,\sigma}}[||e|| > \sigma\sqrt{n}] \leqslant 2^{-n+1}$.*

LWE [23]. Let λ be a positive integer parameter, and $n = n(\lambda)$, $q = q(\lambda)$. Let χ be a distribution over Z. For $m = poly(\lambda)$ samples, the following two distributions are computationally hard to distinguish: (1) $(a_i, a_i^T \mathbf{s} + e) \in Z_q^n \times Z_q$, where $a_i, s \leftarrow Z_q^n$ and $e \leftarrow \chi$. (2) $(a_i, b_i) \in Z_q^n \times Z_q$, where $a_i \leftarrow Z_q^n$ and $b_i \leftarrow Z_q$. For any algorithm $\mathcal{A}$, we define

$$\begin{aligned}
\mathrm{Adv}_{\mathcal{A}}^{LWE_{n,m,q,\chi^m}} = \\
| \Pr[b' = 1 : A^{n \times m} \leftarrow Z_q^{n \times m}, \mathbf{s} \leftarrow Z_q^n, \mathbf{e} \leftarrow \chi^m, b' \leftarrow \mathcal{A}(A, A^T \mathbf{s} + \mathbf{e})] \quad (1) \\
- \Pr[b' = 1 : A^{n \times m} \leftarrow Z_q^{n \times m}, \mathbf{b} \leftarrow Z_q^m, b' \leftarrow \mathcal{A}(A, \mathbf{b})] |.
\end{aligned}$$

Lemma 6. *[23] Let n and prime q be positive integers, and let $m \geqslant 2n \log q$. Then for all but a q^{-n} fraction of all $A \in Z_q^{n \times m}$, the distribution of $u = A^T r$ mod q is statistically close to uniform over Z_q^n, where $r \leftarrow \{0,1\}^m$.*

2.4 Cryptographic Primitives

Public-Key Encryption

Definition 1. *A public key encryption (PKE) consists of three algorithms Gen, Enc and Dec:*

- *Gen(1^λ) takes as input a security parameter 1^λ and outputs a public/secret key pair (pk, sk).*
- *Enc(pk, m) takes as input a public key pk and a message m in the message space $\mathcal{M}$, and outputs a ciphertext c.*
- *Dec(pk, sk, c) takes as input a secret key sk and a ciphertext c, and outputs a message m or a special failure symbol "$\perp \notin \mathcal{M}$".*

Definition 2. *A $PKE = (Gen, Enc, Dec)$ scheme has δ-correctness if*

$$E[\max_{m \in \mathcal{M}} \Pr[Dec(sk, c) \neq m : c \leftarrow Enc(pk, m)]] \leqslant \delta,$$

where the expectation is taken over $(pk, sk) \leftarrow Gen(1^\lambda)$.

Define multiple challenges one-way CPA (mOW-CPA) game and multiple challenges IND-CPA (mIND-CPA) game for a PKE in Figs. 1 and 2, respectively. Single challenge games are very similar to them, except that there is only one challenge ciphertext.

$\mathbf{Game}_{PKE,\mathcal{A}}^{\mathrm{mOW\text{-}CPA}}$

$(pk, sk) \leftarrow Gen(1^\lambda)$
$\mathbf{m}^* \leftarrow \mathcal{M}$
$\mathbf{c}^* \leftarrow Enc(pk, \mathbf{m}^*)$
$m \leftarrow \mathcal{A}(pk, \mathbf{c}^*)$
return $(m \overset{?}{=} m_1^* \vee m \overset{?}{=} m_2^* \vee, \ldots, \vee m \overset{?}{=} m_t^*)$

Fig. 1. mOW-CPA game for PKE.

$\mathbf{Game}_{PKE,\mathcal{A}}^{\mathrm{mIND\text{-}CPA}}$

$(pk, sk) \leftarrow Gen(1^\lambda)$
$(st, \mathbf{m}_1, \mathbf{m}_2) \leftarrow \mathcal{A}_1(pk)$
$b \leftarrow \{0, 1\}$
$\mathbf{c}^* \leftarrow Enc(pk, \mathbf{m}_b)$
$b' \leftarrow \mathcal{A}_2(st, pk, \mathbf{c}^*)$
return $(b' \overset{?}{=} b)$

Fig. 2. mIND-CPA game for PKE.

Definition 3. *A PKE is said to be mOW-CPA secure if* $\Pr[\mathbf{Game}_{PKE,\mathcal{A}}^{mOW\text{-}CPA} = 1] \leq \varepsilon$ *for any probabilistic polynomial time (PPT) adversary, where ε is a negligible function in security parameter λ.*

The mIND-CPA security of PKE can be defined in a similar way.

Key Encapsulation Mechanism

Definition 4. *A key encapsulation mechanism (KEM) consists of three algorithms Gen, Encaps and Decaps:*

- *Gen(1^λ) takes as input a security parameter 1^λ and outputs a public/secret key pair (pk, sk).*
- *Encaps(pk) takes as input a public key pk and outputs a ciphertext c and an ephemeral key $k \in \mathcal{K}$, where $\mathcal{K}$ denotes the KEM's key space.*
- *Decaps(pk, sk, c) takes as input a public key pk, a secret key sk and a ciphertext c, and outputs the ephemeral key k or a special failure symbol "$\perp \notin \mathcal{K}$".*

The multiple challenges mIND-CCA game for a KEM is defined in Fig. 3.

Definition 5. *A KEM is said to be mIND-CCA secure if* $|\Pr[\mathbf{Game}_{KEM,\mathcal{A}}^{mIND\text{-}CCA} = 1] - \frac{1}{2}| \leq \varepsilon$ *for any PPT adversary, where ε is a negligible function in security parameter λ.*

3   Generic Conversion in the Multi-challenge Setting

In this section, we show the security of the $FO^\perp$ transformation for multiple challenges case in QROM. The $FO^\perp$ transformation is presented in Fig. 4. The key space and ciphertext space of the KEM are $\mathcal{K}$ and $\mathcal{C}$, respectively. The message space and random number space of the underlying PKE scheme are $\mathcal{M}$ and $\mathcal{R}$, respectively. $G: \mathcal{M} \to \mathcal{R}$ and $H: \mathcal{M} \times \mathcal{C} \to \mathcal{K}$ are hash functions. $\mathcal{G}$ and $\mathcal{H}$ denote the sets of all Gs and Hs, respectively.

$\text{Game}_{KEM,\mathcal{A}}^{\text{mIND-CCA}}$	$Dec(c)$
$(pk, sk) \leftarrow Gen(1^\lambda)$	if $c = c^* (c \in (c_1^*, \ldots, c_t^*))$
$b \leftarrow \{0, 1\}$	return $\perp$
$(\mathbf{c}^*, \mathbf{K}_0^*) \leftarrow Encaps(pk)$	else $K = Decaps(sk, c)$
$\mathbf{K}_1^* \leftarrow \mathcal{K}$	return K
$b' \leftarrow \mathcal{A}^{Decaps()}(pk, (\mathbf{c}^*, \mathbf{K}_b^*))$	
return $(b' \overset{?}{=} b)$	

Fig. 3. mIND-CPA game for KEM.

$KEM.Gen(1^\lambda)$	$Encaps(pk)$	$Decaps(pk, (sk, s), c)$
$(pk, sk) \leftarrow Gen(1^\lambda)$	$m \leftarrow \mathcal{M}$	$m = Dec(sk, c)$
$s \leftarrow \mathcal{M}$	$c = Enc(pk, m; G(m))$	if $Enc(pk, m; G(m)) = c$
return $((sk, s), pk)$	$K = H(m, c)$	return $K = H(m, c)$
	return (K, c)	else return $K = H(s, c)$

Fig. 4. $FO^\perp[PKE, G, H]$.

Theorem 1. *Suppose PKE with δ-correct is mOW-CPA secure. For any quantum adversary $\mathcal{A}$ against mIND-CCA security, there exists a mOW-CPA adversary $\mathcal{B}$ against PKE , such that*

$$Adv_{KEM,\mathcal{A}}^{\text{mIND-CCA}} \leqslant 2q_H \frac{1}{\sqrt{\mathcal{M}}} + 4q_G\sqrt{\delta} + 2(q_G + q_H)\sqrt{Adv_{PKE,\mathcal{B}}^{\text{mOW-CPA}}},$$

where q_G and q_H are the number of the quantum queries to random oracle G and H, q_D is the classical queries to Decaps oracle.

Proof. The proof is analogous to that of [19] and [25] except that we need to rely on mOW2H lemma. In the proof, random oracle can be simulated by 2q-wise random function according to Lemma 1. We prove the theorem via the following sequence of games, which is also shown in Fig. 5.

GAME G_0. This is the original $\text{Game}_{PKE,\mathcal{A}}^{\text{mIND-CCA}}$.

GAME G_1. $H(s, c)$ in *Decaps* oracle is replaced with $\hat{H}_1$, where $\hat{H}_1 \leftarrow \hat{\mathcal{H}}$ and $\hat{\mathcal{H}}$ denotes the function space of $\hat{H} : \mathcal{C} \rightarrow \mathcal{K}$. Since s is uniformly chosen from message space $\mathcal{M}$, we have

$$|\Pr[G_0 = 1] - \Pr[G_1 = 1]| \leqslant 2q_H \frac{1}{\sqrt{\mathcal{M}}}.$$

due to Lemma 3.

GAME G_2. We change the way that H responds to the adversary $\mathcal{A}$. If $\mathcal{A}$ queries on a valid pair (m, c), i.e. $c = Enc(pk, m; G(m))$. Then response is replaced by $\hat{H}_2(c)$, where $\hat{H}_2 \leftarrow \hat{\mathcal{H}}$.

For a tuple (pk, sk, m), we the following notations:

GAMES $G_0 - G_4$

$(pk, sk, s) \leftarrow Gen; G \leftarrow \mathcal{G}; \hat{H}_1, \hat{H}_2 \leftarrow \hat{\mathcal{H}}; H \leftarrow \mathcal{H}$

$\mathbf{m}^* \leftarrow \mathcal{M};$

$\mathbf{r}^* = G(\mathbf{m}^*) \;//G_0 - G_3$

$\mathbf{r}^* \leftarrow \mathcal{R} \;//G_4$

$\mathbf{c}^* = Enc(pk, \mathbf{m}^*; \mathbf{r}^*); \mathbf{K}_0^* = H(\mathbf{m}^*, \mathbf{c}^*)$

$\mathbf{K}_0^* \leftarrow \mathcal{K}// \; G_4$

$\mathbf{K}_1^* \leftarrow \mathcal{K}$

$b \leftarrow \{0, 1\}$

$b' \leftarrow \mathcal{A}^{|H\rangle, |G\rangle, Dec}(pk, \mathbf{c}^*, \mathbf{K}_b^*)$

return $(b' \overset{?}{=} b)$

$H(m, c)$	$Decaps(c) \;//G_0 - G_2$	$Decaps(c) \;//G_3 - G_4$
if $Enc(pk, m; G(m)) = c \;//G_2 - G_4$	$m = Dec(sk, c)$	return $K = \hat{H}_2(c)$
return $K = \hat{H}_2(c) \;//G_2 - G_4$	if $Enc(pk, m; G(m)) = c$	
else return $H(m, c)$	return $K = H(m, c)$	
	else return	
	$K = H(s, c) \;// \; G_0$	
	$K = \hat{H}_1(c) \;// \; G_1 - G_2$	

Fig. 5. Games G_0–G_4 for the proof of Theorem 1.

– The set of bad randomness

$$R_{bad}(pk, sk, m) = \{r \in \mathcal{R} | Dec(sk, Enc(pk, m; r)) \neq m\}.$$

– The set of good randomness

$$R_{good}(pk, sk, m) = \{r \in \mathcal{R} | Dec(sk, Enc(pk, m; r)) = m\}.$$

– The error rate of randomness for a tuple (pk, sk, m)

$$\delta(pk, sk, m) = \frac{|R_{bad}(pk, sk, m)|}{|\mathcal{R}|}.$$

– The error rate of randomness for a pair (pk, sk)

$$\delta(pk, sk) = \max_{m \in \mathcal{M}} \delta(pk, sk, m).$$

According to Definition 2, we have $\delta = \mathbf{E}[\delta(pk, sk)]$ where the expectation is taken over $(pk, sk) \leftarrow Gen(1^\lambda)$.

We denote $DEnc(\cdot) = Enc(pk, \cdot; G(\cdot))$. Since $DEnc(\cdot)$ is an injective function when there is no bad random numbers, we can see that $\hat{H}_2 \circ DEnc(m)$ and $H(m, c)$ have the same output distribution when input a valid pair (m, c). I.e., given (pk, sk), in **GAME** G_1, the random oracle $G(\cdot)$ outputs bad random numbers with probability $\delta(pk, sk, m)$, and in **GAME** G_2, the random oracle $G'(\cdot)$ always output good random numbers. Define $\hat{G}(\cdot)$ may be $G(\cdot)$ in GAME

$\mathcal{B}^{\|\hat{g}\rangle}, \hat{g} : \mathcal{M} \to \mathcal{R}$	$\hat{G}(m)$
Pick $2q_G$-wise function f	if $\hat{g}(m) = 0$
$(pk, sk) \leftarrow Gen(1^\lambda)$	$\hat{G}(m) = Sample(R_{good}(pk, sk, m); f(m))$
$\mathbf{m}^* \leftarrow \mathcal{M}$	if $\hat{g}(m) = 1$
$\mathbf{r}^* = \hat{G}(\mathbf{m}^*)$	$\hat{G}(m) = Sample(R_{bad}(pk, sk, m); f(m))$
$\mathbf{c}^* = Enc(pk, \mathbf{m}^*; \mathbf{r}^*)$	return $\hat{G}(m)$
$\mathbf{K}_0^* = H(\mathbf{m}^*, \mathbf{c}^*)$	
$\mathbf{K}_1^* \leftarrow \mathcal{K}$	
$b \leftarrow \{0, 1\}$	
$b' \leftarrow \mathcal{A}^{\|H\rangle, \|\hat{G}\rangle, Dec}(pk, \mathbf{c}^*, \mathbf{K}_b^*)$	
return b'	

Fig. 6. $\mathcal{B}^{\|\hat{g}\rangle}$ the proof of Theorem 1.

G_1 or $G'(\cdot)$ in GAME G_2. If for a quantum adversary $\mathcal{A}$ given $\hat{G}(\cdot)$ which can distinguish **GAME** G_1 and **GAME** G_2, there exists a adversary $\mathcal{B}$ which can solve the generic search problem in Lemma 4. The construction of $\mathcal{B}$ is in Fig. 6. The $\mathcal{B}$ is given $\hat{g}$ which may be g or g' in the generic search problem, and constructs $\hat{G}(m)$ for $\mathcal{A}$ as following: The probabilistic algorithm $Sample(\cdot; \cdot)$ can uniformly choose good and bad random numbers, and the randomness space of the $Sample(\cdot; \cdot)$ is $\mathcal{R}_f$. $Sample(\cdot; f(m))$ denotes the deterministic execution of $Sample(\cdot; \cdot)$ using explicitly given randomness $f(m)$, where $f : \mathcal{M} \to \mathcal{R}_f$.

For fixed pair (pk, sk), we have

$$| \Pr[1 \leftarrow \mathcal{B}^{\|g\rangle}|(pk, sk)] - \Pr[1 \leftarrow \mathcal{B}^{\|g'\rangle}|(pk, sk)]|$$
$$= || \Pr[1 \leftarrow \mathcal{A}^{\|G\rangle}|(pk, sk)] - \Pr[1 \leftarrow \mathcal{A}^{\|G'\rangle}|(pk, sk)]|| \qquad (2)$$
$$\leqslant 2q_G \sqrt{\delta(pk, sk)}.$$

By averaging over $(pk, sk) \leftarrow Gen(1^\lambda)$, we have

$$| \Pr[G_1 = 1] - \Pr[G_2 = 1]| \leqslant 2q_G \sqrt{\delta}.$$

GAME G_3. In the G_3, the Dec oracle does not use (sk, s) any more. When $\mathcal{A}$ queries the $Decaps$ oracle on c s.t. $c \notin \{c_1^*, c_2^*, \ldots, c_t^*\}$, it receives $K = \hat{H}_2(c)$. We call a ciphertext c valid if $Enc(pk, Dec(sk, c); G(Dec(sk, c))) = c$, and invalid otherwise, and consider the following two cases.

- **Case 1:** When $\mathcal{A}$ asks a valid c , both Dec oracle in G_2 and G_3 return $\hat{H}_2(c)$.
- **Case 2:** When $\mathcal{A}$ asks a invalid c, $\hat{H}_1(c)$ and $\hat{H}_2(c)$ are returned in G_2 and G_3 respectively. In G_2, the $\hat{H}_1(c)$ is a random function independent of G and H in $\mathcal{A}$'view. In G_3, $\mathcal{A}$ can also access to $\hat{H}_2(c)$ by querying H on a pair (m, c), where $Enc(pk, m; G(m)) = c$. If $\mathcal{A}$ cannot find such m, then $\hat{H}_2(c)$ is random number in $\mathcal{A}$' view. If $\mathcal{A}$ finds such m, then it find a bad random number $G(m)$ for m.

We can see that if $G(m)$ is not a bad random number for each m, the G_2 and G_3 are identical. Using the same analysis in **GAME** G_2, we have

$$|\Pr[G_2 = 1] - \Pr[G_3 = 1]| \leqslant 2q_G\sqrt{\delta}.$$

GAME G_4. In G_4, $\mathbf{r}$ and $\mathbf{K_0}$ are chosen uniformly and independently from $\mathcal{R}$ and $\mathcal{K}$, respectively. Therefore, $\mathcal{A}$ cannot get any information about the bit b. Hence,

$$\Pr[G_4 = 1] = \frac{1}{2}.$$

GAME G_5. In G_5, Lemma 2 is used to bound $|\Pr[G_3 = 1] - \Pr[G_4 = 1]|$. Due to $c_i^* = Enc(pk, m_i^*; G(m_i^*))$ for each (m_i^*, c_i^*) pair, we have $H(m_i^*, c_i^*) = \hat{H}_1(c_i^*)$, where $i \in [t]$. We construct the random oracle $W(m) = (G(m), \hat{H}_1(c))$: $\mathcal{M} \to \mathcal{R} \times \mathcal{K}$, where $c = Enc(pk, m; G(m))$. Then we can write the adversary $\mathcal{A}$ in G_3 or G_4 as $\mathcal{A}^{|W\rangle, Dec}(pk, \mathbf{m}^*, (\mathbf{r}^*, \mathbf{K}_b^*))$, where $b \leftarrow \{0,1\}$. Here, we do not write the $H(\cdot, \cdot)$[3], because it is independent to W and unhelpful for $\mathcal{A}$ to distinguish $(\mathbf{m}^*, W(\mathbf{m}^*))$ from $(\mathbf{m}^*, (\mathbf{r}^*, \mathbf{K}_0^*))$, where $\mathbf{m}^*$, $\mathbf{r}^*$ and $\mathbf{K}_0^*$ are chosen uniformly. Consider the game G_5 shown in Fig. 7. It is obvious that if $\mathcal{A}^{|W\rangle, Decaps}$ takes as input $(pk, \mathbf{m}^*, W(\mathbf{m}^*))$, G_5 perfectly simulates the environment in G_3, and if $\mathcal{A}^{|W\rangle, Decaps}$ takes as input $(pk, \mathbf{m}^*, (\mathbf{r}^*, \mathbf{K}_0^*))$, G_5 perfectly simulates the environment in G_4, where $\mathbf{m}^*$, $\mathbf{r}^*$ and $\mathbf{K}_0^*$ are chosen uniformly.

GAME G_5

$i \leftarrow [q_G + q_H], (pk, sk, s) \leftarrow Gen$

$H \leftarrow \mathcal{H}, G \leftarrow \mathcal{G}, \hat{H}_1 \leftarrow \hat{\mathcal{H}}_1$

$\mathbf{m}^* \leftarrow \mathcal{M}, \mathbf{r}^* \leftarrow \mathcal{R}$

$\mathbf{c}^* = Enc(pk, \mathbf{m}^*; \mathbf{r}^*)$

$(\mathbf{K}_0^*, \mathbf{K}_1^*) \leftarrow \mathcal{K}, b \leftarrow \{0,1\}$

run $\mathcal{A}^{|W\rangle, Decaps}(pk, \mathbf{m}^*, (\mathbf{r}^*, \mathbf{K}_b^*))$ until the i-th query to W

measure the argument m

return $(m \overset{?}{=} m_1^* \vee m \overset{?}{=} m_2^*, \ldots, m \overset{?}{=} m_t^*)$

Fig. 7. $\mathcal{A}^{|W\rangle, Dec}$ for the proof of Theorem 1.

By applying Lemma 2, we have

$$|\Pr[G_3 = 1] - \Pr[G_4 = 1]| \leqslant 2(q_G + q_H)\sqrt{\Pr[G_5 = 1]}.$$

Then we can use the adversary $\mathcal{A}$ in G_5 to construct an adversary $\mathcal{B}$ against the mOW-CPA security of the underlying PKE scheme. $\mathcal{B}$ on input $(pk, \mathbf{c}^*)$ executes as G_5 except running $\mathcal{A}^{|W\rangle, Decaps}$ on input the challenges ciphertexts in the mOW-CPA game. Finally, $\mathcal{B}$ outputs the measurement of the argument m. So we get $\Pr[G_5 = 1] \leqslant \Pr[\mathbf{Game}_{PKE}^{mOW\text{-}CPA} = 1]$. In summary, we have

$$\mathrm{Adv}_{KEM,\mathcal{A}}^{mIND\text{-}CCA} \leqslant 2q_H \frac{1}{\sqrt{\mathcal{M}}} + 4q_G\sqrt{\delta} + 2(q_G + q_H)\sqrt{\mathrm{Adv}_{PKE,\mathcal{B}}^{mOW\text{-}CPA}}.$$

---

[3] The proof of OW2H with redundant oracle can be found in Lemma 3 of [19].

4   Tight Security Proofs for mOW-CPA PKE

In this section, we identify two PKE schemes with tight mOW-CPA security.

4.1   Tight Security of Regev's PKE Scheme

We recall the Regev's PKE scheme [23].

- $Gen(1^\lambda)$: On input security parameter λ, choose a prime q, positive integers n, m, where $m \geqslant 2(n + 1) \log q$, and Gaussian parameters σ. Then choose a matrix $A \in Z_q^{n \times m}$ and a secret $s \in Z_q^n$ randomly, and output the public key $pk = A' = \begin{bmatrix} A \\ \mathbf{b}^T \end{bmatrix}$, where $\mathbf{b} = A^T\mathbf{s} + \mathbf{e}$ and $\mathbf{e} \leftarrow D_{Z^m,\sigma}$, and the secret key $sk = (A, \mathbf{b}, \mathbf{s})$.
- $Enc(pk, m)$: To encrypt a message $m \in \{0,1\}$, one uniformly chooses $r \in \{0,1\}^m$ and outputs the ciphertext

$$c = A'r + (\mathbf{0}, m\lfloor q/2 \rfloor)^T \in Z_q^{n+1}.$$

- $Dec(sk, c)$: To decrypt a ciphertext c with a secret key sk, compute $\omega = (-\mathbf{s}^T, 1) \cdot c$ and outputs 0 if ω is closer to 0 than to $\lceil q/2 \rceil$ modulo q. Otherwise it outputs 1.

Correctness. According to Dec algorithm, we have

$$\begin{aligned} (-\mathbf{s}^T, 1) \cdot c &= (-\mathbf{s}^T, 1) \cdot A' \cdot r + m\lfloor q/2 \rfloor \\ &= \mathbf{e}^T \cdot r + m\lfloor q/2 \rfloor \qquad\qquad (3) \\ &\approx m\lfloor q/2 \rfloor. \end{aligned}$$

In order to make the absolute value less than $q/4$, it suffices to choose appropriate σ and q.

For the longer messages, we can just encrypt each bit of the message using the original Regev's PKE. Note that [22] provides another efficient variant of Regev's PKE for long message. In this paper, we only consider the first method of Regev's PKE for long messages.

Theorem 2. *The Regev's PKE scheme is mOW-CPA secure assuming the hardness of $LWE_{n,m,q,D_{Z^m,\sigma}}$. Namely, for any quantum adversary $\mathcal{A}$ against mOW-CPA security, there exists an adversary B against $LWE_{n,m,q,D_{Z^m,\sigma}}$, such that*

$$\Pr[\mathbf{Game}_{PKE,\mathcal{A}}^{mOW\text{-}CPA} = 1] \leq Adv_B^{LWE_{n,m,q,D_{Z^m,\sigma}}} + q^{-(n+1)} + lt\varepsilon + \frac{t}{2^l},$$

where l is the length of the each message, t is the number of the challenge ciphertexts and ε is a negligible function.

Proof. **GAME** G_0. G_0 is the original mOW-CPA game. The challenger $\mathcal{C}$ runs $Gen(1^\lambda)$ to get the $pk = A' = \begin{bmatrix} A \\ b^T \end{bmatrix}$. Then $\mathcal{C}$ randomly and independently chooses t messages $m_1, m_2, \ldots, m_t$, where the length of each message is l bits. The challenger runs $Enc(pk, m_i)$ to get the corresponding ciphertexts. For each bit b_{ij}, $Enc(pk, b_{ij}) = c_{ij} = A'r + (0, b_{ij}\lfloor q/2 \rfloor)$, where $i \in [t]$, $j \in [l]$ and $r \leftarrow \{0, 1\}^m$. Finally, $\mathcal{C}$ sends the pk and ciphertexts to the adversary $\mathcal{A}$. At the end of the game, $\mathcal{A}$ outputs m. We say $\mathcal{A}$ wins the games if $m = m_i$ for some $i \in [t]$. That is, $\Pr[G_0 = 1] = \Pr[\mathbf{Game}_{PKE,\mathcal{A}}^{\text{mOW-CPA}} = 1]$.

GAME G_1. G_1 is similar to G_0 with the exception that the way of generating **b**. The challenger randomly chooses $\mathbf{b} \leftarrow Z_q^n$. Now we prove that G_0 and G_1 are computationally indistinguishable. Suppose there exists an adversary $\mathcal{A}$ which can distinguish G_0 and G_1. We show how to construct an algorithm $\mathcal{B}$ which can distinguish uniform $(A, \mathbf{b})$ and LWE samples. $\mathcal{B}$ is given $(A, \mathbf{b})$ and sends it to $\mathcal{A}$ as the pk. Then $\mathcal{B}$ generates challenge ciphertexts by the way as in G_0. $\mathcal{B}$ outputs 1 if $\mathcal{A}$ wins the game and outputs 0 otherwise. It's easy to see that if $(A, \mathbf{b}) \leftarrow LWE_{n,m,q,D_{Z^m},\sigma}$ the view of the adversary is identical to that of G_0, and if $(A, \mathbf{b}) \leftarrow Z_q^{n \times m} \times Z_q^m$ the view of the adversary is identical to that of G_1. So we have $|P[G_0 = 1] - P[G_1] = 1| \leqslant \mathrm{Adv}_{\mathcal{B}}^{LWE_{n,m,q,D_{Z^n,\sigma}}}$.

If $(A, \mathbf{b}) \leftarrow Z_q^{n \times m} \times Z_q^m$, the distribution of $A'r$ is statistically ε-close to uniform distribution over Z_q^{n+1}, for all but a $q^{-(n+1)}$ fraction of all $A' \in Z_q^{(n+1) \times m}$ by Lemma 6. Therefore, for any of $A'r$ to encrypt the messages are not random numbers with probability $q^{-(n+1)} + lt\varepsilon$. If all $A'r$ are random numbers, the probability of $\mathcal{A}$ guessing right one of the messages is $\frac{1}{2^l}$. So any adversary $\mathcal{A}$ wins the game with probability at most $q^{-(n+1)} + lt\varepsilon + \frac{t}{2^l}$. We can see

$$\Pr[G_1 = 1] \leqslant q^{-(n+1)} + lt\varepsilon + \frac{t}{2^l}.$$

Finally, we can conclude that

$$\Pr[\mathbf{Game}_{PKE,\mathcal{A}}^{\text{mOW-CPA}} = 1] \leq \mathrm{Adv}_{\mathcal{B}}^{LWE_{n,m,q,D_{Z^m,\sigma}}} + q^{-(n+1)} + lt\varepsilon + \frac{t}{2^l}.$$

4.2   A Variant of Frodo

In this section, we show the mOW-CPA security of a variant of Frodo. Firstly, we review some important results on information theory and lossy mode LWE.

Randomness Extraction. We recall the definition of the min-entropy of a random variable X, and a similar notion called the average min-entropy, as introduced by [12].

Definition 6. *The min-entropy of a random variable X is defined as $H_\infty(X) = -\log(\max_x \Pr[X = x])$. The average min-entropy is defined as $\hat{H}_\infty(X|I) = -\log(E_{i \leftarrow I}[2^{-H_\infty(X|I=i)}])$. The average min-entropy corresponds to the optimal probability of guessing X, given knowledge of I.*

ε-smooth min-entropy introduced by [24] considers all distributions that are ε-close to X, which has highest entropy. The ε-smooth average min-entropy can be defined similarly.

Definition 7. *The ε-smooth min-entropy of a random variable X is defined as: $H_{\infty}^{\varepsilon}(X) = \max_{Y:\triangle(X,Y)\leqslant\varepsilon} H_{\infty}(Y)$, and the ε-smooth average min-entropy is defined as: $\hat{H}_{\infty}^{\varepsilon}(X|I) = \max_{(Y,J):\triangle((X,I),(Y,J))\leqslant\varepsilon} H_{\infty}^{\varepsilon}(Y|J)$.*

Definition 8. *(2-Universal Hash Functions.) A family of functions $\mathcal{H} = \{H : \mathcal{X} \rightarrow [m]\}$ is called a family of 2-universal hash functions, if $\forall x, x' \in \mathcal{X}$ and $x \neq x', \Pr_{H\leftarrow\mathcal{H}}[H(x) = H(x')] \leqslant \frac{1}{m}$.*

Fact 1. Let $q > 2$, and $\mathcal{H} = \{H : Z_q^n \rightarrow Z_q\}$ be a family of hash functions. $H(s)$ is defined as $H(s) = h^T s \mod q$, where $h \in Z_q^n$. Then, $\mathcal{H}$ is a family of 2-universal hash functions.

The leftover hash lemma shows that good randomness can be extracted from a random variable using universal hash functions. Here, we consider whether universal hash functions can be used to extract good randomness from more independent random variables. Next, we prove the multi-leftover hash lemma where the hash function is used in multiple times.

Lemma 7. *(Multi-Leftover Hash Lemma.) Let $X_1, X_2, \ldots, X_t$ be some independent random variables over $\mathcal{X}$. Let $\mathcal{H} = \{H : \mathcal{D} \times \mathcal{X} \rightarrow \{0,1\}^l\}$ be a family of 2-universal hash functions, where $|\mathcal{D}|$ is 2^d and $|\mathcal{X}|$ is 2^e. If for all the random variables X_n, and $n \in [t]$, we have $H_{\infty}(X_n) \geq k$, where $k = l + 2\log\frac{\sqrt{t}}{\varepsilon}$. Then for $U_{\mathcal{D}}$ which is uniform distribution over $\mathcal{D}$ and independent of any X_n,*

$$\triangle((U_{\mathcal{D}}, H(X_1), H(X_2), \ldots, H(X_t)), (U_{\mathcal{D}}, U_{t\times l})) \leq \varepsilon/2,$$

where $U_{t\times l}$ is uniform distribution over $\{0,1\}^{t\times l}$ and independent of $U_{\mathcal{H}}$. That is,

$$\triangle((U_{\mathcal{D}}, H(X_1), H(X_2), \ldots, H(X_t)), (U_{\mathcal{D}}, U_{t\times l})) \leq \frac{1}{2}\sqrt{t2^{l-k}}.$$

The proof of Lemma 7 is similar to that of the original leftover hash lemma and thus omitted.

The leftover hash lemma generalized in [12] and [20] shows a lower bound for average min-entropy or smooth average min-entropy of a random variable, and hash functions can still extract random numbers from the random variable. We consider the case that universal hash functions extract random numbers from more independent random variables given a lower bound for average min-entropy or smooth average min-entropy of those random variables.

Lemma 8. *Let $X_1, X_2, \ldots, X_t$ be independent random variables over $\mathcal{X}$. Let $\mathcal{H}$ be a family of 2-universal hash functions and $H : \mathcal{D} \times \mathcal{X} \rightarrow \{0,1\}^l$, where $|\mathcal{D}|$ is 2^d and $|\mathcal{X}|$ is 2^e. Suppose for a random variables I and all the $n \in [t]$, we have $\hat{H}_{\infty}(X_n|I) \geq k$, where $k = l + 2\log\frac{\sqrt{t}}{\varepsilon}$. Then we have*

$$\triangle((I, U_{\mathcal{D}}, H(X_1), H(X_2), \ldots, H(X_t)), (I, U_{\mathcal{D}}, U_{t\times l})) \leq \varepsilon/2,$$

where $U_\mathcal{D}$ is uniform distribution over $\mathcal{D}$ and $U_{t \times l}$ is uniform distribution over $\{0,1\}^{t \times l}$.

The proof of Lemma 8 is similar to that of Lemma 2.4 in [12] and hence omitted.

Lemma 9. *Let* Y, J_1 *and* J_2 *be random variables. If* $\Delta(J_1, J_2) \leq \varepsilon/2$, *then* $|\hat{H}_\infty(Y|J_1) - \hat{H}_\infty(Y|J_2)| \leq \varepsilon 2^{\hat{H}_\infty(Y|J_1)} + 1$.

The details of the proof can be found in Appendix A.

Lemma 10. *Let* $X_1, X_2, \ldots, X_t$ *be independent random variables over* $\mathcal{X}$. *Let* $\mathcal{H}$ *be a family of 2-universal hash functions and* $H : \mathcal{D} \times \mathcal{X} \to \{0,1\}^l$, *where* $|\mathcal{D}|$ *is* 2^d *and* $|\mathcal{X}|$ *is* 2^e. *Suppose for a random variables* I *and all the* $n \in [t]$, *we have* $\hat{H}^\varepsilon_\infty(X_n|I) \geq k$, *where* $k = l + 2\log \frac{\sqrt{t}}{\varepsilon'} + 2$ *and* $\varepsilon = 2^{-k}$. *Then we have*

$$\Delta((I, U_\mathcal{D}, H(X_1), H(X_2), \ldots, H(X_t)), (I, U_\mathcal{D}, U_{t \times l})) \leq \varepsilon'/2 + (t+1)\varepsilon,$$

where $U_\mathcal{D}$ *is uniform distribution over* $\mathcal{D}$ *and* $U_{t \times l}$ *is uniform distribution over* $\{0,1\}^{t \times l}$.

Proof. Let $\Delta(Y_n, J_n) \leq \varepsilon$ be the random variables such that

$$\max_{\Delta((X_n, I), (Y_n, J_n)) \leq \varepsilon} \hat{H}_\infty(Y_n|J_n).$$

I.e., $\hat{H}_\infty(Y_n|J_n) = \hat{H}^\varepsilon_\infty(X_n|I)$ where $n \in [t]$.

Without loss of generality, we will show that $\hat{H}_\infty(Y_n|J_1) \geq k - 2$ for all the $n \in [t]$. First, we have $\Delta(J_1, J_n) \leq \Delta(J_1, I) + \Delta(I, J_n) \leq 2\varepsilon$. If $\hat{H}_\infty(Y_n|J_1) < k - 2$, $\hat{H}_\infty(Y_n|J_n) - \hat{H}_\infty(Y_n|J_1) \leq 4\varepsilon 2^{\hat{H}_\infty(Y_n|J_1)} + 1 \leq 2$ by Lemma 9. That is,

$$\hat{H}_\infty(Y_n|J_1) \geq \hat{H}_\infty(Y_n|J_n) - 2 \geq k - 2,$$

which contradicts our assumption. According to the Lemma 8, we know that

$$\Delta((J_1, U_\mathcal{D}, H(Y_1), H(Y_2), \ldots, H(Y_t)), (J_1, U_\mathcal{D}, U_{t \times l})) \leq \varepsilon'/2.$$

Since $\Delta((X_n, I), (Y_n, J_n)) \leq \varepsilon$ for all the $n \in [t]$, we have

$$\Delta((I, U_\mathcal{D}, H(X_1), H(X_2), \ldots, H(X_t)), (J_1, U_\mathcal{D}, H(Y_1), H(Y_2), \ldots, H(Y_t))) \leq t\varepsilon,$$

and $\Delta((J_1, U_\mathcal{D}, U_{t \times l}), (I, U_\mathcal{D}, U_{t \times l})) \leq \varepsilon$. Therefore,

$$
\begin{aligned}
&\Delta\left((I, U_\mathcal{D}, H(X_1), H(X_2), \ldots, H(X_t)), (I, U_\mathcal{D}, U_{t \times l})\right) \\
&\leq \Delta\left((I, U_\mathcal{D}, H(X_1), H(X_2), \ldots, H(X_t)), (J_1, U_\mathcal{D}, H(Y_1), H(Y_2), \ldots, H(Y_t))\right) \\
&\quad + \Delta((J_1, U_\mathcal{D}, H(Y_1), H(Y_2), \ldots, H(Y_t)), (J_1, U_\mathcal{D}, U_{t \times l})) \\
&\quad + \Delta((J_1, U_\mathcal{D}, U_{t \times l}), (I, U_\mathcal{D}, U_{t \times l})) \\
&\leq (t+1)\varepsilon + \varepsilon'/2.
\end{aligned}
$$

$$\tag{4}$$

Lossy Mode for LWE [15]. The LWE problem has a lossy mode, where the secret s still leaves high min-entropy given the pair $(A, A^T s + e)$, when A is chosen from lossy mode, a special kind of distribution. What is more, a matrix A chosen from lossy mode is computationally indistinguishable from a uniformly random A under the LWE assumption.

Definition 9. *[15] (SampleLossy.) Let* $\chi = \chi(\lambda)$ *be an efficiently sampled distribution over* Z_q. *The algorithm SampleLossy(n, m, w, χ) is as follow:* $B \leftarrow Z_q^{n \times w}, C \leftarrow Z_q^{w \times m}, F \leftarrow \chi^{n \times m}$ *and output* $A = BC + F$.

Lemma 11. *[15] For any quantum adversary* $\mathcal{A}$ *which can distinguish* $A_0 \leftarrow Z_q^{n \times m}$ *and* $A_1 \leftarrow SampleLossy(n, m, w, \chi)$ *, there exists adversary* $\mathcal{B}$ *against* LWE_{w,m,q,χ^m} *such that* $|\Pr[\mathcal{A}(A_0)] - \Pr[\mathcal{A}(A_1)]| \leqslant nAdv_{\mathcal{B}}^{LWE_{w,m,q,\chi^m}}$.

Although the A_0 is statistically far from A_1, it is easy to show that they are computationally indistinguishable from each other by a standard hybrid argument.

Lemma 12. *[2] Let* $w, n, m, q, \alpha, \beta, \gamma$ *be integer parameters and* χ *distribution (all parameterized by* λ*) such that* $\Pr_{x \leftarrow \chi}[|x| \geqslant \beta] \leqslant negl(\lambda)$ *and* $\alpha \geqslant \beta \gamma n m$. *Let* s *and* e *be random variables distributed according to* $U([-\gamma, \gamma]^n)$ *and* $D_{Z^m, \alpha}$. *Furthermore, let* A *be a matrix sampled by SampleLossy(n, m, w, χ). Then, for any* $\varepsilon \geqslant 2^{-\lambda}$, *we have the following:*

$$H_\infty^\varepsilon(s | A, A^T s + e) \geqslant H_\infty(s) - (w + 2\lambda) \log q - negl(\lambda). \tag{5}$$

Construction. We will show a variant of Frodo with the modification that we change the way of sampling s' and do not need additional error in the ciphertext c_1.

- $Gen(1^\lambda)$: On input security parameter λ, choose a prime q, positive integers n, m, γ, and Gaussian parameters α, σ. Then choose a matrix $A \in Z_q^{n \times m}$ randomly, and output the public key $pk = (A, \mathbf{b})$, where $\mathbf{s} \leftarrow D_{Z^m, \sigma}$, $\mathbf{e} \leftarrow D_{Z^n, \sigma}$ and $\mathbf{b} = A\mathbf{s} + \mathbf{e}$, and the secret key $sk = (A, \mathbf{b}, \mathbf{s})$.
- $Enc(pk, m)$: To encrypt a message $m \in \{0, 1\}$, sample $\mathbf{s}' \leftarrow U[-\gamma, \gamma]^n$ and $\mathbf{e}' \leftarrow D_{Z^m, \alpha}$.
  Then compute the ciphertext as $c_1 = \mathbf{b}^T \mathbf{s}' + m\lfloor q/2 \rfloor$ and $c_2 = A^T \mathbf{s}' + \mathbf{e}'$. Finally output the ciphertext $c = (c_1, c_2) \in Z_q \times Z_q^m$.
- $Dec(sk, c)$: To decrypt a ciphertext $c = (c_1, c_2)$ with a secret key sk, compute $\omega = c_1 - \mathbf{s}^T c_2 \in Z_q$ and outputs 0 if ω is closer to 0 than to $\lceil q/2 \rfloor$ modulo q. Otherwise it outputs 1.

Correctness. According to the Dec algorithm, we have

$$\begin{aligned}\omega = c_1 - \mathbf{s}^T c_2 &= \mathbf{s}^T A^T \mathbf{s}' + \mathbf{e}^T \mathbf{s}' + m\lceil q/2 \rfloor - \mathbf{s}^T A^T \mathbf{s}' - \mathbf{s}^T \mathbf{e}' \\ &= m\lceil q/2 \rfloor + \mathbf{e}^T \mathbf{s}' - \mathbf{s}^T \mathbf{e}'.\end{aligned} \tag{6}$$

The error term is bounded by

$$|\mathbf{e}^T\mathbf{s}' + \mathbf{s}^T\mathbf{e}'| \leqslant |\mathbf{e}^T\mathbf{s}'| + |\mathbf{s}^T\mathbf{e}'|$$

$$\leqslant \gamma\sum_{i=1}^{n}|e_i| + \sum_{j=1}^{m}|s_j e_j'| \tag{7}$$

$$\leqslant \gamma||\mathbf{e}||_1 + ||\mathbf{s}|| \cdot ||\mathbf{e}'||.$$

In order to make the error term less than $q/4$, it suffices to choose appropriate σ and q.

Theorem 3. *The variant of Frodo scheme is mOW-CPA secure assuming the hardness of $LWE_{n,m,q,D_{Z^n,\sigma}}$ and LWE_{w,m,q,χ^m}. Namely, for any quantum adversary $\mathcal{A}$ against mOW-CPA security, there exists an adversary $\mathcal{B}$ against $LWE_{n,m,q,D_{Z^n,\sigma}}$ and $\mathcal{B}'$ against LWE_{w,m,q,χ^m}, such that*

$$\Pr[\mathbf{Game}_{PKE,\mathcal{A}}^{mOW\text{-}CPA} = 1] \leq Adv_{\mathcal{B}}^{LWE_{n,m,q,D_{Z^n,\sigma}}} + nAdv_{\mathcal{B}'}^{LWE_{w,m,q,\chi^m}} + t/2^{\Omega(\lambda)} + t/2^l.$$

Proof. **GAME G_0.** G_0 is the original mOW-CPA game. The challenger $\mathcal{C}$ runs $Gen(1^\lambda)$ to get the $pk = (A, \mathbf{b})$ and $sk = (A, \mathbf{b}, s)$. Then $\mathcal{C}$ randomly and independently chooses t messages $m_1, m_2, \ldots, m_t$, where the length of each message is l bits. The challenger runs $Enc(pk, m_i)$ to get the corresponding ciphertexts $c_i = (c_{i1}, c_{i2})$, where $i \in [t]$. Finally, $\mathcal{C}$ sends the pk and ciphertexts to the adversary $\mathcal{A}$. At the end of the game, $\mathcal{A}$ outputs m. We say $\mathcal{A}$ wins the games if $m = m_i$ for some $i \in [t]$. That is,

$$\Pr[G_0 = 1] = \Pr[\mathbf{Game}_{PKE,\mathcal{A}}^{mOW\text{-}CPA} = 1].$$

GAME G_1. G_1 is similar to G_0 with the exception that the way of generating $\mathbf{b}$. The challenger randomly chooses $\mathbf{b} \leftarrow Z_q^n$. Now we prove that G_0 and G_1 are computationally indistinguishable. Suppose there exists an adversary $\mathcal{A}$ which can distinguish G_0 and G_1. We show how to construct an algorithm $\mathcal{B}$ which can distinguish uniform $(A, \mathbf{b})$ and LWE samples. $\mathcal{B}$ is given $(A, \mathbf{b})$ and sends it to $\mathcal{A}$ as the pk. Then $\mathcal{B}$ generates challenge ciphertexts by the way as in G_0. $\mathcal{B}$ outputs 1 if $\mathcal{A}$ wins the game and outputs 0 otherwise. It's easy to see that if $(A, \mathbf{b}) \leftarrow LWE_{n,m,q,D_{Z^m,\sigma}}$ the view of the adversary is identical to that of G_0, and if $(A, \mathbf{b}) \leftarrow Z_q^{n \times m} \times Z_q^m$ the view of the adversary is identical to that of G_1. So we have $|P[G_0 = 1] - P[G_1] = 1| \leqslant Adv_{\mathcal{B}}^{LWE_{n,m,q,D_{Z^n,\sigma}}}$.

GAME G_2. In this game, we change the way that the matrix A is generated. A is not chosen uniformly but in the lossy mode. In order to show the indistinguishability between G_2 and G_1, we construct an algorithm $\hat{\mathcal{B}}$ which can tell whether $A \leftarrow Z^{n \times m}$ or $A \leftarrow SampleLossy(n, m, w, \chi)$, where χ is a distribution such that $\Pr_{x \leftarrow \chi}[|x| \geqslant \beta] \leqslant negl(\lambda)$ and $\alpha \geqslant \beta\gamma nm$. $\hat{\mathcal{B}}$ is given A and sends $pk = (A, \mathbf{b})$ to $\mathcal{A}$. Then $\hat{\mathcal{B}}$ generates challenge ciphertexts as in G_0. $\mathcal{B}$ outputs 1 if $\mathcal{A}$ wins the game and outputs 0 otherwise. We can see that if $A \leftarrow Z^{n \times m}$ the view of the adversary corresponds to that of G_1, and if $A \leftarrow SampleLossy(n, m, w, \chi)$ the view of the adversary is identical to that of G_2. We have the second inequality

$$|\Pr[G_1 = 1] - P[G_2 = 1]| \leqslant \mathrm{Adv}_{\tilde{\mathcal{B}}}^{Lossy} \leqslant n\mathrm{Adv}_{\mathcal{B}'}^{LWE_{w,m,q,\chi^m}}.$$

by Lemma 11. It remains to show that no adversary has non-negligible chance in winning G_2. For each challenge ciphertext c_i^*, it can be written as

$$c_{i1}^* = \mathbf{b}^T \mathbf{s}_i' + m\lceil q/2 \rceil, c_{i2}^* = A^T \mathbf{s}_i' + \mathbf{e}'.$$

For $\varepsilon = 2^{-k}$, where $k = \log q + \Omega(n)$, we have

$$\begin{aligned}
H_\infty^\varepsilon(\mathbf{s}_i'|A, c_{i2}^* = A^T \mathbf{s}_i' + \mathbf{e}') &\geqslant H_\infty(\mathbf{s}_i') - (w + 2k)\log q \\
&= n\log 2\gamma - (w + 2k)\log q \qquad (8) \\
&\geqslant \log q + \Omega(n).
\end{aligned}$$

where the first inequality follows by Lemma 12 and let $n\log 2\gamma - (w+2k)\log q \geqslant \log q + \Omega(n)$.

Since $\mathbf{s}_i'$ is independent of c_{j2}^* for $j \neq i$, we get

$$H_\infty^\varepsilon(\mathbf{s}_i'|A, c_{12}^*, \ldots, c_{t2}^*) = H_\infty^\varepsilon(\mathbf{s}_i'|A, c_{i2}^*) \geqslant \log q + \Omega(n).$$

Now we can apply Lemma 10 to conclude that $(\mathbf{b}, \mathbf{b}^T \mathbf{s}_1', \ldots, \mathbf{b}^T \mathbf{s}_t')$ is $t/2^{\Omega(n)}$-close to the uniform distribution. Hence, we have $\Pr[G_2 = 1] \leq t/2^{\Omega(\lambda)} + t/2^l$.

Acknowledgements. We would like to thank the anonymous reviewers for their helpful comments. Zhengyu Zhang and Puwen Wei were supported by the National Natural Science Foundation of China (No. 61502276). Haiyang Xue was supported by the National Natural Science Foundation of China (No. 61602473) and the National Cryptography Development Fund (No. MMJJ20170116).

A   Proof of Lemma 9

Proof. We have

$$\begin{aligned}
&|\mathbf{E}_{j \leftarrow J_1}[2^{-H_\infty(Y|J_1=j)}] - \mathbf{E}_{j \leftarrow J_2}[2^{-H_\infty(Y|J_2=j)}]| \\
&= |\sum_j \Pr[J_1 = j]2^{-H_\infty(Y|j)} - \sum_j \Pr[J_2 = j]2^{-H_\infty(Y|j)}| \qquad (9) \\
&\leq \sum_j |\Pr[J_1 = j] - \Pr[J_2 = j]|2^{-H_\infty(Y|j)} \leq \varepsilon.
\end{aligned}$$

The last inequality follows from $\Delta(J_1, J_2) \leq \varepsilon/2$ and the $2^{-H_\infty(Y|j)} \leq 1$. Hence

$$\begin{aligned}
&|\hat{H}_\infty(Y|J_1) - \hat{H}_\infty(Y|J_2)| \\
&= |-\log \mathbf{E}_{j \leftarrow J_1}[2^{-H_\infty(Y|J_1=j)}] + \log \mathbf{E}_{j \leftarrow J_2}[2^{-H_\infty(Y|J_2=j)}]| \\
&= |\log \frac{\mathbf{E}_{j \leftarrow J_2}[2^{-H_\infty(Y|J_2=j)}]}{\mathbf{E}_{j \leftarrow J_1}[2^{-H_\infty(Y|J_1=j)}]}| \leq \log(1 + \frac{\varepsilon}{\mathbf{E}_{j \leftarrow J_1}[2^{-H_\infty(Y|J_1=j)}]}) \qquad (10) \\
&\leq \frac{\varepsilon}{\mathbf{E}_{j \leftarrow J_1}[2^{-H_\infty(Y|J_1=j)}]} + 1 = \varepsilon 2^{\hat{H}_\infty(Y|J_1)} + 1.
\end{aligned}$$

For the last inequality we use the fact that $\log(x + 1) \leq x + 1$.

References

1. Abe, M., David, B., Kohlweiss, M., Nishimaki, R., Ohkubo, M.: Tagged one-time signatures: tight security and optimal tag size. In: Kurosawa, K., Hanaoka, G. (eds.) PKC 2013. LNCS, vol. 7778, pp. 312–331. Springer, Heidelberg (2013). https://doi.org/10.1007/978-3-642-36362-7_20
2. Alwen, J., Krenn, S., Pietrzak, K., Wichs, D.: Learning with rounding, revisited. In: Canetti, R., Garay, J.A. (eds.) CRYPTO 2013. LNCS, vol. 8042, pp. 57–74. Springer, Heidelberg (2013). https://doi.org/10.1007/978-3-642-40041-4_4
3. Ambainis, A., Hamburg, M., Unruh, D.: Quantum security proofs using semi-classical oracles. In: Boldyreva, A., Micciancio, D. (eds.) CRYPTO 2019. LNCS, vol. 11693, pp. 269–295. Springer, Cham (2019). https://doi.org/10.1007/978-3-030-26951-7_10. https://eprint.iacr.org/2018/904
4. Ambainis, A., Rosmanis, A., Unruh, D.: Quantum attacks on classical proof systems: the hardness of quantum rewinding. In: 55th Annual Symposium on Foundations of Computer Science, pp. 474–483. IEEE (2014)
5. Bellare, M., Boldyreva, A., Micali, S.: Public-key encryption in a multi-user setting: security proofs and improvements. In: Preneel, B. (ed.) EUROCRYPT 2000. LNCS, vol. 1807, pp. 259–274. Springer, Heidelberg (2000). https://doi.org/10.1007/3-540-45539-6_18
6. Bellare, M., Rogaway, P.: Random oracles are practical: a paradigm for designing efficient protocols. In: CCS 1993, pp. 62–73. ACM (1993)
7. Blazy, O., Kiltz, E., Pan, J.: (Hierarchical) identity-based encryption from affine message authentication. In: Garay, J.A., Gennaro, R. (eds.) CRYPTO 2014. LNCS, vol. 8616, pp. 408–425. Springer, Heidelberg (2014). https://doi.org/10.1007/978-3-662-44371-2_23
8. Blazy, O., Kakvi, S.A., Kiltz, E., Pan, J.: Tightly-secure signatures from chameleon hash functions. In: Katz, J. (ed.) PKC 2015. LNCS, vol. 9020, pp. 256–279. Springer, Heidelberg (2015). https://doi.org/10.1007/978-3-662-46447-2_12
9. Boneh, D., Dagdelen, Ö., Fischlin, M., Lehmann, A., Schaffner, C., Zhandry, M.: Random oracles in a quantum world. In: Lee, D.H., Wang, X. (eds.) ASIACRYPT 2011. LNCS, vol. 7073, pp. 41–69. Springer, Heidelberg (2011). https://doi.org/10.1007/978-3-642-25385-0_3
10. Bos, J., et al.: Frodo: take off the ring! Practical, quantum-secure key exchange from LWE. In: CCS 2016, pp. 1006–1018. ACM (2016)
11. Chen, J., Wee, H.: Fully, (almost) tightly secure IBE and dual system groups. In: Canetti, R., Garay, J.A. (eds.) CRYPTO 2013. LNCS, vol. 8043, pp. 435–460. Springer, Heidelberg (2013). https://doi.org/10.1007/978-3-642-40084-1_25
12. Dodis, Y., Reyzin, L., Smith, A.: Fuzzy extractors: how to generate strong keys from biometrics and other noisy data. In: Cachin, C., Camenisch, J.L. (eds.) EUROCRYPT 2004. LNCS, vol. 3027, pp. 523–540. Springer, Heidelberg (2004). https://doi.org/10.1007/978-3-540-24676-3_31
13. Gay, R., Hofheinz, D., Kiltz, E., Wee, H.: Tightly CCA-secure encryption without pairings. In: Fischlin, M., Coron, J.-S. (eds.) EUROCRYPT 2016. LNCS, vol. 9665, pp. 1–27. Springer, Heidelberg (2016). https://doi.org/10.1007/978-3-662-49890-3_1
14. Gentry, C., Peikert, C., Vaikuntanathan, V.: Trapdoors for hard lattices and new cryptographic constructions. In: STOC 2008, pp. 197–206. ACM (2008)
15. Goldwasser, S., Kalai, Y., Peikert, C., Vaikuntanathan, V.: Robustness of the learning with errors assumption. In: ICS 2010, pp. 230–240 (2010)

16. Hofheinz, D., Jager, T.: Tightly secure signatures and public-key encryption. In: Safavi-Naini, R., Canetti, R. (eds.) CRYPTO 2012. LNCS, vol. 7417, pp. 590–607. Springer, Heidelberg (2012). https://doi.org/10.1007/978-3-642-32009-5_35

17. Hofheinz, D., Hövelmanns, K., Kiltz, E.: A modular analysis of the Fujisaki-Okamoto transformation. In: Kalai, Y., Reyzin, L. (eds.) TCC 2017. LNCS, vol. 10677, pp. 341–371. Springer, Cham (2017). https://doi.org/10.1007/978-3-319-70500-2_12

18. Hofheinz, D., Koch, J., Striecks, C.: Identity-based encryption with (almost) tight security in the multi-instance, multi-ciphertext setting. In: Katz, J. (ed.) PKC 2015. LNCS, vol. 9020, pp. 799–822. Springer, Heidelberg (2015). https://doi.org/10.1007/978-3-662-46447-2_36

19. Jiang, H., Zhang, Z., Chen, L., Wang, H., Ma, Z.: IND-CCA-secure key encapsulation mechanism in the quantum random oracle model, revisited. In: Shacham, H., Boldyreva, A. (eds.) CRYPTO 2018. LNCS, vol. 10993, pp. 96–125. Springer, Cham (2018). https://doi.org/10.1007/978-3-319-96878-0_4

20. Katsumata, S., Yamada, S., Yamakawa, T.: Tighter security proofs for GPV-IBE in the quantum random oracle model. In: Peyrin, T., Galbraith, S. (eds.) ASIACRYPT 2018. LNCS, vol. 11273, pp. 253–282. Springer, Cham (2018). https://doi.org/10.1007/978-3-030-03329-3_9

21. Libert, B., Joye, M., Yung, M., Peters, T.: Concise multi-challenge CCA-secure encryption and signatures with almost tight security. In: Sarkar, P., Iwata, T. (eds.) ASIACRYPT 2014. LNCS, vol. 8874, pp. 1–21. Springer, Heidelberg (2014). https://doi.org/10.1007/978-3-662-45608-8_1

22. Peikert, C., Vaikuntanathan, V., Waters, B.: A framework for efficient and composable oblivious transfer. In: Wagner, D. (ed.) CRYPTO 2008. LNCS, vol. 5157, pp. 554–571. Springer, Heidelberg (2008). https://doi.org/10.1007/978-3-540-85174-5_31

23. Regev, O.: On lattices, learning with errors, random linear codes, and cryptography. J. ACM **56**, 34 (2009)

24. Renner, R., Wolf, S.: Smooth Renyi entropy and applications. In: ISIT 2004, pp. 233–233. IEEE (2004)

25. Saito, T., Xagawa, K., Yamakawa, T.: Tightly-secure key-encapsulation mechanism in the quantum random oracle model. In: Nielsen, J.B., Rijmen, V. (eds.) EUROCRYPT 2018. LNCS, vol. 10822, pp. 520–551. Springer, Cham (2018). https://doi.org/10.1007/978-3-319-78372-7_17

26. Targhi, E.E., Unruh, D.: Post-quantum security of the Fujisaki-Okamoto and OAEP transforms. In: Hirt, M., Smith, A. (eds.) TCC 2016. LNCS, vol. 9986, pp. 192–216. Springer, Heidelberg (2016). https://doi.org/10.1007/978-3-662-53644-5_8

27. Unruh, D.: Revocable quantum timed-release encryption. In: Nguyen, P.Q., Oswald, E. (eds.) EUROCRYPT 2014. LNCS, vol. 8441, pp. 129–146. Springer, Heidelberg (2014). https://doi.org/10.1007/978-3-642-55220-5_8

28. Wei, P., Wang, W., Zhu, B., Yiu, S.M.: Tightly-secure encryption in the multi-user, multi-challenge setting with improved efficiency. In: Pieprzyk, J., Suriadi, S. (eds.) ACISP 2017. LNCS, vol. 10342, pp. 3–22. Springer, Cham (2017). https://doi.org/10.1007/978-3-319-60055-0_1

29. Zhandry, M.: Secure identity-based encryption in the quantum random oracle model. In: Safavi-Naini, R., Canetti, R. (eds.) CRYPTO 2012. LNCS, vol. 7417, pp. 758–775. Springer, Heidelberg (2012). https://doi.org/10.1007/978-3-642-32009-5_44

Searchable Encryption

Practical, Dynamic and Efficient Integrity Verification for Symmetric Searchable Encryption

Lanxiang Chen[✉] and Zhenchao Chen

College of Mathematics and Informatics, Fujian Normal University,
Fujian Provincial Key Laboratory of Network Security and Cryptology,
Fuzhou, China
lxiangchen@fjnu.edu.cn

Abstract. Symmetric searchable encryption (SSE) has been proposed for years and widely used in cloud storage. It enables individuals and enterprises to outsource their encrypted personal data to cloud server and achieves efficient search. Currently most SSE schemes are working in the semi-honest or curious cloud server model in which the search results are not absolutely trustworthy. Thus, the verifiable SSE (VSSE) schemes are proposed to enable data integrity verification. However, the majority of existing VSSE schemes have their own limitations, such as not supporting dynamics (data updates), working in single-user mode or not practical etc. In this paper, we propose a practical, dynamic and efficient integrity verification method for SSE construction that is decoupled from original SSE schemes. This paper proposed a practical and general SSE (PGSSE for short) scheme to achieve more efficient data integrity verification in comparison with the state-of-the-art schemes. The proposed PGSSE can be applied to any top-k ranked SSE scheme to achieve integrity verification and efficient data updates. Security analysis and experimental results demonstrate that the proposed PGSSE scheme is secure and efficient.

Keywords: Symmetric searchable encryption · Shamir's secret sharing · Merkle Patricia Tree · Integrity verification · Data update

1 Introduction

Accompanying with the explosive growth of cloud computing, data security has become the most important aspect that needs to be guaranteed. To preserve data security and privacy, individuals and enterprises usually encrypt their data before outsourcing them to the remote cloud computing or cloud storage server.

This work was supported by the National Natural Science Foundation of China under Grant No. 61602118, No. 61572010 and No. U1805263, Natural Science Foundation of Fujian Province under Grant No. 2019J01274 and No. 2017J01738.

© Springer Nature Switzerland AG 2019
Y. Mu et al. (Eds.): CANS 2019, LNCS 11829, pp. 163–183, 2019.
https://doi.org/10.1007/978-3-030-31578-8_9

As a result, how to process and search on the encrypted data becomes the critical problem that needs to be resolved. Searchable encryption facilitates search operations on the encrypted data meanwhile it preserves the privacy of users' sensitive data, and thus it has attracted extensive attentions recently.

Since the first searchable encryption scheme [23] proposed in 2000, a large number of related works have emerged in the literature during the last two decades. According to the encryption algorithm adopted in searchable encryption schemes, they can be classified into public key searchable encryption (PKSE) and symmetric searchable encryption (SSE). As there are massive data in a cloud storage, the data usually are encrypted by a symmetric encryption algorithm to guarantee efficiency and data availability.

For a practical SSE scheme, it should satisfy at least the following properties: sublinear search time, compact indexes, supporting ranked search, efficient updates, integrity verification and data security. Unfortunately, none of the existing SSE constructions achieves all these properties at the same time, which has limited their practicability. If an SSE scheme does not support top-k ranked search, the cloud server will return all data files that contain the queried keywords. As the user have no pre-knowledge of the encrypted files, he has to decrypt all these files to further find the most matching files. These will result in unnecessary computing overheads, time consumption and network traffics. Hence, without it, SSE schemes are impractical in the pay-as-you-use cloud computing era. By returning the most related files, ranked search schemes greatly facilitate system practicability.

To enrich the functionalities of SSE, a variety of multi-keyword, multi-user or multiple data owner, dynamic or verifiable SSE schemes have been proposed. However, majority of existing SSE schemes have their own ways of index construction, integrity verification and data updates. A general scheme with more functionalities decoupling from any special constructions is lacking. Motivated by this idea, in this paper, we propose a practical, dynamic and efficient integrity verification method for SSE construction that is decoupled from original SSE schemes. Our work is a one-step forward to the work due to Zhu et al. [31] in terms of top-k ranked search and data update efficiency. The contributions of this work can be summarized as follows.

- We proposed a practical and general integrity verification scheme (PGSSE) with the aid of secret sharing scheme and Merkle Patricia Tree (MPT). Compared with existing SSE schemes, the proposed scheme firstly introduces the secret sharing scheme to SSE to make the general SSE scheme support top-k ranked search.
- Thanks to the secret sharing scheme, users do not need to update the MPT tree but just to update their keys when they update their data without keyword addition. Thus the data updates of the proposed scheme are very efficient.

This paper is arranged as follows. We will discuss related work in Sect. 2. Section 3 gives the preliminaries. The system model and formal definition are presented in Sect. 4. Section 5 describes the details of our PGSSE scheme construction. Section 6 present security and performance evaluation of the proposed scheme. We give a conclusion in Sect. 7.

2 Related Work

In 2000, Song et al. [23] proposed the first searchable encryption scheme which needs to search all encrypted documents in a non-interactive way to check whether a queried keyword is contained or not. For each queried keyword, it has to scan all files and thus the search time was linear in the length of the documents collection. In addition, the proposed scheme is only adapted to single keyword search. In 2003, Goh [11] first proposed to construct index to achieve search and a Bloom filter based index scheme is introduced. He gave a formal security of IND-CKA for SSE and proved the proposed Bloom filter based SSE scheme is IND-CKA secure. The drawback is that the Bloom filters based construction had a possibility of false positives. In 2006, Curtmola et al. [8] proposed two efficient SSE schemes, SSE-1 and SSE-2 with $O(1)$ search time complexity. They gave the formal security definitions for the proposed schemes and utilized broadcast encryption to enable multi-user search in SSE-2. Both schemes only support single-keyword search.

After that, a variety of functionally rich SSE schemes were proposed in the last two decades, including multi-keyword search, top-k ranked search, dynamic data update, verifiable SSE, fuzzy and similarity search etc. As described above, if an SSE scheme does not support top-k ranked search, the cloud server will return all data files containing the queried keywords, which will greatly reduce the practicability of these schemes.

In 2010, Wang et al. [26,27] first proposed to use order-preserving symmetric encryption (OPSE) to achieve a ranked keyword search scheme which protects the privacy of relevance scores. The measure of relevance scores is based on a TF×IDF model. To reduce the amount of information leakage, they proposed to use a one-to-many OPSE scheme to obfuscate the original relevance score distribution. In 2011, Cao et al. [1,2] proposed a privacy-preserving multi-keyword ranked searchable encryption (MRSE) scheme by using "coordinate matching" and "inner product similarity".

To improve the accuracy of search results, almost all multi-keyword SSE schemes [5,9,18,29,30] support top-k ranked search since the ranked SSE scheme has been proposed. However, the SSE constructions described above are static, which means that they did not have the ability to add or delete documents efficiently.

In 2010, Liesdonk et al. [19] proposed the first dynamic SSE scheme which supports a limited number of updates and has a linear search time in the worst case. In 2012, Kamara et al. [15] constructed a dynamic SSE scheme which is an extension of SSE-1. They presented a formal security definition for dynamic SSE. Their scheme is adaptively secure against chosen-keyword attacks (CKA2) and it is also secure in the random oracle model. Then, in 2013, they presented a parallelizable and dynamic sub-linear SSE scheme with the help of multi-core architectures [14]. The search time is about $O(r/p)$ (r and p is the number of documents and cores respectively) for searching a keyword with a logarithmic number of cores. Compared to the SSE scheme in [15], this SSE scheme does not leak the tokens of the keywords contained in an updated document. Their scheme uses a keyword red-black tree (KRB) to construct index that makes updates simple. However, this scheme focuses on the case of single-keyword equality queries only. Since then, some dynamic SSE schemes are presented [3,6,10,12, 21,28].

As the cloud server is not trustable in some circumstances, verifiable SSE (VSSE) [4,7,13,16,17,20,24,25] is proposed to check the integrity of search results and data. In 2012, Kurosawa and Ohtaki [16] first formulate the security of VSSE against active adversaries and proposed a UC-security (abbreviation of Universal Composability) single-keyword VSSE scheme. Their scheme preserves the search results correct even if the server is malicious. Later in 2013, they gave a more efficient VSSE scheme [17] and extended the scheme to dynamic VSSE scheme. In 2015, Sun et al. [24] proposed a dynamic conjunctive keyword VSSE scheme by using bilinear-map accumulator tree. Recently, Jiang et al. [13] proposed a multi-keyword ranked VSSE scheme and a special data structure QSet based on an inverted index. The basic idea is to estimate the least frequent keywords in the query to reduce the search times. The verification is based on a keyword binary vector.

However, all the above works have their own ways of index construction, integrity verification and data updates. A general scheme with more functionalities decoupling from any special constructions is lacking. Recently, Zhu et al. [31] proposed a generic and verifiable SSE scheme (GSSE). It can be adopted to any SSE scheme to provide integrity verification and data updates. They proposed to use the Merkle Patricia Tree (MPT) and incremental hash to construct proof index and develop a timestamp chain to resist data freshness attacks. As MPT is a kind of prefix tree, it is efficient to insert and delete nodes. Hence the GSSE achieves data integrity verification efficiently.

But there is a shortcoming in the proposed GSSE scheme that the user has to get all documents which contain the queried keyword in the document collection to perform the integrity verification. If the queried keyword is common, there could be quite a lot of documents containing the keyword. Many documents returned may be not desired by the user while they will consume a lot of time to search and verify. It also means that the GSSE scheme does not support top-k ranked search.

Comparing to the GSSE scheme, the proposed PGSSE scheme supports top-k ranked search which makes it more practical. As the incremental hash is utilized for integrity verification in GSSE, users have to compute the hash of all queried documents to verify the root of the MPT tree. Thus, the GSSE scheme cannot support top-k ranked search. To overcome this disadvantage, we propose to utilize a secret sharing scheme to replace incremental hash to perform integrity verification. Comparing to GSSE, the PGSSE allows users to perform integrity verification when they only get the top-k documents containing the queried keywords. Meanwhile, we found that the proposed scheme can achieve data updates efficiently.

3   Preliminary

In this section, the notations, MPT and Shamir's secret sharing scheme are revisited.

3.1   Notations

In the following sections, the pseudo-random functions h_1, h_2 and h_3 are defined as $\{0,1\}^* \times \{0,1\}^\lambda \to \{0,1\}^*$. The other notations are described in Table 1.

Table 1. Notations and descriptions

Symbol	Denote
DC	The document collection, including N documents and denoted as $DC = (D_1, D_2, \cdots, D_N)$
C	The encrypted document collection $C = (C_1, C_2, \cdots, C_N)$ stored in the cloud server
I	The encrypted index
W	The keyword dictionary $W = (w_1, w_2, \cdots, w_m)$
m	The number of keywords in W
$DC(w_i)$	The collection of documents that contain keyword w_i
WD_i	The keyword set in the document D_i
Au	The authenticator
$Enc/Dec(\cdot)$	Symmetric encryption/decryption algorithm
Key	The secret key stored by the data owner $Key = \{k_1, k_2, k_3, k_4, (spk, ssk), S, P\}$

For the secret key stored by the data owner, k_1, k_2 and k_3 are used for pseudo-random functions h_1, h_2 and h_3 respectively, k_4 is used for the symmetric encryption algorithm $Enc()$, (spk, ssk) is the public/private key pair, S is a matrix in which each row represents one polynomial's coefficients, P is a set of m arrays.

3.2   Merkle Patricia Tree

Merkle Patricia Tree (MPT) proposed in Ethereum is a mixture of Merkle tree and Patricia tree. It is a kind of prefix tree that has high efficiency in insert and delete operations. There are four kinds of nodes in the MPT, namely null node, leaf node, extension node and branch node. The null node is simple and we use a blank string to represent it. Leaf node (LN) and extension node (EN) are both represented as one key-value pair and those keys are encoded in Hex-Prefix. The keys in extension nodes indicate their descendant nodes' common prefix and their values are their children nodes' hash values. The keys in leaf nodes indicate the rest part except for the common prefix and the values are their own values. Differing from LN and EN, the branch nodes' keys consist of 17 elements in which 16 elements correspond to Hex-Prefix codes. The last element is used only when the search route terminates here in which the value in BN plays the same role as that in LN.

The construction of a MPT is through the "insert" operation which will be demonstrated according to different situations.

(1) Insert to branch node (the current key is empty)

The initial MPT is empty as Fig. 1(a) and a new node with value '223' will be inserted to the MPT. The insert operation directly set the value of the MPT to '223' and get the MPT as Fig. 1(b).

Fig. 1. Insert to branch node (the current key is empty).

(2) Insert to branch node (the current key is not empty)

Assume the initial MPT is Fig. 2(a) and a new node with key-value pair ['a2912', '22'] will be inserted to the branch node. As the descendant of the element 'a' is empty, the "insert" algorithm will create a new leaf node (LN2) to store the rest key '2912' and the value '22'. The new MPT is illustrated as Fig. 2(b).

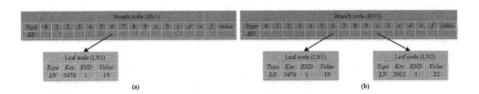

Fig. 2. Insert to branch node (the current key is not empty).

(3) Insert to Extension node

Assume the initial MPT is Fig. 2(b) and a new node with key-value pair ['a2535', '57'] will be inserted to the MPT. As the key 'a2535' has a common prefix 'a2' with LN2, the "insert" algorithm will create an extension node (EN1) whose key is the rest common prefix '2' and a branch node (BN2) whose key '5' and key '9' are linked to newly created leaf nodes with key '35' and '12' respectively. The insertion is completed as Fig. 3.

When searching for a node in the MPT, the "Search" algorithm will start from the root to bottom to check the nodes' key at each level. For example, the user wants to search the node with key 'a2535'. The "search" algorithm will first find the BN1 and then go on to EN1, and finally the path from BN1 to LN2 will be found.

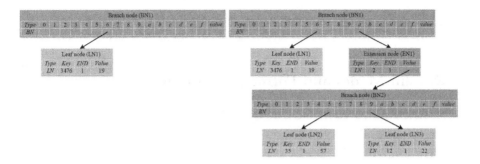

Fig. 3. Insert to extension node.

3.3 Shamir's Secret Sharing Scheme

As pointed out that the incremental hash needs to compute the hash of all queried documents to check the integrity of search results, it is unsuitable for ranked search. With a secret sharing scheme, the pre-defined number of participants can compute the secret without the involvement of all participants. This property can be applied to the ranked search. For the sake of generality, the Shamir's secret sharing scheme is chosen in the proposed PGSSE scheme.

Shamir's secret sharing scheme [22] is a threshold scheme to share a secret in a distributed way. For a (k, n) threshold scheme, a secret is split into n pieces for n participants and any more than $k - 1$ participants can reconstruct the secret (k is the threshold), but the secret cannot be reconstructed with fewer than k pieces. With the feature of dynamics, it can be applied to the ranked search with efficient data updates.

With the dynamics of Shamir's secret sharing scheme, the security can be easily enhanced without changing the secret, and only need to change the polynomial coefficients and construct new shares to the participants.

To construct a (k, n) Shamir's secret sharing scheme, it needs to construct a k-1 degree polynomial $f(x)$ in the finite field GF(q) and the polynomial's constant term is the secret s, where q is a big prime number ($q > n$). Firstly, it randomly generates a k-1 degree polynomial based on GF(q) and set $f(0) = a0 = s$. Then, it randomly selects n different non-zero numbers $(x_1, x_2, ..., x_n)$ and allocates $(x_i, f(x_i))$ to each participant $p_i (0 < i < n)$, where x_i is public and $f(x_i)$ is kept secret.

Then to recover the secret s, it randomly selects k pairs $(x_j, f(x_j))(0 < j < k)$ and utilizes the Lagrange's polynomial interpolation algorithm as Eq. (1) to reconstruct the secret as Eq. (2).

$$f(x) = \sum_{j=1}^{k} f(x_j) \prod_{l=1}^{k} \frac{x - x_l}{x_j - x_l} \bmod q \tag{1}$$

$$s = (-1)^{k-1} \sum_{j=1}^{k} f(x_j) \prod_{l=1}^{k} \frac{x_l}{x_j - x_l} \bmod q \tag{2}$$

4   System Model and Formal Definition

The system model and formal definition are described in this section.

4.1   System Model

The system model is illustrated in Fig. 4. There are three entities, namely data owner, data user and cloud server. Data owner is in charge of constructing index and authenticator. He receives the request from data user and authenticates the data user. After being authenticated, the data user can access cloud server to obtain some search results and he will perform an integrity verification for the search results and the corresponding document data. The cloud server is responsible for storing users' indexes, authenticator and document data. When receiving the token from a data user, the cloud server will make the corresponding proof and authenticator to the data user.

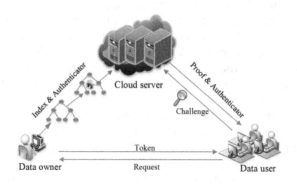

Fig. 4. System model.

4.2   Formal Definition

The proposed PGSSE scheme has seven polynomial-time algorithms.

(1) $Setup(1^\lambda) \rightarrow Key$: It is run by the data owner to setup the scheme. The algorithm takes as input the security parameter λ and outputs the secret Key.

(2) $MPTBuild(Key, W, DC) \rightarrow \{I, Au\}$: It is run by the data owner. It takes as input the Key and keyword dictionary W, and outputs the index and authenticator.

(3) $TokenGen(k_3, Q) \rightarrow \{Token\}$: It is run by the data user. It takes as input k_3 and queried keywords, and outputs the token.

(4) $ProofBuild(I, Token, t_q) \rightarrow \{Proof, Au_q^t, Au_c\}$: It is run by the cloud server. It takes as input index I, the token and the query time t_q, and outputs the corresponding proof and two authenticators Au_q^t and Au_c.

(5) $CheckAu(k_4, Au_q^t, Au_c) \rightarrow \{result\}$: It is run by the data user. It takes as input the key k_4 and two authenticators Au_q^t, Au_c, and outputs a result to indicate whether the root of MPT has been tampered.

(6) $Verify(k_2, k_4, S, P, C_Q, Proof, Token_Q, Au_q^t) \rightarrow \{result\}$: It is run by the data user. It takes as input the k_2 and k_4, the search result C, the authenticator Au_q^t and P, and outputs a result to indicate whether the queried documents have been tampered.

(7) $Update(P, D_j, I) \rightarrow \{P', I'\}$: It is run by the data owner. It takes as input the set P, update document D_j and index I, and outputs the new P' and new I'.

5   Scheme Construction

In this section, the seven polynomial-time algorithms of the proposed PGSSE are detailed respectively. The authenticator of the "CheckAu" algorithm is used to make this scheme to resist the freshness attack on the root of MPT. As it is the same as the scheme in [31] and we will not elaborate on it here.

5.1   Initialization

The "Setup" algorithm will initiate the system parameters and generate all keys. It is executed by data owner and the detailed process is illustrated in Algorithm 1. There are m polynomials and each keyword corresponds to a polynomial. Meanwhile each keyword corresponds to the secret S_{w_i} of the polynomial which is also called node secret. All the node secrets are stored in the MPT. When a data user receives the top-k documents, he/she would try to recover the node secret and execute the integrity verification. The set of P consists of m arrays and each array is in form of ['key' $\rightarrow$ 'value']. This array could help user recover the node secret.

5.2   MPT Building

The MPT building algorithm is performed by the data owner and the detailed procedure is illustrated in Algorithm 2. The index I is the MPT and the "insert" algorithm can refer to Sect. 3.2. When $|DC(w_i)| \geq k$, the node secret will be computed and inserted into MPT, and there will be more than k key-value pairs and the node secret can be recovered through the secret sharing scheme. However, when $|DC(w_i)| < k$, there are less than k key-value pairs in the keyword

---

Algorithm 1. Setup Algorithm

Input: Parameter λ;
Output: $Key = \{k_1, k_2, k_3, k_4, (spk, ssk), S, P\}$.

1 Randomly generates k_1, k_2, k_3, k_4 and q;
2 Generates the public/private key pair (spk, ssk);
3 Compute node secret:
4 **for** $w_i \in W$ **do**
5 $\quad|\quad$ compute $S_{w_i} = h_1(k_1, w_i)$;
6 **end**
7 Generate the matrix $S : S$ is an $m \times k$ matrix, the i-th row is $\{a_{i_1}, a_{i_2}, \cdots, a_{i,k-1}, S_{w_i}\}(i \in [1, m])$ where $\{a_{i_1}, a_{i_2}, \cdots, a_{i,k-1}\}$ are the coefficients of the Shamir's secret sharing polynomials $f_i(x), S_{w_i}$ is the secret of $f_i(x)$;
8 Generate the set P which consists of m arrays. Each array corresponds to a keyword and the keyword w_i's array ($array_i$) is generated as follows:
9 **for** $w_i \in W$ **do**
10 $\quad|\quad$ **for** $D_j \in DC(w_i)$ **do**
11 $\quad|\quad\quad|\quad$ Calculate $key_{D_j} = h_2(k_2, D_j)$;
12 $\quad|\quad\quad|\quad$ Encrypt the key_{D_j} with k_4 and $vx_{D_j} = Enc(k_4, key_{D_j})$;
13 $\quad|\quad\quad|\quad$ Extract the w_i's coefficients in S and build the k-1 degree polynomials $f_i(x)$;
14 $\quad|\quad\quad|\quad$ Calculate $value = f_i(key_{D_j})$;
15 $\quad|\quad\quad|\quad$ Set the $array_i[vx_{D_j}] = value$;
16 $\quad|\quad$ **end**
17 **end**

---

arrays and the node secret cannot be reconstructed by the secret sharing scheme. Hence, the sum of document hash values that can be used to check the integrity of all returned documents is calculated in this situation.

The authenticator Au is used to ensure the freshness of the MPT's root rt that is proposed in [31] and it is generated in Eq. (3), where tp is the timestamp, up_i is the i-th update time point, $Sig(ssk, *)$ is a signature with the private key ssk, $Au_{i,j}$ represents the j-th authenticator in the i-th update interval.

Between a fixed update time point and a query time, more than one data update may happen. Under such circumstances, the cloud server may return the old Au in which tp is after the latest fixed update time point but before the query time. Namely there is at least one data update during this period. To resist this type of freshness attack, Zhu et al. introduce a timestamp-chain mechanism in [31]. In each update interval, it generates a timestamp-chain which is constructed according to Eq. (3) and the last authenticator in the chain also locates at the beginning of the next update interval.

Algorithm 2. MPT Building Algorithm

Input: Key, W, DC;
Output: Index I and authenticator Au.

1 Extract keyword dictionary W from DC ;
2 **for** $w_i \in W$ **do**
3   **if** $|DC(wi)| \geq k$ **then**
4   Compute $T_{w_i} = h_3(k_3, w_i)$;
5   Compute $S_{w_i} = h_1(k_1, w_i)$;
6   Execute $I = I.\text{insert}(T_{w_i}, S_{w_i})$;
7   **else**
8   Compute $T_{w_i} = h_3(k_3, w_i)$;
9   **for** $D_j \in DC(w_i)$ **do**
10   Compute $S_{w_i} = \sum h_1(k_1, D_j)$;
11   Execute $I = I.\text{insert}(T_{w_i}, S_{w_i})$;
12   **end**
13   **end**
14 **end**
15 **end**
16 Generate the authenticator Au as Eq. (3) with k_4 and ssk ;
17 Send the index I and authenticator Au to the cloud server.

$$\begin{cases} \text{xcon}_{i,0} = Enc(k_4, rt_{i,0} \| tp_{i,0}), up_i \leq tp_{i,0} \leq up_{i+1} \\ Au_{i,0} = \big(\text{xcon}_{i,0}, Sig(ssk, \text{xcon}_{i,0})\big) \\ \vdots \\ \text{xcon}_{i,j} = Enc(k_4, rt_{i,j} \| tp_{i,j} \| \text{xcon}_{i,j-1}), up_{i,j-1} \leq tp_{i,j} \leq up_{i+1} \\ Au_{i,j} = \big(\text{xcon}_{i,j}, Sig(ssk, \text{xcon}_{i,j})\big) \\ \vdots \\ \text{xcon}_{i,n} = Enc(k_4, rt_{i,n} \| tp_{i,n} \| \text{xcon}_{i,n-1}), tp_{i,n} = up_{i+1} \\ Au_{i,n} = \big(\text{xcon}_{i,n}, Sig(ssk, \text{xcon}_{i,n})\big) \end{cases} \quad (3)$$

If no data update, the data owner just needs to generate the authenticator with a new timestamp at the fixed update time point. If data update happens, the data owner will generate the new authenticator with a new rt and tp. To check whether the rt is the latest one, the data user just needs to compare whether the tp in Au is before the latest update time.

5.3 Token Generation

This algorithm is run by data user and the procedure is described in Algorithm 3. The *Token* can be regarded as the path from the root of the MPT to the node corresponding to the keyword. The cloud server could find the corresponding keyword in the MPT according to the *Token*.

Algorithm 3. Token Generation Algorithm

    Input: k_3;
    Output: $Token$.
1 **for** $w_i \in Q$ **do**
2 | Compute $T_{w_i} = h_3(k_3, w_i)$;
3 **end**
4 Send the $Token_Q = \{T_{w_1}, T_{w_2}, \cdots, T_{w_c}\}$ to the cloud server.

5.4 Proof Generation

Proof generation algorithm is run by the cloud server and the detailed description is illustrated in Algorithm 4. The checkpoint is the update time point which is closest to the user's query time.

The *proof* is used to provide necessary information for user to reconstruct the root of MPT. If the *Token* sent by data user exists in MPT, the cloud server will generate corresponding *proof* to provide necessary information for user to reconstruct the root of MPT. If the *Token* is not existing in MPT, the server could also generate the *proof*. Namely the server would return *proof* to user no matter whether the keyword exists or not. It will help user to detect whether the server deliberately omits all documents and returns an empty result to evade the integrity verification. In addition, as PGSSE is designed for the multi-keyword SSE scheme, the *proof* is not only a search path but also in the form of a sub-tree.

5.5 Integrity Verification

The integrity verification algorithm is used to check the integrity of search results and the procedure is described in Algorithm 5. According to the value of k, there are corresponding operations to calculate the node secret. According to whether the returned ciphertext and *'remain'* are null for the queried keyword w_i, it performs different operations.

(1) If both the ciphertext and *'remain'* are not null, the data user would look up the $array_i$ and recover the node secret with the help of the returned documents. Then he reconstructs the MPT's root and compare it with rt_q^t decrypted from Au_q^t.
(2) If both the ciphertext and *'remain'* are null, the data user directly reconstructs the MPT's root, and compare it with rt_q^t decrypted from Au_q^t. Only if they are matched, the data user will think that there is no search result for this keyword. Otherwise, the search results must be tampered.
(3) If one of the ciphertext and *'remain'* is null, the search results must be tampered and the verification algorithm return 0.

Algorithm 4. Proof Generation Algorithm

Input: Index $I, Token$ and t_q;
Output: $Proof$, authenticator Au_q^t and Au_c.

1 **for** $w_i \in Q$ **do**
2   Find the search path $route_{w_i} = \{n_1, n_2, \cdots, n_s\}$ according to the $Token$, where $n_i \in \{EN, BN, LN\}, i \in [1, s]$, and n_1 is the root. The sequence of $route$ is the nodes from top to bottom of the MPT. Find the longest search path $route_{w_l}$;
3 **end**
4 **if** T_{w_l} *exists* **then**
5   **for** $l = s - 1$ *to 0* **do**
6   **if** $n_l = BN$ **then**
7              $Proof = Proof \cup V_{n_l}$, where V_{n_l} includes all key-value pairs of the descendant nodes of the BN;
8   **end**
9   **else if** $n_l = EN$ **then**
10              $Proof = Proof \cup V_{n_l}$, where V_{n_l} is the key which is on the search path;
11   **end**
12   **else**
13              $Proof = Proof \cup V_{n_l}$, where V_{n_l} is key-value pair of node n_l;
14   **end**
15   **end**
16 **end**
17 **else**
18   **for** $l = s$ *to 0* **do**
19   Repeat steps from 6-14;
20   **end**
21 **end**
22 Get the Proof generated based on the keyword w_l;
23 **for** $w_j \in Q$ *(except for w_l)* **do**
24   Scan the $Proof$ and find the position of w_j and shade the node secret in the corresponding position;
25 **end**
26 Get the latest authenticator Au_q^t according to the query time and Au_c at the checkpoint;
27 Send $Proof, Au_q^t$ and Au_c to the data user.

5.6 Update Algorithm

The update operations include document addition, modification and deletion. According to whether there is addition or deletion of keywords, there are corresponding operations and it is described in Algorithm 6. If a keyword is newly added, it will insert a new node for this keyword. If there is no keyword addition, the MPT would remain unchanged and it only update the set P. For document deletion, it just needs to refresh the set P and keeps the MPT unchanged.

Algorithm 5. Integrity Verification Algorithm

Input: $k_2, k_4, S, P, C_Q, Proof, Token_Q, Au_q^t$;
Output: $result$.

1 **for** $w_i \in Q$ **do**
2 | Compute $remain_{w_i} = String.match(Token_{w_i}, Proof)$;
3 **end**
4 Let $(rt_q^t, tp_q^t, \mathsf{xcon}) = Dec(k_4, \mathsf{xcon}_q^t)$;
5 **if** $C_Q \neq null \wedge remain_{w_i} \neq null$ **then**
6 | Decrypt the C_Q, and get document collection D_Q;
7 | **if** $|D_Q| > k$ **then**
8 | | **for** $D_j \in DC(w_i)$ **do**
9 | | | Compute $hash_{D_j} = h_2(k_2, D_j)$, and insert it into set M;
10 | | | Get $vx_{D_j} = Enc(k_4, hash_{D_j})$;
11 | | **end**
12 | | **for** $hash_{D_j} \in M$ **do**
13 | | | **if** $array_i[vx_{D_j}] = null$ **then**
14 | | | | return $result=0$;
15 | | | **end**
16 | | | **else**
17 | | | | According to $value_j = array_i[vx_{D_j}]$, there are k $(hash_{D_j}, value_j)$ pairs. Thus utilizing Shamir's secret sharing scheme with S to recover the node secret;
18 | | | **end**
19 | | **end**
20 | **end**
21 | **else if** $|D_Q| < k$ **then**
22 | | **for** $D_j \in D_Q$ **do**
23 | | | Compute $S_{w_i} = \sum h_1(k_1, D_j)$;
24 | | **end**
25 | | Calculate the root rt' according to the LN and $proof$ from bottom to the root;
26 | | **if** $rt' = rt_q$ **then**
27 | | | return $result=1$;
28 | | **end**
29 | | **else**
30 | | | return $result=0$;
31 | | **end**
32 | **end**
33 **end**
34 **else if** $C_Q = null$ and at least one $remain_{w_i}$ doesn't exist($w_i \in Q$) **then**
35 | Calculate the root rt' according to the $proof$ from bottom to the root;
36 | **if** $rt' = rt_q$ **then**
37 | | return $result=1$;
38 | **end**
39 | **else**
40 | | return $result=0$;
41 | **end**
42 **end**
43 **else**
44 | return $result=0$.
45 **end**

Algorithm 6. Update Algorithm

Input: The set P, update document D_j and I;
Output: New set P' and new MPT I'.

1 **if** *the document D_j is added into the DC* **then**
2   **for** $w_i \in D_j$ **do**
3   **if** $w_i \in W$ **then**
4   Calculate $key_{D_j} = h_2(k_2, D_j)$ and $vx_{D_j} = Enc(k_4, key_{D_j})$;
5   Extract the w_i's coefficients in S and rebuild the k-1 degree polynomials $f_i(x)$;
6   Calculate $value = f_i(key_{D_j})$ and set $array_i[vx_{D_j}] = value$ in the set P';
7   **end**
8   **if** $w_i \notin W$ **then**
9   Compute $T_{w_i} = h_3(k_3, w_i)$ and $S_{w_i} = h_1(k_1, w_i)$;
10   Execute $I' = I.\text{insert}(T_{w_i}, S_{w_i})$;
11   Randomly generate the coefficients $(a_{i_1}, a_{i_2}, \cdots, a_{i,k-1})$ for w_i and then insert into S;
12   Generate keyword w_i's array $(array_i)$ for the set P' according to Setup algorithm;
13   **end**
14   **end**
15 **end**
16 **if** *the document D_j is removed from DC* **then**
17   **for** $w_i \in D_j$ **do**
18   Calculate $key_{D_j} = h_2(k_2, D_j)$ and $vx_{D_j} = Enc(k_4, key_{D_j})$;
19   Delete the $array_i[vx_{D_j}]$ in the set P'.
20   **end**
21 **end**

---

6 Security and Performance Evaluation

The proposed PGSSE scheme acts as a general method for any SSE scheme for integrity verification. We needs to guarantee that PGSSE can preserve the data confidentiality and results verifiability for SSE schemes. It means that it does not leaks any useful information about documents and keywords in the verification process and it can be detected if the search results are tampered. The security proof of PGSSE is similar to that of Ref. [31]. To achieve top-k ranked search, we utilize Shamir's secret sharing scheme in PGSSE to replace incremental hash in GSSE of [31]. It will not bring more security risks. Because of space limitation, we omitted the formal security proof.

    The performance of PGSSE includes storage overhead, the time overhead of index building, integrity verification and data updates. We compared the performance of PGSSE with GSSE to better evaluate its efficiency. The configuration of a PC used in experiments is core i5-M480 2.67 GHz CPU, 8 GB memory, and Win10 (64 bit) operation system. The SHA-1, 256-bit AES and 1024-bit RSA is used as the hash function, symmetric encryption/decryption algorithm and signature algorithm respectively. The construction of the MPT is implemented in Java with about 800 lines code.

As the basic index structure of PGSSE is MPT that is same as GSSE, the storage overhead of PGSSE and GSSE is close to each other. As the size of MPT largely depends on the number of keywords in the dictionary, the storage overhead of both PGSSE and GSSE grows linearly with the growth of keywords when set the depth of MPT to be fixed. For 5000 documents in the document collection, the storage overhead is about 17 MB and 15 MB for PGSSE and GSSE respectively. As the PGSSE scheme has to store the coefficients of the Shamir's secret sharing polynomials and the set P of m arrays, the storage overhead of PGSSE is slightly more than that of GSSE.

6.1   MPT Construction

The time overhead of MPT construction in the PGSSE scheme largely depends on the number of document-keyword pairs in the inverted index. When set the depth of MPT and the number of documents to be 5 and 3000 respectively, Fig. 5(a) shows the time of MPT construction of both PGSSE and GSSE grows linearly with the growth of document-keyword pairs. For 30000 document-keyword pairs, the time of MPT construction is about 243 ms and 221 ms for PGSSE and GSSE respectively. When set the depth of MPT and the number of document-keyword pairs to be 5 and 5000 respectively, Fig. 5(b) shows the time of MPT construction of both PGSSE and GSSE grows also with the growth of the number of documents. For 3000 documents, the time of MPT construction is about 142 ms and 126 ms for PGSSE and GSSE respectively. It demonstrates that the time overhead of MPT construction for both schemes is positively correlated with the number of document-keyword pairs and documents. As the more the number of document-keyword pairs, the more the dimension of the Shamir's

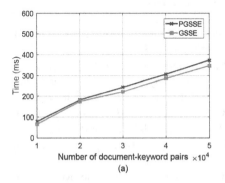
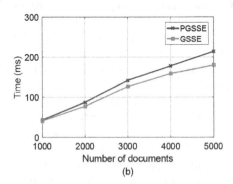

Fig. 5. (a) The time cost of MPT construction with variable number of document-keyword pairs (MPT depth = 5 and the number of documents is $n = 3000$); (b) the time cost of MPT construction with variable number of documents (MPT depth = 5 and the number of document-keyword pairs is 5000).

secret sharing matrix and the more keyword arrays in the set P will be generated in PGSSE, hence the time overhead of PGSSE is a little more than that of GSSE.

6.2 Integrity Verification

The time cost of integrity verification in PGSSE largely depends on the threshold k and the number of queried keywords. The bigger the threshold k is, the more documents will be returned. As a result, the more time will be consumed to construct node secret in the "Verify" algorithm.

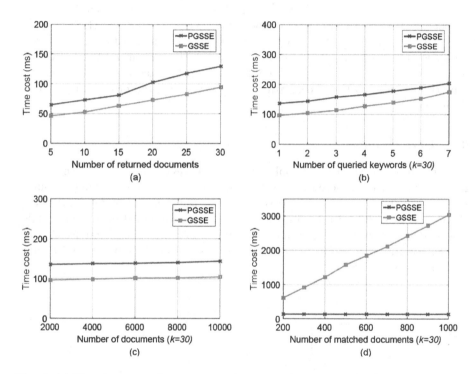

Fig. 6. (a) The time cost of integrity verification with variable k and fixed number of documents $n = 3000$; (b) the time cost of integrity verification with variable number of queried keywords ($n = 3000$ and $k = 30$); (c) the time cost of integrity verification with variable number of documents ($k = 30$); (d) the time cost of integrity verification with variable number of matched documents ($k = 30$).

Assume the number of documents $n = 3000$, Fig. 6(a) shows the time cost of integrity verification of both PGSSE and GSSE grow linearly with k. To return 15 documents, the time of verification is about 81.1 ms and 62.8 ms for PGSSE and GSSE respectively. Assume the number of documents $n = 3000$ and $k = 30$, Fig. 6(b) shows the time cost of integrity verification of both PGSSE and GSSE grow linearly with the number of queried keywords. For 7 queried keywords, the time of verification is about 204.2 ms and 174.6 ms for PGSSE and GSSE respectively. As the time cost of integrity verification is related to the number of returned documents, for $k = 30$, Fig. 6(c) shows the time cost of both PGSSE and GSSE keeps stable with the growth of the number of documents.

The above experimental results show that the efficiency of integrity verification of PGSSE is a bit lower than that of GSSE. In fact, as PGSSE is proposed to improve the practicability of GSSE to enable the ranked top-k search, Fig. 6(d) shows that the efficiency of integrity verification of PGSSE is superior to that of GSSE when the number of matched documents increases sharply. For $k = 30$, if the number of matched documents is 200, the time cost is about 136 ms and 612 ms for PGSSE and GSSE respectively. If the number of matched documents is 1000, the time cost is about 140 ms and 3045 ms for PGSSE and GSSE respectively. This is because of that GSSE has to get all documents which contain the queried keyword in DC to perform the integrity verification. While PGSSE just needs to get the top-k documents to perform the integrity verification, and thus the verification time keeps stable in PGSSE with the growth of matched documents.

6.3   Data Updates

Differing from GSSE, PGSSE only updates the keyword array and the MPT remains unchanged if there is no keyword addition or deletion. Assume the number of documents $n = 3000$, Fig. 7(a) shows that the time cost of data updates of PGSSE keeps stable with the growth of the number of added documents, while it grows linearly with the growth of the number of added documents for that of GSSE. For adding 400 documents, the time cost is about 77 ms and 355 ms for PGSSE and GSSE respectively.

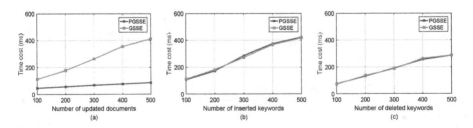

Fig. 7. (a) The update time cost with no keyword insertion and deletion (the number of documents $n = 3000$); (b) the update time cost with keyword insertion ($n = 3000$); (c) the update time cost with keyword deletion ($n = 3000$).

If there is new keywords addition in data updates, it will insert new nodes to the MPT for the new keywords and update the keyword array for PGSSE. While it will also insert new nodes to the MPT for the new keywords and update the incremental hashes of these nodes for GSSE. Figure 7(b) shows that the time cost of data updates of both PGSSE and GSSE grows linearly with the number of added keywords and the time cost is similar. If there is keywords deletion in data updates, it will delete the corresponding nodes of MPT for both PGSSE and GSSE. Figure 7(c) shows that the time cost of data updates of both PGSSE and GSSE grows linearly with the number of deleted keywords and the time cost is also similar.

7   Conclusion

To improve the practicability of existing SSE schemes, we proposed a general and efficient method that provides dynamic and efficient integrity verification for SSE construction that is decoupled from original SSE schemes. The proposed PGSSE overcomes the disadvantages of the GSSE on ranked search. The experimental results demonstrate that PGSSE is greatly superior to GSSE in integrity verification and data updates for top-k ranked search.

References

1. Cao, N., Wang, C., Li, M., Ren, K., Lou, W.: Privacy-preserving multi-keyword ranked search over encrypted cloud data. In: 30th IEEE International Conference on Computer Communications, Joint Conference of the IEEE Computer and Communications Societies, INFOCOM 2011, Shanghai, China, 10–15 April 2011, pp. 829–837 (2011)
2. Cao, N., Wang, C., Li, M., Ren, K., Lou, W.: Privacy-preserving multi-keyword ranked search over encrypted cloud data. IEEE Trans. Parallel Distrib. Syst. **25**(1), 222–233 (2014)
3. Cash, D., et al.: Dynamic searchable encryption in very-large databases: Data structures and implementation. In: 21st Annual Network and Distributed System Security Symposium, NDSS 2014, San Diego, California, USA, 23–26 February 2014
4. Chai, Q., Gong, G.: Verifiable symmetric searchable encryption for semi-honest-but-curious cloud servers. In: Proceedings of IEEE International Conference on Communications, ICC 2012, Ottawa, ON, Canada, 10–15 June 2012, pp. 917–922 (2012)
5. Chen, C., et al.: An efficient privacy-preserving ranked keyword search method. IEEE Trans. Parallel Distrib. Syst. **27**(4), 951–963 (2016)
6. Chen, L., Qiu, L., Li, K., Shi, W., Zhang, N.: DMRS: an efficient dynamic multi-keyword ranked search over encrypted cloud data. Soft Comput. **21**(16), 4829–4841 (2017)
7. Chen, X., Li, J., Weng, J., Ma, J., Lou, W.: Verifiable computation over large database with incremental updates. IEEE Trans. Comput. **65**(10), 3184–3195 (2016)
8. Curtmola, R., Garay, J., Kamara, S., Ostrovsky, R.: Searchable symmetric encryption: improved definitions and efficient constructions. In: Proceedings of the 13th ACM Conference on Computer and Communications Security, CCS 2006, Alexandria, VA, USA, 30 October–3 November 2006, pp. 79–88 (2006)

9. Fu, Z., Ren, K., Shu, J., Sun, X., Huang, F.: Enabling personalized search over encrypted outsourced data with efficiency improvement. IEEE Trans. Parallel Distrib. Syst. **27**(9), 2546–2559 (2016)
10. Gajek, S.: Dynamic symmetric searchable encryption from constrained functional encryption. In: Proceedings of Topics in Cryptology - CT-RSA 2016 - The Cryptographers' Track at the RSA Conference 2016, San Francisco, CA, USA, 29 February–4 March 2016, pp. 75–89 (2016)
11. Goh, E.: Secure indexes. IACR Cryptology ePrint Archive 2003, p. 216 (2003)
12. Hahn, F., Kerschbaum, F.: Searchable encryption with secure and efficient updates. In: Proceedings of the 2014 ACM SIGSAC Conference on Computer and Communications Security, Scottsdale, AZ, USA, 3–7 November 2014, pp. 310–320 (2014)
13. Jiang, X., Yu, J., Yan, J., Hao, R.: Enabling efficient and verifiable multi-keyword ranked search over encrypted cloud data. Inf. Sci. **403**, 22–41 (2017)
14. Kamara, S., Papamanthou, C.: Parallel and dynamic searchable symmetric encryption. In: Financial Cryptography and Data Security - 17th International Conference, FC 2013, Okinawa, Japan, 1–5 April 2013, Revised Selected Papers, pp. 258–274 (2013)
15. Kamara, S., Papamanthou, C., Roeder, T.: Dynamic searchable symmetric encryption. In: The ACM Conference on Computer and Communications Security, CCS 2012, Raleigh, NC, USA, 16–18 October 2012, pp. 965–976 (2012)
16. Kurosawa, K., Ohtaki, Y.: UC-secure searchable symmetric encryption. In: Financial Cryptography and Data Security - 16th International Conference, FC 2012, Kralendijk, Bonaire, 27 Februray–2 March 2012, Revised Selected Papers, pp. 285–298 (2012)
17. Kurosawa, K., Ohtaki, Y.: How to update documents verifiably in searchable symmetric encryption. In: Proceedings of Cryptology and Network Security - 12th International Conference, CANS 2013, Paraty, Brazil, 20–22 November 2013, pp. 309–328 (2013)
18. Li, R., Xu, Z., Kang, W., Yow, K., Xu, C.: Efficient multi-keyword ranked query over encrypted data in cloud computing. Future Gener. Comp. Syst. **30**, 179–190 (2014)
19. van Liesdonk, P., Sedghi, S., Doumen, J., Hartel, P.H., Jonker, W.: Computationally efficient searchable symmetric encryption. In: Proceedings of Secure Data Management, 7th VLDB Workshop, SDM 2010, Singapore, 17 September 2010, pp. 87–100 (2010)
20. Liu, Z., Li, T., Li, P., Jia, C., Li, J.: Verifiable searchable encryption with aggregate keys for data sharing system. Future Gener. Comp. Syst. **78**, 778–788 (2018)
21. Naveed, M., Prabhakaran, M., Gunter, C.A.: Dynamic searchable encryption via blind storage. In: 2014 IEEE Symposium on Security and Privacy, SP 2014, Berkeley, CA, USA, 18–21 May 2014, pp. 639–654 (2014)
22. Shamir, A.: How to share a secret. Commun. ACM **22**(11), 612–613 (1979)
23. Song, D.X., David A. Wagner, Perrig, A.: Practical techniques for searches on encrypted data. In: 2000 IEEE Symposium on Security and Privacy, Berkeley, California, USA, 14–17 May 2000, pp. 44–55 (2000)
24. Sun, W., Liu, X., Lou, W., Hou, Y.T., Li, H.: Catch you if you lie to me: efficient verifiable conjunctive keyword search over large dynamic encrypted cloud data. In: 2015 IEEE Conference on Computer Communications, INFOCOM 2015, Kowloon, Hong Kong, 26 April–1 May 2015, pp. 2110–2118 (2015)
25. Sun, W., et al.: Verifiable privacy-preserving multi-keyword text search in the cloud supporting similarity-based ranking. IEEE Trans. Parallel Distrib. Syst. **25**(11), 3025–3035 (2014)

26. Wang, C., Cao, N., Li, J., Ren, K., Lou, W.: Secure ranked keyword search over encrypted cloud data. In: 2010 International Conference on Distributed Computing Systems, ICDCS 2010, Genova, Italy, 21–25 June 2010, pp. 253–262 (2010)
27. Wang, C., Cao, N., Ren, K., Lou, W.: Enabling secure and efficient ranked keyword search over outsourced cloud data. IEEE Trans. Parallel Distrib. Syst. **23**(8), 1467–1479 (2012)
28. Xia, Z., Wang, X., Sun, X., Wang, Q.: A secure and dynamic multi-keyword ranked search scheme over encrypted cloud data. IEEE Trans. Parallel Distrib. Syst. **27**(2), 340–352 (2016)
29. Yu, J., Lu, P., Zhu, Y., Xue, G., Li, M.: Toward secure multikeyword top-k retrieval over encrypted cloud data. IEEE Trans. Dependable Sec. Comput. **10**(4), 239–250 (2013)
30. Zhang, W., Lin, Y., Xiao, S., Wu, J., Zhou, S.: Privacy preserving ranked multi-keyword search for multiple data owners in cloud computing. IEEE Trans. Comput. **65**(5), 1566–1577 (2016)
31. Zhu, J., Li, Q., Wang, C., Yuan, X., Wang, Q., Ren, K.: Enabling generic, verifiable, and secure data search in cloud services. IEEE Trans. Parallel Distrib. Syst. **29**(8), 1721–1735 (2018)

Multi-owner Secure Encrypted Search Using Searching Adversarial Networks

Kai Chen[1], Zhongrui Lin[2], Jian Wan[2], Lei Xu[1], and Chungen Xu[1(✉)]

[1] School of Science, Nanjing University of Science and Technology, Nanjing, China
[2] School of Computer Science and Engineering, NJUST, Nanjing, China
{kaichen,xuchung}@njust.edu.cn

Abstract. Searchable symmetric encryption (SSE) for multi-owner model draws much attention as it enables data users to perform searches over encrypted cloud data outsourced by data owners. However, implementing secure and precise query, efficient search and flexible dynamic system maintenance at the same time in SSE remains a challenge. To address this, this paper proposes secure and efficient multi-keyword ranked search over encrypted cloud data for multi-owner model based on searching adversarial networks. We exploit searching adversarial networks to achieve optimal pseudo-keyword padding, and obtain the optimal game equilibrium for query precision and privacy protection strength. Maximum likelihood search balanced tree is generated by probabilistic learning, which achieves efficient search and brings the computational complexity close to $\mathcal{O}(\log N)$. In addition, we enable flexible dynamic system maintenance with balanced index forest that makes full use of distributed computing. Compared with previous works, our solution maintains query precision above 95% while ensuring adequate privacy protection, and introduces low overhead on computation, communication and storage.

Keywords: Searchable symmetric encryption · Multi-owner · Ranked search · Searching adversarial networks · Maximum likelihood

1 Introduction

Background and Motivation. In cloud computing, searchable symmetric encryption (SSE) for multiple data owners model (multi-owner model, MOD) draws much attention as it enables multiple data users (clients) to perform searches over encrypted cloud data outsourced by multiple data owners (authorities). Unfortunately, none of the previously-known traditional SSE scheme for MOD achieve secure and precise query, efficient search and flexible dynamic system maintenance at the same time [9]. This severely limits the practical value of SSE and decreases its chance of deployment in real-world cloud storage systems.

Related Work and Challenge. SSE has been continuously developed since it was proposed by Song et al. [12], and *multi-keyword ranked search* over

© Springer Nature Switzerland AG 2019
Y. Mu et al. (Eds.): CANS 2019, LNCS 11829, pp. 184–195, 2019.
https://doi.org/10.1007/978-3-030-31578-8_10

encrypted cloud data scheme is recognized as outstanding [9]. Cao et al. [1] first proposed privacy-preserving multi-keyword ranked search scheme (MRSE), and established strict privacy requirements. They first employed *asymmetric scalar-product preserving encryption* (ASPE) approach [15] to obtain the similarity scores of the query vector and the index vector, so that the cloud server can return the *top-k* documents. However, they did not provide the optimal balance of query precision and privacy protection strength. For better query precision and query speed, Sun et al. [13] proposed MTS with the TF × IDF keyword weight model, where the keyword weight depends on the frequency of the keyword in the document and the ratio of the documents containing this keyword to the total documents. This means that TF × IDF cannot handle the differences between data from different owners in MOD, since each owner's data is different and there is no uniform standard to measure keyword weights. Based on MRSE, Li et al. [8] proposed a better solution (MKQE), where a new index construction algorithm and trapdoor generation algorithm are designed to realize the dynamic expansion of the keyword dictionary and improve the system performance. However, their scheme only realized the linean search efficiency. Xia et al. [16] provided EDMRS to support flexible dynamic operation by using balanced index tree that builded following the bottom-up strategy and "greedy" method, and they used parallel computing to improve search efficiency. However, when migrating to MOD, ordinary balanced binary tree they employed is not optimistic [6]. It is frustrating that the above solutions only support SSE for single data owner model. Due to the diverse demand of the application scenario, such as emerging authorised searchable technology for multi-client (authority) encrypted medical databases that focuses on privacy protection [17,18], research on SSE for MOD is increasingly active. Guo et al. [6] proposed MKRS_MO for MOD, they designed a heuristic weight generation algorithm based on the relationships among keywords, documents and owners (KDO). They considered the correlation among documents and the impact of documents' quality on search results, so that the KDO is more suitable for MOD than the TF × IDF. However, they ignored the secure search scheme in *known background model* [1](a *threat model* that measures the ability of "honest but curious" cloud server [14,20] to evaluate private data and the risk of revealing private information in SSE system). Currently, SSE for MOD is still these challenges: **(1)** comprehensively optimizing query precision and privacy protection is difficult; **(2)** a large amount of different data from multiple data owners makes the data features sparse, and the calculation of high-dimensional vectors can cause "curse of dimensionality"; **(3)** frequent updates of data challenge the scalability of dynamic system maintenance.

Our Contribution. This paper proposes secure and efficient multi-keyword ranked search over encrypted cloud data for multi-owner model based on searching adversarial networks (MRSM_SAN). Specifically, including the following three techniques: **(1) optimal pseudo-keyword padding based on searching adversarial networks (SAN):** To improve the privacy protection strength of SSE is a top priority. Padding random noise into the data [1,8,19] is a current popular method designed to interfere with the analysis and evaluation from cloud server, which protects the document content and keyword information better. However, such an operation will reduce the query precision [1]. In response to this, we creatively use *adversarial learning* [4] to obtain the *optimal probability distribution* for controlling pseudo-keyword padding and the *optimal game equilibrium* for the query precision and the privacy protection strength. This makes query precision exceeds 95% while ensuring adequate privacy protection, which is better than traditional SSE [1,6,8,13,16]; **(2) efficient search based on maximum likelihood search balanced tree (MLSB-Tree):** The construction of the index tree is the biggest factor affecting the search efficiency. If the leaf nodes of the index tree are sorted by maximum probability (the ranking of the index vectors from high to low depends on the probability of being searched), the computational complexity will be close to $\mathcal{O}(\log N)$ [7]. *Probabilistic learning* is employed to obtain MLSB-Tree, which is ordered in a maximum probability. The experimental evaluation shows that MLSB-Tree-based search is faster and more stable compare with related works [6,16]; **(3) flexible dynamic system maintenance based on balanced index forest (BIF):** Using *unsupervised learning* [3,10] to design a fast index clustering algorithm to classify all index into multiple index partitions, and a corresponding balanced index tree is constructed for each index partition, thus all index trees form BIF. Owing to BIF is distributed, it only needs to maintain the corresponding index partition without touching all indexes in dynamic system maintenance, which improves the efficiency of index update operations and introduces low overhead on the computation, communication and storage. In summary, MRSM_SAN increases the possibility of deploying dynamic SSE in real-world cloud storage systems.

Organization and Version Notes. Section 2 describes scheme. Section 3 conducts experimental evaluation. Section 4 discusses our solution. Compared with the preliminary version [2], this paper adds algorithms, enhances security analysis, and conducts more in-depth experimental analysis of the proposed scheme.

2   Secure and Efficient MRSM_SAN

2.1   System Model

The proposed system model consists of four parties, is depicted in Fig. 1. **Data owners** (DO) are responsible for constructing searchable index, encrypting data and sending them to cloud server or trusted proxy; **Data users** (DU) are consumers of cloud services. Based on *attribute-based encryption* [5], once DO authorize DU attributes related to the retrieved data, DU can retrieve the corresponding data; **Trusted proxy** (TP) is responsible for index processing, query and trapdoor generation, user authority authentication; **Cloud server** (CS) provides cloud service, including running authorized access controls, performing searches for encrypted cloud data based on query requests, and returning *top-k* documents to data users. CS is considered "honest but curious" [14,20], so that it is necessary to provide a secure search scheme to protect privacy. Our goal is to protect index privacy, query privacy and keyword privacy in dynamic SSE.

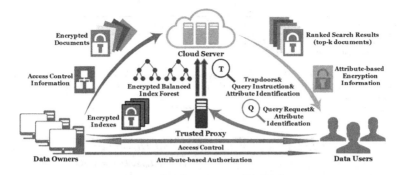

Fig. 1. The basic architecture of MRSM_SAN

2.2   MRSM_SAN Framework

***Setup*:** Based on index clustering results (s index partitions) and privacy requirements in *known background model* [1], TP determines the size N_i of sub-dictionary D_i, the number U_i of pseudo-keyword, sets the parameter $V_i = U_i + N_i$. Thus $V = \{V_1, \ldots, V_s\}$, $U = \{U_1, \ldots, U_s\}$, $N = \{N_1, \ldots, N_s\}$.

***KeyGen*(V):** TP generates key $SK = \{SK_1, \ldots, SK_s\}$, where $SK_i = \{S_i, M_{i,1}, M_{i,2}\}$, $M_{i,1}$ and $M_{i,2}$ are two $V_i \times V_i$-dimensional invertible matrices, S_i is a random V_i-dimensional vector. Symmetric key $\overline{SK}_i = \{S_i, M_{i,1}^{-1}, M_{i,2}^{-1}\}$.

***Extended-KeyGen*(SK_i, Z_i):** For dynamic search [8,16], if Z_i new keywords are added into the i-th sub-dictionary, TP generates a new key $SK_i' = \{S_i', M_{i,1}', M_{i,2}'\}$, where $M_{i,1}'$ and $M_{i,2}'$ are two $(V_i+Z_i) \times (V_i+Z_i)$-dimensional invertible matrices, S_i is new a random $(V_i + Z_i)$-dimensional vector.

***BuildIndex*(F, SK):** To realize secure search in *known background model* [1], TP pads U_i pseudo-keyword into weighted index $\overline{I}$ (associated with document F) to obtain secure index $\widetilde{I}$, and uses $\widetilde{I}$ and SK to generate BIF $\mathcal{F} = \{\tau_1, \ldots, \tau_s\}$ and encrypted BIF $\widetilde{\mathcal{F}} = \{\widetilde{\tau}_1, \ldots, \widetilde{\tau}_s\}$. Finally, TP sends $\widetilde{\mathcal{F}}$ to CS.

***Trapdoor*(Q, SK):** DU sends query request (keywords and their weights) and attribute identification to TP. TP generates query $Q = \{Q_1, \ldots, Q_s\}$ and generates trapdoor $T = \{T_1, \ldots, T_s\}$ using SK. Finally, TP sends T to CS.

***Query*($T, \widetilde{\mathcal{F}}, t, k$):** TP sends query information to CS and specifies t index partitions to be queried. CS performs searches and retrieves *top-k* documents.

2.3   Algorithms for Scheme

1. ***Binary Index Generation*:** DO_i uses algorithm 1 to generate the binary index (vector) I_i for the documents F_i, then sends I_i to TP.
2. ***Fast Index Clustering & Keyword Dictionary Segmentation*:** We employ Algorithm 2 to solve "curse of dimensionality" issue in computing.
3. ***Weighted Index Generation*:** TP exploits the KDO weight model [6] to generate the weight index, as shown in Algorithm 3.

Algorithm 1. *Binary Index Generation*

Input: Document set $F = \{F_1, \ldots, F_m\}$, keyword dictionary $D = \{w_1, \ldots, w_n\}$.
Output: Binary index set $I = \{I_1, \ldots, I_m\}$.
1: **for** $i = 1$, **to** m **do**
2:      Based on **Vector Space Model** [11] and keyword dictionary D, DO_i generates binary index $I_i = \{I_{i,1}, \ldots, I_{i,n_i}\}$ for documents $F_i = \{F_{i,1}, \ldots, F_{i,n_i}\}$, where $I_{i,j}$ is a binary index vector.
3:      **return** Binary index Index I

Algorithm 2. *Fast Index Clustering & Keyword Dictionary Segmentation*

Input: Binary index (vector) I from DO, where $I = \{I_i, \ldots, I_m\}$, $DO = \{DO_i, \ldots, DO_m\}$.
Output: s index partitions, s sub-dictionaries and new binary index $\widehat{I} = \{\widehat{I}_1, \ldots, \widehat{I}_s\}$.
1: *Local Clustering:* For $I = \{I_i, \ldots, I_m\}$, TP uses **Twin Support Vector Machine** [3] to classify the index vectors $\{I_{i,1}, \ldots, I_{i,n_i}\}$ in I_i into 2 clusters (i-th and $(m + i)$-th initial index partition) and obtain the representative vectors for the i-th and the $(m + i)$-th initial index partition. **return** $2m$ initial index partitions and their representative vectors.
2: *Global Clustering:* TP uses **Manhattan Frequency k-Means** [10] algorithm to group all initial index partitions (representative vectors) into s final index partitions. **return** s index partitions.
3: *Keyword Dictionary Segmentation:* According to the obtained s index partitions, the keyword dictionary D is divided into s sub-dictionaries $D_1, \ldots, D_s$ correspondingly. TP obtains new binary index $\widehat{I} = \{\widehat{I}_1, \ldots, \widehat{I}_s\}$, where $I_i = \{\widehat{I}_{i,1}, \ldots, \widehat{I}_{i,M_i}\}$, $\widehat{I}_{i,j}$ is a N_i-dimensional vector. Delete **"public redundancy zero element"** of all index vectors in the same index partition. **return** s sub-dictionaries and new binary index $\widehat{I}$ (low-dimensional) for s index partitions.

Algorithm 3. *Secure Weighted Index Generation*

Input: Binary index $\widehat{I}$ for s index partitions.
Output: Secure Weighted index $\overline{I}$ for s index partitions, i.e. the data type is floating point.
1: *Correlativity Matrix Generation:* Using the corpus to determine the semantic relationship between different keywords and obtain the *correlativity matrix* $S_{N_i \times N_i}$ (symmetric matrix).
2: *Weight Generation:* Based on KDO [6], construct the average keyword popularity AKP about DO. Specifically, calculate AKP_i of DO_i with equation "$AKP_i = (P_i \cdot \widehat{I}_i) \otimes \alpha_i$", where $\widehat{I}_i$ is the index after index clustering, the operator $\otimes$ denotes the product of two vectors corresponding elements, $\alpha_i = (\alpha_{i,1}, \ldots, \alpha_{i,N_i})$, if $|L_i(w_t)| \neq 0$ (the number of documents contain keyword w_t), $\alpha_{i,t} = \frac{1}{|L_i(w_t)|}$, otherwise $\alpha_{i,t} = 0$. Calculate the raw weight information for DO_i, $W_i^{raw} = S_{N_i \times N_i} \cdot AKP_i$, where $W_i^{raw} = (W_{i,1}^{raw}, \ldots, W_{i,N_i}^{raw})$.
3: *Normalized Processing:* Obtain the maximum raw weight of every keyword among different DO, $W_{max} = (W_{i',1}^{raw}, W_{i',2}^{raw}, \ldots)$. Based on the W_{max}, calculate $W_{i,t} = \frac{W_{i,t}^{raw}}{W_{max}[j]}$
4: *Weighted Index Generation:* TP obtains weighted index vector with "$\overline{I}_{i,j} = \widehat{I}_{i,j} \otimes W_i$", where $\overline{I}_{i,j}$ associated with document $F_{i,j}$ corresponds to the i-th index partition ($j \in \{1, \ldots, M_i\}$).
5: *Secure Weighted Index Generation:* TP pads U_i pseudo-keyword into $\overline{I}$ that in the i-th index partition, and obtains secure weighted index $\widetilde{I}$ with high privacy protection strength [16, 19].

Algorithm 4. *MLSB-Tree and BIF Generation*

Input: Secure weighted index $\widetilde{I}$ for s index partitions, randomly generated query vector Q.
Output: MLSB-Tree $\tau_1, \ldots, \tau_s$ for s index partitions and BIF $\mathcal{F}$ for all indexes belong to DO.
1: **for** $i = 1$, **to** s **do**
2:      **for** $j = 1$, **to** M_i **do**
3:           Based on **probabilistic learning**, TP calculates "$Score_{i,j} = \sum_{k=1}^{n} I_{i,j}^T \cdot Q_{i,k}$"; Then, TP sorts $I_{i,1}, \ldots, I_{i,M_i}$ according to $Score_{i,1}, \ldots, Score_{i,M_i}$; Finally, TP follows the bottom-up strategy and generates MLSB-Tree τ_i (balanced tree) with **greedy** method.
4:      **return** MLSB-Tree $\tau_1, \ldots, \tau_s$.
5: **return** BIF $\mathcal{F} = \{\tau_1, \ldots, \tau_s\}$.

Algorithm 5. *Encrypted MLSB-Tree and Encrypted BIF Generation*

Input: BIF $\mathcal{F} = \{\tau_1, \ldots, \tau_s\}$ and key $SK = \{SK_1, \ldots, SK_s\}$, where $SK_i = \{S_i, M_{i,1}, M_{i,2}\}$.
Output: Encrypted BIF $\widetilde{\mathcal{F}} = \{\widetilde{\tau_1}, \ldots, \widetilde{\tau_s}\}$.

1: **for** $i = 1$, to s **do**
2:     TP encrypts MLSB-Tree τ_i with the secret key SK_i to obtain encrypted MLSB-Tree $\widetilde{\tau_i}$.
3:     **for** $j = 1$, to n **do**
4:        TP "splits" vector $u_{i,j}.v$ of $u_{i,j}$ (node of τ_i) into two random vectors $u_{i,j}.v_1$, $u_{i,j}.v_2$.
5:        **if** $S_i[t] = 0$ **then**
6:           $u_{i,j}.v_1[t] = u_{i,j}.v_2[t] = u_{i,j}.v[t]$.
7:        **else**
8:           **if** $S_i[t] = 1$ **then**
9:              $u_i.v_1[t]$ is a random value $\in (0,1)$, $u_{i,j}.v_2[t] = u_{i,j}.v[t] - u_{i,j}.v_1[t]$.
10:        TP encrypts $u_{i,j}.v$ with reversible matrices $M_{i,1}$ and $M_{i,2}$ to obtain "$\widetilde{u_{i,j}.v} = \{\widetilde{u_{i,j}.v_1}$,
      $\widetilde{u_{i,j}.v_2}\} = \{M_{i,1}^T u_{i,j}.v_1, M_{i,2}^T u_{i,j}.v_2\}$", where $u_{i,j}.v_1$ and $u_{i,j}.v_2$ are V_i-length vectors
11:     **return** Encrypted MLSB-Tree $\widetilde{\tau_i}$.
12: **return** Encrypted Encrypted BIF $\widetilde{\mathcal{F}} = \{\widetilde{\tau_1}, \ldots, \widetilde{\tau_s}\}$.

Algorithm 6. *Trapdoor Generation and GDFS$(T, \widetilde{\mathcal{F}}, t, k)$*

Trapdoor Generation

Input: Query vectors $Q = \{Q_1, \ldots, Q_s\}$.
Output: Trapdoor $T = \{T_1, \ldots, T_s\}$.

1: **for** $i = 1$, to s **do**
2:    TP "splits" query vector Q_i into two random vectors $Q_{i,1}$ and $Q_{i,2}$.
3:    **if** $S_i[t] = 0$ **then**
4:      $Q_{i,1}[t]$ is a random value $\in (0,1)$, $Q_{i,2}[t] = Q_i[t] - Q_{i,1}[t]$.
5:    **else**
6:      **if** $S_i[t] = 1$ **then**
7:        $Q_{i,1}[t] = Q_{i,2}[t] = Q_i[t]$, where $t \in \{1, 2, \ldots, N_i\}$.
8:    TP encrypts $Q_{i,1}$ and $Q_{i,2}$ with reversible matrices $M_{i,1}^{-1}$ and $M_{i,2}^{-1}$ to obtain trapdoor "$T_i = \{\widetilde{Q}_{i,1}, \widetilde{Q}_{i,2}\} = \{M_{i,1}^{-1} Q_{i,1}, M_{i,2}^{-1} Q_{i,2}\}$".
9: **return** $T = \{T_1, \ldots, T_s\}$.

GDFS$(T, \widetilde{\mathcal{F}}, t, k)$

Input: Query$(T, \widetilde{\mathcal{F}}, t, k)$.
Output: *top-k* documents.

1: **for** $i = 1$, to s **do**
2:    **if** τ_i is the specified index tree **then**
3:     **if** $u_{i,j}$ is a non-leaf node **then**
4:      **if** $Score(\widetilde{u_{i,j}.v}, T_i) > \lceil \frac{k}{t} \rceil$-th score **then**
5:        GDFS($u_{i,j}$.high-child)
6:        GDFS($u_{i,j}$.low-child)
7:      **else**
8:        **return**
9:     **else**
10:      **if** $Score(\widetilde{u_{i,j}.v}, T_i) > \lceil \frac{k}{t} \rceil$-th score **then**
11:        Update $\lceil \frac{k}{t} \rceil$-th score for i-th index tree τ_i and the ranked search result list for $\widetilde{\mathcal{F}}$.
12: **return** the final *top-k* documents for $\widetilde{\mathcal{F}}$.

---

4. **MLSB-Tree and BIF Generation:** TP uses Algorithm 4 to generate MLSB-Tree $\tau_1, \ldots, \tau_s$ and BIF $\mathcal{F} = \{\tau_1, \ldots, \tau_s\}$.

5. **Encrypted MLSB-Tree and Encrypted BIF Generation.** TP encrypts $\mathcal{F}$ using Algorithm 5 and sends encrypted $\widetilde{\mathcal{F}}$ to CS. τ_i and $\widetilde{\tau_i}$ are isomorphic (i.e. $\tau_i \cong \widetilde{\tau_i}$) [16]. Thus, the search capability of tree is still well maintained.

6. **Trapdoor Generation.** Based on query request from DU, TP generates $Q = \{Q_1, \ldots, Q_s\}$ and $T = \{T_1, \ldots, T_s\}$ using Algorithm 6, and sends T to CS.

7. **Search Process.** (1) *Query Preparation*: DU send query requests and attribute identifications to TP. If validating queries are valid, TP generates trapdoors and initiates search queries to CS. If access control passes, CS performs searches and returns *top-k* documents to DU. Otherwise CS refuses to query. (2) *Calculate Matching Score for Query on MLBS-Treeτ_i*:

$$Score(\widetilde{u_{i,j}.v}, T_i) = \{M_{i,1}^T u_{i,j}.v_1, M_{i,2}^T u_{i,j}.v_2\} \cdot \{M_{i,1}^{-1} Q_{i,1}, M_{i,2}^{-1} Q_{i,2}\} = u_{i,j}^T.v \cdot Q_i$$

(3) *Search Algorithm for BIF*: the greedy depth-first search (GDFS) algorithm for BIF as shown in Algorithm 6.

2.4 Security Improvement and Analysis

Adversarial Learning. Padding random noise into the data [1,8,19] is a popular method to improve security. However, pseudo-keyword padding that follows different probability distributions will reduce query precision to varying degrees [1,8]. Therefore, it is necessary to optimize the probability distribution that controls pseudo-keyword padding. To address this, *adversarial learning* [4] for optimal pseudo-keyword padding is proposed. As shown in Fig. 2. ***Searcher Network $S(\varepsilon)$*** : The search result is generated by taking the random noise ε (the object probability distribution $p(\varepsilon)$) as an input and performing a search, and supplies the search result to the discriminator network $D(x)$. ***Discriminator Network $D(x)$***: The input has an accurate actual result or search result and attempts to predict whether the current input is an actual result or a search result. One of the inputs x is obtained from the real search result distribution $p(x)$, and then one or two are solved. Classify problems and generate scalars ranging from 0 to 1. Finally, in order to reach a balance point which is the best point of the minimax game, $S(\varepsilon)$ generates search results, and $D(x)$ (considered as adversary) considers the probability that $S(\varepsilon)$ produces the accurate real results is 0.5, i.e. it is difficult to distinguish between padding and without-padding, thus it can achieve effective security [19].

Fig. 2. Searching Adversarial Networks

Similar to GAN [4], to learn the searcher's distribution p_s over data x, we define a prior on input noise variables $p_\varepsilon(\varepsilon)$, then represent a mapping to data space as $S(\varepsilon; \theta_s)$, where S is a differentiable function represented by a multi-layer perception with parameters θ_g. We also define a second multi-layer perception $D(x; \theta_d)$ that outputs a single scalar. $D(x)$ represents the probability that x came from the data rather than p_s. We train D to maximize the probability of assigning the correct label to both training examples and samples from S. We simultaneously train S to minimize $\log(1 - D(S(\varepsilon)))$: In other words, D and S play the following two-player minimax game with value function $V(S, D)$:

$$\min_S \max_D V(D, S) = \mathbb{E}_{x \sim p_{data}(x)}[\log D(x)] + \mathbb{E}_{x \sim p_\varepsilon(\varepsilon)}[\log(1 - D(S(\varepsilon)))]$$

Security Analysis. Index confidentiality and query confidentiality: ASPE approach [15] is widely used to generate secure index/query in privacy-preserving keyword search schemes [1,6,8,13,16] and its security has been proven. Since the index/query vector is randomly generated and search queries return only the secure inner product [1] computation results (non-zero) of encrypted index and trapdoor, thus CS is difficult to accurately evaluate the keywords including in the query and matching *top-k* documents. Moreover, confidentiality is further enhanced as the optimal pseudo-keyword padding is difficult to distinguish and the transformation matrices are harder to figure out [15].

Query Unlinkability: By introducing the random value ε (padding pseudo-keyword), the same search requests will generate different query vectors and receive different relevance score distributions [1,16]. The optimal game equilibrium for precision and privacy is obtain by *adversarial learning*, which further improves query unlinkability. Meanwhile, SAN are designed to protect access pattern [19], which makes it difficult for CS to judge whether the retrieved ranked search results come from the same requests.

Keyword Privacy: According to the security analysis in [16], for i-th index partition, aiming to maximize the randomness of the relevance score distribution, it is necessary to obtain as many different $\sum \varepsilon_{\nu_i}$ as possible (where $\nu_i \in \{j | Q_i[j + N_i] = \alpha_i, j = 1, \ldots, U_i\}$; in [16], $\alpha_i = 1$). Assuming each index vector has at least 2^{ω_i} different $\sum \varepsilon_{\nu_i}$ choices, the probability of two $\sum \varepsilon_{\nu_i}$ share the same value is less than $\frac{1}{2^{\omega_i}}$. If we set each $\varepsilon_j \sim U(\mu_i' - \delta_i, \mu_i' + \delta_i)$ (*Uniform distribution*), according to the central limit theorem, $\sum \varepsilon_{\nu_i} \sim N(\mu_i, \sigma_i^2)$ (*Normal distribution*), where $\mu_i = \omega_i \mu_i'$, $\sigma^2 = \frac{\omega_i \delta_i^2}{3}$. Therefore, it can set $\mu_i = 0$ and balance precision and privacy by adjusting the variance σ_i in real-world application. In fact, when $\alpha_i \in [0,1]$ (floating point number), SAN can achieve stronger privacy protection.

3 Experimental Evaluation

We implemented the proposed scheme using Python in Windows 10 operation system with Intel Core i5 Processor 2.40 GHz and evaluated its performance on a real-world data set (academic conference publications provided by IEEE xplore https://ieeexplore.ieee.org/, including 20,000 papers and 80,000 different keywords, 400 academic conferences were randomly selected as data owners DO). All experimental results represent the average of 1000 trials.

Optimal Pseudo-Keyword Padding. The parameters controlling the probability distribution (using SAN to find or approximate) are adjusted to find the optimal game equilibrium for *query precision* P_k (denoted as x) and *rank privacy protection* P_k' (denoted as y) (where $P_k = k'/k$, $P_k' = \sum |r_i - r_i'|/k^2$, k' and r_i are respectively the number of real *top-k* documents and the rank number of document in the retrieved k documents, and r_i' is document's real rank number in the whole ranked results [1]). We choose 95% query precision and 80% rank privacy protection as benchmarks to get the *game equilibrium score* calculation formula: $f(x,y) = \frac{1}{95}x^2 + \frac{1}{80}y^2$ (objective function to be optimized). As shown in Fig. 3, we find the optimal game equilibrium (max $f(x,y) = 177.5$) at $\sigma_1 = 0.05$, $\sigma_2 = 0.08$, $\sigma_3 = 0.12$. The corresponding query precision are: 98%, 97%, 93%. The corresponding rank privacy protection are: 78%, 79%, 84%. Therefore, we can choose the best value of σ to achieve optimal pseudo-keyword padding to satisfy query precision requirement and maximize rank privacy protection.

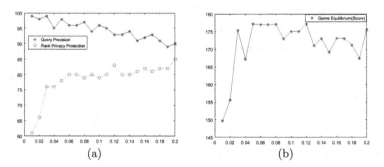

(a)   (b)

Fig. 3. With different choice of standard deviation σ for the random variable ε. {(a) query precision (%) and rank privacy protection (%); (b) game equilibrium (score). explanation for $\sigma \in [0.01, 0.2]$: When σ is greater than 0.2, the weight of the pseudo-keyword may be greater than 1, which violates our weight setting (between 0 and 1), so we only need to find the best game equilibrium point when $\sigma \in [0.01, 0.2]$}.

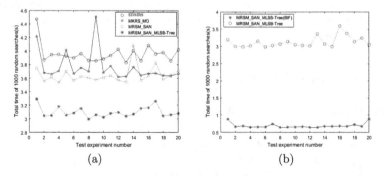

(a)   (b)

Fig. 4. Time cost of query for 1000 random searches in 500 sizes of data set. (a) Comparison of tree-based search efficiency. Since the query is random, the search time fluctuates, which causes the curves in the graph to have intersections; (b) Comparison of MLSB-Tree and BIF search efficiency.

Search Efficiency of MLSB-Tree. Search efficiency is mainly described by query speed, and our experimental objects are index trees that are structured with different strategy: EDMRS [16] (ordinary balanced binary tree), MKRS_MO [6] (grouped balanced binary tree), MRSM_SAN (globally grouped balanced binary tree), MRSM_SAN_MLSB-Tree. We first randomly generate 1000 query vectors, then perform search on each index tree respectively, finally take the results of 20 repeated experiments for analysis. As shown in Fig. 4(a), the query speed and query stability based on MLSB-Tree are better than other index trees. Compared with EDMRS and MKRS_MO, query speed increased by 21.72% and 17.69%. In terms of stability, MLSB-Tree is better than other index trees. (variance of search time(s): 0.0515 [6], 0.0193 [16], **0.0061** [MLSB-Tree]).

Search Efficiency of BIF. As shown in Fig. 4(b), query speed of MRSM_SAN (with MLSB-Tree and BIF) is significantly higher than MRSM_SAN (only with MLSB-Tree), and the search efficiency is improved by 5 times and the stability increase too. This is just the experimental result of 500 documents set with the 4000-dimension keyword dictionary. After the index clustering operation, the keyword dictionary is divided into four sub-dictionaries with a dimension of approximately 1000. As the amount of data increases, the dimension of the keyword dictionary will become extremely large, and the advantages of BIF will become more apparent. In our analytical experiments, the theoretical efficiency ratio before and after segmentation is: $\eta = s \frac{\mathcal{O}(\log N)}{\mathcal{O}(\log N) - \mathcal{O}(\log s)}$, where s is the number of index partitions after fast index clustering, and N is the number of documents included. When the amount of data increases to 20,000, the total keyword dictionary dimension is as high as 80,000. If the keyword sub-dictionary dimension is 1000, the number of index partitions after fast index clustering is 80, the search efficiency will increase by more than 100 times ($\eta = 143$). This will bring huge benefits to large information systems, and our solutions can exchange huge returns with minimal computing resources.

Comparison of Search Efficiency (Larger Data Set). The efficiency of MRSM_SAN (without BIF) and related works [1,6,8,16] are show as Fig. 5(a), and the efficiency of MRSM_SAN(without BIF) and MRSM_SAN(with BIF) are show as Fig. 5(b). It is more notable that the maintenance cost of scheme based on BIF is much lower than the cost of scheme only based on a balanced index tree. When the data owner has added a new document to the cloud server, and TP needs to insert a new index node in the index tree of the cloud server accordingly. If it is only based on an index tree, it must search for at least $\mathcal{O}(\log N)$ times search and at least $\mathcal{O}(\log N)$ times data updates, the total cost is $2\mathcal{O}(\log N)$ (where N is the number of index vectors that contained by the index tree). But BIF is very different, because we group all index vectors into s different index partitions. We assume that the number of index vectors in each partition is equal so we need to spend the same update operation for each partition, which makes the overhead is only $2(\mathcal{O}(\log N) - \mathcal{O}(\log s))$. Moreover, the increase in efficiency is positively correlated with the increase in data volume and data sparsity.

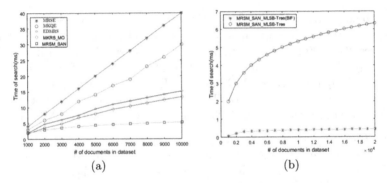

(a)   (b)

Fig. 5. Time cost of query for the same query keywords (10 keywords) in different sizes of data set. {(a) Experimental results show that our solution achieves near binary search efficiency and is superior to other existing comparison schemes. As the amount of data increases, our solution has a greater advantage. It's worth noting that this is just the search performance based on MLSB-Tree; (b) Comparison of MLSB-Tree and BIF. Based on experimental analysis, it concludes that when data volume grows exponentially data features become more sparse, if all index vectors only rely on an index tree to complete the search task, the computational complexity will be getting farther away from $\mathcal{O}(\log N)$. Sparseness of data features makes the similarity between index vectors is mostly close to zero or even equal to zero, which brings trouble to the pairing of index vectors. Moreover, the construction of balanced index tree is not global order, so it is necessary to traverse many nodes in the search, which proves the limitation of *balanced binary tree* [6, 16]. We construct MLSB-Tree with *maximum likelihood method* and *probabilistic learning*. Interestinglythe closer the number of random searches is to infinity, the higher the search efficiency of obtained index tree, this makes the computational complexity of search can converge to $\mathcal{O}(\log N)$}.

4   Discussion

This paper proposes secure and efficient MRSM_SAN, and conducts in-depth security analysis and experimental evaluation. Creatively using *adversarial learning* to find *optimal game equilibrium* for query precision and privacy protection strength and combining traditional SSE with *uncertain control theory*, which opens a door for *intelligent SSE*. In addition, we propose MLSB-Tree, which generated by a sufficient amount of random searches and brings the computational complexity close to $\mathcal{O}(\log N)$. It means that using probabilistic learning to optimize the query result is effective in an *uncertain system* (owner's data and user's queries are uncertain). Last but not least, we implement flexible dynamic system maintenance with BIF, which not only reduces the overhead of dynamic maintenance and makes full use of distributed computing, but also improves the search efficiency and achieves fine-grained search. This is beneficial to improve the availability, flexibility and efficiency of dynamic SSE system.

Acknowledgment. This work was supported by "the Fundamental Research Funds for the Central Universities" (No. 30918012204) and "the National Undergraduate Training Program for Innovation and Entrepreneurship" (Item number: 201810288061). NJUST graduate Scientific Research Training of 'Hundred, Thousand and Ten Thousand' Project *"Research on Intelligent Searchable Encryption Technology"*.

References

1. Cao, N., Wang, C., Li, M., Ren, K., Lou, W.: Privacy-preserving multi-keyword ranked search over encrypted cloud data. IEEE TPDS **25**(1), 222–233 (2014)
2. Chen, K., Lin, Z., Wan, J., Xu, L., Xu, C.: Multi-client secure encrypted search using searching adversarial networks. IACR Cryptology ePrint Archive **2019**, 900 (2019)
3. Chen, S., Cao, J., Huang, Z., Shen, C.: Entropy-based fuzzy twin bounded support vector machine for binary classification. IEEE Access **7**, 86555–86569 (2019)
4. Goodfellow, I.J., et al.: CoRR abs/1406.2661 (2014)
5. Goyal, V., Pandey, O., Sahai, A., Waters, B.: Attribute-based encryption for fine-grained access control of encrypted data. In: ACM CCS 2006, pp. 89–98. ACM (2006)
6. Guo, Z., Zhang, H., Sun, C., Wen, Q., Li, W.: Secure multi-keyword ranked search over encrypted cloud data for multiple data owners. J. Syst. Softw. **137**(3), 380–395 (2018)
7. Knuth, D.E.: The Art of Computer Programming, vol. III, 2nd edn. Addison-Wesley, Boston (1998)
8. Li, R., Xu, Z., Kang, W., Yow, K., Xu, C.: Efficient multi-keyword ranked query over encrypted data in cloud computing. FGCS **30**, 179–190 (2014)
9. Poh, G.S., Chin, J., Yau, W., Choo, K.R., Mohamad, M.S.: Searchable symmetric encryption: designs and challenges. ACM Comput. Surv. **50**(3), 40:1–40:37 (2017)
10. Salem, S.B., Naouali, S., Chtourou, Z.: A fast and effective partitional clustering algorithm for large categorical datasets using a k-means based approach. Comput. Electr. Eng. **68**, 463–483 (2018)
11. Salton, G., Wong, A., Yang, C.: A vector space model for automatic indexing. Commun. ACM **18**(11), 613–620 (1975)
12. Song, D.X., Wagner, D.A., Perrig, A.: Practical techniques for searches on encrypted data. In: IEEE S and P 2000, pp. 44–55. IEEE Computer Society (2000)
13. Sun, W., et al.: Verifiable privacy-preserving multi-keyword text search in the cloud supporting similarity-based ranking. IEEE TPDS **25**(11), 3025–3035 (2014)
14. Wang, C., Wang, Q., Ren, K., Lou, W.: Privacy-preserving public auditing for data storage security in cloud computing. In: IEEE INFOCOM 2010, pp. 525–533. IEEE (2010)
15. Wong, W.K., Cheung, D.W., Kao, B., Mamoulis, N.: Secure KNN computation on encrypted databases. In: ACM SIGMOD 2009, pp. 139–152. ACM (2009)
16. Xia, Z., Wang, X., Sun, X., Wang, Q.: A secure and dynamic multi-keyword ranked search scheme over encrypted cloud data. IEEE TPDS **27**(2), 340–352 (2016)
17. Xu, L., Sun, S., Yuan, X., Liu, J.K., Zuo, C., Xu, C.: Enabling authorized encrypted search for multi-authority medical databases. IEEE TETC (2019). https://doi.org/10.1109/TETC.2019.2905572
18. Xu, L., Xu, C., Liu, J.K., Zuo, C., Zhang, P.: Building a dynamic searchable encrypted medical database for multi-client. Inf. Sci. (2019). https://doi.org/10.1016/j.ins.2019.05.056
19. Xu, L., Yuan, X., Wang, C., Wang, Q., Xu, C.: Hardening database padding for searchable encryption. In: IEEE INFOCOM 2019, pp. 2503–2511. IEEE (2019)
20. Yu, S., Wang, C., Ren, K., Lou, W.: Achieving secure, scalable, and fine-grained data access control in cloud computing. In: IEEE INFOCOM 2010, pp. 534–542. IEEE (2010)

Blockchains

Designing Smart Contract for Electronic Document Taxation

Dimaz Ankaa Wijaya[1,2(✉)], Joseph K. Liu[1], Ron Steinfeld[1], Dongxi Liu[2], Fengkie Junis[3], and Dony Ariadi Suwarsono[3,4]

[1] Monash University, Melbourne, Australia
{dimaz.wijaya,joseph.liu,ron.steinfeld}@monash.edu
[2] Data61, CSIRO, Canberra, Australia
dongxi.liu@data61.csiro.au
[3] Gadjah Mada University, Yogyakarta, Indonesia
fengkie.junis@mail.ugm.ac.id
[4] Directorate General of Taxes, Jakarta, Indonesia
dony.suwarsono@pajak.go.id

Abstract. In recent years, we have seen a massive blockchain adoption in cryptocurrencies such as Bitcoin. Following the success of blockchain in cryptocurrency industry, many people start to explore the possibility of implementing blockchain technology in different fields. In this paper, we propose Smart Stamp Duty, a system which can revolutionize the way stamp duty (document tax) is managed and paid. The proposed Smart Stamp Duty offers significant improvements on the convenience when paying stamp duty by removing the need of physical revenue stamp. At the same time, the blockchain technology also provides the auditability of the recorded data. Smart stamp duty enables the expansion of the existing electronic stamp duty application to retail level. Smart stamp duty also enables the taxpayers to pay the stamp duty of their electronic documents. Our proposed system also enables the taxpayers to convert their electronic documents into physical documents while still maintain the ability to refer the electronic-based stamp duty payments.

Keywords: Blockchain · Smart contract · Electronic document · Taxation

1 Introduction

According to 2018 Indonesia State Budget, taxation is the major source of state revenue, comprising 1.6 quadrillion Rupiah. The number has sharply increased from 1.47 quadrillion Rupiah in 2017 [9–11, 15]. The increase of taxation target in 2018 is due to several assumptions, including automatic exchange of information (AEoI), tax incentives, and improvements on human resources, information technology, and taxation services to the taxpayers [10].

Stamp duty is a state tax levied on specific types of documents. Stamp duty is categorised as "Other Taxes" category in Indonesia's with a total revenue target

© Springer Nature Switzerland AG 2019
Y. Mu et al. (Eds.): CANS 2019, LNCS 11829, pp. 199–213, 2019.
https://doi.org/10.1007/978-3-030-31578-8_11

of 9 trillion Rupiahs [9]. This target is about 0.6% of the overall 2018 revenue, increasing 11% from the 2017 state budget. Despite the small percentage the stamp duty target carries, Directorate General of Taxes (DGT) as the Indonesian Tax Authority looks for revenue improvements in this area [17]. Online transactions should be evaluated as the users might also need to pay stamp duties on the electronic documents, aside from the paper-based documents, involved in the business [21].

1.1   Problem Definition

Indonesian stamp duty is currently only levied on paper-based documents, as defined in the Stamp Duty Act 1985 [13]. The taxpayers need to attach physical revenue stamps, sign them properly, write the dates, and only then the tax payments are deemed valid. The problem with physical stamp is that it is inconvenient to get, because the taxpayers need to buy the physical stamps on brick-and-mortar stores to be attached on their physical documents.

There is also a limited use of computerized stamp duty which is only valid under certain circumstances [16]. To use the computerized stamp duty, the taxpayers need to purchase tax credits. The amount of the tax credits to be purchased is based on their estimated number of documents they will create in the following month. However, not everyone is allowed to use the computerized stamp duty. Only those who create at least one hundred documents per day subject to stamp duty are allowed to use the mechanism.

The existing stamp duty mainly focuses on physical documents; hence, the physical stamp is required. However, as technology advances, computerization changes the way people conduct their business. Electronic documents are more convenient to create, manage, and process compared to physical documents. Although these electronic documents may be owed stamp duty, it is infeasible for the taxpayers to purchase physical stamps and attach them to the electronic documents [5].

In summary, we identify the gap of using computerized stamp duty where individuals or companies that prefer to use electronic documents where these documents are taxable by the stamp duty regulations. Several solutions have been proposed to solve the problem [7, 21]. However, the usability and correctness of the data stored in the proposed solutions are questionable, where the payment proofs are stored in traditional information systems without any proper auditable features by all parties.

1.2   Our Contribution

Considering the current limitation of both physical and electronic stamp duty mechanism, we propose a new blockchain-based electronic stamp duty as an effort to increase the stamp duty revenue by using technological advancement. The system will increase the convenience of the taxpayers to pay the stamp duty when creating taxable documents. The system can be utilised by all taxpayers through information technology devices. The system will also provide a public verifiability to prove that an electronic document has the stamp duty paid. The built-in cryptographic techniques also prevent forgery which happened in the physical stamp duty [8].

We coin the term `Smart Stamp Duty` to refer to our proposed system which utilises smart contract as well as blockchain technology to record information and conduct predetermined business processes automatically based on the user's inputs and program logic. The smart contract is fully auditable, transparent to its stakeholders, and the finality of the data is guaranteed. Moreover, its standard features such as timestamp and digital signature fulfil the requirements of a valid stamp duty. Smart contract is scalable; not only retail users but also wholesale users can use the `Smart Stamp Duty` as long as there exist sufficient applications to pose as user interfaces to communicate with the smart contract.

2   Background

2.1   Indonesian Stamp Duty

Indonesian Stamp Duty Act 1985 determines that there are documents (in the physical form) owed stamp duty [13]. As a proof that the tax is paid, a form of physical stamp is glued on the documents and signed [13]. The cost of the stamp depends on the category of the document and the amount of money printed on the document [13].

2.2   Blockchain

Blockchain was first applied in Bitcoin, a decentralized payment system introduced by Satoshi Nakamoto [12]. The blockchain technology enables users to create transactions without the need of proving their real identities. There is no central authority controlling the system. Instead every user can verify and validate transactions data by using a set of protocols.

Blockchain can be divided into three models based on its type of participants, namely public blockchain, private blockchain, and consortium blockchain [18]. Public blockchain is a type of blockchain which allows everyone to participate at any time and allows anyone to leave whenever they desire. A private blockchain, as its name implies, is a closed system where the participants need specific permissions from a central authority which controls the whole system. Although blockchain is originally intended to remove the central authority, however in some cases the central authority is still required and cannot be removed, for example due to legal requirements.

The last blockchain model is consortium blockchain. It is a combination between public and private blockchain, where there will be several authorities sharing the power equally among them to control the system, however the membership of the authorities is not open as in a public blockchain. Consortium blockchain also applies access control mechanism where not everyone has access to the blockchain.

To replace the role of a central authority, blockchain utilises consensus mechanism to determine the canonical version of the record and rules who may write new information for a given block. Proof-of-Work (PoW) and proof-of-stake (PoS) are two most common consensus mechanisms for public blockchains. Consortium or private blockchain might use different methods of consensus algorithm that meet the system requirement.

2.3   Smart Contract

While Bitcoin's blockchain has a limited set of operation codes (*opcodes*), smart contract platforms such as Ethereum was created to support Turing-complete application in the blockchain [20]. Smart contract is generally an application stored in the blockchain. It receives inputs from the user, runs functions, and generates outputs. The execution of the smart contract codes are generally executed on-chain by the nodes of the network. As it is stored in the blockchain, the codes are permanent and transparent, hence it offers fairness to the participants. A smart contract platform also inherits the same traits as other blockchain systems, where the transaction data is visible and verifiable to anyone.

3   Related Work

3.1   Electronic Stamp Duty

The business model of electronic stamp duty is investigated by [7,21]. The Information and Electronic Transaction Act 2008 has provided a foundation in which the term document can be extended, not only in the physical form but also in electronic form, since both forms are now considered as legal proofs in Indonesian jurisdiction [14]. Before the act was established, all legal proofs must be printed and legalized, but now all electronic information no longer need to be printed in physical form, and if the information is properly handled, it can be a legal proof.

3.2   Blockchain Technology in Taxation

Blockchain does not only show its massive potential in the payment industry, but also other areas such as taxation. Previous studies describe that blockchain technology can mitigate the problem of tax losses due to international trades [1–4]. The blockchain technology has been proposed to modernize Value-Added Tax (VAT) [19]. In this case, the blockchain is utilized to enhance the information openness across multiple authorities and, likewise, to share information between those authorities.

4   Smart Stamp Duty

4.1   Overview

Our system works in the blockchain environment where cryptographic techniques such as digital signature and hash functions exist. We will define the term tokens to refer to electronic money balance owned by a user. The tokens can be used to pay the stamp duty in a smart contract provided by the Tax Authority. Before a user pays the stamp duty, she needs to buy the tokens before paying the stamp duty for each taxable document she creates. The transparency of the system enables the actors of the system to verify and validate all information stored in the blockchain.

4.2   The Blockchain

The blockchain as a shared database which can be used to store the data as well as the smart contracts, and then run the smart contracts based on the input provided by the users. In the proposed system, we use the consortium-based permissioned blockchain model where the Tax Authority along with other trusted participants such as government's auditing bodies can validate new transactions and secure the blockchain.

There will be multiple nodes running the system. These nodes can be managed by multiple participants, for example banks, the tax authority, and the monitoring bodies. Each of these participants keeps a copy of all transactions in the system for auditability and accountability purposes. Although there can be many nodes owned by different participants, the access rights will still need to be set up by the Tax Authority as the system administrator.

4.3   The Smart Contract

The smart contract is a set of permanent codes stored in the blockchain. The smart contract also stores information and runs application logic based on the previously stored codes. The purpose of the smart contract is to make the application logic transparent, auditable, and to let everyone sees information stored in the smart contract. The smart contract of our solution will have several parameters to be set by the owner based on the regulations. The smart contract of our solution is created and owned by the Tax Authority.

4.4   The Roles

We determine several roles involved in the system: the users, the bank, the Tax Authority, and monitoring organisations. Each role has its own credentials to do different activities.

The Taxpayers. The taxpayers are using the system to pay their tax dues, e.g. stamp duty. They interact with the system by using application interfaces provided. The taxpayers interact with the banks or any retail sellers to buy balance by using electronic money, credit cards, bank accounts, or cash. The users produce the documents, pay the stamp duty, and send the documents to other users. The users can also verify that the documents they receive have the tax paid.

The Bank. The bank usually provides financial services to the users. In the proposed system, the bank also provides a gateway to convert the local currency to tokens to access the system. The bank ensures that the exchange system is accessible by the taxpayers whenever they want to buy the tokens to be used to pay the stamp duty. At the end of a predetermined period, the bank reports to the tax authority regarding the amount of money collected from the taxpayers and send all the money to the government's account.

The Tax Authority. The Tax Authority has the authority to determine the tax rate to be paid by the taxpayers. In our system, the Tax Authority controls the blockchain system, including deploying or modifying the smart contracts used to run the system. The Tax Authority could authorize other parties to get involved in the system.

Tax Monitoring Organisations. The tax monitoring organisations help running the system by participating the closed consensus mechanism. In this mechanism, these organisations along with the Tax Authority determine the transactions to be stored in the blockchain. These organisations can have access to the system and the information stored in it.

4.5   The Applications

The applications are the interface between the users and the blockchain/smart contract. The applications used by different actors can be different based on their own authority. The application used by The Tax Authority may have the ability to modify the variables (by using the appropriate private keys), but the users can only add new records.

There can also be some additional applications to conduct the business, such as word extractor which enables the user to create the appropriate Bloom Filter to supply the stamp duty payment. Bloom Filter checker will also be required for validation in case the user prints the electronic documents and the printed documents require verification. However, these additional applications are not discussed in this paper.

4.6   The Token

Tokens are numbers stored in a smart contract and can be moved from an address to other addresses. The tokens are not intended to replace Indonesian national fiat currency as it will violate Indonesian Currency Act 2011. Tokens are used as proofs that the taxpayers have purchased tax credits which can be used to pay stamp duty. The tax credits are in the form of the balance of addresses. Only one token type will be created and available in the `Smart Stamp Duty` system, where the number of tokens can be increased or reduced by the Tax Authority.

4.7   Core Business Process

The core business process is shown in Fig. 1. A simple description for each activity is as follows.

1. User A creates an electronic document (e-document) subject to stamp duty. The electronic document can be in any form (word processor, spreadsheet, raw data, etc).
2. User A purchases tokens from a bank which can be used to pay the stamp duty. The user informs the bank of her "address".
3. The bank credits the tokens to the user's address by sending the correct transaction to the smart contract. The token balance can be checked by User A.
4. User A converts the electronic document into a protected format (i.e. PDF format), then computes the hash value of the e-document. The user then creates a transaction to put the hash value of the document to the smart contract. The transaction will decrease User A's token balance based on the stamp duty paid.
   The original document that has been signed needs to be kept by User A, while a new copy of the original document can be modified to include the receipt of the transaction.
5. The bank routinely provides a report to the Tax Authority regarding the total amount of balance purchased by taxpayers. The report can also be sent automatically or by using a specific interface.
6. The Tax Authority rechecks the report received from the bank to the actual transactions in the smart contract. The process can also be converted into an automatic job.
7. User A sends both the original copy and the paid copy of the electronic document to User B.
8. User B checks whether the stamp duty has been paid by querying the blockchain.

New users can join the system immediately by creating new public key pairs. All transactions can be read by all actors, thus reduce the risk of fraud.

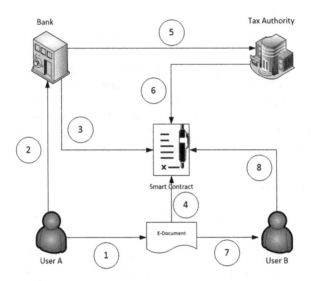

Fig. 1. The core business process of the Smart Stamp Duty.

4.8   The Parameters

Based on the business process we have defined, we then determine the detailed information to be stored in the smart contract. There are 3 types of the data: account parameters, system parameters, and transaction parameters.

Account Parameters. Account parameters are information regarding the token balance of the users. A new account will be created for a new user containing at least the user's public key and an initial token balance (can be zero). The public key cannot be modified while the token balance will always be recalculated by the smart contract based on new transaction parameters.

System Parameters. System parameters are information required to run the system. The smart contract looks up to the system parameters before executing transactions. The smart parameters are determined by the Tax Authority based on the existing regulations. The parameters can always be changed whenever the regulation changes. The system parameters contain at least the following:

- The cost of the stamp.
- The regulation references.

The system parameters can be stored on the same script of the main smart contract or be put on a different script.

Transaction Parameters. Transaction parameters are information regarding stamp duty transaction created by the users. To satisfy the existing regulation on stamp duty, the information items required for a valid transaction are the following:

- The public key of the taxpayer.
- The amount of the tax paid.
- Timestamp of the document.
- Hash value of the document.
- The payer's signature.

In the real world, it might be possible for the taxpayers to create identical documents multiple times, hence they produce an identical hash value. Thus, to minimize the occurrence, it is possible to add unique information into the documents, such as public keys of the taxpayers or their tax file number and the document creation time.

4.9   The Program Logic

Based on the business process and the parameters, it is possible to determine the program logic needed to run the system. All the program logic are run by the smart contract independently.

System Parameter Definition. If the regulation changes, then the Tax Authority can change the system parameter by sending a new transaction. The smart contract checks if all required data is satisfied by the new transaction then stores the new parameter. The system parameters are defined in Sect. 4.8.

Token Transaction. Activities related to tokens are: token creation, token distribution and token sales, and lastly stamp duty payment.

- *Token creation.* As a central controller, Tax Authority is the only party that is given the authority to create tokens. The number of tokens to be created depends on the need. It is also possible to "mint" or create more tokens as required. However, during contract creation, token amount can also be initialized.
- *Token distribution and token sales.* Based on existing regulations, the Tax Authority can distribute the tokens to authorized participants such as banks and post office. The authorized participants can sell the tokens to any secondary markets such as distributors and retailers or sell them directly to end users (taxpayers).
- *Stamp duty payment.* If the user wants to pay the stamp duty, then the user creates a new transaction to do so. The smart contract first checks if there is enough balance in the user's account, checks if required data is supplied. If everything is satisfied, then the smart contract reduces the amount of the stamp duty to be paid in the user's balance before storing the document data in the smart contract. To prove that the user has paid the cost, the tokens will be sent to the Tax Authority's address. The tokens will be accumulated by the Tax Authority which will be used for audit or reconciliation when calculating the real number of tax paid in fiat currency.

Tax Payment Audit. For the purpose of audit, the tokens received by the Tax Authority will not be recirculated. When the audit is finished, then the same number of tokens verified to be the stamp duty revenue could be sent to an unusable address (an address without known private key) such as zero address.

The scope of the audit can be expanded such that auditing bodies can directly assess and monitor the system and evaluate the data in a real time. By using blockchain as a shared ledger, the amount of token that has been sold and the amount of token that has been spent will be easy to calculate.

4.10   Transaction Fee

Assuming the blockchain will run in a permissioned environment, then it is possible to waive transaction fees for all transactions created within the system. It is assumed that the Tax Authority will not abuse the system by flooding the system by sending system parameter-defining transactions. It is also assumed that when creating transactions, the Bank and the User need to have enough balance to pay the stamp duty. Without a proper information supply, the transaction will not be processed and confirmed by the system.

However, without any transaction fee, the system is prone to abuse by participants, for example creating many transactions to their own addresses to slow down the entire system. To avoid the useless transaction flooding, a small transaction fee can be introduced such that creating many transactions will be costly to the attacker. The implementation of transaction fee is beyond the scope of this paper.

4.11   Security Model

We define two possible attacks to the proposed system as follows.

1. An adversary tries to mint new token without authorization.
2. Malicious bank tries to reduce the report provided to Tax Authority to keep the tax money illegally.

In a system where Tax Authority is assumed to be always honest and no security vulnerability is found on the blockchain system and assuming that the cryptographic techniques are secure, then:

1. The probability of an adversary to mint their own token is negligible; a token can only be minted by authorized party, which is Tax Authority.
2. The probability of a malicious bank modifying the report without being detected is also negligible, given the auditability feature of the blockchain.

5   Implementation and Evaluation

We have implemented our solution which can be deployed on Ethereum blockchain. Ethereum-like private blockchain environment such as Quorum[1] is preferred during implementation phase because of the following consideration.

---

[1] https://www.jpmorgan.com/country/US/EN/Quorum.

- A wide number of contributors and software developers that develop and use Ethereum system provide confidence for future improvements.
- Much peer-reviewed research has been conducted on Ethereum in terms of its security. Its security vulnerabilities have been well-discussed and documented. This provides a great assistance in developing secure system.
- Ethereum system has endured severe attacks and exploits since the first launch and proves its durability.

Our solution consists of two parts: the smart contract and the user interface. The smart contract code is available on Github[2]. The user interface was written in Python and it is also available on Github[3].

The smart contract uses a standard `ERC-20` as its foundation which adds the capability of being traded on `ERC-20` token markets due to its compatibility. We call the token `Smart Stamp Duty` (`SSD`). As with other `ERC-20` tokens, our `SSD` is transferrable. Before paying the stamp duty, the user needs to buy the `SSD` token. Aside from `SSD` token, the native token Ether also exists in Ethereum environment. The transaction fee is paid by using Ether, while the amount of the fee is determined by calculating the complexity of the smart contract, expressed by gas. The gas is then converted into Ether by determining a conversion rate called gas cost or gas price. For simplification, the gas cost can be set to 0 Ether such that paying Ether will not be necessary in the implementation.

The main component, class `StampDuty`, contains two important data structures to contain information related to our solution, namely `StampParam` and `PayParam`. Each structure will be further explained.

5.1 Implementing Data Structures

StampParam. `StampParam`, contains information about types of stamp duty. Currently, there are only two stamps available which have different denominations: *"Rp3000"* and *"Rp6000"*[4]. Below is the detail of the `StampParam` structure.

- `StampCode`, to store stamp primary key (unique).
- `StampName`, to store human-readable stamp information.
- `StampPrice`, to store the price of the stamp.
- `RegulationReference`, to store the regulation reference of the stamp.
- `IsActive`, to flag active and inactive stamps.

StampParam is the master table for the stamp duty. Its importance is shown by only allowing the owner (i.e. the Tax Authority) to add and modify the data.

---

[2] https://github.com/sonicskye/smart-stamp-duty.

[3] https://github.com/sonicskye/smart-stamp-duty-ui.

[4] *Rp* stands for Rupiah, the Indonesian national currency.

PayParam. The second structure, `PayParam`, contains stamp duty payment data. A user adds a `PayParam` information every stamp duty paid. The `PayParam` structure is shown below.

- `PayCode`, works as the primary key of the payment (unique).
- `DocHash`, stores the hash value of the document related to the paid stamp.
- `PayIndex`, to store the index of the payment.
- `Payer`, to store the payer address.
- `StampCode`, to refer to the stamp paid by the transaction.
- `BloomFilter`, to store Bloom Filter information for testing.
- `TimeStamp`, to store timestamp in integer format.
- `TimeStampStr`, to store timestamp information in string format.
- `PayerSignature`, to store the payer's signature.

As a unique key, the `PayCode` can be generated by a hash function H of multiple information, such as `DocHash`, `Payer`, `StampCode`, and `Timestamp` when the document is submitted such that:

$$PayCode = H(DocHash\|Payer\|StampCode\|Timestamp)$$

where the symbol ($\|$) is concatenation operation. `PayerSignature` is generated by a signing function $SIGN$ using `Ethereum`'s library under the following formula:

$$PayerSignature = SIGN(DocHash, PrivateKey)$$

where $PrivateKey$ is the private key owned by the payer, associated to the payer's address which is stored in `Payer` variable. Aside from the two data structures, there are also built-in functions to manage the system, including token minting and token burning to control the token supply.

Bloom Filter [6] value of the document is also recorded on the smart contract. It supports content matching mechanism when printing the electronic document into paper-based document. The value `BloomFilter` can be used to verify that the electronic and paper-based versions are identical. When computing the `BloomFilter`, the system will list all words in the document under a parameter set, including number of words. The verification is done by sampling keywords from the document. If the verification passes, then we can conclude that the paper-based document is identical to the electronic version of it.

5.2   Implementing Content Matching

The content matching feature is implemented in our source code. The feature simply tests a set of input strings which comes from a document against a Bloom Filter value taken from the claimed electronic stamp duty. The input can either be done manually or automatically, where the topic is beyond the scope of the paper. The content matching protocol is as follows.

- Determine the document W and its associated Bloom Filter value `BloomFilter` which can be retrieved by querying the smart contract using `PayCode`.
- Extract all distinct words w from the input document W such that $W = \{w_1, w_2, w_3, \ldots, w_n\}$ where n is the number of distinct words.
- For each word w, check whether $w \in$ `BloomFilter`. Store the result in R.
- The result is the percentage of $True$ in R against the value n.
- The expected value is 100%. This result determines that there is a high probability that the document W is identical to the original document. Due to the false positive rate in Bloom Filter, it can never be confirmed that the examined document is identical to the original document.

6 Evaluation

6.1 System Evaluation

We deployed our implementation by using `Truffle` and `Ganache` as development environment in a private Ethereum network. The `PayParam` contains data duplicates such as timestamp and the payer's signature which are actually embedded in the transaction data and in the block where the transactions are included. However, reading the timestamp and signature directly from the block requires enormous computing power. When we tried this, the `Ganache` which we used as a node in our development environment crashed every time the code was tested.

To reduce the computing resources to read the transaction signatures and the block's timestamp. `PayParam` explicitly includes the payer's signature and data creation timestamp such that they can be read directly from the smart contract. As the result of this extra information, the required transaction fee doubled from around 235,000 gas to more than 450,000 gas.

6.2 Security Evaluation

Our system is designed to be implemented in Ethereum-like private blockchain environment where the Tax Authority poses as the central authority to control the whole system. It is assumed that the Tax Authority behaves honestly and never gives the tokens for free to any users and never leaks the private keys to the adversary. It is also assumed that the system does not have severe bugs which can be exploited by the adversary. In this scenario, it is infeasible for the adversary to create SSD tokens without a proper authorisation from the Tax Authority.

7 Conclusion and Future Work

We have described a new blockchain-based electronic stamp duty in the system we call `Smart Stamp Duty` (SSD). The proposed system aims to expand the tax base for the stamp duty as it enables the recording of electronic documents' stamp duty payments. The benefit of our solution is threefold:

1. to increase the convenience for the taxpayers to pay the stamp duty,
2. to increase the state revenue in the stamp duty sector, and
3. to increase the transparency and auditability of the stamp duty mechanisms.

Our solution also enables the format conversion of the document from electronic format into paper-based format where the integrity of both documents can be verified.

For future work, we plan to evaluate the privacy issue of our proposed smart contract system. It is possible that the users' privacy is analysed based on the pattern of the transactions, thus expose risks on their privacy. We will develop mitigation strategies based on our analysis and implement them on the next version of our design. We also plan to evaluate the privacy issues of our system. Although the users will not put their identities on the blockchain, empirical analyses can be conducted to evaluate and discover the users, including their behaviors, locations, or even their real identities.

References

1. Ainsworth, R.T., Alwohaibi, M.: Blockchain, bitcoin, and VAT in the GCC: The missing trader example (2017)
2. Ainsworth, R.T., Alwohaibi, M., Cheetham, M.: The GCC's cryptotaxcurrency, VATCoin (2016)
3. Ainsworth, R.T., Shact, A.: Blockchain (distributed ledger technology) Solves VAT Fraud (2016)
4. Ainsworth, R.T., Viitasaari, V.: Payroll tax & the blockchain (2017)
5. Alaudin, A.: Formulation of stamp duty regulations on e-commerce transactions in Indonesia. Journal of Faculty of Law, Brawijaya University (2016)
6. Bloom, B.H.: Space/time trade-offs in hash coding with allowable errors. Commun. ACM **13**(7), 422–426 (1970)
7. Darono, A.: Study of the electronic stamp duty business model, April 2013
8. Megawati, A.: The police arrested the seller of thousands of fake stamps (2018). https://www.merdeka.com/jakarta/polisi-bekuk-penjual-ribuan-meterai-palsu.html
9. Republic of Indonesia Ministry of Finance. Book ii financial note and the draft 2018 state revenue and expenditure budget (2018). http://www.anggaran.kemenkeu.go.id/content/Publikasi/NK%20APBN/2018%20Buku%20II%20Nota%20Keuangan%20Beserta%20RAPBN%20TA%202019.pdf
10. Republic of Indonesia Ministry of Finance. Information of state revenue and expenditure budget (2018). https://www.kemenkeu.go.id/media/6552/informasi-apbn-2018.pdf
11. Republic of Indonesia Ministry of Finance. State revenue and expenditure budget (2018). https://www.kemenkeu.go.id/apbn2018
12. Nakamoto, S.: Bitcoin: A Peer-to-peer Electronic Cash System (2008). http://bitcoin.org/bitcoin.pdf
13. Republic of Indonesia. Law of the republic of indonesia number 13 of 1985 concerning stamp duty (1985). https://jdih.kemenkeu.go.id/fulltext/1985/13tahun~1985UU.HTM
14. Republic of Indonesia. Law on transactions and electronic information (2008). https://jdih.kemenkeu.go.id/fulltext/2008/11tahun2008uu.htm

15. Republic of Indonesia. Law of the republic of indonesia number 15 of 2017 concerning the state budget for budget year (2018). https://jdih.kemenkeu.go.id/fullText/2017/15TAHUN2017UU.pdf
16. Directorate General of Taxes. Director general of taxes decree number kep-122d/pj./2000 concerning procedure for repayment of stamp duty by affixing customs signs with a computerized system (2000). https://www.ortax.org/ortax/?mod=aturan&page=show&id=1151
17. Salebu, J.B.: Exploring potential revenues to reach tax targets (2018). http://www.pajak.go.id/article/menggali-potensi-penerimaan-untuk-capai-target-pajak-2018
18. Wijaya, D.A., Darmawan, O.: Blockchain from Bitcoin for the World. Jasakom, Jakarta (2017)
19. Wijaya, D.A., Liu, J.K., Suwarsono, D.A., Zhang, P.: A new blockchain-based value-added tax system. In: Okamoto, T., Yu, Y., Au, M.H., Li, Y. (eds.) ProvSec 2017. LNCS, vol. 10592, pp. 471–486. Springer, Cham (2017). https://doi.org/10.1007/978-3-319-68637-0_28
20. Wood, G.: Ethereum: a secure decentralised generalised transaction ledger. Ethereum project yellow paper **151**, 1–32 (2014)
21. Yanwar, H.J., Jati, W.: Extensification of stamp duty on electronic documents (2018). http://www.pajak.go.id/article/menggali-potensi-penerimaan-untuk-capai-target-pajak-2018

PAChain: Private, Authenticated and Auditable Consortium Blockchain

Tsz Hon Yuen$^{(\boxtimes)}$ (iD)

The University of Hong Kong, Pok Fu Lam, Hong Kong
thyuen@cs.hku.hk

Abstract. Blockchain provides a distributed ledger recording a globally agreed, immutable transaction history, which may not be suitable for Fintech applications that process sensitive information. This paper aims to solve three important problems for practical blockchain applications: privacy, authentication and auditability.

Private transaction means that the transaction can be validated without revealing the transaction details, such as the identity of the transacting parties and the transaction amount. Auditable transaction means that the complete transaction details can be revealed by auditors, regulators or law enforcement agencies. Authenticated transaction means that only authorized parties can be involved in the transaction. In this paper, we present a private, authenticated and auditable consortium blockchain, using a number of cryptographic building blocks.

Keywords: Blockchain · Privacy

1 Introduction

Blockchain is a fast-growing field of technology since it is recognized as the core component of Bitcoin [10]. Blockchain can serve as a distributed ledger in a peer-to-peer network to provide publicly verifiable data. Blockchain can be mainly classified into three categories: public, private and consortium blockchain. The *public blockchain* is an open, permissionless system such that every one is allowed to join as a user or miner freely. Bitcoin and most cryptocurrencies falls into this category. Public blockchain is suitable for decentralized network without trusted authority. However, they usually have limited throughput. The *private blockchain* is a closed, permissioned system such that a user must be validated in order to use the system. It is a traditional centralized system with auditability attached. The *consortium blockchain* is a partly private blockchain. The submission of transactions can be performed by many (authorized) users, but the verification of transactions is only permitted by a few predetermined parties. Consortium blockchain provides a higher efficiency (more than 10 K transactions per second (tps)) than the public blockchain, without consolidating power with a single party as the private blockchain. Consortium blockchain is suitable for organizational collaboration.

© Springer Nature Switzerland AG 2019
Y. Mu et al. (Eds.): CANS 2019, LNCS 11829, pp. 214–234, 2019.
https://doi.org/10.1007/978-3-030-31578-8_12

Consortium blockchain received a great interest from the industry due to the similarity with the existing business model, especially in the finance industry. Hyperledger, an umbrella project of open source consortium blockchain, has 130 members including members from the IT industry (e.g., IBM, Intel) and the financial industry (e.g., JP Morgan, American Express).

Privacy in Blockchain. Privacy is important for commercial system, especially in the financial sector where money is transferred from one party to another. No one would like to have his bank account transaction history posted on a public blockchain. We define three key privacy properties that we want to achieve in this paper:

1. Sender Privacy: the sender's identity is not known by any third party and two valid transactions of the same sender should not be linked.
2. Recipient Privacy: the recipient's identity is not known by any third party and two valid transactions of the same recipient should not be linked.
3. Transaction Privacy: the content of the transaction is not known by any third party. General transaction privacy for smart contract is difficult to achieve without using general zero-knowledge proof of circuit or fully homomorphic encryption, which are both not quite practical. In this paper, we only consider the privacy for *transaction amount*.

The above conditions should hold for any third party (including the parties running the consensus algorithm). Cryptocurrencies such as Monero and Zcash offer privacy in the public blockchain. There are also a number of academic and industrial solutions for privacy-preserving consortium blockchain. Details will be discussed in Sect. 3.1.

Our Contributions: Privacy, Authenticated and Auditable in Consortium Blockchain. Auditability is essential for financial blockchain applications and unconditional anonymity may not be desirable. Financial institution has to undertake both internal and external audit for checking if there is any money laundering or terrorist-related activities, under regulations from the government. Auditing is usually performed by sample checking of all transaction records. In case of court order, the institute has to provide the complete information of a particular transaction to the court. For all of the above situations, the privacy of certain transaction has to be revoked if necessary. For simplicity, we denote the party to legally revoke privacy as the *auditor*.

Authentication is important for consortium blockchain in two aspects. First, the consortium companies need to ensure that the user is authenticated to use the system (e.g., he has paid/subscribed for the blockchain service). The consortium companies do not earn from the "mining" process and no new coin is generated from consortium blockchain. Second, authentication is useful for tracing real user identity during the auditing process. If users can transact in the consortium blockchain without registration, then the auditor can only discover the self-generated public key after opening the transaction. The real world identity can only be recovered if the user was registered before and is authenticated during the transaction.

Table 1. Comparison with existing privacy-preserving blockchain schemes

		Sender privacy	Recipient privacy	Transaction privacy	Auditability	Authentication	Efficiency
Public blockchain	Monero/RingCT-based solutions	◐	●	●			◑
	Zcash/zk-SNARK-based solutions	●	●	●			◐
	DAP [8]	●	●	●	●		◐
Consortium/ Private blockchain	R3 Corda, [9], Hyperledger Fabric	◐	◐	◐		●	●
	Fabric experiment	◑	◐	◐	◐	◑	◑
	PRCash [15]	◑	◑	●	◑		◑
	This paper	◑	●	●	●	●	◑

In this paper, we show how to construct a private, authenticated and auditable consortium blockchain: PAChain. We give the sender privacy, recipient privacy and transaction privacy by three separate modules. Auditability is provided for all three modules. Authentication is analyzed for the sender privacy and recipient privacy modules. It allows us to analysis the security of each module clearly. It gives the flexibility for system architects to choose the properties according to the business requirements. Therefore, our construction is suitable to be deployed in real world business use cases. In our construction, we use a number of cryptographic techniques (e.g., anonymous credential, zero-knowledge range proof, additive homomorphic encryption) and modify them for higher efficiency in consortium blockchain. Table 1 gives the comparison of our paper and related works described in Sect. 3.1.

2   PAChain Overview

In this paper, we show the high level overview of how to achieve privacy and auditability in consortium blockchain.

2.1   System Model

In blockchain system, clients can submit transactions (Tx) to nodes running the consensus algorithms. The nodes validate the Tx and add it to a block. In this paper, we extend the system model of Hyperledger Fabric 1.0[1], which is designed for consortium blockchain. Transactions have to be *endorsed* by endorsers and only endorsed Txs can be committed.

---

[1] Hyperledger Fabric Architecture Explained. http://hyperledger-fabric.readthedocs. io/en/latest/arch-deep-dive.html.

There are five types of nodes in PAChain:

1. Client: A client invokes a Tx to the endorsers, and also broadcasts the transaction-proposals to the orderer.
2. Peer: There are two types of peers. *Endorser* checks the validity of the Tx submitted by the client. *Committing peers* commits Txs and maintains the ledger. Note that a peer can play the role of endorser and committing peers at the same time.
3. Orderer: A node running the service of ordering Txs. Consensus algorithm is run between many orderers.
4. Auditor: The auditor can recover the sender identity, recipient identity and/or the Tx amount of any transaction.
5. Certificate Authority(CA): CA issues certificate for the public key of clients. Only authorized party can be involved in a Tx.

UTXO Model for Digital Assets. The model of Unspent Transaction Outputs (UTXO) is used in many blockchain systems, such as Bitcoin. If a user wants to spend his digital assets in a Tx, he has to refer to the specific assets that he wants to spend. If he spends the same asset twice, the verifier will notice it and will reject the Tx. In this paper, we consider the general UTXO model for digital assets and provides anonymity for their transactions[2].

Transaction Workflow. Assume that the client obtains a certificate from the CA. The basic workflow of a transaction (Tx) is as follows:

1. The client creates a signed Tx and sends it to endorser(s) of its choice.
2. Each endorser validates a Tx and produces an endorsement signature.
3. The submitting client collects endorsement(s) for a Tx and broadcasts it to the orderers.
4. The orderers deliver the block of ordered Txs to all peers.
5. The committing peer checks if every ordered Tx is endorsed correctly according to some policy. It also removes double-spending Tx endorsed by different endorsers concurrently (only the first valid Tx is accepted). If the checking is correct, it commits Txs and maintains the state and a copy of the ledger.

The auditor can recover the sender identity, recipient identity and the transaction amount of the Tx if necessary.

2.2   High-Level Description of PAChain

We briefly describe how we can achieve privacy and auditability in our PAChain. Generally speaking, we use the semi-trusted setting of consortium blockchain to set up system parameters and user credentials. Then, we can achieve a more efficient solutions for privacy and auditability, as compared to public blockchain (where all nodes may have Byzantine faults).

---

[2] Hyperledger Fabric 1.0 currently uses the account balance model by default, but it also supports the UTXO model.

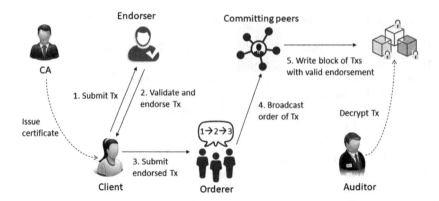

Fig. 1. Nodes and workflow of PAChain

Sender Privacy: It is achieved by the sender Alice using anonymous credential in Step 1 of the workflow. The credential is issued by the endorser of the last corresponding Tx received by Alice, to ensure that she is authenticated. We have to add a tag (which is a deterministic function of the user secret key) to the signature in Step 1 in order to detect double spending. We can take advantage of having some semi-trusted endorsers in our framework, which can act as the group manager of anonymous credential. It provides sender anonymity for all users using the same endorser (see Sect. 6.2). This anonymous credential approach is more efficient the ring signature-based approach in Monero and zk-SNARK-based approach in Zcash.

Recipient Privacy: It is achieved by generating one-time ephemeral key via Diffie-Hellman protocol between the sender and the recipient, in Step 1 of the workflow. However, we face the challenge that only authorized recipients are allowed in consortium blockchain. Therefore, we have to embed the zero-knowledge proof that the recipients are authorized into the generation of one-time ephemeral key (see Sect. 5.2). It is a new security requirement which does not exist in the public blockchain. This requirement is a major difference between PAChain and [8].

Transaction Privacy with Auditability: Our transaction amount privacy is achieved by using additive homomorphic encryption and zero-knowledge proof in Step 1 of the workflow. The proof shows that the Tx output amount is encrypted correctly, falls within a valid range, and the total Tx input and output amount are balanced. The major challenge is that using Paillier encryption with zero-knowledge range proof is not efficient. To be more specific, encrypting 2048-bit plaintext and performing the binary-decomposition range proof is an overkill for 64-bit of transaction amount (see Sect. 4). Therefore, we propose the use of modified ElGamal encryption with signature-based range proof to enhance the efficiency (see Sect. 4.2). The size of the zero knowledge proof is independent to the size of the range. It cannot be used in public blockchain since it requires trusted setup.

Auditability: Auditability for sender and recipient identity can be achieved by encrypting their public keys to the auditor, followed the zero-knowledge proof of the correctness of such encryption in Step 1 of the workflow. We have discussed the auditability of transaction amount above.

Our construction can also be modified to provide fine-grained privacy policy. For example, if the transaction amount M is under \$10,000, then the bank does not need to perform checking for anti-money laundering. As a result, we can leverage our efficient range proof to run the proof of knowledge:

$$PoK\{(M, \text{sender ID, recipient ID}) : 0 \leq M \leq 10,000$$
$$\text{or Encrypt } (M, \text{sender ID, recipient ID}) \text{ to the auditor}\}.$$

2.3   Threat Model

We assume that the system parameters are generated honestly. We consider the following attacker model for privacy:

- The attacker can create malicious client or corrupt any client.
- The peer and CA are assumed to be honest-but-curious: it tries to break privacy passively by recording all inputs, outputs and randomness used, but it still follows the protocols.
- All keys and data used by the orderers are known to the attacker.
- The auditor is assumed to be honest for privacy. Compromising the auditor trivially breaks all privacy[3].

We do not consider network-level privacy issues, such as tracing the sender's IP address or analyzing meta-data in network packets. The correctness of the consensus algorithm is not considered in this paper.

Consider the example of a consortium of banks. It is reasonable to assume that the banks jointly generate system parameters (e.g., by multi-party computation). Each bank acts as a peer and follows the protocol (if it ignores Tx or endorses invalid Tx, it will be discovered by other banks and will be handled by other means outside the blockchain system). However, the bank may be curious to view the Tx details of other banks. Our privacy model captures this scenario. As a result, we allow a larger degree of decentralization by allowing multiple endorsers to validate a Tx, without causing extra privacy leakage.

3   Backgrounds

3.1   Related Works

Public Blockchain. Monero is a cryptocurrency created in 2014 that provides privacy by linkable ring signature, stealth address and Ring Confidential

---

[3] One may argue that it gives too much power for auditor. However in most companies, internal auditor should always be able to control and governance business operations. In some industries, laws require that information must be provided to the court when requested (e.g., anti-money laundering in banks and lawful interception in telecommunication industry).

Transactions [11]. The major disadvantage of Monero is the size of the linkable ring signature, which is proportional to the size of the ring (related to the level of sender anonymity).

Zcash is a cryptocurrency created in 2016 that offers privacy and selective transparency of Txs by using zero-knowledge proofs (zk-SNARK) on special *shielded* transactions [2]. The major disadvantage of Zcash is the large signing key size of 896 MB, long key generation time of 8 min and long signing time of 3 min. As a result, only around 7% of Txs are shielded in Zcash as of March 2017. Recently, Zcash proposed a new Sapling update which reduced the proving time to a few seconds. However, it is still far less efficient than the Monero's approach (and also this paper's approach) which is about 100 ms.

A decentralized anonymous payment (DAP) with the support of accountability is proposed in [8]. They tackle the accountability problem in the public blockchain by using the zk-SNARK approach. Hence, it is also not efficient.

No authentication is provided for all solutions in the public blockchain.

Consortium Blockchain. For consortium blockchain, the major platforms provide limited support for privacy. In R3's Corda and Hyperledger Fabric, Txs are handled by different channels and users can only view Txs in their own channel. Therefore, privacy is maintained by the system's access control policy. The level of privacy is lower than that of Monero and Zcash, which are not affected by system administrators. In addition, Hyperledger Fabric uses *VKey* to provide transaction privacy by symmetric encryption. The problem of key distribution between all parties is a severe challenge for a global deployment of such system. There is an experiment to integrate identity mixer [7] with Fabric[4]. The identity mixer provides a better sender anonymity (by preventing the system administrator to link all transactions from the same user), and provides auditability for the sender identity.

Private Blockchain. Recently, a private industrial blockchain is proposed in [9], where multiple distributed private ledgers are maintained by a subset of stakeholders in the network. Each private ledger is maintained by its stakeholder only. This approach only provides privacy for users outside the private ledger. To provide higher level of privacy, the system must be divided into smaller private ledgers. However, more expensive cross ledger asset transfer is needed if there are some Txs between private ledgers.

PRCash [15] is a centrally-issued cryptocurrency with privacy and auditability. They provide anonymity of the sender and recipient identity. A mixing technique is used to obfuscate the relation between multiple inputs and outputs. However, there are still certain linkability between Txs. Anyone can see that the input of a Tx comes from which previous Tx output. For auditability, PRCash provides prefect transaction amount privacy and no auditability for small amount Tx. For Txs with large amount, it cannot provide privacy for these Txs.

---

[4] https://jira.hyperledger.org/browse/FAB-2005.

3.2   Mathematical Backgrounds

Bilinear Groups. $\mathcal{G}$ is an algorithm, which takes as input a security parameter λ and outputs a tuple $(p, \mathbb{G}_1, \mathbb{G}_2, \mathbb{G}_T, \hat{e})$, where $\mathbb{G}_1$, $\mathbb{G}_2$ and $\mathbb{G}_T$ are multiplicative cyclic groups with prime order p, and $\hat{e} : \mathbb{G}_1 \times \mathbb{G}_2 \rightarrow \mathbb{G}_T$ is a map, which has the following properties: (1) Bilinearity: $\hat{e}(g^a, \hat{g}^b) = \hat{e}(g, \hat{g})^{ab}$ for $\forall g \in \mathbb{G}_1, \hat{g} \in \mathbb{G}_2$ and $\forall a, b \in \mathbb{Z}_p$. (2) Non-degeneracy: There exists $g \in \mathbb{G}_1, \hat{g} \in \mathbb{G}_2$ such that $\hat{e}(g, \hat{g}) \neq 1_{\mathbb{G}}$. (3) Computability: There exists an efficient algorithm to compute $\hat{e}(g, \hat{g})$ for $\forall g \in \mathbb{G}_1, \hat{g} \in \mathbb{G}_2$.

4   Transaction Privacy

One of the challenging part for privacy in blockchain is the confidentiality of the transaction amount. The major difficulty is how to verify the Tx that (1) the total committed input amount is equal to the total committed output amount; (2) all committed amounts fall within a valid range, e.g., from 0 to 2^{64}. This requirement is commonly known as the *confidential transaction*. Theoretically, it can be achieved by combining additive homomorphic commitment with zero-knowledge range proof. One example is Monero [11], which uses Pedersen commitment with 1-out-of-2 ring signature-based binary decomposition range proof.

In PAChain, we require auditability of transaction amount. It is common in business use cases that all Txs can be audited by internal auditors or some external regulators. Therefore, it is necessary to replace additive homomorphic commitment with additive homomorphic encryption in confidential transaction, such that the decryption key is hold by the auditor. However, technical difficulties arise when combining the additive homomorphic Paillier encryption [12] with existing zero-knowledge range proof.

Range Proof with Encryption. Range proof is one essential element in transaction privacy. Without a range proof, the attacker can create a transaction with one input \$0 and two outputs of \$100 and −\$100. The sum of input and output amount is still balanced. If the output amount is encrypted, the attacker can create \$100 out of an input \$0. Therefore range proof is needed to prevent any negative amount or overflow amount.

Problems of Using Paillier Encryption with Range Proof. Paillier encryption [12] is the most common additive homomorphic encryption to date. However, combining Paillier encryption with existing range proof is not trivial in blockchain applications. Therefore, there is no simple and efficient solution for using Paillier encryption with range proof in blockchain.

4.1   Security Model of Transaction Privacy Protocol

We give a high level description on the notion and security model for transaction privacy. Details will be given in the full version of the paper.

A transaction privacy protocol consists of:

- Setup: It outputs the system parameters and the secret key of an auditor.
- TxPrivacySpend: When given some UTXO input amount, UTXO input ciphertexts and some output amount, it outputs some UTXO output ciphertexts and the proof of correctness.
- TxPrivacyVerify: When given some UTXO input ciphertexts, some UTXO output ciphertexts and the proof of correctness, it checks if the proof is correct or not.
- Decrypt: When given a UTXO ciphertext and the secret key of an auditor, it outputs the decrypt amount.

Security Model. We define the security requirements for transaction privacy:

- No output transaction amount is outside the range with a valid proof in TxPrivacySpend.
- The total input transaction amount is equal to the total output transaction amount in TxPrivacySpend.
- No one can learn the transaction amount from the ciphertext, except the auditor.

4.2   Transaction Privacy for PAChain

We give our efficient transaction privacy solution for consortium blockchain. We overcome the problem for effectively combining additive homomorphic encryption and range proof due to two properties: (1) consortium blockchain allows trusted setup; (2) the transaction amount is short.

The first observation is that encrypting a "short" transaction amount (e.g., 51-bit can represent all 21M Bitcoins in terms of satoshi, or 64-bit can represent trillions of dollars with sixth decimals) with Paillier encryption of message space of 2048-bit is superfluous. Therefore, we propose to use the additive homomorphic ElGamal encryption instead. However, decrypting such ciphertext requires the computation of discrete logarithm, which is not feasible for large message space. Hence, we decompose the K-bit transaction amount M into ℓ segment $\mu_0, \ldots, \mu_{\ell-1}$ which are smaller than the message space u by $M = \sum_{j=0}^{\ell-1} \mu_j u^j$. As a result, we encrypt each μ_j by additive homomorphic ElGamal encryption. The small message space of u guarantees efficient decryption. In our implementation, we consider 64-bit transaction amount and the auditor uses a pre-computation table of $(g, g^2, \ldots, g^{u-1})$ for efficient decryption. We can choose $\ell = 4, u = 65536$ and the pre-computation table is about 2 MB. The ciphertext size is 2048-bit, which is still less than the 4096-bit ciphertext size of Paillier encryption.

Another advantage of using ElGamal encryption is the ease to combine with range proof. By using ECC ElGamal encryption, it can be combined with the Boneh-Boyen signature-based range proof [6]. Note that this solution is only suitable for consortium blockchain since the non-interactive version of such range proof requires a trusted setup.

Our Construction. Our transaction privacy (TP) protocol is described below.

- Setup. On input a security parameter 1^λ and the range parameter $R = \mathfrak{u}^\ell$, it generates the bilinear group by $(p, \mathbb{G}_1, \mathbb{G}_2, \mathbb{G}_T, \hat{e}) \leftarrow \mathcal{G}(1^\lambda)$. It randomly picks generators $g, g_0 \in \mathbb{G}_1, \hat{g} \in \mathbb{G}_2$ and $\mathfrak{X} \in \mathbb{Z}_p$. It computes $\hat{\mathfrak{Y}} = \hat{g}^{\mathfrak{X}}$, $\mathfrak{A}_i = g^{\frac{1}{\mathfrak{X}+i}}$ for $i \in [0, \mathfrak{u} - 1]$. Suppose $H : \{0,1\}^* \to \mathbb{Z}_p$ is a collision resistant hash function. In addition, suppose the auditor picks a random secret key $\mathsf{ask}_{\mathsf{tp}}$ in $\mathbb{Z}_p$ and outputs its public key $h_{\mathsf{tp}} = g^{\mathsf{ask}_{\mathsf{tp}}}$. It outputs $\mathsf{param} = (g, g_0, h_{\mathsf{tp}}, \hat{g}, \hat{e}(g, \hat{g}), \hat{\mathfrak{Y}}, \mathfrak{u}, \ell, H, \mathfrak{A}_0, \ldots, \mathfrak{A}_{\mathfrak{u}-1})$.

- TxPrivacySpend. On input n' output transaction amount $M_{\mathsf{out},j}$, n transaction input ciphertext $C_{\mathsf{in},i}$, amount $M_{\mathsf{in},i}$ and randomness $r_{\mathsf{in},i}$, it outputs $\perp$ if $C_{\mathsf{in},i}$ is not a valid ciphertext for $(M_{\mathsf{in},i}, r_{\mathsf{in},i})$ for some $i \in [1, n]$, or $\sum_{i=1}^{n} M_{\mathsf{in},i} \neq \sum_{j=1}^{n'} M_{\mathsf{out},j}$.

For each $M_{\mathsf{out},j}$, it obtains $(C_{\mathsf{out},j}, r_{\mathsf{out},j}, \pi_{\mathsf{enc},j})$ by running the sub-protocol $\mathsf{EncProof}(M_{\mathsf{out},j})$.

EncProof. Suppose that the prover wants to prove some M lies in $[0, \mathfrak{u}^\ell)$. It runs as follows:

1. It first decomposes M into $\mu_k \in [0, \mathfrak{u} - 1]$ such that $M = \sum_{k=0}^{\ell-1} \mu_k \mathfrak{u}^k$.
2. For each μ_k, the prover computes the ElGamal ciphertext $(C_k = g_0^{\mu_k} h_{\mathsf{tp}}^{r_k}$, $B_k = g^{r_k})$ for some random $r_k \in \mathbb{Z}_p$. Denote $C_{\mathsf{enc}} = \{C_k, B_k\}_{k \in [0, \ell-1]}$ and $r_{\mathsf{enc}} = \sum_{k=0}^{\ell-1} r_k \mathfrak{u}^k$.
3. For each μ_k, it proves in zero-knowledge that the encrypted μ_k corresponds to some $\mathfrak{A}_{\mu_k}$:

$$\pi_{\mathsf{enc}} \leftarrow PoK\{(\{\mu_k, r_k, \mathfrak{A}_{\mu_k}\}_{k \in [0, \ell-1]}) :$$
$$\hat{e}(g, \hat{g}) = \hat{e}(\mathfrak{A}_{\mu_k}, \hat{g}^{\mu_k} \cdot \hat{\mathfrak{Y}}) \wedge C_k = g_0^{\mu_k} h_{\mathsf{tp}}^{r_k} \wedge B_k = g^{r_k}\}.$$

Details of the zero-knowledge proof is as follows. The prover randomly picks $v_k, s_k, t_k, \nu \in \mathbb{Z}_p$ for $k \in [0, \ell-1]$ and computes:

$$V_k = \mathfrak{A}_{\mu_k}^{v_k}, \quad a_k = \hat{e}(V_k, \hat{g})^{-s_k} \cdot \hat{e}(g, \hat{g})^{t_k}, \quad E_k = g_0^{s_k} h_{\mathsf{tp}}^{\nu_k}, \quad D_k = g^{\nu_k}.$$

It computes $\tilde{c} = H(\mathsf{param}, \{V_k, a_k, B_k, C_k, D_k, E_k\}_{k \in [0, \ell-1]})$ and for $k \in [0, \ell-1]$:

$$z_{\mu_k} = s_k - \tilde{c}\mu_k, \quad z_{v_k} = t_k - \tilde{c}v_k, \quad z_{r_k} = \nu_k - \tilde{c}r_k.$$

Then $\pi_{\mathsf{enc}} = (\{V_k, z_{\mu_k}, z_{v_k}, z_{r_k}\}_{k \in [0, \ell-1]}, \tilde{c})$.

4. It outputs $(C_{\mathsf{enc}}, r_{\mathsf{enc}}, \pi_{\mathsf{enc}})$.

Observe that $C_{\mathsf{out},j} = \{C_{j,k}, B_{j,k}\}_{k \in [0, \ell-1]}$. For simplicity, denote $C'_{\mathsf{out},j} = \prod_{k=0}^{\ell-1} C_{j,k}^{\mathfrak{u}^k}$ and $B'_{\mathsf{out},j} = \prod_{k=0}^{\ell-1} B_{j,k}^{\mathfrak{u}^k}$. The same definition applies for input ciphertext.

Next, it proves that the total input Tx amount is equal to the total output Tx amount. It is equivalent to know $x_{\mathsf{tp}} = \sum_{j=1}^{n'} r_{\mathsf{out},j} - \sum_{i=1}^{n} r_{\mathsf{in},i}$, such that

$$\prod_{j=1}^{n'} C'_{\mathsf{out},j} / \prod_{i=1}^{n} C'_{\mathsf{in},i} = h_{\mathsf{tp}}^{x_{\mathsf{tp}}}.$$

The zero knowledge proof of x_{tp} is as follows. It picks some random $r_{\mathsf{tp}} \in \mathbb{Z}_p$ and computes $R_{\mathsf{tp}} = h_{\mathsf{tp}}^{r_{\mathsf{tp}}}$, $\tilde{R}_{\mathsf{tp}} = g^{r_{\mathsf{tp}}}$, $\tilde{c}' = H(\mathsf{param}, R_{\mathsf{tp}}, \tilde{R}_{\mathsf{tp}}, \{C_{\mathsf{in},i}\}_{i\in[1,n]}, \{C_{\mathsf{out},j}, \pi_{\mathsf{enc},j}\}_{j\in[1,n']})$, $z_{\mathsf{tp}} = r_{\mathsf{tp}} + \tilde{c}' x_{\mathsf{tp}}$. Denote $\pi_{\mathsf{tp}} = (z_{\mathsf{tp}}, \tilde{c}', \{\pi_{\mathsf{enc},j}\}_{j\in[1,n']})$.

The algorithm outputs $(\{C_{\mathsf{out},j}, r_{\mathsf{out},i}\}_{j\in[1,n']}, \pi_{\mathsf{tp}})$.

- TxPrivacyVerify. On input n Tx input ciphertext $C_{\mathsf{in},i}$, n' Tx output ciphertext $C_{\mathsf{out},j}$ and a proof $\pi_{\mathsf{tp}} = (z_{\mathsf{tp}}, \tilde{c}', \{\pi_{\mathsf{enc},j}\}_{j\in[1,n']})$. For each $\pi_{\mathsf{enc},j}$, it runs the following sub-protocol:

EncVerify. On input the ciphertext $C_{\mathsf{out},j} = \{C_k, B_k\}_{k\in[0,\ell-1]}$ and the proof $\pi_{\mathsf{enc},j} = (\{V_k, z_{\mu_k}, z_{v_k}, z_{r_k}\}_{k\in[0,\ell-1]}, \tilde{c})$, it validates the proof by computing for all $k \in [0, \ell - 1]$:

$$D_k = B_k^{\tilde{c}} g^{z_{r_k}}, \quad E_k = C_k^{\tilde{c}} g_0^{z_{\mu_k}} h_{\mathsf{tp}}^{z_{r_k}}, \quad a_k = \hat{e}(V_k, \hat{\mathfrak{Y}}^{\tilde{c}} \hat{g}^{-z_{\mu_k}}) \cdot \hat{e}(g, \hat{g})^{z_{v_k}}.$$

It outputs 1 if $\tilde{c} = H(\mathsf{param}, \{V_k, a_k, B_k, C_k, D_k, E_k\}_{k\in[0,\ell-1]})$ or 0 otherwise.

It computes $R'_{\mathsf{tp}} = h_{\mathsf{tp}}^{z_{\mathsf{tp}}} \left(\frac{\prod_{i=1}^{n} C'_{\mathsf{in},i}}{\prod_{j=1}^{n'} C'_{\mathsf{out},j}}\right)^{\tilde{c}'}$, $\tilde{R}'_{\mathsf{tp}} = g^{z_{\mathsf{tp}}} \left(\frac{\prod_{i=1}^{n} B'_{\mathsf{in},i}}{\prod_{j=1}^{n'} B'_{\mathsf{out},j}}\right)^{\tilde{c}'}$. It returns 1 if and only if all EncVerify outputs 1, and $\tilde{c}' = H(\mathsf{param}, R'_{\mathsf{tp}}, \tilde{R}'_{\mathsf{tp}}, \{C_{\mathsf{in},i}\}_{i\in[1,n]}, \{C_{\mathsf{out},j}, \pi_{\mathsf{enc},j}\}_{j\in[1,n']})$.

- Decrypt. On input the auditor's secret key $\mathsf{ask}_{\mathsf{tp}}$ and a ciphertext $C_{\mathsf{enc}} = \{C_k, B_k\}_{k\in[0,\ell-1]}$, it computes $g_0^{\mu_k} = \frac{C_k}{B_k^{\mathsf{ask}_{\mathsf{tp}}}}$ for $k \in [0, \ell - 1]$. The auditor uses a pre-computation table containing $(g_0^0, g_0^1, \ldots, g_0^{u-1})$ to find out the value of μ_k. Finally, the auditor recovers $M = \sum_{k=0}^{\ell-1} \mu_k u^k$.

Security of Transaction Privacy. We give the security theorem of our TP protocol. The proofs are given in the full version of the paper.

Theorem 1. *Our TP protocol is sound if the u-Strong Diffie-Hellman (SDH) assumption holds in $(\mathbb{G}_1, \mathbb{G}_2)$ in the random oracle model. Our TP protocol is private if the decisional Diffie-Hellman (DDH) assumption holds in $\mathbb{G}_1$ in the random oracle model. Our TP protocol is balance if the discrete logarithm (DL) assumption holds in $\mathbb{G}_1$ in the random oracle model.*

5   Recipient Privacy

In blockchain, the user address is the hash of his public key, and hence it represents his identity. If we want to preserve the recipient privacy, we can always use a new public key for each Tx. However, this approach is problematic in some consortium blockchain which only allows Txs between authenticated users. It means that all recipient (and sender) address should be authenticated. A straightforward approach is to associate each address with a certificate issued by a CA. The key challenge is how to validate the certificate while hiding the public key/address at the same time.

Previous Works. Dash's PrivateSend is a coin-mixing service based on Coin-Join [13]. Dash requires combining identical input amounts from multiple senders at the time of mixing, and thus it restricts the mixing to only accept certain denominations (e.g. $0.1, $1, $10, etc.). The level of anonymity is related to number of Txs mixed. In Monero, the recipient uses *stealth address* [11], which is a one-time DH-type public key computed from the recipient's public key and some randomness included in the transaction block. The corresponding one-time secret key is only computable by the recipient. In Zcash, it uses the general zk-SNARK to provide zero knowledge for all Tx details including UTXO used [2].

5.1 Security Model of Recipient Privacy Protocol

The formal security notion and security models will be given to the full version of the paper. Roughly speaking, the security requirements for recipient privacy are *soundness* and *anonymity*. It includes:

1. No adversary can be a recipient without credential, even with colluding auditor.
2. No one can learn the identity of the recipient, except the auditor.

5.2 Recipient Privacy for PAChain

Stealth address [11] appears to be the most efficient approach for recipient privacy. However in consortium blockchain, only the recipient's public key is authenticated by the CA, but not the one-time public key. Therefore, the sender additionally needs to show that the one-time public key is computed from an authenticated public key, without revealing the public key itself.

Our Construction. The recipient's certificate is signed by the CA using BBS+ signature [1], which allows efficient zero-knowledge proof. In addition, we encrypt the long-term public key in the zero-knowledge proof, such that the auditor can decrypt the real address (long-term public key) of the recipient.

Our recipient privacy (RP) protocol is described below.

- Setup. On input a security parameter 1^λ, the setup algorithm generates the bilinear group by $(p, \mathbb{G}_1, \mathbb{G}_2, \mathbb{G}_T, \hat{e}) \leftarrow \mathcal{G}(1^\lambda)$. It picks some random generators $g, g_2, g_3, h_2 \in \mathbb{G}_1$ and $\hat{g}_2 \in \mathbb{G}_2$. Suppose $H : \{0,1\}^* \to \mathbb{Z}_p$, $H' : \mathbb{G}_1 \to \mathbb{Z}_p$ are collision resistant hash functions. In addition, suppose the auditor picks a random secret key $\mathsf{ask_{rp}} \in \mathbb{Z}_p$ and outputs its public key $h_{\mathsf{rp}} = g^{\mathsf{ask_{rp}}}$. It outputs the public parameters $\mathsf{param} = (p, \mathbb{G}_1, \mathbb{G}_2, \mathbb{G}_T, \hat{e}, g, h_{\mathsf{rp}}, g_2, g_3, h_2, \hat{g}_2, H, H')$.

- UserKeyGen. The user randomly picks a long-term secret key $x_1, x_2 \in \mathbb{Z}_p$ and computes a long-term public key $Y_1 = g_2^{x_1}, Y_2 = g_2^{x_2}$. It outputs the user key pair $(\mathsf{usk} = (x_1, x_2), \mathsf{upk} = (Y_1, Y_2))$.
- OneTimePkGen. On input $\mathsf{upk} = (Y_1, Y_2)$, the sender randomly picks $r_{tx} \in \mathbb{Z}_p$ and outputs $(R_{tx} = g_2^{r_{tx}}, \mathsf{otpk} = Y_1 g_2^{H'(Y_2^{r_{tx}})})$.
- OneTimeSkGen. On input otpk, R_{tx} and $\mathsf{usk} = (x_1, x_2)$, the recipient computes a one-time secret key $\mathsf{otsk} = x_1 + H'(R_{tx}^{x_2})$, and it outputs otsk if $\mathsf{otpk} = g_2^{\mathsf{otsk}}$.
- CAKeyGen. The CA randomly picks $\beta \in \mathbb{Z}_p$ and computes $\hat{W}_2 = \hat{g}_2^{\beta}$. It outputs the CA key pair $(\mathsf{cask} = \beta, \mathsf{capk} = \hat{W}_2)$.
- CertIssue. On input CA's public key capk and the user long-term public key $\mathsf{upk} = (Y_1, \cdot)$, the user first performs a zero-knowledge proof of discrete logarithm: $x_1 = \log_{g_2} Y_1$. Denote this proof as π_{ca}. After the CA validates π_{ca}, the CA picks some random $s, w \in \mathbb{Z}_p$ and uses his private key $\mathsf{cask} = \beta$ to compute: $F = (h_2 \cdot Y_1 \cdot g_3^s)^{\frac{1}{\beta+w}}$. The CA returns the certificate (F, w, s) to the user.
- RecPrivacySpend. On input $\mathsf{param}, \mathsf{capk} = \hat{W}_2$:

1. The sender with usk decides one or more UTXOs that he wants to spend. For simplicity, assume he picks one UTXO with $(\mathsf{otpk}_s, R_{tx,s})$. He runs $\mathsf{otsk}_s \leftarrow$ **OneTimeSkGen**$(\mathsf{param}, \mathsf{otpk}_s, R_{tx,s}, \mathsf{usk})$. It runs a zero-knowledge proof of discrete logarithm: $\mathsf{otsk}_s = \log_{g_2} \mathsf{otpk}_s$. Denote this proof as π_{otsk}.
2. The sender chooses the set of recipients. For simplicity, assume there is only one recipient with long-term public key $\mathsf{upk}_r = (Y_1, Y_2)$. The sender generates the recipient's one-time public key by running **OneTimePkGen**. The sender obtains $\mathsf{otpk}_r, R_{tx,r}$ and the randomness r_{tx}. Denote $h_{tx} = H'(Y_2^{r_{tx}})$.
3. The sender encrypts Y_1 to the auditor by picking a random $r_{\mathsf{cert}} \in \mathbb{Z}_p$ and computing $C_{\mathsf{rp}} = (C_{\mathsf{cert}} = Y_1 \cdot h_{\mathsf{rp}}^{r_{\mathsf{cert}}}, B_{\mathsf{cert}} = g^{r_{\mathsf{cert}}})$.
4. The sender runs the following proof of knowledge for showing that (1) otpk_r is computed from a public key, (2) the public key has a valid certificate (F, w, s), (3) the public key is encrypted to the auditor:

$$\pi_{\mathsf{rp}} \leftarrow PoK\{(F, w, s, Y_1, h_{tx}, r_{\mathsf{cert}}) : \hat{e}(F, \hat{g}_2^w \cdot \hat{W}_2) = \hat{e}(h_2 \cdot Y_1 \cdot g_3^s, \hat{g}_2)$$
$$\wedge \ \mathsf{otpk}_r = Y_1 g_2^{h_{tx}} \wedge B_{\mathsf{cert}} = g^{r_{\mathsf{cert}}} \wedge C_{\mathsf{cert}} = Y_1 h_{\mathsf{rp}}^{r_{\mathsf{cert}}}\}.$$

The details of the zero knowledge proof π_{rp} is as follows.

(a) **ZKCommit:** It picks some random $\rho, r_\tau, r_w, r_\sigma, r_\rho, r_{\mathsf{cert}}, r_c, r_s \in \mathbb{Z}_p$, computes $\Theta = F^\rho$ and

$$R_{\mathsf{cert},1} = \hat{e}((h_2 \cdot C_{\mathsf{cert}})^{r_\rho} g_3^{r_s} \Theta^{-r_w} h_{\mathsf{rp}}^{-r_\sigma}, \hat{g}_2),$$
$$R_{\mathsf{cert},2} = g^{r_c}, \quad R_{\mathsf{cert},3} = B_{\mathsf{cert}}^{r_\rho} g^{-r_\sigma}, \quad R_{\mathsf{cert},4} = h_{\mathsf{rp}}^{r_c} g_2^{-r_\tau}.$$

(b) **ZKChallenge:** It computes $c = H(\mathsf{CertAuth.mpk}, C_{\mathsf{rp}}, \Theta, R_{\mathsf{cert},1}, R_{\mathsf{cert},2}, R_{\mathsf{cert},3}, R_{\mathsf{cert},4})$.

(c) **ZKResponse:** It computes:

$$z_w = r_w + c \cdot w, \qquad z_\tau = r_\tau + c \cdot h_{tx}, \qquad z_\rho = r_\rho + c \cdot \rho,$$
$$z_c = r_c + c \cdot r_{\mathsf{cert}}, \qquad z_\sigma = r_\sigma + c \cdot r_{\mathsf{cert}} \cdot \rho, \qquad z_s = r_s + c \cdot \rho \cdot s.$$

It outputs $\pi_{\mathsf{rp}} = (c, \Theta, z_\omega, z_\tau, z_\rho, z_c, z_\sigma, z_s)$.

5. Output $\pi_{\mathsf{otsk}}, \mathsf{otpk}_s, \mathsf{otpk}_r, R_{tx,r}, C_{\mathsf{rp}}$ and π_{rp}.

- RecPrivacyVerify. On input $\mathsf{param}, \mathsf{capk}, \pi_{\mathsf{otsk}}, \mathsf{otpk}_s, \mathsf{otpk}_r, R_{tx,r}, C_{\mathsf{rp}}$ and π_{rp}, it outputs 1 if π_{otsk} and π_{rp} are valid zero knowledge proofs.

The details of verifying the zero knowledge proof $\pi_{\mathsf{rp}} = (c, \Theta, z_\omega, z_\tau, z_\rho, z_c, z_\sigma, z_s)$ is as follows.

1. **ZKReconstruct:** Denote $C_{\mathsf{rp}} = (C_{\mathsf{cert}}, B_{\mathsf{cert}})$. It computes:

$$R_{\mathsf{cert},1} = \hat{e}((h_2 \cdot C_{\mathsf{cert}})^{z_\rho} g_3^{z_s} \Theta^{-z_\omega} h_{\mathsf{rp}}^{-z_\sigma}, \hat{g}_2) \cdot \hat{e}(\Theta, \hat{W}_2)^c,$$

$$R_{\mathsf{cert},2} = g^{z_c} B_{\mathsf{cert}}^{-c}, \quad R_{\mathsf{cert},3} = B_{\mathsf{cert}}^{z_\rho} g^{-z_\sigma}, \quad R_{\mathsf{cert},4} = h_{\mathsf{rp}}^{z_c} g_2^{-z_\tau} (\mathsf{otpk}_r / C_{\mathsf{cert}})^c.$$

2. **ZKCheck:** It computes $c' = H(\mathsf{CertAuth.mpk}, C_{\mathsf{rp}}, \Theta, R_{\mathsf{cert},1}, R_{\mathsf{cert},2}, R_{\mathsf{cert},3}, R_{\mathsf{cert},4})$. If $c = c'$, then π_{rp} is a valid zero knowledge proof.

Security of Recipient Privacy. We give the security theorem of our RP protocol. The proofs are given in the full version of the paper.

Theorem 2. *Our RP protocol is sound if the q-SDH assumption holds in $(\mathbb{G}_1, \mathbb{G}_2)$ in the random oracle model, where q is the maximum number of* Issue *oracle query. Our RP protocol is anonymous if the DDH assumption holds in $\mathbb{G}_1$ in the random oracle model.*

6   Sender Privacy

In the UTXO model, the sender has to specify the UTXOs that he wants to use. The UTXOs include the information of the owner's address as well as the transaction amount. The linkage between the current transaction and UTXOs guarantees the validity of the transaction and ensures that there is no double spending. However, this linkage violates the privacy of the sender (no matter the address is used for one time only and the transaction amount is encrypted). It is a dilemma to preserve the transaction correctness and to protect the sender privacy at the same time.

Previous Works. The sender privacy for Dash and Zcash are achieved as the same way as the recipient privacy. In Monero, it uses linkable ring signature (LRS) for hiding the real UTXOs used with other UTXOs (by the anonymity property of LRS), preventing double spending (by the linkability property of LRS) and ensuring transaction correctness (by the unforgeability property of LRS) at the same time [11]. The level of anonymity is related to number of UTXOs (denote as L) included in LRS. However, the number of computation used in signing and the signature size are both $O(L)$. Recently, Sun *et al.* reduced the signature size to $O(1)$ [14], at the price of using trusted setup.

6.1   Security Model of Sender Privacy Protocol

The formal security notion and model for the sender privacy protocol will be given in the full version of the paper. In short, the security requirements for sender privacy are *soundness, unforgeability* and *anonymity*. It includes:

1. No adversary can spend without credential, even with colluding auditor.
2. No adversary can spend money of honest user, even with colluding endorser and auditor.
3. No one can learn the identity of the sender, except the auditor.

6.2   Sender Privacy for PAChain

We give our efficient sender privacy solution for consortium blockchain. By the semi-trusted property of consortium blockchain, we can use the anonymous credential approach to achieve sender privacy. By using the semi-trusted endorser as the group manager (in the honest-but-curious security model), we provide an efficient solution which has the signing time, verification time and signature size independent to the number of UTXO included in the group. At the same time, the sender can be revealed by the auditor. Note that similar to group signature, the endorser (who issued credentials) cannot link the transaction by the credential he issued. Credential is issued to the recipient when the endorser approve the transaction. The endorser does not have any advantage in breaking anonymity in the UTXO model.

Our Construction. Our construction differs from traditional group signatures in two ways: (1) we have to hide both the sender's public key as well as the Tx amount, (2) we have to add a linkability tag to avoid double spending. For the first requirement, we use the BBS group signature [4], since the underlying credential is signed by Boneh-Boyen signature [3], which can be modified to sign on multiple committed values [1]. For the second requirement, we use the tag structure used in most linkable ring signature schemes.

There are two possible constructions: the Tx amount is in plaintext or in ciphertext. In the UTXO model, transaction privacy is required in order to protect sender privacy (otherwise, the attacker can use the Tx amount to link past transactions). In the account-based model, transaction privacy may or may not be needed in the blockchain. For simplicity, we only give the Tx amount ciphertext version here and the plaintext version can be constructed similarly.

- Setup. On input a security parameter 1^λ, the setup algorithm generates the bilinear group by $(p, \mathbb{G}_1, \mathbb{G}_2, \mathbb{G}_T, \hat{e}) \leftarrow \mathcal{G}(1^\lambda)$. It picks some random generators $g, g_1, g_2, u_1, h_s, f \in \mathbb{G}_1$ and $\hat{g}_2 \in \mathbb{G}_2$. Suppose $H : \{0,1\}^* \rightarrow \mathbb{Z}_p$ is a collision resistant hash function. Denote h_{tp} as the public key of the auditor in transaction privacy. Suppose the auditor picks a random secret key ask_{sp} in $\mathbb{Z}_p$ and outputs its public key $h_{\text{sp}} = g^{\text{ask}_{\text{sp}}}$. It outputs the public parameters $\text{param} = (p, \mathbb{G}_1, \mathbb{G}_2, \mathbb{G}_T, \hat{e}, g, g_1, g_2, u_1, h_s, f, h_{\text{tp}}, h_{\text{sp}}, \hat{g}_2, H)$.

- UserKeyGen. The user secret key is $\mathsf{usk} = x_1 \in \mathbb{Z}_p$ and the public key is $\mathsf{upk} = Y_1 = g_2^{x_1}$[5].
- EndorserKeyGen. The endorser randomly picks $\alpha \in \mathbb{Z}_p$ and computes $W_{\mathsf{sp}} = \hat{g}_2^{\alpha}$. It outputs the endorser key pair ($\mathsf{esk} = \alpha, \mathsf{epk} = W_{\mathsf{sp}}$).
- CredIssue. On input endorser public key epk, the user public key Y_1 and the Tx amount ciphertext C, the user first performs a zero-knowledge proof of secret key: $x_1 = \log_{g_2} Y_1$. Denote this proof as π_{ci}. The user sends π_{ci} and π_{tp} to the endorser, where π_{tp} is the zero-knowledge transaction privacy proof (showing the knowledge of (m, r_{tp}) such that $C = g^m h_{\mathsf{tp}}^{r_{\mathsf{tp}}}$).
  After the endorser validates the proofs π_{ci} and π_{tp}, the endorser picks some random $v, z \in \mathbb{Z}_p$ and uses his secret key $\mathsf{esk} = \alpha$ to compute: $A = (h_s \cdot g_1^v \cdot C \cdot Y_1)^{\frac{1}{\alpha+z}}$. The endorser returns the credential $\mathsf{cred} = (A, v, z)$ to the user.
- CredSign. On input param, and private input tuples $x_{\mathsf{in}}', \mathsf{cred}_{\mathsf{in}}, m_{\mathsf{in}}', r_{\mathsf{in}}', Y_{\mathsf{in}}'$ (such that $C_{\mathsf{in}} = g^{m_{\mathsf{in}}'} h_{\mathsf{tp}}^{r_{\mathsf{in}}'}$), it runs the following:

1. It computes the tag for detecting double spending: $T = f^{x_{\mathsf{in}}'}$.
2. It encrypts the public key Y_{in}' to the auditor, by randomly choosing $r_{\mathsf{cred}} \in \mathbb{Z}_p$ and computing $C_{\mathsf{sp}} = (C_{\mathsf{cred}} = Y_{\mathsf{in}}' \cdot h_{\mathsf{sp}}^{r_{\mathsf{cred}}}, B_{\mathsf{cred}} = g^{r_{\mathsf{cred}}})$.
3. It computes the zero knowledge proof π_{sp} for: (1) the credential $\mathsf{cred}_{\mathsf{in}} = (A, v, z)$ corresponds to $Y_{\mathsf{in}}' = g_2^{x_{\mathsf{in}}'}$ and $C_{\mathsf{in}} = g^{m_{\mathsf{in}}'} h_{\mathsf{tp}}^{r_{\mathsf{in}}'}$; (2) $T = f^{x_{\mathsf{in}}'}$; (3) Y_{in}' is encrypted to the auditor.

$$\pi_{\mathsf{sp}} = PoK\{(x_{\mathsf{in}}', m_{\mathsf{in}}', r_{\mathsf{in}}', A, v, z, r_{\mathsf{cred}}) : \hat{e}(A, W_{\mathsf{sp}}\hat{g}_2^z) = \hat{e}(h_s g_1^v g^{m_{\mathsf{in}}'} h_{\mathsf{tp}}^{r_{\mathsf{in}}'} g_2^{x_{\mathsf{in}}'}, \hat{g}_2)$$
$$\wedge T = f^{x_{\mathsf{in}}'} \wedge C_{\mathsf{cred}} = g_2^{x_{\mathsf{in}}'} \cdot h_{\mathsf{sp}}^{r_{\mathsf{cred}}} \wedge B_{\mathsf{cred}} = g^{r_{\mathsf{cred}}}\}.$$

The output signature $\sigma = (\pi_{\mathsf{sp}}, C_{\mathsf{sp}}, T)$. Details of the zero-knowledge proof is shown as follows.

(a) **ZKCommit:** It picks some random $a, r_{\psi}, r_k, r_a, r_b, r_z, r_m, r_r, r_v \in \mathbb{Z}_p$. It computes:

$$S = A \cdot u_1^a, \quad \Xi = g_1^a,$$
$$R_{\mathsf{cred},1} = \hat{e}(u_1^{r_b} S^{-r_z} g_1^{r_k} g^{r_m} h_{\mathsf{tp}}^{r_r} g_2^{r_k}, \hat{g}_2) \cdot \hat{e}(u_1, W_{\mathsf{sp}})^{r_a}, \quad R_{\mathsf{cred},2} = g_1^{r_a},$$
$$R_{\mathsf{cred},3} = \Xi^{r_z} g_1^{-r_b}, \quad R_{\mathsf{cred},4} = g^{r_{\psi}}, \quad R_{\mathsf{cred},5} = g_2^{r_k} h_{\mathsf{sp}}^{r_{\psi}}, \quad R_{\mathsf{cred},6} = f^{r_k}.$$

(b) **ZKChallenge:** It computes $c = H(\mathsf{CredAuth.mpk}, C_{\mathsf{sp}}, T, S, \Xi, R_{\mathsf{cred},1}, R_{\mathsf{cred},2}, R_{\mathsf{cred},3}, R_{\mathsf{cred},4}, R_{\mathsf{cred},5}, R_{\mathsf{cred},6})$.
(c) **ZKResponse:** It computes:

$$z_k = r_k + c \cdot x_{\mathsf{in}}', \qquad z_a = r_a + c \cdot a, \qquad z_z = r_z + c \cdot z,$$
$$z_b = r_b + c \cdot a \cdot z, \qquad z_v = r_v + c \cdot v, \qquad z_m = r_m + c \cdot m_{\mathsf{in}}',$$
$$z_r = r_r + c \cdot r_{\mathsf{in}}', \qquad z_{\psi} = r_{\psi} + c \cdot r_{\mathsf{cred}}.$$

It outputs the proof $\pi_{\mathsf{sp}} = (c, S, \Xi, z_k, z_a, z_z, z_b, z_v, z_m, z_r, z_{\psi})$.

---

[5] This public key Y_1 can be a long term public key if recipient anonymity is not protected in the previous transaction. Otherwise, it can be a one-time public key.

- Verify. On input param, the endorser public keys W_s, a signature $\sigma = (\pi_{sp}, C_{sp} = (C_{cred}, B_{cred}), T)$, it checks the validity of the proof $\pi_{sp} = (c, S, \Xi, z_k, z_a, z_z, z_b, z_v, z_m, z_r, z_\psi)$:

1. **ZKReconstruct:** It computes:

$$R'_{cred,1} = \hat{e}(u_1^{z_b} S^{-z_z} g_1^{z_v} g^{z_m} h_{tp}^{z_r} g_1^{z_k} h_s^c, \hat{g}_2) \cdot \hat{e}(u_1^{z_a} S^{-c}, W_{sp}),$$
$$R'_{cred,2} = g_1^{z_a} \Xi^{-c}, \quad R'_{cred,3} = \Xi^{z_z} g_1^{-z_b}, \quad R'_{cred,4} = g^{z_\psi} B_{cred}^{-c},$$
$$R'_{cred,5} = g_2^{z_k} h_{sp}^{z_\psi} C_{cred}^{-c}, \quad R'_{cred,6} = f^{z_k} T^{-c}.$$

2. **ZKCheck:** It computes $c' = H(\mathsf{CredAuth.mpk}, C_{sp}, T, S, \Xi, R_{cred,1}, R_{cred,2}, R_{cred,3}, R_{cred,4}, R_{cred,5}, R_{cred,6})$.

It outputs 1 if $c = c'$; and outputs 0 otherwise.

- Link. On input param and two tags T_1, T_2 in signatures σ_1, σ_2, such that $T_1 = T_2$, it outputs 1. Otherwise it outputs 0.

- Decrypt. On input a ciphertext (C_{cred}, B_{cred}) and ask_{sp}, it computes $Y' = C_{cred}/B_{cred}^{\mathsf{ask}_{sp}}$.

Security of Sender Privacy. We give the security theorem of our sender privacy (SP) protocol. The proofs are given in the full version of the paper.

Theorem 3. *The SP protocol is sound if the q-SDH assumption holds in $(\mathbb{G}_1, \mathbb{G}_2)$ in the random oracle model, where q is the maximum number of Issue_e oracle query. The SP protocol is unforgeable if the DL assumption holds in $\mathbb{G}_1$ in the random oracle model. The SP protocol is anonymous if the DDH assumption holds in $\mathbb{G}_1$ in the random oracle model.*

7   Performance Analysis

We analyze our PAChain in terms of throughput and latency, two of the most important metrics for analyzing the performance of a blockchain system. The latency of our PAChain is affected by the running time of the modules. The throughput of our PAChain is affected by both the running time of our three modules, and the size of each transaction.

7.1   Transaction Overhead

In this paper, we consider 128-bit security. The transaction amount is represented by a 64-bit positive integer (the same setting as Bitcoin and Monero).

For PAChain's transaction privacy, 64-bit of transaction amount implies that the range $R = 2^{64}$. We can take $\mathfrak{u} = 2^{16} = 65536$, $\ell = 4$. The public parameters for transaction privacy is about 2MB. The size of the ciphertext is 256 bytes. For each transaction output amount, the size of the range proof π_{enc} is 544 bytes [6].

---

[6] A 64-bit range proof by the recent Bulletproof [5] is about 800 bytes.

Table 2. Comparison of privacy-preserving blockchain schemes, for a standard 2-input-2-output transaction.

		Sender privacy	Recipient privacy	Tx privacy	Auditability	Authentication	Tx overhead (bytes)	Sender running time	Verifier running time
Public blockchain	Monero	◑	●	●			12704	300 ms	300 ms
	Zcash	●	●	●			576	120 s	10 ms
Consortium blockchain	Hyperledger Fabric	○	◑	◑		●	628	10 ms	10 ms
	This paper	◑	●	●	●	●	2720	100 ms	100 ms

Table 3. Comparison for transaction privacy for a single output

	Setup time	C_{tp} Enc time	C_{tp} Dec time	π_{tp} Proof time	π_{tp} Verify time
Our scheme	53.8 s	2.8 ms	3.0 ms	27.1 ms	25.6 ms
Paillier encryption	402.6 ms	27.1 ms	7.4 ms		

The size of π_{tp} is 64 bytes plus all π_{enc} for all transaction outputs. For recipient privacy, the size of C_{rp} is 64 bytes, π_{rp} is 256 bytes for each recipient. The block randomness $R_{tx,r}$ is 32 bytes. (The 32 bytes of otpk_r replaces the output address and hence it is not viewed as an overhead). For sender privacy, the size of C_{sp} is 64 bytes, π_{rp} is 352 bytes and T is 32 bytes for each sender.

Considering a classical transaction of 2 inputs and 2 outputs, the overhead for privacy-enhancing consortium blockchain is 2720 bytes. We compare our PAChain with other schemes in Table 2:

- For consortium blockchain (e.g., Fabric or Corda), the classical transaction of 2 inputs and 2 outputs includes 2 ECDSA signatures from two inputs (128 bytes) and two X.509 certificates for 2 outputs' ECDSA public keys (about 500 bytes). The overhead is 628 bytes.
- For the public blockchain Monero, even if we consider the minimum ring size for ring signature as 3 (i.e., the real sender is one-out-of-three public keys. Hence the anonymity is very limited.), the total overhead is 12704 bytes for 2 inputs and 2 outputs.
- For Zcash, all the proofs can be combined to a single 288 bytes zk-SNARK proof. The total proof size becomes 576 bytes. However, the time for generating the proof will be much longer (> 120 s) and it requires a lot of RAM (> 3 GB). It causes a long latency in the blockchain system.

7.2 Module Implementation

We implemented our modules in a server with Intel Core i5 3.4GHz, 8GB RAM, running on Linux. Our implementation is by Golang, using BN256 pairing library.

Transaction Privacy. For transaction privacy, the running time for a single output is shown in Table 3. We compare our scheme with the additive homomorphic Paillier encryption with the same security level. When comparing with the encryption and decryption part only, our scheme is about 9 times and 2 times more efficient than the Paillier encryption. For the prover side, the complete transaction privacy is almost as efficient as a single Paillier encryption. Comparatively, our scheme takes a longer time for Setup, mainly for the generation of system parameters for the range proof.

Recipient Privacy. For recipient privacy for a single output, the Setup time is 4.6 ms, the CertIssue time is 1.4 ms, the Spend Time is 11.2 ms and the Verify time is 10.6 ms.

Sender Privacy. For sender privacy for a single input, the Setup time is 7.8 ms, the CredIssue time is 1.5 ms, the CredSign Time is 15.0 ms and the Verify time is 16.3 ms.

For a standard 2-input 2-output transaction, the total running time of our scheme (achieving all three properties) is 112 ms for the prover and 105 ms for the verifier side.

7.3 Testing Transaction Privacy with Hyperledger Fabric

We integrate the transaction privacy protocol in Hyperledger Fabric 1.0, in order to demonstrate our modulus can be consolidated into real world consortium blockchain. There are a few technical obstacles to implement our scheme. The first obstacle is that Fabric does not support optimization code of BN256 pairing written in C language. It results in > 10 times slower exponentiation and pairing computation. We expect future version of Fabric to allow optimization for pairing-based computation.

The second difficulty is to implement the verification logic into the smart contract (chaincode) of Fabric. We built a complete flow of transaction, including the creation of money (deposit), normal transaction, balance query and the destroy of money (withdraw). The chaincode has 2223 lines of codes. The extra codes for server side and client are 575 lines and 1061 lines respectively. The common module has 823 lines. (Comparatively, the core transaction privacy protocol has 2143 lines of codes.)

Transaction Privacy. In our current implementation for a 2-input 2-output transaction in Hyperledger Fabric 1.0, the signing time is 988 ms and the verification time is 1.35 s. Our implementation shows that other processing time for the transaction packet in negligible when compared to cryptographic operations. We expect that if optimization code of pairing is allowed, the signing and verification time can be about 100 ms.

The consensus algorithm is the current bottleneck of most consortium blockchain systems. The PBFT consensus algorithm used in Hyperledger Fabric 1.0 allows about 2000 transactions per second and has about 1 s of latency. If optimization is allowed in Fabric, our scheme has a running time of 100 ms for both the prover and verifier side, for a standard 2-input 2-output transaction. Therefore, our scheme is practical and will not become the bottleneck of the consortium blockchain system.

8 Conclusion

In this paper, we propose efficient solution for privacy, auditability and authentication in consortium blockchain. We give module solutions for them, so that they can be added to blockchain according to actual business need. We implemented our schemes and they are more efficient than the existing solutions in public blockchain.

References

1. Au, M.H., Susilo, W., Mu, Y.: Constant-size dynamic k-TAA. In: De Prisco, R., Yung, M. (eds.) SCN 2006. LNCS, vol. 4116, pp. 111–125. Springer, Heidelberg (2006). https://doi.org/10.1007/11832072_8
2. Ben-Sasson, E., et al.: Zerocash: decentralized anonymous payments from bitcoin. In: IEEE SP 2014, pp. 459–474. IEEE Computer Society (2014)
3. Boneh, D., Boyen, X.: Short signatures without random oracles. In: Cachin, C., Camenisch, J.L. (eds.) EUROCRYPT 2004. LNCS, vol. 3027, pp. 56–73. Springer, Heidelberg (2004). https://doi.org/10.1007/978-3-540-24676-3_4
4. Boneh, D., Boyen, X., Shacham, H.: Short group signatures. In: Franklin, M. (ed.) CRYPTO 2004. LNCS, vol. 3152, pp. 41–55. Springer, Heidelberg (2004). https://doi.org/10.1007/978-3-540-28628-8_3
5. Bünz, B., Bootle, J., Boneh, D., Poelstra, A., Wuille, P., Maxwell, G.: Bulletproofs: short proofs for confidential transactions and more. In: IEEE SP 2018, pp. 315–334. IEEE (2018). https://doi.org/10.1109/SP.2018.00020
6. Camenisch, J., Chaabouni, R., Shelat, A.: Efficient protocols for set membership and range proofs. In: Pieprzyk, J. (ed.) ASIACRYPT 2008. LNCS, vol. 5350, pp. 234–252. Springer, Heidelberg (2008). https://doi.org/10.1007/978-3-540-89255-7_15
7. Camenisch, J., Mödersheim, S., Sommer, D.: A formal model of identity mixer. In: Kowalewski, S., Roveri, M. (eds.) FMICS 2010. LNCS, vol. 6371, pp. 198–214. Springer, Heidelberg (2010). https://doi.org/10.1007/978-3-642-15898-8_13
8. Garman, C., Green, M., Miers, I.: Accountable privacy for decentralized anonymous payments. In: Grossklags, J., Preneel, B. (eds.) FC 2016. LNCS, vol. 9603, pp. 81–98. Springer, Heidelberg (2017). https://doi.org/10.1007/978-3-662-54970-4_5
9. Li, W., Sforzin, A., Fedorov, S., Karame, G.O.: Towards scalable and private industrial blockchains. In: BCC 2017, pp. 9–14. ACM (2017)
10. Nakamoto, S.: Bitcoin: A peer-to-peer electronic cash system (2009). https://bitcoin.org/bitcoin.pdf
11. Noether, S.: Ring signature confidential transactions for monero. Cryptology ePrint Archive, Report 2015/1098 (2015). http://eprint.iacr.org/
12. Paillier, P.: Public-key cryptosystems based on composite degree residuosity classes. In: Stern, J. (ed.) EUROCRYPT 1999. LNCS, vol. 1592, pp. 223–238. Springer, Heidelberg (1999). https://doi.org/10.1007/3-540-48910-X_16

13. Ruffing, T., Moreno-Sanchez, P., Kate, A.: CoinShuffle: practical decentralized coin mixing for bitcoin. In: Kutyłowski, M., Vaidya, J. (eds.) ESORICS 2014. LNCS, vol. 8713, pp. 345–364. Springer, Cham (2014). https://doi.org/10.1007/978-3-319-11212-1_20

14. Sun, S.-F., Au, M.H., Liu, J.K., Yuen, T.H.: RingCT 2.0: a compact accumulator-based (linkable ring signature) protocol for blockchain cryptocurrency monero. In: Foley, S.N., Gollmann, D., Snekkenes, E. (eds.) ESORICS 2017. LNCS, vol. 10493, pp. 456–474. Springer, Cham (2017). https://doi.org/10.1007/978-3-319-66399-9_25

15. Wüst, K., Kostiainen, K., Capkun, V., Capkun, S.: Prcash: Centrally-issued digital currency with privacy and regulation. In: FC 2019, Cryptology ePrint Archive, Report 2018/412 (2018). https://eprint.iacr.org/2018/412

Senarai: A Sustainable Public Blockchain-Based Permanent Storage Protocol

Dimaz Ankaa Wijaya[1,2(✉)], Joseph Liu[1], Ron Steinfeld[1], Dongxi Liu[2], and Limerlina[3]

[1] Monash University, Melbourne, Australia
{dimaz.wijaya,joseph.liu,ron.steinfeld}@monash.edu
[2] Data61, CSIRO, Canberra, Australia
dongxi.liu@data61.csiro.au
[3] Senarai, Jakarta, Indonesia
merlina@senarai.xyz

Abstract. Storing data in the blockchain brings many benefits. The block-and-chain structure protects the data stored in the blockchain from unauthorised changes. The data will also be duplicated to all nodes which maintain the blockchain. These features guarantee the data immutability and availability. However, such operation is expensive when large size data needs to be stored. We propose Senarai, a protocol which provides an affordable blockchain data storage. Our protocol utilises the fee-less transaction feature provided by Tron. By using this feature, storing an arbitrary size data can be done cheaply. We show that our protocol is sustainable such that after initialisation phase, the permanent data storage capacity is provided for free on daily basis.

Keywords: Storage · Immutability · Senarai · Tron · Blockchain

1 Introduction

Blockchain, which was first introduced in Bitcoin in 2009 by a pseudonym Satoshi Nakamoto [12], has many features that makes it a permanent storage. The tamper-proof ledger system is protected by multiple mechanisms: a block-and-chain structure, cryptographic functions, and data duplications in nodes located in different parts of the world. These nodes communicate to each other by using a peer-to-peer network. The blockchain system protects all recorded information which cannot be modified without a massive effort.

Much research has been done to formulate the best method to store information in a blockchain, mostly in Bitcoin's blockchain. Although the blockchain is not intended to store information not related to bitcoin transactions, people have found secondary blockchain usages other than just storing financial transactions. By storing information in the blockchain, the information becomes permanent

© Springer Nature Switzerland AG 2019
Y. Mu et al. (Eds.): CANS 2019, LNCS 11829, pp. 235–246, 2019.
https://doi.org/10.1007/978-3-030-31578-8_13

such that it cannot be modified easily. The immutability of the blockchain makes it a good solution for problems such as digital notarisation and proof of existence.

Small size data such as hash values of digital information and bigger data such as pictures have been embedded in Bitcoin's blockchain [16]. However, storing information in the blockchain is expensive such that one needs to pay all necessary transaction fees which will increase linearly to the data size [16].

We propose **Senarai**, a sustainable public blockchain-based storage protocol. After a proper initialisation, our protocol does not incur any fee when repeatedly storing arbitrary size data in a blockchain, hence it is sustainable for a long-term usage. To the best of our knowledge, our protocol has the best fee structure compared to existing blockchain-based storage solutions.

2   Background

2.1   Blockchain and Public Blockchain

Blockchain is a ledger where the information stored in the ledger is made public; any parties can read and verify the correctness of all information stored. Blockchain is also a distributed database, where identical copies of the ledger are stored on multiple servers called nodes. The nodes validate new transactions and embed new blocks containing transactions to the existing blockchain. The structure of the blockchain enables the immutability feature, where information that has been validated will be infeasible to tamper with after a certain depth.

2.2   Tron

Tron is a project initiated by Tron Foundation led by Justin Sun [3]. Its native token is called Tronix, abbreviated as TRX. The project aims to compete with Ethereum by scaling the system to allow more transactions to be stored in the blockchain compared to Ethereum. Similar to Ethereum, Tron also supports smart contract creation and interaction, where Ethereum's Solidity programming language is also used in Tron development environment to write the smart contract.

Tron offers free transactions, where each account of at least 24 h old is entitled to 5,000 free "bandwidth points". Each bandwidth point equals to 1 byte of storage [6]. The bandwidth points can be used to create around 20–25 free standard transactions per day, where each transaction requires around 200 bandwidth points [6]. The free bandwidth points allowance is replenished every 24 h [8].

Tron also provides extra bandwidth points to any accounts that "freeze" TRX for at least 72 h in exchange for the extra bandwidth points. The amount of the extra bandwidth points given to the account depends on the number of TRX frozen and total net weight [8]. Each TRX that is frozen for bandwidth points will be awarded bandwidth points by using the following formula: $b = t \times \frac{TNW}{TNL}$ [8], where b is the awarded bandwidth points, and t is the amount of TRX. TNW (Total Network Weight) is the total number of TRX frozen for bandwidth points in the network, while TNL (Total Network Limit) is a predefined constant value of 43,200,000,000 [6].

Whenever a user creates a transaction, the extra bandwidth points owned by the user will be consumed first; if the extra bandwidth points are not sufficient, then the basic allowance will be used. If these points are still insufficient for the transaction fee, then the user needs to pay the transaction fee in TRX. One TRX equals to 100,000 bandwidth points which equals to 100,000 bytes. [8].

3 Related Work

3.1 Bitcoin-Based Storage Systems

Much work has been done to create an efficient storage system on top of Bitcoin [16]. One of the first methods to embed extra data inside Bitcoin transaction is by using OP_RETURN operand. An analysis shows shown that the operand has been used for multiple services such as asset management, document notarisation, or to simply store short messages [2]. While the use of OP_RETURN can only store up to 80 bytes, other methods have been developed to increase the storage capacity.

The use of Pay to Script Hash (P2SH) feature increases the storage capacity where at most 1,400 bytes data can be embedded in a single P2SH transaction scheme with a data efficiency of 85% [19]. Another method is to utilise data dropping operands in Bitcoin, namely OP_DROP and OP_2DROP. If no digital signature is required in the transaction, the data dropping method can achieve up to 94.1% data efficiency [16]. However, this method might be out of control because everyone can try to redeem the transaction. A modified data dropping method which requires a digital signature will give a data efficiency of 88.2% [16].

3.2 Permacoin

Permacoin was proposed to completely turn Bitcoin's popular consensus method, Proof-of-Work (PoW), into a more usable work. This is done by dropping PoW and replacing it with a Proof-of-Retrievability (PoR), where the miners compete to provide data storage instead of computing power [11]. The PoR provides a mechanism to prove that the miners correctly store the data in their storage systems [11]. Although Permacoin solution shows a promising resource utilisation, the solution is currently not compatible with the current Bitcoin protocol, where Bitcoin's PoW is unlikely to be modified into PoR in the near future.

3.3 Ethereum-Based Storage

There are two methods to store arbitrary size data into Ethereum blockchain: either by storing the data in a smart contract or by storing the data directly into a transaction [7]. There are three different storage systems in Ethereum, namely volatile stack, volatile memory, and nonvolatile (permanent) storage [14]. Each of these storage systems has different cost with nonvolatile storage is the most expensive in Ethereum [14].

3.4 Storage Service Cryptocurrencies

InterPlanetary File System (IPFS) is a decentralised network storage which was initially developed by Juan Benet and Protocol Labs [20]. IPFS imitates Bitcoin's network and optimises it to identify duplicates as well as to locate specific files in the nodes connected to its network.

The growing popularity of IPFS initiated cryptocurrency projects providing an identical service: to provide decentralised and/or distributed storage system. Filecoin[1] is the incentive system built on IPFS protocol, while other products such as Sia[2], Storj[3] and IPDB[4] also provide similar storage service [17]. All of the mentioned systems do not provide a permanent storage; the user needs to pay monthly rent for uploading and downloading data from the service. Storj for example, requires the user to pay a monthly fee of US$0.015 per GB data upload and US$0.05 per GB data download [15]. At the time of writing, Filecoin, Storj, and IPDB are still under development.

Although these storage service cryptocurrencies implements blockchain technology, the storage offered by the services is not inside blockchains. The blockchain technology only helps to manage administrative and financial matters such as storage contracts and payments, while the storage management will be done by agents executing the contracts. This enables the services to offer data deletion, which is not possible in a pure blockchain storage solution.

4 Preliminaries

4.1 Resources in Tron

There are two valuable resources in Tron other than tokens or its native currency, TRX. These resources are `bandwidth` and `energy`. Bandwidth is the term to determine the cost to store data in the blockchain, where one byte of data consumes one bandwidth point. Energy is the term to determine the cost to run Tron's smart contract, which is calculated based on CPU time required to run the smart contract. Each microsecond of CPU time consumption costs one energy [9].

To get bandwidth and energy, a user must "freeze" her TRX. Upon freezing the TRX, she must choose whether she would prefer bandwidth or energy in exchange of the frozen TRX. One frozen TRX can only be exchanged with either bandwidth or energy [9]. The frozen TRX can be unfrozen after 72 h.

---

[1] https://filecoin.io.
[2] https://sia.tech.
[3] https://storj.io.
[4] https://ipdb.io.

4.2   Types of Token in Tron

In Tron, token generally means a value that is stored in a smart contract or in the blockchain. There are two types of token in Tron, namely TRC-10 and TRC-20 [10]. TRC-10 is similar to Tron's native currency, TRX, where it is transferrable between accounts. Each account or address in Tron is allowed to create at most one TRC-10 token. One-time cost needs to be paid by the token creator in TRX currency, which currently stands at 1,024 TRX.

TRC-20 is a token built in a smart contract. TRC-20 imitates Ethereum's ERC-20. It is a set of rules in the smart contract development, where every token which follows the rules needs to implement mandatory functions and naming convention. By adopting the standard, an interface can be built for those tokens without any significant changes.

The main difference between TRC-10 and TRC-20 is in their operation requirement. Transferring TRC-10 only requires bandwidth, while executing TRC-20 requires both bandwidth and energy.

5   Senarai

5.1   Overview

Senarai (a Malay language for *list*) is a transaction feeless-based blockchain storage solution. It utilises Tron's free daily bandwidth points allowance given to Tron accounts. Senarai uses the allowance to create transactions that contain payloads. The payloads are fragments of an arbitrary-size data to be stored permanently in the blockchain. The size of the payload is designed such that it does not exceed the maximum free allowance, hence, no transaction fee is required. Senarai has two main phases: **one-time initialisation phase** and **data storage and management phase**. While the first phase requires transaction fees, the latter phase does not.

5.2   Senarai Phases

One-Time Initialisation Phase. There are two steps required upon initialisation. The first step is to generate a new TRC-10 token [10]. The token is used to replace transacted TRX so that no actual coins are involved in the transaction. There is a one-time cost involved in creating TRC-10 token, which is 1,024 TRX [10]. When creating a new TRC-10 token, one can set an *int*64 number of tokens [18] such that it is sufficient enough for a foreseeable future.

The second step is to create as many new accounts as possible. There are two methods to pay the fee: either by freezing TRX for bandwidth points or by paying TRX. However, if one chooses to pay the transaction fee in bandwidth points instead of TRX, the number of TRX to be frozen to get the sufficient bandwidth points vary, depending on the total TRX frozen in the network.

If there is no frozen TRX in the creator's account, then creating a new account in Tron requires 0.1 TRX as a fee f. The maximum daily bandwidth size b

depends on the number of accounts n multiplied by a constant a, where a equals to 5,000 bandwidth points or 5,000 bytes. The formula to get the maximum bandwidth b can be expressed as $b = n \times a$. The cost c for a desired bandwidth b is expressed as $c = b \div a \times f$.

Data Storage and Management Phase. It is assumed that there are n Tron accounts g in a set G, where $g \in G$. The remaining daily bandwidth b' is defined as the maximum bandwidth b subtracted by the bandwidth that has been used. The maximum free transaction size is 5,000 bytes per account. One standard token transaction requires around 285–287 bytes transaction data; therefore, the maximum payload size is 4,715 bytes. To simplify and to give extra space to the transaction data, the maximum payload size will be rounded to a constant p equals to 4,700 bytes. The transaction data constant q becomes 300 bytes. A static account g_s will be the receiver of all related transactions. The protocol for data storage is as follows.

1. Compute the total data size to be stored D in bytes. If $D > b'$ then abort.
2. Compute the number of transactions required $t = CEILING(D \div p)$.
3. Compute the total bandwidth B required, which is the data payload and transaction information, formulated as $B = d + t \times q$. If $B > b'$ then abort.
4. Split the data D into t chunks of p bytes. A data chunk d is at most p bytes long, where $D = \{d_0, d_1, \ldots, d_{t-1}\}$.
5. For each $d_i \in D$:
   (a) Pick an account $g_j \in G$.
   (b) If g_j's remaining free bandwidth allowance is less than the payload d_i, then skip and continue to the next g.
   (c) Create a transaction from g_j to g_s that sends one token and contains the payload d_i.
   (d) If successful, record the **transaction ID** (or **transaction hash**) to the database as well as the chunk index number i.
6. The result of the storing process is the transaction hashes and the chunk indexes which can be used to retrieve the data.

Data Retrieval. Data retrieval can be done in parallel, where multiple data chunks d can be queried from the blockchain at the same time. Once all data chunks are retrieved, the original data D can be reconstructed since the indexes of the chunks are known.

6   Implementation and Evaluation

We have implemented our proposed protocol into a customised wallet where the source code of the wallet is available on Github[5]. The wallet uses a modified

---

[5] https://github.com/sonicskye/senarai.

version of the standard Tron Python API caller library[6]. The communication between the wallet and the Tron system is done by calling APIs provided by Trongrid.io. The local wallet stores private keys, while the API provider conducts all other processes such as signature generation and transaction deployment. The wallet consists of five parts:

- `Account generator`. The code generates new accounts (addresses and private keys) on the local wallet.
- `Account initiator`. The code creates transactions to new accounts by sending them tokens.
- `Remaining bandwidth point calculator`. The code calculates bandwidth points from all initialised accounts.
- `Data storage`. The code stores a local file to the blockchain by splitting the file into chunks of data then embed the data chunks in blockchain transactions.
- `Data retrieval`. The code retrieves the embedded data chunks based on stored transaction ID and join the data chunks to get the original file.

6.1   Experiments

Experiments were conducted on Tron's mainnet[7]. First, we generated a hefty 99 trillion TRC-10 token called `NullCoin`. The `NullCoin` token generation costed 1,024 TRX which was valued around US$20.48 at the time of token creation. An amount of 9,225 TRX worth US$184.50 was frozen in exchange of 54,387 extra daily free bandwidth points[8]. From the given bandwidth points, only 49,387 bandwidth points are usable to create transactions to new accounts (hence, create the accounts), whereas the rest 5,000 bandwidth points can only be used to create transactions to any existing accounts.

By using the given bandwidth points, after several days, 441 new accounts were created. These new accounts are created by sending 100,000 `NullCoin` to each account. All of these accounts are accountable to 2.2 MB of daily free permanent data storage.

6.2   Evaluation

Performance. The wallet was operated on an Ubuntu 18.04.1 LTS 64-bit virtual machine configured with one CPU core and 4 GB RAM. The virtual machine was hosted on a Macbook Pro 2012 Dual Core 2.5 GHz Intel Core i5 and 16 GB RAM connected to a shared wireless ADSL2+ Home Internet connection. The performance evaluation was done on protocol-related operations, namely `account generation`, `account initialisation`, `data storage`, and `data retrieval`. The result of the performance evaluation is as shown in Fig. 1.

---

[6] https://github.com/sonicskye/tron-api-python.

[7] Mainnet or main network is where the tokens have real value. In contrast, testnet or test network provides free zero value tokens that can be used for testing purposes.

[8] The number of free bandwidth points varies depending on the network parameters.

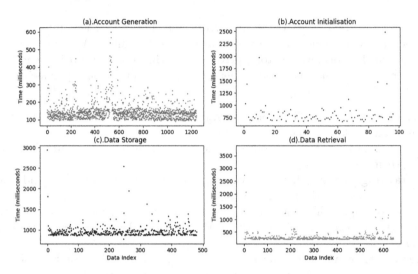

Fig. 1. The required time to finish operations.

Table 1. Evaluation on operation time

Operation	Data count	Lowest time (ms)	Highest time (ms)	Average time (ms)
Account generation	1,240	90.78	600.72	146.69
Account initialisation	97	675.10	2,489.82	862.17
Data storage	480	772.86	2,940.94	964.97
Data retrieval	645	221.36	3,722.77	309.69

The data in Table 1 shows that the `account generation` is the fastest operation in average compared to any other operations, while `data retrieval` comes second. `Account initialisation` and `data storage` requires relatively the same time; however, `data storage` operation needs slightly more time than `account initialisation` operation. The `account initialisation` and `data storage` operations are in the same level, because fundamentally they perform the same task, which is sending transactions to the network and receiving results from the transactions that were sent. The main difference is that the `account initialisation` operation sends much lower data than `data storage` operation. However, the experiment shows that the data size does not make any differences to the performance of the operation.

File storage performance tests[9] were also conducted. The stored data is double the size of the original files due to the inefficiency of the data encoding in the Tron system library, where each byte from the original file will be stored as two characters hexadecimal which is two bytes long. In this case, the maximum size for the data chunk is 2,350 bytes to get an optimum encoded data of 4,700 bytes. The results show that the average time required to store each data chunk

---

[9] The experiments use medium-size royalty-free image files provided by Pixabay.com.

is almost one second, where the total required time depends on the number of data chunks.

The results might be affected by the following factors:

- The Internet connection from the wallet to the API provider. Sending and receiving data are done through the Internet. Considering that the time calculation starts when the wallet sends the data and ends when the wallet receives the feedback, the Internet connection quality may affect the calculated time.
- The workload of the API provider. The raw transaction generation and signature generation are done on the API provider side. This puts a heavy burden to the provider's servers if there are massive requests to be processed.

Although these two factors may affect the results, the performance improvement over these two factors will not be further discussed.

Total Cost and Cost Comparison. The experiments were conducted by using a minimum cost, where there is only fee for creating tokens. The frozen coins were not lost; hence, it can be resold in the future. Although there is a potential loss due to price difference, it will be ignored for simplification. It is possible to get a higher daily storage capacity by spending TRX instead of freezing coins. It is assumed that the preferred capacity of our solution is 2 GB in order to store 1 GB data. At the time of the writing, the market price is US$0.02 per TRX. The cost breakdown for our solution is shown in Table 2.

Table 2. Senarai's cost breakdown for 1 GB storage capacity

Function	No. of operations	Cost (in TRX)	Cost (in US$)
TRC-10 token creation	1	1,024	20.48
Account generation	400,000	0	0.00
Account initialisation	400,000	40,000	800.00
Total cost		41,024	820.48

Cost comparison was also done. Storage cost in Bitcoin and Ethereum were taken from [13]. Adjustments were made to the calculations by applying the most recent cryptocurrency values and transaction fee rates. At the time of the writing, Bitcoin is traded at US$3,733 and Ethereum at US$133. The average transaction cost in Bitcoin is US$0.03 [4], while the average Ethereum gas cost is 2.1 GWei [5]. The cost comparison detail is provided in Table 3.

As shown in Table 3, the one-time cost in our solution does not need to be paid for storing the next gigabytes of data. Assuming that there will be 1 GB data stored each day, the operational cost for 30 days and 1 year (365 days) is shown in Table 4.

As it can be seen in Table 4, our solution does not incur any cost at all. This is because there is no more operational fee involved after the initialisation phase.

Table 3. Cost comparison between solutions

Protocol	Storage cost per GB (in US$)	Remark
Bitcoin	629,777.25	One-time storage
Ethereum	175,634.85	One-time storage
Senarai	820.48	One-time cost

Table 4. Storage cost for a long term usage

Protocol	Monthly cost (in US$)	Annual cost (in US$)
Bitcoin	18,893,317.50	229,868,696.25
Ethereum	5,269,045.57	64,106,721.13
Senarai	0.00	0.00

When the storage becomes enormously big, the cost during the initialisation phase for our solution becomes incredibly small compared to other solutions. This concludes that our solution is way more economical than the existing solutions in the long term.

Payload and Data Efficiency. Our protocol optimises the free 5,000 bandwidth points given by Tron system. To ensure that the transaction size does not exceed the free bandwidth points, a maximum payload size of 4,700 bytes can be stored in each transaction per account, per day. With the transaction data requires around 300 bytes, our protocol has a payload efficiency of 94% with 6% overhead. Table 5 shows the data efficiency comparison between Senarai, Bitcoin, and Ethereum. Senarai's efficiency rate is higher data dropping method in Bitcoin (90.5%), but lower than Ethereum's transaction data storage, where the efficiency can be as high as 99.7%[10].

Table 5. Data efficiency comparison

Method	Max. data per Tx	Data efficiency
Bitcoin's data drop [16]	92,507	90.5%
Ethereum's transaction [16]	44,444[a]	99.7%
Senarai	2,700	94%

[a]The size can vary depending on the block gas limit

---

[10] Based on the transaction created by an Ethereum Stackexchange user in August 2016 which utilised the maximum block gas limit at the time of transaction creation [1]. The data efficiency rate is calculated by comparing the data payload and the raw transaction. The Ethereum block gas limit is scalable, therefore it is possible to increase the efficiency further [1].

However, since the current inefficient encoding scheme requires double the original file size, our protocol's real data efficiency is only 47% of the total transaction size. This low efficiency significantly increases the cost during initialisation phase, but the data storage phase remains free.

7 Conclusion and Future Work

We present `Senarai`, a sustainable blockchain-based storage protocol. Our protocol utilises free storage feature in Tron and expands the storage by creating as many accounts as possible. Our protocol is cost-efficient; the expense is only required during initialisation phase, where as little as US$820.48 will provide a daily free data storage of `1GB`. The sustainability will be maintained such that the used storage will be replenished every 24 h.

For future work, we plan to optimise the encoding scheme in the standard Tron library and to add a compression scheme to improve the current data efficiency. By improving the encoding scheme and incorporating a compression scheme in the protocol, it is expected that more data can be stored in the blockchain without any additional fees.

References

1. Afri: Is there a limit for transaction size? (2016). https://ethereum.stackexchange.com/a/1110
2. Bartoletti, M., Pompianu, L.: An analysis of bitcoin OP_RETURN metadata. In: Brenner, M., et al. (eds.) FC 2017. LNCS, vol. 10323, pp. 218–230. Springer, Cham (2017). https://doi.org/10.1007/978-3-319-70278-0_14
3. BeckyMH: TRON TRX Coin – review of tronix, price, ICO, wallet – bitcoinwiki, October 2017. https://en.bitcoinwiki.org/wiki/TRON
4. BitInfoCharts: Bitcoin median transaction fee historical chart, January 2019. https://bitinfocharts.com/comparison/bitcoin-median_transaction_fee.html
5. Ethgasstation: Eth gas station, January 2019. https://ethgasstation.info/
6. Tron Foundation: Tron protocol release (2018). https://media.readthedocs.org/pdf/tron-wiki/latest/tron-wiki.pdf
7. Manidos: How can i store data in ethereum blockchain, August 2016. https://ethereum.stackexchange.com/a/7886
8. Mao, T.: Bandwidth points (2018). https://developers.tron.network/docs/bandwith
9. Mao, T.: Energy (2018). https://developers.tron.network/docs/energy
10. Mao, T.: Trc10 & trc20 tokens (2018). https://developers.tron.network/docs/trc10-token
11. Miller, A., Juels, A., Shi, E., Parno, B., Katz, J.: Permacoin: repurposing bitcoin work for data preservation. In: 2014 IEEE Symposium on Security and Privacy (SP), pp. 475–490. IEEE (2014)
12. Nakamoto, S.: Bitcoin: a peer-to-peer electronic cash system. Report (2008)
13. Omaar, J.: Forever isn't free: the cost of storage on a blockchain database, July 2017. https://medium.com/ipdb-blog/forever-isnt-free-the-cost-of-storage-on-a-blockchain-database-59003f63e01

14. Palau, A.: Storing on ethereum. Analyzing the costs, July 2018. https://medium.com/coinmonks/storing-on-ethereum-analyzing-the-costs-922d41d6b316
15. rumblestiltskin: My attempt at using storj (2017). https://steemit.com/cryptocurrency/@rumblestiltskin/my-attempt-at-using-storj
16. Sward, A., Vecna, I., Stonedahl, F.: Data insertion in bitcoin's blockchain. Ledger **3**, 1–23 (2018)
17. Terado, T.: What is decentralized storage? July 2018. https://medium.com/bitfwd/what-is-decentralised-storage-ipfs-filecoin-sia-storj-swarm-5509e476995f
18. Tronprotocol: Protobuf protocol, July 2018. https://github.com/tronprotocol/Documentation/blob/master/English_Documentation/TRON_Protocol/TRON_Protobuf_Protocol_document.md
19. Wijaya, D.A.: Extending asset management system functionality in bitcoin platform. In: 2016 International Conference on Computer, Control, Informatics and Its Applications (IC3INA), pp. 97–101. IEEE (2016)
20. Wikipedia: Interplanetary file system, September 2015. https://en.wikipedia.org/wiki/InterPlane-tary_File_System

Cloud Security

Breaking and Fixing Anonymous Credentials for the Cloud

Ulrich Haböck[1]([⊠])(iD) and Stephan Krenn[2](iD)

[1] University of Applied Sciences FH Campus Wien, Vienna, Austria
`ulrich.haboeck@fh-campuswien.ac.at`
[2] AIT Austrian Institute of Technology GmbH, Vienna, Austria
`stephan.krenn@ait.ac.at`

Abstract. In an attribute-based credential (ABC) system, users obtain a digital certificate on their personal attributes, and can later prove possession of such a certificate in an unlinkable way, thereby selectively disclosing chosen attributes to the service provider. Recently, the concept of encrypted ABCs (EABCs) was introduced by Krenn et al. at CANS 2017, where virtually all computation is outsourced to a semi-trusted cloud-provider called wallet, thereby overcoming existing efficiency limitations on the user's side, and for the first time enabling "privacy-preserving identity management as a service".

While their approach is highly relevant for bringing ABCs into the real world, we present a simple attack fully breaking privacy of their construction if the wallet colludes with other users – a scenario which is not excluded in their analysis and needs to be considered in any realistic modeling. We then revise the construction of Krenn et al. in various ways, such that the above attack is no longer possible. Furthermore, we also remove existing non-collusion assumptions between wallet and service provider or issuer from their construction. Our protocols are still highly efficient in the sense that the computational effort on the end user side consists of a single exponentiation only, and otherwise efficiency is comparable to the original work of Krenn et al.

Keywords: Attribute-based credentials · Privacy-preserving authentication · Strong authentication

1 Introduction

Anonymous attribute-based credential systems (ABCs) – first envisioned by Chaum [11,12] and extended in a large body of work [5–10,15,20–22] – are a cryptographic primitive enabling user-centric identity management. In ABC systems, a *user* receives a certificate on his personal data such as name, nationality, or date of birth from an *issuer*. Later, the user can *present* this certificate to *service providers* (or *relying parties*), thereby deciding which attributes to reveal or to keep private, in a way that makes different authentication processes

© Springer Nature Switzerland AG 2019
Y. Mu et al. (Eds.): CANS 2019, LNCS 11829, pp. 249–269, 2019.
https://doi.org/10.1007/978-3-030-31578-8_14

unlinkable to each other. While the service provider receives strong authenticity guarantees on the received attributes, the user's privacy is maintained, even against colluding issuers and service providers.

However, despite of their obvious benefits, ABC systems have not yet found their way into relevant real-world applications. One main reason for this are computational costs, which make them unsuitable for resource-constraint devices.

This drawback was recently addressed by Krenn et al. [18], who proposed a scheme dubbed EABC, where virtually all computations can be outsourced to a semi-trusted *wallet*. The underlying idea was that users get signatures on their attributes, encrypted under some proxy re-encryption [3] scheme, from the issuer, and upload signature and ciphertexts to the wallet, together with a re-encryption key from their own public key to the intended service provider's public key. For presentation, the wallet re-encrypts the ciphertexts of the revealed attributes for the service provider, randomizes the remaining ciphertexts, and attaches a zero-knowledge proof of knowledge of a signature on the underlying ciphertexts. By the privacy property of the proxy re-encryption scheme, the wallet can translate encryptions from users to service providers, without ever learning any information about the underlying plaintexts. However, while solving the efficiency drawbacks of previous ABC systems, the attacker model underlying [18] is unrealistic, as they make very strong non-collusion assumptions between the wallet on the one hand, and service providers or issuers on the other hand. Even worse, we point out a trivial attack which fully breaks privacy in their construction. That is, their security analysis would only hold true if in addition also any collusion between wallet and other users is forbidden, which is clearly unrealistic, e.g., in the case of malicious administrators.

The Attack on Krenn et al. [18]. The fundamental problem of [18] is that for efficiency reasons their construction makes use of bi-directional multi-hop proxy re-encryption schemes. That is, having a re-encryption key $rk_{A \to B}$ that allows a proxy to translates a ciphertext c_A encrypted under pk_A to a ciphertext c_B under pk_B without learning the plaintext, and a re-encryption key $rk_{B \to C}$, the proxy can also translate c_A to c_C under pk_C (multi-hop); furthermore, $rk_{A \to B}$ can efficiently be turned into $rk_{B \to A}$ (bi-directionality).

Assume now that Alice A wants to authenticate herself towards some service provider SP and thus stores $rk_{A \to SP}$ and encryptions c_A of her personal attributes on the wallet. Let the malicious administrator M also sign up for SP and compute $rk_{M \to SP}$. Using the bi-directionality of the proxy re-encryption scheme, this directly gives $rk_{SP \to M}$, and using the multi-hop functionality, M can now translate all of A's ciphertexts for herself, thereby fully breaking Alice's privacy. Even more, because of the concrete choice of the deployed re-encryption scheme, the attacker could even recover Alice's secret key as $sk_A = rk_{A \to SP}^{-1} \cdot sk_{M \to SP} \cdot sk_M^{-1}$ without having to assume a corrupt service provider. Actually, also the secret key of the service provider can be recovered as $sk_{SP} = sk_{M \to SP} \cdot sk_M^{-1}$.

Note that this attack is not specific to the deployed scheme of Blaze et al. [3], but arises in any multi-hop proxy re-encryption scheme that is used for outsourced data sharing application using long-term keys for the relying parties.

Mitigation Strategies. A straightforward solution to this problem might be to replace the deployed proxy re-encryption scheme by a single-hop and/or unidirectional encryption scheme. However, it turns out that the algebraic structures of existing signature and encryption schemes (with the required properties) would then no longer allow for efficient zero-knowledge proofs or knowledge, and the benefits of [18] would dissolve. Very informally speaking, the reason for this is that all such potential schemes would "consume" the one available pairing in the system. Furthermore, the other limitations of [18] (i.e., non-collusion assumptions) would not be addressed by such a modification.

Our Contribution. The main contribution of this paper is to overcome the security limitations of [18] without harming the efficiency of the scheme. That is, we provide an instantiation of an EABC system that does not require any artificial non-collusion assumptions, at the cost of only a single exponentiation on the user's side. Furthermore, in contrast to [18], our system also gives metadata-privacy guarantees in the sense that the wallet only learns the policy for which it is computing the presentation tokens (i.e., which attributes are revealed and which remain undisclosed), but does no longer learn for which service provider it is computing the presentation, such that reliably tracking users becomes virtually impossible. Hiding the presentation policy within a set of policies could be achieved by the techniques of Krenn et al. [18, Sect. 6.1] for a linear overhead.

In a bit more detail, our contribution is multifold.

- Firstly, we replace the static long-term keys used by the service providers in [18] by ephemeral keys which are only used for a single authentication. This is achieved through an interactive key agreement protocol between the two parties, which guarantees freshness of the agreed keys. By this, a malicious administrator can no longer run the attack described above, as the $rk_{A \to SP}$ and $rk_{M \to SP}$ will no longer be bound to the same key of the service provider.
- Next, by using independent keys for the individual user attributes, even a collusion of service provider and wallet may only reveal the information that the user was willing to share with the service provider in any case.
- Thirdly, by replacing the signature scheme deployed in the issuance phase by a blinded version of the same scheme, our construction achieves high unlinkability guarantees even in the case of wallet-issuer collusions. Our blinded version of the structure-preserving signature scheme of Abe et al. [2] may be also of independent interest beyond the scope of this paper.
- Finally, by having a separate identity key that is not stored on the wallet but locally on the user's device, the service provider is guaranteed that the user is actively participating in the protocol. While Krenn et al. [18] considered it undesirable that users need to carry secret key material with them, we believe that having no information stored locally results in unrealistic trust assumptions as there the wallet could impersonate a user towards any service provider that the user ever signed up for.

Outline. This paper is organized as follows. In Sect. 2 we discuss the building blocks of EABC systems, and in particular the schemes needed for our concrete instantiation. Then, in Sect. 3 we give a high-level description of EABC systems, its revised adversary model and security notions. Finally, Sect. 4 presents the concrete EABC instantiation, including security statements, the proofs of which are given in the full version of the paper.

2   Preliminaries

In the following we introduce the necessary background needed in the rest of the paper. In particular, we recap the notions of proxy re-encryption and structure-preserving signatures. We then present a transformation of the AGHO signature scheme [2] into a blinded version, which combines both features, blindness and structure-preservation, needed to efficiently instantiate EABC systems.

2.1   Notation

We denote the security parameter by λ. All probabilistic, polynomial time (PPT) algorithms are denoted by sans-serif letters (A, B, . . .), and their combination in two-party or three party protocols by $\langle A, B \rangle$ and $\langle A, B, C \rangle$, respectively. Whenever we sample a random element m uniformly from a finite set M, we denote this by $m \leftarrow_s M$. We write $\mathbb{Z}_q$ for the integers modulo a prime number q, $\mathbb{Z}_q^*$ for its multiplicative group, and $1/e$ for the modular inverses. We shall make extensive use of non-interactive zero-knowledge proofs of knowledge, where we use the Camenisch-Stadler notation to specify the proof goal. For example,

$$\mathsf{NIZK} \left[(\alpha, \beta, \Gamma) : y_1 = g^\alpha \wedge y_2 = g^\alpha \cdot h^\beta \wedge R = e(\Gamma, H) \right]$$

denotes a non-interactive zero-knowledge proof of knowledge proving knowledge of values α, β, Γ such that the expression on the right-hand side is satisfied. In most situations, extractability of zero-knowledge proofs will be sufficient. However, in a single case we will require simulation-sound extractability [17].

2.2   Anonymous and Re-randomizable Proxy Re-encryption

A proxy re-encryption (PRE) scheme is an asymmetric encryption scheme which allows a third party (the *proxy*) to transform ciphertexts encrypted for one party into ciphertexts encrypted for another one, without learning the underlying plaintext. As in [18], we instantiate our EABC system by using the scheme by Blaze et al. [3] (BBS); the associated issues in [18] are mitigated by a different use of the scheme. It possesses all the security properties needed for proving our system secure, yet it yields algebraic simple relations for encryption, re-encryption and re-randomization, altogether allowing for efficient zero-knowledge proofs of statements which involve these operations.

The BBS scheme consists of six PPT algorithms,

$$\mathsf{PRE}_{BBS} = (\mathsf{Par}, \mathsf{Gen}, \mathsf{Enc}, \mathsf{Dec}, \mathsf{ReKey}, \mathsf{ReEnc}),$$

where $\mathsf{Par}(\lambda)$ outputs the system parameters $pp = (\mathbb{G}, q, g)$, where $\langle g \rangle = \mathbb{G}$ is a group of prime order q. $\mathsf{Gen}(pp)$ generates a key pair (sk, pk) by $sk \leftarrow_\$ \mathbb{Z}_q$ and $pk = (pp, g^{sk})$. Encryption and decryption works as for ElGamal [16], i.e.,

$$\mathsf{Enc}(pk, m) = c = (c_1, c_2) = (g^r, pk^r \cdot m),$$

where $m \in \mathbb{G}$ is the message, $r \leftarrow_\$ \mathbb{Z}_q$, and $\mathsf{Dec}(sk, c) = c_1^{-sk} \cdot c_2$.

Given two key pairs (pk_1, sk_1), (pk_2, sk_2), their re-encryption key $rk = rk_{pk_1 \to pk_2}$ is derived by $\mathsf{ReKey}(sk_1, pk_1, sk_2, pk_2) = sk_1 \cdot sk_2^{-1}$, and

$$\mathsf{ReEnc}(rk, c) = (c_1^{rk}, c_2)$$

transforms a ciphertext $c = (c_1, c_2)$ for pk_1 to one with respect to pk_2.

The relevant properties of the BBS scheme are summarized next.

Proposition 1 ([3]). *Under the DDH assumption in the message space $\mathbb{G}$, the BBS scheme is PRE-IND-CPA secure. That is, it is IND-CPA secure even under knowledge of (polynomially many) re-encryption keys that do not allow the adversary to trivially decrypt the challenge ciphertext.*

Proposition 2 ([18]). *The BBS PRE scheme with re-randomization function* $\mathsf{ReRand}(pk, c) = \mathsf{Enc}(pk, 1) \cdot \mathsf{Enc}(pk, c) = (g^r \cdot c_1, pk^r \cdot c_2), r \leftarrow_\$ \mathbb{Z}_q$, *has the ciphertext re-randomization property. That is, given pk, a message m and its ciphertext c, then the output distribution of $\mathsf{ReRand}(pk, c)$ is computationally indistinguishable from that of $\mathsf{Enc}(pk, m)$.*

Proposition 3 ([18]). *Under the DDH assumption in $\mathbb{G}$, the BBS proxy-re-encryption scheme is* anonymous. *That is, for any PPT adversary Adv there exists a negligible function ν such that*

$$\left| \Pr \left[\begin{array}{l} pp \leftarrow \mathsf{Par}(\lambda); (sk_i, pk_i) \leftarrow \mathsf{Gen}(pp), i \in \{0, 1\}; \\ (m, st) \leftarrow \mathsf{Adv}(pp, pk_1, pk_2); \\ b \leftarrow_\$ \{0, 1\}; b^* \leftarrow \mathsf{Adv}(st, \mathsf{Enc}(pk_b, m)) \end{array} : b^* = b \right] - \frac{1}{2} \right| \le \nu(\lambda)$$

2.3 Structure-Preserving Blind Signatures

The structure-preserving signature scheme of Abe et al. [1,2] is based on asymmetric bilinear groups $(\mathbb{G}, \mathbb{H}, \mathbb{T}, e)$ with the feature that messages, signatures and verification keys consist of elements from $\mathbb{G}$ and/or $\mathbb{H}$, and verification is realized by pairing-product equations over the key, the message and the signature. This allows for efficient zero-knowledge proofs of claims involving the message and the signature, which is why they apply to various cryptographic protocols, e.g.,

[1,14,15,18]. Similarly, our construction relies on the scheme in [2] (AGHO), since it allows to sign vectors of group elements. The AGHO scheme

$$\mathsf{SIG}_{AGHO} = (\mathsf{Par}, \mathsf{Gen}, \mathsf{Sig}, \mathsf{Vf})$$

consists of four PPT algorithms. The setup algorithm Par generates the scheme's parameters $pp = (\mathbb{G}, \mathbb{H}, \mathbb{T}, q, e, G, H)$ which are comprised of groups $\mathbb{G}, \mathbb{H}, \mathbb{T}$ of prime order q, a bilinear mapping $e : \mathbb{G} \times \mathbb{H} \longrightarrow \mathbb{T}$, and their respective generators $G, H, e(G, H)$. $\mathsf{Gen}(pp)$ produces a private-public key pair (sk, vk),

$$sk = (v, (w_i)_{i=1}^l, z) \quad \text{and} \quad vk = (V, (W_i)_{i=1}^l, Z) = (H^v, (H^{w_i}), H^z),$$

where all the secret components v, z, and w_i are randomly sampled from $\mathbb{Z}_q$. Given $m = (g_i)_{i=1}^l$ from $\mathbb{G}^l$, we have that $\sigma = \mathsf{Sig}(sk, m) = \sigma = (R, S, T) \in \mathbb{G} \times \mathbb{G} \times \mathbb{H}$, where

$$R = G^r, \qquad S = G^z \cdot R^{-\cdot v} \cdot \prod_{i=1}^l g_i^{-w_i}, \qquad T = H^{1/r},$$

for $r \leftarrow_s \mathbb{Z}_q^*$. The verification condition of $\sigma = (R, S, T)$ is given by the two bilinear equations $e(S, H) \cdot e(R, V) \cdot \prod_i e(g_i, W_i)$ and $e(R, T) = e(G, H)$.

Theorem 1 ([2]). *In the generic group model, the AGHO signature scheme* $\mathsf{SIG} = (\mathsf{Par}, \mathsf{Gen}, \mathsf{Sig}, \mathsf{Vf})$ *is strongly existentially unforgeable under adaptive chosen message attacks (sEUF-CMA). That is, for every* PPT *adversary* Adv *there exists a negligible function* ν *such that*

$$\Pr\begin{bmatrix} pp \leftarrow \mathsf{Par}(\lambda); (vk, sk) \leftarrow \mathsf{Gen}(pp) & \mathsf{Vf}(vk, (m^*, \sigma^*)) = 1 \wedge \\ (m^*, \sigma^*) \leftarrow \mathsf{Adv}^{\mathsf{Sig}(pp, sk, \cdot)} & : \quad (m^*, \sigma^*) \notin Q \end{bmatrix} \leq \nu(\lambda)$$

where Adv *has access to a signing oracle* $\mathsf{Sig}(pp, sk \cdot)$, *which on input m computes a valid signature σ, adds (m, σ) to the initially empty list Q, and returns σ.*

Blind signatures allow a user to obtain signatures in a way such that both the message as well as the resulting signature remain hidden from the signer. *Restrictive* blind schemes additionally allow the signer to encode information into the message, while still preserving the unlinkability of the resulting message-signature pair to the issuance session. The notion of restrictiveness goes back to Brands [4], and various adaptions have been made since then, e.g., [5,13,19]. In the context of anonymous credentials, and for the first time done in [5], such restricted message is typically a commitment on a value defined by the issuer. As such, we consider a restrictive blind signature scheme

$$\mathsf{BSIG} = (\mathsf{Par}, \mathsf{Gen}, \mathsf{User}, \mathsf{Signer}, \mathsf{Vf})$$

being based on a blind signature scheme and a commitment scheme

$$\mathsf{COM} = (\mathsf{Par}_{COM}, \mathsf{Comm}_{COM}, \mathsf{Vf}_{COM})$$

for values x such that its output is in the message space of the signature. Par, on input the security parameter λ, sets up the scheme's parameters pp, including a compliant setting of COM, and $\mathsf{Gen}(pp)$ generates a private-public key pair (sk, vk). The interactive algorithms User and Signer define the issuance protocol

$$\langle \mathsf{User}(vk, x), \mathsf{Signer}(sk, x) \rangle$$

between a user and a signer with private-public key pair (sk, pk), which on input a commonly agreed value $x \in X$ outputs to the user a certificate (w, com, σ) which consists of a commitment com of x together with an opening w and a valid signature σ on com. The verification $\mathsf{Vf}(vk, x, (w, com, \sigma)) = 1$ is a separate validity check of the commitment com on (x, w) and the signature σ on com.

The notions of unforgeability and blindness adapted to our setting of restrictive blind signatures are as follows.

Definition 1. *A restrictive blind signature scheme* $\mathsf{BSIG} = (\mathsf{Par}, \mathsf{Gen}, \mathsf{User}, \mathsf{Signer}, \mathsf{Vf})$ *is* strongly unforgeable *if for any* PPT *adversary* Adv *there exists a negligible function* ν *such that*

$$\Pr \left[\begin{array}{l} pp \leftarrow \mathsf{Par}(\lambda); \\ (vk, sk) \leftarrow \mathsf{Gen}(pp); Q \leftarrow \emptyset \\ (x^*, (w_i^*, com_i^*, \sigma_i^*)_{i=1}^q) \leftarrow \mathsf{Adv}^{\langle \cdot, \mathsf{Signer}(sk, \cdot)\rangle} \end{array} : \begin{array}{l} q > mult(x^*) \wedge \\ \text{for } 1 \le i \le q \\ \mathsf{Vf}(vk, x^*, w_i^*, (com_i^*, \sigma_i^*)) = 1 \wedge \\ \bigwedge_{j \neq i} (com_j^*, \sigma_j^*) \neq (com_i^*, \sigma_i^*) \end{array} \right] \le \nu(\lambda),$$

where Adv *has access to a signing oracle* $\langle \cdot, \mathsf{Signer}(sk, \cdot) \rangle$ *which logs every successful query* x *in an initially empty list* Q, *and* $mult(x^*)$ *denotes the multiplicity of successful queries with* $x = x^*$.

Definition 2. *A restrictive blind signature scheme* $\mathsf{BSIG} = (\mathsf{Par}, \mathsf{Gen}, \mathsf{User}, \mathsf{Signer}, \mathsf{Vf})$ *satisfies* blindness, *if for any* PPT *adversary* Adv *there exists a negligible function* ν *such that*

$$\left| \Pr \left[\begin{array}{l} pp \leftarrow \mathsf{Par}(\lambda); (vk^*, x_0^*, x_1^*, st) \leftarrow \mathsf{Adv}(pp); \\ ((w_i, com_i, \sigma_i), st) \leftarrow \langle \mathsf{User}(vk, x_i^*), \mathsf{Adv}(st) \rangle, \ i \in \{0, 1\} \\ \text{if } \sigma_0 = \bot \vee \sigma_1 = \bot \text{ then } (\sigma_0, \sigma_1) = (\bot, \bot) \\ b \leftarrow_\$ \{0, 1\}; \ b^* \leftarrow \mathsf{Adv}(st, (com_b, \sigma_b), (com_{1-b}, \sigma_{1-b})) \end{array} : b^* = b \right] - \frac{1}{2} \right| \le \nu(\lambda).$$

Blind AGHO Scheme. Our structure-preserving restrictive blind signature scheme $\mathsf{BSIG}_{AGHO} = (\mathsf{Par}, \mathsf{Gen}, \mathsf{User}, \mathsf{Signer}, \mathsf{Vf})$ is based on the AGHO scheme SIG_{AGHO}, a compatible commitment scheme COM, and two non-interactive extractable zero-knowledge proof systems, to both of which we refer to as NIZK without causing confusion. Par, Gen, and Vf are the corresponding algorithms from SIG_{AGHO}, besides that Par also queries Par_{COM} so that the commitments are elements of the messages space $\mathbb{G}^l$ of the AGHO scheme, and furthermore generates a common reference string for the zero-knowledge proof systems. We stress that our scheme is not merely based on a structure-preserving signature (as the ones from, e.g., [1,14,15]) but is structure-preserving by itself, which is an essential feature for our EABC instantiation.

Definition 3 (Blind AGHO signature on commited values). *The issuance protocol* $\langle \mathsf{User}(vk, x), \mathsf{Signer}(sk, x) \rangle$ *runs between a signer S with AGHO signing keys* $sk = (v, (w_i)_{i=1}^l, z)$, vk, *and a user U who wishes to receive a certificate* (w, com, σ) *on the commonly agreed value x from the signer.*

1. *U computes com on x with opening w by using Comm_{COM}. By our assumption on COM, $m = com$ is from the message space of the signature, i.e. $m = (m_i)_{i=1}^l \in \mathbb{G}^l$.*
2. *U blinds m using a random pad $P = (P_i)_{i=1}^l \leftarrow_\$ \mathbb{G}^l$, and obtains $\overline{m} = (\overline{m}_i)_{i=1}^l = (m_i \cdot P_i^{-1})_{i=1}^l$. It further chooses $e, f \leftarrow_\$ \mathbb{Z}_q^*$, a random decomposition $f = f_1 + f_2$ of f, and sets $\overline{P} = (P_i^e)_{i=1}^l$, $(G_1, G_2, G_3) = (G^e, G^{f_1}, G^{e \cdot f_2})$. U then sends $\overline{m}$, $\overline{P}$, (G_1, G_2, G_3) to S and gives a zero knowledge of well-formedness*

$$\pi_U = \mathsf{NIZK}\Big[(\eta, \varphi_1, \varphi_2, \omega) : G_1^\eta = G \wedge G^{\varphi_1} = G_2 \wedge$$

$$G_1^{\varphi_2} = G_3 \wedge \mathsf{Vf}_{COM}(\overline{m} \cdot \overline{P}^\eta, x, \omega) = 1\Big],$$

using the witnesses $(\eta, \varphi_1, \varphi_2, \omega) = (1/e, f_1, f_2, w)$.
3. *S verifies π_U, and returns $\bot$ if not valid. Otherwise it generates a random decomposition $z = z_1 + z_2$ of its signing key's z, and computes the 'signatures' $\overline{\sigma} = (\overline{R}, \overline{S}_1, \overline{S}_2, \overline{T})$, with $\overline{R} = G^r$, $\overline{T} = H^{1/r}$, $\overline{S}_1 = G^{z_1} \cdot G_2^{-r \cdot v} \cdot \prod_{i=1}^l \overline{m}_i^{-w_i}$, $\overline{S}_2 = G_1^{z_2} \cdot G_3^{-r \cdot v} \cdot \prod_{i=1}^l \overline{P}_i^{-w_i}$, where $r \leftarrow_\$ \mathbb{Z}_q^*$. It then returns $\overline{\sigma}$ to U supplemented by a proof of wellformedness*

$$\pi_S = \mathsf{NIZK}\Big[(\rho, \tau, (\omega_i)_i, \zeta_1, \zeta_2) : \bigwedge_i H^{\omega_i} = W_i \wedge H^{\zeta_1} \cdot H^{\zeta_2} = Z \wedge$$

$$G^\rho = \overline{R} \wedge \overline{T}^\rho = H \wedge V^\rho \cdot H^{-\tau} = 1 \wedge$$

$$G^{\zeta_1} \cdot G_2^{-\tau} \cdot \prod_{i=1}^l \overline{m}_i^{-\omega_i} = \overline{S}_1 \wedge G^{\zeta_2} \cdot G_3^{-\tau} \cdot \prod_{i=1}^l \overline{P}_i^{-\omega_i} = \overline{S}_2\Big],$$

by using the witnesses $(\rho, \tau, (\omega_i)_i, \zeta_1, \zeta_2) = (r, r \cdot v, (w_i)_i, z_1, z_2)$.
4. *U checks if π_S is valid. If so, she outputs $\sigma = (R, S, T)$, where $R = \overline{R}^f$, $S = \overline{S}_1 \cdot \overline{S}_2^{1/e}$, $T = \overline{T}^{1/f}$. (Otherwise she outputs $\bot$).*

Corrrectness of the blind AGHO scheme is straightforward, the proof of the following theorem is given in the full version of the paper.

Theorem 2. *Suppose that both* NIZK *in Definition 3 are extractable. If the commitment scheme* COM *is computationally hiding, then under the DDH assumption in* $\mathbb{G}$ *the restrictive blind signature scheme* BSIG_{AGHO} *satisfies blindness. Furthermore, if* COM *is computationally binding,* BSIG_{AGHO} *is strongly unforgeable in the generic group model.*

3 EABC: High-Level Description

An *encrypted* attribute-based credential (EABC) system, introduced in [18], allows the delegation of selective disclosure to a third party in a privacy-preserving manner by means of proxy re-encryption and redactable signatures.

There are four types of players in an EABC system: *issuers*, *users*, *services*, and the central *wallet*. Each user U holds an *identity key* sk_U proving her identity and which is securely stored on her trusted device (e.g., a smart card or TPM). U engages in an *issuance protocol* with an issuer I to receive an encrypted credential C on certain attributes only readable to her such that C is bound to her identity key sk_U. U further owns an account managed by the wallet W, a typically cloud-based identity provider, to which she uploads all of her encrypted credentials C while not providing it the encryption keys $k(C)$. At any time later, when U wants to access a service S she is asked to attest some of her attributes. To convince S of the requested attributes without revealing any further testified information, U chooses one (or several) of her credentials from her account, selects a subset of attributes contained therein, and instructs W to engage in a *presentation protocol* with S, which serves the latter re-encryptions of the requested attributes together with a proof of validity. In this protocol, the wallet W undertakes (almost) all costly operations while reducing U's effort to a possible minimum, requiring her only to supply the re-encryption keys for the selected attributes, and a proof of her consent (via her sk_U) to the presentation process. This last proof is also how the overall model of EABCs differs from that in [18], where no computation is required on the user's side at all; however, as discussed earlier, we believe this is needed for a realistic attacker scenario, as otherwise the wallet could arbitrarily impersonate the user towards any service provider that the user ever signed up for.

3.1 Formal Definition

An EABC system with attribute space $\mathbb{A}$ is built on a structure-preserving blind signature scheme BSIG in the sense of Sect. 2.3, an anonymous re-randomizable proxy re-encryption scheme PRE (cf. Sect. 2.2) which acts on the message space of BSIG, and two zero-knowledge proof systems to which we both refer as ZKP without causing confusion. Formally, an EABC system

$$\mathsf{EABC} = (\mathsf{Par}, \mathsf{Gen}_I, \mathsf{Gen}_U, \mathsf{Iss}, \mathsf{User}_I, \mathsf{User}_P, \mathsf{Wall}, \mathsf{Serv})$$

consists of the (PPT) algorithms Par, Gen_I, and Gen_U for setup and key generation, and the interactive (PPT) algorithms Iss, User_I, User_P, Wall, Serv which are the components of the issuance and presentation protocol described below. Given the security parameter λ, $\mathsf{Par}\,(\lambda)$ generates the system parameters sp. These are comprised of the parameters for BSIG, PRE, and a common reference string for ZKP. Every user U holds a PRE secret-public key pair (sk_U, pk_U) generated by $\mathsf{Gen}_U(sp) = \mathsf{Gen}_{PRE}(sp)$, her *identity key*, which is used to generate her certificate pseudonyms. On demand, a user repeatedly queries $\mathsf{Gen}_U(sp)$ to generate the encryption keys $k(C)$ of her credentials. Each issuer I is holder of a key pair for the blind signature scheme $(sk_I, vk_I) \leftarrow \mathsf{Gen}_I(sp)$, $\mathsf{Gen}_I = \mathsf{Gen}_{BSIG}$, where sk_I denotes the secret signing key and vk_I its public verification key. Note that, unlike in [18] a service has no permanent key material for the proxy re-encryption scheme. Its PRE key will be an ephemeral one-time key, one for each presentation. Both the issuance and the presentation protocol run over server-side authenticated, integrity protected and confidential connections (associated with some random session identifier sid) and are as follows.

Issuance. The issuance protocol

$$\langle \mathsf{User}_I\,(sid, sk_U, A, vk_I)\,, \mathsf{Iss}\,(sid, A, sk_I[, pk_U])\,\rangle$$

is performed between a user U with identity keys (sk_U, pk_U) and an issuer I with signature key pair (sk_I, vk_I) who is the supplier of the random session identifier sid. Both user and issuer agreed on the unencrypted content, the attributes $A = (a_i)_{i=1}^l \in \mathbb{A}^l$ beforehand. Depending on the type of issuance, it might be mandatory that U authenticates with its identity key, hence we leave it optional whether pk_U is supplied to I or not, denoted by $[, pk_U]$. If successful, the protocol outputs to U an encrypted attribute-based credential (C, vk_I) together with it's (secret) key material $sk = sk(C)$, the latter of which U keeps on her device (and never provides it to a third party). In all other cases, the user receives $\perp$.

Presentation. The presentation protocol is a three party protocol

$$\langle \mathsf{User}_P\,(sid, sk_U, sk(C), D)\,, \mathsf{Wall}\,(sid, C, D, vk_I)\,, \mathsf{Serv}\,(sid, D, vk_I)\,\rangle$$

and involves a user U with identity key sk_U, the wallet W which hosts U's credential (C, vk_I), and a service S, who provides the random session identifier sid. As before, $sk(C)$ is the user's secret key material for C. The user decides the attributes in C to be disclosed to S beforehand, associated with some index subset $D \subseteq \{1, \dots, l\}$. At the end of the protocol the service receives a *presentation* C^* of C, which is comprised of the requested attributes, re-encrypted to a random one-time key sk' of S, together with a proof of validity. The service verifies the proof by help of the issuer's public key vk_I. If valid, the service accepts and decrypts the attributes using sk'. Otherwise it rejects and outputs $\perp$ to both U and W.

3.2 EABC Security Notions

We widen the adversary model from [18] to a setting which does not impose any trust assumption on the wallet. An attacker who controls several players of the

EABC system, i.e. the central wallet, some of its users, service providers and issuers, should not be able to compromise the system in a more than obvious manner. That is, the adversary should not be able to

1. efficiently generate valid presentations which do not match any of the adversary's credentials (*unforgeability*),
2. alter the statement of a presentation successfully without knowledge of the user (*non-deceivability*),
3. learn anything from the encrypted credentials besides the information disclosed under full control of the owner (*privacy*), and
4. distinguish presentations with the same disclosed content when the underlying encrypted credentials are not known to the adversary (*unlinkability*).

All security notions are given in a game-based manner, and we assume server-authenticated, confidential and integrity protected connections between the protocol participants.

Unforgeability for EABC. The *unforgeability experiment* paraphrases a malicious wallet and (adaptively many) malicious users, who altogether try to trick an honest service into accepting a presentation which does not match any of the adversary's queries to an honest issuer. The experiment manages a list L which records the key material of all honest system participants during the entire lifetime of the system, i.e. the honest user's identity keys (pk_U, sk_U) and honest issuer keys (vk_I, sk_I). The list L is also used to log all honest user's credentials $(C, vk_I, sk_U, sk(C))$ under a unique handle h.

At any time the adversary Adv *is given access to all public information contained in L, i.e. the public keys pk_U, vk_I and the handles h, and as wallet W^* it may retrieve the encrypted credentials (C, vk_I) of every handle h contained in L.*

Besides L, the experiment maintains another list Q_{Adv} used for logging all adversaries queries to honest issuers of the system. At first, the experiment initializes the system by running $sp \leftarrow \mathsf{Par}(\lambda)$, setting $L = \emptyset$, $Q_{\mathsf{Adv}} = \emptyset$, and returns sp to the adversary Adv. The adversary may then generate and control adaptively many (malicious) players, and interact with the honest ones by use of the following oracles:

Issuer oracle $\mathsf{I}(vk_I, A\,[, pk_U^*])$. This oracle, on input an issuer's vk_I, attributes $A = (a_i)_i$, and optionally a public identity key pk_U^*, provides the adversary a (stateful) interface to an honest issuer's $\mathsf{Iss}(A, sk_I\,[, pk_U^*])$ in the issuance protocol, provided that vk_I is listed in L. If not, then the oracle generates a fresh pair of issuer keys $(sk_I, vk_I) \leftarrow \mathsf{Gen}_I(sp)$, adds it to L, and returns vk_I to the caller Adv. Whenever the protocol execution is successful from the issuer's point of view, the oracle adds $(vk_I, (a_i)_i\,[, pk_U^*])$ to Q_{Adv}.

User-issuance oracle $\mathsf{U}_I(pk_U, A, vk_I^*)$. This oracle provides the interface to an honest user's $\mathsf{User}_I(sk_U, A, vk_I^*)$ in an adversarily triggered issuance session. If vk_I^* belongs to an honest issuer (being listed in L) the oracle aborts. As above, if pk_U is not in L, the oracle adds fresh $(sk_U, pk_U) \leftarrow \mathsf{Gen}_U(pp)$ to

L, and informs adversary about the new pk_U. Whenever the session yields a valid credential C for the user, the oracle adds $(C, vk_I^*, sk_U, sk(C))$ together with a fresh handle h to L, and outputs (h, C, vk_I^*) to the adversary.

Issuance oracle $\mathsf{UI}(pk_U, A, vk_I)$. This oracle performs a full issuance session between an honest user pk_U and an honest issuer vk_I on the attributes A, logs the resulting credential $(C, vk_I, sk_U, sk(C))$ in L and outputs (C, vk_I), its handle h and the protocol transcript to the caller. Again, if either pk_U or vk_I are not in L the oracle generates the required identities, adds them to L and returns their new public keys before running the issuance.

User-presentation oracle $\mathsf{U}_P(h, D)$. The user-presentation oracle initiates a presentation session for an existing handle h, and provides both interfaces of $\mathsf{User}_P(sk_U, sk(C), D)$, where $(sk_U, sk(C))$ belong to h, to the caller. If the handle h is not listed in L, the oracle aborts.

Eventually the adversary Adv runs a presentation session claiming credentials of some honest, but adversarily chosen vk_I. The experiment is successful if Adv manages to make $\mathsf{Serv}[vk_I]$ accept the presentation but the disclosed attributes $o_S^* = (a_i^*)_{i \in D^*}$ do not correspond to any of the adversary's credentials issued by vk_I, which we denote by $o_S^* \notin Q_{\mathsf{Adv}}|_{D^*}$.

Definition 4. *An EABC system EABC is* unforgeable, *if for any PPT adversary Adv the success probability in the following experiment is bounded by a negligible function in* λ.

Unforgeability Experiment $\mathsf{Exp}_{\mathsf{Adv}}^{\mathrm{forg}}(\lambda)$

$pp \leftarrow \mathsf{Par}(\lambda); L = \emptyset; Q_{\mathsf{Adv}} = \emptyset;$

$(vk_I, st) \leftarrow \mathsf{Adv}(pp),$ *with* vk_I *listed in* L

$\langle o_U^*, o_S^* \rangle \leftarrow \langle \mathsf{Adv}(st), \mathsf{Adv}(st), \mathsf{Serv}(vk_I, D) \rangle$

if $o_U^* \neq \bot \wedge o_S^* \notin Q_{\mathsf{Adv}}|_{D^*}$ **return** *success* **else return** *failed*

In this experiment, $\mathsf{Adv} = \mathsf{Adv}^{\mathsf{I}, \mathsf{U}_I, \mathsf{UI}, \mathsf{U}_P}$ has access to the above defined (interactive) oracles, o_U denotes the service's verdict (output to the user-side), and o_S^* are the disclosed attributes $(a_i^*)_{i \in D^*}$ (output on the service-side).

Non-deceivability of Honest Users. *Non-deceivability* (of honest users) is the infeasibility of successfully altering the presentation goal without being exposed to the honest user. Note that this property is not automatically covered by Definition 4, since such a change of goal might be just between two vk_I-credentials of one and the same user. We formulate this property by means of the *non-deceivability experiment*, which is almost identical to the unforgeability experiment, except that in the last step the adversary Adv opens a presentation session on behalf of an *honest* user for a credential C and index set D chosen by the adversary.

Definition 5. *An EABC system EABC is* non-deceivable towards a user, *if for any PPT adversary Adv the success probability in the following experiment is bounded by a negligible function in* λ.

Non-Deceivability Experiment $\mathsf{Exp}_{\mathsf{Adv}}^{\mathrm{decv}}(\lambda)$

$pp \leftarrow \mathsf{Par}(\lambda); L = \emptyset;$

$(h, D, st) \leftarrow \mathsf{Adv}(pp),$ *such that* h *is listed in* L

Let C, vk_I, sk_U, $sk(C)$, *and* $(a_i)_i$ *belong to* h;

$\langle o_U^*, o_S^* \rangle \leftarrow \langle \mathsf{User}_P(sk_U, sk(C), D), \mathsf{Adv}(st), \mathsf{Serv}(D, vk_I) \rangle$

if $o_U^* \neq \bot \wedge o_S^* \neq (a_i)_{i \in D}$ **return** *success* **else return** *failed*

As in Definition 4, $\mathsf{Adv} = \mathsf{Adv}^{\mathsf{I}, \mathsf{U}_I, \mathsf{UI}, \mathsf{U}_P}$ has access to the oracles described in Sect. 3.2, o_U^* denotes the serivce's verdict (ouput to the user-side), and o_S^* are the disclosed attributes $(a_i^*)_{i \in D^*}$ (output on the service-side).

Privacy for EABC. The adversary's environment in $\mathsf{Exp}_{\mathsf{Adv}}^{\mathrm{priv}}$ is as in the unforgeability experiment from Sect. 3.2. That is, the experiment maintains a list L for the public and secret data of all honest participants, and the adversary is given access to the same honest participant oracles I, U_I, UI, U_P. First, the experiment generates the system parameters pp and a (random) subset $D \subseteq \{1, \ldots, n\}$, and lets the adversary choose one of its issuance keys vk_I^*, two honest (not necessarily different) user identities pk_{U_1}, pk_{U_2} and their queries A_0, A_0 being compliant on D, i.e. $A_1|_D = A_2|_D = (a_i)_{i \in D}$. Then the experiment performs issuance sessions with vk_I^* on A_0 and A_1 (but does not log the resulting credentials C_0 and C_1 in the list L). It chooses a random bit b, tells the adversary C_b and lets Adv participate in a presentation session for C_b as many times as wanted. Based on its experience during all these interactions, the adversary tries to guess the random bit b.

Definition 6. *An EABC system* EABC *satisfies privacy, if for any* PPT *adversary* Adv *the advantage* $\left| \Pr\left[\mathsf{Exp}_{\mathsf{Adv}}^{\mathrm{priv}}(\lambda) = success \right] - \frac{1}{2} \right|$ *in the following experiment is bounded by a negligible function in* λ.

Indistinguishability Experiment $\mathsf{Exp}_{\mathsf{Adv}}^{\mathrm{priv}}(\lambda)$

$pp \leftarrow \mathsf{Par}(\lambda); L = \emptyset; D \leftarrow_\$ 2^{\{1, \ldots, l_{max}\}};$

$(st, vk_I^*, (pk_{U_0}, A_0), (pk_{U_1}, A_1)) \leftarrow \mathsf{Adv}(pp),$

  with $A_0|_D = A_1|_D$ *and* pk_{U_0}, pk_{U_1} *listed in* L

$\langle (C_i, sk(C_i)), st \rangle \leftarrow \langle \mathsf{User}_I(sk_{U_i}, A_i, vk_I^*), \mathsf{Adv}(st) \rangle, i \in \{0, 1\}$

$b \leftarrow_\$ \{0, 1\},$

$b^* \leftarrow \langle \mathsf{User}_P(sk_{U_b}, sk(C_b), D, vk_I^*), \mathsf{Adv}(st, C_b) \rangle$

if $b = b^*$ **return** *success* **else return** *failed*

Again, the adversary $\mathsf{Adv} = \mathsf{Adv}^{\mathsf{I}, \mathsf{U}_I, \mathsf{UI}, \mathsf{U}_P}$ is given access to the oracles as described in Sect. 3.2.

Unlinkability of Presentations. Unlinkability of presentations is the infeasability for a malicious service to link any two presentation sessions with the user or the credentials hidden behind the presentation. Here, the service may collude with issuers (in practice both can be even one and the same entity), but in contrast to the above experiments, the wallet W is assumed to be honest. We express this property by means of the *unlinkability experiment* which essentially is as $\mathsf{Exp}_{\mathsf{Adv}}^{\mathsf{priv}}$ from Sect. 3.2, but the adversary is not given access to the U_P oracle, and it is forbidden to retrieve any credential C from L. In return it is given access to the following honest wallet oracles:

Wallet oracle $\mathsf{W}[h, D]$. This oracle provides the interfaces to an honest wallet's $\mathsf{Wall}[C, D, vk_I]$, where C and vk_I belong to the handle h listed in L. If the handle does not exist, the oracle aborts.

User-wallet oracle $\mathsf{UW}[h, D]$. This oracle, on input the handle h and index subset D, looks up the corresponding credential C and key material sk_U, $sk = sk(C)$ in L and provides the caller the interfaces to the presentation session $\langle \mathsf{User}_P[sk_U, sk(C), vk_I], \mathsf{Wall}[C, D, vk_I], \,.\,\rangle$. As above, if the handle does not exist the oracle aborts.

Definition 7. *An EABC system* EABC *is* unlinkable, *if for any* PPT *adversary* Adv *the advantage* $\left| \Pr\left[\mathsf{Exp}_{\mathsf{Adv}}^{\mathsf{link}}(\lambda) = success \right] - \frac{1}{2} \right|$ *in the following experiment is bounded by a negligible function in* λ.

Unlinbkability Experiment $\mathsf{Exp}_{\mathsf{Adv}}^{\mathsf{link}}(\lambda)$

$pp \leftarrow \mathsf{Par}(\lambda); L = \emptyset;$

$(st, vk_I^*, (pk_{U_0}, A_0), (pk_{U_1}, A_1), D) \leftarrow \mathsf{Adv}(pp),$

  with $A_0|_D = A_1|_D$ and pk_{U_0}, pk_{U_1} listed in L

$\langle (C_i, sk(C_i)), st \rangle \leftarrow \langle \mathsf{User}_I(sk_{U_i}, A_i, vk_I^*), \mathsf{Adv}(st) \rangle, i \in \{0, 1\}$

$b \leftarrow_{\$} \{0, 1\};$

$b^* \leftarrow \langle \mathsf{User}_I(sk_{U_i}, sk(C_b), vk_I^*), \mathsf{Wall}(C_b, D, vk_I^*), \mathsf{Adv}(st) \rangle$

if $b^* = b$ **return** *success* **else return** *failed*

In the experiment the adversary $\mathsf{Adv} = \mathsf{Adv}^{\mathsf{I}, \mathsf{U}_1, \mathsf{UI}, \mathsf{W}, \mathsf{UW}}$ is given access to all the honest-participant oracles from Sect. 3.2, and the above defined honest wallet oracle W and UW.

4 Instantiating EABCs

We instantiate $\mathsf{EABC} = (\mathsf{Par}, \mathsf{Gen}_I, \mathsf{Gen}_U, \mathsf{User}_I, \mathsf{Iss}, \mathsf{User}_P, \mathsf{Wall}, \mathsf{Serv})$ using the structure-preserving blind signature scheme $\mathsf{BSIG} = \mathsf{BSIG}_{AGHO}$ from Sect. 2.3, the anonymous re-randomizable proxy re-encryption scheme $\mathsf{PRE} = \mathsf{PRE}_{BBS}$ by Blaze, Bleumer, and Strauss (BBS, cf. Sect. 2.2), and two non-interactive zero-knowledge proof systems (cf. Sect. 2.1), one of which is simulation extractable. For notational convenience, we shall refer to both as NIZK without causing confusion.

4.1 System Parameters and Key Generation

Given the security parameter λ, a trusted[1] third party generates the system parameters by $sp \leftarrow \mathsf{Par}(\lambda)$, which internally queries Par_{BSIG} and Par_{PRE} in such a way that the message space for BSIG and the ciphertext space of PRE is the same group $\mathbb{G}$ of prime order q. Furthermore, it uses Gen_{ZKP} to set up a common reference string for the NIZK. Specifically, the system parameters are

$$sp = (L, \mathbb{G}, \mathbb{H}, \mathbb{T}, q, e, G, H, g, crs),$$

whereas L is the maximum number of attributes allowed in a credential, $\mathbb{G}$, $\mathbb{H}$, $\mathbb{T}$ are the AGHO pairing groups of prime order q, with bilinear mapping $e : \mathbb{G} \times \mathbb{H} \longrightarrow \mathbb{T}$ and respective generators G, H, and $e(G, H)$, $g \in \mathbb{G}$ is the generator for the BBS encryption scheme, and crs is the common reference string for the NIZK proof systems. We further assume that all attributes $a \in \mathbb{A}$ are represented by group elements from $\mathbb{G}$.

An issuer I's signing key generated by $\mathsf{Gen}_I(sp) = \mathsf{Gen}_{BSIG}(sp)$ consists of the AGHO keys $sk_I = (v, (w_i)_{i=1}^l, z)$ and $vk_I = (V, (W_i)_{i=1}^l, Z)$, where $l \leq L$, and a user's identity key consists of the BBS keys (sk_U, pk_U) generated by $\mathsf{Gen}_U(sp) = \mathsf{Gen}_{PRE}(sp)$.

4.2 Issuance

The issuance protocol is the restrictive blind AGHO signature (Definition 3) based on the commitment

$$\mathsf{Comm}(sk_U, (a_i)_i) = (c_0, (pk_i, c_i)_i),$$

which embodies the encrypted certificate to be signed, being comprised by U's *certificate pseudonym* $c_0 = \mathsf{Enc}_{BBS}(pk_U, 1)$, and the *attribute encryptions* $c_i = \mathsf{Enc}_{BBS}(pk_i, a_i)$ with respect to fresh proxy re-encryption keys $(pk_i)_i$. Although authentication of U is outside the scope of the blind AGHO scheme, we nevertheless integrate it into the issuance protocol by extending the well-formedness proof of the restrictive scheme by a proof of knowing the secret key belonging to pk_U. The protocol runs over a server-authenticated, confidential and integrity protected channel.

Definition 8 (Issuance Protocol). $\langle \mathsf{User}_I(sid, sk_U, (a_i)_{i=1}^l, vk_I), \mathsf{Iss}(sid, pk_U, (a_i)_{i=1}^l, sk_I) \rangle$ *is a protocol between a user U with identity key (sk_U, pk_U) and an issuer I with AGHO keys (sk_I, vk_I). Both user and issuer agreed on the unencrypted content, the attributes $A = (a_i)_{i=1}^l \in \mathbb{A}^l$, beforehand.*

1. *U computes $com = (c_0, (pk_i, c_i)_{i=1}^l)$ by generating a fresh pseudonym $c_0 = (c_{0,1}, c_{0,2}) = \mathsf{Enc}_{BBS}(pk_U, 1)$ and attribute encryptions $c_i = (c_{i,1}, c_{i,2}) = \mathsf{Enc}_{BBS}(pk_i, a_i)$ using a fresh set of attribute keys $(sk_i, pk_i) \leftarrow \mathsf{Gen}_{BBS}(sp)$, $1 \leq i \leq l$. For notational convenience we write $m = (m_{i,j}) = (pk_i, c_{i,1}, c_{i,2})_{i=0}^l$ for com, where we set $(sk_0, pk_0) = (sk_U, 1)$ and $a_0 = 1$.*

---

[1] In practice, the generation of the system parameters can be realized using multiparty techniques.

2. *With the above described commitment scheme, U engages the restrictive blind signing session with I (Definition 3) to receive an encrypted credential*

$$C = \left\{ \left(c_0, (pk_i, c_i)_{i=1}^l \right), \sigma = (R, S, T) \right\},$$

with σ being a valid AGHO signature by I on com, and with $(sk_i)_{i=0}^l$ as the opening of com. Demanding additional authentication of the user U by means of her identity key sk_U, the zero-knowledge proof π_U as described in Definition 3, is explicitly described by

$$\pi_U = \mathsf{NIZK}\left[(\eta, \varphi_1, \varphi_2, (\kappa_i, \lambda_i)_{i=0}^l) : g^{\kappa_0} = pk_U \right.$$
$$\wedge\ G_1^\eta = G \ \wedge\ G^{\varphi_1} = G_2 \ \wedge\ G_1^{\varphi_2} = G_3$$
$$\left. \bigwedge_{i=0}^l G_1^{\lambda_i} = G^{\kappa_i} \wedge \overline{m}_{i,0} \cdot \overline{P}_{i,0}^\eta = g^{\kappa_i} \wedge \overline{m}_{i,2} \cdot \overline{P}_{i,2}^\eta = \overline{m}_{i,1}^{\kappa_i} \cdot \overline{P}_{i,1}^{\lambda_i} \cdot a_i \right],$$

which is bound to the unique session identifier sid. Here, the user U chooses $(\eta, \varphi_1, \varphi_2) = (1/e, f_1, f_2)$, and $(k_i, \lambda_i) = (sk_i, {}^{sk_i}/e)$, $0 \le i \le l$, as witnesses.

Remark 1. In some situations a user might be allowed to stay anonymous towards the issuer. In such case the user's public identity pk_U is not part of the commited x and hence not provided to I, the term $g^{\kappa_0} = pk_U$ in π_U is omitted. In another setting similar to [5,7] a user might be known to I under a pseudonym $P_U = \mathsf{Enc}_{BBS}(pk_U, 1)$ of her. Here, U proves to I that the same secret key sk_U is used in both $P_U = (P_{U,1}, P_{U,2})$ and the certificate pseudonym, replacing $g^{\kappa_0} = pk_U$ by $c_{0,1}^{\kappa_0} = c_{0,2} \wedge P_{U,1}^{\kappa_0} = P_{U,2}$.

4.3   Presentation

The presentation of a credential C, as described in full detail by Definition 10, is essentially based on re-encrypting a re-randomization of C into a selective readable version $\overline{C}$ for the service S, supplemented by two linked zero-knowledge proofs: the computationally costly *presentation proof* π_P, which is performed by the wallet W and which relates the transformed $\overline{C}$ to the original C (the latter, including its signature is hidden from S), and the *ownership proof* π_O on the pseudonym of $\overline{C}$, proving knowledge of the secret identity key belonging to the pseudonym. The first proof is efficiently instantiated by help of the structure-preservation property of the blind AGHO scheme, the ownership proof supplied by the user is a simple proof of knowledge of a single exponent.

For the sake of readability, we gather the establishment of the service's session keys and its corresponding transformation information in a separate subprotocol, the ReKey protocol. Both Protocols from Defintions 9 and 10 run over a server-authenticated, confidential and integrity protected channel, and are associated with the same random session identifier *sid* supplied by the service. Furthermore, the non-interactive proofs (π_O and π_S below) are bound to the context of the presentation, in particular the common public parameters *sid*, vk_I and D.

Definition 9 (ReKey protocol). *This protocol between the user U and the service S is a subprotocol of the presentation protocol from Defintion 10.*

1. *U chooses a random one-time key $sk' \leftarrow_\$ \mathbb{Z}_q$, and forwards sk' and D to S.*
2. *U re-randomizes[2] her pseudonym c_0 by $e \leftarrow_\$ \mathbb{Z}_q$, $\bar{c}_0 = (\bar{c}_{0,1}, \bar{c}_{0,2}) = (c_{0,1}^e, c_{0,2}^e)$ and proves to S that she is in posession of its secret key, by supplying a simulation extractable zero-knowledge proof*

$$\pi_O = \mathsf{NIZK}\left[\kappa : \bar{c}_{0,1}^\kappa = \bar{c}_{0,2}\right],$$

in which she uses $\kappa = sk_U$ as witness.
3. *S verifies π_O, and if valid it keeps $\bar{c}_0$ and sk'. Otherwise, S aborts the protocol. On the user side, U takes the secret attribute keys $(sk_i)_i$ belonging to C and determines $rk_0' = 1/e$ and the re-encryption keys $rk_i' = sk_i/sk'$, $i \in D$.*

Definition 10 (Presentation Protocol). *The presentation protocol of encrypted ABCs $\langle \mathsf{User}_P(sid, sk_U, (sk_i)_{i \in D}), \mathsf{Wall}(sid, C, D, vk_I), \mathsf{Serv}(sid, D, vk_I) \rangle$ is between a user U with identity key sk_U who owns the credential C issued by I, the wallet W, and the service S (the supplier of the random session identifier sid). Here, $D \subseteq \{1, \ldots, l\}$ denotes the index set of the attributes to be disclosed, and $(sk_i)_{i \in D}$ are U's corresponding attribute keys.*

1. *U performs Protocol from Definition 9 with S, and if successful it sends the re-randomized one-time pseudonym $\bar{c}_0$ together with the re-encryption keys rk_0', $(rk_i')_{i \in D}$ and D to W.*
   From now on we proceed similar to [18]:
2. *(Randomization and re-encryption) For $i \in D$, the wallet W re-randomizes the ciphertexts c_i to $\bar{c}_i = \left(c_{i,1} \cdot g^{f_i}, c_{i,2} \cdot pk_i^{f_i}\right)$, with $f_i \leftarrow_\$ \mathbb{Z}_q$. All other attributes are randomized inconsistently, by choosing $v_{i,0}, v_{i,1}, v_{i,2} \leftarrow_\$ \mathbb{Z}_q$ and setting $\overline{pk}_i = pk_i \cdot g^{v_{i,0}}$, $\bar{c}_i = (c_{i,1} \cdot g^{v_{i,1}}, c_{i,2} \cdot g^{v_{i,3}})$ for all $i \notin D$. Using the re-encryption keys $(rk_i')_{i \in D}$ the wallet translates the attributes belonging to $i \in D$ by $d_i = \mathsf{ReEnc}_{BBS}(rk_i', \bar{c}_i) = \left(\bar{c}_{i,1}^{rk_i'}, \bar{c}_{i,2}\right)$.*
3. *(Presentation) W randomizes T by $\overline{T} = T^x$, $x \leftarrow_\$ \mathbb{Z}_q$, forwards $(d_i)_{i \in D}$, $\left(\overline{pk}_i, \bar{c}_i\right)_{i \notin D}$, and $\overline{T}$ to S, and provides a wellformedness proof of these elements via*

$$\pi_P = \mathsf{NIZK}\left[(P, \Sigma, \xi), \eta, (\kappa_i, \gamma_i)_{i \in D}, (\nu_{i,0}, \nu_{i,1}, \nu_{i,2})_{i \notin D} : (1) \wedge (2)\right],$$

which is defined by the relations (1) and (2) below.
4. *S verifies π_P using the verified one-time pseudonym received in Protocol 9. If valid, it decrypts the attributes $(d_i)_{i \in D}$ with its one-time key sk'. (Otherwise S outputs $\perp$).*

---

[2] For the sake of efficiency, U might outsource the re-randomization of its pseudonym to the wallet.

Remark 2. Showing more than one credential is efficiently implemented by merging their ownership proofs to a single NIZK which simultaneously proves knowledge of sk_U on all used pseudonyms, NIZK $[(\kappa) : \bigwedge_C \overline{c}_{0,1}(C)^\kappa = \overline{c}_{0,2}(C)]$.

Equations (1) and (2) mentioned in Protocol 10, state that $(P, \Sigma, \overline{T}^{1/\xi})$ is a valid signature for a quadratic derivative of the above group elements $(\overline{c}_{0,1}, \overline{c}_{0,2})$, $(d_{i,1}, d_{i,2})_{i \in D}$ and $(\overline{pk}_i, \overline{c}_{i,1}, \overline{c}_{i,2})_{i \notin D}$, i.e.

$$
\begin{aligned}
& e(\Sigma, H) \cdot e(\rho, V) \cdot e(\overline{c}_{0,1}, W_{0,1})^\eta \cdot e(\overline{c}_{0,2}, W_{0,2})^\eta \\
& \quad \cdot \prod_{i \in D} e\left(pk', W_{i,0}\right)^{\kappa_i} \cdot e(g, W_{i,1})^{-\gamma_i \cdot \kappa_i} \cdot e(pk', W_{i,2})^{-\gamma_i} \\
& \quad \cdot \prod_{i \notin D} e\left(g, W_{i,0}\right)^{-\nu_{i,0}} \cdot e(g, W_{i,1})^{-\nu_{i,1}} \cdot e(g, W_{i,2})^{-\nu_{i,2}} \\
& = e(G, Z) \cdot \prod_{i \in D} e(d_{i,1}, W_{i,1})^{-1} \cdot e(d_{i,2}, W_{i,2})^{-1} \cdot \\
& \quad \prod_{i \notin D} e(\overline{pk}_i, W_{i,0})^{-1} \cdot e(\overline{c}_{i,1}, W_{i,1})^{-1} \cdot e(\overline{c}_{i,2}, W_{i,2})^{-1}, \quad (1)
\end{aligned}
$$

where V, Z, and $(W_{i,0}, W_{i,1}, W_{i,2})_{i=0}^l$ are the components of the issuers verification key, and

$$
e(P, \overline{T}) \cdot e(G, H)^{-\xi} = 1. \tag{2}
$$

Linearization of the quadratic terms in (1) is accomplished by standard techniques and given in the full version of the paper. An honest prover chooses $(P, \Sigma, \xi) = (R, S, x)$, and uses the parameters from step 2 of Protocol 10, i.e. $\eta = rk_0'$, $(\kappa_i, \gamma_i) = (1/rk_i', rk_i' \cdot f_i)$ for $i \in D$, and $(\nu_{i,0}, \nu_{i,1}, \nu_{i,2}) = (v_{i,0}, v_{i,1}, v_{i,2})$ for all $i \notin D$.

Theorem 3. *Suppose that the AGHO signature scheme is EUF-CMA secure, and that the NIZK from Definition 9 is simulation extractable. Then, under the DDH-assumption in $\mathbb{G}$,*

1. *the proxy re-encryption scheme PRE_{BBS} is PRE-IND-CPA secure, anonymous, and has the ciphertext re-randomization property,*
2. *the structure-preserving blind signature scheme BSIG_{AGHO} is unforgeable and has the blinding property,*

hence our EABC system satisfies unforgeability, non-deceivability, privacy *and* unlinkability *in the sense of Sect. 3.2.*

The proof of Theorem 3 is given in the full version of the paper.

5 Conclusions

In this paper, we pointed out a problem in Krenn et al.'s cloud-based attribute-based credential system [18] by presenting a simple and efficient attack fully breaking the privacy guarantees of their construction in the real world. We then provided a revised, provably secure construction which not only solves this issue, but also reduces the trust assumptions stated in [18] with regards to collusions between the central wallet and other entities in the system. As a building block of potentially independent interest we presented a blind variant of the Abe et al. structure-preserving signature scheme [2].

While we did not provide a concrete implementation of our construction, we expect only very minor performance drawbacks with respect to [18], while correcting all the deficiencies in their work. There, for a security parameter of $\lambda = 112$, all computations on all parties' sides were between 50ms and 440ms when presenting 12 out of 25 attributes. By inspecting the computational efforts needed in our protocol and theirs, one can see only negligible differences, except for the proof of knowledge of a single exponent which is required on the user's side in our construction. However, such computations are efficiently doable, and thus our protocol still provides a significant performance improvement compared to fully locally hosted "conventional" attribute-based credential systems. Finally, we leave a full-fledged implementation, not only of the cryptographic algorithms but of the full system, as open work to demonstrate the real-world applicability of EABCs in general and our construction in particular, and to help ABC systems to finally pave their way into the real world. For this, several approaches can be envisioned, in particular for the presentation protocol, where the optimal choice may depend on external constraints as well as requirements of the specific application domain. Firstly, using the wallet also as a communication proxy, would not require further network anonymisation layers, yet leak metadata to the wallet. Alternatively, by merely outsourcing the computational effort to the wallet and routing the traffic through the user, one could reach the same privacy guarantees as in conventional systems, at the cost of increased bandwidth requirements compared to the first approach; furthermore, the responsibility of transport layer anonymity would be with the user. Finally, an approach close to OpenID Connect could be achieved by combining these two approaches.

Acknowledgements. The first author was partly supported by the "Embedded Lab Vienna for IoT & Security" (ELVIS), funded by the City of Vienna. The second author has received funding from the European Union's Horizon 2020 research and innovation programme under grant agreement no. 830929 ("CyberSec4Europe").

References

1. Abe, M., Fuchsbauer, G., Groth, J., Haralambiev, K., Ohkubo, M.: Structure-preserving signatures and commitments to group elements. In: Rabin, T. (ed.) CRYPTO 2010. LNCS, vol. 6223, pp. 209–236. Springer, Heidelberg (2010). https://doi.org/10.1007/978-3-642-14623-7_12
2. Abe, M., Groth, J., Haralambiev, K., Ohkubo, M.: Optimal structure-preserving signatures in asymmetric bilinear groups. In: Rogaway, P. (ed.) CRYPTO 2011. LNCS, vol. 6841, pp. 649–666. Springer, Heidelberg (2011). https://doi.org/10.1007/978-3-642-22792-9_37

3. Blaze, M., Bleumer, G., Strauss, M.: Divertible protocols and atomic proxy cryptography. In: Nyberg, K. (ed.) EUROCRYPT 1998. LNCS, vol. 1403, pp. 127–144. Springer, Heidelberg (1998). https://doi.org/10.1007/BFb0054122

4. Brands, S.: Untraceable off-line cash in wallet with observers. In: Stinson, D.R. (ed.) CRYPTO 1993. LNCS, vol. 773, pp. 302–318. Springer, Heidelberg (1994). https://doi.org/10.1007/3-540-48329-2_26

5. Brands, S.: Rethinking public key infrastructure and digital certificates - building in privacy. Ph.D. thesis, Eindhoven Institute of Technology (1999)

6. Camenisch, J., Dubovitskaya, M., Haralambiev, K., Kohlweiss, M.: Composable and modular anonymous credentials: definitions and practical constructions. In: Iwata, T., Cheon, J.H. (eds.) ASIACRYPT 2015. LNCS, vol. 9453, pp. 262–288. Springer, Heidelberg (2015). https://doi.org/10.1007/978-3-662-48800-3_11

7. Camenisch, J., Herreweghen, E.V.: Design and implementation of the idemix anonymous credential system. In: Atluri, V. (ed.) ACM CCS 2002, pp. 21–30. ACM (2002). https://doi.org/10.1145/586110.586114

8. Camenisch, J., Lysyanskaya, A.: An efficient system for non-transferable anonymous credentials with optional anonymity revocation. In: Pfitzmann, B. (ed.) EUROCRYPT 2001. LNCS, vol. 2045, pp. 93–118. Springer, Heidelberg (2001). https://doi.org/10.1007/3-540-44987-6_7

9. Camenisch, J., Lysyanskaya, A.: A signature scheme with efficient protocols. In: Cimato, S., Persiano, G., Galdi, C. (eds.) SCN 2002. LNCS, vol. 2576, pp. 268–289. Springer, Heidelberg (2003). https://doi.org/10.1007/3-540-36413-7_20

10. Camenisch, J., Lysyanskaya, A.: Signature schemes and anonymous credentials from bilinear maps. In: Franklin, M. (ed.) CRYPTO 2004. LNCS, vol. 3152, pp. 56–72. Springer, Heidelberg (2004). https://doi.org/10.1007/978-3-540-28628-8_4

11. Chaum, D.: Untraceable electronic mail, return addresses, and digital pseudonyms. Commun. ACM **24**, 84–88 (1981). https://doi.org/10.1145/358549.358563

12. Chaum, D.: Security without identification: transaction systems to make big brother obsolete. Commun. ACM **28**, 1030–1044 (1985). https://doi.org/10.1145/4372.4373

13. Chen, X., Zhang, F., Mu, Y., Susilo, W.: Efficient provably secure restrictive partially blind signatures from bilinear pairings. In: Di Crescenzo, G., Rubin, A. (eds.) FC 2006. LNCS, vol. 4107, pp. 251–265. Springer, Heidelberg (2006). https://doi.org/10.1007/11889663_21

14. Fuchsbauer, G., Hanser, C., Slamanig, D.: Practical round-optimal blind signatures in the standard model. In: Gennaro, R., Robshaw, M. (eds.) CRYPTO 2015. LNCS, vol. 9216, pp. 233–253. Springer, Heidelberg (2015). https://doi.org/10.1007/978-3-662-48000-7_12

15. Fuchsbauer, G., Hanser, C., Slamanig, D.: Structure-preserving signatures on equivalence classes and constant-size anonymous credentials. J. Cryptol. **32**(2), 498–546 (2018). https://doi.org/10.1007/s00145-018-9281-4

16. ElGamal, T.: A public key cryptosystem and a signature scheme based on discrete logarithms. In: Blakley, G.R., Chaum, D. (eds.) CRYPTO 1984. LNCS, vol. 196, pp. 10–18. Springer, Heidelberg (1985). https://doi.org/10.1007/3-540-39568-7_2

17. Groth, J.: Simulation-sound NIZK proofs for a practical language and constant size group signatures. In: Lai, X., Chen, K. (eds.) ASIACRYPT 2006. LNCS, vol. 4284, pp. 444–459. Springer, Heidelberg (2006). https://doi.org/10.1007/11935230_29

18. Krenn, S., Lorünser, T., Salzer, A., Striecks, C.: Towards attribute-based credentials in the cloud. In: Capkun, S., Chow, S.S.M. (eds.) CANS 2017. LNCS, vol. 11261, pp. 179–202. Springer, Cham (2018). https://doi.org/10.1007/978-3-030-02641-7_9

19. Maitland, G., Boyd, C.: A provably secure restrictive partially blind signature scheme. In: Naccache, D., Paillier, P. (eds.) PKC 2002. LNCS, vol. 2274, pp. 99–114. Springer, Heidelberg (2002). https://doi.org/10.1007/3-540-45664-3_7

20. Paquin, C., Zaverucha, G.: U-prove cryptographic specification v1.1 (revision 2). Technical report, Microsoft Corporation, April 2013

21. Ringers, S., Verheul, E., Hoepman, J.-H.: An efficient self-blindable attribute-based credential scheme. In: Kiayias, A. (ed.) FC 2017. LNCS, vol. 10322, pp. 3–20. Springer, Cham (2017). https://doi.org/10.1007/978-3-319-70972-7_1

22. Yang, R., Au, M.H., Xu, Q., Yu, Z.: Decentralized blacklistable anonymous credentials with reputation. In: Susilo, W., Yang, G. (eds.) ACISP 2018. LNCS, vol. 10946, pp. 720–738. Springer, Cham (2018). https://doi.org/10.1007/978-3-319-93638-3_41

Enabling Compressed Encryption
for Cloud Based Big Data Stores

Meng Zhang[1,2], Saiyu Qi[1(✉)], Meixia Miao[3], and Fuyou Zhang[1]

[1] School of Cyber Engineering, Xidian University, Xi'an 710071, China
zhangmeng1575431@163.com, syqi@connect.ust.hk, fuyouzhang@yeah.net
[2] State Key Laboratory of Cryptology, P.O. Box 5159, Beijing 100878, China
[3] National Engineering Laboratory for Wireless Security,
Xi'an University of Posts and Telecommunications, Xi'an 710121, China
miaofeng415@163.com

Abstract. We propose a secure yet efficient data query system for cloud-based key-value store. Our system supports encryption and compression to ensure confidentiality and query efficiency simultaneously. To reconcile encryption and compression without compromising performance, we propose a new encrypted key-value storage structure based on the concept of horizontal-vertical division. Our storage structure enables fine-grained access to compressed yet encrypted key-value data. We further combine several cryptographic primitives to build secure search indexes on the storage structure. As a result, our system supports rich types of queries including key-value query and range query. We implement a prototype of our system on top of Cassandra. Our evaluation shows that our system increases the throughput by up to 7 times and compression ratio by up to 1.3 times with respect to previous works.

Keywords: Encryption · Compression · Key-value store

1   Introduction

With the popularity of cloud computing and the growing demand for big data processing, key-value (KV) store is adopted in many public cloud services to enable efficient and scalable data processing tasks on behalf of users [2,3,16, 32]. However, storing sensitive data on untrusted cloud incurs serious privacy issues [13,14,28,37]. As a result, we need to build a data query system that can efficiently handle big data workloads and guarantee data confidentiality. On the one hand, many big data stores employ compression [4,7,12,24] to significantly increase efficiency, sometimes by up to an order of magnitude. Compression can improve performance gains effectively because it enables servers to fit more data in main memory, thus decreasing the number of accesses to persistent storage. On the other hand, data is often encrypted and the key is preserved at the user side [8,17,18,27,35,38,39] in order to protect data confidentiality. Therefore, an ideal cloud data storage system that aims to protect data confidentiality and preserve efficiency should incorporate both encryption and compression into its design.

© Springer Nature Switzerland AG 2019
Y. Mu et al. (Eds.): CANS 2019, LNCS 11829, pp. 270–287, 2019.
https://doi.org/10.1007/978-3-030-31578-8_15

Unfortunately, many state of arts only support either compression or encryption. The reason is that compression and encryption are not directly compatible. Firstly, since pseudo-random data cannot be compressed, we cannot encrypt data before compressing it. Secondly, if we first compress the data and then encrypt it, we will not be able to perform fine-grained data query, making the system difficult to manage. To the best of our knowledge, only MiniCrypt [41] enables compression and encryption simultaneously. To achieve this goal, MiniCrypt equally divides a key-value table into multiple record groups, with each group containing multiple key-value records. For each record group, MiniCrypt compresses it and then encrypts the compression result. Their observation is that the compression ratio of the entire key-value table is similar to that of a small number of key-value records. However, this approach only supports group-level data access, incurring high communication overhead in data queries.

1.1   Challenge and Contribution

In this paper, we aim to build a secure key-value query system that combines compression and encryption while enabling fine-grained data access. Directly packing multiple key-value records into a record group incurs high communication overhead in data query. When a user queries some columns of a key-value record, the entire record group that contains the record needs to be returned, incurring additional communication overhead. Therefore, how to design an encrypted storage structure that can provide a high compression ratio and support fine-grained data query becomes a challenge.

In response, our system proposes a new encrypted key-value storage structure based on the concept of horizontal-vertical division. Specifically, we first horizontally divide the key-value table into multiple record groups. We then vertically divide each record group into multiple column packets, with each packet containing a column of it. For each record group, we compress and encrypt each column packet of it. This design brings many advantages. Firstly, the design enables fine-grained data access to largely reduce communication overhead of data queries. Suppose that a user wants to query several columns of a key-value record. Different with MiniCrypt, our design only needs to return some column packets of the record group that contain these columns. Secondly, column-wise data compression can lead to a higher compression ratio.

We also build secure search indexes over the encrypted storage structure to support rich types of queries including key-value query and range query. In brief, our contributions are listed as follows:

- We propose an encrypted key-value storage system that guarantees data confidentiality and efficiency.
- We design a new encrypted key-value storage structure based on the concept of horizontal-vertical division to reconcile data compression and encryption. Our design reduces communication overhead of data queries to improve efficiency by avoiding unnecessary column returns and increasing compression ratios.

- We build a secure search index on the encrypted key-value storage structure by using cryptography to support rich queries including KV query and range query.
- We implement a prototype of our system on Cassandra, and deploy it on Aliyun ECS server. We test the communication overhead, throughput, latency, and query cost time. The experimental results show that our system significantly reduces communication overhead of data queries and increases system throughput. Compared with MiniCrypt, our system increases the throughput by up to 7 times.

1.2 Related Work

Encrypted Databases. Currently, the existing key-value stores (e.g., MongoDB [1] and Cassandra [29]) either do not provide encryption or store the secret key on the server side. They cannot resist attacks from the server. However, if the key is stored by the user, the server will not be able to decrypt the data, making data management difficult. In order to support rich queries on encrypted database, many schemes have been proposed [21,30,33–35,40]. Among them, the first encrypted database supporting rich functions is CryptDB [35]. CryptDB uses onion encryption to perform multi-layer encryption in order to meet the requirements of different encryption functions. However, encrypted data brings a lot of storage overhead. Their storage overhead is even several times that of unencrypted data. Yuan et al. [40] implement a scheme for efficient querying on an encrypted key-value storage database. They provide a data partitioning algorithm on the encrypted domain, but since each key-value record has been reconstructed, the storage overhead is large. We test the storage capacity of this scheme, and the storage overhead is more than ten times that of unencrypted data. Macedo et al. [30] implement a NoSQL framework using modular and extensible technology, supporting different cryptographic components for different columns in the database, enabling the database to support richer query methods. These encrypted databases all do not support compression of data, which seriously affects system performance.

Compressed Databases. Previous works [19,20,25,26,36,42] have shown that compression schemes significantly improve the query processing performance of systems. So many databases currently use compression technology to reduce storage overhead and improve system performance [4,5,7,24]. Abadi et al. [4] propose an architecture for a query executor that allows for direct operation on compressed data. It is commonly used in column-oriented database. But it can't be combined well with encryption technology. Binnig et al. [7] propose data structures that support a dictionary compression efficiently. They build a table that maps values to compressed codes (and vice versa). It can combine compression with encryption, but raises storage issues for mapping tables while making data updates complicated. MiniCrypt proposed by Zheng et al. [41] combines compression technology and encryption technology, but due to records group-level compression, it causes unnecessary communication overhead, which in turn reduces system performance. Compression ratios are generally higher in column-stores because consecutive entries in a column are often quite similar to each other, whereas adjacent attributes in a tuple are not [31]. We design a new data storage structure which not only increases the compression ratio, but also reduces communication overhead. The difference between our scheme and MiniCrypt is that our scheme is based on column-wise compression and encryption, which enables access to specific columns. The horizontal-vertical partitioned data structure designed in our scheme is similar to the data structure of RCFile [22]. However, the design goal of RCFile is different with that of our work since it does not ensure data privacy.

OPE and ORE: Agrawal et al. [6] introduce the definition of order-preserving encryption and show how to use order-preserving encryption (OPE) for range query. Boldyreva et al. [9] give the security notion IND-OCPA for order-preserving encryption, and point out that no effective order-preserving scheme can achieve IND-OCPA security. In the same work, they also introduce a weaker security notion (POPF-CCA security). After that, Boldyreva et al. [10] point out that in their OPE scheme, nearly half of the bits in the underlying ciphertexts are leaked. Boneh et al. [11] give a special OPE scheme called order-revealing encryption (ORE) which achieves the IND-OCPA security. In the ORE scheme, a comparison function is used to compare two input ciphertexts and outputs the order. However, it is difficult to implement in practice because of its complex tools. Chenette et al. [15] give a practical ORE scheme and give a definition of security. In the same job, Chenette et al. also give the strategy of converting its ORE scheme to OPE scheme, which achieves the same security as ORE. In order to adapt to our system, we will adopt this scheme of converting ORE to OPE.

1.3 Organization

The rest of this paper is organized as follows. Section 2 introduces the system architecture, threat model and preliminaries. We introduce our proposed storage structure in Sect. 3. We elaborates the system operations in Sect. 4. We give our system security analysis in Sect. 5. We discuss our system implementation and performance evaluation in Sect. 6. At last, we summarize the paper in Sect. 7.

2 Overview

2.1 System Architecture

Our system consists of users and a cloud server. The users are within an organization and share a secret key, which is not available to the cloud server. The cloud server hosts the encrypted key-value store, and the users issue data queries toward the cloud server.

Figure 1 shows the format of a key-value table. We consider that each key-value record in the table is represented as a row indexed by a search key(k). The row consists of multiple columns with each column indexed by a column attribute(*att*). We aim to develop a secure key-value query system that combines compression and encryption while supporting fine-grained data access. Specifically, we aim to achieve the following requirements:

Rich queries: The system should be able to support rich types of queries over the key-value store:

- KV query: Given a search key with (possibly) several column attributes, the cloud server searches the key-value record that matches the key and returns the columns of the record that match the column attributes. For example, given a search key 103 with the column attribute att_1, the cloud server returns 'Jenny'.
- Range query: Given two search keys with (possibly) several column attributes, the cloud server searches the key-value records with search keys that within the range of the two search keys, and returns the columns of these records that match the column attributes. For example, given two search keys 103 and 105 with the column attribute att_1, the cloud server returns ('Jenny', 'Bob', 'Danny').

Data Confidentiality: The cloud server should not be able to get any non-trivial information about its stored data.

Efficiency and Scalability: The system should be able to efficiently process big data workloads and support fast queries without incurring high communication overhead.

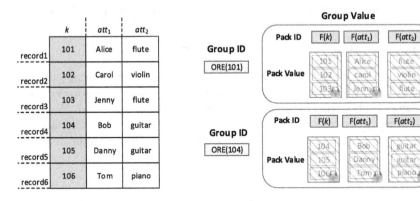

Fig. 1. Key-value table   **Fig. 2.** Storage structure details

2.2   Threat Model

We consider a honest-but-curious threat model. Specifically, we consider the cloud server as a passive adversary to faithfully perform data queries of users but deliberately obtain sensitive information from its hosted key-value store. On the other hand, we assume that the users are trusted and allowed to query the key-value store hosted at the cloud side. As a result, we mainly focus on protecting the confidentiality of the key-value store against a server adversary.

2.3   Preliminaries

In this section, we present some basic cryptographic primitives that are used in this work.

Symmetric Encryption: A symmetric encryption scheme consists of three algorithms: (KGen, Encrypt, Decrypt).

- Sym.KGen(1^λ): The algorithm inputs a security parameter λ, and then outputs a secret key sk.
- Sym.Encrypt(sk, m): The algorithm inputs a secret key sk and a plaintext m, and then outputs a ciphertext c.
- Sym.Decrypt(sk, c): The algorithm inputs a secret key sk and a ciphertext c, and then outputs a plaintext m.

Order-Revealing Encryption: An order-revealing encryption (ORE) scheme on an ordered domain consists of three algorithms: $\Pi =$ (ORE.KGen, ORE.Encrypt, ORE.Compare):

- ORE.KGen(1^λ): The algorithm inputs a security parameter λ, and then outputs a secret key sk.
- ORE.Encrypt(sk, m) $\rightarrow c$: The algorithm inputs the secret key sk and a plaintext m, and then outputs a ciphertext c.
- ORE.Compare(c_1, c_2) $\rightarrow b$: The algorithm inputs two ciphertexts c_1, c_2, and then outputs a bit $b \in \{0, 1\}$

Order-Preserving Encryption: An order-preserving encryption (OPE) scheme [6,9] is an encryption scheme that supports comparisons over encrypted values. An OPE scheme must ensure that $m_1 > m_2 \rightarrow c_1 > c_2$ where $c_1 =$ OPE.Encrypt(m_1) and $c_2 =$ OPE.Encrypt(m_2). For simplicity, an OPE scheme on an ordered domain consists of two algorithms $\Pi =$ (OPE.KGen, ORE.Encrypt).

- OPE.KGen(1^λ): The algorithm inputs a security parameter λ, and then outputs a secret key sk.
- OPE.Encrypt(sk, m): The algorithm inputs the secret key sk and a message m, and then outputs a ciphertext c.

We follow the ORE scheme proposed by Chenette et al. [15] which achieves a simulation-based security notion. In order to better adapt to our database system, we adopt the scheme of ORE conversion to OPE proposed in the same work. The reason is that it supports custom comparator and does not need to change the internal design of the DBMS. And it achieves the same security level as the ORE scheme.

3   Encrypted Key-Value Storage Structure

3.1   Notations

In this section, we present some notations (as shown in Table 1) that are used in this work.

Table 1. Notations

Notations	Meaning
k	Search key of a key-value record
C_k	Search key name of a key-value record
C_{att}	Column attribute name of a key-value record
D_{sort}	The data which is ordered by the search key
RG	Record group
GID	Group ID
k_{min}	The smallest key in a record group
V_c	Column attribute values for each column in a record group
C_{pack}^*	Compressed and encrypted column packet
V_p	Column attribute values in returned packets
V_r	Matched column attribute values in V_p
s	The total number of record groups in result
F	A pseudo-random function

3.2   Basic Idea

We design a new encrypted key-value storage structure based on the concept of horizontal-vertical division. Given the key-value table (as shown in Fig. 1), Fig. 2 shows the details of our storage structure. Specifically, we first horizontally divide the table into multiple record groups. We then vertically divide each record group into multiple column packets with each packet containing a column of it. For each record group, we compress and encrypt individual column packets of it. Each group consists of T key-value records specified by users. Our system supports flexible group size T, which can be adjusted to balance the data compression ratio and query efficiency. For example, if a user specifies $T = 3$, the key-value table can be divided into two record groups. To further improve the compression ratio, a user can select different compression algorithms for different column packets within a record group according to the actual situation (e.g., data type and data distribution).

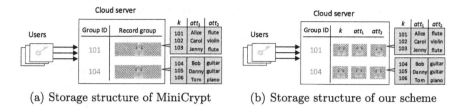

(a) Storage structure of MiniCrypt   (b) Storage structure of our scheme

Fig. 3. Comparison of MiniCrypt and our scheme structure

We construct a secure index on the encrypted key-value storage structure. We construct group IDs for record groups and organize them as an ordered index. A group ID is the smallest search key of the key-value record in the group. We use order-revealing encryption to protect the confidentiality of group IDs while preserving their orders. A column attribute is the attribute of the column compressed in the packet. We use pseudo-random function (PRF) to protect the confidentiality of column attributes. We use the ordered index to support KV query and range query.

3.3   Design Advantages

Figure 3 compares the data storage structures of our system and MiniCrypt [41]. MiniCrypt equally divides a key-value table into multiple record groups and directly compress and encrypt each record group. Instead, we adopt a horizontal-vertical division to further divide record groups into fine-grained column packets. Then we compress and encrypt individual column packets of each record group. Such a design enables our system to support finer-grained data access compared with MiniCrypt. Considering a KV query, for MiniCrypt, even if only one column of the record needs to be returned, the cloud server still needs to return the entire record group that contains the record. For our system, the cloud server only needs to return the matched column packets of the record group that contains the record, avoiding return of unmatched columns.

Moreover, since our system compresses a record group in column-level, the compression ratio is significantly higher than MiniCrypt. Table 2 shows the evaluation results over a key-value table containing a million of key-value records. For a record group, the compression ratios of our system and MiniCrypt increase as the size of the group increases. When the record group is the entire table, the maximum compression ratios (MCRs) of our system and MiniCrypt is 5.3 and 3.92, respectively. We also observe that when increasing the size of a record group to 1000, the compression ratio of our system is even higher than the MCR of MiniCrypt. With higher compression ratio, our system can cache more data in memory to decrease the number of accesses to persistent storage.

Table 2. Our scheme and MiniCrypt compression comparison

Schemes	Max com ratio (MCR)	90% MCR	90% MCR num of records	90% MCR original data size	90% MCR com data size
Our scheme	5.3	4.7	1000	139.7K	29.5K
MiniCrypt	3.92	3.5	300	41.6K	11.6K

4   System Operations

4.1   Construction of Encrypted Key-Value Data Storage

A user preprocesses its key-value table before submitting to the cloud server. First, the user sorts the data in ascending order and then divide it according to the given record group size. For each record group, the user compresses and encrypts each column of it. Algorithm 1 describes the configuration process.

Algorithm 1. PackSetup

Input: parameter λ, record group size: T, the key-value table D;
Output: compressed and encrypted record groups
1: **User:**
2: $(sk_o, sk_e, sk_c) \leftarrow$ KGen (λ);
3: $D_{sort} \leftarrow$ Sort(D, k);
4: $RGs \leftarrow$ Divide(D_{sort}, T)
5: **for** each RG in RGs **do**
6:     $GID \leftarrow$ ORE.Encrypt(k_{min}, sk_o);
7:     **for** each column in RG **do**
8:         $C^*_{pack} \leftarrow$ Sym.Encrypt(Compress$(V_c), sk_e)$;
9:     **end for**
10: **end for**

4.2   Secure KV Query

Since the key-value records are packed and encrypted, users can only query the key-value table in the granularity of column packet. Considering a KV query consisting of a search key k_c and several column attributes $C_{att}[m]$, recall that the ID of a group is the smallest search key of the key-value record within the group. The cloud server first searches the record group with the highest group ID from all the group IDs that are smaller than or equal to k_c. The cloud server then searches the column packets of the record group with column attributes matching $C_{att}[m]$, and finally returns them as query results. This query operation can run efficiently because group IDs are organized as an ordered index, which is stored in memory. Upon receiving the column packets, the user decrypts and decompresses them to filter out the desired values based on $(k_c, C_{att}[m])$. Algorithm 2 presents the overall procedure.

Algorithm 2. KVQuery

Input: private key $\{sk_o, sk_e, sk_c\}$; search key k_c; request column attributes $C_{att}[m]$;
Output: matched column attribute values V_r

1: **User:**
2: $k_c^* \leftarrow$ ORE.Encrypt (k_c, sk_o);
3: **for** each $C_{att|i} \in C_{att}[m]$ **do**
4:   $C_{att|i}^* \leftarrow \mathsf{F}(C_{att|i}, sk_c)$
5: **end for**
6: $C_k^* \leftarrow \mathsf{F}(C_k, sk_c)$
7: **Cloud server:**
8: $C_{pack}^*[m+1] \leftarrow$ select $C_k^*, C_{att|1}^*, C_{att|2}^*, ..., C_{att|m}^*$ from table
9:                   where $GID \leq k_c^*$ order by GID desc limit 1
10: **User:**
11: $V_p \leftarrow$ Sym.Decrypt(Decompress($C_{pack}^*[m+1]$), sk_e)
12: Filter V_p to get V_r by the search key k_c
13: return V_r

4.3 Secure Range Query

Our system supports range query. For a query within a wide range, our system can greatly reduce communication overhead. Considering a range query consisting of two search keys k_{low}, k_{high} and several column attributes $C_{att}[m]$. The cloud server first searches all the record groups with IDs within (k_{low}, k_{high}). After that, if k_{low} is not equal to the smallest record group ID in the result set, the cloud server also needs to get the record groups potentially contains search keys from k_{low} to the smallest record group ID. The cloud server then searches the column packets of these record groups with column attributes matching $C_{att}[m]$, and finally returns them as query results. The user decrypts and decompresses them to filter out the desired values based on $(k_{low}, k_{high}, C_{att}[m])$. The procedure is presented in Algorithm 3.

5 Security Analysis

Our system is designed to protect the confidentiality of user data and provide efficient performance. For performance, we use column-wise compression technology, which not only fits more data in the system memory but also reduces communication overhead. For the confidentiality of data, the system relies on various encryption strategies to ensure.

We use Advanced Encryption Standard (AES) [23] for internal data, and the adversary cannot get any other information about the data from the encrypted data. Classical encryption algorithms that ensure semantic security are not designed for computing. Therefore, we only use these technologies for data transmission and data storage. Our system does not reveal information about the contents of record groups, except for the number of columns and the size of the column packets. We use pseudo-random functions to ensure that the original values of column attributes are not leaked.

Algorithm 3. RanQuery

Input: private key sk_o, sk_c, sk_e; query range $\{k_{low}, k_{high}\}$; request column attributes
   $C_{att}[m]$

Output: matched column attribute values V_r

1: **User:**
2:    $k_{low}^* \leftarrow$ ORE.Encrypt (k_{low}, sk_o);
3:    $k_{high}^* \leftarrow$ ORE.Encrypt (k_{high}, sk_o);
4: **for** each $C_{att|i} \in C_{att}[m]$ **do**
5:        $C_{att|i}^* \leftarrow F(C_{att|i}, sk_c)$
6: **end for**
7: $C_k^* \leftarrow F(C_k, sk_c)$
8: **Cloud server:**
9:    $(C_{pack}^*[m+1][s], GID[s]) \leftarrow$ select $GID, C_k^*, C_{att|1}^*, C_{att|2}^*, ..., C_{att|m}^*$
10:                    from table where $k_{low}^* \leq GID \leq k_{high}^*$
11: **if** $k_{low}^* < GID_{min}$ in $GID[s]$ **then**
12:        $C_{pack}^*[m+1] \leftarrow$ select $C_k^*, C_{att|1}^*, C_{att|2}^*, ..., C_{att|m}^*$ from table
13:                            where $GID \leq k_{low}^*$ order by GID desc limit 1
14:        Add the $C_{pack}^*[m+1]$ to the $C_{pack}^*[m+1][s]$
15: **end if**
16: **User:**
17: $V_p \leftarrow$ Sym.Decrypt(Decompress($C_{pack}^*[m+1][s]$), sk_e)
18: Filter V_p to get V_r by the range (k_{low}, k_{high})
19: **return** V_r

For KV query and range query, we need to compare the record group IDs based on the ordered index. To protect the security of ordered indexes, on the one hand, in the design process, we first partitioned the ordered data by partition key. The partition key is the result of modulo the hash value of the search key, which disrupts the original sequence of the data and makes it order in the respective partition. In addition, the original data is compressed into record groups, and each record group is assigned to a different node according to the partition key, which hides the distance between the plaintext values to some extent. On the other hand, all the record group IDs are encrypted by the ORE scheme proposed by Chenette [15] which is given the security definition. In order to adapt to the common comparator of Cassandra, in our system, we adopt the scheme of ORE conversion to OPE proposed in the same work. It achieves the same security level as ORE.

Once the user uploads the encrypted value and encrypted index to the server node, the size of index, the size of each column pack and the column number of each record will be learned. When querying, the access pattern and search pattern will be exposed, where access pattern indicates the accessed entries and the associations between those columns, and search pattern represents the query token that is repeatedly submitted.

Based on the information we leaked, we consider this situation. When the server monitors multiple queries, it can infer some relevant information by comparing the results of each query. The leak of column packet size makes it vulnerable to inference attacks. By comparing the size of different column packets in each query result, the server can infer the column attributes of the packet, the association between the columns, and so on. Our system enables reducing the information leaked by the size of the packet by padding the encrypted column packets to a tier of a few possible sizes. Our system allows the customer to specify the padding tiers, such as small-medium-large or exponential scale, and pads each pack to the smallest tier value that is at least the pack size.

6   Implementation and Evaluation

We implement our system using Java on an existing key-value storage system, Cassandra [29]. Cassandra is an open source distributed NoSQL database system that supports high scalability. In Cassandra, the unique primary key cannot perform range queries, which acts as a partition key to determine which node the data is located on. Our system makes some minor adjustments to accommodate this design. We use composite key as the primary key in Cassandra. The composite key consists of a partition key and a clustering key. The clustering key is used to sort the data inside the partition. Our design does not modify the internals of Cassandra, but adds several interfaces to the upper layer of Cassandra. For the key-value records, we first calculate hash values for keys, and then perform modular operations on the hash values, so that the data is allocated to M buckets. The number of buckets is determined by the user. The partition key is the result of the modular operation. The clustering key is the record group ID. For the column attribute name, after we perform pseudo-random processing, we can add a letter in front of it to support Cassandra's column attribute rule.

All benchmarks are conducted on Aliyun ECS server with 10 Mbps bandwidth. The Cassandra replication factor is set to 3. All benchmarks use the data warehouse benchmark set TPC-H. TPC-H contains eight basic tables. We use the lineitem table because it contains the most columns. And then we initialize the lineitem table. As a result, the entire data set contains fifteen columns of data. All experiments set record group size to 1000 key-value records.

We compare the performance of our system with MiniCrypt [41]. MiniCrypt adopt records group-level compression method. This scheme has similar security to our system, but our system supports finer-grained access. In order to demonstrate the advantages of our storage structure in a fair way, we use order-revealing

Table 3. Our system and MiniCrypt performance comparison in one KV query with a single column attribute

Schemes	Communication overhead	ORE token generation time	Transmission time	Decrypt and decompress
Our system	4.907K	14 ms	31 ms	3 ms
MiniCrypt	37.275K	14 ms	59 ms	10 ms

encryption to encrypt search keys of both schemes, but when testing throughput and latency, the overhead of order-revealing encryption is not included.

We pre-load 1 GB of data into Cassandra. Our evaluation consists of two groups of experiments: (1) basic evaluation to show that our system takes up less communication overhead; (2) query performance evaluation to demonstrate the performance advantages of our system.

6.1 Basic Evaluation

In this experiment, we measure the communication overhead of MiniCrypt and our system when performing KV queries with different column attributes and range queries with 7 column attributes in different query ranges.

As shown in Fig. 4, when performing KV queries, compared with MiniCrypt, our system results in less communication overhead no matter how many columns are requested. The smaller number of request columns, the less communication overhead of our system, and MiniCrypt remains unchanged.

Our system adopts the concept of horizontal-vertical division so that each column in a record group can be independently accessed. When performing a KV query with a small number of column attributes, it is not necessary to return the entire record group. And the user only needs to decrypt and decompress the returned column packets. What'a more, column-wise compression provides a higher compression ratio because consecutive entries in a column are often quite similar to each other. However, due to the mix of different data types, MiniCrypt does not easily achieve a high compression ratio. At the same time, regardless of how many columns of data are requested, MiniCrypt needs to return the entire record group of data. Therefore, its communication overhead remains unchanged (Fig. 6).

As shown in Fig. 5, when performing range queries, both our system and MiniCrypt may incur the same communication overhead for different query

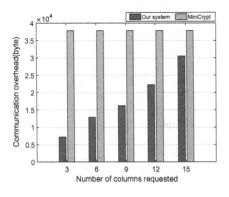

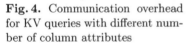

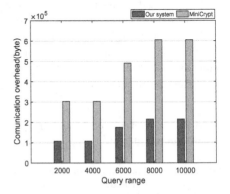

Fig. 4. Communication overhead for KV queries with different number of column attributes

Fig. 5. Communication overhead for range queries with different ranges

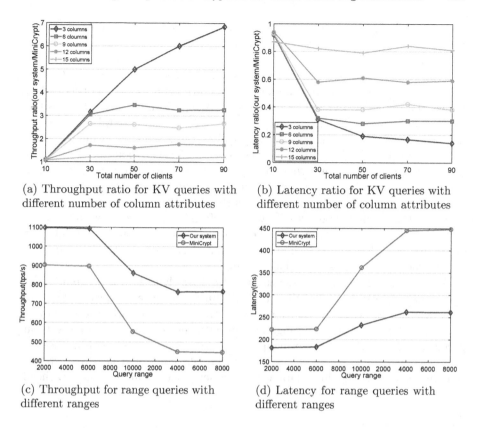

(a) Throughput ratio for KV queries with different number of column attributes

(b) Latency ratio for KV queries with different number of column attributes

(c) Throughput for range queries with different ranges

(d) Latency for range queries with different ranges

Fig. 6. Query performance evaluation.

ranges. The reason is that each record group contains multiple key-value records. The requested data fitted in different query intervals may be included in the same record group. Moreover, we also find that the communication overhead of our system is significantly less than that of MiniCrypt in range queries. The communication overhead of minicrypt is 3 times higher than our system. The reason is that our encrypted key-value storage structure allows fine-grained data access, reducing the amount of data to be returned.

6.2 Query Performance Evaluation

In this experiment, we first test the time cost of each phase of a KV query, the results are shown in Table 3. After that, we compare the throughput ratio and latency ratio of our system and Minicrypt when performing KV queries with different number of columns. At last, we compare the throughput and latency of our system and Minicrypt when performing range queries with 7 column attributes. Experimental results show that our system can achieve higher performance.

Since our system only needs to return the requested column packets when querying a few columns, the communication overhead is greatly reduced, which results in a reduction in transmission time. When performing a KV query with a single column attribute, in our system, the transfer process takes 31 ms and the process of decryption and decompression takes 3 ms. In MiniCrypt, the transmission time takes 59 ms, and the process of decryption and decompression takes 10 ms. Therefore when multiple users concurrently issue KV queries, our system can process more query requests per unit of time, thereby achieving higher throughput. Our system increases the throughput by up to 7 times and decreases the latency by about 80% when performing KV queries with three column attributes. What's more, since the compression ratio of our system is higher, even if issuing a KV query with all the column attributes, the transmission time of our system is still less than that of MiniCrypt.

When performing range queries, we also find that the throughput of our system is significantly higher than that of MiniCrypt in range queries. Our system increases the throughput by up to 1.5 times and decreases the latency by about 33% when performing range queries with seven column attributes. Compared with MiniCrypt, the throughput improvement of our system increases as the query range increases. The reason is that our encrypted key-value storage structure allows fine-grained data access, reducing the communication overhead.

7   Conclusion

We propose an encrypted key-value storage system that combines compression techniques to improve system performance. To support fine-grained access, we build a new key-value storage structure using the concept of horizontal-vertical division. We use order-revealing encryption to encrypt record group IDs to support KV queries and range queries. We implement the prototype of our scheme on Cassandra. Experimental evaluation results show that our system can effectively reduce communication overhead and improve system throughput. In the future work, we plan to explore advanced encryption technologies to support richer queries, such as aggregate query and conditional query. Meanwhile, how to make our system better adapt to the distributed environment, with better scalability will also become our future work.

Acknowledgement. We acknowledge the support from National Natural Science Foundation of China (Nos. 61602363 and 61572382), China 111 Project (No. B16037) and China Postdoctoral Science Foundation (No. 2016M590927).

References

1. Enron email dataset. https://www.cs.cmu.edu/~enron/, Accessed 4 Feb 2019
2. Hbase: the hadoop database, a distributed, scalable, big data store. http://hbase. apache.org. Accessed 4 Feb 2019
3. Redis: an advanced key-value cache and store. http://redis.io/. Accessed 4 Feb 2019
4. Abadi, D., Madden, S., Ferreira, M.: Integrating compression and execution in column-oriented database systems. In: Proceedings of the 2006 ACM SIGMOD International Conference on Management of Data, pp. 671–682. ACM (2006)
5. Agarwal, R., Khandelwal, A., Stoica, I.: Succinct: enabling queries on compressed data. In: 12th USENIX Symposium on Networked Systems Design and Implementation (NSDI 2015), pp. 337–350 (2015)
6. Agrawal, R., Kiernan, J., Srikant, R., Xu, Y.: Order preserving encryption for numeric data. In: Proceedings of the 2004 ACM SIGMOD International Conference on Management of Data, pp. 563–574. ACM (2004)
7. Binnig, C., Hildenbrand, S., Färber, F.: Dictionary-based order-preserving string compression for main memory column stores. In: Proceedings of the 2009 ACM SIGMOD International Conference on Management of Data, pp. 283–296. ACM (2009)
8. Blaze, M.: A cryptographic file system for UNIX. In: Proceedings of the 1st ACM Conference on Computer and Communications Security, pp. 9–16. ACM (1993)
9. Boldyreva, A., Chenette, N., Lee, Y., O'Neill, A.: Order-preserving symmetric encryption. In: Joux, A. (ed.) EUROCRYPT 2009. LNCS, vol. 5479, pp. 224–241. Springer, Heidelberg (2009). https://doi.org/10.1007/978-3-642-01001-9_13
10. Boldyreva, A., Chenette, N., O'Neill, A.: Order-preserving encryption revisited: improved security analysis and alternative solutions. In: Rogaway, P. (ed.) CRYPTO 2011. LNCS, vol. 6841, pp. 578–595. Springer, Heidelberg (2011). https://doi.org/10.1007/978-3-642-22792-9_33
11. Boneh, D., Lewi, K., Raykova, M., Sahai, A., Zhandry, M., Zimmerman, J.: Semantically secure order-revealing encryption: multi-input functional encryption without obfuscation. In: Oswald, E., Fischlin, M. (eds.) EUROCRYPT 2015. LNCS, vol. 9057, pp. 563–594. Springer, Heidelberg (2015). https://doi.org/10.1007/978-3-662-46803-6_19
12. Chang, F., et al.: A distributed storage system for structured data. In: Proceedings of OSDI, pp. 6–8 (2006)
13. Chen, X., Huang, X., Li, J., Ma, J., Lou, W., Wong, D.S.: New algorithms for secure outsourcing of large-scale systems of linear equations. IEEE Trans. Inf. Forensics Secur. 10(1), 69–78 (2015)
14. Chen, X., Li, J., Ma, J., Tang, Q., Lou, W.: New algorithms for secure outsourcing of modular exponentiations. IEEE Trans. Parallel Distrib. Syst. 25(9), 2386–2396 (2014)
15. Chenette, N., Lewi, K., Weis, S.A., Wu, D.J.: Practical order-revealing encryption with limited leakage. In: Peyrin, T. (ed.) FSE 2016. LNCS, vol. 9783, pp. 474–493. Springer, Heidelberg (2016). https://doi.org/10.1007/978-3-662-52993-5_24
16. DeCandia, G., et al.: Dynamo: Amazon's highly available key-value store. In: ACM SIGOPS Operating Systems Review, vol. 41, pp. 205–220. ACM (2007)
17. Escriva, R., Wong, B., Sirer, E.G.: HyperDex: a distributed, searchable key-value store. In: Proceedings of the ACM SIGCOMM 2012 Conference on Applications, Technologies, Architectures, and Protocols for Computer Communication, pp. 25–36. ACM (2012)

18. Goh, E.J., Shacham, H., Modadugu, N., Boneh, D.: SiRiUS: securing remote untrusted storage. In: NDSS, vol. 3, pp. 131–145 (2003)
19. Goldstein, J., Ramakrishnan, R., Shaft, U.: Compressing relations and indexes. In: Proceedings of the 14th International Conference on Data Engineering, pp. 370–379. IEEE (1998)
20. Graefe, G., Shapiro, L.D.: Data compression and database performance. In: Proceedings of the 1991 Symposium on Applied Computing, pp. 22–27. IEEE (1991)
21. Guo, Y., Yuan, X., Wang, X., Wang, C., Li, B., Jia, X.: Enabling encrypted rich queries in distributed key-value stores. IEEE Trans. Parallel Distrib. Syst. **30**(6), 1283–1297 (2018)
22. He, Y., et al.: RCFile: a fast and space-efficient data placement structure in MapReduce-based warehouse systems. In: 2011 IEEE 27th International Conference on Data Engineering, pp. 1199–1208. IEEE (2011)
23. Heron, S.: Advanced encryption standard (AES). Netw. Secur. **2009**(12), 8–12 (2009)
24. Holloway, A.L., Raman, V., Swart, G., DeWitt, D.J.: How to barter bits for chronons: compression and bandwidth trade offs for database scans. In: Proceedings of the 2007 ACM SIGMOD International Conference on Management of Data, pp. 389–400. ACM (2007)
25. Iyer, B.R., Wilhite, D.: Data compression support in databases. In: VLDB, vol. 94, pp. 695–704 (1994)
26. Johnson, T.: Performance measurements of compressed bitmap indices. In: Proceedings of the 25th International Conference on Very Large Data Bases, pp. 278–289. Morgan Kaufmann Publishers Inc. (1999)
27. Kallahalla, M., Riedel, E., Swaminathan, R., Wang, Q., Fu, K.: Plutus: scalable secure file sharing on untrusted storage. In: Fast, vol. 3, pp. 29–42 (2003)
28. Kher, V., Kim, Y.: Securing distributed storage: challenges, techniques, and systems. In: Proceedings of the 2005 ACM Workshop on Storage Security and Survivability, pp. 9–25. ACM (2005)
29. Lakshman, A., Malik, P.: Cassandra: a decentralized structured storage system. ACM SIGOPS Oper. Syst. Rev. **44**(2), 35–40 (2010)
30. Macedo, R., et al.: A practical framework for privacy-preserving NoSQL databases. In: 2017 IEEE 36th Symposium on Reliable Distributed Systems (SRDS), pp. 11–20. IEEE (2017)
31. MacNicol, R., French, B.: Sybase IQ multiplex-designed for analytics. In: Proceedings of the Thirtieth International Conference on Very Large Data Bases, vol. 30, pp. 1227–1230. VLDB Endowment (2004)
32. Ousterhout, J., et al.: The ramcloud storage system. ACM Trans. Comput. Syst. (TOCS) **33**(3), 7 (2015)
33. Pappas, V., et al.: Blind seer: a scalable private DBMS. In: 2014 IEEE Symposium on Security and Privacy, pp. 359–374. IEEE (2014)
34. Poddar, R., Boelter, T., Popa, R.A.: Arx: a strongly encrypted database system. IACR Cryptology ePrint Archive 2016/591 (2016)
35. Popa, R.A., Redfield, C., Zeldovich, N., Balakrishnan, H.: CryptDB: protecting confidentiality with encrypted query processing. In: Proceedings of the Twenty-Third ACM Symposium on Operating Systems Principles, pp. 85–100. ACM (2011)
36. Ray, G., Haritsa, J.R., Seshadri, S.: Database compression: a performance enhancement tool. In: COMAD. Citeseer (1995)
37. Ren, K., Wang, C., Wang, Q.: Security challenges for the public cloud. IEEE Internet Comput. **16**(1), 69–73 (2012)

38. Wang, J., Chen, X., Li, J., Zhao, J., Shen, J.: Towards achieving flexible and verifiable search for outsourced database in cloud computing. Future Gener. Comput. Syst. **67**, 266–275 (2017)
39. Wang, J., Miao, M., Gao, Y., Chen, X.: Enabling efficient approximate nearest neighbor search for outsourced database in cloud computing. Soft. Comput. **20**(11), 4487–4495 (2016)
40. Yuan, X., Wang, X., Wang, C., Qian, C., Lin, J.: Building an encrypted, distributed, and searchable key-value store. In: Proceedings of the 11th ACM on Asia Conference on Computer and Communications Security, pp. 547–558. ACM (2016)
41. Zheng, W., Li, F., Popa, R.A., Stoica, I., Agarwal, R.: MiniCrypt: reconciling encryption and compression for big data stores. In: Proceedings of the Twelfth European Conference on Computer Systems, pp. 191–204. ACM (2017)
42. Zukowski, M., Heman, S., Nes, N., Boncz, P.A.: Super-scalar RAM-CPU cache compression. In: ICDE, vol. 6, p. 59 (2006)

Secret Sharing and Interval Test

Evolving Perfect Hash Families: A Combinatorial Viewpoint of Evolving Secret Sharing

Yvo Desmedt[1,2], Sabyasachi Dutta[3], and Kirill Morozov[4(✉)]

[1] University of Texas at Dallas, Richardson, USA
y.desmedt@cs.ucl.ac.uk
[2] University College London, London, UK
[3] Faculty of Information Science and Electrical Engineering,
Kyushu University, Fukuoka, Japan
saby.math@gmail.com
[4] Department of Computer Science and Engineering, University of North Texas,
Denton, USA
Kirill.Morozov@unt.edu

Abstract. Shamir's threshold secret sharing scheme gives an efficient way to share a secret among n participants such that any k or more of them can reconstruct the secret. For implementation, Shamir's scheme requires a finite field. Desmedt et al. (AsiaCrypt '94) proposed a multiplicative secret sharing scheme over non-abelian groups. In this paper, we extend (non-abelian) multiplicative secret sharing to accommodate unbounded number of participants. We introduce a new combinatorial concept of "evolving" Perfect Hash Families and present a secret sharing scheme as a consequence.

Keywords: Evolving · Perpetual · Perfect hash family · Combinatorial object · Secret sharing

1 Introduction

Threshold secret sharing, independently introduced by Shamir [31] and Blakley [5], is a method to share a secret information among n participants in such a way that any k or more of them can recover the secret but $k - 1$ or less many participants do not have any information about the secret. Shamir's construction of secret sharing requires a finite field for implementation. This puts an inherent limitation—at the same time, extensions of this scheme can work over abelian groups [8, 9, 16].

Y. Desmedt thanks the Jonsson Endowment. He is also an Honorary Professor at University College London.
S. Dutta is grateful to the NICT, Japan for granting a financial support under the NICT International Exchange Program.

© Springer Nature Switzerland AG 2019
Y. Mu et al. (Eds.): CANS 2019, LNCS 11829, pp. 291–307, 2019.
https://doi.org/10.1007/978-3-030-31578-8_16

1.1   Background and Motivation

Homomorphic secret sharing was a popular topic after it was introduced by Benaloh [3]. The fact that Shamir's [31] secret sharing scheme is homomorphic for the addition, is used heavily in secure multiparty computation. In [16], Shamir's secret sharing scheme and its homomorphic property were generalized to work over any abelian group. This result played a key role to obtain the first threshold RSA schemes (see [12] and [10]). So, an algebraic study of the homomorphic properties of secret sharing was natural (see [17]). One of the results of [17] shows that the homomorphic property cannot be achieved when the secret to be shared belongs to a non-abelian group, except when the access structure is trivial.

In the early 1990's the topic of zero-knowledge interactive proof [19] was very popular. In Goldreich et al.'s [18] zero-knowledge proof of graph isomorphism, the prover, knowing a secret, proves that two graphs are isomorphic without divulging the secret. *This secret is in fact an element of S_n, the symmetric group, which is non-abelian.* A threshold cryptographic version was proposed in [13]. It allows t parties (out of n) to share the secret and then jointly co-prove that the two graphs are isomorphic. The main idea was to have a sequential approach used by the t co-provers, this bypassed the fact that the homomorphic property is impossible over non-abelian groups, which would have allowed a parallel execution by the t co-provers. Note that, if the secret would be mapped into a finite field and then one would use Shamir's secret sharing, the threshold proof of [13] would collapse. The way the reconstruction of the secret works starting from the shares was called multiplicative[1] secret sharing [13]. Followup work at Eurocrypt 1996 by Blackburn et al. [4] generalized some of these results. Stinson [33] later in an e-mail pointed out that the results in [4] can be explained using perfect hash families (PHF). Note that the case $t = 2$ in [13] can also be explained using PHFs. Besides the use of PHFs to make multiplicative secret shares, secret shares over non-abelian groups have other properties.

Note also that block ciphers are substitutions and that these form a non-abelian group. An attempt to make threshold block ciphers was made in [6].

Due to results by Barrington [2] one can perform any computation provided one can perform certain operations over S_5, a non-abelian group. This was used in [14] (see also [15] and [7]) to propose a then new approach to perform secure multiparty computation. The work uses shares that belong to, e.g., S_5. The approach in [14,15] were used in 2015 to propose an e-voting scheme using an unconditionally secure MIX operation [11]. Note that a MIX performs a permutation, which is a non-abelian group operation. The use of secret shares over this non-abelian group was essential to obtain the results in [11].

---

[1] Since this requirement corresponds exactly with the reconstruction used in this paper, we do not explain the definition in background. We refer the reader to [13] or [4].

1.2   Challenging Issue and Our Contribution

The above mentioned secret sharing schemes are applicable only when the number of participants is pre-fixed and *finite*. To address the possibility of sharing a secret with an unbounded number of participants, *evolving* secret sharing schemes were introduced by Komargodski, Naor and Yogev [24] and were further improved by [25]. However, both schemes use Shamir's secret sharing and thus work only over finite fields. Then, using [8,9,16], they can be extended to work over abelian groups, but *not* over non-abelian groups.

In this paper, we extend the work of Desmedt et al. [13] by accommodating unbounded number of participants. To achieve our goal we introduce a new combinatorial concept of "evolving" Perfect Hash Families. We explain the underlying combinatorics in Sect. 2.1 and a formal description in Sect. 3. To have an overview of the idea, consider a situation when the domain of a perfect hash function (PHF) family is not known in advance but—according to some application—it may have to be increased in the future. Since PHF is typically stored as a table, it is quite reasonable to keep it small unless the application requires to extend it. We use the term *evolving*,[2] borrowing from the work of Komargodski, Naor and Yogev [24]. We show how to extend the domain of an PHF family adaptively by introducing new partial functions and extending the initial ones. Hereby, we achieve a perpetually evolving PHF, i.e., growing to infinity.

Reflecting back to secret sharing, we recall that PHF has been used to construct secret sharing schemes as shown implicitly by Desmedt et al. in [13] and by Blackburn et al. in [4]. The link between these and PHF was made in [29,33].

Organization of the Paper

Our paper is organized as follows: Sect. 2 describes background in combinatorics, as well as in secret sharing, including the Komargodski-Naor-Yogev scheme. In Sect. 3, we present our model and definitions, including evolving set families and evolving PHF. Some examples that clarify our concept are provided in Sect. 4.2. Section 4 introduces our construction of the perpetually evolving PHF and its proof of correctness. Then, a multiplicative secret sharing scheme over non-abelian quasigroup implied by evolving PHF is sketched in Sect. 5. Finally, Sect. 6 provides concluding remarks and discusses open questions.

2   Preliminaries

Let us fix some notation first. For a natural number n, the set $\{1, \ldots, n\}$ will be denoted as $[n]$. We will write "n-subset" to denote a subset of size n. Given a function f, we will write its restriction to a subset X as $f|_X$. All the logarithms are to the base 2 unless stated otherwise.

---

[2] A similar idea in the context of publishing was introduced by Sloane [32] under the label of "eternal home page".

2.1   Perfect Hash Families

Perfect hash families were first discovered and studied as a part of compiler design. Mehlhorn [28] gives a summary of early work made in this area. PHF have later found applications in the realm of cryptography, for instance, in secret sharing [4], threshold cryptography [30] etc. to name a few.

Definition 1. *A family of functions $\mathcal{F}$ is called an $(N; n, m, w)$-Perfect Hash Family (PHF, in short) if:*

- *each $f \in \mathcal{F}$ is a function from $[n] \longrightarrow [m]$ with $|\mathcal{F}| = N$ and,*
- *for any w-subset $X \subset [n]$, there exists some $g \in \mathcal{F}$ s.t. $g|_X$ is one-to-one.*

Note 1. We emphasize the following two points:

1. An inequality $w \leq m$ follows from the second condition above.
2. If we consider the binary case only, i.e., $m = 2$, then for non-triviality we must have $w = 2$. In other words, for every 2-subset X of the domain, there exists a function f in the family such that $f|_X$ is one-to-one.

Throughout this paper, we will use the following notation related to PHF:

- If the co-domain is of size $m = 2$, the function family is denoted by $\mathcal{F}_2$.
- For $m = 2$, we write the co-domain space as $\{0, 1\}$.
- If $|\mathcal{F}| = N$ then the family is denoted by $PHF(N; n, m, w)$.
- Let $N(n, m, w)$ be the minimum value N for which a $PHF(n, m, w)$ exists.

Results on PHF. Let us survey some useful results on recursive constructions of perfect hash families [1, 27].

Theorem 1 ([27]). *For $PHF(N; n, m, w)$, we have that*

$$N(n, m, w) \geq 1 + N\left(\left\lceil \frac{n}{m} \right\rceil, m, w\right)$$

and it follows that

$$N(n, m, w) \geq \frac{\log n}{\log m}.$$

Corollary 1. *In the setting of the above theorem, for $m = w = 2$, we have that $N(n, 2, 2) \geq \log n$.*

The following stronger lower bound is given by Fredman et al. [20].

Theorem 2 ([20]). *For $PHF(N; n, m, w)$,*

$$N(n, m, w) \geq \frac{\binom{n-1}{w-2} \cdot m^{w-2} \cdot \log(n - w + 2)}{\binom{m-1}{w-2} \cdot n^{w-2} \cdot \log(m - w + 2)}.$$

However, when restricted to the case $m = w = 2$, we get the same lower bound.

Corollary 2. *For* $m = w = 2$, $N(n, 2, 2) \geq \log n$.

The following upper bound was introduced by Mehlhorn [27].

Theorem 3 ([27]). *For* $PHF(N; n, m, w)$, $N(n, m, w) \leq \lceil we^{\frac{w^2}{m}} \ln n \rceil$.

Corollary 3. *In the setting of the above theorem,* $N(n, 2, 2) \leq \lceil 2e^2 \ln n \rceil$ *and thus the behaviour of* N *as a function of* n *varies as* $\Theta(\log n)$ *(for fixed values of* $m = 2$ *and* $w = 2$*).*

Atici et al. [1] focused on the problem of the growth of $N(n, m, w)$ as a function of n. They provided some explicit constructions in which N grows as a polynomial function of $\log n$ (for fixed m and w).

The first recursive construction (based on a difference matrix and a basic PHF) by Atici et al. [1] gives us the following:

Theorem 4 ([1]). *Suppose there is an* $\left(n_0, \binom{w}{2} + 1; 1\right)$*-difference matrix and a* $PHF(N_0; n_0, m, w)$. *Then there exists a* $PHF\left(\left(\binom{w}{2} + 1\right)N_0; n_0^2, m, w\right)$.

Corollary 4. *For* $m = w = 2$, *if there exists an* $(n_0, 2; 1)$*-difference matrix and a* $PHF(N_0; n_0, 2, 2)$, *then there exists a* $PHF(2N_0; n_0^2, 2, 2)$. *Thus, the size of* $\mathcal{F}_2$ *becomes twice to make the domain size quadratic.*

Using an easily constructive special family of difference matrices, Atici et al. [1] gave an iterated result.

Theorem 5 ([1]). *Suppose there is a* $PHF(N_0; n_0, m, w)$ *and suppose that* $\gcd\left(n_0, \binom{w}{2}!\right) = 1$. *Then there exists a* $PHF\left(\left(\binom{w}{2} + 1\right)^j N_0; n_0^{2^j}, m, w\right)$.

Corollary 5. *For* $m = w = 2$, *the condition* $\gcd\left(n_0, \binom{w}{2}!\right) = 1$ *is satisfied. Therefore, from a* $PHF(N_0; n_0, 2, 2)$ *one can construct another PHF with parameters* $(2^j N_0; n_0^{2^j}, 2, 2)$.

Due to another (Kronecker product type) recursive construction by Atici et al. [1] we have the following theorem.

Theorem 6 ([1]). *Suppose that the following exist:*
a $PHF(N_1; n_0 n_1, m, w)$, *a* $PHF(N_2; n_2, n_1, w - 1)$, *and a* $PHF(N_3; n_2, m, w)$.
Then, there exists a $PHF(N_1 N_2 + N_3; n_0 n_2, m, w)$.

Theorem 7 ([1]). $N(n, m, 2) = \left\lceil \frac{\log n}{\log m} \right\rceil$.

Corollary 6. *For* $m = 2$, $N(n, 2, 2) = \lceil \log n \rceil$.

2.2 Secret Sharing

Secret sharing schemes were proposed independently by Shamir [31] and Blakley [5] in 1979. The idea of achieving such a primitive was present in mathematics for quite some time [26] and some combinatorial solutions were proposed. However, inefficiency of these solutions was resolved by [5,31]. They proposed schemes where any k (or more) out of n participants are qualified to recover the secret with $1 < k \leq n$. The resulting access structure is called a (k, n)-*threshold* access structure where k acts as a threshold value for being qualified. Both schemes were fairly efficient in terms of the size of the shares and computational complexity. See also [21,22] for a generalization. In Appendix, we present a basic $(2, 2)$ secret sharing scheme.

The classical secret sharing schemes assume that the number of participants and the access structure is known in advance. Komargodski, Naor, and Yogev [24] introduced evolving secret sharing schemes where the dealer does not know in advance the number of participants that will participate, and moreover there is no upper bound on their number. Thus, the number of participants could be potentially infinite and the access structure may change with time. Komargodski et al. [24] considered the scenario when participants come one by one and receive their share from the dealer; the dealer however cannot update the shares which have already been distributed. They showed that for every evolving access structure there exists a secret sharing scheme where the share size of the tth participant is 2^{t-1}. They also constructed (k, ∞)-threshold evolving secret sharing scheme for constant k in which the share size of the tth participant is $(k - 1) \log t + \mathcal{O}(\log \log t)$. Furthermore, they have provided an evolving 2-threshold scheme which is nearly optimal in the share size of the tth participant viz. $\log t + \mathcal{O}(\log \log t)$.

The main trick that Komargodski et al. used to significantly reduce the share size is introducing the concept of generations. Each generation consists of participants and the size of every generation grows exponentially with time. The sizes of generations are however prefixed depending on the threshold value k. Deployment of Shamir's secret sharing scheme helped to reduce share sizes exponentially. In short, the authors combined the Shamir secret sharing with combinatorics to achieve an efficient construction. In Appendix, we describe (see Fig. 7) a simple example to visualize the main idea behind the construction of $(2, \infty)$ secret sharing proposed by Komargodski, Naor, and Yogev [24]. In Fig. 7, the value s denotes the secret bit, b_i's denote random bits and Shamir$(2, n)(s)$ denotes the corresponding shares of the 2-out-of-n Shamir scheme on a set of n parties.

Later, Komargodski and Paskin-Cherniavsky [25] applied the idea of evolving k-threshold schemes to evolving dynamic threshold schemes and provided a secret sharing scheme in which the share size of the tth participant is $\mathcal{O}(t^4 \log t)$ bits. Furthermore, they showed how to transform evolving threshold secret sharing schemes into robust schemes with the help of algebraic manipulation detection (AMD) codes.

3   Model and Definitions

We begin with the definition of an evolving set. Roughly speaking, points are included in a set from time to time. We consider the case such that this inclusion process gives rise to an at most countable family of finite sets. We denote the evolving sequence of sets by $\{X_n\}_{n\geq 0}$ where X_0 is the basic set. More formally,

Definition 2. *(Evolving family of sets)*
A sequence of sets $\{X_n\}_{n\geq 0}$ is an evolving family of sets if $X_i \subset X_{i+1}$ for all $i \geq 0$, i.e., the family is strictly monotone increasing.
If the sequence is finite, the family is called an evolving *family of sets.*
If the sequence can proceed indefinitely to accommodate infinitely many points then the family is called perpetually-evolving.

To draw the similarity with evolving access structure as defined by Komargodski, Naor, and Yogev [24] we observe that the underlying set of participants in their work is perpetually evolving.

We now define the concept of partial function which plays an instrumental role in our construction. Roughly speaking, a partial function generalizes the concept of a function $f : X \longrightarrow Y$, by not forcing f to map every element of X to some element in Y. Rather, there is a subset X' of X such that f maps every element of f to some element in Y. If $X' = X$, then we sometimes call it a *total* function in order to specify that the domain is not a proper subset of X. Partial functions are recurrent in the theory of computation.

Definition 3 (Partial function). *A rule $f : X \longrightarrow Y$ is called a partial function if there exists a subset $X' \subset X$ such that when restricted to X', $f|_{X'} : X' \longrightarrow Y$ is a (total) function.*

Note 2. We will call a "total function", simply a function when there is no ambiguity.

Definition 4 (Evolving PHF). *Let $\{X_r\}$ be an evolving family of sets, $\{Y_r\}$ be a sequence of sets (which may or may not be evolving) and $\{w_r\}$ be a non-decreasing sequence of positive integers. A sequence of family of partial and total functions $\{\mathcal{F}_r\}$ is called an $(\{X_r\}, \{Y_r\}, w_r)$-evolving PHF if:*

– *each $f \in \mathcal{F}_r$ is a partial/total function from $X_r \longrightarrow Y_r$ and*
– *for any w_r-subset $X' \subset X_r$, there exists $g \in \mathcal{F}_r$ such that the restriction of g on X' is one-to-one.*

Remark 1. We remark that if the family $\{X_r\}$ is perpetually evolving then the corresponding PHF is called *perpetual* PHF. The sequence of co-domains $\{Y_r\}$ need not be evolving. In fact, it can be a constant sequence i.e., $Y_r = Y$ for all r. In addition, the non-decreasing sequence of positive integers $\{w_r\}$ can very well be a constant sequence.

Note 3. We note that for every $r \geq 1$, the collection $\mathcal{F}_r$ is a $(|X_r|, |Y_r|, w_r)$-PHF in a weaker sense as $\mathcal{F}_r$ may contain partial functions.

4   Main Result

In this section, we first describe an algorithm to construct perpetually evolving perfect hash families. Then, through examples, we clarify the rationale behind our construction.

4.1   Construction of Perpetually Evolving PHFs

We introduce Algorithm 1 for constructing perpetually evolving PHF.

We only consider the binary case, i.e., the co-domain is the set $\{0,1\}$ and $w = 2$. A rough sketch of the algorithm is depicted in Fig. 1. In Algorithm 1, we will use the following notations:

- Denote an m-dimensional null vector by $\bar{0}_m$ (written as a column).
- Denote an m-dimensional column vector with all entries equal to 1 by $\bar{1}_m$.
- $\bar{0}_m^T$ and $\bar{1}_m^T$, respectively, denote the transposes of the column vectors.
- After the introduction of rth partial row, the evolved matrix is denoted by $M(r)$.

Algorithm 1. Construction of perpetually evolving PHF

1: **procedure** INIT.
2: Assign $\bar{0}_t$ as the first column, where t is an appropriate value.

3: **procedure** INTRODUCTION OF FIRST PARTIAL ROW
4: Place the remaining $2^t - 1$ columns $\bar{C}_t^1, \ldots, \bar{C}_t^{2^t-1}$ to the right of $\bar{0}_t$.
5: Append a partial row $\bar{0}_{2^t-1}^T$ just below $\bar{C}_t^1, \ldots, \bar{C}_t^{2^t-1}$.
6: Append $\bar{C}_t^1, \ldots, \bar{C}_t^{2^t-1}$ to the right as columns $2^t + 1$ to $2^{t+1} - 1$.
7: Append a partial row $\bar{1}_{2^t-1}^T$ just below the copied columns in Line 6.

8: **procedure** INTRODUCTION OF rth PARTIAL ROW TO THE $(r-1)^{th}$ EVOLVED MATRIX $M(r-1)$
9: Choose the last $\lceil \frac{\alpha}{2} \rceil$ columns $B[1], B[2], \ldots, B[\lceil \frac{\alpha}{2} \rceil]$ of $M(r-1)$, where α denotes the number of columns in the evolved matrix $M(r-1)$.
10: Append $\bar{0}_{\frac{\alpha}{2}}^T$ just below $B[1], B[2], \ldots, B[\lceil \frac{\alpha}{2} \rceil]$.
11: Copy $B[1], B[2], \ldots, B[\lceil \frac{\alpha}{2} \rceil]$ to the right of $M(r-1)$.
12: Append $\bar{1}_{\frac{\alpha}{2}}^T$ just below the columns appended in Line 11.

Theorem 8. *Algorithm 1 admits a perpetually evolving PHF. The number of columns that can be accommodated with the introduction of the rth partial row is exponential in r. To be more precise, if the first column is a t-dimensional $(\bar{0})$ column then the number of new columns that can be accommodated by the* *introduction of rth partial row is $\left\lceil \left(\frac{3}{2}\right)^{r-2} 2^t \right\rceil$. Thus, fixing the value of t, we can see that $\mathcal{O}\left(\left(\frac{3}{2}\right)^r\right)$ new columns can be accommodated by the introduction of rth row.*

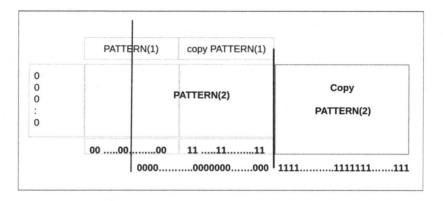

Fig. 1. Pictorial illustration of Algorithm 1.

Proof. We can prove that the algorithm admits a perpetually evolving PHF using induction. More precisely, we will show that after r^{th} partial row the resulting matrix $M(r)$ gives rise to perfect hash family for every positive integer r.

The fact that for $r = 1$, i.e., after the introduction of the first partial row, the resulting matrix $M(1)$ gives rise to an PHF follows easily from the following three observations:

- The first 2^t columns when restricted to the first t entries are nothing but all possible binary columns of length t. Therefore, any two columns in this list differ at least in one position which gives us the required result.
- Any column C from 2 to 2^t (Line 4 of the Algorithm 1) and any column D from $2^t + 1$ to $2^{t+1} - 1$ (Line 6 of the Algorithm 1) differ at the last position because last entry of C is 0 (Line 5 of the Algorithm 1) and last entry of D is 1 (Line 7 of the Algorithm 1).
- Any column from $2^t + 1$ to $2^{t+1} - 1$, when restricted to the first t entries, is also non-zero and hence differ in at least one position from the first column $\bar{0}_t$.

Now to prove the general case, let us suppose that our hypothesis be true for $r-1$, i.e., $M(r - 1)$ gives rise to an PHF. We maintain the same notations as used in the Algorithm 1. Specifically, $B[1], B[2], \ldots, B[\lceil \frac{\alpha}{2} \rceil]$ denote the last $\lceil \frac{\alpha}{2} \rceil$ columns of $M(r-1)$ (Line 10 of the Algorithm 1), where α denotes the number of columns in $M(r - 1)$. They are appended with a zero vector of length $\lceil \frac{\alpha}{2} \rceil$ (Line 10 of the Algorithm 1). The new columns that are accommodated/appended by Lines 11 and 12 in the algorithm contain 1 as the last entry. Therefore, any two columns in the last α columns of $M(r)$ differ in at least one position either because of the last entry or they already differed as columns in $M(r-1)$. The remaining columns in $M(r)$ differ with the newly added columns by our hypothesis when we restrict the newly added columns to their first $t + r - 1$ entries.

It is not very hard to see that the introduction of rth partial row can accommodate $\left\lceil \frac{2^t + (2^t - 1) + 2^t + 3 \cdot 2^{t-1} + 3^2 \cdot 2^{t-2} + \cdots + 3^{r-3}2^{t-r+3}}{2} \right\rceil$ many new columns. A simplification of the above expression results in $\left\lceil (\frac{3}{2})^{r-2}2^t \right\rceil$.   □

A variant of Algorithm 1 consists of modifying Steps 5–7, by placing zeros below the columns $2^{t-1}+1$ up to 2^t and then copy the last 2^{t-1} columns, but having the last row having the values 1. Obviously, orderwise this will make no improvement, but for the first iterations, in this variant, less rows are needed for the first few columns.

4.2   Some Examples

In this section we study some examples to clarify the concept.

Example 1. Let us consider the following finite evolving family:
$\{X_r\} = \{X_0, X_1, X_2, X_3\}$ with $X_0 = \emptyset, X_1 = \{1\}, X_2 = \{1,2\}, X_3 = \{1,2,3\}$;
$\{Y_r\} = \{Y\} = \{\{0,1\}\}$ and $w_1 = 1, w_2 = w_3 = 2$. We note that $\mathcal{F}_0 = \emptyset$.
The evolving family of functions is described below.

- $\mathcal{F}_1 = \{f_1\}$, where $f_1 : \{1\} \longrightarrow \{0,1\}$ defined by $f_1(1) = 0$.
- $\mathcal{F}_2 = \{\boldsymbol{f_1}, f_2\}$, where $\boldsymbol{f_1} : \{1,2\} \longrightarrow \{0,1\}$ defined by $\boldsymbol{f_1}(1) = 0, \boldsymbol{f_1}(2) = 1$. (Here, $\boldsymbol{f_1}$ is actually an extension of f_1 but we will abuse notation from now on to denote the extended function by the function itself and a partial function $f_2 : \{1,2\} \longrightarrow \{0,1\}$ defined by $f_2(2) = 0$.)
- $\mathcal{F}_3 = \{f_1, f_2\}$, where $f_1 : \{1,2,3\} \longrightarrow \{0,1\}$ defined by $f_1(1) = 0, f_1(2) = 1$, $f_1(3) = 1$, and $f_2 : \{1,2,3\} \longrightarrow \{0,1\}$ defined by $f_2(2) = 0, f_2(3) = 1$.

It is easy to check that the above is an example of an evolving PHF. For a better representation, we use Fig. 2 to depict the example.

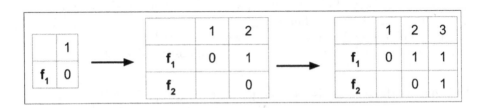

Fig. 2. Evolving PHF from Example 1.

We represent a perfect hash family with co-domain $\{0,1\}$ by a binary matrix where the columns are indexed by the members of the domain and the rows are indexed by (partial or total) functions. The $(i, j)^{th}$ entry of the matrix is the output of ith (partial or total) function with input j. If the function is partial then it is easy to see that the corresponding rows are also partial. Thus total rows correspond to total functions and partial rows correspond to partial functions.

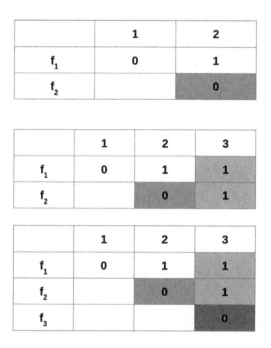

Fig. 3. Introducing a 3rd partial row for accommodating a future 4th column from Example 2. (Color figure online)

Example 2. Taking cue from Example 1, we now want to see whether the process can be continued (in a systematic way) indefinitely giving rise to a perpetually evolving PHF. To accommodate "possible future points" we give an extra entry (0) to the last point/column by defining a new partial row/function (see Figs. 3 and 4). Column 2 is given an extra 0 (defined by f_2) to accommodate a "future" Column 3.

These extra entries are marked with different colors. In Fig. 3, we see that the third column receives a third "extra" entry to ensure that a future fourth column can be added later, if need arise. It is not very hard to observe that continuing this way we can accommodate points/columns indefinitely giving rise to a perpetually evolving PHF. However, this approach is not efficient (see Algorithm 1 for a better approach.

Note 4. From Example 2 we observe the following: (a) points/columns which are added early are of shorter "length", (b) the tth column receives t entries, (c) one can accommodate an unbounded number of columns and (d) adding one new partial row can accommodate only one new column.

	1	2	3	4
f_1	0	1	1	1
f_2		0	1	1
f_3			0	1

Fig. 4. PHF with domain size 4; one new partial row accommodates one new column from Example 2.

	1	2
f_1	0	1
f_2	0	0
f_3		0

	1	2	3	4	5	6	7
f_1	0	1	0	1	1	0	1
f_2	0	0	1	1	0	1	1
f_3		0	0	0	1	1	1

Fig. 5. Seven columns are accommodated using three rows from Example 3.

Example 3 (Gaining efficiency). In this example (see Fig. 5) we see that although each of first two columns get one extra entry as compared to Example 2 but with the 3 rows, we can accommodate upto 7 columns.

Example 4. In this example (see Fig. 6) we present a case where our technique gives better results than the recursive constructions of Atici et al. [1]. Suppose we have a $PHF(N; 3, 2, 2)$. We know that $N(3, 2, 2) = \lceil log\ 3 \rceil = 2$. Applying Corollary 4, we obtain a $PHF(N'; 9, 2, 2)$ with $N' = 4$. Our constructions achieve the same result but the first two columns use less entries than the next ones. Furthermore, we can accommodate one more column viz. 10th column.

5   Impact on Secret Sharing

The connection between perfect hash family and secret sharing is well-established. In this section, we discuss the possibility of achieving non-abelian secret sharing, or more generally, secret sharing schemes over quasi-groups for an evolving access structure. The construction of Komargodski-Naor-Yogev [24] uses combinatorics and applies it to the existing literature on secret sharing over finite fields. However, our construction of perpetual PHF enables us to create

	1	2	3
f_1	0	1	0
f_2	0	0	1
f_3			0
f_4			0

	1	2	3	4	5	6	7	8	9	10
f_1	0	1	0	1	0	1	0	1	0	1
f_2	0	0	1	1	1	1	1	1	1	1
f_3			0	0	0	0	1	1	1	1
f_4			0	0	1	1	0	0	1	1

Fig. 6. Evolving from $PHF(2; 3, 2, 2)$ to $PHF(4; 10, 2, 2)$ via partial rows (from Example 4).

multiplicative sharing schemes for $(2, \infty)$-threshold access structure, in which the secret belongs to a group/quasi-group, which is not necessarily abelian.

Suppose the secret s is an element of a quasigroup $(Q, \cdot)$. Algorithm 2 gives a (non-abelian) secret sharing scheme over the quasigroup Q.

Algorithm 2. Construction of $(2, \infty)$-secret sharing over quasigroup

---

1: **procedure** SHARE GENERATION

2: For every row i of perpetually evolving PHF (described in Algorithm 1):
the dealer assigns a random element $r_i \in Q$ and computes $s \cdot r_i$.

3: Participant j receives r_i or $s \cdot r_i$ or " $-$ " (denoting an empty string),
if the $(i, j)^{th}$ position of the evolving PHF is 0 or 1 or blank respectively.

4: **procedure** RECONSTRUCTION

5: If two participants p_1 and p_2 come together then by construction of the
evolving PHF, there is a row k such that $(k, p_1)^{th}$ and $(k, p_2)^{th}$ entries are
different.

6: They solve the equation $x \cdot r_k = s \cdot r_k$ to retrieve the secret.

---

6   Conclusion

We introduced evolving combinatorial objects, and exemplified this concept with evolving perfect hash family. This construction, in turn, gives rise to an evolving secret sharing scheme, which does not rely on the operations over finite fields, when compared to the scheme by Komargodski, Naor and Yogev [25].

We emphasize that our result paves way for studying evolving combinatorial objects. A natural blueprint is to start from a recursive construction for some combinatorial objects and to develop it into an evolving scheme. Note however that depending on application, different parameters of the object need to be extended. For example, when constructing secret sharing schemes from PHF the domain size matters, but for some applications it could be a co-domain, etc. Another issues to consider is a fine-grained control over parameters. For example, at every step of evolution, our construction doubles the domain size. It is possible that one would want to increase the parameter by some factor, possibly not known in advance.

A natural open question for our future work is to develop evolving PHF for larger co-domains and study the applicability in the realm of secret sharing.

Appendix

$(2, 2)$ Secret Sharing Scheme

- Dealer has a secret bit $s \in \{0, 1\}$;
- Dealer generates a random bit b and computes $s \oplus b$;
- Share of Participant 1 is b and;
- Share of Participant 2 is $s \oplus b$.

It is easy to check that each participant individually has no information about the secret bit s. This is because the share of Participant 1 is independent of s, and Participant 2 holds a one-time pad encryption of it. However, if two participants pool their shares, collaborate then they easily reconstruct the secret bit simply by computing their XOR.

Depiction of $(2, \infty)$ Secret Sharing Scheme

In Fig. 7, we present the share generation process for a $(2, \infty)$ secret sharing scheme.

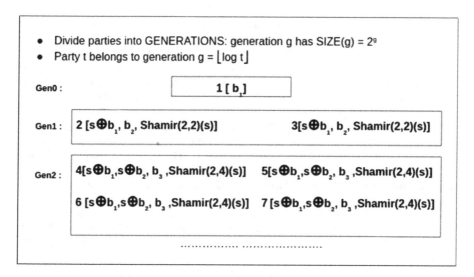

Fig. 7. Shares of parties in an $(2, \infty)$ secret sharing scheme [24]. Entries in the square brackets represent the shares. b_i's are random bits and s the secret bit.

References

1. Atici, M., Magliveras, S.S., Stinson, D.R., Wei, W.-D.: Some recursive constructions for perfect hash families. J. Comb. Des. **4**(5), 353–363 (1996)
2. Barrington, D.A.: Bounded-width polynomial-size branching programs recognize exactly those languages in NC^1. In: STOC 1986, pp. 1–5 (1986)
3. Benaloh, J.C.: Secret sharing homomorphisms: keeping shares of a secret secret (extended abstract). In: Odlyzko, A.M. (ed.) CRYPTO 1986. LNCS, vol. 263, pp. 251–260. Springer, Heidelberg (1987). https://doi.org/10.1007/3-540-47721-7_19
4. Blackburn, S.R., Burmester, M., Desmedt, Y., Wild, P.R.: Efficient multiplicative sharing schemes. In: Maurer, U. (ed.) EUROCRYPT 1996. LNCS, vol. 1070, pp. 107–118. Springer, Heidelberg (1996). https://doi.org/10.1007/3-540-68339-9_10
5. Blakley, G.R.: Safeguarding cryptographic keys. In: AFIPS 1979, pp. 313–317 (1979)
6. Brickell, E., Di Crescenzo, G., Frankel, Y.: Sharing block ciphers. In: Dawson, E.P., Clark, A., Boyd, C. (eds.) ACISP 2000. LNCS, vol. 1841, pp. 457–470. Springer, Heidelberg (2000). https://doi.org/10.1007/10718964_37
7. Cohen, G., et al.: Efficient multiparty protocols via log-depth threshold formulae. In: Canetti, R., Garay, J.A. (eds.) CRYPTO 2013. LNCS, vol. 8043, pp. 185–202. Springer, Heidelberg (2013). https://doi.org/10.1007/978-3-642-40084-1_11
8. Cramer, R., Fehr, S.: Optimal black-box secret sharing over arbitrary Abelian groups. In: Yung, M. (ed.) CRYPTO 2002. LNCS, vol. 2442, pp. 272–287. Springer, Heidelberg (2002). https://doi.org/10.1007/3-540-45708-9_18
9. Cramer, R., Fehr, S., Stam, M.: Black-box secret sharing from primitive sets in algebraic number fields. In: Shoup, V. (ed.) CRYPTO 2005. LNCS, vol. 3621, pp. 344–360. Springer, Heidelberg (2005). https://doi.org/10.1007/11535218_21
10. De Santis, A., Desmedt, Y., Frankel, Y., Yung, M.: How to share a function securely. In: STOC 1994, pp. 522–533 (1994)

11. Desmedt, Y., Erotokritou, S.: Making code voting secure against insider threats using unconditionally secure MIX schemes and human PSMT protocols. In: Haenni, R., Koenig, R.E., Wikström, D. (eds.) VOTELID 2015. LNCS, vol. 9269, pp. 110–126. Springer, Cham (2015). https://doi.org/10.1007/978-3-319-22270-7_7

12. Desmedt, Y., Frankel, Y.: Shared generation of authenticators and signatures. In: Feigenbaum, J. (ed.) CRYPTO 1991. LNCS, vol. 576, pp. 457–469. Springer, Heidelberg (1992). https://doi.org/10.1007/3-540-46766-1_37

13. Desmedt, Y., Di Crescenzo, G., Burmester, M.: Multiplicative non-abelian sharing schemes and their application to threshold cryptography. In: Pieprzyk, J., Safavi-Naini, R. (eds.) ASIACRYPT 1994. LNCS, vol. 917, pp. 19–32. Springer, Heidelberg (1995). https://doi.org/10.1007/BFb0000421

14. Desmedt, Y., et al.: Graph coloring applied to secure computation in non-abelian groups. J. Cryptol. **25**(4), 557–600 (2012)

15. Desmedt, Y., Pieprzyk, J., Steinfeld, R., Wang, H.: On secure multi-party computation in black-box groups. In: Menezes, A. (ed.) CRYPTO 2007. LNCS, vol. 4622, pp. 591–612. Springer, Heidelberg (2007). https://doi.org/10.1007/978-3-540-74143-5_33

16. Desmedt, Y.G., Frankel, Y.: Homomorphic zero-knowledge threshold schemes over any finite Abelian group. SIAM J. Discrete Math. **7**(4), 667–679 (1994)

17. Frankel, Y., Desmedt, Y., Burmester, M.: Non-existence of homomorphic general sharing schemes for some key spaces. In: Brickell, E.F. (ed.) CRYPTO 1992. LNCS, vol. 740, pp. 549–557. Springer, Heidelberg (1993). https://doi.org/10.1007/3-540-48071-4_39

18. Goldreich, O., Micali, S., Wigderson, A.: How to play any mental game. In: STOC 1987, pp. 218–229 (1987)

19. Goldwasser, S., Micali, S., Rackoff, C.: The knowledge complexity of interactive proof systems. SIAM J. Comput. **18**(1), 186–208 (1989)

20. Fredman, M.L., Komlós, J.: On the size of separating systems and families of perfect hash functions. SIAM J. Algebraic Discrete Methods **5**(1), 61–68 (1984)

21. Ito, M., Saito, A., Nishizeki, T.: Secret sharing scheme realizing general access structure. Electron. Commun. Jpn. Part III **72**, 56–64 (1989)

22. Ito, M., Saito, A., Nishizeki, T.: Multiple assignment scheme for sharing secret. J. Cryptol. **6**(1), 15–20 (1993)

23. Krawczyk, H.: Secret sharing made short. In: Stinson, D.R. (ed.) CRYPTO 1993. LNCS, vol. 773, pp. 136–146. Springer, Heidelberg (1994). https://doi.org/10.1007/3-540-48329-2_12

24. Komargodski, I., Naor, M., Yogev, E.: How to share a secret, infinitely. In: Hirt, M., Smith, A. (eds.) TCC 2016. LNCS, vol. 9986, pp. 485–514. Springer, Heidelberg (2016). https://doi.org/10.1007/978-3-662-53644-5_19

25. Komargodski, I., Paskin-Cherniavsky, A.: Evolving secret sharing: dynamic thresholds and robustness. In: Kalai, Y., Reyzin, L. (eds.) TCC 2017. LNCS, vol. 10678, pp. 379–393. Springer, Cham (2017). https://doi.org/10.1007/978-3-319-70503-3_12

26. Liu, C.L.: Introduction to Combinatorial Mathematics. McGraw-Hill, New York (1968)

27. Mehlhorn, K.: On the program size of perfect and universal hash functions. In: FOCS 1982, pp. 170–175 (1982)

28. Mehlhorn, K.: Data Structures and Algorithms, vol. 1. Springer, Heidelberg (1984). https://doi.org/10.1007/978-3-642-69672-5

29. Safavi-Naini, R., Wang, H.: Robust additive secret sharing schemes over Z_m. In: Cryptography and Computational Number Theory. Progress in Computer Science and Applied Logic, vol. 20, pp. 357–368. Birkhauser (2001)
30. Safavi-Naini, R., Wang, H., Lam, K.-Y.: A new approach to robust threshold RSA signature schemes. In: Song, J.S. (ed.) ICISC 1999. LNCS, vol. 1787, pp. 184–196. Springer, Heidelberg (2000). https://doi.org/10.1007/10719994_15
31. Shamir, A.: How to share a secret. Commun. ACM **22**(11), 612–613 (1979)
32. Sloane, N.J.A.: Proposal for an Internet Service: The Eternal Home Page. http://neilsloane.com/doc/eternal.html
33. Stinson, D.: Private communication with Yvo Desmedt, 13 June 1996

Threshold Changeable Ramp Secret Sharing

Fuchun Lin$^{(\boxtimes)}$, San Ling, Huaxiong Wang, and Neng Zeng

School of Physical and Mathematical Sciences, Nanyang Technological University,
Jurong West, Singapore
{fclin,lingsan,HXWang}@ntu.edu.sg
ZENG0106@e.ntu.edu.sg

Abstract. Threshold changeable secret sharing studies the problem of changing the thresholds of a secret sharing scheme after the shares of the initial scheme have been distributed to players. We focus on the most studied scenario of dealer-free threshold increase in the absence of secure channels with an outsider adversary. Previous theoretical works in this scenario only consider an unchanged privacy threshold and define optimal threshold changeable secret sharing schemes as ones meeting the bounds in this case. We highlight increasing the privacy threshold as an independent design goal on top of increasing the reconstruction threshold. We prove new bounds for the above threshold increase scenario with respect to a new privacy threshold that is possibly bigger than the initial privacy threshold. We similarly define an optimal threshold changeable secret sharing scheme as one that achieves equality in all these bounds. A trade-off between the new privacy threshold and the required combiner communication complexity is discovered and new optimal schemes for the case when privacy threshold also increases are identified. These theoretical results put our new construction of threshold changeable secret sharing on a firm ground. Our threshold changeable ramp scheme does not need a priori knowledge of the targeted thresholds to design the protocol and allow the conversion into a ramp scheme with arbitrary new reconstruction thresholds while the privacy threshold grows proportionally as the reconstruction threshold grows. Previous such schemes were only known from lattice-based constructions that use a non-standard privacy definition. Our new schemes are statistical secret sharing schemes that guarantee indistinguishability of shares up to the new privacy threshold.

Keywords: Threshold changeable secret sharing · Communication efficient secret sharing

1 Introduction

Secret sharing, introduced independently by Blakley [1] and Shamir [2], is one of the most fundamental cryptographic primitives. The general goal in secret sharing is to encode a secret $\mathbf{s}$ into a number of *shares* $s_1, \ldots, s_n$ that are distributed

© Springer Nature Switzerland AG 2019
Y. Mu et al. (Eds.): CANS 2019, LNCS 11829, pp. 308–327, 2019.
https://doi.org/10.1007/978-3-030-31578-8_17

among n participants such that only certain *authorized subsets* of the players can reconstruct the secret. An *authorized* subset of players is a set $\mathcal{A} \subset [n]$ such that the shares with indices in $\mathcal{A}$ can collectively be used to reconstruct the secret **s**. On the other hand, $\mathcal{A}$ is an *unauthorized* subset if the knowledge of the shares with indices in $\mathcal{A}$ reveals no information about the secret. A secret sharing is called *perfect* if all the subsets are either authorized or unauthorized. The set of authorized and unauthorized sets define an access structure, of which the most widely used is the so-called *threshold* structure. A threshold secret sharing scheme is defined with respect to an integer parameter r called reconstruction threshold and satisfies the following property: Any set $\mathcal{A} \subset [n]$ with $|\mathcal{A}| < r$ is an unauthorized set and any set $\mathcal{A} \subset [n]$ with $|\mathcal{A}| \geq r$ is an authorized set. When the share lengths are below the secret length, the threshold guarantee that requires all subsets of participants be either authorized, or unauthorized can no longer be attained. Instead, the notion can be relaxed to *ramp* secret sharing which allows some subset of participants to learn some partial information about the secret. A ramp scheme is defined with respect to two threshold parameters, t and r. In a (t, r, n)-ramp scheme, the knowledge of any t shares or fewer does not reveal any information about the secret. On the other hand, any r shares can be used to reconstruct the secret. The subsets of size at least $t + 1$ and at most $r - 1$, may reveal some partial information about the secret. We consider a threshold secret sharing scheme with reconstruction threshold r as a special case of a (t, r, n)-ramp scheme with $t = r - 1$. We denote it an (r, n)-threshold scheme.

A basic application of (r, n)-threshold scheme is for achieving *robustness* of distributed security systems. A distributed system is robust if the system security is guaranteed even against an adversary who eavesdrops up to a certain number of components of the distributed system. In such applications, an (r, n)-threshold scheme can be directly applied. The threshold r of the scheme is determined by a security policy based on an assessment which is a compromise between the value of the protected system and adversary capabilities on one hand, and user convenience and cost on the other hand. As often seen in many applications, the value of the system and adversary capabilities may change over time. This has motivated the study of *threshold changeable secret sharing*.

Changing threshold would be a trivial problem if the dealer is still active and the secret channels for distributing the shares to participants are maintained. Because one can simply start a new run of the protocol after instructing all participants to discard their previous shares. Special threshold change without the dealer assistance has been studied as a *secret redistribution* problem [3,4], where communication among the shareholders through secure channels are allowed. Techniques for changing threshold in [5,6] do not require secure channels but rely on intensive use of broadcast channels.

In this work, we focus on the scenario of dealer-free threshold increase in the absence of secure channels. This is probably the most studied scenario of threshold changeable secret sharing, due to the fact that the setting asks for minimum resources and, with careful designing, provides rather meaningful benefits. A theoretical model was discussed and three efficiency bounds were proved in [4]. The model considers an (r, n)-threshold scheme that, through each shareholder locally applying a publicly known share conversion function to the individual share he/she is holding, can be converted into a new ramp scheme with bigger reconstruction threshold $r' > r$. Of course, to make the idea work, the shareholders are assumed honest and delete the initial shares once the new shares are generated. As it turned out, it is impossible to have the threshold gap remaining 1 while increasing the thresholds, assuming both the initial and target schemes have optimal share size (see [7, Lemma 1]). Remaining unchanged. Somehow counter-intuitively, as the reconstruction threshold increases, the combiner communication complexity for pooling the new shares to reconstruct the secret can be significantly reduced. A threshold scheme with minimum share size that can be converted into such a ramp scheme with minimum share size and minimum combiner communication complexity is defined to be optimal. An explicit construction of such optimal (r, n)-*threshold scheme threshold changeable to* r' was also given in [4] using the geometric construction of Blakley [1]. There are follow up works on threshold changeability that gave new constructions of optimal (or near optimal) threshold changeable secret sharing in this model with various features [8–11]. For example, the packed Shamir scheme construction in [9] is essentially an optimal construction for threshold changeable secret sharing and [10] proposed a variant of the construction that reduces the share size.

Another line of works in this scenario of dealer-free threshold increase in the absence of secret channels, however, concern with the practical applicability rather than achieving theoretical optimality. In the theoretical model, there is an initial reconstruction threshold r and a target new reconstruction threshold r' known a priori to the designer of the protocol, who then tune the construction to achieve the best efficiency. But in practice, one may ask why not simply use a threshold scheme with the bigger reconstruction threshold r' from the beginning, if we already know what will be the "right" threshold value a priori. Note that directly using a standard (r', n)-threshold scheme enjoys the benefit of having threshold gap 1, which no (r, n)-threshold scheme threshold changeable to r' can achieve [7, Lemma 1]. Practical threshold changeability should at least remove the need for a priori knowledge of the new threshold r' and allows the protocol designer to start with a standard (r, n)-threshold scheme without consideration of future threshold increase. An important breakthrough in this line was the lattice-based construction of threshold changeable schemes in [12] (standard Shamir scheme) and its companion paper in [13] (standard Chinese Remainder Theorem scheme). The share conversion functions are realised through deliberately adding structured errors into the initial shares of the standard threshold scheme. The combiner algorithm then adopts a positive application of lattice reduction algorithm to correct the errors. The authors discovered that these constructions not only enjoys the salient feature of removing the need for a priori knowledge of the new reconstruction threshold r', but also allows the privacy threshold to increase proportionally to the increase of reconstruction threshold

r'. Their definition of privacy, however, is not a standard security definition, which might be inherent limitation of the lattice techniques they use.

Our Contributions. Our main contributions in this work are two-fold. On the theory side, we generalise the classical model of (r, n)-*threshold scheme threshold changeable to* r' by allowing the privacy threshold to grow as the reconstruction threshold grows. A direct motivation for this generalisation is of course the break-through constructions of practical threshold changeable secret sharing [12,13], where the new privacy threshold becomes bigger than the initial privacy threshold. We highlight increasing the privacy threshold as an independent design goal that can be investigated on top of the conventional study of increasing the reconstruction threshold. On the construction side, we construct a new threshold changeable secret sharing that has better performance than the lattice-based schemes [12,13] in all aspects.

Modeling. We generalise the (r, n)-*threshold scheme threshold changeable to* r' model to the (t, r, n)-*ramp scheme threshold changeable to* (t', r', n)-*ramp scheme* model. To have succinct notations, we denote it by $(t, r, n) \rightarrow (t', r', n)$-*ramp scheme* model. We generalise the three information-theoretical bounds for measuring the efficiency of threshold changeable secret sharing to cover this general setting. Let $\mathsf{H}(\mathbf{S})$ denote the secret length measured in bits. The initial share spaces $\mathcal{S}_i$, $i \in [n]$ and new share spaces $\mathcal{S}'_i$, $i \in [n]$, are bounded as follows.

1. Bound on initial share size: $\max_{1 \leq i \leq n} \log |\mathcal{S}_i| \geq \frac{\mathsf{H}(\mathbf{S})}{r-t}$;
2. Bound on new share size: $\max_{1 \leq i \leq n} \log |\mathcal{S}'_i| \geq \frac{\mathsf{H}(\mathbf{S})}{r'-t'}$;
3. Bound on combiner communication complexity: for any $I \subset [n]$ with $|I| = r'$,

$$\sum_{i \in I} \log |\mathcal{S}'_i| \geq \frac{r'\mathsf{H}(\mathbf{S})}{r' - t'}.$$

In particular, given a new reconstruction threshold r', the combiner communication complexity bound reveals a trade-off between the new privacy threshold t' and the necessary combiner communication complexity that is required for having privacy threshold t'. The necessary combiner communication complexity decreases as the required privacy threshold decreases, and with the minimum value at the well studied case when $t' = t$. We identify optimal constructions for the case when $r' : r = t' : t$, whose optimality was not known before. Interestingly, the combiner communication complexity in this case remains unchanged during the threshold changes. We leave the quest for optimal constructions with other threshold parameter relations as an interesting open question.

Construction. We make black-box usage of the recent construction of binary secret sharing scheme in [14] and construct a practical threshold changeable secret sharing that does not require a priori knowledge of the future threshold change. We simply use such a binary secret sharing scheme with appropriate choice of parameters and divide the bit-string into n blocks to form n shares. This gives us the initial (t, r, n)-threshold scheme. The share conversion functions are simply deleting an appropriate portion of the initial share. The scheme obtained supports arbitrary reconstruction threshold change and the privacy threshold grows proportionally $t' = \lfloor \frac{tr'}{r} \rfloor$ as the reconstruction threshold grows. When the new thresholds t' and r' satisfy $r' : r = t' : t$, the secret length of our scheme almost matches the bounds in the theoretical model. Moreover, thanks to the statistical secret sharing we use in this construction, the privacy of the obtained scheme is measured by statistical distance between two distributions of any subset of at most t' converted shares corresponding to a pair of secrets, which has significant advantage over the lattice based constructions in [12,13].

Related Works. Wang and Wong [9] initiated the study of minimising the combiner communication complexity through allowing a set of players of size beyond the reconstruction threshold to take part in the reconstruction. Each player apply a *processing function* to his/her own share and communicate the output of the function instead of the complete share. A trade-off between combiner communication complexity and the number of participants involved in the secret reconstruction was shown and the scheme achieving the bound is called optimal communication efficient secret sharing scheme. Note that an optimal communication efficient secret sharing scheme that requires $d > r$ shares in reconstruction naturally yields an optimal threshold changeable secret sharing scheme, where the reconstruction threshold change from r to $r' = d$ and the privacy threshold remains the same ($t' = t$). More concretely, the processing functions of the former serve as the share conversion functions of the latter. An optimal construction of communication efficient secret sharing scheme for a given d was proposed in [9]. In the follow-up works [15,16], constructions that are universally optimal for all possible values of d, $r \le d \le n$ were constructed using packed Shamir secret sharing technique. Since the packed Shamir secret sharing inherently requires large share size, an alternative construction using algebraic geometry codes was proposed in [17] to cope with the share size issue. But since algebraic geometry codes with small alphabet and large length do not give optimal secret sharing schemes, the schemes obtained are only near optimal.

2   Preliminary

The log is to the base 2. Let $\mathbf{X}$ denote a random variable. The Shannon entropy of $\mathbf{X}$ is denoted by $\mathsf{H}(\mathbf{X})$. The mutual information between $\mathbf{X}$ and $\mathbf{Y}$ is given by

$$\mathsf{I}(\mathbf{X}; \mathbf{Y}) = \mathsf{H}(\mathbf{X}) - \mathsf{H}(\mathbf{X}|\mathbf{Y}) = \mathsf{H}(\mathbf{Y}) - \mathsf{H}(\mathbf{Y}|\mathbf{X}),$$

where $\mathsf{H}(\mathbf{Y}|\mathbf{X})$ denotes the conditional Shannon entropy. The *statistical distance* of two random variables (their corresponding distributions) is defined as follows. For $\mathbf{X}, \mathbf{Y} \leftarrow \Omega$,

$$\mathsf{SD}(\mathbf{X}; \mathbf{Y}) = \frac{1}{2} \sum_{\omega \in \Omega} |\Pr(\mathsf{X} = \omega) - \Pr(\mathsf{Y} = \omega)|.$$

Definition 1. *Let $\mathcal{P} = \{P_1, \ldots, P_n\}$ be a group of n participants (a.k.a. share-holders). Let $\mathcal{S}$ be the set of secrets and a secret $\mathbf{s} \in \mathcal{S}$ is denoted in boldface. Let $\mathcal{S}_i$ be the share space of the participant P_i. A secret sharing scheme of n participants is a pair of algorithms: the dealer and the combiner. For a given secret from $\mathcal{S}$ and some random string from $\mathcal{R}$, the dealer algorithm applies the mapping*

$$\mathsf{D} \colon \mathcal{S} \times \mathcal{R} \to \mathcal{S}_1 \times \ldots \times \mathcal{S}_n$$

to assign shares to participants from $\mathcal{P}$. The shares of a subset $\mathcal{A} \subset [n]$ of participants can be input into the combiner algorithm

$$\mathsf{C} \colon \prod_{P_i \in \mathcal{A}} \mathcal{S}_i \to \mathcal{S}$$

to reconstruct the secret.

Definition 2. *A (t, r, n)-ramp scheme is a secret sharing scheme with n participants such that the combiner algorithm always reconstructs the correct secret for any $\mathcal{A} \subset [n]$ of size $|\mathcal{A}| \geq r$ and for any $\mathcal{A} \subset [n]$ of size $|\mathcal{A}| \leq t$, no information about the secret can be learned from pooling their shares together. Moreover if $t < r - 1$, for any $\mathcal{A} \subset [n]$ of size $t < |\mathcal{A}| < r$, neither the combiner algorithm can uniquely reconstruct a secret nor the secret can remain unknown. That is r is the smallest integer such that correct reconstruction is guaranteed and t is the biggest integer such that privacy is guaranteed. We associate a probability with each $\mathbf{s} \in \mathcal{S}$ and obtain a random secret $\mathbf{S} \leftarrow \mathcal{S}$. The share vector obtained from sharing the random secret $\mathbf{S}$ is denoted by*

$$\mathbf{V} \leftarrow \mathcal{S}_1 \times \ldots \times \mathcal{S}_n.$$

Let $\mathsf{H}(\mathbf{S}|\mathbf{V}_{\mathcal{A}})$ denote the entropy of the random variable $\mathbf{S}$ conditioned on the knowledge of the shares held by the participants in $\mathcal{A}$. The above conditions defining a (t, r, n)-ramp scheme can be described information-theoretically as follows.

- *Correctness: $\mathsf{H}(\mathbf{S}|\mathbf{V}_{\mathcal{A}}) = 0$, for any $\mathcal{A} \subset [n]$ of size $|\mathcal{A}| \geq r$;*
- *Privacy: $\mathsf{H}(\mathbf{S}|\mathbf{V}_{\mathcal{A}}) = \mathsf{H}(\mathbf{S})$, for any $\mathcal{A} \subset [n]$ of size $|\mathcal{A}| \leq t$;*
- *Ramp security: $0 < \mathsf{H}(\mathbf{S}|\mathbf{V}_{\mathcal{A}}) < \mathsf{H}(\mathbf{S})$, for any $\mathcal{A} \subset [n]$ of size $t < |\mathcal{A}| < r$.*

The parameter t is called the *privacy threshold* and the parameter r is called the *reconstruction threshold*. The difference of these two thresholds is called the gap and denoted by $g = r - t$. The last item *ramp security* in Definition 2 is void when $t = r - 1$, as there is no integer lying in between t and r. In this case when the gap $g = 1$, we call it a (r, n)-*threshold scheme* for short. A (t, r, n)-ramp scheme is called an *optimal (t, r, n)-ramp scheme* if it has minimum share length and the ramp security is strengthened to $\mathsf{H}(\mathbf{S}|\mathbf{V}_{\mathcal{A}}) = \frac{r-k}{r-t}$, for any $\mathcal{A} \subset [n]$ of size $|\mathcal{A}| = k$ with $t < k < r$.

Definition 3. *A (ε, δ)-statistical (t, r, n)-ramp scheme is a secret sharing scheme with n participants such that the combiner algorithm C correctly reconstructs the secret for any $\mathcal{A} \subset [n]$ of size $|\mathcal{A}| \geq r$ with probability at least $1 - \delta$*

over the randomness of the dealer algorithm D and for any $\mathcal{A} \subset [n]$ of size $|\mathcal{A}| \leq t$, the leakage of information about the secret measured by statistical distance between two distributions of the shares specified by $\mathcal{A}$ corresponding to a pair of secrets is at most ε. For a particular secret $\mathbf{s} \in \mathcal{S}$, let

$$D(\mathbf{s}) \leftarrow \mathcal{S}_1 \times \ldots \times \mathcal{S}_n$$

denote the share vector corresponding to $\mathbf{s}$, which is a random variable distributed on the set $\mathcal{S}_1 \times \ldots \times \mathcal{S}_n$ with randomness from the dealer algorithm D. The above conditions defining a (ε, δ)-statistical (t, r, n)-ramp scheme can be described as follows.

- Correctness: for any $\mathcal{A} \subset [n]$ of size $|\mathcal{A}| \geq r$,

$$\Pr[C(D(\mathbf{s})_{\mathcal{A}}) = \mathbf{s}] \geq 1 - \delta.$$

- Privacy (non-adaptive): for any pair of secrets $\mathbf{s}_0, \mathbf{s}_1 \in \mathcal{S}$ and any $\mathcal{A} \subset [n]$ of size $|\mathcal{A}| \leq t$,

$$SD(D(\mathbf{s}_0)_{\mathcal{A}}; D(\mathbf{s}_1)_{\mathcal{A}}) \leq \varepsilon.$$

Note that when $\varepsilon > 0$, the above defined privacy against a non-adaptive reading adversary is not equivalent to privacy against an adaptive reading adversary (see [14] for an example showing the separation). An adaptive reading adversary first chooses a subset of shares to read and is allowed to choose other (up to the reading budget) shares to read using the knowledge gained in the first reading. In the extreme case, the adaptive adversary choose the set $\mathcal{A}$ one share by one share and base the choice on all previously read values. We only consider non-adaptive reading adversary for statistical secret sharing in this paper, for which we know the following.

Lemma 1 ([14]). *For any $0 \leq \tau < \rho \leq 1$, there is an explicit construction of a family of binary $(\varepsilon(N), \delta(N))$-statistical $(\tau N, \rho N, N)$-ramp schemes against non-adaptive adversary, labeled by the share vector length N, such that $(\varepsilon(N)$ and $\delta(N)$ are both negligible in N and the secret is asymptotically equal to $N(\rho - \tau)$ bits.*

Lemma 1 shows that by relaxing from the error-free definition of secret sharing to allowing some negligible amount of errors $\varepsilon(N)$ and $\delta(N)$, the stringent share size requirement can be overcome and the share size can be reduced to the smallest possible 2. An asymptotic bound on the secret length of such binary statistical secret sharing was shown in [14] and the explicit construction above achieves that bound. By an explicit construction, we mean that both the dealer algorithm and the combiner algorithm run in time polynomial in N.

3   Threshold Changeable Ramp Schemes

We revisit the problem of increasing the thresholds of a secret sharing scheme after the shares are distributed to the shareholders, without further secret communication between the dealer and the shareholders or among shareholders themselves. There might be instructions on how the conversion into a new scheme should be accomplished publicly broadcasted to the shareholders. Each shareholder must generate the new share from the initial share he/she receives, following the instructions. In this work, we consider an outsider adversary. That is we assume that all the shareholders are honest and delete their initial shares after generating the new shares from them. The adversary, who is an outsider, corrupts the shareholders only after the initial shares are successfully deleted. We discuss the necessity of considering an outsider adversary in more details after we formally define our model.

Previous theoretical studies of threshold changeable secret sharing consider a fixed privacy threshold and only increase the reconstruction threshold. That is from a (r, n)-threshold scheme to a $(r - 1, r', n)$-ramp scheme, for $r' > r$. In this work, we do not restrict the target privacy threshold to be equal to the initial privacy threshold. That is from a (r, n)-threshold scheme to a (t', r', n)-ramp scheme, where the target privacy threshold t' can be an integer bigger than the initial privacy threshold $r - 1$. We observe that allowing the privacy threshold to increase as the reconstruction threshold increases is an arguably more natural model. In fact, increasing the privacy threshold can be studied as an independent design goal, on top of increasing the reconstruction threshold, in threshold changeable secret sharing literature. From explicit construction point of view, the lattice-based constructions in [12,13] yield threshold changeable secret sharing schemes whose privacy threshold t' increases proportionally as the reconstruction threshold r' grows. We believe that a thorough study of the model with a changeable privacy threshold is well motivated and will stimulate new constructions of threshold changeable secret sharing.

In the sequal, the results are described in a general language of changing threshold parameters from a (t, r, n)-ramp scheme to a (t', r', n)-ramp scheme, where t is not necessarily equal to $r - 1$. Results stated in this general form enjoys a symmetric look that is not obvious when substituting in $t = r - 1$. Moreover, including the $t < r - 1$ cases into discussion is of independent interest, for example, to a *continuous threshold changeable secret sharing* model, where threshold changes may be activated multiple times before the shares are finally pooled together to reconstruct the secret. According to the impossibility result ([7, Lemma 1]) mentioned above, the obtained scheme after the first change is necessarily a ramp scheme with threshold gap strictly bigger than 1. All the subsequent threshold changes are then between two general ramp schemes. Investigating threshold changeability in the full generality of ramp schemes is then a necessary first step for modelling continuous threshold changeable secret sharing. We call such general schemes (t, r, n)-*ramp schemes threshold changeable to* (t', r', n). To make the notation short, we simply denote $(t, r, n) \rightarrow (t', r', n)$ *ramp schemes.*

3.1   Defining $(t, r, n) \rightarrow (t', r', n)$ Ramp Schemes

Definition 4. *A (t, r, n)-ramp scheme threshold changeable to (t', r', n), or a $(t, r, n) \rightarrow (t', r', n)$ ramp scheme for short, is a (t, r, n)-ramp scheme together* with a set of publicly known share conversion functions

$$h_i \colon \mathcal{S}_i \to \mathcal{S}'_i, i = 1, \ldots, n,$$

and a new combiner algorithm

$$\mathsf{C}' \colon \prod_{P_i \in \mathcal{A}} \mathcal{S}'_i \to \mathcal{S}$$

for a subset $\mathcal{A} \subset [n]$ of participants such that the following properties are satisfied. The share conversion function h_i convert the share $s_i \in \mathcal{S}_i$ of the ith participant P_i into a new share $s'_i = h_i(s_i) \in \mathcal{S}'_i$. The new combiner algorithm C' always reconstructs the correct secret for any $\mathcal{A} \subset [n]$ of size $|\mathcal{A}| \geq r'$ and for any $\mathcal{A} \subset [n]$ of size $|\mathcal{A}| \leq t'$, no information about the secret can be learned from pooling their new shares together. Moreover for any $\mathcal{A} \subset [n]$ of size $t' < |\mathcal{A}| < r'$, neither the combiner algorithm can uniquely reconstruct a secret nor the secret can remain unknown. That is r' is the smallest integer such that correct reconstruction is guaranteed and t' is the biggest integer such that privacy is guaranteed. We associate a probability with each $\mathbf{s} \in \mathcal{S}$ and obtain a random secret $\mathbf{S} \leftarrow \mathcal{S}$. The share vector obtained from sharing the random secret $\mathbf{S}$ and then applying the share conversion functions is denoted by

$$\mathbf{V}' \leftarrow \mathcal{S}'_1 \times \ldots \times \mathcal{S}'_n.$$

Let $\mathsf{H}(\mathbf{S}|\mathbf{V}'_\mathcal{A})$ denote the entropy of the random variable $\mathbf{S}$ conditioned on the knowledge of the new shares held by the participants in $\mathcal{A}$. The above conditions can be described information-theoretically as follows.

- Correctness: $\mathsf{H}(\mathbf{S}|\mathbf{V}'_\mathcal{A}) = 0$, for any $\mathcal{A} \subset [n]$ of size $|\mathcal{A}| \geq r'$;
- Privacy: $\mathsf{H}(\mathbf{S}|\mathbf{V}'_\mathcal{A}) = \mathsf{H}(\mathbf{S})$, for any $\mathcal{A} \subset [n]$ of size $|\mathcal{A}| \leq t'$;
- Ramp security: $0 < \mathsf{H}(\mathbf{S}|\mathbf{V}'_\mathcal{A}) < \mathsf{H}(\mathbf{S})$, for any $\mathcal{A} \subset [n]$ of size $t' < |\mathcal{A}| < r'$.

In Definition 4, the set of share conversion functions $\{h_i | i = 1, \ldots, n\}$ together with the dealer algorithm D of the initial (t, r, n)-ramp scheme in fact define a new dealer algorithm D' through composition of functions.

$$\mathsf{D}' \colon \mathcal{S} \times \mathcal{R} \to \mathcal{S}'_1 \times \ldots \times \mathcal{S}'_n. \tag{1}$$

This new dealer algorithm D' together with the new combiner algorithm C' define a new secret sharing scheme. In this interpretation, a $(t, r, n) \to (t', r', n)$ ramp scheme is a (t, r, n)-ramp scheme equipped with a set of share conversion functions that can transform it into a (t', r', n)-ramp scheme.

The threshold changeable secret sharing as defined in Definition 4 is in the *outsider model*, since the properties of the correctness, privacy and ramp security for the target thresholds t' and r' are defined with respect to the new shares. This makes sense only when all shareholders are honest and follow the protocol deleting the initial shares. If on the contrary, the adversary is among the shareholders and may not follow the protocol, he/she will store the initial share s_i

instead of the new share $s'_i = h_i(s_i)$. One can also consider a *semi-insider* model where the adversary corrupts up to t shareholders before the transformation and $t' - t$ more shareholders after the transformation. The $t' - t$ more shareholders corrupted after the transformation delete the initial shares completely and only store the new shares. The privacy and ramp security conditions in the semi-insider model will be alternatively defined with respect to the conditional entropy $\mathsf{H}(\mathbf{S}|\mathbf{V}_\mathcal{A}, \mathbf{V}'_{\mathcal{B}-\mathcal{A}})$, for any $\mathcal{A} \subset \mathcal{B} \subset [n]$ of size $|\mathcal{A}| \leq t$ and $|\mathcal{B}| \leq t'$. We only briefly discuss the semi-insider model for the special case $t' = t$.

We first state a generalisation of [4, Lemma 1] that intuitively says that by increasing the reconstruction threshold, the gap has to increase as well.

Lemma 2. *Let* Π *be a* (t, r, n) *ramp scheme with privacy threshold changeable to* t' *and reconstruction threshold changeable to* $r' > r$. *Let* Π' *be the resulting* (t', r', n) *ramp scheme. If both* Π *and* Π' *have minimum share size, then* $t' \geq t$ *and* $r' - t' > r - t$.

The proof is given in Appendix A. Lemma 2 shows that the range of achievable new privacy threshold is $t' \in \{t, t+1, \ldots, r' - r + t - 1\}$.

3.2   Bounds and Optimal Schemes

It is then highly interesting to find out what are the theoretical bounds for the full range $t' \in \{t, t + 1, \ldots, r' - r + t - 1\}$ and whether there are optimal constructions. We begin with the information theoretic bounds, which generalise [4, Theorem 2], where only $t' = t$ was treated and was stated for $t = r - 1$ only.

Theorem 1. *Let* Π *be a* $(t, r, n) \rightarrow (t', r', n)$ *ramp scheme, where* $r < r' \leq n$ *and* $t \leq t' < r' - (r - t)$. *Let* $h_i : \mathcal{S}_i \rightarrow \mathcal{S}'_i, i = 1, \ldots, n$, *be its share conversion functions. Then the following three bounds hold.*

1. *Bound on initial share size:* $\max\limits_{1 \leq i \leq n} \log |\mathcal{S}_i| \geq \frac{\mathsf{H}(\mathbf{S})}{r-t}$;

2. *Bound on new share size:* $\max\limits_{1 \leq i \leq n} \log |\mathcal{S}'_i| \geq \frac{\mathsf{H}(\mathbf{S})}{r'-t'}$;

3. *Bound on combiner communication complexity: for any* $I \subset [n]$ *of* $|I| = r'$,

$$\sum_{i \in I} \log |\mathcal{S}'_i| \geq \frac{r' \mathsf{H}(\mathbf{S})}{r' - t'}.$$

Proof. Item 1. follows from the fact that Π is by definition a (t, r, n)-ramp scheme. We now prove Item 2. Consider arbitrary $I = \{i_1, \ldots, i_{|I|}\}$ such that $|I| = r'$. Assume without loss of generality that $|\mathcal{S}'_{i_1}| \leq |\mathcal{S}'_{i_2}| \leq \ldots \leq |\mathcal{S}'_{i_{|I|}}|$. Let $\mathbf{S}$ be uniformly distributed, denote $\mathbf{S}'_I = (\mathbf{S}'_{i_1}, \ldots, \mathbf{S}'_{i_{r'}})$ be the output of the share conversion functions.

$$H(\mathbf{S}'_{i_1}, \ldots, \mathbf{S}'_{i_{r'-t'}}) \overset{(a)}{\geq} H(\mathbf{S}'_{i_1}, \ldots, \mathbf{S}'_{i_{r'-t'}} \mid \mathbf{S}'_{i_{r'-t'+1}}, \ldots, \mathbf{S}'_{i_{r'}})$$

$$\overset{(b)}{=} H(\mathbf{S}'_{i_1}, \ldots, \mathbf{S}'_{i_{r'-t'}} \mid \mathbf{S}'_{i_{r'-t'+1}}, \ldots, \mathbf{S}'_{i_{r'}}) + H(\mathbf{S} \mid \mathbf{S}'_{i_1}, \ldots, \mathbf{S}'_{i_{r'}})$$

$$\overset{(c)}{=} H(\mathbf{S}, \mathbf{S}'_{i_1}, \ldots, \mathbf{S}'_{i_{r'-t'}} \mid \mathbf{S}'_{i_{r'-t'+1}}, \ldots, \mathbf{S}'_{i_{r'}})$$

$$\geq H(\mathbf{S} \mid \mathbf{S}'_{i_{r'-t'+1}}, \ldots, \mathbf{S}'_{i_{r'}})$$

$$\overset{(d)}{=} H(\mathbf{S}),$$

where (a) follows from conditioning reduces entropy, (b) follows from the correctness property in Definition 4, (c) follows from the chain rule and (d) follows from the privacy property in Definition 4. Therefore it follows that

$$\sum_{j=1}^{r'-t'} \log |\mathcal{S}'_{i_j}| \geq H(\mathbf{S}). \tag{2}$$

It then follows from $|\mathcal{S}'_{i_1}| \leq |\mathcal{S}'_{i_2}| \leq \ldots \leq |\mathcal{S}'_{i_{|I|}}|$ that

$$\log |\mathcal{S}'_{i_{r'-t'}}| \geq \frac{H(\mathbf{S})}{r' - t'} \tag{3}$$

and that,

$$\log |\mathcal{S}'_{i_{r'-t'+j}}| \geq \log |\mathcal{S}'_{i_{r'-t'}}| \geq \frac{H(\mathbf{S})}{r' - t'}, j = 1, \ldots, t'. \tag{4}$$

According to Eq. (4), we have

$$\log |\mathcal{S}'_{i_{r'}}| \geq \frac{H(\mathbf{S})}{r' - t'}.$$

Item 2 is obtained since

$$\max_{1 \leq i \leq n} \log |\mathcal{S}'_i| \geq \log |\mathcal{S}'_{i_{r'}}| \geq \frac{H(\mathbf{S})}{r' - t'}.$$

Combining Eqs. (2) and (4) we have,

$$\sum_{j=1}^{r'} \log |\mathcal{S}'_{i_j}| \geq H(\mathbf{S}) + \frac{t' H(\mathbf{S})}{r' - t'} \geq \frac{r' H(\mathbf{S})}{r' - t'}.$$

Item 3 is obtained.

As a result of Theorem 1, we observe a trade-off between the target privacy threshold t' and the combiner communication complexity. The lower bound on the combiner communication complexity for a fixed r' is determined by the target privacy threshold t': the necessary communication complexity decreases as t' decreases. We summarise this observation in the following corollary.

Corollary 1. *Given a fixed r'. Among all choices of target privacy threshold $t' \in \{t, t+1, \ldots, r' - r + t - 1\}$, the necessary communication complexity of $(t, r, n) \rightarrow (t', r', n)$ ramp schemes decreases as t' decreases, with the minimum at the case $t' = t$.*

Corollary 1 shows that the single goal of minimising the combiner commu-nication complexity is optimised when consider the special case $t' = t$. This explains the reason that most of the previous works on threshold changeable secret sharing [8,10,11] were focused on this case. These constructions usually restrict to the parameter setting $t = r - 1$. As mentioned in the Related works, the independent line of works [9,15–17] on minimising the communication com-plexity by pooling more shares than the designed reconstruction threshold (no threshold change is required and usually not restricted to the parameter setting $t = r - 1$) naturally give constructions of threshold changeable ramp schemes for the special case $t' = t$. Another reason that makes the $t' = t$ case special is the fact that explicit schemes resistant to semi-insider attacks can be constructed in this case. See an example construction in Appendix B and the discussion follow-ing it.

In this work, we highlight increasing the privacy threshold as a secondary design goal and ask for constructions that simultaneously achieve the threshold changes and minimise combiner communication complexity.

Definition 5. *A $(t, r, n) \to (t', r', n)$ ramp scheme is called optimal if equality is achieved in all the bounds in Theorem 1.*

An intuitive observation is in place. Using the dealer algorithm D' induced by the initial dealer algorithm D and the share conversion functions $\{h_i\}_{i \in [n]}$ as described in (1), we immediately have the following corollary of Theorem 1.

Corollary 2. *A $(t, r, n) \to (t', r', n)$ ramp scheme is optimal if and only if both its initial scheme with (D, C) and its new scheme with $(\mathsf{D}', \mathsf{C}')$ have minimum share size.*

The optimality for the cases when $t' > t$ was not studied before, although the following construction that achieves optimality is folklore (see e.g. [4]). To distinguish this technique from the packing technique that is commonly used in $t' = t$ case, we call the technique used in this case a *folding* technique, as in the *folded codes.*

Let $\frac{u}{v} = \frac{r}{r'} = \frac{t}{t'}$. Let Π_v be the (tv, rv, nv)-ramp scheme constructed from polynomials. Let $(s_1, \ldots, s_{nv})$ be the share vector of Π_v. Now we fold v shares of Π_v to form a new share and obtain a folded scheme Π that is obvi-ously a (t, r, n)-ramp scheme with optimal share size. The share vector of Π is then $(S_1, \ldots, S_n)$, where $S_i = (s_{(i-1)v+1}, \ldots, s_{iv})$. The dealer algorithm D and combiner algorithm C of Π can be trivially adapted from those of Π_v. We next define a share conversion algorithm $\{h_i\}_{i \in [n]}$ to transform the scheme Π into Π' through dropping the $v - u$ components of each share. More pre-cisely, let $h_i(S_i) = (s_{(i-1)v+1}, \ldots, s_{(i-1)v+u})$. We show that Π' with share vector $(S'_1, \ldots, S'_n)$, where $S'_i = h_i(S_i)$, is a (t', r', n)-ramp scheme with optimal share size. The new combiner algorithm C' is similar to C. We note that any k shares of Π' are corresponding to ku shares of Π_v. We then have any k shares of Π' do not contain any information about the secret if and only if $ku \leq tv$, namely, $k \leq \frac{tv}{u} = t'$. Similarly, any k shares of Π' reconstruct the full secret if and only if if $ku \geq rv$, namely, $k \geq \frac{rv}{u} = r'$. We have shown that the scheme Π together with the transformation algorithm $\{h_i\}_{i \in [n]}$ give an optimal $(t, r, n) \to (t', r', n)$ ramp scheme.

We notice that in the above construction, the parameters satisfy $t' : t = r' : r$. The optimal combiner communication complexity of the transformed scheme in the above construction is the same as that of the original scheme. Indeed, substituting $t' : t = r' : r$ into the combiner communication complexity bound in Theorem 1, we have

$$r' \times \frac{\mathsf{H}(\mathbf{S})}{r' - t'} = r \times \frac{\mathsf{H}(\mathbf{S})}{r - t}.$$

We conclude this section with an interesting open question of constructing optimal $(t, r, n) \to (t', r', n)$ ramp schemes with $t' > t$ and $t' : t \neq r' : r$.

4   Arbitrarily Threshold Changeable Statistical Secret Sharing

From now on, we discuss practical aspects of threshold changeable secret sharing. The biggest drawback of the theoretical model in the previous section is that in practice, it is not natural to assume that the new threshold values t' and r' are known before the distribution of the shares of the initial scheme. On the contrary, the need for change of thresholds is the consequence of unforeseeable development of the application environment. A practical threshold changeable secret sharing should at least allow the users to start with a plain threshold scheme (not a specially built $(t, r, n) \to (t', r', n)$ secret sharing for a pair of fixed t' and r') and at a later stage, when the need for changing into arbitrary new thresholds comes up, be able to implement the change.

4.1   Arbitrarily Threshold Changeable Model

Definition 6. *An arbitrarily threshold changeable (ε, δ)-statistical (t, r, n)-ramp scheme with respect to a set $\mathcal{T} \subset [n] \times [n]$ of new thresholds is a (ε, δ)-statistical (t, r, n)-ramp scheme together with publicly known share conversion functions*

$$h_i^{(t', r')} : \mathcal{S}_i \to \mathcal{S}_i^{(t', r')}, i = 1, \ldots, n, \ (t', r') \in \mathcal{T},$$

and combiner algorithms

$$\mathsf{C}^{(t', r')} : \prod_{P_i \in \mathcal{A}} \mathcal{S}_i^{(t', r')} \to \mathcal{S}, \ (t', r') \in \mathcal{T},$$

for a subset $\mathcal{A} \subset [n]$ of participants such that the following properties are satisfied. The share conversion function $h_i^{(t', r')}$ convert the share $s_i \in \mathcal{S}_i$ of the ith participant P_i into a new share $s_i^{(t', r')} = h_i^{(t', r')}(s_i) \in \mathcal{S}_i^{(t', r')}$. The new combiner algorithm $\mathsf{C}^{(t', r')}$ correctly reconstructs the secret for any $\mathcal{A} \subset [n]$ of size $|\mathcal{A}| \geq r'$ with probability at least $1 - \delta$ over the randomness of the dealer algorithm D and *for any $\mathcal{A} \subset [n]$ of size $|\mathcal{A}| \leq t'$, the leakage of information about the secret measured by statistical distance between the distributions of the subset of new shares specified by $\mathcal{A}$ corresponding to a pair of secrets is at most ε.*

More concretely, for a particular secret $\mathbf{s} \in \mathcal{S}$, let

$$\mathsf{D}^{(t',r')}(\mathbf{s}) \leftarrow \mathcal{S}_1^{(t',r')} \times \ldots \times \mathcal{S}_n^{(t',r')}$$

denote the new share vector of $\mathbf{s}$ (obtained by applying the share conversion functions $\{h_i^{(t',r')}\}_{i \in [n]}$), which is a random variable distributed on the set $\mathcal{S}_1^{(t',r')} \times \ldots \times \mathcal{S}_n^{(t',r')}$ with randomness from the dealer algorithm D. The above conditions defining a (ε, δ)-statistical (t, r, n)-ramp scheme can be described as follows.

– *Correctness: for any $\mathcal{A} \subset [n]$ of size $|\mathcal{A}| \geq r'$,*

$$\Pr[\mathsf{C}^{(t',r')}(\mathsf{D}^{(t',r')}(\mathbf{s})_{\mathcal{A}}) = \mathbf{s}] \geq 1 - \delta.$$

– *Privacy (non-adaptive): for any pair of secrets $\mathbf{s}_0, \mathbf{s}_1 \in \mathcal{S}$ and any $\mathcal{A} \subset [n]$ of size $|\mathcal{A}| \leq t'$,*

$$\mathsf{SD}(\mathsf{D}^{(t',r')}(\mathbf{s}_0)_{\mathcal{A}}; \mathsf{D}^{(t',r')}(\mathbf{s}_1)_{\mathcal{A}}) \leq \varepsilon.$$

The folklore construction described in previous section can be extended to cope with the arbitrary threshold change scenario. But as we will see, the share size required for this construction to support arbitrary threshold change is huge. Recall that, to realise a $(t, r, n) \to (t', r', n)$ ramp scheme, we need to decide on an integer v such that $\frac{u}{v} = \frac{r}{r'}$. Since the choice of v directly affects the share size of the building block Π_v, which is a (tv, rv, nv)-ramp scheme, we want it to be as small as possible. In the situation when r' is not known a priori, one would need to choose an integer v that is a multiple of $n \cdot (n-1) \cdot \ldots \cdot (r+1)$. Using the polynomial based construction of secret sharing, such a (tv, rv, nv)-ramp scheme Π_v requires share space $\mathbb{F}_q$ such that

$$q \geq nv \geq n^2 \cdot (n-1) \cdot \ldots \cdot (r+1).$$

Note that the actual threshold scheme Π is a v-folded version of Π_v and hence will have share size

$$q^v \geq (nv)^v \geq (n^2 \cdot (n-1) \cdot \ldots \cdot (r+1))^{n \cdot (n-1) \cdot \ldots \cdot (r+1)}.$$

In order to keep the share size under control, we sacrifice the perfect privacy ($\varepsilon = 0$) and perfect reconstruction ($\delta = 0$) of the polynomial based construction and settle for an imperfect but almost equally effective (ε, δ)-statistical secret sharing. Through a black-box usage of the explicit construction of binary statistical secret sharing of Lemma 1, we show the following.

Theorem 2. *There is an arbitrarily threshold changeable (ε, δ)-statistical (t, r, n)-ramp scheme with respect to $\mathcal{T} = \{(\lfloor \frac{tr'}{r} \rfloor, r') | r' = r + 1, \ldots, n\}$. In particular, when $t' = \lfloor \frac{tr'}{r} \rfloor = \frac{tr'}{r}$, there is a quantity $\xi_{\varepsilon, \delta} = o(\mathsf{H}(\mathbf{S}))$ such that*

1. *Initial share size:* $\max_{1 \leq i \leq n} \log |\mathcal{S}_i| = \frac{H(\mathbf{S}) + \xi_{\varepsilon,\delta}}{r - t}$;
2. *New share size:* $\max_{1 \leq i \leq n} \log |\mathcal{S}_i^{(t',r')}| = \frac{H(\mathbf{S}) + \xi_{\varepsilon,\delta}}{r' - t'}$;
3. *Combiner communication complexity: for any* $\mathcal{A} \subset [n]$ *of* $|\mathcal{A}| = r'$,

$$\sum_{i \in \mathcal{A}} \log |\mathcal{S}_i^{(t',r')}| = \frac{r'(H(\mathbf{S}) + \xi_{\varepsilon,\delta})}{r' - t'}.$$

Proof. Let $\rho = \frac{r}{n}$ and $\tau = \frac{t}{n}$. Let $\mathbf{c} = (c_1, \ldots, c_N) \in \{0,1\}^N$ be the binary share vector of the binary (ε, δ)-statistical $(N\tau, N\rho, N)$-secret sharing scheme in Lemma 1, where N is a multiple of $n\frac{n!}{r!}$ and is chosen big enough such that the security level (ε, δ) is met. The (t, r, n)-ramp scheme initially shared to n participants is the binary secret sharing scheme with the N-bit share vector folded into n blocks, each block of length $\frac{N}{n}$. More exactly, for $i \in [n]$, the share for the ith participant P_i is

$$s_i = (c_{\frac{(i-1)N}{n} + 1}, \ldots, c_{\frac{iN}{n}}).$$

It is straightforward to see that this gives a (ε, δ)-statistical (t, r, n)-ramp scheme.

We now define the share conversion functions. For any integer $r < r' \leq n$, let $t' = \lfloor \frac{tr'}{r} \rfloor$. For $i = 1, \ldots, n$, let

$$\mathcal{S}_i = \{0,1\}^{\frac{N}{n}}, \quad \mathcal{S}_i^{(t',r')} = \{0,1\}^{\frac{Nr}{nr'}} \tag{5}$$

and

$$h_i^{(t',r')} : \mathcal{S}_i \to \mathcal{S}_i^{(t',r')} : (x_1, \ldots, x_{\frac{N}{n}}) \mapsto (x_1, \ldots, x_{\frac{Nr}{nr'}}),$$

where $\frac{Nr}{nr'}$ is an integer since by construction N is a multiple of $n\frac{n!}{r!}$. The correctness and privacy follows from that of the binary secret sharing scheme naturally.

- Correctness: for any $\mathcal{A} \subset [n]$ of size $|\mathcal{A}| \geq r'$, since the number of bits of $\mathbf{c} = (c_1, \ldots, c_N) \in \{0,1\}^N$ that are contained in these shares is at least $r' \times \frac{Nr}{nr'} = N\rho$, we then have

$$\Pr[C^{(t',r')}(D^{(t',r')}(\mathbf{s})_{\mathcal{A}}) = \mathbf{s}] \geq 1 - \delta.$$

- Privacy (non-adaptive): for any pair of secrets $\mathbf{s}_0, \mathbf{s}_1 \in \mathcal{S}$ and any $\mathcal{A} \subset [n]$ of size $|\mathcal{A}| \leq t'$, since the number of bits of $\mathbf{c} = (c_1, \ldots, c_N) \in \{0,1\}^N$ that are contained in these shares is at least $t' \times \frac{Nr}{nr'} = \lfloor \frac{tr'}{r} \rfloor \times \frac{Nr}{nr'} \leq N\rho$, we have

$$SD(D^{(t',r')}(\mathbf{s}_0)_{\mathcal{A}}; D^{(t',r')}(\mathbf{s}_1)_{\mathcal{A}}) \leq \varepsilon.$$

Finally, in the cases when $\lfloor \frac{tr'}{r} \rfloor = \frac{tr'}{r}$, we have $t' = \frac{tr'}{r}$ and want to show the three equalities. According to Lemma 1, we know there exist a $\xi_{\varepsilon,\delta} = o(H(\mathbf{S}))$ such that

$$H(\mathbf{S}) = (\rho - \tau)N - \xi_{\varepsilon,\delta}.$$

Substituting in $\rho = \frac{r}{n}$ and $\tau = \frac{t}{n}$, we have

$$H(\mathbf{S}) + \xi_{\varepsilon,\delta} = \frac{(r-t)N}{n} \Leftrightarrow \frac{H(\mathbf{S}) + \xi_{\varepsilon,\delta}}{r-t} = \frac{N}{n} = \log|\mathcal{S}_i|. \tag{6}$$

According to (5), we have

$$\frac{\log|\mathcal{S}_i|}{\log|\mathcal{S}_i^{(t',r')}|} = \frac{r'}{r},$$

which together with $t' = \frac{tr'}{r}$ substituting into (6) yields

$$\frac{H(\mathbf{S}) + \xi_{\varepsilon,\delta}}{r' - t'} = \log|\mathcal{S}_i^{(t',r')}|. \tag{7}$$

For any $\mathcal{A} \subset [n]$ of $|\mathcal{A}| = r'$, it follows from (7) that

$$\sum_{i \in \mathcal{A}} \log|\mathcal{S}_i^{(t',r')}| = \frac{r'(H(\mathbf{S}) + \xi_{\varepsilon,\delta})}{r' - t'}.$$

4.2   Comparing with Lattice-Based Construction

The only previously known threshold changeable secret sharing schemes that have this feature of not requiring a priori knowledge of the new thresholds t' and r' were constructed in [12,13]. The privacy and reconstruction threshold pairs (t',r') achieved by our construction is the same as those achieved by the lattice based constructions of [12,13]. Their definition of privacy, however, is not a standard security definition, which might be inherent limitation of the lattice techniques they use. Their construction uses public parameters. Firstly, the privacy their constructions achieved is probabilistic over the randomness of the public parameters. More accurately, with overwhelming probability over the randomness of the public parameters, their system generates a "good" value of public parameter. Secondly, "good" value of public parameter provides the guarantee that any subset of shares up to the privacy threshold leak at most a $\eta H(\mathbf{S})$ bits of entropy, where η can be made as small as one wishes when the secret is uniformly distributed. Note that η here is the fraction of the leakage not the amount of leakage. A small fraction of a long secret can become a huge amount of leakage.

Our construction enjoys a standard indistinguishability fashion of privacy, thanks to the binary statistical secret sharing we used as a black-box. Our construction almost achieves equalities in terms of the three efficiency bounds in Theorem 1[1]. The secret length $H(\mathbf{S})$ is $\xi_{\varepsilon,\delta}$ bits smaller than the bounds in Theorem 1. But the quantity $\xi_{\varepsilon,\delta} = o(H(\mathbf{S}))$ is negligible compare with the secret length. Note that the deficiency here comes in the form of shorter secret, not as leakage of the secret. A shorter secret that remains private is obvious more desirable than a secret that leaks.

---

[1] It should be understood that the bounds in Theorem 1 are derived assuming $\varepsilon = \delta = 0$ and hence is only used here as an indication of being almost optimal.

5   Conclusion

We revisited the (r, n)-*threshold scheme threshold changeable to* r' model, which conventionally considered a fixed privacy threshold (at least in the theoretical model), and highlighted an additional design goal of changing the privacy threshold to t'. We casted the problem in the general setting of changing a (t, r, n)-ramp scheme into a (t', r', n)-ramp scheme and extended the three information-theoretic bounds to cover the general model. We identified a new optimal construction with respect to the new bounds for the case when t' grows proportionally as r' grows. An open question of theoretical interest is whether there are other t' values for which optimal constructions can be found. We also gave a construction that supports arbitrary reconstruction threshold increase without a priori knowledge of the new threshold. The achievable new privacy and reconstruction threshold pairs for our construction is similar to those achieved by the lattice based constructions of [12, 13]. Our construction has the advantage of enjoying a standard definition of privacy that measures leakage in terms of statistical distance.

Acknowledgements. We thank the anonymous reviewers for their comments that improve the presentation of this work. The research is supported by Singapore Ministry of Education under Research Grant MOE2016-T2-2-014(S) and RG133/17 (S).

Appendices

A   Proof of Lemma 2

Proof. For each participant P_i, $i \in [n]$, let the original share of P_i be $\mathbf{S}_i$ and the new share of P_i be $\mathbf{S}'_i$. Here $\mathbf{S}_i$ and $\mathbf{S}'_i$ are random variables, $i \in [n]$. Then the quantity $\mathsf{H}(\mathbf{S}_i)$ is referred to as the size of $\mathbf{P}_i$'s initial share and $\mathsf{H}(\mathbf{S}'_i)$ as the size of P_i's new share.

We first prove $t' \geq t$. Since the new shares are generated from the initial shares through applying deterministic functions, no information can be generated other than those already contained in the initial shares. We then have that any set of t new shares does not contain information about the secret, hence $t' \geq t$.

We next prove $r' - t' > r - t$. Assume by contradiction that we have $r' - t' \leq r - t$. Since Π and Π' both have minimum share size, we have $\mathsf{H}(\mathbf{S}_i) = \mathsf{H}(\mathbf{S})/(r - t)$ and $\mathsf{H}(\mathbf{S}'_i) = \mathsf{H}(\mathbf{S})/(r' - t')$ from [18]. We then have $\mathsf{H}(\mathbf{S}_i) \leq \mathsf{H}(\mathbf{S}'_i)$. Since the conversion function h_i is deterministic, we know that $\mathsf{H}(\mathbf{S}'_i \mid \mathbf{S}_i) = 0$. On the other hand, by the chain rule of mutual information, we have

$$\mathsf{I}(\mathbf{S}_i; \mathbf{S}'_i) = \mathsf{H}(\mathbf{S}_i) - \mathsf{H}(\mathbf{S}_i \mid \mathbf{S}'_i)$$
$$= \mathsf{H}(\mathbf{S}'_i) - \mathsf{H}(\mathbf{S}'_i \mid \mathbf{S}_i).$$

Substituting $\mathsf{H}(\mathbf{S}'_i \mid \mathbf{S}_i) = 0$, we deduce that

$$\mathsf{H}(\mathbf{S}_i \mid \mathbf{S}'_i) = \mathsf{H}(\mathbf{S}_i) - \mathsf{H}(\mathbf{S}'_i) \leq 0,$$

where the inequality follows from the fact that $H(\mathbf{S}_i) \leq H(\mathbf{S}'_i)$. It is obvious that $H(\mathbf{S}_i \mid \mathbf{S}'_i)$ can not be negative. We are left with $H(\mathbf{S}_i \mid \mathbf{S}'_i) = 0$, which means there is a one-to-one correspondence between shares from Π and Π'. That is, the smallest number of new shares that can reconstruct the full secret in Π' must be r, which contradicts the fact that $r' > r$.

B   Semi-insider Secure $(t, r, n) \rightarrow (t', r', n)$ Ramp Scheme

The following construction is a simple adaption of a construction of optimal communication efficient secret sharing [9].

Let $g = r - t$ and $g' = r' - t$. We first parse the secret into v parts: $\mathbf{s}^{(1)} || \ldots || \mathbf{s}^{(v)}$, where each $\mathbf{s}^{(j)} \in \mathbb{F}_q^g$. Now we share $\mathbf{s}^{(1)}$ using a (t, r, n)-ramp scheme $\Pi^{(1)}$ with minimum share size, such as the polynomial based construction. We denote the share vector thus obtained by $(s_1^{(1)}, \ldots, s_n^{(1)})$. Then we share $\mathbf{s}^{(1)} || \mathbf{s}^{(2)}$ using the $(t, r + g, n)$-ramp scheme $\Pi^{(2)}$ with randomness independent from the randomness in the previous step. We denote the share vector thus obtained by $(s_1^{(2)}, \ldots, s_n^{(2)})$. We iterate this process for positive integer $j \leq v$ and share $\mathbf{s}^{(1)} || \ldots || \mathbf{s}^{(j)}$ using the $(t, r + (j - 1)g, n)$-ramp scheme $\Pi^{(j)}$ with randomness independent from the randomness in all previous steps. We denote the share vector thus obtained by $(s_1^{(j)}, \ldots, s_n^{(j)})$. Finally, for $i \in [n]$, we let

$$S_i = (s_i^{(1)}, \ldots, s_i^{(v)})$$

be the share of the ith player and obtain a ramp scheme Π with share vector $(S_1, \ldots, S_n)$.

We now show that Π is a (t, r, n)-ramp scheme with minimum share size. Firstly, the t-privacy follows from the fact that all $\Pi^{(j)}$'s have privacy threshold t and they use independent randomness. Secondly, from any r shares $S_{i_1}, \ldots, S_{i_r}$ of Π, we can extract r shares of $s_{i_1}^{(j)}, \ldots, s_{i_r}^{(j)}$ of $\Pi^{(j)}$ for each $j \in [v]$. Now given r shares $s_{i_1}^{(1)}, \ldots, s_{i_r}^{(1)}$ of $\Pi^{(1)}$, its secret $\mathbf{s}^{(1)}$ can be fully recovered. The knowledge of $\mathbf{s}^{(1)}$ together with r shares $s_{i_1}^{(2)}, \ldots, s_{i_r}^{(2)}$ of $\Pi^{(2)}$ uniquely determine its secret $\mathbf{s}^{(1)} || \mathbf{s}^{(2)}$. By iterating this process, the full secret $\mathbf{s}^{(1)} || \ldots || \mathbf{s}^{(v)}$ can be reconstructed. A dealer algorithm D and a combiner algorithm C for Π can be built from the dealer algorithms $\{\mathsf{D}^{(j)}\}_{j \in [v]}$ and combiner algorithms $\{\mathsf{C}^{(j)}\}_{j \in [v]}$ of $\{\Pi^{(j)}\}_{j \in [v]}$, respectively. Finally, the secret is consist of $g' = vg$ finite field elements while each share of Π is consist of v finite field elements. The scheme Π obviously has the minimum share size.

We next define a share conversion algorithm $\{h_i\}_{i \in [n]}$ to transform the scheme Π into Π' that is a (t, r', n)-ramp scheme. Let

$$h_i(S_i) = s_i^{(v)}.$$

The new combiner algorithm is $\mathsf{C}' = \mathsf{C}^{(v)}$.

We show that Π' with share vector $(S'_1, \ldots, S'_n)$, where $S'_i = h_i(S_i)$, is a (t, r', n)-ramp scheme with minimum share size. This is trivial, since $(S'_1, \ldots, S'_n)$

is just the share vector of $\Pi^{(v)}$, which is a (t, r', n)-ramp scheme with minimum share size by construction.

Let us re-examine the construction above and show security against semi-insider adversary. A share of the packed scheme Π is consist of shares from distinct schemes $\Pi^{(1)}, \ldots, \Pi^{(v)}$ sharing related secrets using independent randomness. One special advantage of this structure is that a subset $\mathcal{A}^{(j_1)}$ of the shares of $\Pi^{(j_1)}$ and a subset $\mathcal{A}^{(j_2)}$ of the shares of $\Pi^{(j_2)}$ for $j_1 \neq j_2$ are independent if one subset is of size at most t. This means that even if at most t shareholders do not erase their original shares of Π after the transformation from Π into Π' through applying the transformation algorithm $\{h_i\}_{i \in [n]}$, the dishonestly kept at most t shares of Π contribute the same amount of information as the transformed partial shares to the transformed scheme Π', since the dishonestly kept extra partial content of the original shares are independent of the share vectors of Π'.

References

1. Blakley, G.R., et al.: Safeguarding cryptographic keys. In: Proceedings of the National Computer Conference, vol. 48 (1979)
2. Shamir, A.: How to share a secret. Commun. ACM **22**(11), 612–613 (1979)
3. Desmedty, Y., Jajodiay, S.: Redistributing secret shares to new access structures and its applications (1997)
4. Martin, K.M., Pieprzyk, J., Safavi-Naini, R., Wang, H.: Changing thresholds in the absence of secure channels. In: Pieprzyk, J., Safavi-Naini, R., Seberry, J. (eds.) ACISP 1999. LNCS, vol. 1587, pp. 177–191. Springer, Heidelberg (1999). https://doi.org/10.1007/3-540-48970-3_15
5. Blundo, C., Cresti, A., De Santis, A., Vaccaro, U.: Fully dynamic secret sharing schemes. In: Stinson, D.R. (ed.) CRYPTO 1993. LNCS, vol. 773, pp. 110–125. Springer, Heidelberg (1994). https://doi.org/10.1007/3-540-48329-2_10
6. Barwick, S.G., Jackson, W.-A., Martin, K.M.: Updating the parameters of a threshold scheme by minimal broadcast. IEEE Trans. Inf. Theory **51**(2), 620–633 (2005)
7. Martin, K.M., Safavi-Naini, R., Wang, H.: Bounds and techniques for efficient redistribution of secret shares to new access structures. Comput. J. **42**(8), 638–649 (1999)
8. Maeda, A., Miyaji, A., Tada, M.: Efficient and unconditionally secure verifiable threshold changeable scheme. In: Varadharajan, V., Mu, Y. (eds.) ACISP 2001. LNCS, vol. 2119, pp. 403–416. Springer, Heidelberg (2001). https://doi.org/10.1007/3-540-47719-5_32
9. Wang, H., Wong, D.S.: On secret reconstruction in secret sharing schemes. IEEE Trans. Inf. Theory **54**(1), 473–480 (2008)
10. Zhang, Z., Chee, Y.M., Ling, S., Liu, M., Wang, H.: Threshold changeable secret sharing schemes revisited. Theoret. Comput. Sci. **418**, 106–115 (2012)
11. Jia, X., Wang, D., Nie, D., Luo, X., Sun, J.Z.: A new threshold changeable secret sharing scheme based on the Chinese Remainder Theorem. Inf. Sci. **473**, 13–30 (2019)
12. Steinfeld, R., Pieprzyk, J., Wang, H.: Lattice-based threshold changeability for standard Shamir secret-sharing schemes. IEEE Trans. Inf. Theory **53**(7), 2542–2559 (2007)

13. Steinfeld, R., Pieprzyk, J., Wang, H.: Lattice-based threshold-changeability for standard CRT secret-sharing schemes. Finite Fields Appl. **12**(4), 653–680 (2006)
14. Lin, F., Cheraghchi, M., Guruswami, V., Safavi-Naini, R., Wang, H.: Secret sharing with binary shares. In: 10th Innovations in Theoretical Computer Science Conference (ITCS 2019). Schloss Dagstuhl-Leibniz-Zentrum fuer Informatik (2018)
15. Huang, W., Langberg, M., Kliewer, J., Bruck, J.: Communication efficient secret sharing. IEEE Trans. Inf. Theory **62**(12), 7195–7206 (2016)
16. Bitar, R., El Rouayheb, S.: Staircase codes for secret sharing with optimal communication and read overheads. IEEE Trans. Inf. Theory **64**(2), 933–943 (2017)
17. Martínez-Peñas, U.: Communication efficient and strongly secure secret sharing schemes based on algebraic geometry codes. IEEE Trans. Inf. Theory **64**(6), 4191–4206 (2018)
18. Blundo, C., De Santis, A., Vaccaro, U.: Efficient sharing of many secrets. In: Enjalbert, P., Finkel, A., Wagner, K.W. (eds.) STACS 1993. LNCS, vol. 665, pp. 692–703. Springer, Heidelberg (1993). https://doi.org/10.1007/3-540-56503-5_68

Client-Aided Two-Party Secure Interval Test Protocol

Hiraku Morita[✉] and Nuttapong Attrapadung

AIST, Tokyo, Japan
{hiraku.morita,n.attrapadung}@aist.go.jp

Abstract. Secure interval test protocol checks if an integer is within some interval in a privacy-preserving manner. A natural application is geological location hiding, where we can check whether a person is in a certain territory without revealing any information. In addition, secure interval test protocol enables us to do arithmetic over private values with rounding errors. Therefore, it allows servers to obtain an approximation of a complicated function.

In this work, we present an efficient secure interval test protocol that checks whether a shared value is within the range of two plain values. We also show that the interval test protocol can be used as a building block to construct protocols with richer functionality such as the approximation of exponential functions or logarithmic functions.

Our protocol is constructed in the *client-aided model*, which is briefly mentioned in some previous work on constructing practical MPC frameworks such as SecureML (S&P'17), in which any number of clients can not only create shares of their inputs but also generate some necessary correlated randomness used in the online phase and distribute them to servers. Such correlated randomness generated by clients serves efficient protocols since servers don't have to jointly generate randomness by themselves, which can avoid costly computation/communication.

In this paper, we improve the state-of-the-art secure interval test protocol by Nishide and Ohta (PKC'07) based on a secret sharing scheme. We use the client-aided model and tree-based techniques, which contribute to reducing communication rounds. Our proposed protocol has only 4 communication rounds regardless of the bit length of inputs. This is about 3 times fewer rounds than existing protocols. Using the proposed protocol, we further introduce a secure look-up table technique that can be utilized to securely compute some richer functions.

Keywords: Two-party computation · Client-server model · Client-aided model · Secure interval test · GMW secret sharing

1 Introduction

Multi-party computation (MPC) is one of the most promising cryptographic primitives, where parties can securely compute a function on their input. It

© Springer Nature Switzerland AG 2019
Y. Mu et al. (Eds.): CANS 2019, LNCS 11829, pp. 328–343, 2019.
https://doi.org/10.1007/978-3-030-31578-8_18

is useful in privacy-preserving data processing such as secure data mining and secure out-sourced computation. In general, MPC enables N parties to jointly compute a function f on their secret values. Formally, each party i has his own secret x_i for $i = 1, \ldots, N$ to jointly compute $F = f(x_1, \ldots, x_N)$ without leaking any information on x_i. Security of MPC guarantees that any party i will not obtain any information on input x_j of any other party $j \neq i$ except what can be derived from the output.

We focus on a secret sharing based MPC [12] in this paper. Compared to garbled circuit based approaches [20] and (fully) homomorphic encryption based approaches, secret sharing based MPC requires less computation and smaller bandwidth, while it generally requires more communication rounds. Therefore, one of the motivations for previous work has been constructing efficient protocols that have fewer communication rounds.

Secure Interval Test Protocols. In this paper, we introduce a secure interval test protocol. Similar to secure comparison protocols, secure interval test protocol itself is useful and also a good building block to construct fruitful secure computations. For instance, interval test protocols can be applied to statistical analysis, data classification, and machine learning.

There are two types of secure interval test protocols, which depend on the types of input. One takes as input a shared value and a plain interval, while another takes as input a shared value and a shared interval in order to check if the value is within the interval. We focus on the former type of secure interval test protocol since it is useful to construct secure protocols with richer functions.

Secure interval test protocol has been inefficient and could be a bottleneck for applications. One of the reasons is that secure interval test protocols essentially need bit-wise comparisons to check if a shared value is larger than the tight lower bound and smaller than the tight upper bound, while applications such as machine learning need to be processed arithmetically since they use various mathematical functions. In a secret sharing scheme, a bit-decomposition protocol for a bit length n requires $\log n$ communication rounds in order to convert an arithmetic shared value into a bit-wise shared value.

However, Damgård *et al.* [7] proposed an innovative method to execute a bit-decomposition protocol in constant rounds, which is applied to construct a constant round secure comparison protocol. Their proposed protocol can be defined under any secret sharing scheme that satisfies linearity and contains constant round multiplications. Nishide and Ohta [16] constructed an interval test protocol, equality check protocol, and comparison protocol in constant rounds without using a bit-decomposition technique. Their proposed protocols were still not very practical since the protocols required more than 10 rounds.

In this paper, we improve the interval test protocol of Nishide and Ohta by assuming the client-aided client-server two-party model and by using tree-based techniques. Details are as follows.

MPC in Client-Server Model. As MPC is getting to be a necessary technique for practical secure applications, the client-server model is drawing attention for its suitability to real-world applications. Such a model is focused on not only by recent research such as Araki *et al.* [1] but also by business applications such as the Sharemind system by Cybernetica. This model captures the situation where there are N servers and an arbitrary number of clients (say, t clients). Each client has their own input x_i for $i \in \{1, \dots, t\}$, and secret-shares it to N servers. The servers jointly compute a function f on these inputs x_i to return $f(x_1, \dots, x_t)$ to the clients in secure manner. This model can also be considered to be $(N+t)$-party MPC that N parties don't have their input while the other t parties do. The above model fits real-world services that users (clients) have their input and ask servers to jointly compute a function. Here, clients who use the service just provide their inputs and wait until they obtain outputs. Therefore, no clients need to take part in heavy computation.

Client-Aided Client-Server Model. As Mohassel-Zhang [13] mentioned, in the client-aided client-server model, clients not only provide input but also can generate and secret-share correlated randomness to servers. Such correlated randomness is used by N servers to make secure computation efficient. The only downside of this model is the restriction that any server is not allowed to collude with any client since, otherwise, it would break security. However, we can assume that no servers will collude with any client because there is no incentive to do so to damage their reputation. The security notion in the semi-honest model follows from the standard notation of private computation [11].

1.1   Contribution

In this paper, we construct an efficient secure interval test protocol (Sect. 3) in the client-aided model where there are two servers and any number of clients. Our proposed protocol is, to the best of our knowledge, the most efficient protocol in the two-party setting. The state-of-the-art secure interval test protocol by Nishide and Ohta [16] in the multi-party setting has 11 rounds[1]. However, our secure interval test protocol only needs 4 rounds, which is about three times faster.

We further provide an explanation of how to construct a certain secure look-up table protocol and construct a secure computation of the function $f(x) = \lfloor \log x \rfloor$ as an example that uses such a look-up table (Sect. 4).

Our Technique. Our proposed protocol is based on the secure interval test protocol by Nishide and Ohta [16]. First of all, we highlight the main idea of Nishide and Ohta's protocol. In their protocol, parties collaboratively generate an arithmetic share and boolean share of randomness r. To test if the (arithmetically) shared form of a value $[\![x]\!]$ is within a public interval $[y_L, y_R]$, the shared value x is masked by the arithmetically shared randomness r and revealed ($c = x + r$). Using the revealed value c and the original interval, the protocol obtains a new interval $[r_{\text{low}}, r_{\text{high}}]$. Then, we can consider the original interval test protocol turns to be the protocol that checks if the boolean shared value r is within

---

[1] According to [16], their protocol takes 13 rounds including 2 rounds for generating randomness.

the new interval. Now, bit-wise secure comparison protocol is used *twice* as a building block to construct the secure interval test protocol.

We follow the above idea that checks whether the randomness is within the new interval. We further reduce communication rounds in the client-aided two-party model as follows:

- P_1 generates randomness r by using a (pre-distributed) share of 0, while P_2 computes $c = x + r$ by just using the corresponding share of 0 without any communication. Therefore, in our IntvlTest protocol, only P_1 knows the value r and only P_2 knows the value c and the new interval $[r_{low}, r_{high}]$. Since each r and $[r_{low}, r_{high}]$ is possessed by P_1 and P_2 in the clear, respectively, we first construct an interval test protocol that takes *plain inputs*, PlainIntvlTest.
- Note that [16] is based on any linear secret sharing scheme defined over a field. On the other hand, our proposed protocol is based on a specific secret sharing scheme, the so-called 2-out-of-2 secret sharing scheme. This is the basic MPC protocol introduced by Goldreich *et al.* [12] and called a two-party GMW-style secret sharing scheme. By focusing on the two-party secret sharing scheme, we can use tree-based techniques to construct efficient sub-protocols.
- We construct PlainIntvlTest protocol by using tree-based techniques in a similar manner to PlainLessThan protocol by [14]. The tree-based technique has two advantages: (1) it enables parallel computations, and (2) each computation can be executed by using the 3-rounds PlainEqual by [14]. Note that it is comparatively easy to generate the correlated randomness used in PlainEqual protocol in the client-aided model.

1.2 Related Work

Multi-party computation has been studied ever since Yao [20] proposed the millionaire's problem. As the millionaire's problem is a secure comparison protocol, secure comparison protocols, in particular, have been and will continue to be widely studied [4–10, 16–18, 20].

As Veugen *et al.* [19] pointed out, secret sharing based secure comparison protocols [6, 10, 16] have advantages in computational cost over the other protocols based on garbled circuits or homomorphic encryptions. A secure interval test protocol based on the secret sharing scheme also has the same advantage over others since the protocol is essentially similar to secure comparison protocols.

Techniques to construct a secret sharing based constant-round secure comparison protocol [7, 14, 16] can be applied to construct a constant-round secure interval test protocol. Nishide and Ohta [16] proposed a secret sharing based constant-round secure interval test protocol as well as a secure comparison protocol. Our protocol is loosely based on their idea.

Nergiz *et al.* [15] proposed a secure interval test protocol constructed from a secure set intersection protocol using homomorphic encryption. Catrina and de Hoogh [6] provided an efficient secure comparison protocol, which can be applied to construct an efficient secure interval test protocol. However, it offers only statistical security We would like to highlight that our protocol does not rely on heavy cryptographic computations such as homomorphic encryption, garbled circuits, and oblivious transfer, and it offers perfect security.

2 Preliminaries

In this section, we introduce some notation and some known techniques.

2.1 Notation

Let n be a positive integer and p be a prime of n-bit. That is, $n = \lceil \log_2 p \rceil$. Note that $x \leq p - 1 < 2^n - 1$ for $x \in \mathbb{F}_p$. We assume that a proposition $(y_L \leq x \leq y_R)$ denotes 1 if $y_L \leq x \leq y_R$, and 0 otherwise. Let $[y_L, y_R]$ denote a set of elements from $\mathbb{F}_p$ that are equal or larger than y_L and equal or less than y_R, which is called the *interval*. Just to stress the two values y_L and y_R, we also write it in a 2-dimensional vector (y_L, y_R). Through this paper, P_1 and P_2 represent servers, and P_3 represents a client. Since our protocol is based on the two-party GMW-style (arithmetic) secret sharing scheme, the shares between two parties are represented as follows: Let $[\![x]\!]$ denote that a secret x is shared between two parties. That is, $[\![x]\!] = ([\![x]\!]_1, [\![x]\!]_2)$, where $[\![x]\!]_j$ is a share of x that a party P_j for $j \in \{1, 2\}$ owns. We use binary logarithm that has base 2. A value x can be represented in bit-wise form as follows: $x = x_{n-1} \| x_{n-2} \| \dots \| x_0$, where $x_i \in \{0, 1\}$ and $x = \sum_{i=0}^{n-1} 2^i \cdot x_i$ for $i \in [0, n-1]$.

Let $\mathbb{T}_{2^n}$ be a complete binary tree whose leaves correspond to integers from 0 to $2^n - 1$. Let $\mathbb{S}_{2^n}$ denote the all nodes of a tree $\mathbb{T}_{2^n}$ and let $w_{i,j}$ denote a node of j-th from the left at the i-th layer (leaves are at the bottom written as 0-th layer and roots are at the top written as n-th layer), where (i, j) satisfies $i \in [0, n]$ and $j \in [0, 2^{n-i} - 1]$.

2.2 GMW-Style Secret Sharing Scheme

In the two-party GMW-style secret sharing scheme [12], a party P_1 has $[\![x]\!]_1 = r$ and a party P_2 has $[\![x]\!]_2 = x - r$ as arithmetic shares of $x \in \mathbb{Z}_{2^n}$, where $r \xleftarrow{\$} \mathbb{Z}_{2^n}$. Suppose that $\mathsf{Share}(\cdot)$ is the algorithm that produces such shares. Thus, $([\![x]\!]_1, [\![x]\!]_2) \leftarrow \mathsf{Share}(x)$.

For $[\![x]\!] = ([\![x]\!]_1, [\![x]\!]_2)$ and $[\![y]\!] = ([\![y]\!]_1, [\![y]\!]_2)$, secure addition $[\![x]\!] + [\![y]\!]$ is computed as follows: Each party P_i computes $[\![x]\!]_i + [\![y]\!]_i$ locally. Moreover, secure subtraction will be $[\![x]\!]_i - [\![y]\!]_i$ and multiplication by a public value c will be $c \cdot [\![x]\!]_i$. Subtraction of public value, $[\![x]\!] - s$, will be as follows: P_1 does nothing, while P_2 computes $[\![x]\!]_2 - s$ locally.

2.3 Multiplication Triple

The secure multiplication protocol in the GMW-style secret sharing scheme uses pre-computed correlated randomness, so-called multiplication triples (MT) [3] such as $[\![c]\!] = [\![a]\!] \cdot [\![b]\!]$ in order to make the protocol efficient. Standard multi-party computation needs to prepare such MTs by executing homomorphic encryption or oblivious transfer which have large offline communication costs.

On the other hand, the client-aided two-party computation [13] we use in this paper allows clients to generate MTs such as $a = a_1 + a_2, b = b_1 + b_2, c = c_1 + c_2$ to whatever extent is needed and send the shares to the servers. This is done by just selecting two randomnesses a, b, computing $c = a \cdot b$, generating shares of a, b, c, and sending them to servers, which can remove the communication cost between servers while generating MTs.

3   Secure Interval Test and Sub-protocols

The secure interval test protocol has two variations depending on the types of input as follows: (1) that takes as input a shared value $[\![x]\!]$ and a plain interval (c_L, c_R), and (2) that takes as input a shared value $[\![x]\!]$ and a shared interval $([\![c_L]\!], [\![c_R]\!])$. There is a trivial construction for both types of secure interval test protocol using secure comparison protocol as sub-protocol as follows:

$$\mathsf{IntvlTest}([\![x]\!], (c_L, c_R)) \overset{\text{def}}{=} \mathsf{LessThan}([\![x]\!] - c_L, c_R - c_L) \qquad (1)$$

$$\mathsf{IntvlTest}([\![x]\!], ([\![c_L]\!], [\![c_R]\!])) \overset{\text{def}}{=} \mathsf{LessThan}([\![x]\!] - [\![c_L]\!], [\![c_R]\!] - [\![c_L]\!]), \qquad (2)$$

where $\mathsf{LessThan}$ takes as input two values and outputs 1 in shared form if the first parameter is strictly less than the second one, and outputs 0 in shared form otherwise. Here, the second parameter can be either a shared value or a plain value. We note that using the $\mathsf{LessThan}$ protocol in [14] makes the above protocols execute in 5 rounds. We focus on reducing communication rounds of the former type of $\mathsf{IntvlTest}$ protocol since it is useful in practice. We present some fruitful applications in Sect. 4. As previous work by Nishide and Ohta [16] has shown, the former type of $\mathsf{IntvlTest}$ protocol can be constructed in a more efficient way without using secure comparison protocol.

Our client-aided two-party secure interval test protocol improves the secure interval test protocol in the multi-party setting introduced by Nishide and Ohta [16] by adjusting to the two-server setting. The model we use captures the situation in which there are 2 servers that run MPC and any number of clients who generate correlated randomness and provide inputs. For simplicity, we will assume that there is only one client unless stated otherwise. Namely, our proposed protocol is constructed under the model where there exist 2 servers (P_1 and P_2) who run MPC and a client (P_3). We note that this model can be considered "an unorthodox 3-party setting".

As Table 1 shows, we succeed in reducing the number of communication rounds compared with [16]. Our proposed protocol requires only 4 rounds while the protocol in [16] requires 11 rounds. This is also 1 round fewer than the trivial construction explained above. Our protocol requires transmitting more data than [16] but since the number of communication rounds matters for execution

Table 1. Comparison of round and communication complexity of IntvlTest

	Ours	Using LessThan in [14]	[16]
Round	4	5	11
Comm.	$\mathcal{O}(n^2)$	$\mathcal{O}(n^2)$	$\mathcal{O}(\log_2 n)$
# of servers	2	2	≥ 2

time, we can estimate that our protocol is about 3 times faster than the current state-of-the-art protocol [16]. We note that the setting in [16] and ours are slightly different, that is, [16] works in the multi-party setting while ours works in the two-server setting (with clients).

In Sect. 3.1, we present our secure interval test protocol, IntvlTest. Then in Sect. 3.2, we present a sub-protocol PlainIntvlTest used to construct our target protocol, IntvlTest.

3.1 Secure Interval Test Protocol

Suppose that P_1 and P_2 have shares of x. For a (public) plain interval (c_L, c_R) known by both P_1 and P_2, we propose a secure interval test protocol that checks if $c_L \leq x \leq c_R$. More formally, the protocol runs as $[\![\delta]\!] \leftarrow \mathsf{IntvlTest}([\![x]\!], (c_L, c_R))$, where the inputs are a share of $[\![x]\!]$ and a plain interval $(c_L, c_R) \in \mathbb{F}_p^2$ for $c_L < c_R$ and the output is a share of $\delta = (c_L \leq x \leq c_R)^2$.

Construction of IntvlTest Protocol. We describe the algorithm of our secret sharing based secure interval test protocol IntvlTest in Algorithm 1. In this algorithm, we only describe the operations of servers (P_1 and P_2). We don't describe client's (P_3's) operation since it just works to generate correlated randomness and provides input before starting online phase.

Correctness/Security. If P_1 chose randomness r and sent $[\![x]\!]_1 + r$ to P_2, then P_2 would be able to obtain $c \leftarrow x + r$ by computing $[\![x]\!]_2 + ([\![x]\!]_1 + r)$. This needs one round of communication, which can be reduced as follows: For pre-distributed shares of 0 (i.e., $[\![0]\!] = r_1 + r_2$), instead of choosing randomness, P_1 sets a value $r \overset{\$}{\leftarrow} -([\![x]\!]_1 + r_1) \bmod p$, while P_2 sets $c \leftarrow [\![x]\!]_2 + r_2 \bmod p$ ($= x + r \bmod p$) as in step 1 in Algorithm 1. After step 1, only P_1 has the randomness r, and only P_2 has the masked value $c = x + r$. The protocol to check whether x is within the interval (c_L, c_R) can be rewritten as the protocol to check whether r is within the new interval $(r_\text{low}, r_\text{high})$ that depends on the value c.

Since the value c is known only by P_2, only P_2 can determine the new interval $(r_\text{low}, r_\text{high})$ depending on c. On the other hand, P_1 does not know the value c, and he just puts randomness r as input to PlainIntvlTest.

As shown in Fig. 1, we first suppose the case of $c_L \leq c \leq c_R$ (Step 2-6). For $r_\text{low} \leftarrow c - c_L + 1$ and $r_\text{high} \leftarrow c + p - c_R - 1$, if $r_\text{low} \leq r \leq r_\text{high}$, then $x < c_L$ or

---

[2] We note that the output of the interval test protocol in [16] is shares of $\delta = (c_L < x < c_R)$.

Algorithm 1. 2-party IntvlTest Protocol

Functionality: $\llbracket d \rrbracket \leftarrow$ IntvlTest($\llbracket x \rrbracket, (c_L, c_R)$).

Input: Arithmetic shared value $\llbracket x \rrbracket$ over $\mathbb{F}_p$ and public integers $c_L, c_R \in \mathbb{F}_p$, where $c_L \leq c_R$.

Auxiliary Input: Shares of 0 such that $r_1 + r_2 = 0$.

Output: Arithmetic shared value $\llbracket d \rrbracket$ over $\mathbb{F}_p$, where $d = 1$ if $c_L \leq x \leq c_R$, and $d = 0$, otherwise.

1: P_1 sets $r \leftarrow -(\llbracket x \rrbracket_1 + r_1) \bmod p$ and P_2 sets $c \leftarrow \llbracket x \rrbracket_2 + r_2 \bmod p$ $(= x + r \bmod p)$
2: **if** $c_L \leq c \leq c_R$ **then**
3: $\qquad$ P_2 computes $r_{\text{low}} \leftarrow c - c_L + 1, r_{\text{high}} \leftarrow c + p - c_R - 1$
4: $\qquad$ Compute $\llbracket d' \rrbracket \leftarrow$ PlainIntvlTest($r, (r_{\text{low}}, r_{\text{high}})$)
5: $\qquad$ Set $\llbracket d' \rrbracket \leftarrow \llbracket d' \rrbracket - 1$ (Note: only P_2 works)
6: $\qquad$ Set $\llbracket d \rrbracket \leftarrow \llbracket d' \rrbracket^2$
7: **else**
8: $\qquad$ **if** $c_R < c$ **then**
9: $\qquad\qquad$ P_2 computes $r_{\text{low}} \leftarrow c - c_R, r_{\text{high}} \leftarrow c - c_L$
10: $\qquad$ **else** ($i.e.$, $c < c_L$)
11: $\qquad\qquad$ P_2 computes $r_{\text{low}} \leftarrow c + p - c_R, r_{\text{high}} \leftarrow c + p - c_L$
12: $\qquad$ $\llbracket d' \rrbracket \leftarrow$ PlainIntvlTest($r, (r_{\text{low}}, r_{\text{high}})$)
13: $\qquad$ Set $\llbracket d \rrbracket \leftarrow \llbracket d' \rrbracket^2$
14: Output $\llbracket d \rrbracket$

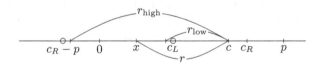

Fig. 1. Case of $c_L \leq c \leq c_R$

$c_R < x$ is satisfied. Therefore, we flip the result of PlainIntvlTest($r, (r_{\text{low}}, r_{\text{high}})$) by computing $\llbracket d' \rrbracket \leftarrow \llbracket d' \rrbracket - 1$ and $\llbracket d \rrbracket \leftarrow \llbracket d' \rrbracket^2$ (Step 5, 6) to satisfy $c_L \leq x \leq c_R$. Readers might wonder why it is not just $\llbracket d \rrbracket \leftarrow 1 - \llbracket d' \rrbracket$. The reason is that if P_1, who did not know the value c, noticed that he was jointly flipping $\llbracket d' \rrbracket$, he would have known some information about c. Thus, P_1 is supposed to perform exactly the same operation regardless of c, while P_2 subtracts a public value 1 from his share in order to flip the value d' in this case. After Step 5, the value d' would be 0 or $-1 \pmod p$. Therefore, two parties compute the square to obtain a required value at Step 6.

As shown in Fig. 2, we now suppose the case of $c_R < c$ (Step 8, 9, 12, 13). For $r_{\text{low}} \leftarrow c - c_R$ and $r_{\text{high}} \leftarrow c - c_L$, if $r_{\text{low}} \leq r \leq r_{\text{high}}$, then $c_L \leq x \leq c_R$ is satisfied. Therefore, two parties compute PlainIntvlTest($r, (r_{\text{low}}, r_{\text{high}})$) to obtain a required value. We add Step 13 to keep the P_1's process consistent with Step 6. Note that Step 13 does not change the value at Step 12.

As shown in Fig. 3, we suppose the case of $c < c_L$ (Step 10-13). For $r_{\text{low}} \leftarrow c + p - c_R$ and $r_{\text{high}} \leftarrow c + p - c_L$, if $r_{\text{low}} \leq r \leq r_{\text{high}}$, then $c_L \leq x \leq c_R$ is

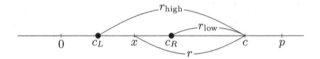

Fig. 2. Case of $c_R < c$

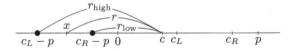

Fig. 3. Case of $c < c_L$

satisfied. Therefore, two parties compute a square of $\mathsf{PlainIntvlTest}(r, (r_{\text{low}}, r_{\text{high}}))$ and output it as in the previous case.

Since transmitting values are masked by randomness and we just use a $\mathsf{PlainIntvlTest}$ protocol as a sub-protocol, security of $\mathsf{IntvlTest}$ protocol holds straightforwardly.

Round Complexity. Since our proposed protocol, $\mathsf{IntvlTest}$, has a $\mathsf{PlainIntvlTest}$ protocol that is 3-round and one secure multiplication protocol, it requires 4 rounds, which is fewer than the $\mathsf{IntvlTest}$ protocol using $\mathsf{LessThan}$ protocol of [14] and the $\mathsf{IntvlTest}$ protocol proposed in [16] as shown in Table 1.

3.2 Plain Interval Test Protocol (PlainIntvlTest)

Suppose that only P_1 has x and only P_2 has an interval (y_L, y_R). We describe $\mathsf{PlainIntvlTest}$ protocol that checks if $y_L \leq x \leq y_R$. More formally, the protocol executes as $[\![\delta]\!] \leftarrow \mathsf{PlainIntvlTest}(x, (y_L, y_R))$, where it takes $x \in \mathbb{F}_p, (y_L, y_R) \in \mathbb{F}_p^2$ as input and outputs a share of $\delta = (y_L \leq x \leq y_R)$.

Approach. We construct $\mathsf{PlainIntvlTest}$ based on tree-based techniques in a similar manner to [2,14]. $\mathsf{PlainIntvlTest}$ protocol takes plaintexts from two servers as input, which allows servers to encode each plaintext to a tree structure. Such a data representation enables two servers to run secure computation in parallel.

Now, we use two encoding techniques. One is called range encoding, and the other is called point encoding [2]. In our construction, using $\mathsf{PlainEqual}$ protocol that checks the equality of two plaintexts is crucial. $\mathsf{PlainIntvlTest}$ checks n equality at the same time. In this paper, we use a secret sharing based constant-round $\mathsf{PlainEqual}$ protocol introduced in [14], which makes the protocol efficient.

Point/Range Encoding. Point encoding works in the same way as that of [14]. On the other hand, range encoding is slightly different from that of [14]. Range encoding, in this paper, adds dummy nodes if necessary.

First, we review some basic notation for tree structures. A value $x \in \mathbb{F}_p$ corresponds to a node $w_{0,x}$. For u, v such as $0 \leq u \leq v \leq 2^n - 1$, we assume that

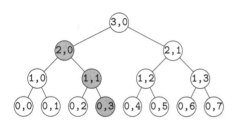

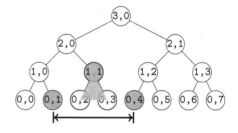

Fig. 4. Point Encoding when $x = 3$. (Here, $\mathsf{pointEnc}(3) = \{(0,3),(1,1),(2,0)\}$)

Fig. 5. Range Encoding when $y_L = 1, y_R = 4$. (Here, $\mathsf{rangeEnc}([1,4]) = \{(0,1),(0,4),(1,1),(1,7),(2,7),(2,7)\}$ including dummy nodes.)

Algorithm 2. PlainIntvlTest Protocol

Functionality: $[\![\delta]\!] \leftarrow \mathsf{PlainIntvlTest}(x,(y_L,y_R))$, such that $\delta = (y_L \leq x \leq y_R)$.
Input: Cleartext $x \in \mathbb{F}_p$ from P_1, $y_L, y_R \in \mathbb{F}_p$ from P_2, where $y_L \leq y_R$.
Output: Arithmetic shared value $[\![\delta]\!]$ over $\mathbb{F}_p$.
1: P_1 sets $\{(i,p_i)\} \leftarrow \mathsf{pointEnc}(x)$;
2: P_2 sets $\{\ldots,(i,r_i),(i,r_i'),\ldots\} \leftarrow \mathsf{rangeEnc}([y_L,y_R])$
3: $[\![d_i]\!] \leftarrow \mathsf{PlainEqual}(p_i,r_i) + \mathsf{PlainEqual}(p_i,r_i')$ for all $i \in [0,n-1]$.
4: $[\![\delta]\!] \leftarrow \sum_{i=0}^{n-1}[\![d_i]\!]$.
5: **return** $[\![\delta]\!]$.

---

an interval $R = [u,v]$. For any interval R, a node $w_{i,j} \in \mathbb{S}_{2^n}$ is called a cover node of R if all descendant leaf nodes $w_{i,j}$ are in R. We write such a set of nodes as $\mathsf{cover}(R)$. For $w_{i,j} \in \mathbb{S}_{2^n}$ such as $(i,j) \neq (n,0)$, $\mathsf{parent}(w_{i,j})$ denotes a parent node of $w_{i,j}$.

Next, we show definitions of range encoding and point encoding. Range encoding is as follows: For an interval $R = [u,v]$ such as $0 \leq u \leq v \leq 2^n - 1$, define

$$\mathsf{rangeEnc}(R) := \{(i,a_i) \in \mathbb{S}_{2^n} \mid (i,a_i) \in \mathsf{cover}(R),$$
$$\mathsf{parent}(i,a_i) \notin \mathsf{cover}(R)\}.$$

Moreover, while $\mathsf{rangeEnc}(R)$ has less than two nodes at each layer j, it adds a dummy node $(j, 2^n - 1)$.

For point $x \in [0, p-1]$, $\mathsf{pointEnc}(x)$ denotes a set of all ancestor nodes of a node $(0,x)$ in a tree $\mathbb{T}_{2^n}$. Note that this includes the node $(0,x)$ itself. We show point encoding in Fig. 4 and range encoding in Fig. 5. These encoding methods have the following properties: For any interval R and any point $x \in \mathbb{F}_p$, if $x \in R$ then $|\mathsf{rangeEnc}(R) \cap \mathsf{pointEnc}(x)|$ is 1, while it is 0 if $x \notin R$.

PlainIntvlTest **Protocol.** We describe our PlainIntvlTest protocol in Algorithm 2.

Correctness/Security. Assume that $y_L \leq x \leq y_R$. Then, one of the elements in $\mathsf{pointEnc}(x)$ should be equal to one of the elements in $\mathsf{rangeEnc}([y_L,y_R])$. All

---

Algorithm 3. PlainEqual Protocol [14]

Functionality: $[\![\delta]\!] \leftarrow \mathsf{PlainEqual}(x, y)$, where $\delta = 1$ if $x = y$; $\delta = 0$, otherwise.
Input: Cleartexts $x, y \in \mathbb{F}_p$
Output: Arithmetic shared value $[\![\delta]\!]$
  1: Parse $x = x_{n-1} \parallel x_{n-2} \parallel \ldots \parallel x_0$.
  2: Parse $y = y_{n-1} \parallel y_{n-2} \parallel \ldots \parallel y_0$.
  3: P_1 sets $[\![x_i]\!]_1 \leftarrow x_i$ and $[\![y_i]\!]_1 \leftarrow 0$ for $i \in [0, n-1]$.
  4: P_2 sets $[\![x_i]\!]_2 \leftarrow 0$ and $[\![y_i]\!]_2 \leftarrow y_i$ for $i \in [0, n-1]$.
  5: Compute $[\![v_i]\!] \leftarrow (1 - [\![x_i]\!] - [\![y_i]\!])^2$ for $i \in [0, n-1]$.
  6: Compute $[\![\delta]\!] \leftarrow \mathsf{AND}^*([\![v_0]\!], [\![v_1]\!], \ldots [\![v_{n-1}]\!])$.
  7: **return** $[\![\delta]\!]$.

---

equality checks are executed at the same time and the final output is OR of all the results. Since at most one of the value d_i can be one, the OR operation here can be done by just using secure addition which is locally processed.

Assume now that $x < y_L$ or $y_R < x$. As long as one uses the above encoding methods, there are no matched nodes between a set $\mathsf{pointEnc}(x)$ and a set $\mathsf{rangeEnc}([y_L, y_R])$. Note that our definition of range encoding further contains some steps to add dummy nodes for later use, which is slightly different from that of [2,14]. After the first run of $\mathsf{rangeEnc}$, we assign a node $(j, 2^n - 1)$ as a dummy node until each layer j has two nodes.

By adding such dummy nodes, we guarantee that there are always two equality checks at each layer, so that no information about an interval of P_2 will be leaked to P_1. Note that an equality check between a node in $\mathsf{pointEnc}(x)$ and a dummy node in $\mathsf{pointEnc}$ always outputs 0.

Security holds straightforwardly since $\mathsf{rangeEnc}$ and $\mathsf{pointEnc}$ are locally executable and only PlainEqual is used as a sub-protocol.

Round Complexity. As PlainEqual protocol at Step 3 of Algorithm 2 requires 3 rounds as we show in the next subsection, PlainIntvlTest protocol requires 3 rounds.

3.3 Equality Check Protocol with Plain Input (**PlainEqual**)

Our proposed IntvlTest protocol uses PlainIntvlTest protocol as its sub-protocol. Then, PlainIntvlTest protocol uses secure equality check protocol with plain input (so-called PlainEqual) introduced in [14] that allows PlainIntvlTest to be constant-rounds. PlainEqual takes two plain values and outputs whether the both values are equivalent or not. PlainEqual is constructed using constant-round multi-fan-in AND protocol AND* that takes several inputs and outputs AND of all of them. We show the algorithm of PlainEqual protocol briefly. Refer to [14] for more details.

Table 2. Look-up table

$x \in$	$f(x)$
(a_1, a_2)	y_1
(a_3, a_4)	y_2
(a_5, a_6)	y_3
(a_7, a_8)	y_4
(a_9, a_{10})	y_5
$\vdots$	$\vdots$
(a_{2t-1}, a_{2t})	y_t

Table 3. Look-up table for function $f(x) = \lfloor \log x \rfloor$

$x \in$	$f(x)$
$\{1\}$	0
$(2, 3)$	1
$(4, 7)$	2
$(8, 15)$	3
$(16, 31)$	4
$\vdots$	$\vdots$
$(2^{n-1}, p - 1)$	$n - 1$

AND^* protocol used in Algorithm 3 is executed as follows:

$$\mathsf{AND}^*(\llbracket v_0 \rrbracket, \llbracket v_1 \rrbracket, \dots, \llbracket v_{n-1} \rrbracket)$$
$$= \begin{cases} \llbracket 0 \rrbracket & \text{if } \sum_{i=0}^{n-1} v_i \in [0, n-1] \\ \llbracket 1 \rrbracket & \text{if } \sum_{i=0}^{n-1} v_i = n. \end{cases}$$

AND^* protocol introduced in [14] requires only 2 rounds. Therefore, $\mathsf{PlainEqual}$ protocol requires only 3 rounds since it has secure multiplication once at Step 5 and AND^* protocol at Step 6.

4 Applications Using IntvlTest and PlainIntvlTest

In this section, we show some applications using IntvlTest and PlainIntvlTest as sub-protocols.

4.1 Secure Look-Up Table

We construct a certain type of secure look-up table by using IntvlTest protocol. Given a Table 2 of function f that has separated intervals as its index, say $(a_1, a_2), (a_3, a_4), (a_5, a_6), \dots, (a_{2t-1}, a_{2t})$ for $a_1 < a_2 < a_3 < \cdots < a_{2t}$ shown in the left-hand column, we can compute $f(x)$ in shared form by using the values in the right-hand column and IntvlTest protocol as follows:

$$\llbracket f(x) \rrbracket \leftarrow \mathsf{IntvlTest}(\llbracket x \rrbracket, (a_1, a_2)) \times y_1 + \mathsf{IntvlTest}(\llbracket x \rrbracket, (a_3, a_4)) \times y_2$$
$$+ \mathsf{IntvlTest}(\llbracket x \rrbracket, (a_5, a_6)) \times y_3 + \mathsf{IntvlTest}(\llbracket x \rrbracket, (a_7, a_8)) \times y_4$$
$$\vdots$$
$$+ \mathsf{IntvlTest}(\llbracket x \rrbracket, (a_{2t-1}, a_{2t})) \times y_t.$$

Algorithm 4. IIT Protocol (Type 1)

Functionality: $[\![\delta]\!] \leftarrow \mathsf{IIT}([\![\![a]\!], [\![b]\!]], [c, d])$, where $\delta = 1$ if $[a, b] \cap [c, d] \neq \emptyset$; $\delta = 0$, otherwise.

Input: Shared interval $[[\![a]\!], [\![b]\!]]$ and public plain interval $[c, d]$

Output: Arithmetic shared value $[\![\delta]\!]$

1: Compute $[\![t']\!] \leftarrow \mathsf{LessThanEQ}([\![a]\!], d)$
2: Compute $[\![t'']\!] \leftarrow \mathsf{LessThanEQ}(c, [\![b]\!])$
3: Compute $[\![t]\!] \leftarrow [\![t']\!] \cdot [\![t'']\!]$
4: **return** $[\![t]\!]$.

Only one of the outputs from IntvlTest protocols should be 1 in shared form and thus the corresponding value y_i alone remains in shared form, which is required.

As a simple example, we construct a function $f(x) = \lfloor \log x \rfloor, x \in \mathbb{F}_p \setminus \{0\}$ that computes a rounding down of logarithm on x. Table 3 shows input and output of the function. Now, we use Table 3 to securely compute a function $f(x)$ as follows: Two servers jointly compute the following equation.

$$
\begin{aligned}
[\![f(x)]\!] \leftarrow & \mathsf{IntvlTest}([\![x]\!], (2, 3)) \times 1 + \mathsf{IntvlTest}([\![x]\!], (4, 7)) \times 2 \\
& + \mathsf{IntvlTest}([\![x]\!], (8, 15)) \times 3 + \mathsf{IntvlTest}([\![x]\!], (16, 31)) \times 4 \\
& \vdots \\
& + \mathsf{IntvlTest}([\![x]\!], (2^{n-1}, p-1)) \times (n-1).
\end{aligned}
$$

Note that this look-up table technique only uses several IntvlTest protocols in parallel, and multiplication of public values and secure addition, which can be executed locally without communication. Therefore, the above secure look-up table protocol requires only 4 rounds.

4.2 Secure Interval Intersection Test Protocol

Secure interval test protocol can be extended to secure interval *intersection* test protocol in an analogous manner, which we denote IIT. We consider two types of IIT protocols that takes as input (i) a shared interval and a plain interval (type 1), and (ii) two shared intervals (type 2), in order to check if the two intervals overlap.

In Algorithm 4, we show an IIT protocol of type 1, where we use LessThanEQ defined below as a sub-routine:

$$
\begin{aligned}
\mathsf{LessThanEQ}([\![x]\!], y) &\leftarrow \mathsf{IntvlTest}([\![x]\!], (0, y)) \\
\mathsf{LessThanEQ}(y, [\![x]\!]) &\leftarrow \mathsf{IntvlTest}([\![x]\!], (y, p-1)).
\end{aligned}
$$

Round complexity of this protocol is 5 since we use 4-round LessThanEQ and one secure multiplication as sub-routines.

Algorithm 5. IIT Protocol (Type 2)

Functionality: $[\![\delta]\!] \leftarrow \mathsf{IIT}([[\![a]\!], [\![b]\!]], [[\![c]\!], [\![d]\!]])$, where $\delta = 1$ if $[a, b] \cap [c, d] \neq \emptyset$; $\delta = 0$, otherwise.

Input: Shared intervals $[[\![a]\!], [\![b]\!]]$ and $[[\![c]\!], [\![d]\!]]$

Output: Arithmetic shared value $[\![\delta]\!]$

1: Compute $[\![t']\!] \leftarrow \mathsf{LessThanEQ}'([\![a]\!], [\![d]\!])$
2: Compute $[\![t'']\!] \leftarrow \mathsf{LessThanEQ}'([\![c]\!], [\![b]\!])$
3: Compute $[\![t]\!] \leftarrow [\![t']\!] \cdot [\![t'']\!]$
4: **return** $[\![t]\!]$.

Algorithm 6. PII Protocol (Type 1)

Functionality: $[[\![e]\!], [\![f]\!]] \leftarrow \mathsf{PII}([[\![a]\!], [\![b]\!]], [c, d])$, where $[e, f] = [a, b] \cap [c, d]$.

Input: Shared interval $[[\![a]\!], [\![b]\!]]$ and public plain interval $[c, d]$

Output: Shared interval $[[\![e]\!], [\![f]\!]]$

1: Compute $[\![t']\!] \leftarrow \mathsf{LessThanEQ}([\![a]\!], c)$
2: Compute $[\![e]\!] \leftarrow [\![t']\!] \cdot c + (1 - [\![t']\!]) \cdot [\![a]\!]$
3: Compute $[\![t'']\!] \leftarrow \mathsf{LessThanEQ}([\![b]\!], d)$
4: Compute $[\![f]\!] \leftarrow [\![t'']\!] \cdot [\![b]\!] + (1 - [\![t'']\!]) \cdot d$
5: **return** $[[\![e]\!], [\![f]\!]]$.

In Algorithm 5, we show an IIT protocol of type 2, where we use $\mathsf{LessThanEQ}'$ defined below as a sub-routine:

$$\mathsf{LessThanEQ}'([\![x]\!], [\![y]\!]) \leftarrow 1 - \mathsf{LessThan}([\![y]\!], [\![x]\!])$$

for 5-round $\mathsf{LessThan}$ defined in [14].

Straightforward round complexity of this protocol would be 6 since we use 5-round $\mathsf{LessThanEQ}'$ and one secure multiplication as sub-routines. However, the secure multiplication can merge into multi-fan-in AND protocol used in $\mathsf{LessThan}$, which makes the type 2 IIT protocol 5 rounds in total.

4.3   Private Interval Intersection

Analogous to the private set intersection, private *interval* intersection protocol takes two intervals and outputs the intersection of intervals in a privacy-preserving manner, which we denote PII protocol. Similar to IIT protocols, we consider two types of PII protocols that take as input (i) a shared interval and a plain interval (type 1), and (ii) two shared intervals (type 2), in order to output the intersection of these two intervals.

In Algorithm 6, we show a PII protocol of type 1, where we use $\mathsf{LessThanEQ}$ defined in Sect. 4.2 as a sub-routine. The round complexity of this type 1 PII protocol is 5 rounds since we use 4-round $\mathsf{LessThanEQ}$ (Step 1, 3) and secure multiplications (Step 2, 4), where Step 1, 2 and Step 3, 4 can be done in parallel.

In Algorithm 7, we show a PII protocol of type 2, where we use $\mathsf{LessThanEQ}'$ defined in Sect. 4.2 as a sub-routine. The round complexity of this type 2 PII

Algorithm 7. PII Protocol (Type 2)

Functionality: $[[e]], [[f]] \leftarrow \text{PII}([[a]], [[b]]], [[c]], [[d]]])$, where $[e, f] = [a, b] \cap [c, d]$.
Input: Shared intervals $[[a]], [[b]]]$ and $[[c]], [[d]]]$
Output: Shared interval $[[e]], [[f]]]$
1: Compute $[[t']] \leftarrow \text{LessThanEQ}'([[a]], [[c]])$
2: Compute $[[e]] \leftarrow [[t']] \cdot [[c]] + (1 - [[t']]) \cdot [[a]]$
3: Compute $[[t'']] \leftarrow \text{LessThanEQ}'([[b]], [[d]])$
4: Compute $[[f]] \leftarrow [[t'']] \cdot [[b]] + (1 - [[t'']]) \cdot [[d]]$
5: **return** $[[e]], [[f]]]$.

protocol is 5 rounds since we use 5-round LessThanEQ' (Step 1, 3) and secure multiplications (Step 2, 4), where secure multiplication can merge into the multi-fan-in AND protocol used in LessThanEQ'.

Acknowledgement. This work was supported by JST CREST JPMJCR19F6.

References

1. Araki, T., Furukawa, J., Lindell, Y., Nof, A., Ohara, K.: High-throughput semi-honest secure three-party computation with an honest majority. In: CCS 2016, pp. 805–817 (2016). https://doi.org/10.1145/2976749.2978331
2. Attrapadung, N., Hanaoka, G., Kiyomoto, S., Mimoto, T., Schuldt, J.C.N.: A taxonomy of secure two-party comparison protocols and efficient constructions. In: PST 2017, pp. 215–224 (2017). https://doi.org/10.1109/PST.2017.00033
3. Beaver, D.: Efficient multiparty protocols using circuit randomization. In: Feigenbaum, J. (ed.) CRYPTO 1991. LNCS, vol. 576, pp. 420–432. Springer, Heidelberg (1992). https://doi.org/10.1007/3-540-46766-1_34
4. Blake, I.F., Kolesnikov, V.: Strong conditional oblivious transfer and computing on intervals. In: Lee, P.J. (ed.) ASIACRYPT 2004. LNCS, vol. 3329, pp. 515–529. Springer, Heidelberg (2004). https://doi.org/10.1007/978-3-540-30539-2_36
5. Blake, I.F., Kolesnikov, V.: Conditional encrypted mapping and comparing encrypted numbers. In: Di Crescenzo, G., Rubin, A. (eds.) FC 2006. LNCS, vol. 4107, pp. 206–220. Springer, Heidelberg (2006). https://doi.org/10.1007/11889663_18
6. Catrina, O., de Hoogh, S.: Improved primitives for secure multiparty integer computation. In: Garay, J.A., De Prisco, R. (eds.) SCN 2010. LNCS, vol. 6280, pp. 182–199. Springer, Heidelberg (2010). https://doi.org/10.1007/978-3-642-15317-4_13
7. Damgård, I., Fitzi, M., Kiltz, E., Nielsen, J.B., Toft, T.: Unconditionally secure constant-rounds multi-party computation for equality, comparison, bits and exponentiation. In: Halevi, S., Rabin, T. (eds.) TCC 2006. LNCS, vol. 3876, pp. 285–304. Springer, Heidelberg (2006). https://doi.org/10.1007/11681878_15
8. Damgård, I., Geisler, M., Krøigaard, M.: Homomorphic encryption and secure comparison. IJACT 1(1), 22–31 (2008). https://doi.org/10.1504/IJACT.2008.017048
9. Damgard, I., Geisler, M., Kroigard, M.: A correction to 'efficient and secure comparison for on-line auctions'. Int. J. Appl. Crypt. 1(4), 323–324 (2009)

10. Garay, J., Schoenmakers, B., Villegas, J.: Practical and secure solutions for integer comparison. In: Okamoto, T., Wang, X. (eds.) PKC 2007. LNCS, vol. 4450, pp. 330–342. Springer, Heidelberg (2007). https://doi.org/10.1007/978-3-540-71677-8_22

11. Goldreich, O.: The Foundations of Cryptography - Volume 2, Basic Applications. Cambridge University Press, Cambridge (2004)

12. Goldreich, O., Micali, S., Wigderson, A.: How to play any mental game or a completeness theorem for protocols with honest majority. In: STOC 1987, pp. 218–229 (1987). https://doi.org/10.1145/28395.28420

13. Mohassel, P., Zhang, Y.: SecureML: a system for scalable privacy-preserving machine learning. In: S&P 2017, pp. 19–38 (2017). https://doi.org/10.1109/SP.2017.12

14. Morita, H., Attrapadung, N., Teruya, T., Ohata, S., Nuida, K., Hanaoka, G.: Constant-round client-aided secure comparison protocol. In: Lopez, J., Zhou, J., Soriano, M. (eds.) ESORICS 2018. LNCS, vol. 11099, pp. 395–415. Springer, Cham (2018). https://doi.org/10.1007/978-3-319-98989-1_20

15. Nergiz, A.E., Nergiz, M.E., Pedersen, T., Clifton, C.: Practical and secure integer comparison and interval check. In: PASSAT 2010, pp. 791–799 (2010)

16. Nishide, T., Ohta, K.: Multiparty computation for interval, equality, and comparison without bit-decomposition protocol. In: Okamoto, T., Wang, X. (eds.) PKC 2007. LNCS, vol. 4450, pp. 343–360. Springer, Heidelberg (2007). https://doi.org/10.1007/978-3-540-71677-8_23

17. Schoenmakers, B., Tuyls, P.: Practical two-party computation based on the conditional gate. In: Lee, P.J. (ed.) ASIACRYPT 2004. LNCS, vol. 3329, pp. 119–136. Springer, Heidelberg (2004). https://doi.org/10.1007/978-3-540-30539-2_10

18. Veugen, T.: Encrypted integer division and secure comparison. Int. J. Appl. Cryptol. **3**(2), 166–180 (2014). https://doi.org/10.1504/IJACT.2014.062738

19. Veugen, T., Blom, F., de Hoogh, S.J.A., Erkin, Z.: Secure comparison protocols in the semi-honest model. J. Sel. Top. Signal Process. **9**(7), 1217–1228 (2015). https://doi.org/10.1109/JSTSP.2015.2429117

20. Yao, A.C.: How to generate and exchange secrets (extended abstract). In: FOCS 1986, pp. 162–167. IEEE Computer Society (1986). https://doi.org/10.1109/SFCS.1986.25

LWE

Improved Multiplication Triple Generation over Rings via RLWE-Based AHE

Deevashwer Rathee[1]([⊠]), Thomas Schneider[2]([⊠]), and K. K. Shukla[1]

[1] Department of Computer Science and Engineering,
Indian Institute of Technology (BHU) Varanasi, Varanasi, India
{deevashwer.student.cse15,kkshukla.cse}@iitbhu.ac.in
[2] Department of Computer Science,
Technische Universität Darmstadt, Darmstadt, Germany
schneider@encrypto.cs.tu-darmstadt.de

Abstract. An important characteristic of recent MPC protocols is an input-independent setup phase in which most computations are offloaded, which greatly reduces the execution overhead of the online phase where parties provide their inputs. For a very efficient evaluation of arithmetic circuits in an information-theoretic online phase, the MPC protocols consume Beaver multiplication triples generated in the setup phase. Triple generation is generally the most expensive part of the protocol, and improving its efficiency is the aim of our work.

We specifically focus on computation over rings of the form $\mathbb{Z}_{2^\ell}$ in the semi-honest model and the two-party setting, for which an Oblivious Transfer (OT)-based protocol is currently the best solution. To improve upon this method, we propose a protocol based on RLWE-based Additively Homomorphic Encryption. Our experiments show that our protocol is more scalable, and it outperforms the OT-based protocol in most cases. For example, we improve communication by up to 6.9x and runtime by up to 3.6x for 64-bit triple generation.

Keywords: Secure two-party computation · Beaver multiplication triples · Ring-LWE · Additively Homomorphic Encryption

1 Introduction

Secure multi-party computation (MPC) allows a set of distrusting parties to jointly compute a function on their inputs while keeping them private from one another. There is a multitude of MPC protocols such as [9,10,15] that allow secure evaluation of arithmetic circuits, which form the basis of many privacy-preserving applications. An important characteristic of many of the recent MPC protocols is an input-independent setup phase in which most computations are offloaded, which greatly reduces the execution overhead of the online phase where parties provide their inputs. The idea is to compute Beaver multiplication triples

© Springer Nature Switzerland AG 2019
Y. Mu et al. (Eds.): CANS 2019, LNCS 11829, pp. 347–359, 2019.
https://doi.org/10.1007/978-3-030-31578-8_19

[3] in the setup phase, and then use them to evaluate arithmetic circuits very efficiently in an information-theoretic online phase, without using any cryptographic operations. In light of their significance on the overall runtime of the protocol, the main focus of this work is efficient generation of such triples in the semi-honest setting.

In the malicious model and the multi-party setting, the first to employ RLWE-based Somewhat Homomorphic Encryption (SHE) for triple generation were [9] in 2012. Their major source of efficiency was the packing method from [22]. In 2016, this method was replaced by an Oblivious Transfer (OT)-based method by Keller et al. [15]. Later in 2017, SHE emerged again with the Overdrive methodology [16]. These protocols were designed to generate triples over a finite field which can only be used to support finite field arithmetic in the online phase. In 2018, Cramer et al. [7] proposed an OT-based protocol that generates triples over rings of the form $\mathbb{Z}_{2^\ell}$. Designing protocols over rings is useful in a lot of applications since it greatly simplifies implementation of comparisons and bitwise operations, which are inefficient to realize with finite field arithmetic. Apart from this, using ring-based protocols also implies that we can leverage some special tricks that computers already implement to make integer arithmetic very efficient. In 2019, Orsini et al. [18] presented a more compact solution based on SHE and argued that it is more efficient than the OT-based protocol of [7]. Concurrently, Catalano et al. [5] used the Joye-Libert homomorphic cryptosystem [14] to improve upon the communication of [7] particularly for larger choices of ℓ.

Our Contributions. In this paper, we consider the semi-honest model and the two-party setting, for which the current best method for generating triples over rings is the OT-based approach of [10]. Taking inspiration from the changing trend in the malicious model, we propose a protocol based on RLWE-based Additively Homomorphic Encryption (RLWE-AHE) that improves upon the OT-based solution. In the process, we analyze the popular approaches for triple generation using AHE and adapt them to using state-of-the-art RLWE-AHE and our scenario. We also argue why the approach taken in [18] does not provide the most efficient solution in our semi-honest setting. Our experiments show that our protocol is more scalable, and it outperforms the OT-based protocol in most cases. For example, we improve communication over [10] by up to 6.9x and runtime by up to 3.6x for 64-bit triple generation.

2   Preliminaries

2.1   Notation

We denote the players as P_0 and P_1. κ denotes the symmetric security parameter, σ the statistical security parameter, and λ the computational security parameter. $\langle x \rangle$ is a shared value of $x \in \{0,1\}^\ell$, which is a pair of ℓ-bit shares $(\langle x \rangle_0, \langle x \rangle_1)$, where the subscript represents the party that holds the share. A vector of shares is represented in bold face e.g. $\langle \mathbf{x} \rangle$, and multiplication, denoted by $\cdot$, is performed component-wise on it. To represent an element x being sampled uniformly at random from G, we use the notation $x \leftarrow_\$ G$. Assignment modulo 2^ℓ is denoted by $\leftarrow_\ell$.

Functionality $\mathcal{F}_{\mathsf{Triple}}^n$: Sample values $\mathbf{a}_0, \mathbf{a}_1, \mathbf{b}_0, \mathbf{b}_1, \mathbf{r} \leftarrow_\$ (\mathbb{Z}_{2^\ell})^n$. Output tuples $(\mathbf{a}_0, \mathbf{b}_0, (\mathbf{a}_0 + \mathbf{a}_1) \cdot (\mathbf{b}_0 + \mathbf{b}_1) + \mathbf{r})$ and $(\mathbf{a}_1, \mathbf{b}_1, -\mathbf{r})$ to P_0 and P_1, respectively, where arithmetic is performed modulo 2^ℓ.

Fig. 1. Functionality for generating Beaver multiplication triples.

2.2 Problem Statement

A Beaver multiplication triple [3] is defined as the tuple $(\langle a \rangle, \langle b \rangle, \langle c \rangle)$ satisfying:

$$(\langle a \rangle_0 + \langle a \rangle_1) \cdot (\langle b \rangle_0 + \langle b \rangle_1) \equiv (\langle c \rangle_0 + \langle c \rangle_1) \bmod 2^\ell.$$

Our aim is to construct a two-party protocol that securely realizes the $\mathcal{F}_{\mathsf{Triple}}^n$ functionality which is defined in Fig. 1.

2.3 Security Model

Our protocol is secure against a semi-honest and computationally bounded adversary. This adversary tries to learn information from the messages it sees during the protocol execution, without deviating from the protocol.

2.4 Ring-LWE-Based Additively Homomorphic Encryption (RLWE-AHE)

We use an IND-CPA secure AHE scheme with the following 5 algorithms:

- $\mathsf{KeyGen}(1^\lambda) \to (\mathsf{pk}, \mathsf{sk})$: Key Generation is a randomized algorithm that outputs the key pair $(\mathsf{pk}, \mathsf{sk})$, with public key pk and secret key sk. We consider a single key pair $(\mathsf{pk}, \mathsf{sk})$ throughout the entire paper.
- $\mathsf{Enc}(\mathsf{pk}, \mathbf{m}) \to \mathsf{ct}$: Encryption is a randomized algorithm that takes a vector $\mathbf{m} \in (\mathbb{Z}_p)^n$ as input, where n depends on scheme parameters m and p (cf. Sect. 4.1), along with pk, and outputs a ciphertext ct. We assume that all ciphertexts in the following description of the scheme are encrypted with public key pk.
- $\mathsf{Dec}(\mathsf{sk}, \mathsf{ct}) \to \mathbf{m}$: Decryption takes the secret key sk and a ciphertext ct, and outputs the plaintext $\mathbf{m} \in (\mathbb{Z}_p)^n$.

- Add(pk; ct_1, ct_2) $\rightarrow$ ct': Addition takes as input two ciphertexts ct_1, ct_2 and the public key pk, and outputs a ciphertext ct' such that Dec(sk, ct') = $m_1 + m_2 \in (\mathbb{Z}_p)^n$, where addition is performed component-wise. This algorithm is also denoted by the $\oplus_{pk}$ operator.
- ScalarMult(pk; ct, s) $\rightarrow$ ct': Given inputs ciphertext ct and scalar s, and the public key pk, scalar-multiplication outputs a ciphertext ct' such that Dec(sk, ct') = Dec(sk, ct) $\cdot$ s $\in (\mathbb{Z}_p)^n$, where multiplication is performed component-wise. This algorithm is also denoted by the $\odot_{pk}$ operator.

Possible instantiations of RLWE-based schemes that satisfy the description above are [4,11]. These schemes are IND-CPA secure, and their security relies on the Decision RLWE assumption [17]. We assume that the parameters of the scheme have been chosen to be large enough to allow evaluation of the circuit for our triple generation protocol and accommodate the extra noise added to prevent leakage through ciphertext noise (cf. Sect. 4.3).

3   Previous Works

The previous approaches for generating multiplication triples in the semi-honest model are based on AHE and OT. Initially, Beaver triples were generated using AHE schemes such as Paillier [19] and DGK [8]. However, the authors in [10] showed that the OT-based generation method greatly outperforms the AHE-based generation, and is currently the best method. In this section, we summarize both approaches. Although the protocols based on AHE are much slower, they are the basis for our proposed protocol.

3.1   AHE-Based Generation

Case I - $2^\ell|p$. Figure 2 describes a well-known protocol for generating triples using AHE [20]. This protocol generates multiplication triples in $\mathbb{Z}_{2^\ell}$, using an AHE scheme with plaintext modulus p, and it works if and only if $2^\ell|p$. This is due to the fact that the AHE scheme implicitly reduces the underlying plaintext modulo p. We can use the DGK cryptosystem [8] since it uses a 2-power modulus.

Case II - $2^\ell \nmid p$. We start by choosing r from an interval such that $d = \langle a \rangle_0 \cdot \langle b \rangle_1 + \langle b \rangle_0 \cdot \langle a \rangle_1 + r$ does not overflow the bound p. This affects the security of the protocol as we no longer have information theoretic security provided by uniform random masking by r. To get around this issue, we resort to "smudging" [1], where we get statistical security of σ-bits by sampling r from an interval that is by factor 2^σ larger than the upper bound on magnitude of the expression $v = \langle a \rangle_0 \cdot \langle b \rangle_1 + \langle b \rangle_0 \cdot \langle a \rangle_1$. Since the upper bound on v is $2^{2\ell+1}$, we sample r from $\mathbb{Z}_{2^{2\ell+\sigma+1}}$. Consequently, the plaintext modulus p has to be of bitlength $2\ell + \sigma + 2$. This prevents the overflow and provides statistical security of σ-bits [20]. We can instantiate this case with the Paillier cryptosystem [19], whose plaintext modulus is the product of two distinct primes.

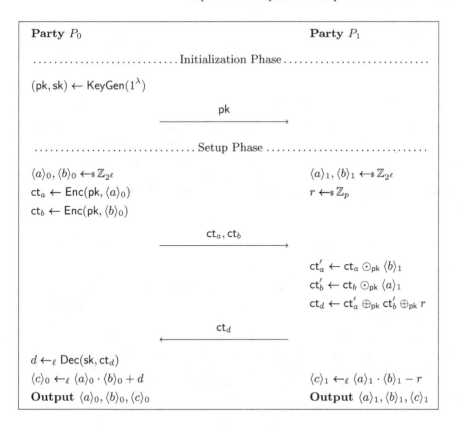

Fig. 2. $\Pi_{\mathsf{BasicTripleAHE}}$: Basic Beaver Triple Generation using AHE.

3.2 OT-Based Generation

The feasibility result for triple generation over $\mathbb{Z}_{2^\ell}$ using Oblivious Transfer was given in [13], and it was shown in [10] that it is the currently best method for triple generation in the semi-honest setting. This protocol facilitates the triple generation by allowing secure computation of the product of two secret values. The amortized complexity of generating a triple in $\mathbb{Z}_{2^\ell}$ using OT-based generation is 2ℓ Correlated-OT (C-OT) over $(\ell+1)/2$-bit strings (cf. [10]). The protocol uses state-of-the-art C-OT extension (cf. [2]) that requires $\kappa + \ell$-bit communication per C-OT on ℓ-bit strings.

4 RLWE-Based Generation

In Sect. 3.1, we described two cases, namely $2^\ell | p$ and $2^\ell \nmid p$, and presented a protocol for both of them. While we can build a protocol based on our RLWE-AHE scheme that follows a similar design as in Sect. 3.1 for both cases, the two protocols are not equally efficient. In this section, we analyze these differences and show that the protocol for $2^\ell \nmid p$ is more efficient. Before comparing the cases, we detail two optimizations and a security consideration that are crucial for our analysis.

4.1   Batching Optimization

Using a RLWE-AHE scheme, we can generate many triples at the cost of generating one by leveraging the ciphertext packing technique described in [22]. For a prime p, we can encrypt a maximum of $n = \phi(m)/\mathrm{ord}_{\mathbb{Z}_m^*}(p)$ plaintexts $m_i \in \mathbb{Z}_p$ in a single ciphertext. The operations performed on a ciphertext are applied to all the slots of the underlying plaintext in parallel. As a result, in a single run of the protocol, we can generate n triples.

4.2   CRT Optimization

Using a very large plaintext modulus p results in inefficient instantiations since a larger p leads to a larger ciphertext modulus to contain the noise growth. Therefore, we use the CRT optimization to split the plaintext modulus p into e distinct primes p_i of equal bitlength such that $p = \prod_{i=1}^{i=e} p_i$ for some $e \in \mathbb{Z}$. We create e different instances of the cryptosystem for each p_i, and the whole protocol is performed for each instance. The plaintexts produced after decryption are combined using the Chinese Remainder Theorem (CRT) (with precomputed tables) to get the output in $\mathbb{Z}_p$. This technique also has the advantage that it can be parallelized in a straightforward manner.

4.3   Leakage Through Ciphertext Noise

The ciphertexts of RLWE-based schemes have noise associated with them, whose distribution gets skewed on performing homomorphic operations on the ciphertext. This can lead to potential leakage through noise, and reveal private information input by the evaluator to the key owner. A solution to this problem, called the noise flooding technique, was proposed in [12]. This technique involves adding a statistically independent noise from an interval B' much larger than B, assuming that the ciphertext noise is bounded by B at the end of the computation. Specifically, this is done by publicly adding an encryption of zero with noise taken uniformly from $[-B', B']$ such that $B' > 2^\sigma B$, to provide statistical security of σ bits. We denote the encryption with noise from an interval $p \cdot 2^\sigma$ times larger than the normal encryption as Enc'.

4.4   Parameter Selection

The plaintext modulus p determines the protocol to be used as described in Sect. 3.1. After determining p, we can determine the other parameters to maximize efficiency as follows:

Case I - $2^\ell | p$: This approach was recently considered in [18] for the malicious model. In order to generate authenticated triples in $\mathbb{Z}_{2^\ell}$, the authors required Zero Knowledge Proofs of Knowledge (ZKPoKs) and triples to be generated in $\mathbb{Z}_{2^{\ell+s}}$ to prevent a malicious adversary from modifying the triples with error probability $2^{-s+\log s}$. However in the semi-honest setting, the adversaries can not deviate from the protocol. Hence we do not require ZKPoKs, and computing triples in $\mathbb{Z}_{2^\ell}$ suffices. We start by choosing m to be a prime like in [18] to ensure a better underlying geometry. Given that d is the order of 2 in $\mathbb{Z}_m^*$, we get $n = \phi(m)/d$ slots, each of which embeds a d-degree polynomial (cf. [22]). In order to utilize the higher coefficients of the polynomial embedded in each slot, we employ the packing method from [18] to achieve a maximum utilization of $\phi(m)/5$ slots. Despite this significant optimization, most of the slots are wasted. Moreover, since p is a power of 2, we can not use the CRT optimization.

Case II - $2^\ell \nmid p$: Here, we choose m to be a power of 2 for efficiency reasons described in [6], and big enough to provide security greater than 128-bits. Accordingly, we choose a prime plaintext modulus p of $2\ell + \sigma + 2$ bits that satisfies $p \equiv 1 \bmod m$, thereby maximizing the number of slots to $\phi(m)$. A concern of inefficiency here is that now our plaintext modulus is much larger than it was in the previous case. However, using the CRT optimization, we can split the plaintext modulus into e distinct primes p_i and get e instances of the cryptosystem with similar parameter lengths as in the previous case. A run of the protocol will require e times more computation and communication, but we can use the maximum number of slots. An important consideration here is that while we will have similar plaintext modulus and ciphertext modulus bitlengths, taking a 2-power m might result in an *at most* twice as large n than is required for 128-bit security. However with increasing n, the communication and computation increase only linearly and quasi-linearly respectively, and the number of triples generated increase linearly as well. Therefore, the amortized communication remains the same and the amortized computation increases *at most* by a factor of $\Delta = (\log(n) + 1)/\log(n)$, which is small for the minimum value of n typically required to maintain security (for $n = 4096$, $\Delta = 1.08$).

Conclusion: A single run of the protocol for Case I requires $e = (2\ell + \sigma + 2)/\ell$ times more computation and communication than Case II. However, the protocol for Case II requires *at least* 5 runs of the protocol to generate the same number of triples. Hence, considering $\sigma = 40$-bits and with the exception of small values of ℓ ($\ell \leq 15$), Case II is more efficient. Although we conclude that Case I could be better for smaller ℓ, we have implemented the protocol just for Case II because SEAL [6], currently the most efficient publicly available library that satisfies the description of our RLWE-AHE scheme, only supports 2-power cyclotomics.

4.5   Our Final Protocol

Our final protocol is given in Fig. 3. In the protocol, we have shown an initialization phase for the generation of n triples. However, arbitrary many triples can be generated following a single initialization phase (involving a single key-pair).

As discussed above, we have used the parameters for Case II with $2^\ell \nmid p$. Rather than drowning the ciphertext noise with a fresh encryption of zero with extra noise, we combine it with the step of adding r, and simply add a fresh encryption of r with extra noise. The advantage of using RLWE-AHE for generating triples is not only efficiency (cf. Sect. 5); we also get post-quantum security, unlike the OT-based approach which heavily relies on OT extension for efficiency.

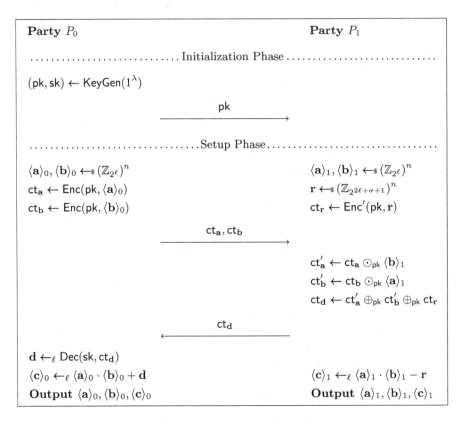

Fig. 3. $\Pi_{\mathsf{TripleRLWE}}$: Beaver Triple Generation using RLWE-AHE. Enc′ denotes encryption with extra noise (cf. Sect. 4.3) and n denotes the number of plaintext slots (cf. Sect. 4.1).

Theorem 1. *The $\Pi_{\mathsf{TripleRLWE}}$ protocol (cf. Fig. 3) securely computes the $\mathcal{F}^n_{\mathsf{Triple}}$ functionality (cf. Fig. 1) in the presence of semi-honest adversaries, providing statistical security against a corrupted P_0 and computational security against a corrupted P_1.*

Proof. We first show that the output of the functionality $\mathcal{F}^n_{\mathsf{Triple}}$ and the output of the protocol $\Pi_{\mathsf{TripleRLWE}}$ are identically distributed. Then we construct a simulator for each corrupted party that outputs a view consistent with the output of the functionality.

Output Distribution. The functionality chooses the shares $\langle \mathbf{a} \rangle_i, \langle \mathbf{b} \rangle_i$ uniformly at random for $i \in \{0, 1\}$, as do the parties P_0 and P_1 in the protocol, which makes them identically distributed in both cases. Let $\mathbf{u} = (\langle \mathbf{a} \rangle_0 + \langle \mathbf{a} \rangle_1) \cdot (\langle \mathbf{b} \rangle_0 + \langle \mathbf{b} \rangle_1) \mod 2^\ell$ and $\mathbf{v} = \langle \mathbf{a} \rangle_1 \cdot \langle \mathbf{b} \rangle_1 \mod 2^\ell$. The functionality sets $\langle \mathbf{c} \rangle_0 = \mathbf{u} + \mathbf{r} \mod 2^\ell$ and $\langle \mathbf{c} \rangle_1 = -\mathbf{r} \mod 2^\ell$ for some $\mathbf{r} \leftarrow_{\$} (\mathbb{Z}_{2^\ell})^n$, while the parties compute $\langle \mathbf{c} \rangle_0 = \mathbf{u} + (\mathbf{r}^* - \mathbf{v}) \mod 2^\ell$ and $\langle \mathbf{c} \rangle_1 = -(\mathbf{r}^* - \mathbf{v}) \mod 2^\ell$ for some $\mathbf{r}^* \leftarrow_{\$} (\mathbb{Z}_{2^{2\ell+\sigma+1}})^n$. Since $2^\ell \mid 2^{2\ell+\sigma+1}$, $\mathbf{t} = \mathbf{r}^* - \mathbf{v} \mod 2^\ell$ is uniformly distributed in $(\mathbb{Z}_{2^\ell})^n$ and the joint distribution of $\langle \mathbf{c} \rangle_0$ and $\langle \mathbf{c} \rangle_1$ is identically distributed in the ideal functionality and the protocol. Hence, the output is identically distributed in both scenarios.

Corrupted P_0. The Simulator S_0 receives $(\langle \mathbf{a} \rangle_0, \langle \mathbf{b} \rangle_0, \langle \mathbf{c} \rangle_0)$ as input. It chooses a uniformly random tape ρ for P_0, and uses this tape to run $(\mathsf{pk}, \mathsf{sk}) \leftarrow \mathsf{KeyGen}(1^\lambda)$. It then uses independent randomness to sample a uniformly random $\mathbf{d}^* \in (\mathbb{Z}_{2^{2\ell+\sigma+1}})^n$ such that $\mathbf{d}^* \equiv \langle \mathbf{c} \rangle_0 - \langle \mathbf{a} \rangle_0 \cdot \langle \mathbf{b} \rangle_0 \mod 2^\ell$, and to encrypt $\mathbf{d}^*$ with extra noise. Its output is $(\rho, \mathsf{ct}_{\mathbf{d}^*} = \mathsf{Enc}'(\mathsf{pk}, \mathbf{d}^*))$.

$\mathsf{ct}_{\mathbf{d}^*}$ is statistically indistinguishable from $\mathsf{ct}_{\mathbf{d}}$ received by P_0 in the protocol. This follows from the fact that $t = v + r$ and r^*, where $v \in \mathbb{Z}_{2^{2\ell+1}}$ and $r, r^* \leftarrow_{\$} \mathbb{Z}_{2^{2\ell+\sigma+1}}$, are statistically $2^{-\sigma}$ indistinguishable [20]. Therefore, the underlying plaintexts are statistically indistinguishable. From a similar argument, the ciphertexts are also statistically indistinguishable (cf. Sect. 4.3). Hence, the output distributions are identical and the corresponding views are statistically indistinguishable, implying that the joint distribution of party P_0's view $(\rho, \mathsf{ct}_{\mathbf{d}})$ and the protocol output is statistically indistinguishable in the ideal and the real execution.

Corrupted P_1. The Simulator S_1 receives $(\langle \mathbf{a} \rangle_1, \langle \mathbf{b} \rangle_1, \langle \mathbf{c} \rangle_1)$ as input. It chooses a uniformly random tape ρ for P_1. It then uses independent randomness to run $(\mathsf{pk}, \mathsf{sk}) \leftarrow \mathsf{KeyGen}(1^\lambda)$, and to perform encryptions on a vector of zeros (denoted by $\mathbf{0}^n$) using pk. Its output is $(\rho, \mathsf{pk}, \mathsf{ct}_{\mathbf{a}} = \mathsf{Enc}(\mathsf{pk}, \mathbf{0}^n), \mathsf{ct}_{\mathbf{b}} = \mathsf{Enc}(\mathsf{pk}, \mathbf{0}^n))$.

The computational indistinguishability of the view follows from the IND-CPA security of the AHE scheme (cf. Sect. 2.4), because the distinguisher doesn't have access to the randomness used to generate the key-pair $(\mathsf{pk}, \mathsf{sk})$ with which $\mathsf{ct}_{\mathbf{a}}$ and $\mathsf{ct}_{\mathbf{b}}$ were encrypted. Hence the joint distribution of party P_1's view $(\rho, \mathsf{pk}, \mathsf{ct}_{\mathbf{a}}, \mathsf{ct}_{\mathbf{b}})$ and the protocol output is computationally indistinguishable in the ideal and the real execution. □

5   Implementation Results

In this section, we compare the performance of our RLWE-based method (cf. Sect. 4) with the OT-based method (cf. Sect. 3.2) for generating Beaver multiplication triples.

Experimental Setup. Our benchmarks were performed on two servers, each equipped with an Intel Core i9-7960X @ 2.8 GHz CPU with 16 physical cores and 128 GB RAM. We consider triple generation for bitlenghts $\ell \in \{8, 16, 32, 64\}$.

We have used the Microsoft SEAL library v3.1 [6] to implement the RLWE-based method $\Pi_{\mathsf{TripleRLWE}}$, and the OT-based method Π_{TripleOT} is implemented in ABY library [10]. In all experiments, we have set the symmetric security parameter to $\kappa = 128$, and the statistical security parameter to $\sigma = 40$. The computational security parameter λ for the RLWE-AHE scheme has been chosen to get security of at least 128-bits (see the full version of our paper [21] for concrete parameters).

We run the benchmarks for three network settings (bandwidth, latency): LAN10 (10 Gbps, 0.5 ms RTT), LAN1 (1 Gbps, 0.5 ms RTT), and WAN (100 Mbps, 50 ms RTT). In each setting, we performed experiments for $N \in \{2^{15}, 2^{16}, \ldots, 2^{22}\}$ triples and $T \in \{2, 8, 32\}$ threads.

Results and Analysis. We give the amortized (over generating $N = 2^{20}$ triples) runtimes in Table 1 and the communication in Table 2 to compute one Beaver multiplication triple using RLWE-AHE and OT for bitlengths $\ell \in \{8, 16, 32, 64\}$. Extensive plots of the results of our experiments are given in the full version of our paper [21] and we give some highlights in Fig. 4. The results of our experiments can be summarized as follows:

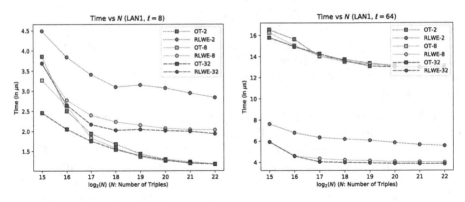

Fig. 4. Performance plots showing amortized runtime (over generating N triples) to compute one ℓ-bit Beaver multiplication triple in the LAN1 scenario. The legend entries represent the method and the number of threads T used.

1. RLWE-AHE requires less communication than OT, and the difference grows with increasing ℓ. For $\ell = 64$, the improvement factor over OT is 6.9x.
2. RLWE-AHE requires more computation than OT for smaller bitlengths, since OT has a smaller runtime than RLWE-AHE in the LAN10 setting where communication is not a bottleneck.
3. RLWE-AHE is faster than OT for larger bitlengths due to lower computation and communication requirements, achieving speedup of 3.6x for $\ell = 64$ in the WAN setting.

Table 1. Amortized runtime (in μs) for generating one ℓ-bit Beaver multiplication triple with T threads in the LAN10, LAN1, and WAN setting. A total of $N = 2^{20}$ triples are generated. Smallest values are marked in bold.

Setting	ℓ	$T = 2$			$T = 8$			$T = 32$		
		OT	RLWE	Impr.	OT	RLWE	Impr.	OT	RLWE	Impr.
LAN10	8	**0.92**	2.36	0.39x	**0.35**	0.70	0.51x	**0.24**	0.51	0.47x
	16	**1.74**	2.38	0.73x	**0.56**	0.69	0.81x	**0.39**	0.50	0.77x
	32	3.35	**2.37**	1.41x	0.99	**0.68**	1.46x	0.75	**0.49**	1.51x
	64	6.53	**4.61**	1.41x	1.89	**1.30**	1.46x	1.61	**0.80**	2.01x
LAN1	8	**1.30**	3.07	0.42x	**1.27**	2.07	0.61x	**1.28**	2.02	0.64x
	16	**2.64**	3.08	0.85x	2.56	**2.09**	1.22x	2.58	**1.99**	1.29x
	32	5.55	**3.07**	1.81x	5.53	**2.34**	2.36x	5.49	**2.24**	2.45x
	64	13.14	**5.85**	2.25x	13.09	**4.06**	3.23x	13.03	**3.88**	3.35x
WAN	8	20.48	**20.02**	1.02x	**19.33**	25.11	0.77x	**20.14**	22.90	0.88x
	16	31.10	**20.39**	1.53x	32.66	**26.11**	1.25x	28.98	**23.83**	1.22x
	32	60.81	**23.85**	2.55x	60.22	**26.42**	2.28x	61.25	**26.44**	2.32x
	64	140.48	**39.34**	3.57x	138.54	**45.20**	3.07x	140.79	**41.57**	3.39x

Table 2. Amortized communication (in Bytes) for generating one ℓ-bit Beaver multiplication triple. Smallest values are marked in bold.

ℓ	OT	RLWE	Impr.
8	272	**224**	1.21x
16	576	**224**	2.57x
32	1280	**256**	5.00x
64	3072	**448**	6.85x

4. OT is faster than RLWE-AHE for smaller bitlengths in most cases. For instance, OT is better than RLWE-AHE in all cases for $\ell = 8$ in the LAN1 setting (cf. Fig. 4a).

5. Due to less communication, the improvement factor in runtime of RLWE-AHE over OT increases with decreasing network performance.

6. RLWE-AHE benefits more from multi-threading than OT for faster networks. For $\ell = 64$ in the LAN1 setting, the improvement factor increases from 2.25x to 3.35x as we move from 2 to 32 threads. When communication is the bottleneck, multi-threading does not benefit either method.

7. OT benefits more from increasing N in general, and the gains are more prominent for smaller ℓ. For $\ell = 8$ (resp. 64) in the LAN1 setting and $T = 2$ threads, the performance of OT improves by 3.27x (resp. 1.25x) as we increase N from 2^{15} to 2^{22}, compared to a performance improvement by 1.58x (resp. 1.37x) for RLWE-AHE.

8. The performance of RLWE-AHE saturates for a smaller N as compared to OT. For instance, in Fig. 4, the performance of RLWE-AHE saturates at

$N = 2^{18}$ for $\ell = 8$ (resp. $N = 2^{17}$ for $\ell = 64$), while the performance of OT saturates at $N = 2^{21}$ (resp. $N = 2^{19}$).

9. Overall, our RLWE-based method is a better option for most practical cases. It is faster in almost all scenarios for the WAN setting, while even in the LAN10 setting, the performance improvement is significant for larger bitlengths. However, for smaller bitlengths such as $\ell = 8$, the OT-based method is more suitable even in the WAN setting.

Acknowledgements. This work was co-funded by the DFG as part of project E4 within the CRC 1119 CROSSING and project A.1 within the RTG 2050 "Privacy and Trust for Mobile Users", and by the BMBF and the HMWK within CRISP.

References

1. Asharov, G., Jain, A., López-Alt, A., Tromer, E., Vaikuntanathan, V., Wichs, D.: Multiparty computation with low communication, computation and interaction via threshold FHE. In: Pointcheval, D., Johansson, T. (eds.) EUROCRYPT 2012. LNCS, vol. 7237, pp. 483–501. Springer, Heidelberg (2012). https://doi.org/10.1007/978-3-642-29011-4_29

2. Asharov, G., Lindell, Y., Schneider, T., Zohner, M.: More efficient oblivious transfer and extensions for faster secure computation. In: ACM CCS (2013)

3. Beaver, D.: Efficient multiparty protocols using circuit randomization. In: Feigenbaum, J. (ed.) CRYPTO 1991. LNCS, vol. 576, pp. 420–432. Springer, Heidelberg (1992). https://doi.org/10.1007/3-540-46766-1_34

4. Brakerski, Z., Gentry, C., Vaikuntanathan, V.: (Leveled) fully homomorphic encryption without bootstrapping. In: Innovations in Theoretical Computer Science Conference (2012)

5. Catalano, D., Raimondo, M.D., Fiore, D., Giacomelli, I.: Mon$\mathbb{Z}_{2^k}$a: fast maliciously secure two party computation on $\mathbb{Z}_{2^k}$. Cryptology ePrint Archive, Report 2019/211 (2019)

6. Chen, H., Laine, K., Player, R.: Simple encrypted arithmetic library - SEAL v2.1. In: Brenner, M., et al. (eds.) FC 2017. LNCS, vol. 10323, pp. 3–18. Springer, Cham (2017). https://doi.org/10.1007/978-3-319-70278-0_1. https://github.com/microsoft/SEAL

7. Cramer, R., Damgård, I., Escudero, D., Scholl, P., Xing, C.: SPD$\mathbb{Z}_{2^k}$: Efficient MPC mod 2^k for dishonest majority. In: Shacham, H., Boldyreva, A. (eds.) CRYPTO 2018. LNCS, vol. 10992, pp. 769–798. Springer, Cham (2018). https://doi.org/10.1007/978-3-319-96881-0_26

8. Damgård, I., Geisler, M., Krøigård, M.: Homomorphic encryption and secure comparison. Int. J. Appl. Crypt. **1**(1), 22–31 (2008)

9. Damgård, I., Pastro, V., Smart, N., Zakarias, S.: Multiparty computation from somewhat homomorphic encryption. In: Safavi-Naini, R., Canetti, R. (eds.) CRYPTO 2012. LNCS, vol. 7417, pp. 643–662. Springer, Heidelberg (2012). https://doi.org/10.1007/978-3-642-32009-5_38

10. Demmler, D., Schneider, T., Zohner, M.: ABY - a framework for efficient mixed-protocol secure two-party computation. In: NDSS (2015). https://encrypto.de/code/ABY

11. Fan, J., Vercauteren, F.: Somewhat practical fully homomorphic encryption. Cryptology ePrint Archive, Report 2012/144 (2012)

12. Gentry, C.: Fully homomorphic encryption using ideal lattices. In: STOC (2009)

13. Gilboa, N.: Two party RSA key generation. In: Wiener, M. (ed.) CRYPTO 1999. LNCS, vol. 1666, pp. 116–129. Springer, Heidelberg (1999). https://doi.org/10.1007/3-540-48405-1_8

14. Joye, M., Libert, B.: Efficient cryptosystems from 2^k-th power residue symbols. In: Johansson, T., Nguyen, P.Q. (eds.) EUROCRYPT 2013. LNCS, vol. 7881, pp. 76–92. Springer, Heidelberg (2013). https://doi.org/10.1007/978-3-642-38348-9_5

15. Keller, M., Orsini, E., Scholl, P.: MASCOT: faster malicious arithmetic secure computation with oblivious transfer. In: ACM CCS (2016)

16. Keller, M., Pastro, V., Rotaru, D.: Overdrive: making SPDZ great again. In: Nielsen, J.B., Rijmen, V. (eds.) EUROCRYPT 2018. LNCS, vol. 10822, pp. 158–189. Springer, Cham (2018). https://doi.org/10.1007/978-3-319-78372-7_6

17. Lyubashevsky, V., Peikert, C., Regev, O.: On ideal lattices and learning with errors over rings. In: Gilbert, H. (ed.) EUROCRYPT 2010. LNCS, vol. 6110, pp. 1–23. Springer, Heidelberg (2010). https://doi.org/10.1007/978-3-642-13190-5_1

18. Orsini, E., Smart, N.P., Vercauteren, F.: Overdrive2k: efficient secure MPC over $\mathbb{Z}_{2^k}$ from somewhat homomorphic encryption. Cryptology ePrint Archive, Report 2019/153 (2019)

19. Paillier, P.: Public-key cryptosystems based on composite degree residuosity classes. In: Stern, J. (ed.) EUROCRYPT 1999. LNCS, vol. 1592, pp. 223–238. Springer, Heidelberg (1999). https://doi.org/10.1007/3-540-48910-X_16

20. Pullonen, P., Bogdanov, D., Schneider, T.: The design and implementation of a two-party protocol suite for sharemind 3. Technical report, CYBERNETICA Institute of Information Security (2012)

21. Rathee, D., Schneider, T., Shukla, K.K.: Improved multiplication triple generation over rings via RLWE-based AHE. Cryptology ePrint Archive, Report 2019/577 (2019)

22. Smart, N.P., Vercauteren, F.: Fully homomorphic SIMD operations. Designs Codes Crypt. **71**(1), 57–81 (2014)

Fractional LWE: A Nonlinear Variant of LWE

Gerald Gavin[(✉)] and Stephane Bonnevay

Laboratory ERIC, University of Lyon, Lyon, France
{gerald.gavin,stephane.bonnevay}@univ-lyon1.fr

Abstract. Many cryptographic constructions are based on the famous problem LWE [Reg05]. In particular, this cryptographic problem is currently the most relevant to build FHE [GSW13,BV11]. In [BV11], encrypting x consists of randomly choosing a vector c satisfying $\langle s, c \rangle = x + \mathsf{noise} \pmod{q}$ where s is a secret size-n vector. While the vector sum is a homomorphic operator, such a scheme is intrinsically vulnerable to lattice-based attacks. To overcome this, we propose to define c as a pair of vectors (u, v) satisfying $\langle s, u \rangle / \langle s, v \rangle = x + \mathsf{noise} \pmod{q}$. This simple scheme is based on a new cryptographic problem intuitively not easier than LWE, called Fractional LWE (FLWE). While some homomorphic properties are lost, the secret vector s could be hopefully chosen shorter leading to more efficient constructions. We extensively study the hardness of FLWE. We first prove that the decision and search versions are equivalent provided q is a *small* prime. We then propose lattice-based cryptanalysis showing that n could be chosen logarithmic in $\log q$ instead of polynomial for LWE.

1 Introduction

Many cryptographic constructions are based on the famous problem *Learning with Errors* (LWE) [Reg05]. Cryptographic work over the past decade has built many primitives based on the hardness of LWE. Today, LWE is known to imply essentially everything you could want from crypto (apart from a few notable exceptions as obfuscation). In particular, this cryptographic problem is currently the most relevant to build FHE [GSW13,BV11]. LWE is known to be hard based on certain assumptions regarding the worst-case hardness of standard lattice problems such as GapSVP and SVP and no quantum attacks against this problem are known.

Typically, LWE deals with a secret vector $s \in \mathbb{Z}_q^n$ and an example w of LWE is a randomly chosen size-n vector satisfying[1] $\langle s, w \rangle = e \pmod{q}$ with $e \ll q$ being a randomly chosen noise value. The problem LWE consists of recovering s from a polynomial number of examples. This problem is equivalent to solve a SVP (Shortest Vector Problem) on a lattice of dimension n. The hardness of LWE holds ensuring that n is chosen sufficiently large, i.e. $\Omega(\log q)$.

---

[1] $\langle s, c \rangle$ denoting the scalar product between s and c.

© Springer Nature Switzerland AG 2019
Y. Mu et al. (Eds.): CANS 2019, LNCS 11829, pp. 360–371, 2019.
https://doi.org/10.1007/978-3-030-31578-8_20

We propose here a nonlinear variant of LWE, called Fractional LWE (FLWE), hopefully less vulnerable to lattice-based attacks. For concreteness, an example of this new problem is a pair of randomly chosen vectors $w = (u, v)$ satisfying

$$\langle s, u \rangle / \langle s, v \rangle = e \pmod{q}$$

This problem does not intuitively seem easier than LWE and the same security level could be hopefully guaranteed in smaller dimension n. But can we quantify this? The main purpose of this paper is to extensively study this problem. Similarly to LWE, we reduce the search version to the decisional one (consisting distinguishing between m examples of FLWE and m randomly chosen vectors) provided q is a small prime (see Sect. 2). Then, we mainly propose two classes of lattice-based attacks. A typical lattice-based attack of the first class exploits the following equation

$$\langle s, u \rangle \cdot \langle s, v \rangle^{q-2} = e \pmod{q}$$

Indeed, by expanding the right term and by sampling sufficiently many examples w_i, the noise values e_i and thus s can be recovered by solving a SVP. However, this attack fails by choosing q sufficiently large. The first class of lattice-based attacks is a generalization of this attack (see Sect. 3.4). We formally prove that this class does not contain any efficient attack for any choice of n provided q is sufficiently large.

We then consider a second class of lattice-based attacks exploiting polynomial equations between noise values (see Sect. 3.5). As the expanded representation size of the involved polynomials exponentially grows with n, it suffices to choose (provided the noise level is large enough to ensure that the noise values cannot be guessed with non-negligible probability)

$$n = \Omega(\log \log q)$$

(instead of $n = \Omega(\log q)$ for LWE [LL15]) to ensure the inefficiency of these attacks.

In Sect. 4, we develop a very simple large plaintext encryption scheme whose security relies on FLWE. Typically, an encryption of $x \in \{0, \ldots, \xi - 1\}$ with $\xi \approx 2^\lambda$ is a pair of vectors $c = (u, v)$ satisfying

$$\langle s, u \rangle / \langle s, v \rangle = x + e\xi \pmod{q}$$

where e is uniform over $\{0, \ldots, \xi - 1\}$. We show that this encryption scheme is significantly more efficient that (large domain) LWE-based schemes to evaluate very short arithmetic circuits assuming the hardness of FLWE with $n = \Omega(\log \log q)$.

However, the homomorphic capabilities of our scheme are very limited due to the ciphertext expansion. Indeed, the ciphertext size polynomially (but not exponentially as we may intuitively think) grows with the number of arithmetic operations restricting evaluation to very short-size arithmetic circuits. However, very small arithmetic circuits can be evaluated very efficiently making this scheme relevant for some (cloud) applications. Can we concretely compare homomorphic performance of our scheme with the ones of LWE? In appearance, LWE seems better because one homomorphic addition only requires $O(n)$ while one homomorphic addition/multiplication requires $O(n^2)$ for our scheme. Nevertheless, one homomorphic multiplication also requires $O(n^2)$ for LWE meaning that these two schemes are equivalent in the worst case. Furthermore, n might be chosen significantly smaller in our scheme. This gives hope to improve existing LWE-based homomorphic encryption schemes. For instance, relinearization technics used in [BGV14] could be perhaps adapted to our scheme in order to overcome the ciphertext expansion.

Because of the lack of space, most of the proofs were omitted. They can be found in [GB19].

Notation. *We use standard Landau notations. Throughout this paper, we let λ denote the security parameter: all known attacks against the cryptographic scheme under scope should require $2^{\Omega(\lambda)}$ bit operations to mount.*

Given two vectors $\boldsymbol{a} = (a_0, \ldots, a_n)$ and $\boldsymbol{b} = (b_0, \ldots, b_n)$, $\boldsymbol{a} \odot \boldsymbol{b} \stackrel{\text{def}}{=} (c_{ij})_{n \geq i \geq j \geq 0}$ with $c_{ii} = a_i b_i$ and $c_{ij} = a_i b_j + a_j b_i$ if $i > j$.

Definition 1. *A rational function $\phi = \phi'/\phi''$ is said to be polynomial-degree if ϕ', ϕ'' are both polynomial-degree polynomials.*

Remark 1. The number $M(n, m)$ of n-variate monomials of degree m is equal to $\binom{m + n - 1}{n - 1}$. Fixing n, $M(n, m) = O(m^{n-1})$.

2 Fractional LWE

For positive integer n and $q \geq 2$, a vector $\boldsymbol{s} \in \{1\} \times \mathbb{Z}_q^n$ and a probability distribution χ on $\mathbb{Z}_q$, let $A_{\boldsymbol{s}, \chi}$ be the distribution obtained by choosing at random a noise term $e \leftarrow \chi$ and two vectors $\boldsymbol{u}, \boldsymbol{v} \leftarrow \mathbb{Z}_q^{n+1}$ satisfying $\langle \boldsymbol{s}, \boldsymbol{u} \rangle / \langle \boldsymbol{s}, \boldsymbol{v} \rangle = e$ and outputting $(\boldsymbol{u}, \boldsymbol{v})$. For concreteness, $(\boldsymbol{u}, \boldsymbol{v})$ can be chosen as follows: $e \leftarrow \chi$, $(u_1, \ldots, u_n, v_0, \ldots, v_n)$ uniform over $\mathbb{Z}_q^{2n+1}$ and $u_0 := e \cdot \langle \boldsymbol{s}, \boldsymbol{v} \rangle - \sum_{i=1}^{n} s_i u_i$. Moreover, if $\langle \boldsymbol{s}, \boldsymbol{v} \rangle = 0$ (this happens with probability $1/q$) then this process is started again.

Definition 2. *For an integer $q = q(n)$, a distribution ψ over $\{1\} \times \mathbb{Z}_q^n$ and an error distribution $\chi = \chi(n)$ over $\mathbb{Z}_q$, the learning with errors problem FLWE$_{n,m,q,\chi,\psi}$ is defined as follows: given m independent samples from $A_{\boldsymbol{s}, \chi}$ where $\boldsymbol{s} \leftarrow \psi$, output $\boldsymbol{s}$ with non-negligible probability.*

The decision variant of the FLWE problem, denoted by DFLWE$_{n,m,q,\chi,\psi}$ is to distinguish (with non-negligible advantage) m samples chosen according to $A_{\boldsymbol{s}, \chi}$ from m samples chosen according to the uniform distribution over $\mathbb{Z}_q^{n+1} \times \mathbb{Z}_q^{n+1}$.

As done for LWE, we propose a reduction from FLWE to DFLWE ensuring that q is a prime polynomial in λ. The proof of the following proposition is largely inspired by the reduction from LWE to DLWE found in [Reg05]. However, we also need that the number m of samples is not too large, i.e. $m = O(q)$.

Lemma 1 *(Search to Decision). Assuming $m = O(q)$, there is a probabilistic polynomial-time reduction from solving $FLWE_{n,m,q,\chi,\psi}$ with overwhelming probability to solving $DFLWE_{n,m,q,\chi,\psi}$ with overwhelming probability provided q is a small prime (polynomial in λ).*

Challenging issues remain unresolved. For instance, can Lemma 1 be extended to large primes q or can worst-case be reduced to average?

3   Analysis of FLWE

3.1   Probability Distributions χ, ψ

An example of $FLWE_{n,m,q,\chi,\psi}$ is a pair of vectors $\boldsymbol{w} = (\boldsymbol{u}, \boldsymbol{v})$ satisfying $\langle \boldsymbol{s}, \boldsymbol{u} \rangle / \langle \boldsymbol{s}, \boldsymbol{v} \rangle = e \pmod{q}$ where $\boldsymbol{s} \leftarrow \psi$ and $e \leftarrow \chi$. To simplify the analysis, we will only consider probability distributions χ which ensure that noise values e cannot be guessed.

Typically, χ refers to the uniform probability distribution over $\{0, \dots, \xi - 1\}$ and ψ refers to the uniform probability distribution over $\{1\} \times \mathbb{Z}_q^n$,

$$\xi \approx 2^\lambda < q$$
$$q \approx 2^{\delta\lambda}$$

3.2   Problem Statement

Let $\boldsymbol{s}^* \leftarrow \psi$ and let $\boldsymbol{w}_1 = (\boldsymbol{u}_1, \boldsymbol{v}_1), \dots, \boldsymbol{w}_m = (\boldsymbol{u}_m, \boldsymbol{v}_m)$ be m examples of $FLWE_{n,m,q,\chi,\psi}$ drawn according to $A_{\boldsymbol{s}^*,\chi}$.

By rewriting the equations $\langle \boldsymbol{s}, \boldsymbol{u}_i \rangle / \langle \boldsymbol{s}, \boldsymbol{v}_i \rangle = e_i \pmod{q}$, we get the following polynomial system $\mathcal{F} = 0$ whose $(s_1 = s_1^*, \dots, s_n = s_n^*, x_1 = e_1, \dots, x_m = e_m)$ is a solution

$$\begin{cases} (u_{10} - x_1 v_{10}) + (u_{11} - x_1 v_{11})s_1 + \dots + (u_{1n} - x_1 v_{1n})s_n = 0 \\ \dots \\ (u_{m0} - x_m v_{m0}) + (u_{m1} - x_m v_{m1})s_1 + \dots + (u_{mn} - x_m v_{mn})s_n = 0 \end{cases} \tag{1}$$

Let $X \subset \mathbb{Z}_q^{m+n}$ be the solution set of $\mathcal{F} = 0$. Throughout this section, $I_\mathcal{F}$ refers to the ideal generated by the family of polynomials $\mathcal{F} \subset \mathbb{Z}_q[S_1, \dots, S_n, X_1, \dots, X_m]$ and I_X refers to the ideal of polynomials which are zero over X. By construction, $I_\mathcal{F} \subseteq I_X$ but it is well-known that the converse is not true in general.

This system is clearly underdefined (n variables can be freely chosen) and hence $\boldsymbol{s}^*$ cannot be recovered without taking into account the shortness of the variables e_i. Szepieniec et al. [SP17] have conjectured that this problem called *Short Solutions to Nonlinear Systems of Equations (SSNE)* is difficult. They identified two types of attacks (algebraic and lattice-based attacks).

3.3 Algebraic Attacks

It is well-known that solving polynomial systems is $\mathcal{NP}$-hard. (ensuring that the degree of the polynomials is at least 2). To solve such systems, we classically compute a (lexicographic order) Groebner basis [BKW93] of $I_{\mathcal{F}}$ which consists of a set of univariate polynomials: this new set of (univariate) polynomial equations can be solved with Berkelamp's algorithm [BRS67]. Although the complexity of the best known algorithm to compute Groebner basis is (at least double) exponential, it is difficult to evaluate their running-time in practice. It mainly depends on the number of variables and the degree of the polynomials.

Nevertheless, this purely algebraic method cannot be applied here because the system $\mathcal{F} = 0$ (1) is underdefined[2]. Some polynomials, exploiting the fact that $\mathbb{Z}_q$ is a finite field or that $e_i \in \{0, \dots, \xi - 1\}$, can be added to $\mathcal{F}$ in order to overdefine the system, i.e. $x_i^q - x_i$, $s_i^q - s_i$ or $\prod_{k \in \{0, \dots, \xi\}} (x_i - k)$. However, the degree of these polynomials is large[3] making Groebner basis computations surely impracticable.

Finally, hybrid attacks consisting of guessing some variables in order to overdefine the system is not relevant here because q, ξ are assumed to be large.

3.4 A First Class of Lattice-Based Attacks

Typically, an example w of LWE satisfies $\langle s, w \rangle = e$ meaning that LWE is natively a lattice problem. Indeed, by considering sufficiently many examples $w_1, \dots, w_t$, the vector noise $(e_1, \dots, e_t)$ and thus s can be recovered by solving a SVP over the lattice spanned by the n vectors $\alpha_i = (w_{1i}, \dots, w_{ti})$.

An example of FLWE is pair of vectors $w = (u, v)$ s.t. $\langle s, u \rangle / \langle s, v \rangle = e$. By using $x^{-1} = x^{q-2} \pmod{q}$, we get the polynomial equation $\langle s, u \rangle \cdot \langle s, v \rangle^{q-2} = e$ (mod q) leading to a lattice-based attack. To highlight this, consider the case $q = 5$ and $n = 1$, i.e. $s = (1, s)$. In this case, $(u_1 + su_2)(v_1 + sv_2)^3 = e$. By developing the right term, we get[4]

$$\sum_{i=0}^{4} s^i p_i(u, v) = e$$

where p_i is a degree-4 polynomial. It follows that s can be recovered by solving a SVP over a small dimension lattice. However, choosing a large prime q (exponential in λ) ensures that the dimension of the lattice is exponential. Nevertheless, one can imagine more efficient attacks based on the same idea. This section aims at formally proving the non-existence of such attacks.

Let us imagine that the attacker is able to recover functions $\varphi_1, \dots, \varphi_\gamma$ such that there are constants (indexed by s) $a_1, \dots, a_\gamma \in \mathbb{Z}_q$ and a function ε satisfying

$$a_1 \cdot \varphi_1(w) + \cdots + a_\gamma \cdot \varphi_\gamma(w) = \varepsilon(w)$$

---

[2] For instance $s_1, \dots, s_n$ can be chosen arbitrarily.

[3] not polynomial in the security parameter λ.

[4] $u_1 v_1^3 + s(u_2 v_1^3 + 3u_1 v_1^2 v_2) + s^2(3u_2 v_1^2 v_2 + 3u_1 v_1 v_2^2) + s^3(u_1 v_2^3 + 3u_2 v_1 v_2^2) + s^4(u_2 v_2^3) = e$.

where $\varepsilon(\boldsymbol{w}) \ll q$. Note that this equality holds with $a_i = s_i$, $\varphi_i(\boldsymbol{w}) = w_i$ and $\varepsilon(\boldsymbol{w}) = e_i$ if $\boldsymbol{w}$ is a LWE example. By sampling sufficiently many instances $\boldsymbol{w}_1, \ldots, \boldsymbol{w}_t$, the coefficients $a_1, \ldots, a_\gamma$ can be recovered by solving an approximate-SVP. This is a relevant attack if $\boldsymbol{s}$ can be derived from the knowledge of $\varepsilon(\boldsymbol{w}_1), \ldots, \varepsilon(\boldsymbol{w}_t)$. This attack can be identified to the tuple $(\varphi_1, \ldots, \varphi_\gamma, \varepsilon)$. This is formally encapsulated in the following definition where the functions $\varphi_1(\boldsymbol{w}), \ldots, \varphi_\gamma(\boldsymbol{w})$ are rational and where $\varepsilon(\boldsymbol{w}) = p(e)$, p being a polynomial.

Definition 3. *Let $(\varphi_1, \ldots, \varphi_\gamma)$ be a (polynomial-size) tuple of polynomial-degree rational functions (see Definition 1) and let p be a non-constant polynomial-degree polynomial. We say that $(\varphi_1, \ldots, \varphi_\gamma, p)$ belongs to the class $\mathcal{C}$ if there exist functions $a_1, \ldots, a_t$ satisfying*

$$a_1(\boldsymbol{s}) \cdot \varphi_1(\boldsymbol{w}) + \ldots + a_\gamma(\boldsymbol{s}) \cdot \varphi_\gamma(\boldsymbol{w}) = p(e) \qquad (2)$$

with non-negligible probability over the choices of $\boldsymbol{s}, \boldsymbol{w}$.

By considering sufficiently many examples $\boldsymbol{w}_i$ and by assuming that p is a small-degree polynomial with small coefficients, i.e. $p(e) \ll q$, the rational functions $\varphi_1, \ldots, \varphi_\gamma$ satisfying (2) can be used to recover $p(e_1), \ldots, p(e_t)$ and thus (hopefully) $e_1, \ldots, e_t$ and then $\boldsymbol{s}$.

Theorem 1. $\mathcal{C}$ *is empty[5] for any $n \geq 1$.*

3.5 Equations Between Noise Values

The second way to investigate $\mathcal{F} = 0$ (1) consists of exploiting the fact that the noise values are relatively small w.r.t. q. However, $s_1^*, \ldots, s_n^*$ are not short and they should be eliminated in order to obtain a system of equations only dealing with $x_1, \ldots, x_n$. In other words, we are looking for polynomials $\phi \in I_X \cap \mathbb{Z}_q[X_1, \ldots, X_m]$. The computational methods to achieve this generally consists of searching polynomials in $\phi \in I_{\mathcal{F}} \cap \mathbb{Z}_q[X_1, \ldots, X_m]$.

Case $n = 1$. Let $\boldsymbol{s} = (1, s)$ and let $\boldsymbol{w} = (\boldsymbol{u}, \boldsymbol{v})$ and $\boldsymbol{w}' = (\boldsymbol{u}', \boldsymbol{v}')$ be two instances of FLWE. We can eliminate s by extracting s from the equations $\langle \boldsymbol{s}, \boldsymbol{u} \rangle = e \langle \boldsymbol{s}, \boldsymbol{v} \rangle$ and $\langle \boldsymbol{s}, \boldsymbol{u}' \rangle = e' \langle \boldsymbol{s}, \boldsymbol{v}' \rangle$, i.e. $s = (ev_1 - u_1)(u_2 - ev_2)^{-1} = (e'v_1' - u_1')(u_2' - e'v_2')^{-1}$ (mod q) leading to the equation

$$u_1 u_2' - u_1' u_2 + e(v_1 u_2' - v_2 u_1') + e'(u_1 v_2' - v_1' u_2) + ee'(v_1 v_2' - v_1' v_2) = 0 \qquad (3)$$

This equation can be seen as a three-variate linear equation having a short solution (e, e', ee'). It is well-known that such a solution can be recovered by considering a dimension-4 lattice[6]. We will investigate the case $n > 1$ in next sections. In particular, we will see that the size of the linear combinations that we obtain by eliminating $s_1, \ldots, s_n$ exponentially grows with n. It follows that n could be chosen logarithmic in λ instead of polynomial for LWE.

---

[5] There does not exist any lattice-based attack satisfying Definition 3.
[6] However, by choosing $\delta = 1$, this attack fails because $ee' \gg q$.

Recovering a Short Integer Solution in Linear Systems. Let q be a large prime, let $\boldsymbol{x}^* = (x_1^*, \cdots, x_\ell^*)$ be a randomly chosen *short* vector and let $\mathcal{A} \in \mathbb{Z}_q^{t \times \ell}$ with $t \leq n$ be a randomly chosen matrix such that $\mathcal{A}\boldsymbol{x}^* = 0$. Our problem simply consists of recovering $\boldsymbol{x}^*$ only given $\mathcal{A}$. This problem looks like a generalization of the Subset Sum Problem but it does not fit to the famous problem SIS (Short Integer Solution) (which is equivalent to SVP on $\mathcal{L}^\perp(\mathcal{A})$) because we want to specifically recover $\boldsymbol{x}^*$ instead of an arbitrary short solution in SIS[7]. Unlike SIS, the smaller is the number of rows t, the harder is our problem. Indeed, if t is too small [8] then many short solutions - even shorter than $\boldsymbol{x}^*$ - could exist. Conversely, by increasing t, *smaller* equations can be found with gaussian eliminations, i.e. equations dealing with $\ell - t + 1$ variables which could be obtained and solved considering dimension-$(\ell - t + 1)$ lattices. More generally, the solution set of $\mathcal{A}\boldsymbol{x} = 0$ is a q-ary[9] dimension-ℓ euclidean lattice $\mathcal{L}$ spanned by at least $\ell - t$ dimension-ℓ (linearly independent) vectors[10] $\boldsymbol{x}_1, \ldots, \boldsymbol{x}_{\ell-t}$ (being solutions of the system). In order to reduce the lattice dimension, these vectors could be truncated ensuring that the truncated vector $\boldsymbol{x}^*$ can be still considered as small in the lattice spanned by the truncated vectors $\boldsymbol{x}_1, \ldots, \boldsymbol{x}_{\ell-t}$. However, more than $\ell - t + 1$ components should be kept ($\ell - t$ is surely not enough because $\mathcal{L} = \mathbb{Z}^{\ell-t}$ in this case). It follows that dimension-$(d \geq \ell - t + 1)$ lattices should be considered. Hence, ensuring that $\ell - t$ is not *too large*, short solutions can be recovered by applying a lattice basis reduction algorithm over $\mathcal{L}$, e.g. LLL or BKZ. Let us try to quantify it.

It is well-known that SVP is a $\mathcal{NP}$-hard (under some conditions) problem and lattice basis reduction algorithms only recover approximations of the shortest vector within a factor[11] γ^d (with $\gamma \approx 1.01$ for the best known polynomial-time algorithms [MR09]). While this approximation may be sufficient to solve SVP on some lattices, it is ensured that $\boldsymbol{x}^*$ cannot be recovered provided[12] $\gamma^d \geq q\sqrt{d}$ and hence (provided $(\log q - \log \log q) \log \gamma \geq 1$)

$$d \geq \ell - t + 1 \geq (\log q + \log \log q)/\log \gamma \qquad (4)$$

Indeed, the euclidean norm of any vector of $\mathbb{Z}_q^d$ is smaller than $q\sqrt{d}$. Consequently, satisfying (4) ensures that any solution of $\mathcal{A}\boldsymbol{x} = 0$ can be potentially output. As the number of solutions of $\mathcal{A}\boldsymbol{x} = 0$ is large, it can be assumed that $\boldsymbol{x}^*$ is output with negligible probability.

---

[7] Unlike our problem, some columns of $\mathcal{A}$ can be removed in SIS (meaning that some components of the searched solution are set to 0) reducing the dimension of the considered lattice. Obviously, if too many columns are removed then short solutions do not exist meaning that a compromise should be done (see [MR09]).

[8] typically $t < \ell/r$ according to gaussian estimations.

[9] meaning that $q\mathbb{Z}^\ell \subset \mathcal{L}$, see [MR09].

[10] and vectors belonging to $q\mathbb{Z}^\ell$.

[11] γ^d for a full rank dimension-d lattice.

[12] The norm of any vector belonging to $\mathbb{Z}_q^d$ is smaller than $q\sqrt{d}$.

Applying It to Our Scheme. Contrarily to LWE-based encryption, eliminating $s_1, \ldots, s_n$ from $\mathcal{F} = 0$ gives nonlinear equations (as observed in the case $n = 1$ (3)) between the variables $x_1, \ldots, x_{n+1}$. This is the major difference with LWE.

We first easily check that there do not exist equations between less than n variables. The most natural way to get equations between $n + 1$ variables is to consider the n first equations of $\mathcal{F} = 0$ as a linear system where the variables are $s_1, s_2, \ldots, s_n$. By doing this, each variable s_i can be expressed as a ratio p_i/p_0 of two degree-n polynomials defined[13] over $x_1, \ldots, x_n$. By injecting these equations in the $(n+1)^{th}$ equation of $\mathcal{F} = 0$, we get an equation between the variables $x_1, \ldots, x_{n+1}$ of degree $n + 1$, i.e. we obtain a polynomial $\phi \in I_{\mathcal{F}} \cap \mathbb{Z}_q[X_1, \ldots, X_{n+1}]$ defined by

$$\phi(x_1, \ldots, x_{n+1}) \stackrel{\text{def}}{=} \sum_{i=0}^{n} (u_{n+1,i} - x_{n+1}v_{n+1,i})p_i(x_1, \ldots, x_n) = 0 \qquad (5)$$

We obviously obtain the same polynomial ϕ by permuting the $(n+1)$ first rows of $\mathcal{F}$. In addition,

$$\phi(x_1, \ldots, x_{n+1}) = \sum_{e \in \{0,1\}^{n+1}} a_e x_1^{e_1} \cdots x_{n+1}^{e_{n+1}}$$

where a_e are degree-$(n+1)$ polynomials defined over $\boldsymbol{w}_1, \ldots, \boldsymbol{w}_{n+1}$. By considering each monomial of ϕ as a variable, we get an linear equation that could lead to lattice-based attacks. However, one could reasonably think that $a_e = 0$ with negligible probability (over the choice of $\boldsymbol{w}_1, \ldots, \boldsymbol{w}_{n+1}$) implying that the number of monomials of ϕ is exponential. Nevertheless, we cannot a priori exclude the possibility to recover smaller equations. The following lemma establishes the non-existence of such equations.

Lemma 2. *Let ϕ be the polynomial defined in Eq. 5. We have,*

1. *ϕ has more than $(1 - 1/\xi - n/q) \cdot 2^{n+1}$ monomials in mean[14].*
2. *Any non-null multiple φ of ϕ has more monomials than ϕ.*
3. *With overwhelming probability (see footnote 14), any polynomial $\varphi \in I_X \cap \mathbb{Z}_q[X_1, \ldots, X_{n+1}]$ s.t. $\deg \varphi < \frac{q}{2(n+1)}$ is a multiple of ϕ.*

By corollary, $I_{\mathcal{F}} \cap \mathbb{Z}_q[X_1, \ldots, X_{n+1}]$ is generated (see footnote 14) by ϕ and any non-null polynomial $\varphi \in I_{\mathcal{F}} \cap \mathbb{Z}_q[X_1, \ldots, X_{n+1}]$ has more than[15] 2^{n+1} monomials. What about polynomials $\varphi \in I_{\mathcal{F}} \cap \mathbb{Z}_q[X_1, \ldots, X_m]$?

---

[13] Consider the $n \times n$ matrix $M = [(u_{ij} - x_i v_{ij})_{1 \le i,j \le n}]$, the vector $\boldsymbol{t} = (u_{i0} - x_i v_{i0})_{1 \le i \le n}$ and the matrix M_j equal to M where the jth column is replaced by $-\boldsymbol{t}$. Solving $\mathcal{F} = 0$ as a linear system gives $s^i = \det M_i / \det M$. It follows that the polynomials $p_i = \det M_i$ and $p_0 = \det M$ have 2^n monomials $x_1^{e_1} \cdots x_n^{e_n}$ where $0 \le e_1, \ldots, e_n \le 1$.

[14] randomness coming from the choice of $\mathcal{F}$, i.e. $\boldsymbol{w}_1, \ldots, \boldsymbol{w}_m$.

[15] a quantity exponentially close to 2^{n+1}.

One can reasonably think that the number of monomials grows with the number of involved variables implying that any $\varphi \in I_{\mathcal{F}} \cap \mathbb{Z}_q[X_1, \ldots, X_m]$ has at least 2^{n+1} monomials. To get such a general result, Lemma 2 should be extended. We did not succeed in proving such a result while we obtained some partial and/or informal results. Details can be found in [GB19].

Conjecture 1. With overwhelming probability (see footnote 14), any non-null polynomial $\varphi \in I_{\mathcal{F}} \cap \mathbb{Z}_q[X_1, \ldots, X_m]$ has more than 2^{n+1} monomials.

Let us now consider a set of t polynomials $\phi_1, \ldots, \phi_t \in I_{\mathcal{F}} \cap \mathbb{Z}_q[X_1, \ldots, X_m]$. Let ℓ denote the number of monomials involved in this set of polynomials. Hence, by considering each monomial as a variable, we get a system $\mathcal{A}\boldsymbol{x}^* = 0$ with t equations and ℓ variables. Without loss of generality, it can be assumed that these equations are linearly independent (otherwise it suffices to remove linearly dependent equations). According to the previous section, short solutions could be found by applying lattice basis reduction algorithms. However, assuming that Conjecture 1 is true, it is ensured that $\ell - t + 1$ is larger than 2^{n+1}. Indeed, if it is not the case, polynomials $\varphi \in I_{\mathcal{F}} \cap \mathbb{Z}_q[X_1, \ldots, X_m]$ containing less than $\ell - t + 1 < 2^{n+1}$ monomials can be obtained by gaussian eliminations. Consequently, according to (4), it suffices that $\ell - t + 1 \geq 2^{n+1} \geq (\log q + \log \log q)/\log \gamma$ to ensure that $\mathcal{A}\boldsymbol{x}^* = 0$ cannot be solved by using lattice basis reduction algorithms. Thus, n can be chosen as follows:

$$\begin{aligned} n &\geq \log(\log q + \log \log q) - \log \log \gamma - 1 \\ &\geq \log \log q - \log \log \gamma \\ &\approx \log \lambda + \log \delta - \log \log \gamma \end{aligned}$$

For instance, one can choose $n = \log \delta + 13$ for $\gamma = 1.01$, $\lambda = 100$.

The monomials were assumed to be small relatively to q. However, it is not the case provided

$$n \geq \delta$$

This ensures the inefficiency of such lattice-based attacks. This suggests that n can be fixed independently of the security parameter λ.

3.6   Discussion

In this section, we investigated the hardness of $\mathsf{FLWE}_{n,m,q,\chi,\psi}$ (and $\mathsf{FDLWE}_{n,m,q,\chi,\psi}$). Our security analysis deals with probability distributions χ ensuring that noise values cannot be guessed with non-negligible probability. Typically, χ is the uniform probability distribution over a set $\{0, \ldots, \xi - 1\}$ with $2^\lambda \approx \xi < q$. Our analysis suggests that $\mathsf{FLWE}_{n,m,q,\chi,\psi}$ is hard ensuring that $n \geq \log \log q - \log \log \gamma$ or $n \geq \log q / \log \xi$.

Let us consider now smaller noise levels. Our analysis remains relevant except that some noise values can be guessed. Assume for instance that $\xi \approx 2^{10}$ and $q \approx \xi^\delta$. At most 10 ($= (\lambda = 100)/\log \xi$) noise values can be guessed, one can reasonably think that it suffices to choose n larger than $10 + \log \log q - \log \log \gamma \approx 27 + \log \delta$ (assuming $\gamma = 1.01$).

4   A Somewhat Homomorphic Private-Key Encryption

Let λ be a security parameter, let ξ be a λ-bit prime and let q be a $(2\delta+1)\lambda$-bit prime with $\delta \geq 1$. Throughout this section, χ refers to the uniform distribution over $\{0,\ldots,\xi-1\}$. Note that this set is also the plaintext domain.

Definition 4. *The functions* KeyGen, Encrypt, Decrypt *are defined as follows:*

- KeyGen(λ,ξ,q). *Let n be indexed by λ,q. The uniform probability over $\{1\}\times\mathbb{Z}_q^n$ is denoted by ψ and let $\boldsymbol{s} \leftarrow \psi$.*

$$K = \{\boldsymbol{s}\} \; ; pp = \{q,\xi\}$$

- Encrypt($K,pp, x \in \{0,\ldots,\xi-1\}$). *Let $e \leftarrow \chi$ and let $\overline{x} = x + e\xi$. Output a pair $\boldsymbol{c} = (\boldsymbol{u},\boldsymbol{v}) \in \mathbb{Z}_q^{n+1} \times \mathbb{Z}_q^{n+1}$ of two randomly chosen vectors[16] satisfying*

$$\langle \boldsymbol{s},\boldsymbol{u}\rangle \cdot \langle \boldsymbol{s},\boldsymbol{v}\rangle^{-1} = \overline{x} \pmod q$$

- Decrypt($K,pp, \boldsymbol{c} = (\boldsymbol{u},\boldsymbol{v})$). *Output $x = \langle \boldsymbol{s},\boldsymbol{u}\rangle \cdot \langle \boldsymbol{s},\boldsymbol{v}\rangle^{-1} \bmod q \bmod \xi$*

In the rest of the paper, it will be assumed that $pp = \{q,\xi\}$ is public. We remark that $\boldsymbol{c}$ and $a\boldsymbol{c}$ are encryptions of the same value for any $a \in \mathbb{Z}_q^*$.

4.1   Homomorphic Properties

Let $\boldsymbol{c} = (\boldsymbol{u},\boldsymbol{v})$ and $\boldsymbol{c}' = (\boldsymbol{u}',\boldsymbol{v}')$ be *fresh* encryptions (output by Encrypt) of respectively x and x'. Similarly to LWE-based encryption schemes, this scheme has natural homomorphic properties coming from the following equalities

$$\frac{\langle \boldsymbol{s},a\boldsymbol{u}\rangle}{\langle \boldsymbol{s},\boldsymbol{v}\rangle} = a\overline{x} \qquad\qquad \frac{\langle \boldsymbol{s},\boldsymbol{u}+a\boldsymbol{v}\rangle}{\langle \boldsymbol{s},\boldsymbol{v}\rangle} = \overline{x}+a$$

$$\frac{\langle \boldsymbol{s},\boldsymbol{u}\rangle\langle \boldsymbol{s},\boldsymbol{u}'\rangle}{\langle \boldsymbol{s},\boldsymbol{v}\rangle\langle \boldsymbol{s},\boldsymbol{v}'\rangle} = \overline{x}\overline{x}' \qquad\qquad \frac{\langle \boldsymbol{s},\boldsymbol{u}\rangle\langle \boldsymbol{s},\boldsymbol{v}'\rangle + \langle \boldsymbol{s},\boldsymbol{u}'\rangle\langle \boldsymbol{s},\boldsymbol{v}\rangle}{\langle \boldsymbol{s},\boldsymbol{v}\rangle\langle \boldsymbol{s},\boldsymbol{v}'\rangle} = \overline{x}+\overline{x}'$$

It follows that vectors[17] $(\boldsymbol{u}\odot\boldsymbol{v}'+\boldsymbol{u}'\odot\boldsymbol{v}, \boldsymbol{v}\odot\boldsymbol{v}')$ and $(\boldsymbol{u}\odot\boldsymbol{u}', \boldsymbol{v}\odot\boldsymbol{v}')$ are encryptions of respectively $x+x'$ and xx' under the key $K_2 = (s_is_j)_{n\geq i\geq j\geq 0}$ with $s_0 = 1$. This process can be naturally iterated. However, the noise exponentially grows with the homomorphic multiplications limiting evaluation to degree-δ polynomials. Moreover, the ciphertext size grows with the number of homomorphic operations m. Nevertheless, it is important to notice that this growth is *only* polynomial and not exponential. Indeed, the size of K_m is equal to the number of degree-m monomials defined over $n + 1$ variables. According to Remark 1, this number is in $O(m^n)$. While this growth strongly limits the homomorphic capabilities, short arithmetic circuits representing degree-δ polynomials could be efficiently evaluated provided n is small enough.

---

[16] For instance, one can randomly choose $\boldsymbol{u}, v_1,\ldots, v_{n-1}, e$ and adjust v_n in order to satisfy the equality.

[17] Recall that given two vectors $\boldsymbol{a} = (a_0,\ldots,a_n)$ and $\boldsymbol{b} = (b_0,\ldots,b_n)$, $\boldsymbol{a} \odot \boldsymbol{b} \stackrel{\text{def}}{=} (c_{ij})_{n\geq i\geq j\geq 0}$ with $c_{ii} = a_ib_i$ and $c_{ij} = a_ib_j + a_jb_i$ if $i > j$.

4.2   Security Analysis

As expected, FLWE can be almost straightforwardly reduced to the security of our scheme.

Proposition 1. *Let m be the number of requests to the encryption oracle done by the CPA attacker. Our scheme is IND-CPA secure assuming the hardness of $DFLWE_{n,m,q,\chi,\psi}$.*

4.3   Efficiency

Proposition 1 and the analysis of FLWE suggests that our scheme is IND-CPA secure assuming either $n \geq \log(2\delta + 1) + \log \lambda - \log \log \gamma$ or $n \geq 2\delta + 1$. For instance, one can choose $n = 9$ for $\delta = 4$ with $\xi \approx 2^{100}$ and $q \approx 2^{900}$. Such parameters lead to a scheme able to evaluate degree-4 polynomials and the ratio (fresh) ciphertext size/plaintext size is close to $\frac{900}{100} \times (9 + 1) \times 2 = 180$. More generally, this ratio is

$$O\left(\delta(\log \delta + \log \lambda - \log \log \gamma)\right)$$

by choosing $n = \log(2\delta + 1) + \log \lambda - \log \log \gamma$.

Let us propose a comparison with a simple (large plaintext) LWE-based encryption where a ciphertext is a vector $c \in \mathbb{Z}_q^n$ satisfying $\langle s, c \rangle = x + e\xi$. Even if we consider the smallest noise level (for instance $e \leftarrow \{0, 1\}$), q should be at least a $\delta\lambda$-bit prime to ensure correctness of degree-δ polynomial evaluation. Moreover, in such schemes, it is required that $n = \Omega(q)$ leading to a ratio ciphertext size/plaintext size in $\Omega(\delta^2 \lambda)$. This shows that our scheme significantly outperforms LWE-based schemes in the evaluation of short arithmetic circuits.

5   Perspectives

We proposed a new cryptographic primitive derived from LWE, called Fractional LWE. Our analysis suggests that n could be chosen logarithmic in $\log q$ instead of polynomial for LWE (FLWE). We then propose a very simple private-key homomorphic encryption based on this problem. This large plaintext encryption scheme achieves good efficiency to evaluate very short arithmetic circuits. Nevertheless, a part of our security analysis is subject to Conjecture 1. While formal and experimental results are proposed in favor of Conjecture 1, we did not manage to formally prove it. In our opinion, this conjecture represents a nice algebraic challenge and its proof would be a great step in the security analysis of our scheme. More fundamentally, the existence of reductions from classical cryptographic problems (LWE, SVP,...) should be investigated. In parallel, it is interesting to wonder whether some LWE-based cryptographic primitives can be improved with our scheme. The most natural one would be an efficient (somewhat) homomorphic encryption by introducing relinearization technics to reduce the ciphertext expansion.

References

[BGV14] Brakerski, Z., Gentry, C., Vaikuntanathan, V.: (Leveled) fully homomorphic encryption without bootstrapping. TOCT **6**(3), 13:1–13:36 (2014)

[BKW93] Becker, T., Kredel, H., Weispfenning, V.: Gröbner Bases: A Computational Approach to Commutative Algebra. Springer, London (1993). https://doi.org/10.1007/978-1-4612-0913-3

[BRS67] Berlekamp, E.R., Rumsey, H., Solomon, G.: On the solution of algebraic equations over finite fields. Info. Control **10**(6), 553–564 (1967)

[BV11] Brakerski, Z., Vaikuntanathan, V.: Efficient fully homomorphic encryption from (standard) LWE. In: Proceedings of the 2011 IEEE 52nd Annual Symposium on Foundations of Computer Science, FOCS 2011, pp. 97–106, Washington DC, USA (2011). IEEE Computer Society

[GB19] Gavin, G., Bonnevay, S.: Fractional LWE: a nonlinear variant of LWE. Cryptology ePrint Archive, Report 2019/902 (2019). https://eprint.iacr.org/2019/902

[GSW13] Gentry, C., Sahai, A., Waters, B.: Homomorphic encryption from learning with errors: conceptually-simpler, asymptotically-faster, attribute-based. In: Canetti, R., Garay, J.A. (eds.) CRYPTO 2013. LNCS, vol. 8042, pp. 75–92. Springer, Heidelberg (2013). https://doi.org/10.1007/978-3-642-40041-4_5

[LL15] Laine, K., Lauter, K.: Key recovery for LWE in polynomial time. Cryptology ePrint Archive, Report 2015/176 (2015). https://eprint.iacr.org/2015/176

[MR09] Micciancio, D., Regev, O.: Lattice based cryptography. In: Bernstein, D.J., Buchmann, J., Dahmen, E. (eds.) Post-Quantum Cryptography, pp. 147–191. Springer, Berlin (2009). https://doi.org/10.1007/978-3-540-88702-7_5

[Reg05] Regev, O.: On lattices, learning with errors, random linear codes, and cryptography. In: Proceedings of the 37th Annual ACM Symposium on Theory of Computing, Baltimore, MD, USA, May 22–24 2005, pp. 84–93 (2005)

[SP17] Szepieniec, A., Preneel, B.: Short solutions to nonlinear systems of equations. In: Kaczorowski, J., Pieprzyk, J., Pomykała, J. (eds.) NuTMiC 2017. LNCS, vol. 10737, pp. 71–90. Springer, Cham (2018). https://doi.org/10.1007/978-3-319-76620-1_5

Encryption, Data Aggregation, and Revocation

Private Data Aggregation over Selected Subsets of Users

Amit Datta[1](✉), Marc Joye[2](✉), and Nadia Fawaz

[1] Snap Inc., Santa Monica, CA, USA
amit.datta@snap.com
[2] OneSpan, Brussels, Belgium
marc.joye@onespan.com

Abstract. Aggregator-oblivious (AO) encryption allows the computation of aggregate statistics over sensitive data by an untrusted party, called aggregator. In this paper, we focus on exact aggregation, wherein the aggregator obtains the exact sum over the participants. We identify three major drawbacks for existing exact AO encryption schemes—no support for dynamic groups of users, the requirement of additional trusted third parties, and the need of additional communication channels among users. We present privacy-preserving aggregation schemes that do not require any third-party or communication channels among users and are exact and dynamic. The performance of our schemes is evaluated by presenting running times.

Keywords: Data aggregation · Privacy · Aggregator obliviousness

1 Introduction

Data aggregation is the process of compiling data from multiple entities or databases into one location with the intent of preparing a combined dataset. The entity performing the aggregation is referred to as the *aggregator*. Data aggregation has many applications in surveys, e-voting, data analytics, etc. However, a major concern with data aggregation is that the data handled in most of these settings are sensitive, or confidential, or may be correlated with sensitive information. This gives rise to data privacy concerns, particularly when the aggregator is *untrusted*.

Consider the simple example of a salary survey, where an aggregator is interested in computing the average salary of a group of users. A user may consider her individual salary as private information that she does not wish to reveal to the aggregator. However, if an aggregation protocol allows the aggregator to compute the average salary without requiring access to any individual's salaries in the clear, then the user may be willing to participate in the protocol.

Similarly, in the health sector, patient health information is protected by legal frameworks, such as Health Insurance Portability and Accountability Act

© Springer Nature Switzerland AG 2019
Y. Mu et al. (Eds.): CANS 2019, LNCS 11829, pp. 375–391, 2019.
https://doi.org/10.1007/978-3-030-31578-8_21

(HIPAA), which regulates the use and disclosure of protected health information, including medical records and payment history, held by covered entities like health-care providers and health insurances. A network of healthcare providers or hospitals may be interested in computing aggregate statistics over their joint populations of patients for research purposes, or for disease prevention and control. However legal frameworks may not allow them to share individual patient records in the clear. In such a scenario, the network would like to leverage a privacy-preserving aggregation protocol which would allow them to compute the aggregate statistics without access to data in the clear. Many businesses also leverage user data analytics, for operations management, marketing and sales campaigns, or to provide personalized services to their customer base, especially in the age of digital economy. Businesses that operate in different regions of the world, such as the USA and the European Union (EU), may face different privacy regulations in each location. Such businesses may see their ability to transfer user data across their different corporate locations or with their global partners restricted, as illustrated by the decision by the EU Court of Justice to invalidate the EU-US Safe Habor agreement, on which thousands of companies had relied for the transatlantic transfer of user data since 2000 [5]. Limitations on data transfer in turn generates the need for new privacy-preserving aggregation techniques to perform aggregate statistics over user data across locations, without transferring individual user data in the clear.

Aggregator Obliviousness. The notion of *aggregator obliviousness* (AO) [17] was introduced to allow an aggregator to receive encrypted data from users and compute the aggregate and *nothing else*. With an AO encryption scheme, the aggregator gains no additional knowledge other than what is evident from the aggregate itself. Formal definitions for AO are given in Sect. 2.

Suppose that there is a population of n users, denoted by $\{1, \ldots, n\}$, as well as a designated entity called the aggregator. Let $x_{i,\tau}$ denote the private data of user i, where $i \in \mathcal{S}$ for a subset $\mathcal{S} \subseteq \{1, \ldots, n\}$ of users, and with tag τ, where the tag is an identifier of the aggregation (τ may for example indicate the time period at which the aggregate is computed). The goal of the aggregator is to compute the aggregate value

$$\Sigma_\tau = \sum_{i \in \mathcal{S}} x_{i,\tau}$$

in an *oblivious way*, that is, without having access to the private data $x_{i,\tau}$ in the clear. For simplicity, we focus on the aggregate sum throughout the paper. Extensions to operations and statistics beyond the simple sum are presented in Sect. 5.

Certain aggregation schemes make use of additional communication channels among users to perform the aggregation. These channels are used to exchange information among users and can enhance schemes. However, such channels may not exist between users, and even when they do exist, such protocols have very high communication overheads. So, a protocol would ideally not make use of communication channels among users.

Several current schemes require the presence of one or more additional *third parties* to carry out the protocol. These third-parties are assumed to be non-colluding with the aggregator and their participation often adds desirable properties (like dynamism or fault tolerance) to the protocol. However, such third parties are often difficult to find in practice, thereby rendering such schemes less practical. For such settings, it is desirable to develop a scheme which does not require an additional third-party.

In many existing AO encryption schemes, the group of users that are aggregated is static. The scheme is setup for a given set of users and if any user joins or leaves, fresh keys need to be distributed to all the users. For many applications, it is desirable to have a *dynamic* scheme, which allows users to easily join or leave a group. With a dynamic scheme, the aggregator can carry out the aggregation over any subset or superset of users without much hassle; in particular, without having to redistribute keys to every user involved in the computation.

The lack of fault tolerance is another drawback present in many aggregation schemes. An aggregation scheme is said *fault-tolerant* when the aggregator is still able to compute an aggregate even if one or several users fail to report their share.

From the above discussion, it turns out that an ideal AO encryption scheme should be one where the users send their encrypted share in a single communication step and, at the same time, must enjoy the properties of being dynamic and fault-tolerant—without additional non-colluding third parties. No such scheme does exist as far are *exact* aggregates are concerned. This is not surprising as the above sought-after features are incompatible all together. In particular, such an exact AO encryption cannot be fault tolerant. Indeed, when there are no extra parties involved in the protocol, the users have to send their encrypted contributions directly to the aggregator. Hence, if the scheme is supposed to be fault-tolerant, the aggregator should be able to compute the aggregate sum in the clear over all participating users but also over all but one participating users, thereby obtaining the private data of the victim user by subtracting the two aggregate sums.

Our Contributions. This paper considers the weaker notion of *selective* fault-tolerance. In this setting, the subset of users over which the aggregate is to be computed can be dynamically chosen but must be known to the users beforehand. A useful application is the ability to leave out persistently failing users by making use of the dynamic nature of the scheme.

More specifically, we present two dynamic AO encryption schemes, which do not require any third party or communication among users. In these schemes, each user can participate in the protocol with the knowledge of the other users' identities. Just by knowing the identities, a user can derive a key to perform the encryption. This key has special properties that allow the aggregator to decrypt the aggregate when all users in a selected subset encrypt with their corresponding keys. We implement the schemes and evaluate their performance.

Related Work. A number of candidates for AO encryption schemes have been proposed, including [1,2,7,8,11,13–17]. However most of them are not dynamic. Among the ones that can support dynamic joins and leaves, [7] requires n^2 messages exchanged among users at each time period, [1,14] need each user to store n keys, which can be impractical for a large set of users, and [2,8,11,15] assume the presence of one or more third parties. Schemes like [1,11,16,17] provide a different notion of privacy: *Differential Privacy* (DP), which makes the final aggregate noisy.

In this paper, we focus on exact aggregation, without addition of any noise. However, our schemes can be modified to return noisy sums, for instance through differential-privacy techniques as was done, e.g., in [17]. Jawurek *et al.* [12] provide a fantastic survey for different techniques used for privacy-preserving aggregation.

2   Definitions

In this section, we formally define relevant notions for privacy-preserving aggregation. We first formally define Aggregator Obliviousness and then briefly describe the scheme by Shi *et al.*

2.1   Aggregator-Oblivious Encryption

The definition of *aggregator-oblivious encryption* was introduced in [17]. Here, we extend their definition to cover a broader class of encryption schemes and use cases.

Definition 1. *An* aggregator-oblivious encryption scheme *is a tuple of algorithms,* (Setup, KeyGen, Enc, AggrDec), *defined as:*

Setup(1^κ) *On input a security parameter κ, the* Setup *algorithm generates the system parameters* param, *a master secret key* msk, *and the aggregation key* sk_0. *Parameters* param *are made public and key* sk_0 *is given to the aggregator.*

KeyGen(param, msk, i) *Given a user's identifier i, the* KeyGen *algorithm produces the private key* sk_i *for user i as* $sk_i =$ KeyGen(param, msk, i).

Enc(param, $sk_i, \tau, \mathcal{S}_\tau, x_{i,\tau}$) *Given the private input $x_{i,\tau}$ with tag τ of user $i \in \mathcal{S}_\tau$ and the private key sk_i, user i applies this algorithm to produce the ciphertext* $c_{i,\tau} =$ Enc(param, $sk_i, \tau, \mathcal{S}_\tau, x_{i,\tau}$).

AggrDec(param, $sk_0, \tau, \mathcal{S}_\tau, \{c_{i,\tau}\}_{i \in \mathcal{S}_\tau}$) *Upon receiving the ciphertexts $c_{i,\tau}$ $\forall i \in \mathcal{S}_\tau$ associated with tag τ, the aggregator obtains the aggregate value*

$$\Sigma_\tau = \sum_{i \in \mathcal{S}_\tau} x_{i,\tau}$$

as $\Sigma_\tau =$ AggrDec(param, $sk_0, \tau, \mathcal{S}_\tau, \{c_{i,\tau}\}_{i \in \mathcal{S}_\tau}$) *using the aggregation key* sk_0.

Remark 1. Three modifications were made to the original definition:

1. In the original definition, the aggregation is always computed over the entire set of users: $\mathcal{S}_\tau = \{1, \ldots, n\}$. This set is fixed and is part of the system parameters. In our case, we allow for subsets $\mathcal{S}_\tau$ of users that may change depending on the value of τ.
2. The setup algorithm is now divided in two sub-algorithms: the Setup algorithm itself that outputs the system parameters and the aggregation key and the KeyGen algorithm that returns the private keys for each user. The original definition corresponds to the case where the master secret key is the vector containing all user's private keys, $\mathsf{msk} = (\mathsf{sk}_1, \ldots, \mathsf{sk}_n)$, and the KeyGen algorithm simply returns the i-th component of msk as the private key of user i.
3. Finally, in the original definition, tag τ explicitly refers to a time period. We use the more generic term of tag, which serves as a unique identifier for the aggregation instance.

2.2 Aggregator Obliviousness

The security notion of *aggregator obliviousness* (AO) requires that the aggregator cannot learn, *for each tag* τ, anything more than the aggregate value Σ_τ from the individual encrypted values. If there are corrupted users (i.e., users sharing their private information with the aggregator), the notion requires that the aggregator gets no additional information about the private values of the honest users beyond what is evident from the final aggregate value and the private values of the corrupted users. Furthermore, in our setting, we assume that each user encrypts only *one* value for a given tag.

More formally, AO security is defined by the following game between a challenger and an attacker.

Setup The challenger runs the Setup algorithm and gets param, msk and sk_0. It also runs KeyGen to obtain the encryption key sk_i of each user i. The challenger gives param to the attacker.

Queries In the first phase, the attacker can submit queries that are answered by the challenger. The attacker can make two types of queries:
1. <u>Encryption queries:</u> The attacker submits $(i, \tau, \mathcal{S}_\tau, x_{i,\tau})$ for a pair (i, τ) with $i \in \mathcal{S}_\tau$ and gets back the encryption of $x_{i,\tau}$ with tag τ under key sk_i;
2. <u>Compromise queries:</u> The attacker submits i and receives the private key sk_i of user i; if $i = 0$, the attacker receives the aggregation key sk_0.

Challenge The attacker chooses a target tag $\tau^\star$. Let $\mathcal{U}^\star \subseteq \{1, \ldots, n\}$ be the whole set of users for which, at the end of the game, no encryption queries associated to tag $\tau^\star$ have been made and no compromise queries have been made. The attacker chooses a subset $\mathcal{S}_{\tau^\star} \subseteq \mathcal{U}^\star$ and two different series of triples

$$\langle (i, \tau^\star, x_{i,\tau^\star}^{(0)}) \rangle_{i \in \mathcal{S}_{\tau^\star}} \quad \text{and} \quad \langle (i, \tau^\star, x_{i,\tau^\star}^{(1)}) \rangle_{i \in \mathcal{S}_{\tau^\star}},$$

that are given to the challenger. Further, if the aggregator capability sk_0 is compromised at the end of the game and $\mathcal{S}_{\tau^\star} = \mathcal{U}^\star$, it is required that

$$\sum_{i \in \mathcal{S}_{\tau^\star}} x_{i,\tau^\star}^{(0)} = \sum_{i \in \mathcal{S}_{\tau^\star}} x_{i,\tau^\star}^{(1)}.$$

Guess The challenger chooses at random a bit $b \in \{0,1\}$ and returns the encryption of $\langle x_{i,\tau^*}^{(b)} \rangle_{i \in \mathcal{S}_{\tau^*}}$ to the attacker.

More queries In the second phase, the attacker can make more encryption queries and compromise queries. Note that since $\mathcal{S}_{\tau^*} \subseteq \mathcal{U}^*$, the attacker cannot submit an encryption query $(i, \tau^*, \cdot, \cdot)$ with $i \in \mathcal{S}_{\tau^*}$ or a compromise query i with $i \in \mathcal{S}_{\tau^*}$.

Outcome At the end of the game, the attacker outputs a bit b' and wins the game if and only if $b' = b$ (i.e., if it correctly guessed the bit b). As usual, $\mathcal{A}$'s advantage is defined to be

$$\mathbf{Adv}^{AO}(\mathcal{A}) := 2 \left| \Pr[b' = b] - 1/2 \right|$$

where the probability is taken over the random coins of the game according to the distribution induced by Setup and over the random coins of the attacker.

Definition 2. *An encryption scheme* (Setup, KeyGen, Enc, AggrDec) *is said to meet the* AO *security notion if no probabilistic polynomial-time attacker $\mathcal{A}$ can win the above* AO *security game with an advantage $\mathbf{Adv}^{AO}(\mathcal{A})$ that is non-negligible in the security parameter.*

2.3 Example Scheme

As an illustration, we briefly review the scheme proposed by Shi *et al.* [17] for achieving aggregator-obliviousness. This notion is met under the DDH assumption [4,10] in the random oracle model. This scheme will serve as a building block for our final scheme.

Setup(1^κ) On input a security parameter κ, a trusted dealer generates a group $\mathbb{G} = \langle g \rangle$ of prime order q for which the DDH assumption holds. It also defines a cryptographic hash function $H \colon \{0,1\}^* \to \mathbb{G}$, viewed as a random oracle. Finally, it generates n random elements $s_1, \ldots, s_n \in \mathbb{Z}/q\mathbb{Z}$ and lets $s_0 = -\sum_{i=1}^n s_i \pmod{q}$, where n denotes the total number of users. The system parameters are param $= \{\mathbb{G}, g, q, H\}$, the master secret key is msk $= (s_1, \ldots, s_n)$, and the aggregation key is $\mathsf{sk}_0 = s_0$. Parameters param are made public and sk_0 is given to the aggregator.

KeyGen(param, msk, i) On input user's identifier i, the KeyGen algorithm returns the private key $\mathsf{sk}_i = s_i$ for user i, where s_i is the i-th component of msk.

Enc(param, $s_i, \tau, \{1, \ldots, n\}, x_{i,\tau}$) User $i \in \{1, \ldots, n\}$ encrypts a value $x_{i,\tau}$ with tag τ using the private key s_i to get the ciphertext

$$c_{i,\tau} = g^{x_{i,\tau}} H(\tau)^{s_i}.$$

AggrDec(param, $s_0, \tau, \{1, \ldots, n\}, \{c_{i,\tau}\}_{1 \leqslant i \leqslant n}$) Upon receiving all the $c_{i,\tau}$'s (with $i \in \{1, \ldots, n\}$) associated with tag τ, the aggregator first computes

$$V_\tau = H(\tau)^{s_0} \prod_{1 \leqslant i \leqslant n} c_{i,\tau}$$

and then obtains the aggregate value $\Sigma_\tau = \sum_{1 \leqslant i \leqslant n} x_{i,\tau} \pmod{q}$ by computing the discrete logarithm of V_τ w.r.t. basis g.

The scheme works because the aggregation key is defined as $s_0 = -\sum_{1 \leqslant i \leqslant n} s_i$. The aggregator is so able to remove the masking expression $H(\tau)^{\sum_{1 \leqslant i \leqslant n} s_i} = H(\tau)^{-s_0}$ by using the aggregation key s_0. Indeed, the aggregation step computes $V_\tau = H(\tau)^{s_0} g^{\sum_{1 \leqslant i \leqslant n} x_{i,\tau}} H(\tau)^{\sum_{1 \leqslant i \leqslant n} s_i} = g^{\Sigma_\tau}$, the discrete logarithm of which yields Σ_τ.

3 Dynamic AO Encryption

In this section, we describe a dynamic aggregation scheme. With a dynamic scheme, the aggregator can carry out the aggregation over any subset or superset of users without needing to perform the full Setup. To aggregate over any subset of users, the aggregator only needs to convey information about the composition of the subset to all the users. To add a new user to the set $\mathcal{S}_\tau$, only KeyGen needs to be carried out for that user, after which the aggregation can be carried out as before.

To devise a dynamic AO scheme, we begin by building on top of the scheme due to Shi *et al.* As seen in Sect. 2.3, the key trick behind that scheme is $\mathsf{sk}_0 = -\sum_i \mathsf{sk}_i$. Shi *et al.* achieve this by having a trusted dealer generate and send keys satisfying this property to the users and the aggregator. To apply Shi *et al.*'s scheme, as it is, on a subset of users, the trusted dealer has to regenerate keys for each subset and send them over to the users. Thus, the scheme would require multiple interactions with the trusted dealer. In settings where the key is burned onto a hardware component (like set-top boxes), such a scheme becomes inapplicable.

In the proposed schemes, we give users the ability to compute such keys on their own. A trusted dealer provides all users and the aggregator with an initial secret key sk_i. If a user knows the identities of the other users are in the desired subset $\mathcal{S}_\tau$, then she can compute the subset-key $s_{i,\tau}$ for that subset such that $s_{0,\tau} = -\sum_{i \in \mathcal{S}_\tau} s_{i,\tau}$. We use a combination of pairings and identity-based encryption so that users can compute the subset key from only the identities of the other users in the given subset. There is no need to redistribute any keys. Kursawe *et al.* [14] use a similar technique, but they combine pairings and public-key encryption. The primary advantage in our schemes is that there is no need for a public-key infrastructure and users do not need to remember or verify public keys. Thus, we obtain more versatile aggregation schemes over dynamic subsets of users.

3.1 Key Ingredient

To provide the ability to users to compute their own subset keys, we use identity-based encryption (IBE) and bilinear type-1 pairings. In an IBE scheme, the public key of a user is a unique identifying information about the user. This allows other users to send encrypted values with just the knowledge of the identity of the recipient. A bilinear type-1 pairing is a symmetric map $e \colon \hat{G} \times \hat{G} \to \hat{G}_T$, where $\hat{G}, \hat{G}_T$ are cyclic groups of order p and the function e satisfies bilinearity: $e(\hat{g}^a, \hat{g}^b) = e(\hat{g}, \hat{g})^{ab}$, where $\langle \hat{g} \rangle = \hat{G}$ and $a, b \in \mathbb{Z}/p\mathbb{Z}$.

In our setting, we assume that i is the unique identifier for user i. The trusted dealer sends the secret key $\mathsf{sk}_i = J(i)^{\mathsf{msk}}$ to user i, where msk is the master secret key of the trusted dealer and $J \colon \mathbb{Z} \to \hat{G}$ is a publicly known cryptographic hash function. Given the identities of users in a subset $\mathcal{S}_\tau$, a user $i \in \mathcal{S}_\tau$ derives on her own the corresponding subset key from $\mathcal{S}_\tau$ and sk_i as

$$s_{i,\tau} = \sum_{\substack{k \in \mathcal{S}_\tau \cup \{0\} \\ k < i}} \mathfrak{H}(K_{i,k}) - \sum_{\substack{k \in \mathcal{S}_\tau \\ k > i}} \mathfrak{H}(K_{i,k}) \pmod{T}$$

with

$$K_{i,k} = e(J(k), \mathsf{sk}_i)$$

and $\mathfrak{H} \colon \hat{G}_T \to \mathbb{Z}/T\mathbb{Z}$ is hash function (typically, $T = q$ a prime, or $T = 2^\ell$).

We assume that the aggregator's identifier is $i = 0$. For a subset $\mathcal{S}_\tau$, the aggregator also derives the matching subset key $s_{0,\tau}$ in a similar manner, which results in

$$s_{0,\tau} = \sum_{\substack{k \in \mathcal{S}_\tau \cup \{0\} \\ k < 0}} \mathfrak{H}(K_{0,k}) - \sum_{\substack{k \in \mathcal{S}_\tau \\ k > 0}} \mathfrak{H}(K_{0,k}) \pmod{T}$$

$$= - \sum_{k \in \mathcal{S}_\tau} \mathfrak{H}(K_{0,k}) \pmod{T}.$$

Property. We now show that the above construction satisfies the following property

$$\sum_{i \in \mathcal{S}_\tau} s_{i,\tau} = -s_{0,\tau} \pmod{T}$$

By construction, variable $K_{i,k}$ is symmetric. Indeed, we have

$$K_{i,k} = e(J(k), \mathsf{sk}_i) = e(\mathsf{sk}_i, J(k)) = e(J(i)^{\mathsf{msk}}, J(k))$$
$$= e(J(i), J(k)^{\mathsf{msk}}) = e(J(i), \mathsf{sk}_k) = K_{k,i}.$$

In a way similar to [9, Proposition 1], letting $m_\tau = \min_{i \in \mathcal{S}_\tau} i$ and $M_\tau = \max_{i \in \mathcal{S}_\tau} i$, it is then readily verified that:

$$\sum_{i \in \mathcal{S}_\tau} s_{i,\tau} = \sum_{i \in \mathcal{S}_\tau} \left[\sum_{\substack{k \in \mathcal{S}_\tau \cup \{0\} \\ k < i}} \mathfrak{H}(K_{i,k}) - \sum_{\substack{k \in \mathcal{S}_\tau \\ k > i}} \mathfrak{H}(K_{i,k}) \right]$$

$$= \sum_{i \in \mathcal{S}_\tau} \left[\mathfrak{H}(K_{i,0}) + \sum_{\substack{k \in \mathcal{S}_\tau \\ k < i}} \mathfrak{H}(K_{i,k}) - \sum_{\substack{k \in \mathcal{S}_\tau \\ k > i}} \mathfrak{H}(K_{k,i}) \right]$$

$$= -s_{0,\tau} + \sum_{\substack{i \in \mathcal{S}_\tau \\ i \neq m_\tau}} \sum_{\substack{k \in \mathcal{S}_\tau \\ k < i}} \mathfrak{H}(K_{i,k}) - \sum_{\substack{i \in \mathcal{S}_\tau \\ i \neq M_\tau}} \sum_{\substack{k \in \mathcal{S}_\tau \\ k > i}} \mathfrak{H}(K_{k,i})$$

$$= -s_{0,\tau} + \sum_{\substack{i \in \mathcal{S}_\tau \\ i \neq m_\tau}} \sum_{\substack{k \in \mathcal{S}_\tau \\ k < i}} \mathfrak{H}(K_{i,k}) - \sum_{\substack{k \in \mathcal{S} \\ k \neq m_\tau}} \sum_{\substack{i \in \mathcal{S}_\tau \\ i < k}} \mathfrak{H}(K_{k,i})$$

$$= -s_{0,\tau} \pmod{T}.$$

This was the key property used by Shi *et al.*'s scheme, but instead of having a key dealer generate keys which satisfy this property, we now have a mechanism to generate keys having this characteristic. With this building block in place, we are now ready to present our new schemes.

3.2 Proposed Schemes

We detail the description of our schemes using the building block we describe in the previous section. The Enc phase now has an additional step wherein the subset keys are computed by users.

In both schemes, user i receives an initial secret key sk_i, while the aggregation key sk_0 is given to the aggregator. Given the identities of the users in the subset and the secret key, each user computes the subset key $s_{i,\tau}$, which satisfies the property $s_{0,\tau} = -\sum_{i \in \mathcal{S}_\tau} s_{i,\tau}$, where $s_{0,\tau}$ is the aggregator's subset key. With the subset key, each user encrypts her value and sends the ciphertext to the aggregator. Upon receiving ciphertexts from all the users in the subset, the aggregator uses the subset key $s_{0,\tau}$ to recover the aggregate value. The correctness of the schemes follows from the correctness of the building block since $s_{0,\tau} = -\sum_{i \in \mathcal{S}_\tau} s_{i,\tau}$. Details of both schemes are provided in the following sections.

Scheme I

Setup(1^κ) On input a security parameter κ, a trusted dealer generates the system parameters param $= \{\mathbb{G}, g, q, \hat{\mathbb{G}}, \hat{\mathbb{G}}_T, e, H, \mathfrak{H}, J\}$ where $\mathbb{G} = \langle g \rangle$ is a group of order q, $\hat{\mathbb{G}}$ and $\hat{\mathbb{G}}_T$ are two groups of order p with a bilinear type-1 pairing $e \colon \hat{\mathbb{G}} \times \hat{\mathbb{G}} \to \hat{\mathbb{G}}_T$, and $H \colon \{0,1\}^* \to \mathbb{G}$, $\mathfrak{H} \colon \hat{\mathbb{G}}_T \to \mathbb{Z}/q\mathbb{Z}$ and $J \colon \mathbb{Z} \to \hat{\mathbb{G}}$ are cryptographic hash functions. The trusted dealer also generates a master secret key $\mathsf{msk} \in \mathbb{Z}/p\mathbb{Z}$ and computes the aggregation key $\mathsf{sk}_0 = [J(0)]^{\mathsf{msk}} \in \hat{\mathbb{G}}$. Parameters param are made public and key sk_0 is given to the aggregator.

KeyGen(param, msk, i) Given user's identifier i, the KeyGen algorithm returns $\mathsf{sk}_i = [J(i)]^{\mathsf{msk}} \in \hat{\mathbb{G}}$, where msk is the master secret key of the trusted dealer.

Enc(param, $\mathsf{sk}_i, \tau, \mathcal{S}_\tau, x_{i,\tau}$) $\mathcal{S}_\tau$ represents the subset of users among whom the aggregation is to be computed. For a private input $x_{i,\tau} \in \mathbb{Z}/q\mathbb{Z}$ with tag τ, provided that $i \in \mathcal{S}_\tau$, user i computes the ciphertext

$$c_{i,\tau} = g^{x_{i,\tau}} H(\tau)^{s_{i,\tau}} (\in \mathbb{G})$$

where

$$s_{i,\tau} = \sum_{\substack{k \in S_\tau \cup \{0\} \\ k < i}} \mathfrak{H}(K_{i,k}) - \sum_{\substack{k \in S_\tau \\ k > i}} \mathfrak{H}(K_{i,k}) \pmod{q}$$

with $K_{i,k} = e(J(k), \mathsf{sk}_i) \in \hat{\mathbb{G}}_T$. The user sends $c_{i,\tau}$ to the aggregator through *any* channel.

AggrDec(param, $\mathsf{sk}_0, \tau, S_\tau, \{c_{i,\tau}\}_{i \in S_\tau}$) Upon receiving all the ciphertexts $c_{i,\tau}$ with $i \in S_\tau$, all with tag τ, the aggregator computes in $\mathbb{G}$

$$V_\tau := H(\tau)^{s_{0,\tau}} \cdot \prod_{i \in S_\tau} c_{i,\tau} = g^{\sum_{i \in S_\tau} x_{i,\tau}}$$

where $s_{0,\tau} = -\sum_{k \in S_\tau} \mathfrak{H}(K_{0,k}) \pmod{q}$ with $K_{0,k} = e(J(k), \mathsf{sk}_0)$. The aggregator obtains the sum $\Sigma_\tau := \sum_{i \in S_\tau} x_{i,\tau} \pmod{q}$ by computing the discrete logarithm of V_τ with respect to g.

In [3], the authors show how the use of two hash functions lead to a much tighter security reduction. The same observation readily applies here.

Scheme II

Setup(1^κ) On input a security parameter κ, a trusted dealer generates the system parameters param $= \{\hat{\mathbb{G}}, \hat{\mathbb{G}}_T, e, \ell, \{\mathfrak{H}_\tau\}_\tau, J\}$ where $\hat{\mathbb{G}}$ and $\hat{\mathbb{G}}_T$ are two groups of order p with a bilinear type-1 pairing $e : \hat{\mathbb{G}} \times \hat{\mathbb{G}} \to \hat{\mathbb{G}}_T$, ℓ is a length upper-bounding the bit-length of the private inputs and their sums, $\{\mathfrak{H}_\tau\}_\tau : \hat{\mathbb{G}}_T \to \mathbb{Z}/2^\ell\mathbb{Z}$ is a family of keyed cryptographic hash functions, and $J : \mathbb{Z} \to \hat{\mathbb{G}}$ is a cryptographic hash function. The trusted dealer also generates a master secret key msk $\in \mathbb{Z}/p\mathbb{Z}$ and computes the aggregation key $\mathsf{sk}_0 = [J(0)]^{\mathsf{msk}} \in \hat{\mathbb{G}}$. Parameters param are made public and key sk_0 is given to the aggregator.

KeyGen(param, msk, i) Given user's identifier i, the KeyGen algorithm returns $\mathsf{sk}_i = [J(i)]^{\mathsf{msk}} \in \hat{\mathbb{G}}$, where msk is the master secret key of the trusted dealer.

Enc(param, $\mathsf{sk}_i, \tau, S_\tau, x_{i,\tau}$) S_τ represents the subset of users among whom the aggregation is to be computed. For a private input $x_{i,\tau} \in \mathbb{Z}/2^\ell\mathbb{Z}$ with tag τ, provided that $i \in S_\tau$, user i forms the ciphertext

$$c_{i,\tau} = x_{i,\tau} + s_{i,\tau} (\in \mathbb{Z}/2^\ell\mathbb{Z})$$

where

$$s_{i,\tau} = \sum_{\substack{k \in S_\tau \cup \{0\} \\ k < i}} \mathfrak{H}_\tau(K_{i,k}) - \sum_{\substack{k \in S_\tau \\ k > i}} \mathfrak{H}_\tau(K_{i,k}) \pmod{2^\ell}$$

with $K_{i,k} = e(J(k), \mathsf{sk}_i) \in \hat{\mathbb{G}}_T$. The user sends $c_{i,\tau}$ to the aggregator through *any* channel.

$\mathsf{AggrDec}(\mathsf{param}, \mathsf{sk}_0, \tau, \mathcal{S}_\tau, \{c_{i,\tau}\}_{i \in \mathcal{S}_\tau})$ Upon receiving all the ciphertexts $c_{i,\tau}$ with $i \in \mathcal{S}_\tau$, all with tag τ, the aggregator computes in $\mathbb{Z}/2^\ell \mathbb{Z}$

$$\Sigma_\tau := s_{0,\tau} + \sum_{i \in \mathcal{S}_\tau} c_{i,\tau} = \sum_{i \in \mathcal{S}_\tau} x_{i,\tau} \pmod{2^\ell}.$$

4 Performance

In this section, we present performance results from implementations of both our schemes. The principal computational difference between our schemes and prior schemes is the additional time required for the computation of the subset key described in Sect. 3.1. This is the most expensive step computationally as it requires a user to compute n pairings, where n is the subset size.

For Scheme I, this computation of the subset key is required only when subsets change. As long as aggregates are computed over the same subsets, in spite of the aggregations being with different tags, there is no need to recompute subset-keys. However, for Scheme II, the subset keys cannot be reused in spite of having the same subsets. For every aggregation with a different tag τ, users must compute fresh subset keys. But if user i has access to the components $K_{i,k}$ for every $k \in \mathcal{S}_\tau$ (e.g., by storing them), this computation is very fast as it only involves evaluations of keyed hash functions.

For the second step of the encryption (i.e., the encryption itself) or for the decryption, Scheme II is far more efficient than Scheme I. This is expected since Scheme II does not require costly operations like exponentiations and discrete-log computations, instead relies solely on modular additions. The computational efficiency is clear from Table 1 of timing comparisons.

All computations described in this section were performed on a 2.6 GHz Intel Core i7 processor with 8 GB 1600 MHz DDR3 RAM.

Subset Keys

For the computation of subset keys, we use Tate pairing with an embedding degree of 2 on a supersingular curve to compute pairings required for subset-key computation. The $\mathrm{GF}(p)$ elliptic curve is assumed to be of the form $y^2 = x^3 + Ax + B \mod p$, where $A = -3$ and $B = 0$. A low embedding degree (2 in our case) has an adverse impact on performance, but timings can be improved by using an embedding degree of 3 as shown by Teruya *et al.* [18].

To compute the subset key, a user has to compute a pairing for every other user in the subset. Thus, the time required for computing the subset key is linear in the size of the subset. The rate of this linear increase (slope) depends on how big the pairings are. We compare the rate of increase in computation times using pairings of size 512, 1179, and 1536 bits. The slope is steeper as the size of pairing increases. With a 512-bit pairing, it takes nearly 1.5 s to compute the subset key when there are 400 users in a subset. Thus, depending on how often users have to compute the subset keys, the size of each subset can be decided. For example, if users are expected to report values once every few minutes, subsets can easily have about 16,000 users. However, if users are to report values every few seconds, then subsets need to be much smaller (less than 300). In Fig. 1, we show how the computation time changes with the number of users and the size of the pairing.

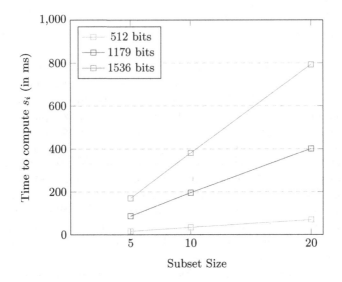

Fig. 1. Variation in subset key computation time with subset size and size of pairing. For a given pairing size, the time is linear in the subset size.

Schemes I and II

We present the computation times of both our schemes in Table 1. The running times are with a margin of error at 95% confidence, computed with 100 samples for 2^{12} users, for both the schemes. For the first scheme, we use a 200-bit curve similar to the one used in [3]. The timings presented here do not include the times required to compute the subset keys; we discuss them in the previous section. The computation time of AggrDec does not include the time to compute the subset key of the aggregator $s_{0,\tau}$. The hashing of tags $(H(\tau))$ is performed using SHA-512.

Table 1. Comparison of running times for Schemes I & II

	Enc		AggrDec	Result recovery
	$H(\tau)$	$c_{i,\tau}$		
Scheme I	0.22(±0.01)	2.40(±0.06)	27.1(±0.38)	112(±11.73)
Scheme II	–	0.003(±0.0)	0.15(±0.01)	–

5   Extensions and Applications

The AO encryption schemes proposed in Sect. 3 can be extended to allow the aggregator to compute other statistics beyond the simple sum $\sum_i x_{i,\tau}$. In this section, we present extensions and example applications of dynamic AO encryption. These applications have in common the following: an aggregate value needs to be computed over a selected subset of a population, while not releasing individual user values in the clear. A major advantage of the schemes proposed in this paper is that they allow to compute aggregate values over any selected subsets of the population, without the need for redistributing keys for every new subset of selected users considered.

Hereafter, we will recurrently use the application example of surveys, polls, or electronic voting. In this case, the subset of users $\mathcal{S}_\tau$ on which the aggregate is computed is the set of users who are eligible to take the survey or poll, or to vote in the election. The tag τ is a unique identifier associated with each survey or election, that determines the subset of users over which the aggregate will be computed. For instance, the subset $\mathcal{S}_\tau$ can have associated eligibility criteria, such as demographic information (citizenship, age, location, ...). The individual user value $x_{i,\tau}$ represents the vote of user i, or her answers to the poll or to the survey. The dynamic schemes proposed in this paper allow to conduct multiple surveys or elections over different subsets of the population without the need to redistribute keys each time.

5.1   Weighted Sums

The proposed schemes can be extended to support weighted sums $\sum_i a_i\, x_{i,\tau}$ for some predetermined individual weights $a_i \in \mathbb{Z}$ known to each user. Particular cases of weighted sums include the simple sum $\sum_{i \in \mathcal{S}_\tau} x_{i,\tau}$ where $a_i = 1$ for all $i \in \mathcal{S}_\tau$, and the sample mean $\bar{X}_\tau := \frac{1}{|\mathcal{S}_\tau|} \sum_{i \in \mathcal{S}_\tau} x_{i,\tau}$ where $a_i = \frac{1}{|\mathcal{S}_\tau|}$ for all $i \in \mathcal{S}_\tau$.

Application. Consider the simple example of a salary survey, where an aggregator is interested in computing the average salary of a subset $\mathcal{S}_\tau$ of the population. User $i \in \mathcal{S}_\tau$ may consider her individual salary $x_{i,\tau}$ as private sensitive information that she does not wish to disclose in the clear. However, she may be willing to take part in the survey provided only the average salary $\Sigma_\tau = \frac{1}{|\mathcal{S}_\tau|} \sum_{i \in \mathcal{S}_\tau} x_{i,\tau}$ over all users in $\mathcal{S}_\tau$ is released in the clear, but not the individual user values, which is the promise of AO encryption.

5.2 k-th Order Moments

If the aggregator wishes to evaluate the k-th order sample moment $m_{k,\tau} := \frac{1}{|\mathcal{S}_\tau|} \sum_{i \in \mathcal{S}_\tau} (x_{i,\tau})^k$, each user encrypts $(x_{i,\tau})^k$. For instance, if the aggregator wishes to evaluate the sample mean (1st order moment) and the sample 2nd order moment, each user has to encrypt both $x_{i,\tau}$ and $(x_{i,\tau})^2$. Applying the basic scheme, the aggregator then gets $\sum_{i \in S_\tau} x_{i,\tau}$ and $\sum_{i \in S_\tau} (x_{i,\tau})^2$, from which the sample mean and 2nd order moment are obtained as

$$\bar{X}_\tau := \frac{1}{|S_\tau|} \sum_{i \in S_\tau} x_{i,\tau} \quad \text{and} \quad m_{2,\tau} := \frac{1}{|S_\tau|} \sum_{i \in S_\tau} (x_{i,\tau})^2.$$

Similarly, the sample variance (2nd order central moment) can be obtained as $\sigma_\tau^2 := \frac{1}{|S_\tau|} \sum_{i \in S_\tau} ((x_{i,\tau})^2 - \bar{X}_\tau^2)$. Note that the unbiased sample variance can be obtained using Bessel's correction, as $\sigma_\tau^2 := \frac{1}{|S_\tau|-1} \sum_{i \in S_\tau} ((x_{i,\tau})^2 - \bar{X}_\tau^2)$.

5.3   Vector Aggregation and Histograms

The schemes can also be extended to support vector aggregation, where each user i holds a vector $x_{i,\tau}$ of some length L. Vector aggregation can be used to compute at once the aggregates of L different variables of interest, or to compute histograms. Using the schemes to encrypt the vectors and then perform vector aggregation is more efficient than encrypting and aggregating each of the L values of interest separately. This can be done via a standard batching technique, that is, by packing the L components of the vector into a single ciphertext.

Application. Consider the example of a survey, where an aggregator is interested in computing both the average height and weight of a subset S_τ of the population. For each user i, $x_{i,\tau} = [h_i, w_i]$ is a vector of length $L = 2$ containing the user's height and weight. The vector aggregate $\bar{X}_\tau = \frac{1}{|S_\tau|} \sum_{i \in S_\tau} x_{i,\tau}$ produces a vector of the same length as $x_{i,\tau}$ which contains the average height and weight.

Application [Histogram]. Consider a survey or an election where users are asked to make a choice among several possible values. In that case, a user's vote or answer to a survey question can be encoded as a binary vector of length the number of possible values among which the user has to chose. Each entry of the vector is a 0 or a 1 indicating whether the user chose this value or not. For instance, if a user is asked to choose one candidate among 3 candidates for an election, then vector $x_{i,\tau}$ will be of length 3, the three entries of the vector representing candidates to the election. An example answer would be $x_{i,\tau} = [0, 1, 0]$, which means that the user voted for the second candidate. The result of the election can be obtained by aggregating the individual user vectors over the set of eligible voters S_τ: $\Sigma_\tau = \sum_{i \in S_\tau} x_{i,\tau}$ is a histogram of the votes, i.e., a vector of the same length as $x_{i,\tau}$ which contains the counts over each entries. In the previous election example, Σ_τ would be of length 3, where each entries contain the vote counts for each candidate.

In the more advanced case of surveys or election ballots with multiple questions, the vector binarization procedure presented in the previous example can be extended. A user's election ballot or answers to the full survey can be encoded as a binary vector of length [number of questions × number of possible answers to each question], and the position of each '1' in the vector indicates the user's selected answers to each question. For instance, if a user is asked to choose one candidate among 3 candidates for an election, and to answer a question with 5 choices, then vector $x_{i,\tau}$ will be of length $3 + 5 = 8$, the first three entries of the vector representing candidates to the election, and the next 5 entries representing the possible answers to the question. An example user ballot would be $x_{i,\tau} = [0, 1, 0, 1, 0, 0, 0, 0]$, which means that the user voted for the second candidate, and chose the first possible answer to the question. The result of the election can be obtained by aggregating the individual user vectors over the set of eligible voters $\mathcal{S}_\tau$: $\Sigma_\tau = \sum_{i \in \mathcal{S}_\tau} x_{i,\tau}$ is a vector of the same length as $x_{i,\tau}$ which contains the counts over each entries. In the previous election example, Σ_τ would be of length 8, the first 3 entries containing the vote counts for each candidate, and the next 5 entries containing the counts that teach possible answer to the question obtained. Note that questions allowing the choice of multiple simultaneous answers are possible, for instance $x_{i,\tau} = [0, 1, 0, 1, 0, 0, 1, 0]$ could be a possible ballot where answers 1 and 4 to the question were simultaneously selected.

Last, ensuring the validity of each individual user inputs $x_{i,\tau}$, to avoid that users game the election or survey, is a problem beyond the scope of this paper. It could be addressed by having a voting interface that restricts the format in which users provide individual ballot inputs rather than allowing a user to return any vector containing any values, or by using zero-knowledge proofs.

5.4   Statistics – Bootstrap Sampling

The bootstrap is a general method for assessing statistical accuracy, that relies on sampling with replacement [6]. Consider a training dataset containing n datapoints $\{x_1, \ldots, x_n\}$ on which an analyst would like to fit a model, or from which he would like to compute an estimate. For instance, the analyst may be interested in computing an estimate of the population mean and assessing its accuracy. The analyst generates B subsets of users of size n, called bootstrap samples. Tag τ encodes each bootstrap sample. Then the analyst computes the sample mean for each bootstrap sample of size n using AO encryption without needing to redistribute keys for each bootstrap sample set. Using the B sample means, the analyst can assess the accuracy of the sample mean estimate.

References

1. Ács, G., Castelluccia, C.: I have a DREAM! (DiffeRentially privatE smArt Metering). In: Filler, T., Pevný, T., Craver, S., Ker, A. (eds.) IH 2011. LNCS, vol. 6958, pp. 118–132. Springer, Heidelberg (2011). https://doi.org/10.1007/978-3-642-24178-9_9
2. Barthe, G., Danezis, G., Grégoire, B., Kunz, C., Zanella-Béguelin, S.: Verified computational differential privacy with applications to smart metering. In: 26th IEEE Computer Security Foundations Symposium (CSF 2013), pp. 287–301. IEEE Press (2013). https://doi.org/10.1109/CSF.2013.26
3. Benhamouda, F., Joye, M., Libert, B.: A new framework for privacy-preserving aggregation of time-series data. ACM Trans. Inf. Syst. Secur. **18**(3) (2016). https://doi.org/10.1145/2873069

4. Boneh, D.: The decision Diffie-Hellman problem. In: Buhler, J.P. (ed.) ANTS 1998. LNCS, vol. 1423, pp. 48–63. Springer, Heidelberg (1998). https://doi.org/10.1007/BFb0054851

5. Court of Justice of the European Union: The court of justice declares that the commission's US safe harbour decision is invalid. Press Release No 117/15, Judgment in Case C-362/14 Maximillian Schrems v Data Protection Commissioner, October 2015. http://curia.europa.eu/jcms/upload/docs/application/pdf/2015-10/cp150117en.pdf

6. Efron, B.: Bootstrap methods: another look at the jackknife. Ann. Stat. **7**(1), 1–26 (1979). http://www.jstor.org/stable/2958830

7. Erkin, Z., Tsudik, G.: Private computation of spatial and temporal power consumption with smart meters. In: Bao, F., Samarati, P., Zhou, J. (eds.) ACNS 2012. LNCS, vol. 7341, pp. 561–577. Springer, Heidelberg (2012). https://doi.org/10.1007/978-3-642-31284-7_33

8. Garcia, F.D., Jacobs, B.: Privacy-friendly energy-metering via homomorphic encryption. In: Cuellar, J., Lopez, J., Barthe, G., Pretschner, A. (eds.) STM 2010. LNCS, vol. 6710, pp. 226–238. Springer, Heidelberg (2011). https://doi.org/10.1007/978-3-642-22444-7_15

9. Hao, F., Zieliński, P.: A 2-round anonymous veto protocol. In: Christianson, B., Crispo, B., Malcolm, J.A., Roe, M. (eds.) Security Protocols 2006. LNCS, vol. 5087, pp. 202–211. Springer, Heidelberg (2009). https://doi.org/10.1007/978-3-642-04904-0_28

10. Hellman, M.E., Diffie, W.: New directions in cryptography. IEEE Trans. Inf. Theory **22**(6), 644–654 (1976). https://doi.org/10.1109/TIT.1976.1055638

11. Jawurek, M., Kerschbaum, F.: Fault-tolerant privacy-preserving statistics. In: Fischer-Hübner, S., Wright, M. (eds.) PETS 2012. LNCS, vol. 7384, pp. 221–238. Springer, Heidelberg (2012). https://doi.org/10.1007/978-3-642-31680-7_12

12. Jawurek, M., Kerschbaum, F., Danezis, G.: SoK: privacy technologies for smart grids - a survey of options. Technical report MSR-TR-2012-119, Microsoft Research, Cambridge, UK, November 2012. https://www.microsoft.com/en-us/research/publication/privacy-technologies-for-smart-grids-a-survey-of-options/

13. Joye, M., Libert, B.: A scalable scheme for privacy-preserving aggregation of time-series data. In: Sadeghi, A.-R. (ed.) FC 2013. LNCS, vol. 7859, pp. 111–125. Springer, Heidelberg (2013). https://doi.org/10.1007/978-3-642-39884-1_10

14. Kursawe, K., Danezis, G., Kohlweiss, M.: Privacy-friendly aggregation for the smart-grid. In: Fischer-Hübner, S., Hopper, N. (eds.) PETS 2011. LNCS, vol. 6794, pp. 175–191. Springer, Heidelberg (2011). https://doi.org/10.1007/978-3-642-22263-4_10

15. Leontiadis, I., Elkhiyaoui, K., Molva, R.: Private and dynamic time-series data aggregation with trust relaxation. In: Gritzalis, D., Kiayias, A., Askoxylakis, I. (eds.) CANS 2014. LNCS, vol. 8813, pp. 305–320. Springer, Cham (2014). https://doi.org/10.1007/978-3-319-12280-9_20

16. Rastogi, V., Nath, S.: Differentially private aggregation of distributed time-series with transformation and encryption. In: Elmagarmid, A.K., Agrawal, D. (eds.) 2010 ACM SIGMOD International Conference on Management of Data, pp. 735–746. ACM Press (2010). https://doi.org/10.1145/1807167.1807247

17. Shi, E., Chan, T.H.H., Rieffel, E.G., Chow, R., Song, D.: Privacy-preserving aggregation of time-series data. In: Network and Distributed System Security Symposium (NDSS 2011). The Internet Society (2011). https://www.ndss-symposium. org/wp-content/uploads/2017/09/shi.pdf
18. Teruya, T., Saito, K., Kanayama, N., Kawahara, Y., Kobayashi, T., Okamoto, E.: Constructing symmetric pairings over supersingular elliptic curves with embedding degree three. In: Cao, Z., Zhang, F. (eds.) Pairing 2013. LNCS, vol. 8365, pp. 305–320. Springer, Cham (2014). https://doi.org/10.1007/978-3-319-04873-4_6

Achieving Efficient and Verifiable Assured Deletion for Outsourced Data Based on Access Right Revocation

Yuting Cheng[1], Li Yang[1(✉)], Shui Yu[2], and Jianfeng Ma[1]

[1] Xidian University, Xi'an, China
`yangli@xidian.edu.cn`
[2] University of Technology, Sydney, Australia

Abstract. With the growing use of cloud storage facilities, outsourced data security becomes a major concern. However, assured deletion for outsourced data, as an important issue for users, but received less attention in academia and industry. Most of traditional deletion solutions require specific data organization forms or storage media, and are not applicable for outsourced data. Moreover, existing access control schemes for cloud which used ciphertext-policy attribute-based encryption (CPABE), focused on fine-grained access control, and completely ignored data deletion. In this paper, we aim to design an effective data deletion scheme that can be applied to any CPABE built on linear secret sharing-scheme. However, the challenge is how to maintain the traits of traditional CPABE while implementing a universal deletion method.

To address this challenge, we propose a policy graph to describe relationships among users, policies, attributes, and files and introduce a new deletion concept for CPABE: when all users are unauthorized for a file, we say that the file is deleted. Then, we extend an efficient and verifiable deletion scheme on a CPABE. Specifically, we give an effective method to select key attributes and update the relevant parts of ciphertext so that all users become unauthorized. Furthermore, we verify the cipher update performed by third-party server through merkle trees. We also demonstrate its universality and prove the security under q-BDHE assumption. Finally, the performance evaluation and simulation results reveal that our solution achieves better performance compared with other schemes.

Keywords: Access policy deletion · Ciphertext update · Verifiable assured deletion · Access right revocation

1 Introduction

During the past decade, we have seen an unprecedented growth in the use of cloud technology. Technological devices are being used in schools, government institutes, business sectors, and even hospitals to store data and retrieve it instantly when desired. Unfortunately, the convenience of technology also brings with it

© Springer Nature Switzerland AG 2019
Y. Mu et al. (Eds.): CANS 2019, LNCS 11829, pp. 392–411, 2019.
https://doi.org/10.1007/978-3-030-31578-8_22

some drawbacks. Outsourced data, especially sensitive user information, are at a risk of threat.

Ciphertext-policy attribute-based encryption (CPABE) is considered one of the most promising technologies to achieve flexible and fine-grained data access control without relying on a third party in cloud storage systems [14]. Many security issues in cloud computing has been fully studied in existing CPABE schemes, such as attribute revocation, ciphertext update, policy hiding, and efficient encryption/decryption. However, assured outsourced data deletion, as an important issue, received less attention in academia and industry.

Why is data deletion important in a CPABE scheme? Consider the following scenario, (1) the data owner encrypts a file using a CPABE scheme, and then stores the ciphers in a semi-trusted cloud server, (2) then he/she requests the server to delete the ciphertexts. In the real world, the cloud server simply releases the logical relationship between blocks, any authorized user can still access the deleted data with the help of the cloud server [6]. Then, he/she decrypts these ciphers with his/her secret key to access files that have been deleted. As we can see, the lack of specific data deletion operations in CPABE may lead to data leakage. Therefore, it's necessary to design extra deletion methods for every CPABE scheme to ensure that all users cannot decrypt deleted data.

Assured deletion has been extensively studied as an important direction of privacy protection. There are two ways to perform secure deletion: (1) overwrite the data [6] or destroy the physical media; (2) encrypt data or the whole media and use the first method or one of the other specified methods to securely delete the key [1]. The first method is not applicable for the cloud-based model because of the popularity of the virtualized storage models in the cloud computing [2]. The second method includes time-based and policy-based. The latter achieves more flexible deletion than the former [13]. Policy-based deletion schemes can be broadly divided into two types: one is based on other encryption technologies, the other is based on CPABE. However, the former cannot provide flexible access control for outsourced data in cloud computing. To the best of our knowledge, only the scheme in [19] is based on CPABE. The basic idea of this scheme is to add a virtual deletion attribute for each access policy and each user's attribute set, and then delete files by revoking the deletion attribute. Unfortunately, we must re-encrypt all data with a new access policy if we want to extend this scheme to an existing CPABE. Moreover, the overhead of file deletion is very high because a data owner is required to perform multiple bilinear pairing operations.

As mentioned above, most of existing deletion schemes don't achieve fine-grained access control, while most of the CPABE schemes completely ignore data deletion. In this paper, we would like to propose an assured deletion method without losing the characteristics of the attribute encryption scheme, such as fine-grained access control, flexible user rights management. However, the challenge is that how to maintain the traits of traditional CPABE while implementing a generic deletion method. That is, the deletion method can be applied to other CPABE schemes. We translate this challenge into the problem how to establish the relationship between attribute revocation, policy deletion and data deletion, which is talked in Sect. 3.

In the CPABE-based outsourcing scenario, files are encrypted by symmetric technology, such as AES. Then, the data key is protected by CPABE. Therefore, data deletion is to delete data keys. It can be achieved while all authorized users cannot correctly recover the data key of deleted files. in order to realize deletion, we make all users' attribute set unauthorized. The main contributions of this work can be summarized as follows.

- We propose a policy graph that defines the relationship among users, policies, attributes, and files. A protected class is considered the smallest access unit. It can support to simultaneously decrypt or delete a group of files. In order to efficiently delete data, we also define a key attribute set for each policy.
- We extend an efficient deletion method on the traditional CPABE. It converts data key deletion into policy deletion. The semi-honest cloud performs ciphers update upon receiving a deletion request. We conduct an operation to select a minimal key attribute set so that the server only updates essential attributes.
- We present a verify method based on Merkle Tree. The authority calculates the Merkle root of the updated ciphertext components as long as the server updates ciphers. Finally, it can return a verify result to the data owner. This method can efficiently verify ciphertext update.
- We give a security proof under decisional q-parallel bilinear Diffie-Hellman (q-BDHE) assumption, and universality analysis of our deletion method. We also analyse the performance of user-side encryption/decryption, and policy deletion. We also simulate our solution using the library in the github [7]. These results show that our scheme not only extends deletion in the CPABE scheme but also is more efficient than schemes in [16,19].

The remaining of this paper is organized as follows. Section 2 gives the background. Section 3 shows the definition of the policy graph, the relationship between attribute revocation and file deletion, and security requirements. Section 4 gives the system model and security model. Section 5 gives the construction of our scheme. Section 6 gives some analysis. Section 7 shows performance analysis and experimental results. Section 8 gives the related works.

2   Preliminaries

2.1   Access Policy

Let $P = \{P_1, P_2, \cdots, P_n\}$ be the attribute universe. The set $\mathbb{A} \subset 2^{\{P_1, P_2, \cdots, P_n\}}$ is monotone if $\forall B, C \in \mathbb{A}$: if $B \in \mathbb{A}$ and $B \subset C$, then $C \in \mathbb{A}$. Access structure is the monotone set $\mathbb{A}$ consisting of all non-empty subsets of P, that is $\mathbb{A} \subset 2^{\{P_1, P_2, \cdots, P_n\}}/\{\emptyset\}$. The sets in $\mathbb{A}$ are called authorized set, and the sets not in $\mathbb{A}$ are called unauthorized set. In the CPABE scheme, only the secret key corresponding to the authorized set can decrypt the ciphertext correctly.

2.2   Linear Secret-Sharing Schemes

A secret-sharing scheme Π over a set of attributes P is called linear (over $\mathbb{Z}_p$) if:

- The shares for each attribute form a vector over $\mathbb{Z}_p$.
- There exists a matrix an M with l rows and n columns called the share generating matrix for Π. For all $i = 1, ...,$ the i-th row of M we let the function ρ defined the attribute labeling row i as $\rho(i)$. When we consider the column vector $v = (s, r_2, ..., r_n)$, where $s \in \mathbb{Z}_p$ is the secret to be shared, and $r_2, ..., r_n \in \mathbb{Z}_p$ are randomly chosen, then $M \cdot v^T$ is the vector of l shares of the secret s according to Π. The share $M_i \cdot v^T$ belongs to the attribute $\rho(i)$.

It is shown in [14] that every linear secret sharing-scheme according to the above definition also enjoys the linear reconstruction property, defined as follows: Suppose that Π is an LSSS for the access structure $\mathbb{A}$. Let $S \in \mathbb{A}$ be any authorized set, and let $I \subset \{1, 2, 3, \cdots, \}$ be defined as $I = i : \rho(i) \in S \subset \{1, 2, 3, \cdots, \}$. Then, there exists constants $\{w_i \in \mathbb{Z}_p | i \in I\}$ such that, if $\{\lambda_i\}$ are valid shares of any secret s according to Π, then $\sum_{i \in I} w_i \lambda_i = s$. Furthermore, it is proved in [14] that these constants w_i can be found in time polynomial in the size of the share-generating matrix $M_{l \times n}$. In our scheme, we encrypt the data key with the share-generating matrix $M_{l \times n}$ which is related with the access policy.

2.3   Merkle Hash Tree

A Merkle Hash tree [9] is a tree in which every leaf node is labelled with the hash of a data block and every non-leaf node is labelled with the cryptographic hash of the labels of its child nodes. Hash trees allow efficient and secure verification of the contents of large data structures. A Merkle tree is recursively defined as a binary tree of hash lists where the parent node is the hash of its children, and the leaf nodes are hashes of the original data blocks. In this paper, we use the merkle tree to efficiently validate the ciphertext update in attribute revocation and file deletion.

2.4   Crypographic Assumption

Waters [14] gave the definition of the decisional q-parallel Bilinear Diffie-Hellman Exponent (q-parallel BDHE) problem as follows. Chooses a group $\mathbb{G}$ of a prime order p. Let $a, s, b_1, b_2, \cdots, b_q \in \mathbb{Z}_p$ be chosen at random and g be a generator of $\mathbb{G}$. Then

$$\boldsymbol{y} = g, g^s, g^a, \cdots, g^{a^q}, , g^{a^{q+2}}, \cdots, g^{a^{2q}}$$

$$\forall 1 \leq j \leq q, g^{sb_j}, g^{\frac{a}{b_j}}, \cdots, g^{\frac{a^q}{b_j}}, , g^{\frac{a^{q+2}}{b_j}}, \cdots, g^{\frac{a^{2q}}{b_j}}$$

$$\forall 1 \leq k, j \leq q, k \neq j, g^{as\frac{b_k}{b_j}}, \cdots, g^{a^q s \frac{b_k}{b_j}}.$$

The advantage of an adversary $\mathcal{A}$ to distinguish a valid tuple $e(g, g)^{a^{q+1}s}$ from a random element $R \in \mathbb{G}_T$ is as follows,

$$Adv_A = |Pr[\mathcal{A}(\boldsymbol{y}, e(g, g)^{a^{q+1}s}) = 0] - Pr[\mathcal{A}(\boldsymbol{y}, R) = 0]|$$

where a bilinear pairing $e : \mathbb{G} \times \mathbb{G} \rightarrow \mathbb{G}_T$.

3   Problem Description

In this section, we present a policy graph describing the relationship among users, files, policies, attributes, and illustrate the mapping between attribute revocation and policy deletion. Finally, we give three security goals.

3.1   Policy Graph

We define the policy graph suitable for our scheme. In the following description, & represents the predicate AND, and | represents the predicate OR.

Protected Class. In a corporate environment, if a file is a minimum access unit, each file will possess a data key, which causes inconvenience to each user in key management. In this paper, we treat files protected by the same access policy as a minimal access unit. A protected class is used to describe the logical group of files. In other words, files under the same protected class have the same data key. For example, in Fig. 1, File01, File02 and File03 possess the same access policy *Att01* & (*Att02* | *Att04*) and the same data key. A minimal access unit is also the smallest deletion unit.

Definition of Policy Graph. Let $G(V, E)$ denote a policy graph, as shown in Fig. 1. V is the set of all nodes, and it includes two kinds of nodes, source nodes (attribute nodes) and interior nodes (protected class nodes). The indegree of a source node is 0. E is the set of all edges. There are also two types of edges, edges from a source node to an interior node and edges from an interior node to an interior node. Each protected class node represents a unique user access policy that is a logical combination of multiple attribute nodes. When a user is in the user list of an attribute node, the node is true for the user. Therefore, the value of a protected class node for a user depends on the access policy of this protected class and the user's attribute set. Note that the value of a protected class node is always consistent with that of its access policy. For example, the access policy of the node p_3 is *Att03* | *Att05*. For *User01* and *User02*, p_3 is *true*. But for *User03* and *User04*, p_3 is *false*. Importantly, when a protected class node is *true* for a user, he/she can access the correct data key. Otherwise, the data key is invisible to the user.

The Key Attribute Set of a Policy. An attribute set is a key attribute set of a policy if and only if all attributes in the set are False, p is always False. Let KAS_p be a key attribute set of the policy p. If p is *Att01* & *Att02*, KAS_p will be one of the following sets {*Att01*},{*Att02*}, {*Att01*, *Att02*}. If p is *Att01* | *Att02*, KAS_p will be {*Att01*, *Att02*}. For an |, if and only if all child nodes are false, the policy is false. For an &, we only need to set one of child nodes to false. There are multiple sets of key attributes for a complex access policy, depending on the type of gates that makes up the policy. In this paper, we only focus on one of the smallest key attribute sets. The reason is given in Sect. 5.3.

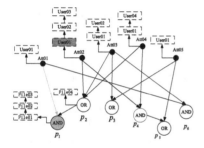

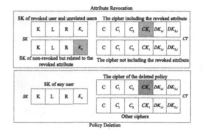

Fig. 1. A policy graph $G(V, E)$ **Fig. 2.** ARPD

3.2 Attribute Revocation and Policy Deletion

Attribute revocation means an attribute of a user is revoked. That is, the revoked user's access right to each policy is updated. In Fig. 1, when we revoke the attribute $Att02$ of the user $User01$, the access policy of p_4 changes from true to false for $User01$ but access rights of other users related to $Att02$, like $User02$, $User03$, are not affected. Yang et al. [17] proposed an effective attribute revocation scheme. As shown in Fig. 2, while an attribute of a user is revoked, other users containing the revoked attribute will update their secret key and ciphertexts related with this revoked attribute will also be updated. SK is user secret key and CT is data key ciphertext. Their generation details are defined in Sect. 5.

In this paper, the deletion of a policy means the deletion of a data key. After a policy is deleted, the policy is false for all users. That is, all users' access rights to the deleted policy are revoked. For example, if p_1 is deleted, then all files protected by p_1 will be inaccessible. p_1 is always false for any user. Therefore, our main deletion idea is to ensure that the value of the deleted policy is always false for any user. According to Sect. 3.1, if we let all attributes in the set KAS of the deleted policy be false, the policy will be inaccessible (false) for any user. For example, if we delete p_1 and $\{Att01\}$ belongs to KAS_{p_1}, we will make $Att01$ false for any user. We achieve a policy deletion method by only updating the deleted policy's ciphertext components related to attributes in the KAS. We give the schematic in Fig. 2. The marked part indicates the part that was updated and ARPD means attribute revocation and policy deletion.

3.3 Security Goals

We distinguish users from three aspects: (1) Is he/she an authorized user? (2) Registration time (bounded by the time for policy deletion). (3) Have she/he visited the correct data key before deletion? According to this, we believe that when a user accesses a ciphertext, he/she always needs to download the latest ciphertext from the cloud server instead of storing previous keys and keys' ciphertexts. We consider three specific security goals as followings:

- Confidentiality of data and keys. It is expected that the data are well protected when stored and processed in the cloud and the corresponding secret keys will not be disclosed to malious users.
- Secure fine-grained access control. This requirement is used to ensure that unauthorized users are always unable to access files. Of course, it also can resist the collusion attack of unauthorized users.
- Assured deletion. These kinds of users cannot gain any knowledge of files protected by the deleted policy. The data owner can verify the data deletion.

4   System Model

In this section, we propose the system model, including the parties and their functions. Then, we give the framework of our scheme to introduce the input and output of each algorithm. Finally, we give the security model.

4.1   Overview of System Model

Our system is composed of five participants: center server (CS), the authority (AA), the public cloud server (PBS), the data owners (DO) and the data consumers (DU), as shown in Fig. 3. In order to achieve higher security, we require that the user identity private key and the user decryption key be generated by two different agencies, CS and AA.

The CS is a trusted server in the system. In the real world, it's usually provided by the government. In the system initialization, it runs *CSetup* for system parameter generation and user registration.

The AA is also a trusted server in the system. It is mainly responsible for issuing secret keys, and assistance for policy deletion. The AA runs *AASetup* for generating attribute parameter. Then, it runs *SKeyGen* to produce secret keys for legal users. In th policy deletion phase, *DeleteKeyGen* is executed. It also performs *Verify* after a cipher is updated.

The PBS is a semi-honest (curious but honest) server. It stores files' and data keys' ciphertexts. In the decryption, it runs *TkeyGen* to assist users to decrypt. It updates the cipher of related data key so as to delete files protected by the same protected class.

The DO encrypts files and data keys. The DO first queries the AA whether the policy has been registered. If so, he/she should execute *TkeyGen* and then *Decrypt* to get the data key, and encrypts files with this key. Otherwise, the AA records this policy and the DO encrypts files and the data key.

Each DU is assigned to a secret key by the AA. With the help of PBS, the DU runs *Decrypt* for decryption in a constant time.

4.2   Definition of Our Scheme Framework

Our assured deletion scheme contains four phases: system initialization, encryption and decryption, policy deletion, and verification of cipher updates. In the Fig. 3, ① - ⑥ refer to the main parameters of *CSetup*, *SKeyGen*, *Encrypt*, *TKeyGen*, *DeleteKeyGen*, and *Verify*, respectively.

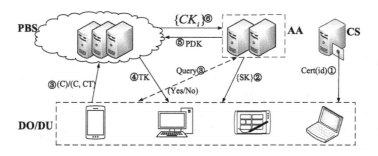

Fig. 3. System model

System Initialization

- $CSetup(1^{\lambda}) \rightarrow (msk, sp, (sk, vk), \{id, UPK_{id}, USK_{id}, Cert(id)\})$: This algorithm takes the security parameter λ. It outputs the master key msk, the system parameter sp, a secret and verificative key pair (sk, vk). It generates a series of parameters for each user, such as a global unique user identity id, a user-independent public and private key pair UPK_{id}, USK_{id}, and the certificate $Cert(id)$.
- $AASetup(U) \rightarrow (ASK, APK, \{SK_{a_i}, PK_{a_i} | a_i \in U\})$: This algorithm takes the attribute set U containing all attributes as input. It outputs the attribute authorization secret key ASK and the attribute authorization public key APK. For each attribute a_i in U, it generates an attribute key pair (SK_{a_i}, PK_{a_i}).
- $SKeyGen(Cert(id), sp, ASK, AT_{id}, \{PK_{a_i}\}) \rightarrow (SK_{id})$: It takes as inputs the user certificate $Cert(id)$, the system parameter sp, the attribute authorization secret key ASK, a user's attribute set AT_{id} and a series of attribute public keys $\{PK_{a_i}\}$. It outputs the secret key SK_{id}.

Encryption and Decryption

- $Encrypt(Cert(id), sp, APK, policy, f, k, (M, \rho), \{PK_{a_i}\}) \rightarrow (C, CT)$: This algorithm takes as inputs the certificate $Cert(id)$, the system parameter sp, the attribute authorization public key APK, the user access policy $policy$, a file f, a data key k, the access structure (M, ρ), and a series of attribute public keys $\{PK_{a_i}\}$. It outputs the file ciphertext C, the data key ciphertext CT.
- $TKeyGen(CT, SK_{id}, UPK_{id}) \rightarrow (TK_{id})$: This algorithm takes as inputs the data key cipher CT, the secret key SK_{id}, and the identity public key UPK_{id}. It outputs the token key TK_{id}.
- $Decrypt(C, CT, TK_{id}, USK_{id}) \rightarrow (f)$: This algorithm takes as inputs the file ciphertext C, the data key ciphertext CT, the token key TK_{id}, and the identity secret key USK_{id}. It outputs the plaintext f.

Policy Deletion

- $DeleteKeyGen(Cert(id), ASK, \{SK_{a_j} | a_j \in KAS_p\}, KAS_p) \to (\{PDK_{a_j} | a_j \in KAS_p\})$: This algorithm takes as inputs the user certificate $Cert(id)$, the attribute authorization secret key ASK, the related attribute secret keys $\{SK_{a'_j}\}$, and the key attribute set KAS_p for the policy p to be deleted. It outputs the policy deletion key $\{PDK_{a_j}\}$.
- $CTDelete(CT, \{PDK_{a_j}\}) \to (CT'')$: This algorithm takes as inputs the data key ciphertext CT and the policy deletion key $\{PDK_{a_j}\}$. It outputs the data key ciphertext CT'', which cannot be decrypted by any user.

Verification of Cipher Updates

- $Verify(CT, \{PDK_{a_j}\}) \to (true/false)$: This algorithm takes as inputs the data key ciphertext CT and the policy deletion key $\{PDK_{a_j}\}$. It outputs true or false.

4.3   Security Model

We define the security model of our scheme by the following game between a challenger $\mathcal{C}$ and an adversary $\mathcal{A}$. $\mathcal{A}$ can query for all secret keys that cannot be used directly to decrypt the challenge ciphertext.

Init. $\mathcal{A}$ determines the access structure $\mathbb{A}^*$ of the deleted policy p^*, meaning the data key encrypted by $\mathbb{A}^*$ is deleted.

Setup. The $CSetup$ and $AASetup$ are run to generate a series of system parameters. Then, $\mathcal{C}$ sends the public keys to $\mathcal{A}$.

Phase 1. $\mathcal{A}$ queries secret key by sending (id, SA_{id}) to $\mathcal{C}$, where SA_{id} is a set of attributes. $\mathcal{C}$ gives the secret key SK_{id} to $\mathcal{A}$.

Challenge. $\mathcal{A}$ submits two equal length messages m_0 and m_1. $\mathcal{C}$ flips a random coin b, and encrypts m_b under $\mathbb{A}^*$. Then, the ciphertext CT^* is sent to $\mathcal{A}$.

Phase 2. The adversary repeats the phase 1.

Guess. The adversary outputs a guess b^* of b.

The advantage of an adversary in this game is defined as $|Pr[b^* = b] - \frac{1}{2}|$.

Definition 2 A scheme is secure if all polynomial time adversaries have at most a negligible advantage in the above game, that is to say, $|Pr[b^* = b] - \frac{1}{2}|$ is negligible.

5   Construction of Our Scheme

In this section, we give the details of our scheme. Let $\mathbb{G}$ and $\mathbb{G}_T$ be the multiplicative groups with the same prime order p and $e : \mathbb{G} \times \mathbb{G} \to \mathbb{G}_T$ be the bilinear map. Let the generator of $\mathbb{G}$ be g. Let $H : \{0,1\}^* \to \mathbb{G}$ be a hash function such that the security is in the random oracle.

5.1 System Initialization

The CS runs $CSetup$. It randomly chooses a from the group $\mathbb{Z}_p$ and sets the master key $msk = a$, the system parameter $sp = g^a$ and a secret and verification key pair (sk, vk). Then the CS sends sp and vk to the AA. When a user initiates a registration request, the AA assigns him a global unique identity id. Then, it randomly chooses $u_{id}, z_{id} \in \mathbb{Z}_p$ to compute a user identity public key $UPK_{id} = g^{u_{id}}$ and a user identity secret key $USK_{id} = z_{id}$. Next, it computes a certificate $Cert(id) = En_{sk}(id, u_{id}, g^{1/z_{id}})$ with public key encryption mechanism and sends $\{id, UPK_{id}, USK_{id}, Cert(id)\}$ to the user. Only the parameter UPK_{id} can be public.

In $AASetup$, the AA generates the authorized secret key $ASK = (\alpha, \beta, \gamma)$, where α, β, γ are randomly selected from the group $\mathbb{Z}_p$. Then, it generates the authorized public key as follows: $APK = (g^{1/\beta}, g^{\gamma/\beta}, e(g, g)^\alpha)$. For each attribute a_i in the attribute set U, it produces the attribute secret key $SK_{a_i} = v_{a_i}$ and the attribute public key $PK_{a_i} = (g^{v_{a_i}} H(a_i))^\gamma$, where v_{a_i} is randomly selected from the group $\mathbb{Z}_p$. The APK, all PK_{a_i} and sp can be public.

In $SKeyGen$, the AA produces the user secret key for each legal user according to his/her attribute set. At first, it uses the verification key vk to decrypt the user's certificate $Dec_{vk}(Cert(id))$. If the DU is illegal, the AA does not respond to the user's request. In contrast, it computes the secret key as follows:

$$SK_{id} = (K_{id} = g^{\alpha/z_{id}} g^{au_{id}} g^{\frac{t_{id}}{\beta}}, L_{id} = (g^{1/z_{id}})^{\beta t_{id}}, R_{id} = g^{t_{id}},$$
$$\forall a_i \in AT_{id}, K_{id,a_i} = g^{\beta \gamma t_{id}/z_{id}} (g^{v_{a_i}} H(a_i))^{\gamma(\beta u_{id} + \gamma)})$$

where t_{id} is a random number in $\mathbb{Z}_p$.

5.2 Encryption and Decryption

In the encryption phase, the DO sends a query and the user access policy p to the AA. The AA sends a confirmation whether the corresponding data key has already existed. If it existed, the owner performs the $TKeyGen$ and $Decrypt$ to get the data key k and performs a symmetric encryption. Otherwise, the owner randomly produces the data key k and runs $Encrypt$ to encrypt the file f and the data key k. The DO computes the data key ciphertext CT as follows:

$$CT = (C = ke(g, g)^{\alpha s}, C_1 = g^s, C_2 = g^{\frac{s}{\beta}}$$
$$\forall i = 1 to l : CK_i = g^{a\lambda_i} \cdot (PK_{\rho(i)})^{-r_i}, DK_{1,i} = g^{\frac{r_i}{\beta}}, DK_{2,i} = g^{-\frac{\gamma r_i}{\beta}})$$

In the decryption phase, the DU sends his/her secret key to the PBS and requests the file ciphertext C. If the DU is authorized, the PBS computes the token key by using $TKeyGen$ as follows and sends the token key TK_{id} to the DU. The calculation by the PBS is as follows. First,

$$e(C_1, K_{id})e(C_2, R_{id})^{-1} = e(g, g)^{\frac{s\alpha}{z_{id}}} e(g, g)^{asu_{id}}$$

Then,

$$e(CK_i, UPK_{id})e(DK_{1,i}, K_{id,\rho(i)})e(DK_{2,i}, L_{id}PK_{\rho(i)}) = e(g,g)^{a\lambda_i u_{id}}$$

Finally,

$$TK = \frac{e(g,g)^{\frac{\alpha s}{z_{id}}} \cdot e(g,g)^{sau_{id}}}{\prod_{\rho(i) \in AT_{id}}(e(g,g)^{a\lambda_i u_{id}})^{w_i}} = \frac{e(g,g)^{\frac{\alpha s}{z_{id}}} \cdot e(g,g)^{sau_{id}}}{e(g,g)^{au_{id}\sum_{\rho(i) \in AT_{id}}\lambda_i w_i}}$$

If AT_{id} is an authorized set of the policy contained in the ciphertext, the token key TK_{id} will be $e(g,g)^{\frac{\alpha s}{z_{id}}}$. The user decrypts the data key ciphertext CT as follows:

$$k = \frac{CK}{TK^{USK_{id}}} = \frac{ke(g,g)^{\alpha s}}{e(g,g)^{\frac{\alpha s}{z_{id}} z_{id}}}.$$

Otherwise, TK_{id} will be $\perp$.

5.3 Policy Deletion

According to Sect. 3, we know that as long as we set all attributes which are included in a key attribute set of a deleted policy to false, all users become unauthorized for this policy. We update the deleted policy's data key ciphertext's components that are related to the attributes in the key attribute set, so that all users cannot correctly decrypt the ciphertext. In the file deletion phase, the AA first produces the deletion key by running $DeleteKeyGen$ for the deleted policy. Then, the PBS runs $CTDelete$ for the data key cipher of the deleted policy.

In $DeleteKeyGen$, the AA searches a minimal key attribute set of the deleted policy by running Algorithm 1. Why do we need to use an extra operation to find the smallest set instead of directly using all attributes as a key attribute set? Because, when the number of attributes in the key attribute set is reduced, the number of the ciphertext component to be updated is also reduced. What's more, the cipher update time far exceeds the execution time of the search operation.

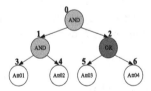

Fig. 4. A special policy

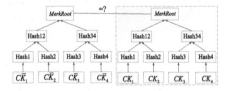

Fig. 5. Verification

How do we choose the smallest key attributes set? For a simple policy, $Att01$ & $Att02$ or $Att01 \mid Att02$, we can easily determine a minimum key attributes set. As shown in Sect. 3.1, as long as it is the &(AND) gate we only need to randomly set one of its child nodes false. For a complex policy, we represent it as a binary tree. All non-leaf nodes are gate nodes, representing an access policy, and all leaf nodes are attribute nodes. We assume that the minimum key attribute set of a leaf node is composed only by itself. The key attribute set of a policy is the minimum key attribute set of the root node of its corresponding binary tree. As shown in Fig. 4, the policy is $(Att01$ & $Att02)$ & $(Att03 \mid Att04)$. For the node 1, its key attribute set is $\{Att01\}$ or $\{Att02\}$. For the node 2, its key attribute set is $\{Att03, Att04\}$. For the root node 0, its key attribute set is $\{Att01\}$ or $\{Att02\}$ or $\{Att03, Att04\}$. Therefore, the minimum key attribute set for this policy is $\{Att01\}$ or $\{Att02\}$. Obviously, the minimum key attribute set of a non-leaf node is the union of the smallest key attribute sets of its child nodes or the smaller one. We present a recursive algorithm $SelKAS$ to implement the selection of a minimum key attribute set for a binary tree.

Then, it computes the deletion key PDK_{a_i} for each key attribute belonging to KAS.

$$\forall a_i \in KAS_p : PDK_{a_i} = \beta d_i,$$

where d_i is a random number in $\mathbb{Z}_p$.

In $CTDelete$, the PBS deletes the data key ciphertext as follows.

$$CT' = (C' = C, C'_1 = C_1, C'_2 = C_2, DK'_{1,i} = DK_{1,i}, DK'_{2,i} = DK_{2,i}$$
$$\forall i = 1tol, if(\rho(i) \notin KAS_{p_i}) : CK_i' = CK_i,$$
$$if(\rho(i) \in KAS_{p_i}) : CK_i' = CK_i \cdot DK_{2,i}^{PDK_{\rho(i)}})$$

We only update the data key ciphertext of p. As we can say, the policy is always False for each DU. No one can obtain the data key of p, which means all files protected by this user access policy are deleted.

5.4   Verification of Cipher Updates

The DO submits a verification request to AA after performing a policy dele- tion operation. Figure 5 shows a verification diagram, where the deleted policy includes four attributes. In the Fig. 5, $Hash_i$ is the hash value of ciphertext component CK_i and $Hash_{ij}$ is the hash value of $Hash_i \| Hash_j$. The process in the dashed box needs to be performed while generating the initial cipher. As long as a data key ciphertext is generated, the AA computes the initial merkle root for all cipher components $\{CK_i\}$ of the policy. Let $MerkRoot$ be the result of $Merkle(\{CK_i\})$.

As long as a data key cipher is deleted, the AA requests the related ciphertext components and recovers the cipher component as following:

$$if(\rho(i) \notin KAS_{p_i}) : C\bar{K}_i = CK_i,$$
$$if(\rho(i) \in KAS_{p_i}) : C\bar{K}_i = CK_i/DK_{2,i}^{PDK_{\rho(i)}}.$$

The AA computes $MerkRoot' = Merkle(\{C\bar{K}_i\})$. It outputs true while $MerkRoot' = MerkRoot$. The DO who initiated the deletion request is con- vinced that the ciphertext has been changed.

Algorithm 1. Select Key Attributes of the deleted policy **SelKAS**

Input: the node of binary tree *treeNode*;
Output: the key attributes set *KAS*;
 1: Create a key attribute set $KAS = \emptyset$;
 2: **if** *treeNode.isLeaf()* **then**
 3:     *KAS.add(treeNode)*
 4:     **return** *KAS*
 5: **end if**
 6: $KASL = SelKAS(treeNode.Lchild)$
 7: $KASR = SelKAS(treeNode.Rchild)$
 8: **if** *treeNode.value* $=$ ”*AND*” **then**
 9:     **if   then**$KASL.size() > KASR.size()$
10:         $KAS = KASR$
11:     **else**
12:         $KAS = KASL$
13:     **end if**
14: **else**
15:     **if** *treeNode.value* $=$ ”*OR*” **then**
16:         $KAS = KASL \cup KASR$
17:     **end if**
18: **end if**
19: **return** *KAS*

---

6   Universality Analysis and Security Proof

In this section, we give the universality analysis and the security proof.

6.1   Universality Analysis

Our scheme can be applied to all CPABE schemes built on the LSSS access structure. The reason why an authorized user can decrypt correctly is that the parameters corresponding to the user attributes match that of the access structure in the ciphertext. While updating parameters related to attributes in the KAS, we break the match between them so that all users can't decrypt correctly during the decryption phase.

6.2   Security Proof

The proof of the first two security goals are similar to [14]. The third goal is proved as follows.

Theorem 1. *Suppose the decisional q-parallel BDHE assumption holds in group* $\mathbb{G}$ *and* $\mathbb{G}_T$, *there is no* $\mathcal{A}$ *who can break the security of the proposed protocol with a nonnegligible advantage.*

Proof. The challanger C randomly chooses $\mu \leftarrow_R \{0,1\}$. If μ is 0, then $Z = e(g,g)^{a^{q+1}s}$. If μ is 1, $Z \leftarrow_R \mathbb{G}_T$. Then, it set $T = (\bar{y}, Z)$. After receiving T, B performs the following games.

Init. The adversary A outputs the challenge access structure $(M^*_{l^* \times n^*}, \rho^*)$.

Setup. B runs CSetup and AASetup and gives g to A. Then it also randomly chooses $\alpha, \beta, \gamma \in \mathbb{Z}_p$, and sets $\alpha = \alpha' + a^{q+1}$ by letting $e(g,g)^\alpha = e(g^a, g^{a^q})e(g,g)^{\alpha'}$.

The B programs the random oracle H as follows. It records every pair $(x, H(x))$, and returns the same value for the same x. Let X be the set of indices i, such that $\rho^*(i) = x$. Then, it runs the oracle as

$$H(x) = g^{d_x} \prod_{i \in X} g^{\frac{a M^*_{i,1}}{b_i} \frac{a^2 M^*_{i,2}}{b_i} \cdots \frac{a^{n^*} M^*_{i,n^*}}{b_i}}.$$

If X is $\emptyset$, $H(x) = g^{d_x}$, where d_x is a random value.

The B generates $APK = (e(g,g)^\alpha, g^{1/\beta}, g^{\gamma/\beta})$. Public attribute keys are produced as follows.

$$PK_x = (g^{d_x + v_x} \prod_{i \in X} g^{\frac{a M^*_{i,1}}{b_i} \frac{a^2 M^*_{i,2}}{b_i} \cdots \frac{a^{n^*} M^*_{i,n^*}}{b_i}})^\gamma.$$

B issues an identity id to A and chooses two random number u'_{id}, z_{id}. Then, $USK_{id} = z_{id}, UPK_{id} = g^{u'_{id}} g^{\frac{-a^q}{z_{id}}}$ by setting $u_{id} = u'_{id} - (\frac{a^q}{z_{id}})$. (USK_{id}, UPK_{id}) is sent to A.

Phase 1. A makes queries for secret key by submitting (id, SA_{id}). B computes a secret key SK_{id} for the set SA_{id} as follows. Note that SA_{id} is an collection of arbitrary attributes. A can obtain the key.

B finds a vector $\boldsymbol{w} = (w_1, w_2, \cdots, w_{n^*}) \in \mathbb{Z}_p^{n^*}$ in which $w_1 = -1$, and $\boldsymbol{w} \cdot M^*_i = 0$ for all i while $\rho(i)^* \in SA_{id}$.

Then, it chooses a random value $r \in \mathbb{Z}_p$ and lets t as $t = r + w_1 a^q + w_2 a^{q-1} + \cdots + w_n a^{q-(n^*-1)}$. It sets

$$L = g^{\frac{r\beta}{z_{id}}} \prod_{i=1}^{n^*} (g^{a^{q+1-i}})^{\frac{w_i \beta}{z_{id}}}, R = g^r \prod_{i=1}^{n^*} (g^{a^{q+1-i}})^{w_i},$$

$$K = g^{\frac{\alpha'}{z_{id}}} g^{au'_{id}} g^{\frac{r}{\beta}} \prod_{i=1}^{n^*} (g^{a^{q+1-i}})^{\frac{w_i}{\beta}}.$$

Then, if x is used in the access policy

$$K_x = L^\gamma PK_x^{\beta u'_{id}} PK_x^\gamma (g^{a^q})^{\frac{-\beta\gamma(v_x + d_x)}{z_{id}}} \prod_{i \in X} \prod_{j=1}^{n^*} (g^{\frac{a^{q+1+j}}{b_i}})^{-\beta\gamma M^*_{i,j}}.$$

If x is not used in the access policy $K_x = L^\gamma GPK_{id}^{\beta\gamma(v_x + d_x)} g^{\gamma^2(v_x + d_x)}$.

Challenge. $\mathcal{A}$ submits two messages of equal length, m_0 and m_1. $\mathcal{B}$ flips a coin b. It computes $C = m_b T e(g^s, g^{\alpha'})$, $C_1 = g^s$ and $C_2 = g^{\frac{s}{\beta}}$. The difficult part C_i is to simulate the values since this contains terms that must be canceled out. However, the simulator can choose the secret splitting, such that these can be canceled out. Intuitively, the simulator will choose random $y_2', \cdots, y_{n^*}'$ and share s using the vector $\boldsymbol{v} = (s, sa + y_2', sa^2 + y_3', \cdots, sa^{n^*-1} + y_{n^*}')$, and

$$\lambda_i = sM_{i,1}^* + \sum_{j=2}^{n^*}(sa^{j-1} + y_j')M_{i,j}^*.$$

It also chooses random values $r_1', \cdots, r_l'$.

Then, it sets $DK_{1,i} = (g^{r_i'}g^{sb_i})^{\frac{1}{\beta}}$, $DK_{2,i} = (g^{r_i'}g^{sb_i})^{-\frac{\gamma}{\beta}}$. Let R_i be the set of the other indices that map to the same attribute as row $i, \forall k \in R_i, k \neq i, \rho^*(i) = \rho^*(k)$, for all $i = 1, 2, \cdots, n^*$. The $\mathcal{B}$ generates a key attribute set KAS of the deleted policy. if $\rho^*(i) \in KAS$,

$$C_i = ((g^{v_{\rho^*(i)}+v_i'}H(\rho^*(i)))^\gamma)^{r_i'}(\prod_{j=2}^{n^*}(g^{aM_{i,j}^*y_j^*}))(g^{-b_i s\gamma})^{v_{\rho^*(i)}+v_i'+d_{\rho^*(i)}}$$

$$(\prod_{k \in R_i}\prod_{j=1}^{n^*}(g^{a^j s(\frac{b_i}{b_k})})^{\gamma M_{k,j}^*}),$$

where v_i' is a random value. Otherwise,

$$C_i = (PK_{\rho^*(i)})^{r_i'}(\prod_{j=2}^{n^*}(g^{aM_{i,j}^*y_j^*}))(g^{-b_i s\gamma})^{v_{\rho^*(i)}+d_{\rho^*(i)}}(\prod_{k \in R_i}\prod_{j=1}^{n^*}(g^{a^j s(\frac{b_i}{b_k})})^{\gamma M_{k,j}^*}).$$

Phase 2. Then, it repeats the phase 1.

Guess. $\mathcal{A}$ will eventually output a guess b^* of b. If $b^* = b$, $\mathcal{B}$ outputs $\mu = 0$ to show that $T = e(g, g)^{a^{q+1}s}$. Otherwise, it outputs $\mu = 1$ to indicate that it believes T is a random group element in $\mathbb{G}_T$. When T is a tuple, $\mathcal{B}$ gives a perfect simulation so we have that $Pr[\mathcal{B}(\boldsymbol{y}, T = e(g, g)^{a^{q+1}s} = 0)] = \frac{1}{2} + Adv_{\mathcal{A}}$.

When T is a random group element, the message m_b is completely hidden from the adversary and we have $Pr[\mathcal{B}(\boldsymbol{y}, T = R) = 1] = \frac{1}{2}$.

Thus, the total advanrange for $\mathcal{A}$ in this $q - parallel$ $BDHE$ game is negligible.

Privacy-Preserving Guarantee. In our scheme, PBS, CS, and AA can only get one of the user's keys (the secret key and the identity secret key). However, only knowing the above two keys can correctly decrypt the ciphertext.

7   Performance Analysis

In this section, we analyze the performance of our scheme and deletion scheme in [19] and some CPABE schemes in [8,11,15–18]. We also further compare our scheme with schemes in [16,19]. Because Yang's scheme in [17] did not support backward security [5] when the revoked user colludes with non-revoked users. And in the remaining CPABE schemes, the scheme in [16] has better encryption performance.

7.1   Comprehensive Analysis

We compare our scheme with previous works in terms of the user-side efficiency and some features, data deletion and the verification of updating ciphers, as shown in Table 1. In our scheme, we achieve efficient encryption by translating key encryption into key decryption and offload part of the key decryption operation to the cloud. In the deletion phase, we use hash and exponential operation rather than time-consuming pair. In general, our scheme not only realizes the efficient operation of the DU, but also improves the shortcomings of existing schemes in [8,11,15–19].

Table 1. Comparisions with previous related schemes

Scheme	User-side efficient		Data deletion	Verifiable cipher update
	Encryption	Decryption		
[8,15,16]	×	×	×	×
[11,17]	×	✓	×	×
[19]	×	×	✓	✓
Ours	✓	✓	✓	✓

7.2   Implementation and Evaluation

We perform experiments on a Windows with an Intel Core $i7$ CPU at 3.40 GHz and 8.00 GB RAM. We conduct implementation by the project *CloudCrypto* based on the jpbc library [3]. While testing the costs in different schemes, we use a prime-order bilinear group with 160-bit and the base field size is 512-bit. In the verification phase, we use SHA-256 to generate the merkle root.

We implement three schemes, including our scheme, Yang's scheme in [16] and Yu's scheme in [19]. We simulate all algorithms and measure the performance of encryption, decryption and deletion. All algorithms are executed 100 times. The results are the average of total time. The number of users' attributes ranges from 1 to 24. Let $n_{f,p}$ be the number of files protected by a policy.

Encryption Efficiency. We test the cases where $n_{f,p}$ is 1, 2, and 3, as shown in Fig. 6(a). When $n_{f,p}$ is 1, the average encryption time of a file changes from 60.267 ms to 558.13 ms as the number of attributes possessed by the policy increased, and when $n_{f,p}$ is greater than 1, the average encryption time reduces rapidly. When $n_{f,p}$ is 2 and 3, the average encryption time is 30.882 ms to 279.7 ms and 20.838 ms to 186.678 ms. The encryption time in [19] ranges from 66.757 ms to 468.413 ms, and the time in [16] is from 62.743 ms to 656.604 ms. It's obvious that our solution shows better encryption performance than that in [16,19] when $n_{f,p}$ is greater than 1. Specifically, we convert the user-side encryption overhead to the outsourced decryption. We also test the cost time of the secret key generation phase of the three scenarios. The secret key in Yu's contains less information, so this scheme has the best secret key generation efficiency. Compared with the scheme in Yang's, ours always takes an extra 20 ms or so. The reason is that we always need to perform an exponential operation more.

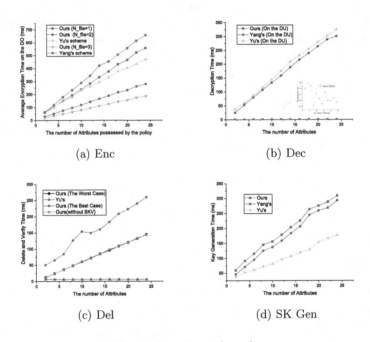

(a) Enc   (b) Dec

(c) Del   (d) SK Gen

Fig. 6. Experimental results

Decryption Efficiency. Our decryption operation is divided into two parts: the generation of token key by the PBS, and then the final encryption by the DU. As shown in Fig. 6(b), a large amount of decryption overhead is offloaded to the PBS, the cost on the DU is almost close to 0 (0.608 ms–0.793 ms). In [19], the DU takes a lot of time (36.042 ms–324.793 ms) for decryption. In [16], decryption time on each DU is from 25.042 ms to 300.793 ms. What stands out in the Fig. 6(b) is that our scheme achieves more efficient decryption.

Policy Deletion Efficiency. We test the total cost of the deletion phase including the policy deletion key generation, the cipher update and the verification, and analyze the best case and the worst case. "Best" means the key attribute set only includes one element. "Worst" is that the key attribute set contains all attributes in the deleted policy. Yu's [19] deletion is also given in the Fig. 6(c). Note that the horizontal axis is the number of attributes of a deleted policy. It's apparent that even in the worst case, our scheme is also more efficient than Yu's in [19]. Moreover, we also test the case while we perform deletion without the operation $SelKAS$, where the overhead is always close to the worst case. In the Fig. 6(c), SKV is $SeleKAS$. Therefore, we can effectively reduce the number of updated ciphertext components and improve the deletion efficiency by performing the selection of key attributes.

8 Related Work

Assured deletion has been extensively studied in a large number of works. These studies are divided into two parts: (1) erase the data [6] or destroy the physical media; (2) encrypt data and then make data key irrecoverable [1].

8.1 Erase the Data

The schemes [6] based on erasing depend heavily on the physical storage. They are only suitable for the deletion of local data. Unfortunately, due to the popularity of the virtualized storage medias in the cloud computing, these schemes are no longer feasible.

8.2 Secure Deletion via Encryption

The encryption-based deletion methods are to implement encryption key deletion. They can be divided into three categories: (1) based on the erasable memory [1,10], (2) time-based [12]and (3) policy-based [2,13,19]. First case mentioned above is to store the key in erasable memory. Obviously, it is difficult and impractical for outsourcing scenarios. The second methods are to generate a time key in the encryption phase. When the time key expired, the corresponding file will not be decrypted correctly. However, we need to know the deletion time of all files in advance. Fortunately, policy-based outsourced data deletion overcomes the shortcomings of the first two solutions.

Tang et al. [13] first proposed a policy-based deletion scheme, emphasizing outsourced data deletion based on multiple key management, but ignoring the flexible access control of data which is an important feature for outsourcing scenario. Cachi [2] et al. then proposed a secret sharing-based deletion scheme. The key is inaccessible by setting the recovery threshold and updating the key component, but the update of the key component still relies on the erasable memory.

With the development of access control schemes, Waters et al. [14] firstly proposed the ABE, especially, CPABE. It is considered to be a promising solution in the outsourcing scenario because of its flexible access control. However, most of attribute encryption schemes focus on user right management, key leakage tracking, lightweight encryption and decryption and so on [4,8,11,16,17]. Assured deletion for outsourced data, as an important issue for users, but received less attention in academia and industry.

Only one paper extends data deletion in traditional CPABE. Yu et al. [19] proposed an efficient deletion scheme that added a deletion attribute for each policy associated with ciphertexts, and the value of the deletion attribute could directly determine the true value of the user policy. File deletion is achieved by canceling the deletion attribute. However, this scheme adds an deletion attribute when encrypting and decrypting, which leads to many additional overhead. In particular, when a user encrypts multiple files, the overhead of adding an attribute is not to be underestimated. In addition, a user needs to perform multiple pair operations in the verification phase, which is inefficient.

Therefore, we want to conduct a deletion scheme based on CPABE, without adding an additional attribute to achieve high efficiency on the user side.

9   Conclusion

In this paper, we study the privacy threat of imcomplete data deletion in CPABE schemes. We propose a novel deletion concept for CPABE: when all users' access right to a file are revoked, we believe that the file was deleted. We also introduce an improved CPABE scheme based on protected class, and realize the user-side efficient encryption. After that, we extend a deletion method by revoking all users' rights. At the same time, in order to improve the efficiency of deletion, we propose a key attribute selection method to reduce the number of attributes that need to be deleted. Finally, we propose a Merkle tree-based verification method for each update so as to ensure that the PBS always stores the latest ciphertext version. Our implementation shows that ours is more efficient than previous works.

Acknowledgement. This work was supported by the National Natural Science Foundation of China (61671360, 61672415,61672413), the National Key Research and Development Project of China (2016YFB0801805), the Key Program of NSFC-Tongyong Union Foundation under Grant (U1636209), and the Key Program of NSFC Grant (U1405255).

References

1. Boneh, D., Lipton, R.J.: A revocable backup system. In: Proceedings of the 6th USENIX Security Symposium, San Jose, CA, USA, July 22–25 1996 (1996)
2. Cachin, C., Haralambiev, K., Hsiao, H., Sorniotti, A.: Policy-based secure deletion. In: 2013 ACM SIGSAC Conference on Computer and Communications Security, CCS 2013, Berlin, Germany, 4–8 November 2013, pp. 259–270 (2013)
3. Caro, A.D., Iovino, V.: JPBC: Java pairing based cryptography. In: Proceedings of the 16th IEEE Symposium on Computers and Communications, ISCC 2011, Kerkyra, Corfu, Greece, 28 June 28–1 July 2011, pp. 850–855 (2011)

4. Chen, Y., Sun, W., Zhang, N., Zheng, Q., Lou, W., Hou, Y.T.: A secure remote monitoring framework supporting efficient fine-grained access control and data processing in IoT. In: Beyah, R., Chang, B., Li, Y., Zhu, S. (eds.) SecureComm 2018. LNICST, vol. 254, pp. 3–21. Springer, Cham (2018). https://doi.org/10.1007/978-3-030-01701-9_1

5. Hong, J., Xue, K., Li, W.: Comments on "DAC-MACS: effective data access control for multiauthority cloud storage systems"/security analysis of attribute revocation in multiauthority data access control for cloud storage systems. IEEE Trans. Inf. Forensics Secur. 10(6), 1315–1317 (2015)

6. Joukov, N., Papaxenopoulos, H., Zadok, E.: Secure deletion myths, issues, and solutions. In: Proceedings of the 2006 ACM Workshop on Storage Security and Survivability, StorageSS 2006, Alexandria, VA, USA, 30 October 2006, pp. 61–66 (2006)

7. Liu, W.: Cloudcrypto. https://github.com/liuweiran900217/CloudCrypto

8. Liu, Z., Duan, S., Zhou, P., Wang, B.: Traceable-then-revocable ciphertext-policy attribute-based encryption scheme. Future Gener. Comput. Syst. 93, 903–913 (2019)

9. Merkle, R.C.: A digital signature based on a conventional encryption function. In: Pomerance, C. (ed.) CRYPTO 1987. LNCS, vol. 293, pp. 369–378. Springer, Heidelberg (1988). https://doi.org/10.1007/3-540-48184-2_32

10. Mitra, S., Winslett, M.: Secure deletion from inverted indexes on compliance storage. In: Proceedings of the Second ACM Workshop on Storage Security and Survivability, pp. 67–72. ACM (2006)

11. Ning, J., Cao, Z., Dong, X., Liang, K., Ma, H., Wei, L.: Auditable σ-time outsourced attribute-based encryption for access control in cloud computing. IEEE Trans. Inf. Forensics Secur. 13(1), 94–105 (2018)

12. Perlman, R.J.: File system design with assured delete. In: Proceedings of the Network and Distributed System Security Symposium, NDSS 2007, San Diego, California, USA, 28th February–2nd March 2007 (2007)

13. Tang, Y., Lee, P.P.C., Lui, J.C.S., Perlman, R.J.: Secure overlay cloud storage with access control and assured deletion. IEEE Trans. Dependable Sec. Comput. 9(6), 903–916 (2012)

14. Waters, B.: Ciphertext-policy attribute-based encryption: an expressive, efficient, and provably secure realization. In: Catalano, D., Fazio, N., Gennaro, R., Nicolosi, A. (eds.) PKC 2011. LNCS, vol. 6571, pp. 53–70. Springer, Heidelberg (2011). https://doi.org/10.1007/978-3-642-19379-8_4

15. Yang, K., Jia, X.: Expressive, efficient, and revocable data access control for multiauthority cloud storage. IEEE Trans. Parallel Distrib. Syst. 25(7), 1735–1744 (2014)

16. Yang, K., Jia, X., Ren, K.: Attribute-based fine-grained access control with efficient revocation in cloud storage systems. In: 8th ACM Symposium on Information, Computer and Communications Security, ASIA CCS 2013, Hangzhou, China, 08–10 May 2013, pp. 523–528 (2013)

17. Yang, K., Jia, X., Ren, K., Zhang, B.: DAC-MACS: effective data access control for multi-authority cloud storage systems. In: Proceedings of the IEEE INFOCOM 2013, Turin, Italy, 14–19 April 2013, pp. 2895–2903 (2013)

18. Yang, K., Jia, X., Ren, K., Zhang, B., Xie, R.: DAC-MACS: effective data access control for multiauthority cloud storage systems. IEEE Trans. Inf. Forensics Secur. 8(11), 1790–1801 (2013)

19. Yu, Y., Xue, L., Li, Y., Du, X., Guizani, M., Yang, B.: Assured data deletion with fine-grained access control for fog-based industrial applications. IEEE Trans. Ind. Inform. 14(10), 4538–4547 (2018)

Integer Reconstruction Public-Key Encryption

Houda Ferradi[2](✉) and David Naccache[1]

[1] Département d'informatique de l'ENS, ENS, CNRS, PSL University, 45 rue d'Ulm,
75230 Paris Cedex 05, France
{houda.ferradi,david.naccache}@ens.fr
[2] Department of Computing, The Hong Kong Polytechnic University (PolyU),
Hung Hom, Hong Kong

Abstract. In [AJPS18], Aggarwal, Joux, Prakash & Santha described an elegant public-key encryption (AJPS-1) mimicking NTRU over the integers. This algorithm relies on the properties of Mersenne primes instead of polynomial rings.

A later ePrint [BCGN17] by Beunardeau et al. revised AJPS-1's initial security estimates. While lower than initially thought, the best known attack on AJPS-1 still seems to leave the defender with an exponential advantage over the attacker [dBDJdW17]. However, this lower exponential advantage implies enlarging AJPS-1's parameters. This, plus the fact that AJPS-1 encodes only a single plaintext bit per ciphertext, made AJPS-1 impractical. In a recent update, Aggarwal et al. overcame this limitation by extending AJPS-1's bandwidth. This variant (AJPS-ECC) modifies the definition of the public-key and relies on error-correcting codes.

This paper presents a different high-bandwidth construction. By opposition to AJPS-ECC, we do not modify the public-key, avoid using error-correcting codes and use backtracking to decrypt. The new algorithm is *orthogonal* to AJPS-ECC as both mechanisms may be concurrently used *in the same ciphertext* and cumulate their bandwidth improvement effects. Alternatively, we can increase AJPS-ECC's information rate by a factor of 26 for the parameters recommended in [AJPS18].

The obtained bandwidth improvement and the fact that encryption and decryption are reasonably efficient, make our scheme an interesting post-quantum candidate.

Keywords: KEM · Efficiency improvement · MERS assumption · Implementation

1 Introduction

The public-key encryption schemes that are mostly used today are RSA [RSA78] and ElGamal [ElG85]. Their security are based on the problems of factoring large composite integers or computing discrete logarithms.

H. Ferradi—This work was done while the first author was at NTT Research - Tokyo.

© Springer Nature Switzerland AG 2019
Y. Mu et al. (Eds.): CANS 2019, LNCS 11829, pp. 412–433, 2019.
https://doi.org/10.1007/978-3-030-31578-8_23

However, in [Sho97], Shor published an algorithm that quantum computers can use to solve both discrete logarithms and factoring in polynomial time. Therefore, it is important to construct new encryption schemes which are to remain secure even after the advent of quantum computers.

For this purpose, in 2016 the National Institute of Standards and Technology (NIST) has initiated a competition to select post-quantum cryptographic algorithms that are supposed to resist future quantum computers [NIS17].

One promising candidate selected for such quantum-resistant encryption scheme is the AJPS cryptosystem [AJPS17a, AJPS18, AJPS17b], introduced by Aggarwal, Joux, Prakash, and Santha, which is an interesting alternative to the well-established NTRU cryptosystem. The security of AJPS encryption relies on the hardness of *Mersenne Low Hamming Combination* (MERS) problem [AJPS17a, AJPS18]: Given a Mersenne prime $p = 2^n - 1$ (where n is prime), samples of the $\mathsf{MERS}_{h,n}$ distribution are constructed as $(a, b = as + e)$, where $a \in_R \mathbb{Z}_p$, the secret s and the error e are chosen uniformly at random from the elements in $\mathbb{Z}_p$ of Hamming weight h. The decisional version of the MERS assumption states that no efficient adversary can distinguish the $\mathsf{MERS}_{h,n}$ distribution from a uniform distribution over $\mathbb{Z}_p^2$.

Despite the efficiency benefit of its reliance on Mersenne primes. The cryptosystem introduced in [AJPS18], however, remains inefficient because of the constraint that $n = \Theta(h^2)$.

In this paper, we present new KEM schemes for enhancing the information rate of [AJPS18] by a factor of 26, this $\mathcal{O}(1)$ improvement is nonetheless significant in practice.

1.1   Related Works

Lattice-based cryptography is the most popular candidate for post-quantum security. Its security relies on the hardness of basis reduction and other related problems in random lattices, like Learning with Errors (LWE) based cryptosystems [Reg06], Ring-LWE based cryptosystems [LPR10] and NTRU [HPS98].

Concurrently to the above, AJPS's Mersenne post-quantum cryptosystems [AJPS18] belong to the NTRU family.

1.2   Organization of the Paper

Section 2, overviews basic notions and notations, key-encapsulation mechanisms and backtracking. Section 3 recalls the MERS problem and its hardness. Section 4 reviews the original AJPS scheme and its variants. In Sect. 5 we propose our KEM schemes, which are the Bivariate KEM and Trivariate KEM and prove their security. Section 7, provideq an instantiation for the backtracking used in our KEMs. Finally, we conclude the paper in Sect. 9.

2   Preliminaries

2.1   Notation

We denote by $\|x\|$ the Hamming weight of an n-bit string x, which is the total number of 1's in x. Let $\mathfrak{H}_{n,h}$ be the set of all n-bit strings of Hamming weight h and $\{0,1\}^n$ the set of all n-bit strings.

Let $\mathbb{Z}_p$ be the integer ring modulo p, where $p = 2^n - 1$ is a Mersenne prime. We have the following property:

Lemma 1. *Let $x, y \in \mathbb{Z}_p$, then the following properties hold:*
Property 1: $\|x + y \pmod{p}\| \le \|x\| + \|y\|$

Property 2: $\|x \cdot y \pmod{p}\| \le \|x\| \cdot \|y\|$

Property 3: $x \ne 0^n \Rightarrow \| -x \pmod{p}\| = n - \|x\|$

The proof can be found in [AJPS18].

2.2   Key-Encapsulation Mechanism Syntax

A key-encapsulation mechanism (KEM) consists in four algorithms: $\Pi = ($Setup, KeyGen, Encap, Decap$)$.

- Setup: The set up algorithm takes as input a security parameter λ and outputs a public parameter pp.
- KeyGen: The key generation algorithm takes as input a public parameter pp and outputs a public-key pk and a secret key sk.
- Encap: The encapsulation algorithm takes as input a public key pk and outputs a ciphertext C and key K.
- Decap: The decapsulation algorithm takes as input a ciphertext C and sk and outputs $\perp$ or a key K.

Definition 1. *We say that $\Pi = ($Setup, KeyGen, Encap, Decap$)$ has (1-λ)-correctness if for any $($pk, sk$)$ generated by KeyGen, we have that:*

$$\Pr\big[\mathsf{Decap}(\mathsf{sk}, \mathsf{C}) = \mathsf{K} : (\mathsf{C}, \mathsf{K}) \leftarrow \mathsf{Encap}(\mathsf{pk})\big] \ge 1 - \lambda.$$

2.3   Backtracking Techniques

For solving constraint satisfaction problems, there are basically three main approaches: backtracking, local search, and dynamic programming. In this work, we consider backtracking (also known as *depth-first search*), which consists of searching every possible combination in order to solve an optimization problem. To that end, backtracking proceeds in three steps: it picks a solution as a sequence of choices to the first sub-problem, then recursively attempts to resolve other sub-problems based on the solution of the first sequence of choices, then it returns the best solution found. We refer the reader to [Knu] for further introductory background.

3   The MERS Assumption: Notation and Definitions

Aggarwal et al. introduced a new assumption [AJPS18] mimicking NTRU over integers, this assumption relies on the properties of Mersenne primes in the ring $\mathbb{Z}_p$ instead of polynomial rings $\mathbb{Z}_q[x]/(x^n - 1)$. Their conjecture is based on the observation that given any number $a \in \mathbb{Z}_p$ we obtain this property: if we multiply a by any number $b = 2^x$ where $x \in [0, n-1]$, then the result $c = a \cdot b$ is just a cyclic shift.

The security of our scheme is based on the following assumptions:

3.1   MERS Assumption

For two integers $4h^2 < n$ and for n-bit Mersenne prime $p = 2^n - 1$, and for integer $s \in \mathbb{Z}_p$, we define a distribution $\mathsf{MERS}_{s,n,h}$ as follows: choose $r \leftarrow \{0,1\}^n$ and $b \leftarrow \mathfrak{H}_{n,h}$, return $(r, r \cdot s + b \bmod p)$. We also define a uniform distribution U as follows: choose $r \leftarrow \{0,1\}^n$ and $b \leftarrow \{0,1\}^n$, return $(r, b \bmod p)$.

Definition 2 ((Decisional) Mersenne Low Hamming Combination Assumption (MERS Assumption) [AJPS18]). *For two positive integers $4h^2 < n$ and for an adversary $\mathcal{A}$, we introduce the $\mathsf{MERS}_{n,h}$ advantage as the quantity:*

$$\mathsf{Adv}^{\mathsf{MERS}_{h,n}}(\mathcal{A}) = \left| \Pr[\mathcal{A}^{\mathsf{MERS}_{s,n,h}()} \Rightarrow \texttt{True}] - \Pr[\mathcal{A}^{U()} \Rightarrow \texttt{True}] \right|,$$

where $s \xleftarrow{\$} \mathfrak{H}_{n,h}$. We say that the $\mathsf{MERS}_{n,h}$ problem is (t, q, ϵ)-hard if for all attackers $\mathcal{A}$ with time complexity t, making at most q queries, we have $\mathsf{Adv}^{\mathsf{MERS}}(\mathcal{A}) \leq \epsilon$.

3.2   RMERS Assumption

Definition 3 (Mersenne Low Hamming Ratio Search Assumption (RMERS) [AJPS18]). *Given an n-bit Mersenne prime $p = 2^n - 1$, $4h^2 < n$ and an integer $H \in \mathbb{Z}_p$, find $F, G \in \mathfrak{H}_{n,h}$ such that:*

$$H = \frac{F}{G} \bmod p$$

3.3   Hardness of the MERS Problem

MEET-IN-THE-MIDDLE ATTACK. de Boer et al. [dBDJdW17] presented a meet-in-the-middle attack for solving the MERS problem. Their classical and quantum attacks run in respective times:

$$\tilde{O}\left(\sqrt{\binom{n-1}{h-1}} \right) \quad \text{and} \quad \tilde{O}\left(\sqrt[3]{\binom{n-1}{h-1}} \right)$$

LLL-ATTACK. [BCGN17, dBDJdW17] present an LLL-based algorithms [LLL82] for solving the RMERS, whose Turing and quantum running times are respectively $O(2^{2h})$ and $O(2^h)$. To date, this is the most efficient known approach for solving RMERS. Assuming a quantum computer using Grover's algorithm, one obtains a quadratic speedup over [dBDJdW17] (i.e. $O(2^h)$). Therefore, as per [AJPS18], our scheme is secure when $h = \lambda = 256$.

PRIMALITY OF n. [AJPS17a, AJPS18] recommend $p = 2^n - 1$ and n to be primes to avoid an attack on composite n.

4 The Original AJPS Cryptosystem

4.1 The AJPS-1 Encryption Scheme

The original AJPS-1 encryption scheme based on MERS assumption [AJPS17a], is defined by the following sub-algorithms:

- Setup(1^λ) → pp. Chooses the public parameters pp $= \{n, h\}$ so that $p = 2^n - 1$ is an n-bit Mersenne prime achieving some λ-bit security level.
- KeyGen(pp) → {sk, pk}. Picks $\{F, G\} \in_R \mathfrak{H}_{n,h}^2$ and returns:

$$\begin{cases} \mathsf{sk} & \leftarrow G \\ \mathsf{pk} & \leftarrow H = F/G \bmod p \end{cases}$$

- Enc(pp, pk, $m \in \{0, 1\}$) → C. Picks $\{A, B\} \in_R \mathfrak{H}_{n,h}^2$, and computes:

$$C \leftarrow (-1)^m (AH + B) \bmod p$$

- Dec(pp, sk, C) → $\{\bot, 0, 1\}$, computes $d = \|GC \bmod p\|$ and returns:

$$\begin{cases} 0 & \text{if } d \leq 2h^2, \\ 1 & \text{if } d \geq n - 2h^2, \\ \bot & \text{otherwise} \end{cases}$$

The intuition behind the decryption formula is the observation that when $m = 0$ we get:

$$W = GC = G(AH + B) = FA + GB \Rightarrow W \text{ is of low Hamming weight}$$

The security of this cryptosystem is based on the (decisional) MERS problem introduced before.

To increase bandwidth, Aggarwal et al. introduced the AJPS-ECC variant described hereafter.

4.2   The **AJPS-ECC** Encryption Scheme

In the second variant AJPS-ECC [AJPS18, AJPS17b] aims to extend AJPS-1's bandwidth, while requiring an ancillary error correction scheme $\{\mathcal{D}, \mathcal{E}\}$.

AJPS-ECC is formally defined by the following sub-algorithms:

- Setup$(1^\lambda) \to$ pp. As in Sect. 4.1.
- KeyGen(pp) $\to$ {sk, pk}. Picks $\{F, G\} \in_R \mathfrak{H}^2_{n,h}$, $R \in_R \{0,1\}^n$ and returns:

$$\begin{cases} \text{sk} & \leftarrow F \\ \text{pk} & \leftarrow \{R, T\} = \{R, FR + G \bmod p\} \end{cases}$$

- Enc(pp, pk, $m \in \{0,1\}^\lambda) \to C$. Picks $\{A, B_1, B_2\} \in_R \mathfrak{H}^3_{n,h}$ and computes the ciphertext:

$$C = \begin{cases} C_1 & \leftarrow AR + B_1 \bmod p \\ C_2 & \leftarrow (AT + B_2 \bmod p) \oplus \mathcal{E}(m) \end{cases}$$

- Dec(pp, sk, $C) \to \{\bot, m\}$ returns:

$$\mathcal{D}((FC_1 \bmod p) \oplus C_2).$$

For the sake of clarity, we keep the definition of C_1 unchanged but slightly depart from [AJPS17a]'s original formulae by modifying the definitions of T and C_2 as follows:

$$T \leftarrow FR - G \ \bmod p$$
$$C_2 \leftarrow (AT - B_2 \bmod p) \oplus \mathcal{E}(m)$$

To understand the intuition behind Dec consider the quantity $W = FC_1 - C_2$ corresponding to the particular case $\mathcal{E}(m) = 0$:

$$W = FAR + FB_1 - AT + B_2 = FAR + FB_1 - A(FR - G) + B_2 = FB_1 + GA + B_2$$

As before, we see that $d = \|W\|$ is low. This means that the noise attached to $\mathcal{E}(m)$ after the clean-off operation $(FC_1 \bmod p) \oplus C_2$ is low and thus surmountable by the error-correcting code $\{\mathcal{E}, \mathcal{D}\}$.

The security of this cryptosystem is based on the (decisional) MERS problem. We refer the reader, again, to [AJPS18] for further details about this cryptosystem and the parameter choices allowing successful decryption and sufficient security. Sticking only to the core idea, we purposely omit the hashing and re-encryption tests performed during the key de-encapsulation process.

5 Proposed Schemes

5.1 Overview of Our Approach

In this section we will describe two AJPS-ECC variants based on the new idea of *Randomness Reconstruction*.

Our idea departs from AJPS-1 in a direction orthogonal to the above.

We set by design $m = 0$ in AJPS-1 or $\mathcal{E}(m) = 0$ in AJPS-ECC and attempt to recover *the randomness*[1] into which information (encapsulated keys and/or plaintext information) will be embedded.

The intuition is that the receiver might be able to recover the randomness if parameters are properly chosen *using his extra knowledge* of G, F and knowing that, in addition, the unknown randomness has a low Hamming weight.

We hence focus the rest of this paper on methods for solving equations of the following forms:

$$W = Fx + Gy \quad \text{or} \quad W = Fx + Gy + z \bmod p$$

Where all parameters[2] and unknowns are *randomly* chosen in $\mathfrak{H}_{n,h}$ and where a solution $\{x, y\}$ or $\{x, y, z\}$ *is known to exist.*

We do not introduce any modifications in Setup and KeyGen, nor do we modify pp or sp[3]. We thus focus on the encapsulation (encryption) and on the de-encapsulation (decryption) processes only, in KEM and PKE respectively.

In a non-KEM version, a plaintext m encoded in the unknowns $(x, y$ or $x, y, z)$ can be directly recovered upon decryption. Such an encryption mode must however be protected against active attacks using padding and randomization that we do not address here.

Remark 1. It is tempting but inadvisable to create dependencies between the variables F, G and/or the unknowns x, y, z. Consider an AJPS-ECC where $m \in_R \{0, 1\}^\lambda$ and $\{A, B_2\} \in_R \mathfrak{H}_{n,h}^2$ but where $B_1 \leftarrow \mathcal{H}(m)$ is obtained by hashing m into $\mathfrak{H}_{n,h}$. Given m, *anybody* can re-compute B_1 and algebraically infer A, B_2. We hence see in this example that A, B_2 do not add extra entropy as security solely rests upon m.

We carefully distinguish between *security bandwidth* and *information rate*. An idea, unexplored in AJPS-ECC, may exploit Remark 1 to transport more plaintext information in $\{C_1, C_2\}$ *without adding extra security*. To encrypt a τ-bit message μ, pick a key $m \in_R \{0, 1\}^\lambda$ and encrypt $c \leftarrow \mathcal{F}_m(\mu)$ using a block cipher $\mathcal{F}$. Set $B_1 \leftarrow \mathcal{H}(m)$. Let $\mathcal{M}$ be any invertible public mapping $\mathcal{M} : \{0, 1\}^\tau \rightarrow \mathfrak{H}_{n,h}^2$. Encode: $\{A, B_2\} \leftarrow \mathcal{M}(c)$ and form $\{C_1, C_2\}$ using AJPS-ECC. To decrypt, recover m using error-correction, recompute B_1, algebraically recover $\{A, B_2\}$ and retrieve the plaintext μ by:

---

[1] A, B or A, B_1, B_2.

[2] Except W.

[3] Note that in AJPS-1/ECC given G one can compute F and vice versa.

$$\mu \leftarrow \mathcal{F}_m^{-1}(\mathcal{M}^{-1}(A, B_2))$$

$$= \mathcal{F}_m^{-1}\left(\mathcal{M}^{-1}(\frac{C_1 - B_1}{R} \bmod p, C_2 - \frac{(C_1 - B_1)T}{R} \bmod p)\right)$$

$$= \mathcal{F}_m^{-1}\left(\mathcal{M}^{-1}(\frac{C_1 - \mathcal{H}(m)}{R} \bmod p, C_2 - \frac{(C_1 - \mathcal{H}(m))T}{R} \bmod p)\right)$$

Because in AJPS-ECC$\{n, h\} = \{756839, 256\}$ the potential encoding capacity of $\mathcal{M}$ can be relatively high:

$$2 \log_2 \binom{756839}{256} = 6631 \text{ bits}$$

This increases AJPS-ECC's information rate by a factor of 26. Again, proper message padding may be necessary to resist active attacks. We stress, again, that this does not increase security bandwidth but information rate only. An attacker guessing m will determine μ.

5.2   The Bivariate KEM

Our Bivariate Cryptosystem $\Pi_1 = (\text{Setup}, \text{KeyGen}, \text{Encap}, \text{Decap})$ is defined as follows:

- Setup$(1^\lambda) \rightarrow$ pp: As in Sect. 4.1.
- KeyGen(pp) $\rightarrow \{\text{sk}, \text{pk}\}$ are identical to AJPS-1.
- Encap(pp, pk) $\rightarrow C$. Picks $\{A, B\} \in_R \mathfrak{H}_{n,h}^2$ and computes:

$$C \leftarrow AH + B \bmod p$$

- Decap(pp, sk, C) $\rightarrow \{\bot, \{A, B\}\}$ returns:

$$\{A, B\} \leftarrow \text{Solve}_{x,y}[GC = Fx + Gy \bmod p]$$

If $\{A, B\} \neq \bot$ use $\{A, B\}$ as KEM entropy for further encryption.

Example: Let μ be a plaintext and $\mathcal{R}$ a redundancy function. Compute $m \leftarrow \mathcal{R}(\mu, \rho)$ where ρ is random. A typical KEM[4] is shown here:

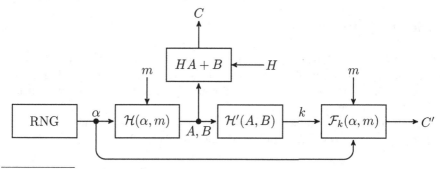

---

[4] Where $\mathcal{H}, \mathcal{H}'$s are hash functions and $\mathcal{F}$ is a block-cipher.

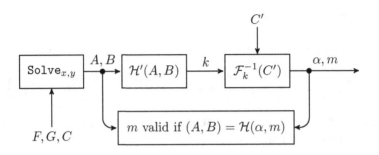

Retrieve m from $\mu \leftarrow \mathcal{R}^{-1}(m)$.

5.3 The Trivariate KEM

Our Trivariate Cryptosystem $\Pi_2 = (\mathsf{Setup}, \mathsf{KeyGen}, \mathsf{Encap}, \mathsf{Decap})$ is defined as follows:

- $\mathsf{Setup}(1^\lambda) \rightarrow \mathsf{pp}$: As in Sect. 4.1
- $\mathsf{KeyGen}(\mathsf{pp}) \rightarrow \{\mathsf{sk}, \mathsf{pk}\}$ are identical to AJPS-ECC but with the modified formula $T \leftarrow FR - G \bmod p$.
- $\mathsf{Encap}(\mathsf{pp}, \mathsf{pk}) \rightarrow C$. Picks $\{A, B_1, B_2\} \in_R \mathfrak{H}_{n,h}^3$ and computes the ciphertext:

$$C = \begin{cases} C_1 & \leftarrow AR + B_1 \bmod p \\ C_2 & \leftarrow AT - B_2 \bmod p \end{cases}$$

- $\mathsf{Decap}(\mathsf{pp}, \mathsf{sk}, C) \rightarrow \{\perp, \{A, B_1, B_2\}\}$ returns:

$$\{A, B_1, B_2\} \leftarrow \mathsf{Solve}_{x,y,z} [FC_1 - C_2 = Fy + Gx + z \bmod p]$$

If $\{A, B_1, B_2\} \neq \perp$ use $\{A, B_1, B_2\}$ as KEM entropy for further encryption.

As noted before, the trivariate version may accommodate in the encryption formula an *independent* $\mathcal{E}(m)$ and thus cumulate the bandwidth improvements due to both mechanisms. This requires that the $\{n, h\}$ values of both schemes coincide and the enforcement of the condition $n \leq 16h^2$, not addressed here. We conjecture that such a meeting point exists.

6 Security Proof

Just as RSA, our encryption process is deterministic (i.e., requires no nonce) if the message is encoded in A, B. As such, unless we use a proper padding before encryption, the encryption function itself cannot provide "native" indistinguishability against chosen plaintext attacks.

We nonetheless provide a proof that breaking our KEM is equivalent to the solving the (search) RMERS Problem, defined in Sect. 3.2.

Theorem 1. *Inverting the* KEM *described in Sect. 5.2 with parameter h is as hard as solving* RMERS *Problem for parameter h/2.*

Proof. We will show that any attack algorithm $\mathcal{A}(C, H, h, p) = \{A, B\}$ extracting the exchanged key $\{A, B\}$ without resorting to the secret key can be used to solve the RMERS for parameters $h/2, p$. There is hence a loss in the reduction of a factor of 2 in the security parameter h.

Assume that $\mathcal{A}(C, H, h, p)$ exists.

We note that $\mathcal{A}$ resolves the equation $C = AH + B \bmod p$ without resorting to the secret elements F, G.

What happens if we invoke $\mathcal{A}(0, -H \bmod p, h, p)$?

In such a case $\mathcal{A}$ will return A, B such that:

$$0 = -AH + B \bmod p \text{ that is: } H = \frac{B}{A} = \frac{F}{G} \bmod p$$

$\mathcal{A}$ has thus solved the RMERS for parameters h, p.

This is, however, insufficient as $\mathcal{A}$ may refuse to solve the equation $C = AH + B \bmod p$ when $C = 0$. We will hence mask the input C so that $\mathcal{A}$ could not refuse to process it.

To do so, we sacrifice reduction tightness to force $\mathcal{A}$ to solve arbitrary target instances H_t of the RMERS for parameters $h/2, p$.

- Generate two random numbers $r_A, r_B \in_R \mathfrak{H}^2_{n,h/2}$.
- Form the quantity:
$$C = H_t r_A - r_B \bmod p$$

- Invoke $\mathcal{A}(C, -H_t \bmod p, h, p)$

$\mathcal{A}$ will return an A, B such that:

$$C = -AH_t + B = H_t r_A - r_B \bmod p$$

In other words:

$$H_t = \frac{B + r_B}{A + r_A} = \frac{F}{G} \bmod p$$

$\mathcal{A}$ was hence instrumented to solve a RMERS instance of parameter $h/2$, as required (Fig. 1).

$\square$

The proof extends, *mutatis mutandis*, to the trivariate KEM of Sect. 5.3.

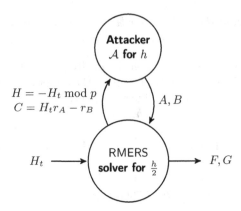

Fig. 1. Security reduction: Turning an attacker into an RMERS solver.

7   Instantiating $\texttt{Solve}_{x,y}$ Using Backtracking

This section explains how to instantiate $\texttt{Solve}_{x,y}$. The routine $\texttt{Solve}_{x,y,z}$ is obtained *mutatis mutandis*.

The intuition behind $\texttt{Solve}_{x,y}$ is the following: assume that we are given the quantity $W = GC = AF + BG \bmod p$ where $\|W\| \cong 2h^2$. Because multiplication modulo p is (somewhat) weight-preserving, we can test the hypothesis that the i-th bit of A is equal to one by looking at the quantity Δ:

$$\Delta = \|W\| - \|W - 2^i F \bmod p\|$$

Intuitively, a good guess should result in a weight decrease of $\simeq h$ whereas a wrong guess should re-blur W by triggering random carry propagations. Evidently, because there may be false positives during this process, we must be able to backtrack. To reduce the false positive error probability, n must be large enough with respect to h. The exact same idea applies to $\texttt{Solve}_{x,y,z}$.

7.1   Prerequisites and Subroutines

We start by introducing three necessary prerequisites.

The Ancillary Function $\texttt{Confirm}$: Our algorithms require an ancillary function $\texttt{Confirm}$-ing a candidate solution $\{x, y\}$. e.g. given a candidate x, $\texttt{Confirm}(x)$ may solve $GC = Fx + Gy \bmod p$ for y and return $\{y, \textsf{True}\}$ if $\|x\| = \|y\| = h$. Because in some cases several solutions may exist, a simpler implementation may just compare $\mathcal{H}(x, y)$ to a confirmation digest τ provided with the ciphertext and return $\{y, \textsf{True}\}$ if the purported solution hashes into τ. If $\mathcal{H}(x, y) \neq \tau$ then $\texttt{Confirm}(x)$ returns $\{\bot, \textsf{False}\}$. Note that using a confirmation digest would not satisfy the standard indistinguishability security requirement for KEMs.

Table 1. Backtracking success chances for $\{t, h, n\} = \{1, 72, 19937\}$. 50 decryption simulations per entry.

Γ	50	51	52	53	54	55	56	57	58	59	60
Probability	24%	20%	26%	30%	32%	20%	22%	18%	14%	9%	4%

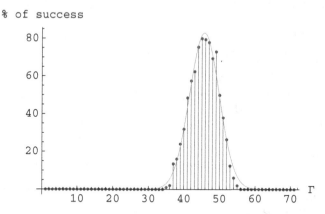

Fig. 2. Backtracking success chances for $\{t, h, n\} = \{1, 65, 19937\}$. 200 decryption simulations per entry. Fitted with $82.922 \exp(-(\Gamma - 45.7122)^2/34.6815)$

Determining the Backtracking Aperture Γ: Backtracking is parametrized by a constant Γ controlling the aperture of the exhausting process (i.e. the marginal tolerance allowing to exclude a search path from further investigation). Simulations indicate that for any given $\{n, h\}$ there is a $\Gamma_{optimal}$ value minimizing the failure probability. We did not attempt a formal analysis of the dependency between $\{n, h\}$ and $\Gamma_{optimal}$ but estimated $\Gamma_{optimal}$ for various $\{n, h\}$ pairs using simulations as shown in Table 1 and Fig. 2.

7.2 The Backtracking Algorithms

The deterministic backtracking algorithm $\mathcal{B}_1$ subtracts left-shifted Fs from W to obtain candidate ws having smaller and smaller weights. $\mathcal{B}_1$ maintains a set of integers R containing the bit positions of x discovered so far. The deterministic algorithm is called by $\{A, B\} \leftarrow \mathcal{B}_1(W, \emptyset, 0)$ and the randomized version is called by $\mathcal{B}_2(W, \emptyset, 0, \phi_j)$ where ϕ_js are random permutations of $\mathbb{Z}_k$. Code is available from the authors upon request.

ALGORITHM 1
Backtracking $\mathcal{B}_1(w, R, e)$

Input: w, R, e. The values $G, F, h, n, p = 2^n - 1, C$ are global and invariant.
Output: $\{x, y\} = \{A, B\}$ such that $W = CG = Fx + Gy \bmod p$ or Failure.

> if $e = n$ then return Failure
> else
> > if $\#R = h$ then
> > > $x \leftarrow \sum_{i \in R} 2^i$
> > > $\{s, y\} \leftarrow \texttt{Confirm}(x)$
> > > if s then return $\{x, y\}$
> > $\overline{w} \leftarrow w - 2^e F \bmod p$
> > if $|\|\overline{w}\| - \|w\| + h| \leq \Gamma$ then
> > > $\mathcal{B}_1(\overline{w}, R \cup \{e\}, e + 1)$
> > else
> > > $\mathcal{B}_1(w, R, e + 1)$

ALGORITHM 2
Backtracking $\mathcal{B}_2(w, R, e, \phi)$

Input: w, R, e, ϕ. The values $G, F, h, n, p = 2^n - 1, C$ are global and invariant.
Output: $\{x, y\} = \{A, B\}$ such that $W = CG = Fx + Gy \bmod p$ or Failure.

> if $e = n$ then return Failure
> else
> > if $\#R = h$ then
> > > $x \leftarrow \sum_{i \in R} 2^{\phi(i)}$
> > > $\{s, y\} \leftarrow \texttt{Confirm}(x)$
> > > if s then return $\{x, y\}$
> > $\overline{w} \leftarrow w - 2^{\phi(e)} F \bmod p$
> > if $|\|\overline{w}\| - \|w\| + h| \leq \Gamma$ then
> > > $\mathcal{B}_2(\overline{w}, R \cup \{e\}, e + 1, \phi)$
> > else
> > > $\mathcal{B}_2(w, R, e + 1, \phi)$

We conjecture that working with a fixed Γ during the entire backtracking process handicaps the algorithm. When the process starts the weight of W is high, hence the probability to strike-out h bits by subtraction is high. However as subtractions make w sparser aperture should intuitively decrease. It may hence make sense to explore algorithms in which the constant Γ_{optimal} is replaced by a function $\Gamma(\|w\|, \|\overline{w}\|, n, h)$.

Best Candidate Search: $\mathcal{B}_1$ and $\mathcal{B}_2$ explore all the paths starting by an a priori promising Δ. However, $\mathcal{B}_1$ and $\mathcal{B}_2$ do not explore the most promising paths first. A more complex backtracking strategy ($\mathcal{B}_3$) trying with priority the paths starting by a Δ as close as possible to h was developed as well (information available from the authors upon request). We do not include this algorithm here for the sake of concision.

Dealing with Decoding Failures: Because we may discard seemingly uninteresting (but actually promising) exploration paths, backtracking may fail to decode W. As it seems complex to formally compute the algorithm's success probability, we estimated it by simulation. To deal with decryption failures we re-attempt backtracking after index randomization i.e. pick t random permutations $\{\phi_0, \ldots, \phi_{t-1}\}$ of $\mathbb{Z}_k$ and re-run $\mathcal{B}_2(W, \emptyset, 0, \phi_j)$ t times hoping that at least one of the t runs will succeed[5]. A more brutal approach consists in sending t encapsulated keys to increase the probability exponentially. This (conjectured) exponential probability gain only handicaps the information rate by a constant factor[6]. A simple idea for (conjectured) squaring the failure probability consists in trying to backtrack on A and, upon failure, re-launch the algorithm to backtrack on B.

Information Leakage from Decryption Failures: Because decryption may fail, a possible cryptanalysis (that we did not investigate) might be to analyze, possibly adaptively, the ciphertexts causing failures and thereby extract information on $\{F, G\}$. We do not regard this as a major problem for the following reasons:

- Failure is highly dependent on the backtracking algorithm chosen by the receiver. The backtracking procedures that we give here are one possibility amongst many.
- An empirical protection consists in randomizing the backtracking process, e.g., assume all the ϕ_i to be *randomly drawn per decryption and secret*.
- Another protection is to purposely fail decryption with some probability ϵ to prevent the cryptanalyst from identifying true failures. Note that the random tape used to simulate false failures must be derived from a fixed secret and *the ciphertext itself* to avoid replays and majority votes.

Protection Against Side-Channel Attacks: It is reasonable to assume that, like most encryption schemes, the algorithms described in this paper are vulnerable to timing and side-channel attacks, an aspect that we did not investigate here.

---

[5] Note that $\mathcal{B}_1(W, \emptyset, 0) = \mathcal{B}_2(W, \emptyset, 0, \mathrm{ID})$.

[6] Link $B_0, \ldots, B_{t-1}$ in a way allowing the recovery of all the B_i if one of them is known (e.g. define $B_i = \mathcal{F}_i(\text{seed})$ where $\mathcal{F}_k(m)$ is a block-cipher encrypting into $\mathfrak{H}_{n,h}$). Use the A_i to transport entropy or information. One successful decryption reveals the seed $\Rightarrow$ open all the B_is $\Rightarrow$ all the t information containers A_i.

7.3 Eccentric Reconstruction Strategies

Backtracking might be improved in a variety of ways. As examples, we are introducing in this section a few research ideas that we did not explore in detail.

Idea 7.3.1: Brittle Encryption Formulae. We may modify the bivariate encryption formula to $C \leftarrow HA + 3B$ and enforce by design that H, A, B do not contain the binary sequence 11. This means that the bit positions representing $3B$ will be "colored" by a pattern 11 making their isolation and identification easier. If $n \gg h$ we may even attempt to brutally reset all the isolated ones in W and divide the result by $3G$ to directly obtain B. For the trivariate version one may use:

$$C = \begin{cases} C_1 & \leftarrow AR + B_1 \mod p \\ C_2 & \leftarrow AT - 3B_2 \mod p \end{cases}$$

resulting in the decryption formula $W = FC_1 - C_2 = FB_1 + GA + 3B_2$. Here as well, we banish the pattern 11 from G, F, A, B_1, B_2. We may thus attempt to identify in W the binary patterns 11, hinting the probable presence of B_2 to ease decoding. Note that the pattern 11 may result naturally from the multiplication, the addition or the reduction and hence mislead the decoder (backtrack). Similarly, an 11 due to $3B_2$ may disappear due to addition (backtrack). Note that marking B with 11s makes backtracking more efficient as this increases the SNR. e.g. if we replace each 1 in B by a 1111 we increase overall weight of W to $5h^2$ but a correct guess will cause a weight decrease of $\simeq 4h$ instead of h. In other words, while requiring a larger n, this improves the SNR:

$$\text{from } \text{SNR} = \frac{h}{2h^2} = \frac{1}{2h} \text{ to } \text{SNR} = \frac{4h}{5h^2} = \frac{4}{5h}$$

Idea 7.3.2: Dye Tracing. In hydrogeology, *dye tracing* is a technique for tracking various flows using dye added to the water source. In other words, dye tracing uses dye as a flow tracer. It is an evolution of the ages-known float tracing method, which consists of throwing a buoyant object into a waterflow to see where it emerges. To simulate the effect of dye tracing, we inject into F's digits a few low-weight binary patterns and track their appearance in W. For instance (toy example), generate an F of weight $h - 10$ not containing any of the ten sequences ℓ_i:

11	101	111	1001	1011	1101	10001	11001	10101	10011

randomly insert those ten ℓ_is into F's blank spaces (insert each ℓ_i once, this will increase the weight of F to $h - 10 + \sum \|\ell_i\| = h - 10 + 26 = h + 16$ and the weight of W to $\simeq 2h^2 + 16h$). To retrieve A, isolate the 10 dyes tracers in W and use majority voting on bit offsets to infer the probable positions of A's bits.

Idea 7.3.3: Demodulation. We can attempt to "travel back in time" and infer $\omega = FA + GB \in \mathbb{Z}$ from W, or at least estimate the probability that a candidate bit in W originates from the number's pre-reduced upper half. Given $\omega \in \mathbb{Z}$ decryption[7] is immediate because:

$$A = \omega F^{-1} \bmod G = (W \bmod p) F^{-1} \bmod G$$

To demodulate W we work modulo $p = 2^n - 3$ that "colors" the folded MSBs by turning them into LSB 11s. The process is error-prone[8] but *actually works* for parameters that are large enough. We implemented the idea very brutally, by simply translating each 11 in W into a 1 in the MSB of ω without taking any further precautions. 100 demodulation attempts for $\{n = 6 \times 10^7, h = 55\}$ resulted in 29 successes. Although n is huge, the resulting information rate is not "that" catastrophic as we can pack:

$$2 \log_2 \binom{6 \times 10^7}{55} = 2356 \text{ plaintext bits into the ciphertext.}$$

In other words, each plaintext bit claims 25461 ciphertext bits and is successfully transmitted with probability 29%.

While $h = 55$ is not very large and $n = 6 \times 10^7$ is extremely large, our simulation shows that it is definitely possible to make ingredients meet at workable parameter combinations. We conjecture that with proper analysis and refined demodulation strategies k might be reduced by at least two orders of magnitude. It may also be possible to work modulo $2^n - \pi$ with a more distinguishable color $\pi \neq 3$ despite an extra weight due to a more complex π. $\pi = -1$ is interesting as well as -1 turns folded bits into long chains of 1s.

Note 1. (important) One of the features preventing lattice-based attacks in AJPS-1/ECC is the emergence of parasitic short vectors due to working modulo $2^n - 1$ (Sect. 5.1. [AJPS17a][9]). We did not evaluate the impact of $\pi \neq 1$ on the number and the norm of parasitic short vectors *and hence on security.*

Idea 7.3.4: Pattern Identification. Another idea consists in exploiting the fact that $W = AF + BG \bmod p$ will naturally contain binary sequences of the form:

$$v_\ell = \underbrace{0, \ldots, 0}_{\ell \text{ zeros}} |1| \underbrace{0, \ldots, 0}_{\ell+1 \text{ zeros}}$$

Let m be an ℓ-bit encapsulated key[10] and define $C = m(AH + B) \bmod p$ we get:

$$CG = m(AF + GB) = mW = m \times (w'|v_\ell|w) = u'|m|u \bmod p$$

---

[7] Take F, G coprime in Setup.

[8] Again, "natural" 11s may be already present in the LSBs of ω, $11 + 01$ may destroy an 11, $10 + 01$ may create fake 11s etc.

[9] ePrint version 20170530:072202.

[10] We consider m to be beyond exhaustive search, typically 160 bits.

m can thus be read[11] on $CG \bmod p$. It remains to identify m. To do so, we can generate $\{A, B\} \leftarrow \mathcal{H}(m)$ and hence confirm proper decryption using n re-hashings and re-encryptions. This workload may be considerably reduced by sacrificing a few bits of m, e.g. 16 bits, to display a specific pattern (e.g. 0xFFFF) allowing a quick identification of m. This divides the number of hashings and re-encryptions by 2^{16}. As a numerical example, $\{n, h\} = \{75 \times 10^4, 100\}$ corresponds to an $\ell \simeq 200$. If we sacrifice 20 bits devoting them to an identification pattern we can hope to decapsulate a $\simeq 160$-bit key using one re-encryption only.

ℓ has a low variance as it is essentially determined by a max-min over the differences between the positions s_i of the bits equal to one in W:

$$\ell \cong \max_i \min(s_i - s_{i-1}, s_{i+1} - s_i)$$

For a subtle technical reason ℓ is actually higher than this crude estimate. To understand why we refer the reader to Fig. 3 where we illustrate the expected distribution of $\hbar = 2h^2$ bits amongst n potential positions. The least significant 1-bit ◦ is expected to appear at γ where:

$$\gamma = \frac{\hbar}{(n-1)^{\hbar}} \int_0^{n-1} x \, (n - 1 - x)^{\hbar - 1} \, dx = \frac{n-1}{\hbar + 1}$$

Similarly, the most significant 1-bit position • is expected at $\simeq \hbar\gamma$. The reason why two other points are singled-out by ◦ and • will be clarified later. Now, because arithmetics modulo p wrap everything that overflows 2^n on 2^0 the actual gap between • and ◦ is not γ but 2γ. This is illustrated in Fig. 4. In other words, the primary formula for ℓ should be corrected to:

$$\ell \cong \max \left(\underline{\alpha}, \overline{\alpha}, \max_i \min(s_i - s_{i-1}, s_{i+1} - s_i) \right)$$

Where:

$$\underline{\alpha} = \min([\circ, \bullet], [\bullet, \circ]) \overset{\mathrm{u}}{=} [\bullet, \circ] \simeq \gamma$$

and

$$\overline{\alpha} = \min([\circ, \bullet], [\bullet, \circ]) \overset{\mathrm{u}}{=} [\bullet, \circ] \simeq \gamma$$

$$
\begin{array}{ccccccccc}
n-1 & \hbar\gamma \;\; (\hbar-1)\gamma & & & & & 3\gamma & 2\gamma & \gamma & 0
\end{array}
$$

Fig. 3. Dots show the expected positions of $\hbar$ bits picked randomly amongst n positions. Here $\gamma = \frac{n-1}{\hbar+1}$.

---

[11] Note that reading is circular i.e. wrapping around $CG \bmod p$.

Where $\overset{u}{=}$ denotes "usually (or frequently) equal to". We therefore see that wrapping due to modular arithmetic has an unexpected favorable effect on ℓ.

Pattern identification is somewhat homomorphic but with a very fast increasing noise: encode a zero as $m = 01$, a one as $m = 11$ and read the plaintext on the most significant bit of that encoding.

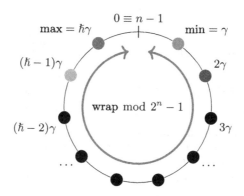

Fig. 4. The interval $[\min, \max] = [\,\circ\,, \bullet\,]$ of size 2γ created by wrapping $\mathrm{mod}\,p$.

Idea 7.3.5: Prime Embedding. A variant of the above, mostly of theoretical interest, is the following: because there is a number of natural leading and tailing zeros in $\eta = AF + GB \bmod p$ we can encode $C = m(AH + B) \bmod p$, recover $m(AF + GB) \bmod p$ and, provided that m is short enough, hope that η is small enough to get[12] $\omega = m\eta \in \mathbb{Z}$. It remains to extract m from ω. To do so, pick m as a product of, say 64-bit random primes. The receiver can pull-out those primes from ω using ECM factorization[13]. When a candidate m was formed, confirm it using hashing and re-encryption as before. $\{n, h\} = \{2.5 \times 10^6, 100\}$ gives an average expected margin of $\simeq 240$ bits for encoding m. The main problem with this variant would be the high variance in the size of m embeddable into the ciphertext which would make decryption uncertain.

Note 2. (research note) To the above we add two ideas that *we conjecture to be insecure* (by opposition to the previous ideas that we do not conjecture to be secure)

Variant 7.3.6. Correlated As: To ease backtracking we wish to give the decoder several Δs generated from the same B. Generate t independent keys

---

[12] The possibly rotated.

[13] Accidental similar-size prime factors may come from $AF + GB$ as well, but those are few and hence easy to filter.

$\{F_i, G_i, H_i\}$ (possibly modulo different p_is). Pick $t-1$ public random permutations $\phi_1, \ldots, \phi_{t-1}$ of $\mathbb{Z}_n$. Generate $\{B_0, \ldots, B_{t-1}, A_0\} \in_R \mathfrak{H}_{n,h}^{t+1}$. Let

$$A_0 = \sum_{i=0}^{n-1} 2^i a_i$$

and form t ciphertexts $\{C_0, \ldots, C_{t-1}\}$:

$$C_j = A_j H_j + B_j \bmod p_j \quad \text{where} \quad A_j = \sum_{i=0}^{n-1} 2^{\phi_j(i)} a_i$$

Simultaneous backtracking on the A_js will reveal more information per bit guess to the decoder.

Variant 7.3.7. Correlated Hs: Set G and define $H_i = F_i/G$ for $i \geq 1$. We illustrate the idea with two H_is. Encrypt $C = A_0 H_0 + A_1 H_1 + B$. We see that $W = GC = F_0 A_0 + F_1 A_1 + BG$. Linking A_0 and A_1 as in note 9.1 we see that the SNR[14] in the case of a successful guess increases:

$$\text{from} \quad \text{SNR} = \frac{h}{2h^2} = \frac{1}{2h} \quad \text{to} \quad \text{SNR} = \frac{2h}{3h^2} = \frac{2}{3h}$$

This modifies the complexity assumption as well.

Note 3. (a broken variant) We close this paper by attracting the reader's attention to the broken variant given in the appendix, that we mention as a target for fixing.

8   Security and Parameter Sizes

Brittle encryption, dye tracing, demodulation, pattern identification and prime embedding are only illustrative research directions that we consider interesting or curious but that we do not claim nor conjecture to be secure. This work did not cover the security of the proposed constructions and focused on the textbook modes in which data is encoded and decoded. Parameter sizes were not recommended and numerical examples are given for illustrative purposes.

A careful trade-off must be established between ① the security, ② the backtracking failure probability and ③ the efficiency of the various Solve processes. So far, simulations indicate that there seem to be ways to practically satisfy those three constraints at once.

---

[14] Note that as backtracking proceeds the SNR improves. In this paper SNR stands for the SNR at the beginning of the backtracking process.

9 Concluding Remarks

In this paper, we have introduced a new KEM schemes improving the AJPS cryptosystem and its variant. They are related to the hardness of MERS problem. Since our constructions rely on newly introduced assumptions, further cryptanalytic efforts are demanded in order to get more confidence about their exact security.

Acknowledgments. The authors thank Waïss Azizian, Sarah Houdaigoui and Quốc Tún Lê for the development and the simulation of different backtracking strategies.

A Appendix: A Broken Scheme Target for Repair

While designing the algorithms in this paper we broke the following variant that we mention here as a fixing target. Let R be a secret n-bit number of the form: $R \leftarrow \text{random}|v_\ell|\text{random}$ where v_ℓ is defined as in Pattern identification, and define the auxiliary public-key $L \leftarrow R/G \bmod p$. Note that this modifies the complexity assumption on which the scheme rests.

Encrypt by $C \leftarrow AH + B + Lm \bmod p$ and decrypt by $W \leftarrow GC = AF + GB + Rm \bmod p$. This offers a (noisy) visibility window on m allowing to extract and error-correct m.

The problem here is the equation defining L from which G can be revealed using LLL. It is unclear if this can be fixed using modifications in the encryption process and/or in parameter sizes. It is also interesting to determine if secure H-less variants[15] can be designed.

Assuming that the above could be repaired, the following is for readers fond of dangerous games.

A dangerous game, that we do not recommend, reduces noise in the visibility window. If A, B, G, F are generated in a biased way by shifting more Hamming weight into the MSBs and the LSBs as shown in Fig. 5, then the result of the multiplication modulo p of two such numbers results in a number of the form shown in Figs. 6 and 7. Those densities are illustrated in Figs. 8 and 9. This reduces the noise in the reading window and allows an easier recovery of m. It must be stressed that this increases the vulnerability of the public-key to the partition attacks of [BCGN17] as the attacker can better zoom on the information-rich part of F and G. This might be compensated by a higher h. Note that weight shifting does not necessarily need to be similar in all four variables A, B, F, G and that several weight shifting schemes are circularly equivalent because of the multiplication modulo p.

---

[15] i.e. where the sender encrypts by $C \leftarrow Lm + B \bmod p$ this is insecure) or a similar trick.

weight = $h/4 + \Delta$	weight = $h/2 - 2\Delta$	weight = $h/4 + \Delta$

Fig. 5. Unbalanced weight of A, B, F, G before multiplication.

weight $\cong$ $h^2/4 + 2\Delta^2$	weight $\cong$ $h^2/2 - 4\Delta^2$	weight $\cong$ $h^2/4 + 2\Delta^2$

Fig. 6. Unbalanced weight of AF mod p and BG mod p after multiplication.

weight $\cong$ $h^2/2 + 4\Delta^2$	weight $\cong$ $h^2 - 8\Delta^2$	weight $\cong$ $h^2/2 + 4\Delta^2$

Fig. 7. Unbalanced weight of $AF + BG$ mod p after addition.

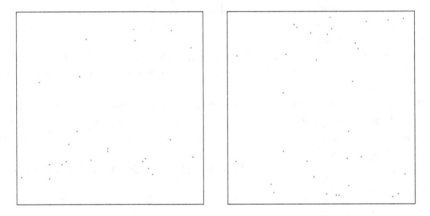

Fig. 8. Unbalanced A and F for $\{n, h, \Delta\} = \{2^{14}, 32, 6\}$.

Fig. 9. Unbalanced AF mod p for $\{n, h, \Delta\} = \{2^{14}, 32, 6\}$. Note the relatively lower dot density at the middle of the diagram.

References

[AJPS17a] Aggarwal, D., Joux, A., Prakash, A., Sántha, M.: A New Public-Key Cryptosystem via Mersenne Numbers. Cryptology ePrint Archive, Report 2017/481 (2017). http://eprint.iacr.org/2017/481

[AJPS17b] Aggarwal, D., Joux, A., Prakash, A., Sántha, M.: Mersenne-756839. Technical report, National Institute of Standards and Technology (2017). https://csrc.nist.gov/projects/post-quantum-cryptography/round-1-submissions

[AJPS18] Aggarwal, D., Joux, A., Prakash, A., Santha, M.: A new public-key cryptosystem via Mersenne numbers. In: Shacham, H., Boldyreva, A. (eds.) CRYPTO 2018. LNCS, vol. 10993, pp. 459–482. Springer, Cham (2018). https://doi.org/10.1007/978-3-319-96878-0_16

[BCGN17] Beunardeau, M., Connolly, A., Géraud, R., Naccache, D.: On the Hardness of the Mersenne Low Hamming Ratio Assumption. Cryptology ePrint Archive, Report 2017/522 (2017). http://eprint.iacr.org/2017/522

[dBDJdW17] de Boer, K., Ducas, L., Jeffery, S., de Wolf, R.: Attacks on the AJPS Mersenne-Based Cryptosystem. Cryptology ePrint Archive, Report 2017/1171 (2017). https://eprint.iacr.org/2017/1171

[ElG85] ElGamal, T.: A public key cryptosystem and a signature scheme based on discrete logarithms. In: Blakley, G.R., Chaum, D. (eds.) CRYPTO 1984. LNCS, vol. 196, pp. 10–18. Springer, Heidelberg (1985). https://doi.org/10.1007/3-540-39568-7_2

[HPS98] Hoffstein, J., Pipher, J., Silverman, J.H.: NTRU: a ring-based public key cryptosystem. In: Buhler, J.P. (ed.) ANTS 1998. LNCS, vol. 1423, pp. 267–288. Springer, Heidelberg (1998). https://doi.org/10.1007/BFb0054868

[Knu] Knuth, D.E.: The Art of Computer Programming. Addison-Wesley, Boston (1968)

[LLL82] Lenstra, A.K., Lenstra, H.W., Lovász, L.: Factoring polynomials with rational coefficients. Mathematische Annalen **261**(4), 515–534 (1982)

[LPR10] Lyubashevsky, V., Peikert, C., Regev, O.: On ideal lattices and learning with errors over rings. In: Gilbert, H. (ed.) EUROCRYPT 2010. LNCS, vol. 6110, pp. 1–23. Springer, Heidelberg (2010). https://doi.org/10.1007/978-3-642-13190-5_1

[NIS17] NIST. Post Quantum Crypto Project, May 2017. http://csrc.nist.gov/groups/ST/post-quantum-crypto/. Accessed 19 May 2017

[Reg06] Regev, O.: Lattice-based cryptography. In: Dwork, C. (ed.) CRYPTO 2006. LNCS, vol. 4117, pp. 131–141. Springer, Heidelberg (2006). https://doi.org/10.1007/11818175_8

[RSA78] Rivest, R.L., Shamir, A., Adleman, L.: A method for obtaining digital signatures and public-key cryptosystems. Commun. ACM **21**(2), 120–126 (1978)

[Sho97] Shor, P.W.: Polynomial-time algorithms for prime factorization and discrete logarithms on a quantum computer. SIAM J. Comput. **26**(5), 1484–1509 (1997)

Distributed Multi-authority Attribute-Based Encryption Using Cellular Automata

Ankit Pradhan, Kamalakanta Sethi$^{(\boxtimes)}$, Shrohan Mohapatra,
and Padmalochan Bera

Indian Institute of Technology, Bhubaneswar, India
{ap36,ks23,sm32,plb}@iitbbs.ac.in

Abstract. Cellular automata (CA) has attracted the attention of research communities for its applications in the design of symmetric and public-key cryptosystems. The strength of cellular automata lies in its inherent data parallelism, which can help accelerate access control mechanisms, and its information scrambling capabilities, which can enhance the security of the system. Also, the cryptosystems designed using CA do not involve number-theoretic methodologies that incur large computational overhead like traditional cryptosystems. However, existing CA-based cryptosystems encompass a limited set from the set of all possible transition rules indicating the existence of CA cryptosystems which are possibly unbreakable but have not been explored sufficiently. Thus, they have not yet been considered for applications involving fine-grained access control for heterogeneous access to the data. In this paper, we propose a secure distributed multi-authority attribute-based encryption using CA, which has potential applications in cloud systems. Our cryptosystem adopts the concept of multi-authority attribute-based access control where the encryption and attribute distribution use reversible CA, and policy satisfiability is achieved by Turing-complete CA in a distributed environment. We illustrate the practical usability of our proposed cryptosystem, in terms of efficiency and security, by extensive experimental results.

Keywords: Cellular automata · Reversible cellular automata ·
Turing-complete cellular automata · Attribute-based access control ·
Multi-authority access control · Cloud system

1 Introduction

A cellular automaton can be defined as an automaton consisting of a set of states, a grid with its initial content, and a transition function [1]. The state of each cell is an element of a pre-defined finite set. The configuration of a cellular automaton is said to be the content of the grid at any instance of time. The future state of a cell depends on its current state and that of its neighbours via a

© Springer Nature Switzerland AG 2019
Y. Mu et al. (Eds.): CANS 2019, LNCS 11829, pp. 434–456, 2019.
https://doi.org/10.1007/978-3-030-31578-8_24

mapping, known as the transition function or the rule of the cellular automaton. A cellular automaton is said to be reversible [18] if and only if there exists a bijective mapping from the set of all possible configurations of a grid to itself. A cellular automaton is said to be Turing-complete [2] if it can emulate any Turing machine, i.e., the functionality within the configurations behaves in the same way as an equivalent Turing machine would do.

Cellular automata have been analyzed for producing interesting informational variations [12,13] from simple rules, which aroused a lot of expectations on their computational power and universality. One-dimensional CA of rule 110 has been proven to be Turing-complete [3], with a polynomial-time emulation [4], but finally has been found to be in NC (Nick's class) [21] in terms of algorithmic complexity. This means if an algorithm takes $T(n)$ time to run on a Turing machine of input size n, then the same algorithm may take $O\big(T(n)^4 \, log(T(n))\big)$ time to run on an equivalent rule 110 CA [4]. Two-dimensional CA is proven to be more efficient in terms of emulation [5], but are still slower in practical applications. The fact that all the cells of the grid change its state simultaneously implies high data-level parallelism that can help accelerate, as well as assess, certain important mechanisms.

Cellular automata is used for designing efficient symmetric-key [18] as well as public-key cryptosystems [20]. The inherent nature of the information produced by the temporal evolution of CA introduces a significant amount of pseudo-randomness [13], and allows flexibility in information scrambling, making it secure. But such traditional systems do not support fine-grained access control, which is one of the essential requirement for cloud systems.

Cloud systems allow users to store data on remote servers in a secure, and reliable manner. As the servers are not completely trusted, traditional encryption methods encrypt the data and distribute keys to the user before storing the data on remote servers. Although these methods provide secure access control, key management is a difficult task for data owners when more users are added to the system. Data owners have to stay online to distribute keys to new users periodically as per their requirement. Also, there are multiple copies of ciphertext for the same data which incurs large storage overhead on the server. Several methods deliver key management and distribution task from the data owner to remote servers under the assumption that these servers are fully trusted. However, remote servers cannot be fully trusted, and thereby, these methods cannot be applied to cloud storage. Attribute-based encryption (ABE) [7] is a promising technique to provide fine-grained access control on encrypted data. But recent implementations of such systems [26] requires a large number of arithmetic computations such as pairing, repeated exponentiation, etc., which in turn result in computation overhead to the system.

In traditional ABE cryptosystems, computations such as pairing, repeated exponentiation, etc., incur a lot of computational overhead, thus rendering them inefficient for practical applications. On the other hand, cellular automata allow highly scaling levels of data parallelism and scramble the information efficiently but do not provide fine-grained access control, which is a primary requirement in large-scale cloud systems. Also, security concerns become even more serious when the access control needs to be distributed among various attribute authorities. This motivates us to design a multi-authority attribute-based cryptosystem using CA.

In this paper, we present a novel CA-based public-key cryptosystem where encryption and decryption are realized using reversible cellular automata, and policy satisfiability is performed using Turing-complete cellular automata. Our proposed cryptosystem has the following features:

1. Turing completeness is highly rare in CA in comparison with the reversibility principle. Proving Turing-completeness is still a venture not completely explored. So using our proposed CA-based cryptosystem, it is very less likely for the attacker to guess the policy satisfiability of the system which uses Turing-complete CA.
2. Reversibility in CAs adds a lot of non-linearity to the encryption and decryption ends, thus preventing linear attacks.
3. Even though the process of attribute authorization is sequentially before the encryption, they do not affect the overall result and are independent. So our cryptosystem is portable across different topologies of distributed authorities.
4. A distinguishable feature of a CA is that all cells of its change their states simultaneously at a single instance of time. This implies a very high level of data parallelism, which can be used to improve the speed of the system for large message blocks sent by a larger number of users.

The paper is structured as follows. In Sect. 2, we present some important related works that are relevant and motivational to the design of our proposed cryptosystem. In Sect. 3, a brief background on first order CA is presented, along with some important features of CAs that form the cornerstone to our cryptosystem. In Sect. 4, we define, describe, exemplify, and illustrate our cryptosystem in terms of its security against some prevalent attacks. We present some experimental results, in Sect. 5, that show the efficacy of our cryptosystem. Finally, we conclude our work in Sect. 6.

2   Related Work

Cellular automata (CA) were first suggested to be used as efficient pseudorandom number generators (PRNG) which is used in stream ciphers [27] and block ciphers [23]. One-dimensional elementary CA with periodic boundary conditions, such as rule 30 [12], were the first suggestions but were cracked through chosen plaintext attacks [13]. Having its vulnerabilities in some periodic patterns, second-order, block, and programmable CA [14] were also suggested as a means to vaccinate itself from such attempts of pattern recognition, but were again cracked employing several layers of linear attacks [15]. Adding some hybridization in terms of transition rules and periodicity, yet another PRNG was proposed by Tomassini [16], which has so far been resistant enough against attacks. A lot of other generic realizations such as S-boxes in DES, AES, etc., some Boolean function based public-key cryptosystems have also been deeply explored and discussed [17]. But these CA-based cryptosystems consider a limited set of transition rules of all the possible ones. Reversible CA have been suggested to be used in symmetric key cryptosystems [18], and public key cryptosystems [20]. Also, there have been recent works on non-linear CA to improve the randomness property [24]. But such primitive systems do not inherently provide any fine-grained access control mechanism, which is a strong necessity in large-scale cloud systems. Such fine-grained access control mechanisms have been incorporated in encryption systems that have been discussed below, but do not use CA anywhere.

Sahai and Waters (2005) [7] first proposed attribute-based encryption (ABE) for supporting secure fine-grained threshold-based access control through secret sharing. Later Goyal et al. [8] introduced Key-Policy Attribute-Based Encryption (KP-ABE) and Bethencourt et al. [9] proposed Ciphertext-Policy Attribute-Based Encryption (CP-ABE). The dichotomy between KP-ABE and CP-ABE is clarified by the parties who have access control. While, in KP-ABE schemes, the access control lies with the system users, in CP-ABE schemes, the access control lies with the data owners [11].

Multiple studies and papers addressing various issues in attribute-based encryption have surfaced since 2005. The domain of ABE schemes consists of designs having large universe constructions with high expressiveness and based on the prime-order setting. With the introduction of decentralization in ABE by M. Chase in 2007 [28], other issues like user and attribute revocation [10] and traceability for malicious user identification [29] have been proposed and improved upon [25]. Most of the previous attribute-based constructions are computationally expensive due to multiple pairing and exponentiation operations involved in their algorithms. This motivates our current research, along with the future scope of CA-based systems in emerging fields such as DNA and molecular computing.

In this work, we aim to improve upon the security of the cryptosystems for cloud storage by making use of reversible CAs [18] in the attribute-based encryption process. The Turing completeness of Game of Life allows us to make the attribute checking process encapsulated in an envelope of homogeneity, in the sense that all the encryption-decryption process happens through the cellular automaton. This way, we inculcate a fine-grained access control mechanism at a significantly low mathematical complexity and computational cost, thus overcoming the disadvantage of traditional ABE schemes. While symmetric and public-key encryption systems do not provide the fine-grained access control, identity-based cryptosystems [19] involve a lot of complicated arithmetic, rendering it to be practically unusable. Also, this should help the shift of paradigm of the use of cellular automaton as a PRNG to that as a fundamental and reliable model of encryption. While PRNG models focused on expanding keys using fixed rules to generate relatively untraceable information, such models allow flexibility in the rules as well. The use of the main key repository using a dictionary data structure adds to the flexibility of the systems.

3   Background

In this section, we formally define and describe a generic first-order CA. We illustrate the definition by some classic examples of Turing-complete CA, followed by an explanation of reversible CAs. Finally, we state some important properties that will help us understand the design of the cryptosystem.

3.1   First-Order Cellular Automata

Any n-dimensional r-neighbourhood first-order cellular automata can be defined as a 4-tuple (S, Q_t, N, f) where $S \subseteq \mathbb{N}$, known as the set of states; Q_t is an $l_1 \times l_2 \cdots \times l_n$ matrix for an instance of time t, where $\forall k \in \{1, 2, \ldots n\}$ $\forall i_k \in \{1, 2, \ldots l_k\}$, $Q_t[i_1, i_2, \ldots i_n] \in S$. N is an $r \times n$ matrix known as the neighbourhood matrix. Finally, $f : S^{r+1} \to S$, known as the transition function. Here the future state of a cellular automaton is defined in the following way:

$$\forall k \in \{1, 2, \ldots n\} \quad \forall i_k \in \{1, 2, \ldots l_k\},$$

where

$$h_m = Q_t[\{((i_k + N[m, k]) \mod l_k) \quad \forall k \in \{1, 2, \ldots l_m\}\}] \quad \forall m \in \{1, 2, \ldots r\}$$

A cellular automaton is said to be elementary if $n = 1, r = 2, S = \{0, 1\}, N = (1, -1)^T$. The decimal equivalent of the binary number formed by an ordered sequence of the images of the function f of an elementary cellular automaton is said to be its 'rule', or the 'Wolfram totalistic rule'. An example running of rule 110 cellular automaton is graphically illustrated in Fig. 1 [4]. The transition function $f_{110} : S^3 \to S$ for rule 110 cellular automaton is given below in the form of a boolean expression:

$$f_{110}(A, B, C) = \overline{A}(B + C) + A(B \oplus C)$$

Throughout the paper, we refer the sequence $Q_1, Q_2, Q_3, Q_4 \ldots$ as the temporal evolution of a cellular automaton, the matrix Q_1 as the seed of the machine.

Local patterns observed in class 4 CA with specific initial configurations are seemingly scrambled in future configurations, but universality in computation requires some special features in these local patterns. Rule 110 is Turing-complete [4], but the equivalence is not efficient enough. This is just one side of the scenario that the use of Turing-complete CA has never occurred specifically in any kind of cryptosystem. Though the Turing-completeness scrambles the information, it is not guaranteed that one can retrace back the original seed from there.

The initial usage of CA in the design of cryptosystems was pertinent and inclined to that of PRNGs in stream ciphers, a classical example of rule 30 [12]. When such cryptosystems were broken [13], then second-order block CA were used [14] and these were again broken [15]. Tomassini's cryptosystem [16] is yet another PRNG, which has not been broken. Several other CA-based cryptosystems [17] have been proposed.

Another classical example of a Turing-complete two-dimensional CA is shown in Fig. 2, known as "Conway's Game of Life" [5,32], whose transition function is described as follows:

1. Any live cell (i.e., a cell with state '1') with fewer than two live neighbours dies (i.e., the cell attains the state '0'), as if by under-population.
2. Any live cell with two or three live neighbours lives on to the next generation.
3. Any live cell with more than three live neighbours dies, as if by overpopulation.
4. Any dead cell with exactly three live neighbours becomes a live cell, as if by reproduction.

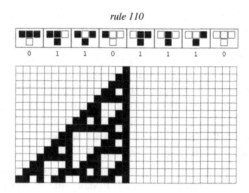

Fig. 1. A demonstration of the rule 110 cellular automaton [4]. The first row shows the transition rules. It is noteworthy of the binary number formed of the outputs, whose decimal equivalent is rule 110. The subsequent grid shows the temporal evolution, where the initial seed consists of a single one and rest all zeroes.

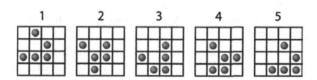

Fig. 2. A demonstration of the Game Of Life cellular automaton [5]. The first grid shows the seed, and the subsequent grids show the temporal evolution of the same. This is known as a 'glider'.

Conway's Game Of Life has been shown to be Turing complete, as it is possible to emulate AND, OR, and NOT gates. It has been shown how to place appropriate p30 gliders and other associated gliders in the seed of the two-dimensional CA [5], which enables us to allow the interpretation of all possible Boolean logic.

This will be used in securing the fine-grained access control mechanism provided by our design of the cryptosystem, described in a later section.

A reversible cellular automata is one where every temporal evolution (or **configuration**) leads to a unique seed in the next time switch of the cellular automata. Thus, if $d = l_1l_2l_3\cdots l_n$, and $k = |S|$ in the above definition and as is defined in [6], there are $k^d!$ possibilities of the $(k^d)^{k^d}$ to get a reversible cellular automaton. Since we will be using these kinds of CAs for encryption and decryption, as described in Sect. 4.3, this poses a huge problem for the adversary to know the key space. Reversible CAs were explored in design of symmetric and public-key cryptosystem [18,20].

3.2   Some Important Observations

The essential properties of a CA that supports the design of our proposed cryptosystem are as follows:

1. **Tracing back the previous states of a temporal evolution of a CA is in NP:** Given a previous configuration of a CA and its corresponding transition rules, we can compute the future configuration of a CA using a deterministic Turing machine in polynomial time. Thus verifying the solution to the posed problem is in P. So the problem is in NP.

2. **Game of Life is Turing-complete and hence is non-linear:** Turing-completeness has already been proved in [5]. Also, to find chaotic patterns that lead to Turing-completeness, as shown in [4] and [5], the transition function should not be generating any such mathematical sequence. The proof is led by the following contradiction: say the 'computationally universal' transition function F allows a generation of some sequence S in its configuration C. Since the automaton is 'computationally universal', one must be able to generate another fundamentally different sequence S' starting from the same seed. Without loss of generality, S' will definitely reflect upon at least one different configuration C' in one of its temporal evolutions. This implies that this might have started from a transition function F'. Thus, the temporal evolution of a Turing-complete CA is not mathematically tractable, and thus consequently non-linear.

3. **Checking if a 1-dimensional CA rule is reversible or not is in NP:** Suppose there is a decision problem 'Given a CA rule, is the mapping from a set of all possible configurations C to the future ones bijective?'. We claim that this decision problem is in NP. Consider the verification version of the problem: Given a bijective mapping $f_c : C \to C$ and a given CA rule R, can f_c be generated by the temporal evolution of R? This can be solved in linear time with respect to the size of C. This proves our claim. There have been solutions proposed [22] but are not polynomial-time in the worst case. It has also been proven that deciding whether an n-dimensional CA is reversible for $n > 1$ is Turing-undecidable [6].

4. **Only a small fraction of all possible CAs are actually reversible:** As discussed in Sect. 3.1, the expected fraction of reversible CAs of k states

and d cells is $\frac{k^{d!}}{(k^d)^{k^d}}$. For example, if $k = 2$ and $d = 16$, the probability that a CA is reversible is $\frac{k^{d!}}{k^{d}{}^{k^d}} = 7.659 \times 10^{-28460}$. This shows a typical value of the fraction of all possible CAs that are reversible. So, it is extremely difficult for the adversary to realize the key space of the proposed cryptosystem. This strengthens the security of the proposed solution as will be seen in Sect. 4.3.

5. **All reversible CAs are non-linear, i.e. no reversible CA are linear:** We will again prove this by contradiction. Let us consider a bijective mapping $f_c : C \to C$ of a linear CA, where configuration $C = \{c_1, c_2, c_3 \cdots c_d\} \in \{0, 1\}^d$, d being the number of cells in the configuration. Since the transition function $f_c : C \to C$ is linear, there exist $a_1, a_2, a_3 \cdots a_r \in \{0, 1\}$ such that a cell c'_i in the future configuration is given by,

$$c'_i = a_1 c_{i-\lfloor \frac{r}{2} \rfloor} \oplus a_2 c_{i-\lfloor \frac{r}{2} \rfloor + 1} \oplus \cdots \oplus a_i c_i \oplus \cdots a_r c_{i+\lfloor \frac{r}{2} \rfloor} \tag{1}$$

Say for the number of neighbours $r \geq 2$, there exist two numbers e, f such that $i - \lfloor \frac{r}{2} \rfloor \leq e, f \leq i + \lfloor \frac{r}{2} \rfloor$ such that $a_e = a_f$, so in two distinct initial configurations where one has $c_e = c_f = 0$ and another has $c_e = c_f = 1$ will lead to same c'_i. Thus f_c is not bijective, leading to a contradiction. Thus no reversible CA is linear.

Such essential, remarkable, and annotated observations have been taken into account for proving the security of our proposed cryptosystem against some prevalent attacks. Various aspects of our proposed cryptosystem are described in the section below.

4 Proposed Cryptosystem

In this section, we present the details of our cryptosystem along with the architectural design. The working of the cryptosystem is illustrated with an example. We finally use the important properties of CAs to substantiate the security of our cryptosystem.

4.1 Operational Flow of Our Cryptosystem

1. We define a function $T_1 : I \to \Theta$ to map attributes to their corresponding authorities. Here I is the set of attributes, and Θ is the set of all authorities. It can be noted that each attribute is associated with only one authority, but a single authority can possess many attributes. Also, two small positive integers N_1 and N_2 are chosen for the Central Authority and the attribute authorities respectively.
2. Data owner DO sends a prototype access structure A^1 and attribute list S to the Central Authority (CAuth), which is fully trusted.

---

[1] A prototype access structure is a Boolean representation of the access policy which is to be satisfied by the user's attributes in order to access the data.

3. Each user U submits its credentials to CAuth and obtains the associated set of attributes based on them.

4. CAuth also possesses a **tokenizing cellular automata**[2] CA_{token} that works on some chosen reversible CA rule R_{token}.

5. For each attribute $i \in S$, CAuth computes the **Intermediate Token** (IT) $i_1 = T_2(i)$, where $T_2 : I \rightarrow I_1$, I_1 being the set of all such possible intermediate tokens. The computation is as follows: the attribute i is fed as the seed of the tokenizing CA and is run for N_1 steps until the intermediate token i_1 is obtained.

6. The central authority CAuth and the attribute authorities exchange certain special messages for obtaining the final tokens from the intermediate tokens. These special messages are in the form of 5-tuples $(\theta_1, \theta_2, \theta_3, i_1, k)$ where

   (a) θ_1 is the ID of the nearest attribute authority from where the 5-tuple will start travelling from and reach back.

   (b) θ_2 is the ID of the attribute authority that is presently going to receive the 5-tuple.

   (c) θ_3 is the ID of the attribute authority that has the attribute associated with the intermediate token i_1 and converts it into final token.

   (d) i_1 could be either intermediate or final token.

   (e) k is a Boolean value, which is True if the i_1 is a final token.

7. CAuth submits an initialized set $S_1 = \{(\theta_n, \theta_n, \theta, i_1, False) : \theta = T_1(i), i_1 = T_2(i) \quad \forall i \in S\}$ to the nearest authority server θ_n.

8. All the attribute authorities are connected to each other by means of an arbitrary topology G. Each authority is associated with some identifier (ID) $\theta_A \in \Theta$ and all agree upon some routing algorithm A_r. For all 5-tuples $(\theta_n, \theta_2, \theta_3, i_1, k)$ coming to a server with ID θ_2, the following cases are considered, and whichever case is satisfied by the 5-tuple, the corresponding procedure is followed.

   (a) If $\theta_2 = \theta_3$ and the Boolean variable k is False, that means this server is the destination server that has the authority for the given attribute. So the server computes the Final Token(FT) $i_2 \in I_2$, I_2 being the set of all FTs, from the IT i_1 by feeding it as a seed to the tokenizing CA running with same reversible rule R_{token} for N_2 steps. Then it consults its own routing table to find the next hop attribute authority server with ID θ_4 which can route this tuple to the one with ID θ_n. Finally, it replaces the 5-tuple $(\theta_n, \theta_2, \theta_3, i_1, k)$ with $(\theta_n, \theta_4, \theta_n, i_1, True)$ and submits it to attribute authority with ID θ_4.

   (b) If $\theta_2 \neq \theta_3$ and the Boolean variable k is False, that means the tuple still has the IT and it has not reached the concerning attribute authority server. So this server consults its routing table to find the next hop attribute authority server with ID θ_4 which can route this tuple to the one with ID θ_3. Then it replaces the 5-tuple $(\theta_n, \theta_2, \theta_3, i_1, k)$ with $(\theta_n, \theta_4, \theta_3, i_1, k)$ and submits it to attribute authority with ID θ_4.

---

[2] Tokenising CA is a CA that converts an attribute into its intermediate token.

(c) If $\theta_2 \neq \theta_3 = \theta_n$ and the Boolean variable k is True, that means the tuple has the FT, but it has not reached the nearest attribute authority server to CAuth. So this server consults its routing table to find the next hop attribute authority server with ID θ_4 which can route this tuple to the one with ID θ_n. Then it replaces the 5-tuple $(\theta_n, \theta_2, \theta_n, i_1, k)$ with $(\theta_n, \theta_4, \theta_n, i_1, k)$ and submits it to attribute authority with ID θ_4.

(d) If $\theta_2 = \theta_3$ and the Boolean variable k is True, that means $\theta_2 = \theta_3 = \theta_n$, the algorithm A_r has terminated and the tuple containing the FT has finally reached back to nearest attribute authority server. So the set $S_2 = \{i_2 : i_2 = T_2(i_1) \forall (\theta_n, \theta_2, \theta_3, i_1, k) \in S_1'\}$, $T_2 : I_1 \to I_2$, is submitted back to CAuth. Here I_2 is the set of all such possible final tokens and S_1' is the modified set of 5-tuples after the termination of the routing algorithm A_r. CAuth waits until the complete set S_2 is received to prevent any replay attacks.

CAuth ensures that all the equality checks performed in the above algorithm are secured and cannot be compromised by an external attacker.

9. Meanwhile, the central authority CAuth waits for a fixed amount of time for the response from the authority server θ_n. If it does not receive any response, it aggressively sends a dummy packet to θ_n. Then attribute authority θ_n replies with the same dummy packet after it has cleared all of its dirtied buffers.

10. For all FTs $i_2 \in S_2$, CAuth computes the original $i, T_2(i) = i_2$ by passing i_2 as a seed to **detokenizing CA**[3] that uses reverse tokenizing CA rule R_{token}^{-1} for $N_1 + N_2$ steps. Then CAuth calculates the policy P representing the prototype access structure A by considering the Boolean min-terms. Finally, CAuth sends P to the key allocation repository (KAR). CAuth controls KAR and periodically monitors it.

11. KAR maintains a one-to-one mapping $M_1 : \mathbb{P}_> \to \mathbb{R}_{CA}$ where $\mathbb{P}_>$ is the set of all policies and $\mathbb{R}_{CA}$ is the set of all reversible rules. KAR submits $R_{encrypt} = M_1(P)$ to the data owner DO.

12. DO chooses a random number N as a public key. It also has a specific CA known as **encryption CA**[4], there data D is fed as its seed. The CA is run N times with rule $R_{encrypt}$, to obtain ciphertext C. DO then sends the 2-tuple $<C, N>$ to the cloud.

13. At the decryption end, the user U places its own attribute set A' to the first Game of Life (GoL) that generates the policy $P_u \in \mathbb{P}_>$ that it satisfies.

14. The user submits P_u to the KAR, which replies back with $R_{decrypt} = M_1(P_u)$. Then $R_{decrypt}$ is fed as the seed to the second GoL which returns reverse CA rule $R_{decrypt}^{-1}$.

15. U now feeds C to the **decryption CA**[5] that runs on rule $R_{decrypt}^{-1}$ for N steps. If $R_{decrypt} = R_{encrypt}$, it means that U satisfied the policy and it

---

[3] Detokenizing CA is a CA that converts a final token into its actual attribute.

[4] Encryption CA is a CA that is responsible for encryption of a message into its ciphertext.

[5] Decryption CA is a CA that is responsible for converting the ciphertext back to its original message.

will obtain the message M successfully from the decryption CA. Else, the policy satisfaction has been unsuccessful; therefore, the decryption is also unsuccessful.

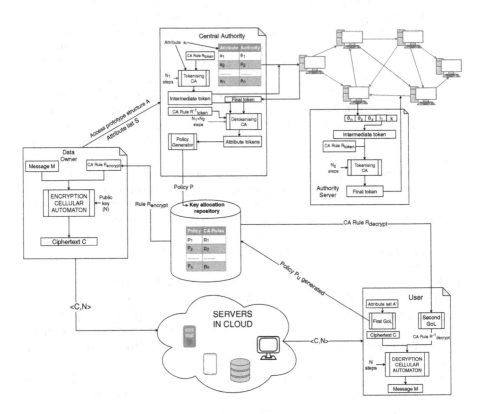

Fig. 3. Architectural design of the proposed cryptosystem.

The following is a description of the encryption-decryption algorithm for our proposed cryptosystem. The pictorial representation of the system is given in Fig. 3.

Now, we discuss the proof of correctness, a proven upper bound on the number of users and some comments on the security of our proposed cryptosystem.

Proof of Correctness: Say for a transition function $f : S^{r+1} \to S$ following a CA rule R beginning with message M_1, ends up in a sequence $M_1, M_2, M_3, \cdots M_N$ after $N = N_1 + N_2$ steps of temporal evolution. Here, $C = M_N$ is the final ciphertext, and $M_2, M_3, \cdots M_{N-1}$ are the intermediate texts during the temporal evolution of encryption cellular automaton. For every intermediate ciphertext M_i, $f_R(M_i) = M_{i+1} \implies f_{R^{-1}}(M_{i+1}) = M_i$, where R^{-1} is the reverse of the CA rule R. So if the attributes satisfy the policy generated from first GoL, the ith step of encryption that corresponds to the stage

$f_R(M_i) = M_{i+1}$ is directly mapped to the $(N-1-i)^{th}$ step of decryption, i.e., $f_{R^{-1}}(M_{i+1}) = M_i$. This allows the original message retrieved from the ciphertext successfully. A similar logic holds good for the case of the generation of intermediate and final tokens from the attributes and vice-versa. There, the attribute behaves like the message, and the first few steps of encryption are carried out at the central authority to generate the intermediate token, and the rest at the respective attribute authorities. This completes the proof of correctness of our cryptosystem.

4.2   A Working Example

Suppose a data owner DO want to send a message "ThisIsACryptoSystem" to all the users satisfying a policy C_p with a first-order cellular automaton with the following parameters:

1. Set of states $S = \{t, h, i, s, a, c, r, y, p, o, e, m, T, H, I, S, A, C, R, Y, P, O, E, M\}$, $k = |S| = 24$
2. A neighbourhood of size $r = 3$
3. A grid of size $d = 19$ (which is actually the length of the message in this case)

We assume a typical reversible CA rule $R_{encrypt}$ that represents a bijective mapping f_{encode} of the set S of these 24 letters to itself, as shown in Table 1, corresponds to C_p. We are choosing $N_1 = 7$ and $N_2 = 3$ for this example.

Table 1. The function $f_{encode} : S \rightarrow S$ shown in tabular form

m_i	$f_{encode}(m_i)$	m_i	$f_{encode}(m_i)$
t	T	h	t
i	A	s	C
a	R	c	I
r	Y	y	i
p	a	o	P
e	H	m	O
T	y	H	s
I	E	S	r
A	c	C	R
R	M	Y	e
P	S	O	o
E	m	M	p

Let there be three attributes in C_p, namely a_1, a_2 and a_3. Let a_1 be encoded as $(03af)_{16}$, a_2 as $(8316)_{16}$, and a_3 as $(941e)_{16}$. Suppose $C_p = a_1 \bar{a}_2 a_3 + \bar{a}_1 a_2 \bar{a}_3$. Let the rule of the tokenizing CA be Wolfram rule 18. For attribute a_1, the

Table 2. The steps of generating the intermediate token in the tokenizing CA.

Step 0 0000001110101111
Step 1 1000010000000000
Step 2 0100101000000001
...   ...
Step 7 1011010100000010

tokenizing CA performs the $N_1 = 7$ steps to obtain the intermediate token $(b502)_{16}$ as shown in Table 2. This step is known as **tokenization**.

Similarly the intermediate tokens for attributes a_2 and a_3 are $(1a83)_{16}$ and $(2a8c)_{16}$ respectively. Now when the nearest authority server θ_n is sent the set S_1 containing the authority IDs and the intermediate tokens, first the corresponding authority IDs are matched. Let us assume that the inherent routing algorithm ensures that the tuples in S_1 attributes a_1, a_2 and a_3 reach the respective authorities θ_1, θ_2 and θ_3 respectively. Since all attribute authorities agree to use the same rule 18 to encode the ITs, the server with ID θ_1 generates the final token for a_1 as $(0231)_{16}$ using $N_2 = 3$ steps of CA computation as shown in Table 3.

Table 3. The steps of generating the final token at the attribute authority.

Step 0 1011010100000010
Step 1 0000000010000100
Step 2 0000000101001010
Step 3 0000001000110001

Similarly, the final tokens for a_2 and a_3 are $(a900)_{16}$ and $(1150)_{16}$ respectively. Now after the collection of final tokens by the server with ID θ_n, the attributes need to retrieved back using rule 18R [33] once the modified version of set S_1 is sent back to the central authority. For example, the code for attribute a_1, i.e. $(03af)_{16}$ is retrieved back from its final token $(0231)_{16}$ by applying rule 18R in the detokenizing CA for $N_1 + N_2 = 10$ steps, as shown in Table 4.

This step is called as **detokenization**, which happens in a similar way for attributes a_2 and a_3 also.

Now the attributes are combined in the policy generator to form a policy C_p resembling the prototype access structure A. Suppose the access policy is given as $A = a_1\bar{a}_2a_3 + \bar{a}_1a_2\bar{a}_3$. Let every Boolean variable x is expressed as logic '1', and its negation $\bar{x}$ as logic '0'. Then, the min-terms $a_1\bar{a}_2a_3$ and $\bar{a}_1a_2\bar{a}_3$ are expressed as $(101)_2 = (5)_{10}$ and $(010)_2 = (2)_{10}$. We express C_p as a single

Table 4. The steps of retrieving the attribute back from the final token at the central authority.

Step 0	0000001000110001
Step 1	0000000101001010
Step 2	0000000010000100
...	...
Step 10	0000001110101111

number $2^2 + 2^5 = 4 + 32 = 36$. This number represents the policy. Likewise, there are 2^{2^v} representative policies for v number of attributes. Now C_p is sent to the key allocation repository, and corresponding CA rule $R_{encrypt}$ is chosen and sent back to DO.

The data owner DO chooses a random number $N = 5$ as its public key, and performs five steps of encryption, as shown in Table 5.

Table 5. The steps of encryption

Step 0	ThisIsACryptoSystem
Step 1	ytACECcRYiaTPriCTHO
...	...
Step 5	IAmaPaOhCEycesEacMr

The message obtained after 5th step of temporal evolution is the final ciphertext. Along with the public key and the list of users, the following ciphertext is uploaded to the cloud - $<$ "IAmaPaOhCEycesEacMr", $5 >$. Now suppose the user Alice (that has the attributes to satisfy the policy C_p) logs in to the system to access this ciphertext. Firstly, she provides her attribute set to the first GoL. Our cryptosystem implements a multiplexer logic to convert this attribute set into the required policy $C_p = 36$. The CA rule R_p is extracted from C_p in the key allocation repository and passed to 2nd GoL. The inverse rule $R_{encrypt}^{-1}$ is computed in the second GoL, that yields the function f_{encode}^{-1}. The Table 6 shows the function f_{encode}^{-1} to compute different characters in plaintext.

Now the decryption CA applies five steps, shown in Table 7, on the ciphertext with inverse mapping f_{encode}^{-1}. After 5th step of temporal evolution of the inverse CA, Alice gets the original message.

Suppose another user Bob satisfies policy $C_p' = 38$. The first GoL, here, would generate C_p' and send it to KAR. Suppose C_p' is mapped to a CA rule R_p' that represents a bijective mapping $f_{decode}^{-1} : S \rightarrow S$. The function f_{decode} as shown in Table 8.

KAR replies with rule R_p' to Bob. In his second GoL, the inverse CA rule $(R_p')^{-1}$ is generated. In this case, the decryption CA performs five steps, shown in

Table 6. The function $f^{-1}_{encode} : S \rightarrow S$ shown in tabular form

m_i	$f^{-1}_{encode}(m_i)$	m_i	$f^{-1}_{encode}(m_i)$
A	i	C	s
E	I	H	e
I	c	M	R
O	m	P	o
R	C	S	P
T	t	Y	r
a	p	c	A
e	Y	h	a
i	y	m	E
o	O	p	M
r	S	s	H
t	h	y	T

Table 7. The steps of decryption

Step 0 IAmaPaOhCEycesEacMr

Step 1 ciEpopmasITAYHIpARS

. . .   . . .

Step 5 ThisIsACryptoSystem

Table 8. The function $f_{decode} : S \rightarrow S$ shown in tabular form

m_i	$f^{-1}_{encode}(m_i)$	m_i	$f^{-1}_{encode}(m_i)$
A	I	C	s
E	i	H	e
I	m	M	R
O	c	P	o
R	t	S	r
T	C	Y	P
a	p	c	A
e	Y	h	a
i	O	m	E
o	y	p	M
r	S	s	T
t	h	y	H

Table 9, on the ciphertext $<$ "IAmaPaOhCEycesEacMr",5 $>$ with the mapping f_{decode} to obtain the message, which is incorrect.

Table 9. The steps of wrong decryption

Step 0 IAmaPaOhCEycesEacMr

Step 1 mIEpopcasiHAYTipARS

. . .   . . .

Step 5 cOAhYhEtIoiHCIhipS

This completes the illustration of the operational flow of our cryptosystem. We note that there are N steps taken in encryption and decryption, N_1 steps for the intermediate token generation, N_2 steps for the final token generation and $N_1 + N_2$ steps of original attribute generation. Taking into consideration the constant time of the routing algorithm, the overall time complexity of our proposed cryptosystem is $O(N + N_1 + N_2)$. And for a certain instance of encryption and decryption, the parameters N, N_1 and N_2 are fixed. Thus our cryptosystem takes constant time for both encryption and decryption. This is clearly visible in the experimental results obtained in Sect. 5.

4.3   Security of the Proposed Cryptosystem

The following are some of the proofs of security of the systems, presenting some possible brute force attacks, and the time complexity of the most intelligent possible scheme for the corresponding attack.

1. **Resistance to brute force attacks to discover the underlying CA rule being used.** As mentioned in Sect. 3.1, for a CA with k states, r neighbours and grid of size d, there are $(k^d)^{k^d}$ possibilities of the mapping of the configurations, and k^{k^r} possibilities of transition functions. So, the most intelligent interception through the brute force attack may scale as
   $$\Omega\left((k^d)^{k^d} + k^{k^r}\right),$$ which will be extremely time consuming. For example, for $k = 2, d = 20, r = 6$, the value of $(k^d)^{k^d}$ is approximately $(2.06 \times 10^6)^{10^6}$ and that of k^{k^r} is approximately 1.8446×10^{19}. These numbers are simply a rough estimate of how many instructions the adversary needs to carry out for this brute force attack.

2. **Resistance to brute force attack on the Game of Life CA.** Here, we discuss the case where the adversary attacks the Game of Life CA that formulates the satisfying policy. The adversary here tries to guess the policies of each and every user. Since 3-SAT is in NP-complete, guessing the policy will be highly time-consuming. The most intelligent brute force attack on the Game of Life CA assuming attribute set of cardinality A and n users, will be $\Omega(exp(n * 2^{2^A}))$, as the number of all policies using the complete attribute set is 2^{2^A}, which will also be highly scaling. For example, if $n = 100, A = 5$, $exp(n * 2^{2^A}) \approx 73 \times 2^{10^{11}}$.

3. **Resistance to linear attacks.** As seen in Sect. 3.2, both the Turing complete and reversible CAs are non-linear, and again looking at the pragmatic magnitudes, prominent attacks on linear systems such as differential, linear and interpolation attacks will be super-exponentially scaling with time. For example, if there are k states and d cells in the grid, the most intelligent linear attack will demand the computation of d coefficients using d equations, with choosing d messages among the k^d possibilities, rendering $\binom{k^d}{d}$ choices, with $O(d^3)$ test time per choice (by means of Gaussian elimination).

Thus the time required for such an attack for a first-order cellular automaton becomes $\Omega\left(\binom{k^d}{d}d^3\right)$. This complexity keeps on scaling with increasing order of linearity. For instance, if $k = 2, d = 5$, an approximate value of $\binom{k^d}{d}d^3$ is 6.95×10^{240}. This itself tells us how strong our system is against linear attacks.

4. **Resistance to attacks on attribute authorities.** Now the attribute authorities exchange intermediate tokens and final tokens among each other. So the most intelligent attacks on the attribute authorities will be a brute-force attack to check which of all the possible CA rules are reversible or not. As concluded from Sect. 3.2, the fraction of reversible CA rules are $\frac{(k^d)!}{k^{dk^d}}$. So the time complexity of a maximal probabilistic brute force attack to check whether a rule is reversible or not is $\Omega(exp((k^d)!))$, which is also high. For instance, if $k = 2, d = 10$, then $exp((k^d)!) \approx 43 \times 2^{10^{2639}}$.

5. **Resistance to attacks by malicious attribute authorities.** It might be a case where some malicious attribute authorities might not forward some of the attributes to the corresponding authorities. In that case, the nearest attribute authority to the central authority is not able to send the final tokens, and the central authority stops waiting after its timer has gone off. The dummy packet it sends cautions the authority that the process of forwarding has gone wrong at some point. So it waits for some time, replies with the dummy packet and the entire process restarts. Since the nearest authority waits for some time, the network is released of its traffic, authorities have their buffers cleared, and the system successfully comes back to its original state.

5   Experimental Results

In this section, we evaluate the performance of our cryptosystem in various scenarios. The important metric of consideration here is the time complexity of encryption and decryption algorithms. The experiments illustrated here are of programmatic nature, i.e., software-enabled mechanisms have been used. The testing environment is set up in Python and Wolfram languages, where standard libraries are used wherever required, and linear operations happen via BLAS (Basic Linear Algebra Subsystem) packages [31].

5.1   Complexity of Encryption and Decryption

The time complexity of the encryption and decryption in such an environment is governed by the size of the hardware, the number of steps (N_1 and N_2) in tokenization, the random public key parameter N and the network topology among the attribute authorities. For the sake of simplicity of analysis, we show the effect assuming the hardware is as large as the input message, however it can be easily generalized easily to large size messages. Figure 4 shows the variation of the encryption and decryption time with the length of the message. Figure 5 shows the same with the number of steps in intermediate tokenization (N_1) and Fig. 6 with that of final tokenization (N_2). Figure 7 shows the variation with the public key parameter N. All these results were obtained by considering the attribute authorities in a logical mesh topology. Also, in each of these plots, the variation with respect to one parameter has been shown having the other parameters held constant. Wherever required, the public key parameter N, and the parameters N_1 and N_2 have been held constant at 60. Also, as and when necessary, the message length has been fixed at 1.5 MB. Figure 8 shows a comparison of the same across different kinds of topology with fixed message length, parameters N_1 and N_2, and public key parameter N. Comparison has been made among mesh, star, tree, ring, and hypercube topology among eight attribute authorities. Figure 4 shows that the practical time complexity is invariant of the length of the message, as the definition of CA itself allows the entire plaintext to be accommodated wholly in a single go. Figures 5, 6 and 7 show a direct linear relationship between the speed and the number of steps taken in tokenization, and also with the public key parameter which substantiates the time complexity of $O(N + N_1 + N_2)$ derived in Sect. 4.2. Figure 8 reveals that the decryption time is invariant of the routing algorithm or the topology, but the encryption time is dependent on the same. For example, the encryption time is maximum for tree and ring topology (as the diameter of the graph is the highest) and minimum for the star topology (as the diameter of the graph is 2). The diameter of a graph is defined as the maximum possible length of the shortest path between all pairs of vertices.

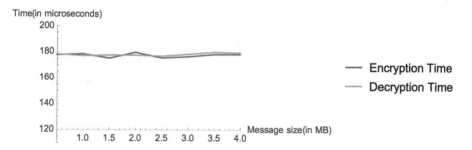

Fig. 4. The plot of encryption and decryption time (in μs) against the message length l (in MB).

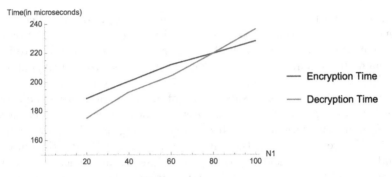

Fig. 5. The plot of encryption and decryption time (in μs) against number of steps N_1 in intermediate token generation.

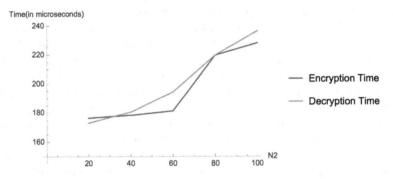

Fig. 6. The plot of encryption and decryption time (in μs) against number of steps N_2 in final token generation.

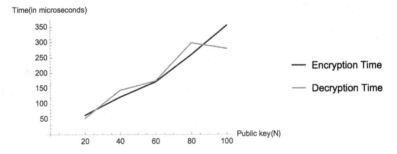

Fig. 7. The plot of encryption and decryption time (in μs) against public key parameter N.

5.2   Comparison with Existing Cryptosystems

A comparison of the performance of our proposed cryptosystem has been made with an existing efficient statically-secure large-universe multi-authority attribute-based cryptosystem [30]. Figures 9 and 10 clearly show that our cryptosystem performs much better than the existing systems in terms of encryption

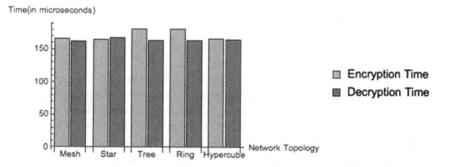

Fig. 8. The plot showing the variation of encryption and decryption time across different kinds of topology of the attribute authorities.

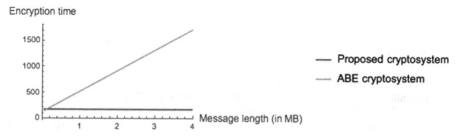

Fig. 9. The plot showing the variation of encryption time with the message length (in MB) of both the proposed cryptosystem and the RW-scheme [30]. The blue curve shows the measure of time in microseconds whereas the red curve shows the same in millisecond. (Color figure online)

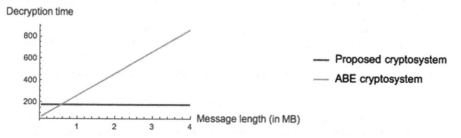

Fig. 10. The plot showing the variation of decryption time with the message length (in MB) of both the proposed cryptosystem and the RW-scheme [30]. The blue curve shows the measure of time in microseconds whereas the red curve shows the same in millisecond. (Color figure online)

and decryption times. This is because existing systems, such as the Rouselakis - Waters scheme (RW-scheme) [30], have linear time complexity. On the other hand, our cryptosystem has constant time complexity in terms of message length. Here the encryption and decryption times consider the processes in totality, i.e., the time required for pairing, key generation, etc.

6   Conclusion

In this paper, we propose a distributed multi-authority attribute-based encryption scheme for large-scale cloud systems using reversible and Turing-complete cellular automata. This system uses reversible CAs for distributed attribute authorization, encryption and decryption and Turing-complete CAs for policy validation. We have shown the security of our system in encryption, decryption, and attribute authorization. The system is designed a way where the attribute authorization and encryption-decryption mechanisms are independent of each other, which makes the system adaptable to any topology of attribute authority network. Also, we have shown the efficacy of the system in terms of requiring lesser hardware and accommodating the unrestricted number of users. We believe that improvement in distributed computational hardware in the future will reduce the cost of CAs.

References

1. Schiff, J.: Introduction to Cellular Automata, 1st edn. Wiley, Hoboken
2. Wolfram, S.: Computation theory of cellular automata. Commun. Math. Phys. **96**, 15–57 (1984). https://doi.org/10.1007/BF01217347
3. Cook, M.: Universality in elementary cellular automata. Complex Syst. **15**(1), 1–40 (2004)
4. Neary, T., Woods, D.: P-completeness of cellular automaton rule 110. In: Bugliesi, M., Preneel, B., Sassone, V., Wegener, I. (eds.) ICALP 2006. LNCS, vol. 4051, pp. 132–143. Springer, Heidelberg (2006). https://doi.org/10.1007/11786986_13
5. Rennard, J.P.: Implementation of logical functions in the game of life. In: Adamatzky, A. (ed.) Collision-Based Computing, pp. 491–512. Springer, London (2002). https://doi.org/10.1007/978-1-4471-0129-1_17
6. Kari, J.: Reversible cellular automata. In: Proceedings of the 9th International Conference on Developments in Language Theory, DLT 2005, pp. 57–68 (2005)
7. Sahai, A., Waters, B.: Fuzzy identity-based encryption. In: Cramer, R. (ed.) EURO-CRYPT 2005. LNCS, vol. 3494, pp. 457–473. Springer, Heidelberg (2005). https://doi.org/10.1007/11426639_27
8. Goyal, V., Pandey, O., Sahai, A., Waters, B.: Attribute-based encryption for fine-grained access control of encrypted data. In: Proceedings of the 13th ACM Conference on Computer and Communications Security - CCS 2006, pp. 89–98 (2006). https://doi.org/10.1145/1180405.1180418
9. Bethencourt, J., Sahai, A., Waters, B.: Ciphertext-policy attribute-based encryption. In: Proceedings of the 2007 IEEE Symposium on Security and Privacy, pp. 321–334 (2007). https://doi.org/10.1109/SP.2007.11
10. Lewko, A., Sahai, A., Waters, B.: Revocation systems with very small private keys. In: Proceedings of the 2010 IEEE Symposium on Security and Privacy, pp. 273–285 (2010). https://doi.org/10.1109/SP.2010.23
11. Waters, B.: Ciphertext-policy attribute-based encryption: an expressive, efficient, and provably secure realization. In: Catalano, D., Fazio, N., Gennaro, R., Nicolosi, A. (eds.) PKC 2011. LNCS, vol. 6571, pp. 53–70. Springer, Heidelberg (2011). https://doi.org/10.1007/978-3-642-19379-8_4

12. Wolfram, S.: Random sequence generation by cellular Automata. In: Advances in Applied Mathematics, vol. 7, no. 2, pp. 123–169 (1986). https://doi.org/10.1016/0196-8858(86)90028-X

13. Meier, W., Staffelbach, O.: Analysis of pseudo random sequences generated by cellular automata. In: Davies, D.W. (ed.) EUROCRYPT 1991. LNCS, vol. 547, pp. 186–199. Springer, Heidelberg (1991). https://doi.org/10.1007/3-540-46416-6_17

14. Nandi, S., Kar, B.K., Chaudhuri, P.P.: Theory and applications of cellular automata in cryptography. IEEE Trans. Comput. **43**(12), 1346–1357 (1994). https://doi.org/10.1109/12.338094

15. Paterson, K., Blackburn, S., Murphy, S.: Comments on theory and applications of cellular automata in cryptography. IEEE Trans. Comput. **46**, 637–638 (1997). https://doi.org/10.1109/12.589245

16. Tomassini, M., Perrenoud, M.: Cryptography with cellular automata. Appl. Soft Comput. **1**(2), 151–160 (2001). https://doi.org/10.1016/S1568-4946(01)00015-1

17. Santos, T.: Cellular automata and cryptography. In: Dissertacao de Mestrado apresentada a Faculdade de Ciencias da Universidade do Porto em Ciencia de Computadores (2014)

18. Bouchkaren, S., Lazaar, S.: A fast cryptosystem using reversible cellular automata. Int. J. Adv. Comput. Sci. Appl. **5**(5), 207–210 (2014)

19. Sakai, R., Kasahara, M.: ID-based cryptosystems with pairing on elliptic curves. IACR Cryptology ePrint Archive, vol. 03 (2003)

20. Guan, P.: Cellular automaton public-key cryptosystem. Complex Syst. **1**, 51–57 (1987)

21. Arora, S., Barak, B.: Computational Complexity: A Modern Approach, 1st edn, pp. 109–111. Cambridge University Press, Cambridge (2007)

22. Amoroso, S., Patt, Y.: Decision procedures for surjectivity and injectivity of parallel maps for tesselation structures. J. Comput. Syst. Sci. **6**(5), 448–464 (1972). https://doi.org/10.1016/S0022-0000(72)80013-8

23. Seredynski, M., Bouvry, P.: Block cipher based on reversible cellular automata. In: Proceedings of the 2004 Congress on Evolutionary Computation (IEEE Cat. No. 04TH8753), Portland, OR, USA, vol. 2, pp. 2138–2143 (2004). https://doi.org/10.1109/CEC.2004.1331161

24. Maiti, S., Roy Chowdhury, D.: Achieving better security using nonlinear cellular automata as a cryptographic primitive. In: Ghosh, D., Giri, D., Mohapatra, R.N., Savas, E., Sakurai, K., Singh, L.P. (eds.) ICMC 2018. CCIS, vol. 834, pp. 3–15. Springer, Singapore (2018). https://doi.org/10.1007/978-981-13-0023-3_1

25. Wei, J., Liu, W., Hu, X.: Secure and efficient attribute-based access control for multiauthority cloud storage. IEEE Syst. J. **12**(2), 1731–1742 (2018). https://doi.org/10.1109/JSYST.2016.2633559

26. Zhang, K., Li, H., Ma, J., Liu, X.: Efficient large-universe multi-authority ciphertext-policy attribute-based encryption with white-box traceability. In: Science China Information Sciences, pp. 2895–2903 (2017). https://doi.org/10.1007/s11432-016-9019-8

27. de Souza Brito, A., Soares, S.S.R.F., Villela, S.M.: Metaheuristics in the project of cellular automata for key generation in stream cipher algorithms. In: 2018 IEEE Congress on Evolutionary Computation (CEC), Rio de Janeiro, pp. 1–8 (2018). https://doi.org/10.1109/CEC.2018.8477658

28. Chase, M.: Multi-authority attribute based encryption. In: Vadhan, S.P. (ed.) TCC 2007. LNCS, vol. 4392, pp. 515–534. Springer, Heidelberg (2007). https://doi.org/10.1007/978-3-540-70936-7_28

29. Hinek, M.J., Jiang, S., Safavi-Naini, R., Shahandashti, S.F.: Attribute-based encryption with key cloning protection. Cryptology ePrint Archive, Report 2008/478 (2008). https://doi.org/10.1504/IJACT.2012.045587

30. Rouselakis, Y., Waters, B.: Efficient statically-secure large-universe multi-authority attribute-based encryption. In: Böhme, R., Okamoto, T. (eds.) FC 2015. LNCS, vol. 8975, pp. 315–332. Springer, Heidelberg (2015). https://doi.org/10.1007/978-3-662-47854-7_19

31. Sloot, P.M.A.: BLAS: basic linear algebra subsystems. In: Psychometrika, January 1993

32. Steinhart, E.: An Introduction to Conway's The Game of Life, in YouTube. www.youtube.com/watch?v=ouipbDkwHWA

33. Toffoli, T., Margolus, N.: Invertible cellular automata. Physica D **45**, 229–253 (1990)

Generic Construction of Bounded-Collusion IBE via Table-Based ID-to-Key Map

Kyu Young Choi$^{(\boxtimes)}$, Eunkyung Kim$^{(\boxtimes)}$, Hyojin Yoon,
Dukjae Moon, and Jihoon Cho

Security Research Team, Samsung SDS, Inc., Seoul, Korea
{ky12.choi,ek41.kim,hj1230.yoon,dukjae.moon,jihoon1.cho}@samsung.com

Abstract. In this paper, we give a new generic construction of bounded-collusion identity-based encryption (BC-IBE) scheme from public key encryption (PKE) scheme. Our construction significantly reduces the number of public parameters to $O(t)$, where t is a collusion parameter. Especially, we present the *first* construction in which the number of public parameters is independent from the total number of identities in the system. To achieve this, we propose a novel table-based ID-to-key map, and a method of deriving key pair from two (public and secret) parameter tables by using a cryptographic hash function. We provide a security proof of our construction in random oracle model.

1 Introduction

In identity-based encryption (IBE) schemes [12], a user's identity itself serves as a public key, and the corresponding secret key is generated by a trusted key generation center. After the first efficient construction of Boneh and Franklin's pairing-based IBE scheme [2], it has been remarkably improved in terms of security and efficiency of IBE schemes [1,3,11].

On the other hand, Dodis et al. [5] considered a variant of IBE scheme, bounded-collusion IBE (BC-IBE) scheme which is secure against any probabilistic polynomial time adversary who can obtain at most t secret keys $sk_1, \cdots, sk_t$. They presented a generic construction of BC-IBE scheme from any semantic secure PKE scheme. Unfortunately, the size of ciphertext in the induced BC-IBE scheme is significantly larger than that of the underlying PKE scheme, but their result is still valuable in that BC-IBE is a stepping stone for a generic reduction from PKE to IBE. In 2012, Goldwasser et al. [8] showed how to construct BC-IBE from PKE with key homomorphism while maintaining ciphertext size. Their security proof of BC-IBE scheme depends on the fact that the underlying PKE satisfies *"linear hash proof property"* inspired by the paradigm of hash proof systems [4]. In 2014, Tessaro and Wilson [13] presented a generic construction of BC-IBE from PKE with key homomorphism and malleability property, in order to apply more PKE schemes. However, the size of public parameters

© Springer Nature Switzerland AG 2019
Y. Mu et al. (Eds.): CANS 2019, LNCS 11829, pp. 457–469, 2019.
https://doi.org/10.1007/978-3-030-31578-8_25

is increased to $O(t^2 \log |ID|)$ from the previous result $O(t \log |ID|)$, where t is the collusion parameter and $|ID|$ is the total number of identities the scheme supports.

An (BC-)IBE scheme consists of four algorithms (Setup, Extract, Enc, Dec): Enc and Dec are usual encryption and decryption algorithms, and Setup generates public parameter pp and master secret key msk. Extract takes pp, msk, and a user's identity ID as input, and outputs the user's secret key sk_{ID}.

In both of [8] and [13], Setup algorithm generates N key pairs $\{(sk_i, pk_i)\}_{i \in [N]}$ of a PKE scheme where $N = O(t \log |ID|)$ in [8] and $N = O(t^2 \log |ID|)$ in [13], respectively. We have $pp = (pk_1, \cdots, pk_N)$ and $msk = (sk_1, \cdots, sk_N)$. The extraction algorithm of [8] is described as follow: for an identity ID and a hash function H, compute $f(H(ID)) = (id_1, \cdots, id_N)$ where f is a function satisfying the linear hash proof property and each id_i is an integer. The output of Extract is the user's secret key computed by $sk_{ID} = \sum_{i=1}^{N} id_i \cdot sk_i$. Note that by the homomorphic property the corresponding public key will be $pk_{ID} = \prod_{i=1}^{N} pk_i^{id_i}$. In [13], they made use of a "(N, s)-cover-free map" $\phi : [N] \to 2^{[N]}$ where $2^{[N]}$ denotes the collection of all the subsets of $[N]$. Extract algorithm takes ID, and computes $sk_{ID} = \sum_{i \in \phi(ID)} sk_i$ and $pk_{ID} = \prod_{i \in \phi(ID)} pk_i$. The (N, s)-cover-free map ϕ guarantees that there always exists an element of $[N]$ contained in exactly one subset among $\{\phi(ID_1), \cdots, \phi(ID_{t-1})\}$ for t distinct identities $ID_1, \cdots, ID_t$, i.e., the set $\phi(ID_t) \setminus \cup_{i=1}^{t-1} \phi(ID_i)$ is nonempty. Note that if we think msk and pp as 1-dimensional tables of length N then Extract chooses s element from the table to compute key pairs. The table size N cannot be reduced since this is intrinsic to guarantee the security from the property of the cover-free map ϕ.

Our Construction. We take a similar approach but in a different way: Instead of using cover-free maps, we set msk and pp as 2-dimensional tables. Given a security parameter λ, take two integers $u, v \in \mathbb{Z}$ satisfying $uv = \lambda$. Then, for a cryptographic hash function $H : \{0,1\}^* \to \{0,1\}^\lambda$, we interpret the hash value $H(ID)$ as a vector of length v with elements in $\{0, 1, \cdots, 2^u - 1\}$, i.e., $H(ID) = h_1 \| h_2 \| \cdots \| h_v$. Then, the key generation algorithm generates secret/public keys by the following rule

$$sk_{ID} = \sum_{j=1}^{v} sk_{(h_j, j)}, \text{ and } pk_{ID} = \prod_{j=1}^{v} pk_{(h_j, j)},$$

i.e., for each j-th column in Table 1, we take (h_j, j)-th component and then compute the sum and product for secret and public keys, respectively.

We show that, with the above identity-to-key assignments, one can generically convert PKE with key homomorphism and additional properties into BC-IBE with collusion parameter $t \approx 2^u \cdot v$ where $uv = \lambda$. In Table 2, we summarize the required assumptions on the underlying PKE schemes. The additional requirement of our construction, "*power of message-and-key*", can be easily satisfied by PKE schemes of our interest. Note that the table size $N = 2^u v$ varies with the choice of (u, v). For example, the possible minimum table size is $N = 2\lambda$ where $u = 1$ and $v = \lambda$ and the maximum table size is $N = 2^\lambda$ where $u = \lambda$ and

Table 1. Master secret table msk and public parameter table pp with 2^u rows and v columns.

$\mathsf{sk}_{(0,1)}$	$\mathsf{sk}_{(0,2)}$	$\cdots$	$\mathsf{sk}_{(0,v)}$
$\mathsf{sk}_{(1,1)}$	$\mathsf{sk}_{(1,2)}$	$\cdots$	$\mathsf{sk}_{(1,v)}$
$\vdots$	$\vdots$		$\vdots$
$\mathsf{sk}_{(2^u-1,1)}$	$\mathsf{sk}_{(2^u-1,2)}$	$\cdots$	$\mathsf{sk}_{(2^u-1,v)}$

$\mathsf{pk}_{(0,1)}$	$\mathsf{pk}_{(0,2)}$	$\cdots$	$\mathsf{pk}_{(0,v)}$
$\mathsf{pk}_{(1,1)}$	$\mathsf{pk}_{(1,2)}$	$\cdots$	$\mathsf{pk}_{(1,v)}$
$\vdots$	$\vdots$		$\vdots$
$\mathsf{pk}_{(2^u-1,1)}$	$\mathsf{pk}_{(2^u-1,2)}$	$\cdots$	$\mathsf{pk}_{(2^u-1,v)}$

$v = 1$. To guarantee the storage efficiency, We encourage the use of (u, v) such that $u \ll v$. We also stress that the table size N in fact determines the collusion parameter $t \approx N$.

Table 2. Comparison with previous constructions for BC-IBE. (t is the collusion parameter and $|\mathcal{ID}|$ is the total number of identities in the system. #pp is the number of public system parameters. Note that linear hash proof is strictly stronger assumption than semantic security.)

Construction	PKE assumptions	Security	#pp
GLW12 [8]	- Linear hash proof	Semantic	$\mathcal{O}(t \log|\mathcal{ID}|)$
	- Key homomorphism		
TW14 [13]	- Semantic-secure PKE	Selective	$\mathcal{O}(t^2 \log|\mathcal{ID}|)$
	- Key homomorphism		
	- Semantic-secure PKE	Semantic	
	- Key homomorphism		
	- Weak multi-key malleability		
	- Semantic-secure PKE	Semantic	
	- Multi-key malleability		
Ours	- Semantic-secure PKE	Semantic	$\mathcal{O}(t)$
	- Key homomorphism		
	- Weak multi-key malleability		
	- Power of message-and-key		

2 Preliminaries

KEY HOMOMORPHISMS. Throughout the paper, we focus on PKE = (KeyGen, Enc, Dec), where the secret and public keys are elements of groups $(\mathbb{G}_{sk}, +)$ and $(\mathbb{G}_{pk}, \cdot)$, respectively. For convenience and ease of distinction, we will use additive notation for the group $\mathbb{G}_{sk}$ of secret keys and multiplicative notation for the group $\mathbb{G}_{pk}$ of public keys.

Definition 1 (Secret-key to Public-key Homomorphism [13]**).** *We say that PKE admits a secret-key to public-key homomorphism (key homomorphism, in short), if there exist a map* $\mu : \mathbb{G}_{sk} \to \mathbb{G}_{pk}$ *such that for all* $sk, sk' \in \mathbb{G}_{sk}$, *it holds that* $\mu(sk + sk') = \mu(sk) \cdot \mu(sk')$ *and further, for all* $(sk, pk) \leftarrow$ KeyGen, *it holds that* $pk = \mu(sk)$.

For example, discrete logarithm problem (DLP)-based encryption schemes have a key pairs $(pk, sk) = (g^a, a)$ where $\mathbb{G}_{pk} = G$ is a cyclic group of prime order p and $\mathbb{G}_{sk} = \mathbb{Z}_p$. Then they allow a secret-key to public-key homomorphism $\mu : \mathbb{Z}_p \to G$ defined by $\mu(a) = g^a$.

As stated in [13], the definition of key homomorphism does not require that every group element $sk \in \mathbb{G}_{sk}$ is a valid secret key. Instead, we say that μ satisfies v-*correctness* if for any $k \leq v$ valid secret keys $sk_1, \cdots, sk_k$ output by KeyGen, the correctness condition holds for the derived secret key $sk = sk_1 + \cdots + sk_k$, i.e., we have for all message m, Dec$(sk, Enc(\mu(sk), m)) = m$ with high probability.

IND-ID-CPA SECURITY FOR BC-IBE. The security model of IBE must allow the adversary to obtain the secret keys of identities $ID_1, ..., ID_n$ of his choice (via secret key extraction queries), where n is a polynomial number of queries. On the other hand, the security model of BC-IBE must allow the adversary to obtain secret keys associated with at most $t(< n)$ identities, where t is a threshold parameter for collusion resistance. The IND-ID-CPA security notion [8] for BC-IBE is defined in term of the following game between a challenger $\mathcal{C}$ and an adversary $\mathcal{A}$:

Setup: $\mathcal{C}$ chooses a security parameter and obtains the public parameters and master secret key by running the setup algorithm. It gives the public parameters to $\mathcal{A}$.

Phase 1: $\mathcal{C}$ initializes a counter to be 0. When $\mathcal{A}$ issues secret key extraction queries for various identities, $\mathcal{C}$ increments its counter for each query. If the resulting counter is $\leq t$, $\mathcal{C}$ responds by running key generation algorithm to generate a secret key for the requested identity. If the counter is $> t$, it does not respond to the query.

Challenge: $\mathcal{A}$ outputs two equal-length messages m_0, m_1 and an identity ID^* that is not queried in any private key extraction query in Phase 1. $\mathcal{C}$ flips a coin $b \in \{0, 1\}$, encrypts m_b to identity ID^* using the encryption algorithm, and responds the ciphertext to $\mathcal{A}$.

Phase 2: $\mathcal{A}$ issues more secret key extraction queries for various identities not equal to ID^*. $\mathcal{C}$ responds as in Phase 1. Note that the same counter is employed, so that only t total queries in the game are answered with secret keys.

Guess: Finally, $\mathcal{A}$ outputs a guess $b' \in \{0, 1\}$. $\mathcal{A}$ wins the game if $b' = b$.

The advantage of $\mathcal{A}$ in the above game is defined as Adv$_{\mathcal{A}} = |$Pr$[b' = b] - 1/2|$. We say that a bounded-collision IBE scheme with parameter t is IND-ID-CPA secure if any adversary $\mathcal{A}$ has a negligible advantage in this game.

3   A New Construction of BC-IBE

In this section, we present how to assign a user's identity to its secret key which is an important building block for our generic construction. Once a secret key is assigned to the user, we can use the secret-key to public-key homomorphism to obtain the corresponding public key, and then encryption and decryption algorithms of our IBE scheme remain the same as the encryption and decryption algorithms of the base PKE scheme.

To fix the idea, let us denote PKE = (KeyGen, Enc, Dec) be a public-key encryption scheme. Assume the scheme has a secret-key to public-key homomorphism $\mu : \mathbb{G}_{sk} \to \mathbb{G}_{pk}$.

3.1   Table-Based Identity to Secret Key Map

In this section, we propose a new map ϕ which maps an arbitrary ID into the corresponding secret/public key which is a group element. We then present a generic construction of BC-IBE from PKE with key homomorphism based on our identity map ϕ. Our ID-to-key map ϕ will depend on a table with random group elements: we first apply a hash function to an ID string of arbitrary length and then the corresponding key will be a subset sum (or product etc.) of group elements related to the hash value of ID.

Let λ be a security parameter, and $(\mathbb{G}, +)$ be a group with additive notation for the simplicity. We first construct a table T of random group element as follows:

(i)   Pick two integers $u, v \in \mathbb{Z}^+$ satisfying $uv = \lambda$ and let $N = 2^u v$ be the table size.
(ii) Choose N distinct group elements $x_{(i,j)} \in \mathbb{G}$ for $0 \leq i < 2^u$ and $1 \leq j \leq v$.
(iii) Construct a table T with 2^u rows and v columns. Set $x_{(i,j)}$ to be the (i,j)-th entry of T.

ID-to-key Map(λ). Given the above table T, a cryptographic hash function $H : \{0,1\}^* \to \{0,1\}^\lambda$, and a group $(\mathbb{G}, +)$, we now define the ID-to-key map $\phi_T : \{0,1\}^* \to \mathbb{G}$ into the group $\mathbb{G}$. For each ID $\in \{0,1\}^*$, we do the following:
1. Compute $H(\text{ID}) \in \{0,1\}^\lambda$.
2. Split the hash string $H(\text{ID})$ into v substrings with u-bit length. Let h_j be the j-th substring for $1 \leq j \leq v$

$$H(\text{ID}) = h_1 \| h_2 \| \cdots \| h_v.$$

   Note that each substring $h_j \in \{0,1\}^u$ can be seen as an integer h_j with $0 \leq h_j < 2^u$.
3. For all $j = 1, 2, \cdots, v$, keep the (h_j, j)-th entry $x_{(h_j,j)}$ from the table T.
4. Output $\phi_T(\text{ID}) \in \mathbb{G}$ defined by

$$\phi_T(\text{ID}) = \sum_{j=1}^{v} x_{(h_j,j)}.$$

This completes the description of our ID-to-key map ϕ_T into the group $\mathbb{G}$ with respect to the table T.

Theorem 1. *Let $\lambda = uv$ be a security parameter and $N = 2^u v$ be the table size. Let ϕ_T be the ID-to-key map into the group $\mathbb{G}$ with respect to the table T. Then, given samples $\{(\mathsf{ID}_i, \phi_T(\mathsf{ID}_i))\}_{1 \leq i \leq \ell-1}$ for $\ell \leq N - 1 - \frac{\lambda}{\log_2 p}$, the probability of constructing the next sample $(\mathsf{ID}_\ell, \phi_T(\mathsf{ID}_\ell))$ for $\mathsf{ID}_\ell \notin \{\mathsf{ID}_i\}_{1 \leq i \leq \ell-1}$ is $1 - \prod_{k=1}^{\ell} \left(1 - \frac{k-1}{2^\lambda}\right)$.*

Lemma 1. *Let $\lambda = uv$ and $N = 2^u v$. For $i = 1, 2, \cdots, \ell$, let $B_i \in \{0,1\}^{2^u \times v}$ be a binary matrix over $\mathbb{Z}_p$ with p prime such that each column has Hamming weight 1. If $\ell \leq N - 1 - \frac{\lambda}{\log_2 p}$, then $\{B_i\}_{1 \leq i \leq \ell}$ are linearly independent with probability at least $\prod_{k=0}^{\ell-1} \left(1 - \frac{k}{2^\lambda}\right)$.*

Proof. Let E_k be the event that $\{B_i\}_{1 \leq i \leq k}$ are linearly independent, and I_k be the event that B_k is linearly independent to $\{B_i\}_{1 \leq i \leq k-1}$. Then by definition it holds that $I_k \cap E_{k-1} = E_k$ for all k and we have

$$\Pr[E_\ell] = \Pr[I_\ell | E_{\ell-1}] \Pr[E_{\ell-1}] = \cdots = \prod_{k=1}^{\ell} \Pr[I_k | E_{k-1}]$$

where $\Pr[E_0] = 1$. Consider the opposite probability $\Pr[I_k^{\mathsf{C}} | E_{k-1}] = 1 - \Pr[I_k | E_{k-1}]$ that B_k is linearly dependent to $\{B_i\}_{1 \leq i \leq k-1}$ given that $\{B_i\}_{1 \leq i \leq k-1}$ are linearly independent. This event is divided into the following two cases:

(i) $B_k = B_i$ for some $1 \leq i \leq k - 1$, or
(ii) $a \cdot B_k = \sum_{i=1}^{k-1} a_i \cdot B_i$ for some $a \in \mathbb{Z}_p^*$, $a_i \in \mathbb{Z}_p$ and at least two of a_i's are nonzero.

The first case occurs with probability $\frac{k-1}{2^\lambda}$ since the number of possible B_k's is $(2^u)^v = 2^\lambda$ and the number of all possible B_i's is $(k - 1)$. To estimate the probability of the second case, we observe that the portion of matrices of the form $a \cdot B_k$ in $\mathbb{Z}_p^{2^u \times v}$ is $(p - 1) \cdot \frac{2^\lambda}{p^N}$. Since there are p^{k-1} possible vectors of the form $\sum_{i=1}^{k-1} a_i \cdot B_i$ for given $\{B_i\}_{1 \leq i \leq k-1}$, we estimate the number of nontrivial linear combinations $a \cdot B_k = \sum_{i=1}^{k-1} a_i \cdot B_i$ by

$$(p - 1) \cdot p^{k-1} \cdot \frac{2^\lambda}{p^N} < \frac{2^\lambda}{p^{N-k}}$$

which is less than 1 by assumption $k \leq \ell < n - \frac{\lambda}{\log_2 p}$. Thus, the second case barely occurs, and hence we can say that $\Pr[I_k^{\mathsf{C}} | E_{k-1}] = \frac{k-1}{2^\lambda}$ and $\Pr[E_\ell] = \prod_{k=1}^{\ell} \left(1 - \frac{k-1}{2^\lambda}\right)$. $\square$

Proof (Proof of Theorem 1). One useful way to see the ID-to-key map $\phi_T(\mathsf{ID})$ is to introduce a binary matrix $B_{\mathsf{ID}} \in \{0,1\}^{2^u \times v}$ defined by

$$B_{\mathsf{ID}}[i,j] = \begin{cases} 1, & \text{if } i = h_j \\ 0, & \text{otherwise} \end{cases}$$

where $H(\mathsf{ID}) = h_1 \| h_2 \| \cdots \| h_v$. Then we have

$$\phi_T(\mathsf{ID}) = \sum_{j=1}^{v} \sum_{i=0}^{2^u-1} B_{\mathsf{ID}}[i,j] \cdot x_{(i,j)} = \sum_{j=1}^{v} \sum_{i=0}^{2^u-1} (B_{\mathsf{ID}} * T)[i,j],$$

where $B_{\mathsf{ID}} * T$ denotes component-wise multiplication. Let B_i be the corresponding binary matrix for ID_i for $1 \leq i \leq \ell - 1$, then $\phi_T(\mathsf{ID}_i)$ is computed by adding all components of $B_i * T$. If one can find ID_ℓ such that $B_\ell = \sum_{i=1}^{\ell-1} a_i \cdot B_i$ for some $a_i \in \mathbb{Z}_p$, then one can also compute $\phi_T(\mathsf{ID}_\ell)$ from $\sum_{i=1}^{\ell-1} a_i \cdot \phi_T(\mathsf{ID}_i)$ since we have $B_\ell * T = \left(\sum_{i=1}^{\ell-1} a_i \cdot B_i \right) * T = \sum_{i=1}^{\ell-1} a_i \cdot (B_i * T)$. However, it occurs with probability $\approx 1 - \prod_{k=0}^{\ell-1} \left(1 - \frac{k}{2^\lambda} \right)$ by Lemma 1.   □

The above probability implies that, for limited numbers of queries, there is little chance that linear dependence occurs. Hence it is hard to find a valid secret key from the queries, and thus it leads to *bounded-collusion* security of our scheme. For instance, at $\lambda = 256, u = 8$ and $v = 32$, the probability $P_\ell (= 1 - \Pr[E_\ell])$ is $\leq 2^{-230}$ if $\ell \leq 8190$, is ≈ 1 otherwise. Hence, if there exists a BC-IBE scheme using the ID-to-key map with the above parameters, we can say the scheme is "8190-bounded collusion IBE scheme".

3.2   Construction

Given a PKE=(KeyGen, Enc, Dec) scheme with key homomorphism, we now construct a BC-IBE scheme by using the method for ID-to-key mapping presented above.

IBE.Setup(λ). To generate public parameters pp and master key msk, run as
   follows:
   1. Pick two integers $u, v \in \mathbb{Z}^+$ such that $uv = \lambda$.
   2. Generate $N(= 2^u v)$ key-pairs $(sk_{(i,j)}, pk_{(i,j)})$ by running KeyGen algorithm N times, where $0 \leq i < 2^u$ and $1 \leq j \leq v$.
   3. Generate two tables T_{sk}, T_{pk} with 2^u rows and v columns using the N key-pairs.
   4. Construct two ID-to-key maps $\phi_{T_{sk}}$ and $\phi_{T_{pk}}$.
   5. Output msk $= T_{sk}$ and pp $= T_{pk}$.
IBE.KeyGen(msk, pp, ID). Given an identity ID, the secret key is $sk_{\mathsf{ID}} = \phi_{T_{sk}}(\mathsf{ID})$.
IBE.Enc(pp, ID, m). To encrypt message m under an identity ID, compute $pk_{\mathsf{ID}} = \phi_{T_{pk}}(\mathsf{ID})$ and then set the ciphertext to be ct $=$Enc(pk_{ID}, m).
IBE.Dec(pp, ct, sk_{ID}). To decrypt the ciphertext ct using the secret key sk_{ID},
   output m $=$ Dec(sk_{ID}, ct).

For the following security analysis, we define a following new property which we call *"power of message-and-key"*. The property states that given an integer v and a ciphertext ct of a message m under a known public key pk where ct is used a random value r, we can generate a new ciphertext ct' of the message m^v under the public key pk^v where ct' is used the same random value r of the given ciphertext ct. As we know, the message encapsulation part in an ElGamal encryption consists of multiplication of a shared secret pk^r and a message m, where pk is a public key and r is an ephemeral key (or random secret value). This ElGamal type encryption scheme satisfies the property by computing ct_2^v, where ct_2 is a ciphertext corresponding to the message encapsulation. We will show for give a concrete example based on DDH and LWE assumptions in Sect. 4.

Definition 2 (Power of Message-and-Key). *We say that a PKE with a secret-key to public-key homomorphism allows* power of message-and-key *computation if there exists an algorithm* MKPower *that takes a ciphertext $ct = \mathsf{Enc}(pk, m; r)$ and an integer v and returns a new ciphertext $ct' = \mathsf{Enc}(pk^v, m^v; r)$ where the same random value r is used.*

Another useful property that we will use is the *"weak multi-key malleability"* of PKE from [13] which describes the property that one can transform a ciphertext under a public key into another ciphertext under a product of some public keys including the original public key. Here the product of public keys is also a public key if PKE admits a key homomorphism.

Definition 3 (Weak Multi-key Malleability [13]). *For an integer v, we say that PKE is weakly v-key malleable if there exists an efficient algorithm* Simulate *such that for all messages m and all sk, sk', the probability distributions D_0 and D_1 are computationally indistinguishable: for any key pairs $(sk, pk), (sk', pk') \leftarrow$* KeyGen$(\lambda)$,

(i) $D_0 = \{((pk, pk'), sk', ct_0) : ct_0 \leftarrow \mathsf{Enc}(pk \cdot pk', m)\}$
(ii) $D_1 = \{((pk, pk'), sk', ct_1) : ct \leftarrow \mathsf{Enc}(pk, m), ct_1 \leftarrow \mathsf{Simulate}(ct, (pk, pk'), sk')\}$.

To prove the security of our construction, we define Γ_{PKE} as a PKE scheme which satisfies the above two properties, and $\Delta_{\mathsf{BC\text{-}IBE}}$ as a BC-IBE scheme generated by our generic construction based on Γ_{PKE}.

Theorem 2. *Let H be a random oracle. Let ϵ_ϕ be an advantage against the ID-to-key map ϕ_T. Suppose there exists an* IND-ID-CPA *adversary $\mathcal{A}$ against $\Delta_{\mathsf{BC\text{-}IBE}}$. Suppose $\mathcal{A}$ makes at most q_H queries to the H. Then, there exists an adversary $\mathcal{B}$ against the Γ_{PKE}. Concretely,*

$$\mathsf{Adv}_{\mathcal{B}, \Gamma_{\mathsf{PKE}}}^{\mathsf{IND\text{-}CPA}}(\lambda) \geq \frac{1}{q_H} \cdot \mathsf{Adv}_{\mathcal{A}, \Delta_{\mathsf{BC\text{-}IBE}}}^{\mathsf{IND\text{-}ID\text{-}CPA}}(\lambda) + \epsilon_\phi.$$

Proof. Let $\mathcal{A}$ be an active adversary that gets an advantage in attacking the IND-ID-CPA security of the BC-IBE scheme $\Delta_{\mathsf{BC\text{-}IBE}}$. $\mathcal{A}$ can get the advantage by following two cases: one is breaking the ID-to-key map ϕ_T, namely computing the secret key of a target identity directly, and the other is attacking the scheme $\Delta_{\text{BC-IBE}}$ without breaking the map ϕ_T. Roughly speaking, if $t \leq N - 2$, the advantage ϵ_ϕ for the first case is negligible by Theorem 1, where t is a parameter for collusion resistance and $N(= 2^u v)$ is a table size.

For the second case, we show how to build an algorithm $\mathcal{B}$ that uses $\mathcal{A}$ to break the IND-CPA security of PKE scheme Γ_{PKE}. (Note that, in the following proof, we will generate a public table that (randomly selected) half of the table is computed using given public key for $\mathcal{B}$ and the other half is using inverse of the public key. This is to satisfy the following three conditions: one is to behave like random oracle, second is to respond for any private key extraction queries, and the last is to connect a public key given to $\mathcal{B}$ to $\mathcal{A}$'s challenged identity.) Without loss of generality, we assume that any extraction queries are preceded by H queries. On input pk^*, $\mathcal{B}$ outputs $\mathsf{b}^* \in \{0,1\}$ by interacting with $\mathcal{A}$ as follows:

Setup. $\mathcal{B}$ picks $u, v \in \mathbb{Z}^+$ such that $uv = \lambda$ and generates $n(= 2^u v)$ key-pairs $(sk_{(i,j)}, pk_{(i,j)})$ by running KeyGen algorithm of Γ_{PKE}, where $i \in [0, 2^u - 1]$ and $j \in [1, v]$. $\mathcal{B}$ sets $\mathcal{RI} = \{0, 1, ..., 2^u - 1\}$ and $\mathcal{CI} = \{0, 1, ..., v\}$. $\mathcal{B}$ selects v random subsets RI_j^+ of $\mathcal{RI}$ such that $|RI_j^+| = 2^{u-1}$. $\mathcal{B}$ sets $RI_j^- = \mathcal{RI} \setminus RI_j^+$, where $j \in [1, v]$ and $|RI_j^+| = |RI_j^-| = 2^{u-1}$. To generate a public table $T_{pk'}$, $\mathcal{B}$ then sets

$$pk'_{(i,j)} = \begin{cases} pk^* pk_{(i,j)} & \text{if } i \in RI_j^+, \\ (pk^*)^{-1} pk_{(i,j)} & \text{if } i \in RI_j^-, \end{cases}$$

where $i \in [0, 2^u - 1]$ and $j \in [1, v]$. $\mathcal{B}$ guesses $\theta \in [1, q_H]$ such that $\mathcal{A}$ outputs ID^* for θ'th H query as a target identity for the challenge phase. To avoid collision and consistently respond to these queries, $\mathcal{B}$ maintains a list $\mathcal{L}_H = \{\langle \mathsf{ID}, H(\mathsf{ID}) = h_1 || h_2 || \cdots || h_v \rangle\}$ which is empty initially. $\mathcal{B}$ then gives the public parameters $\mathsf{pp} = T_{pk'}$ to $\mathcal{A}$.

Phase 1. $\mathcal{A}$ issues H and extraction queries. $\mathcal{B}$ responds as follows:

H-queries: When $\mathcal{A}$ makes H query ID, $\mathcal{B}$ does the following:

- If the query ID is θ'th H query (we let $\mathsf{ID} = \mathsf{ID}^\theta$ at this point), $\mathcal{B}$ chooses v random values $h_j^\theta \in RI_j^+$ where $j \in \{1, 2, ..., v\}$. $\mathcal{B}$ adds $\langle \mathsf{ID}^\theta, h_1^\theta || h_2^\theta || \cdots || h_v^\theta \rangle$ to $\mathcal{L}_H$ and responds with $H(\mathsf{ID}^\theta) = h_1^\theta || h_2^\theta || \cdots || h_v^\theta$. Note that the public value pk_{ID^θ} for the $H(\mathsf{ID}^\theta)$ is $(pk^*)^v pk_{(h_1^\theta, 1)} pk_{(h_2^\theta, 2)} \cdots pk_{(h_v^\theta, v)}$. (Note that, although the values h_j^θ are picked in the fixed sets RI_j^+, the simulate for θ'th H query acts like a random oracle because the subsets RI_j^+ are selected randomly in each game.)
- Otherwise, $\mathcal{B}$ selects a random subset CI^+ of $\mathcal{CI}$ such that $|CI^+| = v/2$. $\mathcal{B}$ sets $CI^- = \mathcal{CI} \setminus CI^+$, where $|CI^+| = |CI^-| = v/2$. For j from 1 to v, $\mathcal{B}$ chooses random $h_j \in \mathcal{RI}$ as follows:

$$h_j \in RI_j^+, \quad \text{if } j \in CI^+,$$
$$h_j \in RI_j^-, \quad \text{if } j \in CI^-.$$

$\mathcal{B}$ adds $\langle \mathsf{ID}, h_1 || h_2 || \cdots || h_v \rangle$ to $\mathcal{L}_H$ and responds with $H(\mathsf{ID}) = h_1 || h_2 || \cdots || h_v$. Note that, the value $pk_{\mathsf{ID}} = (pk^*)^{v/2}(pk^*)^{-v/2}pk_{(h_1,1)}pk_{(h_2,2)} \cdots pk_{(h_v,v)} = pk_{(h_1,1)}pk_{(h_2,2)} \cdots pk_{(h_v,v)}$.

<u>Extraction queries:</u> When $\mathcal{A}$ makes a secret key extraction query on ID, if $\mathsf{ID} = \mathsf{ID}^\theta$, $\mathcal{B}$ aborts the simulation. Otherwise, $\mathcal{B}$ finds the tuple $\langle \mathsf{ID}, h_1 || h_2 || \cdots || h_v \rangle$ in $\mathcal{L}_H$. $\mathcal{B}$ picks the $sk_{(h_j,j)}$ for each substring h_j. $\mathcal{B}$ then responds to $\mathcal{A}$ with $\mathsf{sk}_{\mathsf{ID}} = sk_{(h_1,1)} + sk_{(h_2,2)} + \cdots + sk_{(h_v,v)}$. (Note that, according to the security model of BC-IBE, $\mathcal{A}$ can obtain at most t secret keys.)

Challenge. $\mathcal{A}$ outputs two messages $\mathsf{m}_0, \mathsf{m}_1 \in \mathcal{M}$ and an identity ID^*. If $\mathsf{ID}^* \neq \mathsf{ID}^\theta$, $\mathcal{B}$ abort the simulation. Otherwise, $\mathcal{B}$ sets $(\mathsf{m}_0' = \mathsf{m}_0^{1/v}, \mathsf{m}_1' = \mathsf{m}_1^{1/v})$, forwards $(\mathsf{m}_0', \mathsf{m}_1')$ to the IND-CPA game, and obtains a challenge ciphertext ct. $\mathcal{B}$ then computes $ct' = \mathsf{MKPower}(ct, v)$, $pk' = \prod_{j=1}^{v} pk_{(h_j^\theta, j)}$, and $sk' = \sum_{j=1}^{v} sk_{(h_j^\theta, j)}$, where ct' is a ciphertext on the message $\mathsf{m}_\mathsf{b}(= (\mathsf{m}_\mathsf{b}')^v)$ of Γ_{PKE} by $\mathsf{MKPower}$ algorithm. $\mathcal{B}$ generates $\mathsf{ct}^* = \mathsf{Simulate}(ct', pk', sk')$ and gives ct^* to $\mathcal{A}$. (Note that, since the ciphertext ct is for pk^*, the algorithms $\mathsf{MKPower}$ and $\mathsf{Simulate}$ are used to modify ct to ct^* which is a cipertext on m_b for pk_{ID^θ}.)

Phase 2. $\mathcal{A}$ issues more H and extraction queries. $\mathcal{B}$ responds as in Phase 1.

Guess. $\mathcal{A}$ outputs a guess $\mathsf{b} \in \{0, 1\}$. $\mathcal{B}$ finally outputs $\mathsf{b}^* = \mathsf{b}$.

The probability that $\mathcal{B}$ correctly guesses θ is $1/q_H$. If the advantage of $\mathcal{A}$ is ϵ_{IBE}, then the advantage of $\mathcal{B}$ is at least $\frac{\epsilon_{\mathsf{IBE}}}{q_H} + \epsilon_\phi$ as required. $\qquad\square$

Note that, an IND-ID-CPA secure BC-IBE scheme generated from our construction can transform into an IND-ID-CCA secure BC-IBE scheme in the random oracle model by using the Fujisaki-Okamoto transformation [6].

4   Instantiation: BC-IBE from PKEs

4.1   DDH-Based Construction

We present a simple instantiation of our generic construction based on ElGamal encryption as follows: $(sk, pk = g^{sk}) \leftarrow \mathsf{KeyGen}(\mathbb{G}, q, g)$; $(g^r, m \cdot pk^r) \leftarrow \mathsf{Enc}(pk, m)$; $m = ct_2 \cdot ct_1^{-sk} \leftarrow \mathsf{Dec}(sk, (ct_1, ct_2))$, where $\mathbb{G}$ is a group with prime order q, g is a generator of $\mathbb{G}$, $m(\in \mathbb{G})$ is a message, and r is a random number in $\mathbb{Z}_q$. As we know, the ElGamal encryption scheme is IND-CPA secure under the DDH assumption and satisfies the following properties:

- *Secret-key to public-key homomorphism* $\mu : \mathbb{Z}_q \to \mathbb{G}$, *where* $\mu(x) = g^x$.
- *Computability of power of message-and-key:* $\mathsf{MKPower}((ct_1, ct_2), v) \to (ct_1, ct_2^v)$.
- *Weak key malleability:* $ct^* = \mathsf{Simulate}((ct_1, ct_2), pk', sk') = (ct_1, ct_2 \cdot ct_1^{sk'})$.

Eventually, it can construct the ElGamal encryption scheme to a BC-IBE scheme which is IND-ID-CPA secure under the DDH assumption by Theorem 2 (Fig. 1).

IBE.Setup($\lambda, \mathbb{G}, q, g$)	IBE.KeyGen(msk, ID)	IBE.Enc(pp, ID, m)
$sk_{(i,j)} \xleftarrow{\$} \mathbb{Z}_q$, $pk_{(i,j)} \leftarrow g^{sk_{(i,j)}}$	$sk_{\mathsf{ID}} \leftarrow \phi_{T_{sk}}(\mathsf{ID})$	$pk_{\mathsf{ID}} \leftarrow \phi_{T_{pk}}(\mathsf{ID})$
$(i,j)^{th}$ entry of $T_{sk} \leftarrow sk_{(i,j)}$	Return sk_{ID}	$r \xleftarrow{\$} \mathbb{Z}_q^*$
$(i,j)^{th}$ entry of $T_{pk} \leftarrow pk_{(i,j)}$		$ct_1 \leftarrow g^r$, $ct_2 \leftarrow \mathsf{m} \cdot pk_{\mathsf{ID}}^r$
$\mathsf{pp} \leftarrow (q, \mathbb{G}, g, T_{pk})$	IBE.Dec(pp, ct, sk_{ID})	Return $ct = (ct_1, ct_2)$
$\mathsf{msk} \leftarrow T_{sk}$	$\mathsf{m} \leftarrow ct_2 \cdot ct_1^{sk_{\mathsf{ID}}}$	
Return (pp, msk)	Return m	

Fig. 1. The BC-IBE scheme based on ElGamal encryption

4.2 LWE-Based Construction

The learning with errors (LWE) problem is one of the most important foundations of lattice-based cryptography which is quantum-resistant. A majority of candidates submitted to Post-Quantum Cryptography Standardization project of NIST are lattice-based cryptography, and its security is fundamentally based on the hardness of LWE problem. The LWE problem was first introduced by Regev [9,10] and has been being studied in a plenty number of papers afterwards.

Here we give a simplified description of the LWE problem for ease of understanding. See [9,10] for more details. The LWE problem is essentially to distinguish a special distribution from the uniform distribution. Let $n \geq 1$, $p \geq 2$ and $m \geq n \log p$ be parameters, and χ be an error distribution over $\mathbb{Z}_p^m$ sampling short elements. We further assume that p is prime. For $A \xleftarrow{\$} \mathbb{Z}_p^{m \times n}$, $s \xleftarrow{\$} \mathbb{Z}_p^n$, $e \xleftarrow{\$} \chi$, and $b \xleftarrow{\$} \mathbb{Z}_p$, the LWE assumption says that the following two distributions are statistically indistinguishable for secret s and e:

$$(A, As + e) \stackrel{s}{\approx} (A, b).$$

Our generic construction can be applied to obtain an LWE-based BC-IBE scheme from PKE in [7] which is a dual encryption scheme of Regev [9,10]. Let us briefly review the PKE scheme of [7]. $(sk = s, pk = (A, b = As)) \leftarrow \mathsf{KeyGen}(n, m, p)$; $(r^T A + e, r^T b + e' + m\lfloor \frac{p}{2} \rfloor) \leftarrow \mathsf{Enc}(pk, m)$; $m' = ct_2 - ct_1 \cdot sk$, $1 \leftarrow \mathsf{Dec}(sk, (ct_1, ct_2))$ if m' is close to $p/2$, 0 otherwise, where $sk = s \in \mathbb{Z}_p^n$, $A \in \mathbb{Z}_p^{m \times n}$, e and e' are sampled from proper error distributions, and $r \in \mathbb{Z}_p^m$ is the random vector for random encryption (Fig. 2).

- *Secret-key to public-key homomorphism* $\mu : \mathbb{Z}_p^n \rightarrow \mathbb{Z}_p^m$, defined by $\mu(s) = A \cdot s$ for a fixed matrix $A \in \mathbb{Z}_p^{m \times n}$.
- *Computability of power of message-and-key*: $\mathsf{MKPower}((ct_1, ct_2), v) \rightarrow (ct_1, v \cdot ct_2)$.
- *Weak key malleability*: $ct^* = \mathsf{Simulate}((ct_1, ct_2), pk', sk') = (ct_1, ct_2 + ct_1 \cdot sk')$.

IBE.Setup(λ, n, m, p)	IBE.KeyGen(msk, ID)	IBE.Enc(pp, ID, m)
$A \xleftarrow{\$} \mathbb{Z}_p^{m \times n},\ \boldsymbol{s}_{(i,j)} \xleftarrow{\$} D_{\mathbb{Z}^m, r}$	$sk_{\mathsf{ID}} \leftarrow \phi_{T_{sk}}(\mathsf{ID})$	$pk_{\mathsf{ID}} \leftarrow \phi_{T_{pk}}(\mathsf{ID})$
$A\boldsymbol{s}_{(i,j)} \in \mathbb{Z}_p^m$	Return sk_{ID}	$\boldsymbol{r}^T \xleftarrow{\$} \mathbb{Z}_p^m$
$(i,j)^{th}$ entry of $T_{sk} \leftarrow \boldsymbol{s}_{(i,j)}$		$e \xleftarrow{\$} \chi,\ e' \xleftarrow{\$} \chi'$
$(i,j)^{th}$ entry of $T_{pk} \leftarrow A\boldsymbol{s}_{(i,j)}$		$ct_1 \leftarrow \boldsymbol{r}^T A + e$
pp $\leftarrow (n, m, p, A, T_{pk})$	IBE.Dec(pp, ct, sk_{ID})	$ct_2 \leftarrow \boldsymbol{r}^T pk_{\mathsf{ID}} + e' + m\lfloor \frac{p}{2} \rfloor$
msk $\leftarrow T_{sk}$	Return 0 or 1 by inspecting	Return (ct_1, ct_2)
Return(pp, msk)	$ct_2 - ct_1 sk_{\mathsf{ID}}$	

Fig. 2. The BC-IBE scheme based on the GPV scheme [7]

References

1. Boneh, D., Boyen, X.: Efficient selective-id secure identity-based encryption without random oracles. In: Cachin, C., Camenisch, J.L. (eds.) EUROCRYPT 2004. LNCS, vol. 3027, pp. 223–238. Springer, Heidelberg (2004). https://doi.org/10.1007/978-3-540-24676-3_14

2. Boneh, D., Franklin, M.: Identity-based encryption from the weil pairing. In: Kilian, J. (ed.) CRYPTO 2001. LNCS, vol. 2139, pp. 213–229. Springer, Heidelberg (2001). https://doi.org/10.1007/3-540-44647-8_13

3. Chen, L., Cheng, Z.: Security proof of Sakai-Kasahara's identity-based encryption scheme. In: Smart, N.P. (ed.) Cryptography and Coding 2005. LNCS, vol. 3796, pp. 442–459. Springer, Heidelberg (2005). https://doi.org/10.1007/11586821_29

4. Cramer, R., Shoup, V.: Universal hash proofs and a paradigm for adaptive chosen ciphertext secure public-key encryption. In: Knudsen, L.R. (ed.) EUROCRYPT 2002. LNCS, vol. 2332, pp. 45–64. Springer, Heidelberg (2002). https://doi.org/10.1007/3-540-46035-7_4

5. Dodis, Y., Katz, J., Xu, S., Yung, M.: Key-insulated public key cryptosystems. In: Knudsen, L.R. (ed.) EUROCRYPT 2002. LNCS, vol. 2332, pp. 65–82. Springer, Heidelberg (2002). https://doi.org/10.1007/3-540-46035-7_5

6. Fujisaki, E., Okamoto, T.: Secure integration of asymmetric and symmetric encryption schemes. J. cryptol. **26**(1), 80–101 (2013)

7. Gentry, C., Peikert, C., Vaikuntanathan, V.: Trapdoors for hard lattices and new cryptographic constructions. In: Proceedings of the Fortieth Annual ACM Symposium on Theory of Computing, pp. 197–206. ACM (2008)

8. Goldwasser, S., Lewko, A., Wilson, D.A.: Bounded-collusion IBE from key homomorphism. In: Cramer, R. (ed.) TCC 2012. LNCS, vol. 7194, pp. 564–581. Springer, Heidelberg (2012). https://doi.org/10.1007/978-3-642-28914-9_32

9. Regev, O.: On lattices, learning with errors, random linear codes, and cryptography (2005)

10. Regev, O.: On lattices, learning with errors, random linear codes, and cryptography. J. ACM (JACM) **56**(6), 34 (2009)

11. Sakai, R., Kasahara, M.: Id based cryptosystems with pairing on elliptic curve

12. Shamir, A.: Identity-based cryptosystems and signature schemes. In: Blakley, G.R., Chaum, D. (eds.) CRYPTO 1984. LNCS, vol. 196, pp. 47–53. Springer, Heidelberg (1985). https://doi.org/10.1007/3-540-39568-7_5

13. Tessaro, S., Wilson, D.A.: Bounded-collusion identity-based encryption from semantically-secure public-key encryption: generic constructions with short ciphertexts. In: Krawczyk, H. (ed.) PKC 2014. LNCS, vol. 8383, pp. 257–274. Springer, Heidelberg (2014). https://doi.org/10.1007/978-3-642-54631-0_15

A *t*-out-of-*n* Redactable Signature Scheme

Masayuki Tezuka$^{(\boxtimes)}$, Xiangyu Su, and Keisuke Tanaka

Tokyo Institute of Technology, Tokyo, Japan
`tezuka.m.ac@m.titech.ac.jp`

Abstract. A redactable signature scheme allows removing parts of a signed message without invalidating the signature. Currently, the need to prove the validity of digital documents issued by governments and enterprises is increasing. However, when disclosing documents, governments and enterprises must remove privacy information concerning individuals. A redactable signature scheme is useful for such a situation.

In this paper, we introduce the new notion of the *t*-out-of-*n* redactable signature scheme. This scheme has a signer, *n* redactors, a combiner, and a verifier. The signer designates *n* redactors and a combiner in advance and generates a signature of a message *M*. Each redactor decides parts that he or she wants to remove from the message and generates a piece of redaction information. The combiner collects pieces of redaction information from all redactors, extracts parts of the message that more than *t* redactors want to remove, and generate a redacted message.

We consider the one-time redaction model which allows redacting signatures generated by the signer only once. We formalize the one-time redaction *t*-out-of-*n* redactable signature scheme, define security, and give a construction using the pairing based aggregate signature scheme in the random oracle model.

Keywords: Redactable signature scheme · Aggregate signature scheme · Shamir's secret sharing · Bilinear map

1 Introduction

1.1 A Redactable Signature Scheme

Recently, due to the development of IoT devices, the number of electronic data is steadily increasing. It is indispensable for future information society to make use of these data. When we use data, it is important to prove that the data has not been modified in any way. A digital signature enables a verifier to verify the authenticity of *M* by checking that σ is a legitimate signature on *M*. However,

A part of this work was supported by NTT Secure Platform Laboratories, JST OPERA JPMJOP1612, JST CREST JPMJCR14D6, JSPS KAKENHI JP16H01705, JP17H01695.

© Springer Nature Switzerland AG 2019
Y. Mu et al. (Eds.): CANS 2019, LNCS 11829, pp. 470–489, 2019.
https://doi.org/10.1007/978-3-030-31578-8_26

in our real-world scenario, when we use data, the confidential information should be deleted from the original data. A digital signature cannot verify the validity of a message with parts of the message removed.

A redactable signature scheme (RSS) is a useful cryptographic scheme for such a situation. This scheme consists of a signer, a redactor, and a verifier. A signer signs a message M with a secret key sk and generates a valid signature σ. A redactor who can become anyone removes some parts of a signed message from M, generate a redacted message M', and updates the corresponding signature σ' without the secret key sk. A verifier still verifies the validity of the signature σ' on message M' using pk.

An idea of a redactable signature scheme was introduced by Steinfeld, Bull, and Zheng [21] as a content extraction signature scheme (CES). This scheme allows generating an extracted signature on selected portions of the signed original document while hiding removed parts of portions. Johnson, Molnar, Song, and Wagner [11] proposed a redactable signature scheme (RSS) which is similar to a content extraction signature scheme.

Security. Security of a redactable signature scheme was argued in many works. Brzuska, Busch, Dagdelen, Fischlin, Franz, Katzenbeisser, Manulis, Onete, Peter, Poettering, and Schröder [3] formalized three security notions of a redactable signature for tree-structured messages in the game-based definition.

- Unforgeability: Without the secret key sk it is hard to generate a valid signature σ' on a message M' except to redact a signed message (M, σ).
- Privacy: Except for a signer and redactors, it is hard to derive any information about removed parts of the original message M from the redacted message M'.
- Transparency: It is hard to distinguish whether (M, σ) directly comes from the signer or has been processed by a redactor.

Derler, Pöhls, Samelin, and Slamanig [4] gave a general framework of a redactable signature scheme for arbitrary data structures and defined its security definitions.

Additional Functionalities. Following additional functionalities for a redactable signature scheme were proposed.

- Disclosure control [5,6,8,9,13–16,19]: Miyazaki, Iwamura, Matsumoto, Sasaki, Yoshiura, Tezuka, and Imai [16] proposed the disclosure control. The signer or intermediate redactors can control to prohibit further redactions for parts of the message.
- Identification of a redactor [7,10]: Izu, Kanaya, Takenaka, and Yoshioka [7] proposed the redactable signature scheme called "Partial Information Assuring Technology for Signature" (PIATS). PIATS allows a verifier to identify the redactor of the signed message.
- Accountability [18]: Pöhls and Samelin proposed an accountable redactable signature scheme that allows deriving the accountable party of a signed message.

- Update and Marge [12,17]: Lim, Lee, and Park [12] proposed the redactable signature scheme where a signer can update signature by adding new parts of a message. Moreover, Pöhls and Samelin [17] proposed the updatable redactable signature scheme that can update a signature and marge signatures derived from the same signer.

1.2 Motivation

Consider the case where a citizen requests the signed secret document disclosure to the government. To disclose the secret signed document, the government must remove sensitive data from it. A decision of deletion for confidential information of a document is performed by multiple officers in the government meeting.

One of the simple solutions is that the signer of the secret document gives the signing key sk to the meeting chair. The chair takes a vote on removing sensitive information and removes it from the secret document and signed it using sk. However, if the meeting chair is malicious, it is risky for the secret document signer to give the meeting chair a signing key sk. Therefore, the secret document signer wants to avoid giving a signing key sk to others.

If we try to adapt the original RSS on this situation, we suffer from the following problem. RSS allows anyone to redact message parts and even removes the necessary information. Moreover, a malicious chair can redact message parts form the signed document regardless of the decision of the officers.

1.3 Our Contributions

We introduce the new notion of t-out-of-n redactable signature scheme to overcome this problem. This scheme is composed of a signer, n redactors, a combiner, and a verifier. The signer designates n redactors and a combiner, generates a key pair (pk, sk) and redactor's secret key $\{rk[i]\}_{i=1}^{n}$ and sends rk[i] to the redactor i. Then signer decides parts of a message that redaction is allowed, signs the message, and sends its signature to n redactor and a combiner. Each redactor i selects parts of the signed message that he or she wants to remove, generates a piece of redaction information RI_i, and sends it to the combiner. The combiner collects all redaction information $\{RI_i\}_{i=1}^{n}$, extracts signed message parts which at least t redactors want to remove using $\{RI_i\}_{i=1}^{n}$, generates the redactable signature. The verifier can verify the validities of signatures.

Now, we reconsider applying the t-out-of-n redactable signature scheme to the above redaction problem. Let the secret document signer be a signer of the t-out-of-n redactable signature scheme, officers be redactors, and the meeting chair be a combiner. The secret document signer does not have to give the signing key sk to the chair. Our t-out-of-n redactable signature only allows the chair to redact parts of message which at least t officers wants to remove.

We consider the one-time redaction model which allows redacting signed message only one time for each signature and gives the unforgeability, privacy, and transparency security of the t-out-of-n redactable signature scheme in the one-time redaction model. Also, we give a concrete construction of the *t*-out-of-*n* redactable signature scheme which satisfies the unforgeability, privacy, and transparency security.

Our construction is based on the (t, n)-Shamir's secret sharing scheme and the redactable signature scheme proposed by Miyazaki, Hanaoka, and Imai [14] which use the aggregate signature scheme proposed by Boneh, Gentry, Lynn, and Shacham [1] based on the BLS signature scheme [2]. Our technical point is to adapt (t, n)-Shamir's secret share scheme and compute Lagrangian interpolation at the exponent part of the group element to reconstruct information for the redaction. Security of our scheme is based on the computational co-CDH assumption in the random oracle model.

2 Preliminaries

Let 1^λ be the security parameter. A function $f(k)$ is negligible in k if $f(\lambda) \leq 2^{-\omega(\log \lambda)}$. PPT stands for probabilistic polynomial time. For strings m and r, $|m|$ is the bit length of m and $m||r$ is the concatenation of m and r. For a finite set S, $\#S$ denotes the number of elements in S, $s \xleftarrow{\$} S$ denotes choosing an element s from S uniformly at random. $y \leftarrow \mathcal{A}(x)$ denotes that an algorithm $\mathcal{A}$ outputs y for an input x.

2.1 Bilinear Map

Let $\mathcal{G}$ be a bilinear group generator that takes as an input a security parameter 1^λ and outputs the descriptions of multiplicative groups $(q, \mathbb{G}_1, \mathbb{G}_2, \mathbb{G}_T, e, g_1, g_2)$ where $\mathbb{G}_1$, $\mathbb{G}_2$, and $\mathbb{G}_T$ are groups of prime order q, e is an efficient, non-degenerating bilinear map $e : \mathbb{G}_1 \times \mathbb{G}_2 \rightarrow \mathbb{G}_T$, g_1, and g_2 are generators of the group $\mathbb{G}_1$ and $\mathbb{G}_2$ respectively, and ϕ is a computable isomorphism from $\mathbb{G}_2$ to $\mathbb{G}_1$ with $\phi(g_2) = g_1$.

1. Bilinear: for all $u \in \mathbb{G}_1$, $v \in \mathbb{G}_2$ and $a, b \in \mathbb{Z}$, then $e(u^a, v^b) = e(u, v)^{ab}$.
2. Non-degenerate: $e(g_1, g_2) \neq 1_{\mathbb{G}_T}$.

Definition 1 (Computational co-Diffie-Hellman Problem). *For a groups* $\mathbb{G}_1 = \langle g_1 \rangle$, $\mathbb{G}_2 = \langle g_2 \rangle$ *of prime order* q, *define* $\mathsf{Adv}^{\text{co-CDH}}_{\mathbb{G}_1, \mathbb{G}_2, \mathcal{A}}$ *of a PPT adversary* $\mathcal{A}$ *as*

$$\mathsf{Adv}^{\text{co-CDH}}_{\mathbb{G}_1, \mathbb{G}_2, \mathcal{A}} = \Pr\left[\mathcal{A}(g_2, g_2^\alpha, h) = h^\alpha \middle| \alpha \xleftarrow{\$} \mathbb{Z}_q, h \leftarrow \mathbb{G}_1\right],$$

where the probability is taken prover the randomness of $\mathcal{A}$ *and the random selection of* (α, h). *The computational co-Diffie-Hlleman* (co-CDH) *assumption is that for all adversaries* $\mathcal{A}$, *the* $\mathsf{Adv}^{\text{co-CDH}}_{\mathbb{G}_1, \mathbb{G}_2, \mathcal{A}}$ *is negligible in* λ.

2.2   Secret Sharing

In order to construct a t-out-of-n redactable signature scheme, we use the (t, n)-Shamir's secret sharing scheme [20]. The (t, n)-secret sharing scheme is composed of a dealer and n users. The dealer decides a secret s, computes secret shares $\{s_i\}_{i=1}^{n}$, and gives the secret share s_i to the user i. If any t of n secret shares or more shares are collected, we can reconstruct the secret s from them. While, with less than t secret shares, we cannot recover the secret s.

Shamir's Secret Sharing Scheme. We refer to the (t, n)-shamir's secret sharing scheme.

1. The dealer chooses the secret $s \in \mathbb{Z}$ and sets $a_0 \leftarrow s$.
2. The dealer chooses $a_1, \cdots, a_{t-1} \in \{0, \cdots, p - 1\}$ independently at random and gets the polynomial $f(X) = \sum_{i=0}^{t-1} a_i X^i$.
3. The dealer computes $f(i)$, sets $s_i \leftarrow (i, f(i))$, and sends the secret share s_i to the user i.

If we collect t or more secret shares, we can reconstruct the secret s by the Lagrange interpolation. Let $J \subset \{1, \cdots, n\}$ and $|J| = t$. If we have secret shares $\{s_j\}_{j \in J} = \{(j, f(j))\}_{j \in J}$, we can compute $s = \sum_{i \in J} \left(s_i \prod_{j \in J, j \neq i} j(j - i)^{-1} \right)$.

3   A t-out-of-n Redactable Signature Scheme

We explain the outline of our proposed t-out-of-n redactable signature scheme in the one-time redaction model. A t-out-of-n redactable signature scheme in the one-time redaction model (t, n)-RSS is a signature scheme that has a signer, n redactors, a combiner, and a verifier. The signer designates n redactors and the combiner.

The signer selects a threshold t and the number of redactors n. Then, he or she runs key generation algorithm and gets $(\mathsf{pk}, \mathsf{sk}, \{\mathsf{rk}[i]\}_{i=1}^{n})$. The pk is published and the redactor's key $\mathsf{rk}[i]$ is sent to the redactor i.

The signer signs a message M with an admissible description ADM which represents parts of the message that redactors cannot remove from the message M. In the processing of the signing, a random document ID (DID) is added to the message M, then the signature σ is generated. $(M, \mathsf{ADM}, \mathsf{DID}, \sigma)$ generated by the signer is sent to n redactors and the combiner.

Each redactor i checks whether DID has never been seen before. If he or she has seen it, then aborts. Also, if the signature is invalid, then aborts. Otherwise, he or she selects parts of the message that he or she wants to remove and makes the redaction information RI_i and sends it to the combiner. The protocol works only once for DID which redactors have not seen before.

The combiner collects pieces of redaction information $\{\mathsf{RI}_i\}_{i=1}^{n}$. From $\{\mathsf{RI}_i\}_{i=1}^{n}$, the combiner extracts parts which at least t redactor want to remove. Finally, the combiner outputs the redacted message M', ADM, DID, and its updated valid signature σ'.

The signature is verified using the signer's public key pk. In the verification, it is possible to prove the validity of the $(M, \mathsf{ADM}, \mathsf{DID}, \sigma)$ made by a legitimate signer or redacted by the redaction protocol for that signature while keeping redactors anonymity.

3.1   A t-out-of-n Redactable Signature Scheme for Set

In this paper, we focus on the t-out-of-n redactable signature scheme in the one-time redaction model for set. In the following, we assume that a message M is a set and use following notations. An admissible description ADM is a set containing all elements which must not be redacted.

A modification instruction MOD is a set containing all elements which a redactor want to redact from M. $\mathsf{ADM} \preceq M$ means that ADM is a valid description. (i.e., $\mathsf{ADM} \cap M = \mathsf{ADM}$.) $\mathsf{MOD} \stackrel{\mathsf{ADM}}{\preceq} M$ means that MOD is valid redaction description respect to ADM and M. (i.e., $\mathsf{MOD} \cap \mathsf{ADM} = \emptyset \wedge \mathsf{MOD} \subset M$.) A redaction $M' \stackrel{\mathsf{MOD}}{\longleftarrow} M$ would be $M' \leftarrow M \backslash \mathsf{MOD}$. In the following definition, we explicit ADM and DID in the syntax.

Definition 2. *A t-out-of-n redactable signature scheme in the one-time redaction model (t, n)-RSS Π is composed of four components (KeyGen, Sign, Redact, Verify).*

KeyGen : *A key generation algorithm is a randomized algorithm that a signer runs. Given a security parameter 1^λ, a threshold t and the number of redactors n, return a signer's public key pk, a signer's secret key sk, and redactor's secret keys $\{\mathsf{rk}[i]\}_{i=1}^n$.*

Sign : *A signing algorithm is a randomized algorithm that a signer runs. Given a signer's secret key sk, a message M and an admissible description ADM, return a message M, an admissible description ADM, a document ID (DID), and a signature σ.*

Redact : *A redact protocol is a 1-round interactive protocol between the combiner and n redactors. Each redactor i generates redaction information RI_i and sends to the combiner. The combiner collects all redaction informations $\{\mathsf{RI}_i\}_{i=1}^n$ and finally outputs the redacted signature $(M', \mathsf{ADM}, \mathsf{DID}, \sigma')$. We describe the protocol as follows:*

  - *Given an input $(M, \mathsf{ADM}, \mathsf{DID}, \sigma)$ from the signer, each redactor i selects a modification instruction MOD_i and runs a redact information algorithm RedInf with $(\mathsf{pk}, \mathsf{rk}[i], M, \mathsf{ADM}, \mathsf{DID}, \sigma, \mathsf{MOD}_i, \mathbb{L}_i^{t-1})$. $\mathbb{L}_i^{t-1}$ is the list which stores on DID sent from the signer. It is used for t-th input of the RedInf by redactor i and $L^0 = \emptyset$. In the processing in RedInf, if DID is previous input to RedInf then redactor i stop interacting with a combiner. Otherwise, output the redact information RI_i and the updated list $\mathbb{L}_i^t$. Each redactor i sends RI_i to the combiner.*
  - *The combiner runs a deterministic threshold redact algorithm ThrRed with $(\mathsf{pk}, M, \mathsf{ADM}, \mathsf{DID}, \sigma, \{\mathsf{RI}_i\}_{i=1}^n)$ as an input. In the algorithm ThrRed, MOD is*

derived from $\{RI_i\}_{i=1}^n$ *and it redacts a message* M *based on* MOD. ThrRed *outputs a redacted message* M', ADM, DID *and the updated signature* σ'. *Finally, the combiner outputs* $(M', \text{ADM}, \text{DID}, \sigma')$ *as an output of* Redact *protocol.*

Verify : *A verification algorithm is a deterministic algorithm. Given an input* $(\text{pk}, M, \text{ADM}, \text{DID}, \sigma)$, *return either 1 (Accept) or 0 (Reject).*

Correctness. We require the correctness that all honestly computed and redacted signatures are accepted.

Definition 3 (Correctness). *A* t-*out-of-*n *redactable signature scheme in the one-time redaction model* (t, n)-RSS Π *is correct,* $\forall \lambda \in \mathbb{N}$, $\forall k \in \mathbb{N}$, $1 \leq \forall i \leq n$, $\forall M_0^k$, $\forall \text{ADM}^k \preceq M_0^k$, $\forall(\text{pk}, \text{sk}, \{rk[i]\}_{i=1}^n) \leftarrow \text{KeyGen}(1^\lambda, t, n)$,

$\forall(M_0, \text{ADM}^k, \text{DID}^k, \sigma_0^k) \leftarrow \text{Sign}(\text{sk}, M_0^k, \text{ADM}^k)$, $\forall \text{MOD}_i^k \overset{\text{ADM}^k}{\preceq} M_0^k$,
$(RI_i^k, \mathbb{L}_i^k) \leftarrow \text{RedInf}(\text{pk}, rk[i], M_0^k, \text{ADM}^k, \text{DID}^k, \sigma_0^k, \text{MOD}_i^k, \mathbb{L}_i^{k-1})$,
$(M_1^k, \text{ADM}^k, \text{DID}, \sigma_1^k) \leftarrow \text{ThrRed}(\text{pk}, M_0^k, \text{ADM}^k, \text{DID}^k, \sigma_0^k, \{RI_i^k\}_{i=1}^n)$,
we require the following.
If $\text{DID}^k \notin \mathbb{L}_t^{k-1}$ *for all* $t \in \{0, 1\}$, $\text{Verify}(\text{pk}, M_t^k, \text{ADM}^k, \text{DID}^k, \sigma_t^k) = 1$.
If $\text{DID}^k \in \mathbb{L}_i^{k-1}$, $\text{Verify}(\text{pk}, M_0^k, \text{ADM}^k, \text{DID}^k, \sigma_0^k) = 1$.

3.2   Security of a t-out-of-n Redactable Signature Scheme

We give the security notion of unforgeability, privacy, and transparency for a redactable signature scheme in the one-time redaction model.

Unforgeability. Unforgeability requires that without a signer's secret key sk, it should be infeasible to compute a valid signature σ' on $(M', \text{ADM}, \text{DID})$ except to redact a signed message $(M, \text{ADM}, \text{DID}, \sigma)$ even if $t - 1$ redactors keys are corrupted.

Definition 4 (Unforgeability). *The unforgeability against redactors security of a* t-*out-of-*n *redactable signature scheme in the one-time redaction model* (t, n)-RSS Π *is defined by the following unforgeability game between a challenger* $\mathcal{C}$ *and a PPT adversary* $\mathcal{A}$.

1. $\mathcal{C}$ *generates key pairs* $(\text{pk}, \text{sk}, \{rk[i]\}_{i=1}^n)$ *using* $\text{KeyGen}(1^\lambda, t, n)$, *and gives* pk *to an adversary* $\mathcal{A}$.
2. $\mathcal{A}$ *is given access (throughout the entire game) to a sign oracle* $\mathcal{O}^{\text{Sign}}(\cdot, \cdot)$ *such that* $\mathcal{O}^{\text{Sign}}(M, \text{ADM})$, *returns* $(M, \text{ADM}, \text{DID}, \sigma) \leftarrow \text{Sign}(\text{sk}, M, \text{ADM})$.
3. $\mathcal{A}$ *is given access (throughout the entire game) to a redact oracle* $\mathcal{O}^{\text{Redact}}(\cdot, \cdot, \cdot, \cdot, \cdot)$. $\mathcal{O}^{\text{Redact}}$ *is defined as follows:*
    For an u-*th query* $(M, \text{ADM}, \text{DID}, \sigma, \text{MOD})$:
    1. $(RI_i, \mathbb{L}_i^u) \leftarrow \text{RedInf}(\text{pk}, rk[i], M, \text{ADM}, \text{DID}, \sigma, \text{MOD}, \mathbb{L}_i^{u-1})$ *for* $i = 1, ..., n$.
    2. $(M', \text{ADM}, \text{DID}, \sigma') \leftarrow \text{ThrRed}(\text{pk}, M, \text{ADM}, \text{DID}, \sigma, \{RI_i\}_{i=1}^n)$.
    3. *Return* $(M', \text{ADM}, \text{DID}, \sigma')$.

4. $\mathcal{A}$ is given up to $t-1$ times access (throughout the entire game) to a corrupt oracle $\mathcal{O}^{\mathsf{Corrupt}}(\cdot)$, where $\mathcal{O}^{\mathsf{Corrupt}}(i)$ outputs a $\mathsf{rk}[i]$ of a redactor i.
5. $\mathcal{A}$ outputs $(M^*, \mathsf{ADM}^*, \mathsf{DID}^*, \sigma^*)$.

A *t*-out-of-*n* redactable signature scheme in the one-time redaction model (t,n)-RSS Π satisfies the unforgeability security if for all PPT adversaries $\mathcal{A}$, the advantage $\mathsf{Adv}_{\Pi,\mathcal{A}}^{\mathsf{Uf\text{-}}(t,n)\text{-}\mathsf{RSS}} = \Pr[\mathsf{Verify}(\mathsf{pk}, M^*, \mathsf{ADM}^*\mathsf{DID}^*, \sigma^*) = 1 \wedge (M^*, \mathsf{ADM}^*, \mathsf{DID}^*) \notin (Q_{\mathsf{Sign}} \cup Q_{\mathsf{Redact}})]$ is negligible in λ.

Here, q_s is the total number of queries to $\mathcal{O}^{\mathsf{Sign}}$, (M_i, ADM_i) is an *i*-th input for $\mathcal{O}^{\mathsf{Sign}}$, $(M^i, \mathsf{ADM}^i, \mathsf{DID}^i, \sigma^i)$ is an *i*-th output of $\mathcal{O}^{\mathsf{Sign}}$ and $Q_{\mathsf{Sign}} := \bigcup_{i=1}^{q_s}\{(M^i, \mathsf{ADM}^i, \mathsf{DID}^i)\}$. Also, q_r is the total number of queries to $\mathcal{O}^{\mathsf{Redact}}$, $(M^i, \mathsf{ADM}^i, \mathsf{DID}^i, \sigma^i, \mathsf{MOD}^i)$ is an *i*-th input for $\mathcal{O}^{\mathsf{Redact}}$, $(M'^i, \mathsf{ADM}^i, \mathsf{DID}^i, \sigma'^i)$ is an *i*-th output of $\mathcal{O}^{\mathsf{Redact}}$ and $Q_{\mathsf{Redact}} := \bigcup_{i=1}^{q_r}\{(M'^i, \mathsf{ADM}^i, \mathsf{DID}^i)\}$.

Privacy. Privacy requires that except for a signer, n redactors, and a combiner, it is infeasible to derive information on redacted message parts when given a message-ADM-DID-signature pair.

Definition 5 (Privacy). *The privacy of a t-out-of-n redactable signature scheme in the one-time redaction model (t,n)-RSS Π is defined by the following weak privacy game between a challenger $\mathcal{C}$ and a PPT adversary $\mathcal{A}$.*

1. $\mathcal{C}$ generates key pairs $(\mathsf{pk}, \mathsf{sk}, \{\mathsf{rk}[i]\}_{i=1}^n)$ using $\mathsf{KeyGen}(1^\lambda, t, n)$, and gives pk to an adversary $\mathcal{A}$.
2. $\mathcal{A}$ is given access (throughout the entire game) to a sign oracle $\mathcal{O}^{\mathsf{Sign}}(\cdot, \cdot)$ such that $\mathcal{O}^{\mathsf{Sign}}(M, \mathsf{ADM})$, returns $(M, \mathsf{ADM}, \mathsf{DID}, \sigma) \leftarrow \mathsf{Sign}(\mathsf{sk}, M, \mathsf{ADM})$.
3. $\mathcal{A}$ is given access (throughout the entire game) to a redact oracle $\mathcal{O}^{\mathsf{Redact}}(\cdot, \cdot, \cdot, \cdot, \cdot)$. $\mathcal{O}^{\mathsf{Redact}}$ is defined as follows:
   For an u-th query $(M, \mathsf{ADM}, \mathsf{DID}, \sigma, \mathsf{MOD})$:
   Let w be the number of queries to $\mathcal{O}^{\mathsf{LoRredact}}$ when $\mathcal{A}$ makes an u-th query to $\mathcal{O}^{\mathsf{Redact}}$.
   1. $(\mathsf{RI}_i, \mathbb{L}_i^{u+2w}) \leftarrow \mathsf{RedInf}(\mathsf{pk}, \mathsf{rk}[i], M, \mathsf{ADM}, \mathsf{DID}, \sigma, \mathsf{MOD}, \mathbb{L}_i^{u+2w-1})$ for $i = 1, ..., n$.
   2. $(M', \mathsf{ADM}, \mathsf{DID}, \sigma') \leftarrow \mathsf{ThrRed}(\mathsf{pk}, M, \mathsf{ADM}, \mathsf{DID}, \sigma, \{\mathsf{RI}_i\}_{i=1}^n)$.
   3. Return $(M', \mathsf{ADM}, \mathsf{DID}, \sigma')$.
4. $\mathcal{A}$ is given access (throughout the entire game) to a left-or-right redact oracle $\mathcal{O}^{\mathsf{LoRredact}}(\cdot, \cdot, \cdot, \cdot, \cdot, \cdot)$. $\mathcal{O}^{\mathsf{LoRredact}}$ is defined as follows:
   For an w-th query $(M^0, \mathsf{ADM}^0, \mathsf{MOD}^0, M^1, \mathsf{ADM}^1, \mathsf{MOD}^1)$:
   Let u be the number of queries to $\mathcal{O}^{\mathsf{Redact}}$ when $\mathcal{A}$ makes an w-th query to $\mathcal{O}^{\mathsf{LoRredact}}$.
   1. Compute $(M^c, \mathsf{ADM}^c, \mathsf{DID}^c, \sigma^c) \leftarrow \mathsf{Sign}(\mathsf{sk}, M^c, \mathsf{ADM}^c)$ for $c \in \{0, 1\}$.
   2. For $i = 1, \cdots n$, compute

   $$(\mathsf{RI}_i^0, \mathbb{L}^{u+2w-1}) \leftarrow \mathsf{RedInf}(\mathsf{pk}, \mathsf{rk}[i], M^0, \mathsf{ADM}^0, \mathsf{DID}^0, \sigma^0, \mathsf{MOD}^0, \mathbb{L}_i^{u+2w-2})$$
   $$(\mathsf{RI}_i^1, \mathbb{L}^{u+2w}) \leftarrow \mathsf{RedInf}(\mathsf{pk}, \mathsf{rk}[i], M^1, \mathsf{ADM}^1, \mathsf{DID}^1, \sigma^1, \mathsf{MOD}^1, \mathbb{L}_i^{u+2w-1}).$$

3. For $i = 1, ..., n$, compute

$$(M^{c\prime}, \mathsf{ADM}^c, \mathsf{DID}^c, \sigma^{c\prime}) \leftarrow \mathsf{ThrRed}(\mathsf{pk}, M^c, \mathsf{ADM}^c, \mathsf{DID}^c, \sigma^c, \{\mathsf{RI}_i^c\}_{i=1}^n).$$

4. If $M^{0\prime} \neq M^{1\prime} \vee \mathsf{ADM}_0 \neq \mathsf{ADM}_1$, return $\perp$.
5. Return $(M^{b\prime}, \mathsf{ADM}^b, \mathsf{DID}^b, \sigma^{b\prime})$. ($b$ is chosen by $\mathcal{C}$ in step 1.)
5. $\mathcal{A}$ outputs b^*.

A t-out-of-n redactable signature scheme in the one-time redaction model (t, n)-RSS Π satisfies the privacy security if for all PPT adversaries $\mathcal{A}$, the following advantage $\mathsf{Adv}_{\Pi, \mathcal{A}}^{\mathsf{Priv}\text{-}(t,n)\text{-}\mathsf{RSS}} = |\Pr[b = b^*] - 1/2|$ is negligible in λ.

Transparency. Transparency requires that except for a signer, n redactors, and a combiner, it is infeasible to distinguish whether a signature directly comes from the signer or has been redacted by redactors.

Definition 6 (Transparency). *The privacy of a t-out-of-n redactable signature scheme in the one-time redaction model (t, n)-RSS Π is defined by the following weak privacy game between a challenger $\mathcal{C}$ and a PPT adversary $\mathcal{A}$.*

1. $\mathcal{C}$ chooses a bit $b \xleftarrow{\$} \{0, 1\}$, generates key pairs $(\mathsf{pk}, \mathsf{sk}, \{\mathsf{rk}[i]\}_{i=1}^n)$ using $\mathsf{KeyGen}(1^\lambda, t, n)$, and gives pk to an adversary $\mathcal{A}$.
2. $\mathcal{A}$ is given access (throughout the entire game) to a sign oracle $\mathcal{O}^{\mathsf{Sign}}(\cdot, \cdot)$ such that $\mathcal{O}^{\mathsf{Sign}}(M, \mathsf{ADM})$, returns $(M, \mathsf{ADM}, \mathsf{DID}, \sigma) \leftarrow \mathsf{Sign}(\mathsf{sk}, M, \mathsf{ADM})$.
3. $\mathcal{A}$ is given access (throughout the entire game) to a redact oracle $\mathcal{O}^{\mathsf{Redact}}(\cdot, \cdot, \cdot, \cdot, \cdot)$. $\mathcal{O}^{\mathsf{Redact}}$ is defined as follows:
   For an u-th query $(M, \mathsf{ADM}, \mathsf{DID}, \sigma, \mathsf{MOD})$:
   Let w be the number of queries to $\mathcal{O}^{\mathsf{Sign/Redact}}$ when $\mathcal{A}$ makes an u-th query to $\mathcal{O}^{\mathsf{Redact}}$.
   1. $(\mathsf{RI}_i, \mathbb{L}_i^{u+2w}) \leftarrow \mathsf{RedInf}(\mathsf{pk}, \mathsf{rk}[i], M, \mathsf{ADM}, \mathsf{DID}, \sigma, \mathsf{MOD}, \mathbb{L}_i^{u+2w-1})$ for $i = 1, ..., n$.
   2. $(M', \mathsf{ADM}, \mathsf{DID}, \sigma') \leftarrow \mathsf{ThrRed}(\mathsf{pk}, M, \mathsf{ADM}, \mathsf{DID}, \sigma, \{\mathsf{RI}_i\}_{i=1}^n)$.
   3. Return $(M', \mathsf{ADM}, \mathsf{DID}, \sigma')$.
4. $\mathcal{A}$ is given access (throughout the entire game) to a sign or redact oracle $\mathcal{O}^{\mathsf{Sign/Redact}}(\cdot, \cdot, \cdot)$. $\mathcal{O}^{\mathsf{Sign/Redact}}$ is defined as follows:
   For an w-th query $(M, \mathsf{ADM}, \mathsf{MOD})$:
   Let u be the number of queries to $\mathcal{O}^{\mathsf{Redact}}$ when $\mathcal{A}$ makes an w-th query to $\mathcal{O}^{\mathsf{Sign/Redact}}$.
   1. Compute $(M, \mathsf{ADM}, \mathsf{DID}_0, \sigma) \leftarrow \mathsf{Sign}(\mathsf{sk}, M, \mathsf{ADM})$.
   2. For $i = 1, \ldots n$, compute

   $$(\mathsf{RI}_i, \mathbb{L}_i^{u+2w-1}) \leftarrow \mathsf{RedInf}(\mathsf{pk}, \mathsf{rk}[i], M, \mathsf{ADM}, \mathsf{DID}^0, \sigma, \mathsf{MOD}, \mathbb{L}_i^{u+2w-2}).$$

   3. Compute $(M', \mathsf{ADM}, \mathsf{DID}^0, \sigma^0) \leftarrow \mathsf{ThrRed}(\mathsf{pk}, M, \mathsf{ADM}, \mathsf{DID}^0, \sigma, \{\mathsf{RI}_i\}_{i=1}^n)$.
   4. Compute $(M', \mathsf{ADM}, \mathsf{DID}^1, \sigma^1) \leftarrow \mathsf{Sign}(\mathsf{sk}, M', \mathsf{ADM})$.
   5. For $i = 1, \ldots n$, $\mathbb{L}_i^{u+2w} \leftarrow \mathbb{L}_i^{u+2w-1} \cup \{\mathsf{DID}^1\}$.

  6. Return $(M', \mathsf{ADM}, \mathsf{DID}^b, \sigma^b)$.

5. $\mathcal{A}$ *outputs* b^*.

A *t*-out-of-*n* *redactable signature scheme in the one-time redaction model* (t, n)-RSS Π *satisfies the transparency security if for all PPT adversaries* $\mathcal{A}$, *the following advantage* $\mathsf{Adv}_{\Pi\mathcal{A}}^{\mathsf{Tran}\text{-}(t,n)\text{-}\mathsf{RSS}} = |\Pr[b = b^*] - 1/2|$ *is negligible in* λ.

Theorem 1. *If t-out-of-n redactable signature scheme in the one-time redaction model* (t, n)-RSS Π *satisfies transparency, then it satisfies privacy.*

We prove **Theorem 1** in a similar way of [3,4]. We will describe the proof of **Theorem 1** in the full version of this paper.

4 Our *t*-out-of-*n* Redactable Signature Scheme

In this section, we give a concrete construction of *t*-out-of-*n* redactable signature scheme in one-time redaction model (t, n)-RSS Π_1. Let ℓ, d be polynomials in λ, $(q, \mathbb{G}_1, \mathbb{G}_2, \mathbb{G}_T, e, g_1, g_2) \leftarrow \mathcal{G}(1^\lambda)$, $H : \{0,1\}^* \rightarrow \mathbb{G}_1$ a hash function, and M a message having a set data structure (i.e., $M = \{m_1, ..., m_\ell\}$) and $\#M \le \ell$.

$\mathsf{KeyGen}(1^\lambda, t, n)$: Given a security parameter 1^λ, a threshold value t, and the number of redactors n, the PPT algorithm KeyGen works as follows:

1. Choose $\tilde{x} \xleftarrow{\$} \mathbb{Z}_q$, compute $\tilde{y} \leftarrow g_2^{\tilde{x}}$, and set $(\mathsf{pk}_{\mathsf{Fix}}, \mathsf{sk}_{\mathsf{Fix}}) \leftarrow (\tilde{y}, \tilde{x})$.
2. Choose $a_0, a_1, \cdots, a_{t-1} \xleftarrow{\$} \mathbb{Z}_q$ independently at random and gets the polynomial $f(X) = \sum_{i=0}^{t-1} a_i X^i$.
3. For $i = 0$ to n, compute $x_i \leftarrow f(i)$, $y_i \leftarrow g_2^{f(i)}$.
4. Set $(\mathsf{pk}_{\mathsf{Agg}}, \mathsf{sk}_{\mathsf{Agg}}) \leftarrow (y_0, x_0)$, $\mathsf{rk}[i] \leftarrow (i, x_i)$ for all $i \in [n]$.
5. Set $(\mathsf{pk}, \mathsf{sk}) \leftarrow ((\mathsf{pk}_{\mathsf{Fix}}, \mathsf{pk}_{\mathsf{Agg}}, t, n), (\mathsf{sk}_{\mathsf{Fix}}, \mathsf{sk}_{\mathsf{Agg}}))$.
6. Return $(\mathsf{pk}, \mathsf{sk}, \{\mathsf{rk}[i]\}_{i=1}^n)$.

$\mathsf{Sign}(\mathsf{sk}, M, \mathsf{ADM})$: Given a signer's secret key sk, a message M, and ADM (In this scheme, ADM represent a set containing all blocks which must not be redacted.), the PPT algorithm Sign works as follows:

1. Parse sk as $(\mathsf{sk}_{\mathsf{Fix}}, \mathsf{sk}_{\mathsf{Agg}})$.
2. If $\mathsf{ADM} \not\preceq M$, (i.e., $\mathsf{ADM} \cap M \ne \mathsf{ADM}$.) then abort.
3. Choose document ID $\mathsf{DID} \xleftarrow{\$} \{0,1\}^d$.
4. Compute $h_{\mathsf{ADM}} \leftarrow H(\mathsf{DID}\|\mathrm{ord}(\mathsf{ADM}))$.
   $\mathrm{ord}(\mathsf{ADM})$ denotes a lexicographic ordering to the elements in ADM.
5. For $m_j \in M$, compute $h_{m_j} \leftarrow H(\mathsf{DID}\|m_j)$.
6. Compute $\sigma_{\mathsf{Fix}} \leftarrow h_{\mathsf{ADM}}^{\mathsf{sk}_{\mathsf{Fix}}}$.
7. Compute $\sigma_{\mathsf{ADM}} \leftarrow h_{\mathsf{ADM}}^{\mathsf{sk}_{\mathsf{Agg}}}$, $\sigma_{m_j} \leftarrow h_{m_j}^{\mathsf{sk}_{\mathsf{Agg}}}$ for $m_j \in M$.
8. Compute $\Sigma_{\mathsf{agg}} \leftarrow \sigma_{\mathsf{ADM}} \cdot \prod_{m_j \in M} \sigma_{m_j}$.
9. Set $\sigma \leftarrow (\sigma_{\mathsf{Fix}}, \Sigma_{\mathsf{agg}})$.
10. Return $(M, \mathsf{ADM}, \mathsf{DID}, \sigma)$.

Redact : Redact is an interactive protocol between the combiner and n redactor. The combiner interacts with the n redactors and finally outputs the redacted signature. Given a tuple $(M, \mathsf{ADM}, \mathsf{DID}, \sigma)$ to n redactors and the combiner from the signer, the interactive protocol works as follows:

1. Each redactor i selects a modifiction instruction MOD_i. Let $\mathbb{L}_i$ be the list which stores DIDs, $\mathbb{L}_i^0 = \emptyset$, and $\mathbb{L}_i^{t-1}$ the list which used in the input of t-th running of the PPT algorithm RedInf by the redactor i.
   The redactor i runs $\mathsf{RedInf}(\mathsf{pk}, \mathsf{rk}[i], M, \mathsf{ADM}, \mathsf{DID}, \sigma, \mathsf{MOD}_i, \mathbb{L}_i^{t-1})$.
   $\mathsf{RedInf}(\mathsf{pk}, \mathsf{rk}[i], M, \mathsf{ADM}, \mathsf{DID}, \sigma, \mathsf{MOD}_i, \mathbb{L}_i^{t-1})$:
   1. Parse pk as $(\mathsf{pk}_{\mathsf{Fix}}, \mathsf{pk}_{\mathsf{Agg}}, t, n)$ and σ as $(\sigma_{\mathsf{Fix}}, \Sigma_{\mathsf{agg}})$.
   2. If $\mathsf{DID} \in \mathbb{L}_i^{t-1}$ then abort.
   3. Update $\mathbb{L}_i^t \leftarrow \mathbb{L}_i^{t-1} \cup \{\mathsf{DID}\}$.
   4. Check $\mathsf{MOD}_i \overset{\mathsf{ADM}}{\preceq} M$. (i.e., $\mathsf{MOD}_i \cap \mathsf{ADM} = \emptyset \wedge \mathsf{MOD}_i \subset M$.)
   5. Compute $h_{\mathsf{ADM}} \leftarrow H(\mathsf{DID}\|\mathrm{ord}(\mathsf{ADM}))$.
      $\mathrm{ord}(\mathsf{ADM})$ denotes a lexicographic ordering to the elements in ADM.
   6. For $m_j \in M$, compute $h_{m_j} \leftarrow H(\mathsf{DID}\|m_j)$.
   7. If $e(\sigma_{\mathsf{Fix}}, g_2) \neq e(h_{\mathsf{ADM}}, \mathsf{pk}_{\mathsf{Fix}})$ then abort.
   8. If $e(\Sigma_{\mathsf{agg}}, g_2) \neq e(h_{\mathsf{ADM}}, \mathsf{pk}_{\mathsf{Agg}}) \cdot \prod_{m_j \in M} e(h_{m_j}, \mathsf{pk}_{\mathsf{Agg}})$ then abort.
   9. For $m_j \in \mathsf{MOD}_i$, compute $\mathsf{RI}_{i,m_j} \leftarrow h_{m_j}^{\mathsf{rk}[i]}$.
   10. For $m_j \notin \mathsf{MOD}_i$, set $\mathsf{RI}_{i,m_j} \leftarrow \emptyset$.
   11. Set a redaction information RI_i of redactor i as $\mathsf{RI}_i \leftarrow \{\mathsf{RI}_{i,m_j}\}_{m_j \in M}$
   12. Output $(\mathsf{RI}_i, \mathbb{L}_i^t)$.
   For one DID, redactor i runs RedInf only once. This can be done by introducing a table $\mathbb{L}_i$.
2. Each redactor i sends (i, RI_i) to the combiner.
3. The combiner collects all n redaction information $\{\mathsf{RI}_i\}_{i=1}^n$.
4. The combiner runs the PPT algorithm $\mathsf{ThrRed}(\mathsf{pk}, M, \mathsf{ADM}, \mathsf{DID}, \sigma, \{\mathsf{RI}_i\}_{i=1}^n)$.
   $\mathsf{ThrRed}(\mathsf{pk}, M, \mathsf{ADM}, \mathsf{DID}, \sigma, \{\mathsf{RI}_i\}_{i=1}^n)$:
   1. Parse pk as $(\mathsf{pk}_{\mathsf{Fix}}, \mathsf{pk}_{\mathsf{Agg}}, t, n)$ and σ as $(\sigma_{\mathsf{Fix}}, \Sigma_{\mathsf{agg}})$.
   2. Parse RI_i as $\{\mathsf{RI}_{i,m_j}\}_{m_j \in M}$.
   3. For $m_j \in M$, define $\mathsf{RI}_{m_j} = \{\mathsf{RI}_{i,m_j}\}_{i=1}^n$.
   4. Define $\mathsf{MOD} = \{m_j | m_j \in M \wedge \#\mathsf{RI}_{m_j} \geq t\}$
   5. For $m_j \in \mathsf{MOD}$, define $\mathsf{InRI}_{m_j} \leftarrow \{i \in \mathbb{N} | \{\mathsf{RI}_{i,m_j}\} \neq \emptyset\}$.
   6. For $m_j \in \mathsf{MOD}$, choose subset $J_{m_j} \subset \mathsf{InRI}_{m_j}$ such that $\#J_{m_j} = t$.
   7. For $m_j \in \mathsf{MOD}$, compute $\sigma_{m_j} \leftarrow \prod_{i \in J_{m_j}} (\mathsf{RI}_{i,m_j})^{\gamma_{i,J_{m_j}}}$,
      where $\gamma_{i,J_{m_j}} = \prod_{j \in J_{m_j}, j \neq i} j(j-i)^{-1}$.
   8. Compute $\sigma_{\mathsf{MOD}} \leftarrow \prod_{m_j \in \mathsf{MOD}} \sigma_{m_j}$, $\Sigma'_{\mathsf{agg}} \leftarrow \Sigma_{\mathsf{agg}}/\sigma_{\mathsf{MOD}}$.
   9. Set $M' \leftarrow M \backslash \{\mathsf{MOD}\}$, $\sigma' \leftarrow (\sigma_{\mathsf{Fix}}, \Sigma'_{\mathsf{agg}})$.
   10. Return $(M', \mathsf{ADM}, \mathsf{DID}, \sigma')$.
5. The combiner outputs $(M', \mathsf{ADM}, \mathsf{DID}, \sigma')$.

Verify$(\mathsf{pk}, M, \mathsf{ADM}, \mathsf{DID}, \sigma)$: Given a tuple $(\mathsf{pk}, M, \mathsf{ADM}, \mathsf{DID}, \sigma)$, the PPT algorithm Verify works as follows:

1. Parse pk as $(\mathsf{pk}_{\mathsf{Fix}}, \mathsf{pk}_{\mathsf{Agg}}, t, n)$ and σ as $(\sigma_{\mathsf{Fix}}, \Sigma_{\mathsf{agg}})$.

2. If $\mathsf{ADM} \cap M \neq \mathsf{ADM}$, return 0.
3. Compute $h_{\mathsf{ADM}} \leftarrow H(\mathsf{DID} \| \mathrm{ord}(\mathsf{ADM}))$.
   $\mathrm{ord}(\mathsf{ADM})$ denotes a lexicographic ordering to the elements in ADM.
4. For $m_j \in M$, compute $h_{m_j} \leftarrow H(\mathsf{DID} \| m_j)$.
5. If $e(\sigma_{\mathsf{Fix}}, g_2) \neq e(h_{\mathsf{ADM}}, \mathsf{pk}_{\mathsf{Fix}})$, return 0
6. If $e(\Sigma_{\mathsf{agg}}, g_2) = e\left(h_{\mathsf{ADM}}, \mathsf{pk}_{\mathsf{Agg}}\right) \cdot \prod_{m_j \in M} e(h_{m_j}, \mathsf{pk}_{\mathsf{Agg}})$, return 1. Otherwise output 0.

Correctness. If $(M, \mathsf{ADM}, \mathsf{DID}, \sigma)$ is honestly generated by the Sign and has not been processed by the Redact protocol, $\mathsf{Verify}(M, \mathsf{ADM}, \mathsf{DID}, \sigma) = 1$ always holds. If $(M, \mathsf{ADM}, \mathsf{DID}, \sigma)$ is honestly generated the Sign and $(M', \mathsf{ADM}, \mathsf{DID}, \sigma')$ is honestly redacted from $(M, \mathsf{ADM}, \mathsf{DID}, \sigma)$ by Redact protocol, $(M', \mathsf{ADM}, \mathsf{DID}, \sigma')$ passes the verification in the Verify. Therefore, our construction of t-out-of-n redactable signature scheme in the one-time redaction model satisfies correctness.

4.1   Security of Our t-out-of-n Redactable Signature Scheme

Theorem 2. *In the random oracle model, if the computational co-Diffie-Hellman problem assumption holds, then our proposed t-out-of-n redactable signature scheme in the one-time redaction model (t, n)-RSS Π_1 satisfies the unforgeability property.*

Here, to explain the outline of the proof, we introduce new notations. Let q_s be the total number of queries from an adversary to $\mathcal{O}^{\mathsf{Sign}}$, (M_i, ADM_i) an i-th input for $\mathcal{O}^{\mathsf{Sign}}$, $(M^i, \mathsf{ADM}^i, \mathsf{DID}^i, \sigma^i)$ the i-th output of $\mathcal{O}^{\mathsf{Sign}}$. We denote

$$Q_{\mathsf{Sign}} := \bigcup_{i=1}^{q_s} \{(M^i, \mathsf{ADM}^i, \mathsf{DID}^i)\}, \quad Q_{\mathsf{Sign}}^{\mathsf{AD}} := \bigcup_{i=1}^{q_s} \{(\mathsf{ADM}^i, \mathsf{DID}^i)\}.$$

Also, let q_r be the total number of queries from an adversary to $\mathcal{O}^{\mathsf{Redact}}$, $(M^i, \mathsf{ADM}^i, \mathsf{DID}^i, \sigma^i, \mathsf{MOD}^i)$ an i-th input for $\mathcal{O}^{\mathsf{Redact}}$, $(M'^i, \mathsf{ADM}^i, \mathsf{DID}^i, \sigma'^i)$ the i-th output of $\mathcal{O}^{\mathsf{Redact}}$. We denote

$$Q_{\mathsf{Redact}} := \bigcup_{i=1}^{q_r} \{(M'^i, \mathsf{ADM}^i, \mathsf{DID}^i)\}, \quad Q_{\mathsf{Redact}}^{\mathsf{AD}} := \bigcup_{i=1}^{q_r} \{(\mathsf{ADM}^i, \mathsf{DID}^i)\}.$$

We assume the following three types of PPT adversaries that breaks the unforgeability security in our proposed scheme.

- An adversary $\mathcal{A}_1$ that outputs a forgery $(M^*, \mathsf{DID}^*, \mathsf{ADM}^*, \sigma^*)$ such that $(\mathsf{ADM}^*, \mathsf{DID}^*) \notin (Q_{\mathsf{Sign}}^{\mathsf{AD}} \cup Q_{\mathsf{Redact}}^{\mathsf{AD}})$.
- An adversary $\mathcal{A}_2$ that outputs a forgery $(M^*, \mathsf{DID}^*, \mathsf{ADM}^*, \sigma^*)$ which satisfies $(\mathsf{ADM}^*, \mathsf{DID}^*) \in (Q_{\mathsf{Sign}}^{\mathsf{AD}} \cup Q_{\mathsf{Redact}}^{\mathsf{AD}})$. Moreover, there is $\tilde{M}$ such that $(\tilde{M}, \mathsf{ADM}^*, \mathsf{DID}^*) \in (Q_{\mathsf{Sign}} \cup Q_{\mathsf{Redact}})$.

– An adversary $\mathcal{A}_3$ that outputs a forgery $(M^*, \mathsf{DID}^*, \mathsf{ADM}^*, \sigma^*)$ which satisfies $(\mathsf{ADM}^*, \mathsf{DID}^*) \in (Q_{\mathsf{Sign}}^{\mathsf{AD}} \cup Q_{\mathsf{Redact}}^{\mathsf{AD}})$. Moreover, there are no $\tilde{M}$ such that $(\tilde{M}, \mathsf{ADM}^*, \mathsf{DID}^*) \in (Q_{\mathsf{Sign}} \cup Q_{\mathsf{Redact}})$ and $\tilde{M} \not\subseteq M$.

To prove the theorem, for each $\mathcal{A}_i$, we consider a sequential of games from the original unforgeability game to game which is directly related to solving a co-CDH problem. Then, We construct $\mathcal{B}_i$ which breaking the co-CDH assumption using $\mathcal{A}_i$. $\mathcal{B}_1$ breaks the co-CDH assumption using the forgery σ_{Fix}^*. In the case of $\mathcal{B}_2$ and $\mathcal{B}_3$, they use the forgery Σ_{agg}^* to break the co-CDH assumption. One difference between $\mathcal{B}_2$ and $\mathcal{B}_3$ is how to program the hash value.

Case 1. We consider an adversary $\mathcal{A}_1$ that can generate a valid forgery with ϵ_{uf1} against our proposal redactable signature scheme. Let $\mathbf{Game}_{1-0}$ be the original unforgeability game in a redactable signature scheme and $\mathbf{Game}_{1-5}$ be directly related to solving the computational co-Diffie-Hellman problem. Define $\mathsf{Adv}_{\mathcal{A}_1}[\mathbf{Game}_{1-X}]$ as the advantage of an adversary $\mathcal{A}_1$ in $\mathbf{Game}_{1-X}$.

– $\mathbf{Game}_{1-0}$: Original unforgeability game in a redactable signature scheme.

$$\mathsf{Adv}_{\mathcal{A}_1}[\mathbf{Game}_{1-0}] = \epsilon_{\mathsf{uf1}}$$

– $\mathbf{Game}_{1-1}$: We change a key generation algorithm KeyGen in Step 1.
   Choose $\tilde{x} \xleftarrow{\$} \mathbb{Z}_q$, $\tilde{r} \xleftarrow{\$} \mathbb{Z}_q$ and compute $u \leftarrow g^{\tilde{x}}$, $\tilde{y} \leftarrow g_2^{\tilde{x}+\tilde{r}}$.
   Set $(\mathsf{pk}_{\mathsf{Fix}}, \mathsf{sk}_{\mathsf{Fix}}) \leftarrow (\tilde{y}, \tilde{x}+\tilde{r})$.

– $\mathbf{Game}_{1-2}$: We change a setting of the random oracle $\mathcal{O}^H$. Fix $h \xleftarrow{\$} \mathbb{G}_2$ and let $\mathbb{T}$ be a table that maintains a list of tuples $\langle v, w, b, c \rangle$ as explain below. We refer to this list for the query to $\mathcal{O}^h$. The initial state of $\mathbb{T}$ is empty. For queries $v^{(i)}$ to $\mathcal{O}^H$:
   - If $\langle v^{(i)}, w^{(i)}, \cdot, \cdot \rangle$ (Here, '$\cdot$' represents an arbitrary value) already appears in $\mathbb{T}$, then return $w^{(i)}$.
   - Choose $s^{(i)} \xleftarrow{\$} \mathbb{Z}_q$.
   - Flip a biased coin $c^{(i)} \in \{0,1\}$ such that $\Pr[c^{(i)} = 0] = 1 - 1/(q_s+1)$ and $\Pr[c^{(i)} = 1] = 1/(q_s+1)$.
   - If $c^{(i)} = 0$, compute $w^{(i)} = \phi(g_2)^{b^{(i)}}$.
   - If $c^{(i)} = 1$, compute $w^{(i)} = h \cdot \phi(g_2)^{b^{(i)}}$.
   - Insert $\langle v^{(i)}, w^{(i)}, s^{(i)}, c^{(i)} \rangle$ in $\mathbb{T}$ and return $w^{(i)}$.

– $\mathbf{Game}_{1-3}$: We modify the signing algorithm Sign in Step 4 as follows:
   - Set $v^{(0)} \leftarrow (\mathsf{DID}\|\mathsf{ord}(\mathsf{ADM}))$.
   - Query $v^{(0)}$ to $\mathcal{O}^H$. We assume $\langle v^{(0)}, w^{(0)}, b^{(0)}, c^{(0)} \rangle$ to be the tuple in $\mathbb{T}$ for $v^{(0)}$.
   - If $c^{(0)} = 1$, return $\perp$ and abort.

– $\mathbf{Game}_{1-4}$: We modify the signing algorithm Sign in Step 6 as follows:
   - Compute $\sigma_{\mathsf{Fix}} \leftarrow \phi(u)^{b^{(0)}} \cdot \phi(g_2)^{\tilde{r}b^{(0)}}$.
   (A signature σ_{Fix} can be generated without a knowledge of $\mathsf{sk}_{\mathsf{Fix}}$.)

– $\mathbf{Game}_{1-5}$: We receive a valid forgery $(M^*, \mathsf{ADM}^*\mathsf{DID}^*, \sigma^*)$ from the adversary $\mathcal{A}_1$, we operate as follows:

- Set $v^{(0)} \leftarrow (\mathsf{DID}^* \| \mathrm{ord}(\mathsf{ADM}^*))$.
- Query $v^{(0)}$ to $\mathcal{O}^H$. We assume $\langle v^{(0)}, w^{(0)}, s^{(0)}, c^{(0)} \rangle$ to be the tuple in $\mathbb{T}$ for each $v^{(0)}$.
- If $c^{(0)} = 0$, then abort.

Lemma 1. *The following equation holds.*

$$\mathsf{Adv}_{\mathcal{A}_1}[\mathbf{Game}_{1-1}] = \mathsf{Adv}_{\mathcal{A}_1}[\mathbf{Game}_{1-0}].$$

Since the distribution of $(\mathsf{pk}_{\mathsf{Fix}}, \mathsf{sk}_{\mathsf{Fix}})$ in $\mathbf{Game}_{1-0}$ and $\mathbf{Game}_{1-1}$ are same.

Lemma 2. *If H is the random oracle model, the following eqauation holds.*

$$\mathsf{Adv}_{\mathcal{A}_1}[\mathbf{Game}_{1-2}] = \mathsf{Adv}_{\mathcal{A}_1}[\mathbf{Game}_{1-1}]$$

Since the distribution of outputs of $\mathcal{O}^H$ in $\mathbf{Game}_{1-1}$ and $\mathbf{Game}_{1-2}$ are identical.

Lemma 3. *The following inequality holds.*

$$\mathsf{Adv}_{\mathcal{A}_1}[\mathbf{Game}_{1-3}] \geq (1 - 1/(q_s + 1))^{q_s} \times \mathsf{Adv}_{\mathcal{A}_1}[\mathbf{Game}_{1-2}].$$

Since the probability that each signing query does not abort at least $1 - 1/(q_s + 1)$.

Lemma 4. *The following equation holds.*

$$\mathsf{Adv}_{\mathcal{A}_1}[\mathbf{Game}_{1-4}] = \mathsf{Adv}_{\mathcal{A}_1}[\mathbf{Game}_{1-3}].$$

Since outputs of Sign in $\mathbf{Game}_{1-3}$ and $\mathbf{Game}_{1-4}$ are same.

Lemma 5. *The following inequality holds.*

$$\mathsf{Adv}_{\mathcal{A}_1}[\mathbf{Game}_{1-5}] \geq (1/(q_s + 1)) \times \mathsf{Adv}_{\mathcal{A}_1}[\mathbf{Game}_{1-4}].$$

Since the probability that the forged signature satisfies $c^{(0)} = 1$ at least $1/(q_s+1)$.

To summarize **Lemma 1** to 5, the following holds.
(In the following equation, e represents the Napier's constant.)

$$\mathsf{Adv}_{\mathcal{A}_1}[\mathbf{Game}_{1-5}] \geq (1 - 1/(q_s + 1))^{q_s} \times (1/(q_s + 1)) \times \mathsf{Adv}_{\mathcal{A}_1}[\mathbf{Game}_{1-0}]$$
$$\geq (1/e) \times (1/(q_s + 1)) \times \mathsf{Adv}_{\mathcal{A}_1}[\mathbf{Game}_{1-0}]$$

Now we construct the algorithm $\mathcal{B}_1$ which breaking the computational co-Diffie-Hellman assumption using the algorithm $\mathcal{A}_1$. The operation of $\mathcal{B}_1$ for the input co-Diffie-Hellman problem instance (g_2, g_2^α, h^*) is changed to h to h^* and u to g_2^α in $\mathbf{Game}_{1-5}$. Suppose $\mathcal{B}_1$ does not abort receiving a forgery $(M^*, \mathsf{ADM}^*, \mathsf{DID}^*, \sigma^*)$ from $\mathcal{A}_1$.

$\mathcal{B}_1$ parses σ^* as $(\sigma_{\mathsf{Fix}}^*, \Sigma_{\mathsf{agg}}^*)$, sets $v^{(0)} \leftarrow (\mathsf{DID}^* \| \mathrm{ord}(\mathsf{ADM}^*))$ and computes $w^{(0)} \leftarrow h^* \cdot \phi(g_2)^{b^{(0)}}$. Since $(M^*, \mathsf{ADM}^*, \mathsf{DID}^*, \sigma^*)$ is valid and $\mathsf{pk}_{\mathsf{Fix}} = g_2^{\alpha + \tilde{r}}$, $e(\sigma_{\mathsf{Fix}}^*, g_2) = e((w^{(0)})^{\alpha + \tilde{r}}, g_2)$ holds. It implies that $\sigma_{\mathsf{Fix}}^* = (w^{(0)})^{\alpha + \tilde{r}} = (h^* \cdot \phi(g_2)^{b^{(0)}})^{\alpha + \tilde{r}}$. Therefore, $\mathcal{B}_1$ computes $(h^*)^\alpha = \sigma_{\mathsf{Fix}}^* \cdot (\phi(u)^{b^{(0)}} \cdot (h^*)^{\tilde{r}} \cdot \phi(g_2)^{\tilde{r} b^{(0)}})^{-1}$

and outputs the solution $(h^*)^\alpha$ of the computational co-Diffie-Hellman problem instance (g_2, g_2^α, h^*).

Let $\epsilon_{\text{co-cdh}}$ is the probability that $\mathcal{B}_1$ break the computational co-Diffie-Hellman assumption. We can bound the probability $\epsilon_{\text{co-cdh1}} \geq \text{Adv}_{\mathcal{A}_1}[\textbf{Game } 1 - 5]$ and $\epsilon_{\text{co-cdh1}} \geq (1/e) \times (1/q_s + 1) \times \epsilon_{\text{uf1}}$ holds. (e represents the Napier's constant.) Hence, if ϵ_{uf1} is non-negligiable in λ, $\mathcal{B}_1$ breaks the computational co-Diffie-Hellman assumption with non-negligiable in $\epsilon_{\text{co-cdh1}}$.

Case 2. We consider an adversary $\mathcal{A}_2$ that can generate a valid forgery with ϵ_{uf2} against our proposal redactable signature scheme. Let $\textbf{Game}_{2-0}$ be the original unforgeability game in a redactable signature scheme and $\textbf{Game}_{2-6}$ be directly related to solve the computational co-Diffie-Hellman problem. Define $\text{Adv}_{\mathcal{A}_2}[\textbf{Game}_{2-X}]$ as the advantage of an adversary $\mathcal{A}_2$ in $\textbf{Game}_{2-X}$.

– $\textbf{Game}_{2-0}$: Original unforgeability game in a redactable signature scheme.

$$\text{Adv}_{\mathcal{A}_2}[\textbf{Game}_{2-0}] = \epsilon_{\text{uf2}}$$

– $\textbf{Game}_{2-1}$: We change a setting of $\mathcal{O}^{\text{Redact}}$.

We introduce a table $\mathbb{L}^t$ that store DIDs and $\mathbb{L}^0 = \emptyset$.

For a t-th query $(M, \text{ADM}, \text{DID}, \sigma, \text{MOD})$ to $\mathcal{O}^{\text{Redact}}$:

- Parse pk as $(\text{pk}_{\text{Fix}}, \text{pk}_{\text{Agg}}, t, n)$ and σ as $(\sigma_{\text{Fix}}, \Sigma_{\text{agg}})$.
- If $\text{DID} \in \mathbb{L}^{t-1}$, then abort.
- Set $\mathbb{L}^t \leftarrow \mathbb{L}^{t-1} \cup \{\text{DID}\}$.
- If $\text{MOD} \not\subseteq M \vee \text{MOD} \cap \text{ADM} \neq \emptyset$, then abort.
- Compute $h_{\text{ADM}} \leftarrow H(\text{DID}\|\text{ord}(\text{ADM}))$.
  ord(ADM) denotes a lexicographic ordering to the elements in ADM.
- For $m_j \in M$, compute $h_{m_j} \leftarrow H(\text{DID}\|m_j)$.
- If $e(\sigma_{\text{Fix}}, g_2) \neq e(h_{\text{ADM}}, \text{pk}_{\text{Fix}})$, then abort.
- If $e(\Sigma_{\text{agg}}, g_2) \neq e(h_{\text{ADM}}, \text{pk}_{\text{Agg}}) \cdot \prod_{m_j \in M} e(h_{m_j}, \text{pk}_{\text{Agg}})$, then abort.
- For $m_j \in \text{MOD}$, compute $\sigma_{m_j} \leftarrow H(\text{DID}\|m_j)^{\text{sk}_{\text{Agg}}}$.
- Compute $\sigma_{\text{MOD}} \leftarrow \prod_{m_j \in \text{MOD}} \sigma_{m_j}$, $\Sigma'_{\text{agg}} \leftarrow \Sigma_{\text{agg}}/\sigma_{\text{MOD}}$.
- Set $M' \leftarrow M\backslash\text{MOD}$, $\sigma' \leftarrow (\sigma_{\text{Fix}}, \Sigma'_{\text{agg}})$.
- Return $(M', \text{ADM}, \text{DID}, \sigma')$.

(Redactions are done using sk_{Agg} instead of using $\{\text{rk}[i]\}_{i=1}^n$.)

– $\textbf{Game}_{2-2}$: We change settings of KeyGen and $\mathcal{O}^{\text{Corrupt}}$.

- We change a key generation algorithm KeyGen in Step 2 to 6.
  * Choose $x \xleftarrow{\$} \mathbb{Z}_q$, $r \xleftarrow{\$} \mathbb{Z}_q$, compute $u \leftarrow g^x$, $y \leftarrow g_2^{x+r}$.
  * Set $\text{pk}_{\text{Agg}} \leftarrow y$, $\text{sk}_{\text{Agg}} \leftarrow x + r$.
  * Return $(\text{pk}, \text{sk}) \leftarrow ((\text{pk}_{\text{Fix}}, \text{pk}_{\text{Agg}}, t, n), (\text{sk}_{\text{Fix}}, \text{sk}_{\text{Agg}}))$.
  (Redactor's keys $\{\text{rk}[i]\}_{i=1}^n$ are not generated in the KeyGen.)
- We change the setting of $\mathcal{O}^{\text{Corrupt}}$ as follows:
  Let CR is a list to store a redactor's key information $(i, \text{rk}[i])$
  For a query i to $\mathcal{O}^{\text{Corrupt}}$,
  * If $(i, \text{rk}[i])$ already appears in CR, then return $\text{rk}[i]$.

* Choose $f(i) \xleftarrow{\$} \mathbb{Z}_q$, set $CR \leftarrow CR \cup \{(i, f(i))\}$.
* Return $\mathsf{rk}[i] \leftarrow (i, f(i))$.

- **Game$_{2-3}$**: We change a setting of the random oracle $\mathcal{O}^H$. Fix $h \xleftarrow{\$} \mathbb{G}_2$ and let $\mathbb{T}$ be a table that maintains a list of tuples $\langle v, w, b, c \rangle$ as explain below. We refer to this list for the query to $\mathcal{O}^h$. The initial state of $\mathbb{T}$ is empty. For queries $v^{(i)}$ to $\mathcal{O}^H$:

  * If $\langle v^{(i)}, w^{(i)}, \cdot, \cdot \rangle$ (Here, '$\cdot$' represents an arbitrary value) already appears in $\mathbb{T}$, then return $w^{(i)}$.
  * Choose $s^{(i)} \xleftarrow{\$} \mathbb{Z}_q$.
  * Flip a biased coin $c^{(i)} \in \{0, 1, 2\}$ such that such that $\Pr[c^{(i)} = 1] = 1 - 1/((\ell+1)(q_s + q_r) + 1)$, $\Pr[c^{(i)} = 1] = 1/(2(\ell+1)(q_s + q_r) + 2)$, $\Pr[c^{(i)} = 2] = 1/(2(\ell+1)(q_s + q_r) + 2)$.
  * If $c^{(i)} = 0$, compute $w^{(i)} = \phi(g_2)^{b^{(i)}}$.
  * If $c^{(i)} = 1$, compute $w^{(i)} = h \cdot \phi(g_2)^{b^{(i)}}$.
  * If $c^{(i)} = 2$, compute $w^{(i)} = h^{-1} \cdot \phi(g_2)^{b^{(i)}}$.
  * Insert $\langle v^{(i)}, w^{(i)}, s^{(i)}, c^{(i)} \rangle$ in $\mathbb{T}$ and return $w^{(i)}$.

- **Game$_{2-4}$**: We modify the signing algorithm Sign in Step 6 as follows:
  * Set $v^{(0)} \leftarrow (\mathsf{DID}\|\mathsf{ord}(\mathsf{ADM}))$, $v^{(j)} \leftarrow (\mathsf{DID}\|m_j)$ $(1 \leq j \leq \#M)$.
  * Query $v^{(j)}$ $(0 \leq j \leq \#M)$ to $\mathcal{O}^H$. We assume $\langle v^{(j)}, w^{(j)}, b^{(j)}, c^{(j)} \rangle$ to be the tuple in $\mathbb{T}$ for each $v^{(j)}$ $(1 \leq j \leq \#M)$.
  * If $c^{(0)} = 2$, $c^{(1)} = 1$, $c^{(j)} = 0$ $(2 \leq \forall j \leq \#M)$ or $c^{(j)} = 0$ $(0 \leq \forall j \leq \#M)$, go to Step 6 of Sign. Otherwise return $\bot$ and abort.

- **Game$_{2-5}$**: We modify the signing algorithm Sign in Step 7, 8 as follows:
  * If $c^{(0)} = 2$, $c^{(1)} = 1$, $c^{(j)} = 0$ $(2 \leq \forall j \leq \#M)$,
    * Compute $\sigma_{\mathsf{ADM}m_1} \leftarrow \phi(u)^{b^{(0)}+b^{(1)}} \cdot \phi(g_2)^{r(b^{(0)}+b^{(1)})}$.
    * For all $m_j \in M \backslash \{m_1\}$, compute $\sigma_{m_j} \leftarrow \phi(u)^{b^{(j)}} \cdot \phi(g_2)^{rb^{(j)}}$.
    * Compute $\Sigma_{\mathsf{agg}} \leftarrow \sigma_{\mathsf{ADM}m_1} \cdot \prod_{m_j \in M \backslash \{m_1\}} \sigma_{m_j}$.
  * If $c^{(j)} = 0$ $(0 \leq \forall j \leq \#M)$,
    * Compute $\sigma_{\mathsf{ADM}} \leftarrow \phi(u)^{b^{(0)}} \cdot \phi(g_2)^{rb^{(0)}}$.
    * For all $m_j \in M$, compute $\sigma_{m_j} \leftarrow \phi(u)^{b^{(j)}} \cdot \phi(g_2)^{rb^{(j)}}$.
    * Compute $\Sigma_{\mathsf{agg}} \leftarrow \sigma_{\mathsf{ADM}} \cdot \prod_{m_j \in M} \sigma_{m_j}$.

(By above modification, a signature Σ_{agg} can be generated without a knowledge of the $\mathsf{sk}_{\mathsf{Agg}}$.)

- **Game$_{2-6}$**: We change a setting of $\mathcal{O}^{\mathsf{Redact}}$.
  * Parse pk as $(\mathsf{pk}_{\mathsf{Fix}}, \mathsf{pk}_{\mathsf{Agg}}, t, n)$ and σ as $(\sigma_{\mathsf{Fix}}, \Sigma_{\mathsf{agg}})$.
  * If $\mathsf{DID} \in \mathbb{L}^{t-1}$, then abort.
  * Set $\mathbb{L}^t \leftarrow \mathbb{L}^{t-1} \cup \{\mathsf{DID}\}$.
  * If $\mathsf{MOD} \not\subseteq M \vee \mathsf{MOD} \cap \mathsf{ADM} \neq \emptyset$, then abort.
  * Set $v^{(0)} \leftarrow (\mathsf{DID}\|\mathsf{ord}(\mathsf{ADM}))$, $v^{(j)} \leftarrow (\mathsf{DID}\|m_j)$ $(1 \leq j \leq \#M)$.
  * Query $v^{(j)}$ $(0 \leq j \leq \#\mathsf{MOD})$ to $\mathcal{O}^H$. We assume $\langle v^{(j)}, w^{(j)}, b^{(j)}, c^{(j)} \rangle$ to be the tuple in $\mathbb{T}$ for each $v^{(j)}$ $(1 \leq j \leq \#\mathsf{MOD})$.
  * If $e(\sigma_{\mathsf{Fix}}, g_2) \neq e(w^{(0)}, \mathsf{pk}_{\mathsf{Fix}})$, then abort.
  * If $e(\Sigma_{\mathsf{agg}}, g_2) \neq \prod_{0 \leq j \leq \#M} e(w^{(j)}, \mathsf{pk}_{\mathsf{Agg}})$, then abort.
  * If $c^{(j)} = 0$ $(\forall m_j \in \mathsf{MOD})$, go to next step. Otherwise return $\bot$ and abort.

- For all $m_j \in \mathsf{MOD}$, compute $\sigma_{m_j} \leftarrow \phi(u)^{b^{(j)}} \cdot \phi(g_2)^{rb^{(j)}}$.
- Compute $\sigma_{\mathsf{MOD}} \leftarrow \prod_{m_j \in \mathsf{MOD}} \sigma_{m_j}$, $\Sigma'_{\mathsf{agg}} \leftarrow \Sigma_{\mathsf{agg}}/\sigma_{\mathsf{MOD}}$.
- Set $M' \leftarrow M\backslash\mathsf{MOD}$, $\sigma' \leftarrow (\sigma_{\mathsf{Fix}}, \Sigma'_{\mathsf{agg}})$.
- Return $(M', \mathsf{ADM}, \mathsf{DID}, \sigma')$.

(Redactions can be done without the knowledge of the $\mathsf{sk}_{\mathsf{Agg}}$.)

- **Game$_{2-7}$:** We receiving the output forgery $(M^*, \mathsf{ADM}^*, \mathsf{DID}^*, \sigma^*)$ from the adversary $\mathcal{A}_3$,
  - Set $v^{(0)} \leftarrow (\mathsf{DID}^* \| \mathrm{ord}(\mathsf{ADM}^*))$, $v^{(j)} \leftarrow (\mathsf{DID} \| m_j^*)$ $(1 \leq j \leq \#M^*)$.
  - Query $v^{(j)}$ $(0 \leq j \leq \#M^*)$ to $\mathcal{O}^H$. We assume $\langle v^{(j)}, w^{(j)}, s^{(j)}, c^{(j)} \rangle$ to be the tuple in $\mathbb{T}$ for each $v^{(j)}$ $(0 \leq j \leq \#M^*)$.
  - If $c^{(0)} = 1$ and $c^{(j)} = 0$ $(1 \leq j \leq \#M^*)$, then accept. Otherwise reject and abort.

Lemma 6. *The following equation holds.*

$$\mathsf{Adv}_{\mathcal{A}_2}[\mathbf{Game_{2-1}}] = \mathsf{Adv}_{\mathcal{A}_2}[\mathbf{Game_{2-0}}].$$

Since outputs of $\mathcal{O}^{\mathsf{Redact}}$ in **Game$_{2-0}$** and **Game$_{2-1}$** are same.

Lemma 7. *The following equation holds.*

$$\mathsf{Adv}_{\mathcal{A}_2}[\mathbf{Game_{2-2}}] = \mathsf{Adv}_{\mathcal{A}_2}[\mathbf{Game_{2-1}}].$$

To simplify the discussion, let $\mathcal{A}_2$ get $\mathsf{rk}[i], \ldots, \mathsf{rk}[t-1]$ from $\mathcal{O}^{\mathsf{Corrupt}}$. In [**Game$_{2-1}$**], the following equation holds.

$$V \begin{pmatrix} a_0 \\ a_1 \\ a_2 \\ \vdots \\ a_{t-1} \end{pmatrix} = \begin{pmatrix} f(0) \\ f(1) \\ f(2) \\ \vdots \\ f(t-1) \end{pmatrix} \quad \text{where } V = \begin{pmatrix} 1 & 0 & 0 & \cdots & 0 \\ 1 & 1 & 1 & \cdots & 1 \\ 1 & 2 & 2^2 & \cdots & 2^{t-1} \\ \vdots & \vdots & \vdots & \cdots & \vdots \\ 1 & t-1 & (t-1)^2 & \cdots & (t-1)^{t-1} \end{pmatrix}.$$

Since V is the Vandermonde matrix, V is the regular matrix. Distributions of $(a_0, a_1, \cdots, a_{t-1})$ and $(f(0), f(1), \ldots, f(t-1))$ are identical. Therefore, distributions of $(\mathsf{sk}_{\mathsf{Agg}}, \mathsf{rk}[1], \ldots, \mathsf{rk}[t-1])$ in [**Game$_{2-1}$**] and [**Game$_{2-2}$**] are same.

Lemma 8. *If H is the random oracle model, the following equation holds.*

$$\mathsf{Adv}_{\mathcal{A}_2}[\mathbf{Game_{2-3}}] = \mathsf{Adv}_{\mathcal{A}_2}[\mathbf{Game_{2-2}}].$$

Since the distribution of outputs of $\mathcal{O}^H$ in **Game$_{2-3}$** and **Game$_{2-2}$** is identical.

Lemma 9. *The following inequality holds.*

$$\mathsf{Adv}_{\mathcal{A}_2}[\mathbf{Game_{2-4}}] \geq (1 - 1/((\ell+1)(q_s + q_r) + 1))^{(\ell+1)q_s} \times \mathsf{Adv}_{\mathcal{A}_2}[\mathbf{Game_{2-3}}].$$

Since the probability that each signing query does not abort at least $(1 - 1/((\ell+1)(q_s + q_r) + 1))^{(\ell+1)}$.

Lemma 10. *The following equation holds.*

$$\mathsf{Adv}_{\mathcal{A}_2}[\mathbf{Game}_{2-5}] = \mathsf{Adv}_{\mathcal{A}_2}[\mathbf{Game}_{2-4}].$$

Since outputs of Sign in $\mathbf{Game}_{2-5}$ and $\mathbf{Game}_{2-4}$ are same.

Lemma 11. *The following inequality holds.*

$$\mathsf{Adv}_{\mathcal{A}_2}[\mathbf{Game}_{2-6}] \geq (1 - 1/((\ell+1)(q_s + q_r) + 1))^{(\ell+1)q_r} \times \mathsf{Adv}_{\mathcal{A}_2}[\mathbf{Game}_{2-5}].$$

Since the probability that each redaction query does not abort at least $(1 - 1/((\ell+1)(q_s + q_r) + 1))^{(\ell+1)}$.

Lemma 12. *The following inequality holds.*

$$\mathsf{Adv}_{\mathcal{A}_2}[\mathbf{Game}_{2-7}]$$

$$\geq \frac{\left(\frac{1}{2(\ell+1)(q_s+q_r)+2}\right)^2}{\left(1 - \frac{1}{(\ell+1)(q_s+q_r)+1}\right)^2 + \left(\frac{1}{2(\ell+1)(q_s+q_r)+2}\right)^2} \times \mathsf{Adv}_{\mathcal{A}_2}[\mathbf{Game}_{2-6}]$$

$$= (1/(4(\ell+1)^2(q_s+q_r)^2 + 1)) \times \mathsf{Adv}_{\mathcal{A}_2}[\mathbf{Game}_{2-6}].$$

Since an output $(M^*, \mathsf{ADM}^*, \mathsf{DID}^*, \sigma^*)$ satisfies $(c^{(0)}, c^{(1)}) = (0, 0)$ or $(2, 1)$.
To summarize **Lemma 6** to **12**, the following holds.
(In the following equation, e represents the Napier's constant.)

$$\mathsf{Adv}_{\mathcal{A}_2}[\mathbf{Game}_{2-7}] \geq (1 - 1/((\ell+1)(q_s + q_r) + 1))^{(\ell+1)(q_s+q_r)}$$
$$\times 1/(4(\ell+1)^2(q_s+q_r)^2 + 1) \times \mathsf{Adv}_{\mathcal{A}_2}[\mathbf{Game}_{2-0}]$$
$$\geq (1/e) \times (1/(4(\ell+1)^2(q_s+q_r)^2 + 1)) \times \mathsf{Adv}_{\mathcal{A}_2}[\mathbf{Game}_{2-0}]$$

Now we construct the algorithm $\mathcal{B}_2$ which breaking the computational co-Diffie-Hellman assumption using the algorithm $\mathcal{A}_2$. The operation of $\mathcal{B}_2$ for the input co-Diffie-Hellman problem instance (g_2, g_2^α, h^*) is changed to h in $\mathbf{Game}_{2-7}$ to h^* and u to g_2^α.

Suppose $\mathcal{B}_2$ do not abort receiving a forgery $(M^*, \mathsf{ADM}^*, \mathsf{DID}^*, \sigma^*)$ from $\mathcal{A}_2$. $\mathcal{B}_3$ parses σ^* as $(\sigma^*_{\mathsf{ADM}^*}, \Sigma^*_{\mathsf{agg}})$, sets $v^{(j)} \leftarrow (\mathsf{DID}^* \| m_j^*)$ $(1 \leq j \leq \#M^*)$, and computes $w^{(1)} \leftarrow h \cdot \phi(u)^{b^{(1)}} \cdot \phi(g_2)^{rb^{(1)}}$, $w^{(j)} \leftarrow \phi(u)^{b^{(j)}} \cdot \phi(g_2)^{rb^{(j)}}$ $(2 \leq j \leq \#M^*)$. Then $\mathcal{B}_3$ computes $\sigma^*_{m_1^*} \leftarrow \Sigma^*_{\mathsf{agg}} / \prod_{j=2}^{\#M^*} \sigma_{m_j}$. Since $(M^*, \mathsf{ADM}^*, \mathsf{DID}^*, \sigma^*)$ is valid signature and $\mathsf{pk}_{\mathsf{Agg}} = g_2^{\alpha+r}$, $e(\sigma^*_{m_1^*}, g_2) = e\left((w^{(1)})^{\alpha+r}, g_2\right)$ holds. It implies that $\sigma^*_{m_1^*} = (w^{(1)})^{\alpha+r} = (h^* \cdot \phi(g_2)^{b^{(1)}})^{\alpha+r}$. Therefore, $\mathcal{B}_3$ computes $(h^*)^\alpha = \sigma^*_{m_1^*} \cdot (\phi(u)^{b^{(1)}} \cdot (h^*)^r \cdot \phi(g_2)^{rb^{(1)}})^{-1}$ and outputs the solution $(h^*)^\alpha$ of the computational co-Diffie-Hellman problem instance (g_2, g_2^α, h^*).

Let $\epsilon_{\mathsf{co-cdh2}}$ is the probability that $\mathcal{B}_2$ break the computational co-Diffie-Hellman assumption. We can bound the probability $\epsilon_{\mathsf{co-cdh2}} \geq \mathsf{Adv}_{\mathcal{A}_2}[\mathbf{Game}_{2-7}]$ and $\epsilon_{\mathsf{co-cdh2}} \geq (1/e) \times (1/(4(\ell+1)^2(q_s+q_r)^2 + 1)) \times \epsilon_{\mathsf{uf2}}$ holds. (e represents the Napier's constant.) If ϵ_{uf2} is non-negligiable in λ, $\mathcal{B}_2$ breaks the computational co-Diffie-Hellman assumption with non-negligiable in $\epsilon_{\mathsf{co-cdh2}}$. We will describe the proof of **Case 3** in the full version of this paper.

Theorem 3. *Our proposed t-out-of-n redactable signature scheme in the one-time redaction model* (t, n)*-RSS* Π_1 *satisfies the transparency.*

We will describe the proof of **Theorem** 3 in the full version of this paper. By **Theorem** 1 and **Theorem** 3, our proposed scheme satisfies the privacy.

5 Conclusion

In this paper, we introduce the new notion of t-out-of-n RSS. Our proposed model supports only the one-time redaction model which allows redacting signed message only one time for each signature. Our construction Π_1 does not satisfy the unforgeability in a model that allows redacting signed message many times. For example, $M = \{m_1, m_2, m_3\}$ and ADM $= \emptyset$, an adversary who does the following operation generates a valid forgery in a multiple redactions model.

1. Given pk from $\mathcal{C}$.
2. Query (M, ADM) to $\mathcal{O}^{\text{Sign}}$ and get $(M, \text{ADM}, \text{DID}, \sigma)$.
3. Let $\text{MOD}^1 = \{m_1\}$ $\text{MOD}^2 = \{m_2\}$. Query $(M, \text{ADM}, \text{DID}, \sigma, \text{MOD}^1)$ to $\mathcal{O}^{\text{Redact}}$ and get $(M', \text{ADM}, \text{DID}, \sigma')$ and query $(M', \text{ADM}, \text{DID}, \sigma', \text{MOD}^2)$ to $\mathcal{O}^{\text{Redact}}$ and get $(M'', \text{ADM}, \text{DID}, \sigma'')$.
4. Parse σ as $(\sigma_{\text{Fix}}, \Sigma_{\text{agg}})$, σ' as $(\sigma'_{\text{Fix}}, \Sigma'_{\text{agg}})$, and σ'' as $(\sigma''_{\text{Fix}}, \Sigma''_{\text{agg}})$.
5. Compute $\sigma_{m_1} \leftarrow \Sigma_{\text{agg}} \cdot (\Sigma'_{\text{agg}})^{-1}$, $\Sigma^*_{\text{agg}} \leftarrow \sigma_{m_1} \cdot \Sigma''_{\text{agg}}$.
6. Set $M^* \leftarrow \{m_1, m_3\}$, $\sigma^* \leftarrow (\sigma_{\text{Fix}}, \Sigma^*_{\text{agg}})$ and output $(M^*, \text{DID}, \text{ADM}, \sigma^*)$.

Giving a construction of RSS in the multiple redactions model is the interesting futures work.

References

1. Boneh, D., Gentry, C., Lynn, B., Shacham, H.: Aggregate and verifiably encrypted signatures from bilinear maps. In: Biham, E. (ed.) EUROCRYPT 2003. LNCS, vol. 2656, pp. 416–432. Springer, Heidelberg (2003). https://doi.org/10.1007/3-540-39200-9_26
2. Boneh, D., Lynn, B., Shacham, H.: Short signatures from the weil pairing. In: Boyd, C. (ed.) ASIACRYPT 2001. LNCS, vol. 2248, pp. 514–532. Springer, Heidelberg (2001). https://doi.org/10.1007/3-540-45682-1_30
3. Brzuska, C., et al.: Redactable signatures for tree-structured data: definitions and constructions. In: Zhou, J., Yung, M. (eds.) ACNS 2010. LNCS, vol. 6123, pp. 87–104. Springer, Heidelberg (2010). https://doi.org/10.1007/978-3-642-13708-2_6
4. Derler, D., Pöhls, H.C., Samelin, K., Slamanig, D.: A general framework for redactable signatures and new constructions. In: Kwon, S., Yun, A. (eds.) ICISC 2015. LNCS, vol. 9558, pp. 3–19. Springer, Cham (2016). https://doi.org/10.1007/978-3-319-30840-1_1
5. Haber, S., et al.: Efficient signature schemes supporting redaction, pseudonymization, and data deidentification. In: Proceedings of the 2008 ACM Symposium on Information, Computer and Communications Security, ASIACCS 2008, Tokyo, Japan, 18–20 March 2008, pp. 353–362 (2008)

6. Izu, T., Izumi, M., Kunihiro, N., Ohta, K.: Yet another sanitizable and deletable signatures. In: 25th IEEE International Conference on Advanced Information Networking and Applications Workshops, WAINA 2011, Biopolis, Singapore, 22–25 March 2011, pp. 574–579 (2011)
7. Izu, T., Kanaya, N., Takenaka, M., Yoshioka, T.: PIATS: a partially sanitizable signature scheme. In: Qing, S., Mao, W., López, J., Wang, G. (eds.) ICICS 2005. LNCS, vol. 3783, pp. 72–83. Springer, Heidelberg (2005). https://doi.org/10.1007/11602897_7
8. Izu, T., Kunihiro, N., Ohta, K., Sano, M., Takenaka, M.: Sanitizable and deletable signature. In: Chung, K.-I., Sohn, K., Yung, M. (eds.) WISA 2008. LNCS, vol. 5379, pp. 130–144. Springer, Heidelberg (2009). https://doi.org/10.1007/978-3-642-00306-6_10
9. Izu, T., Kunihiro, N., Ohta, K., Sano, M., Takenaka, M.: Yet another sanitizable signature from bilinear maps. In: Proceedings of the the Forth International Conference on Availability, Reliability and Security, ARES 2009, Fukuoka, Japan, 16–19 March 2009, pp. 941–946 (2009)
10. Izu, T., Kunihiro, N., Ohta, K., Takenaka, M., Yoshioka, T.: A sanitizable signature scheme with aggregation. In: Dawson, E., Wong, D.S. (eds.) ISPEC 2007. LNCS, vol. 4464, pp. 51–64. Springer, Heidelberg (2007). https://doi.org/10.1007/978-3-540-72163-5_6
11. Johnson, R., Molnar, D., Song, D., Wagner, D.: Homomorphic signature schemes. In: Preneel, B. (ed.) CT-RSA 2002. LNCS, vol. 2271, pp. 244–262. Springer, Heidelberg (2002). https://doi.org/10.1007/3-540-45760-7_17
12. Lim, S., Lee, E., Park, C.: A short redactable signature scheme using pairing. Secur. Commun. Netw. **5**(5), 523–534 (2012)
13. Ma, J., Liu, J., Wang, M., Wu, W.: An efficient and secure design of redactable signature scheme with redaction condition control. In: Au, M.H.A., Castiglione, A., Choo, K.-K.R., Palmieri, F., Li, K.-C. (eds.) GPC 2017. LNCS, vol. 10232, pp. 38–52. Springer, Cham (2017). https://doi.org/10.1007/978-3-319-57186-7_4
14. Miyazaki, K., Hanaoka, G., Imai, H.: Digitally signed document sanitizing scheme based on bilinear maps. In: Proceedings of the 2006 ACM Symposium on Information, Computer and Communications Security, ASIACCS 2006, Taipei, Taiwan, 21–24 March 2006, pp. 343–354 (2006)
15. Miyazaki, K., Hanaoka, G., Imai, H.: Invisibly sanitizable digital signature scheme. IEICE Trans. **91-A**(1), 392–402 (2008)
16. Miyazaki, K., et al.: Digitally signed document sanitizing scheme with disclosure condition control. IEICE Trans. **88-A**(1), 239–246 (2005)
17. Pöhls, H.C., Samelin, K.: On updatable redactable signatures. In: Boureanu, I., Owesarski, P., Vaudenay, S. (eds.) ACNS 2014. LNCS, vol. 8479, pp. 457–475. Springer, Cham (2014). https://doi.org/10.1007/978-3-319-07536-5_27
18. Pöhls, H.C., Samelin, K.: Accountable redactable signatures. In: 10th International Conference on Availability, Reliability and Security, ARES 2015, Toulouse, France, 24–27 August 2015, pp. 60–69 (2015)
19. Samelin, K., Pöhls, H.C., Bilzhause, A., Posegga, J., de Meer, H.: Redactable signatures for independent removal of structure and content. In: Ryan, M.D., Smyth, B., Wang, G. (eds.) ISPEC 2012. LNCS, vol. 7232, pp. 17–33. Springer, Heidelberg (2012). https://doi.org/10.1007/978-3-642-29101-2_2
20. Shamir, A.: How to share a secret. Commun. ACM **22**(11), 612–613 (1979)
21. Steinfeld, R., Bull, L., Zheng, Y.: Content extraction signatures. In: Kim, K. (ed.) ICISC 2001. LNCS, vol. 2288, pp. 285–304. Springer, Heidelberg (2002). https://doi.org/10.1007/3-540-45861-1_22

Signature, ML, Payment, and Factorization

A Framework with Randomized Encoding for a Fast Privacy Preserving Calculation of Non-linear Kernels for Machine Learning Applications in Precision Medicine

Ali Burak Ünal[1]([✉]), Mete Akgün[2]([✉]), and Nico Pfeifer[1,3]

[1] Methods in Medical Informatics, Department of Computer Science,
University of Tuebingen, Tuebingen, Germany
{uenal,pfeifer}@informatik.uni-tuebingen.de
[2] Translational Bioinformatics, University Hospital Tuebingen, Tuebingen, Germany
akguen@informatik.uni-tuebingen.de
[3] Statistical Learning in Computational Biology,
Max Planck Institute for Informatics, Saarbrücken, Germany

Abstract. For many diseases it is necessary to gather large cohorts of patients with the disease in order to have enough power to discover the important factors. In this setting, it is very important to preserve the privacy of each patient and ideally remove the necessity to gather all data in one place. Examples include genomic research of cancer, infectious diseases or Alzheimer's. This problem leads us to develop privacy preserving machine learning algorithms. So far in the literature there are studies addressing the calculation of a specific function privately with lack of generality or utilizing computationally expensive encryption to preserve the privacy, which slows down the computation significantly. In this study, we propose a framework utilizing randomized encoding in which four basic arithmetic operations (addition, subtraction, multiplication and division) can be performed, in order to allow the calculation of machine learning algorithms involving one type of these operations privately. Among the suitable machine learning algorithms, we apply the oligo kernel and the radial basis function kernel to the coreceptor usage prediction problem of HIV by employing the framework to calculate the kernel functions. The results show that we do not sacrifice the performance of the algorithms for privacy in terms of F1-score and AUROC. Furthermore, the execution time of the framework in the experiments of the oligo kernel is comparable with the non-private version of the computation. Our framework in the experiments of radial basis function kernel is also way faster than the existing approaches utilizing integer vector homomorphic encryption and consequently homomorphic encryption based solutions, which indicates that our approach has a potential for application to many other diseases and data types.

Keywords: Privacy preserving machine learning · Randomized encoding · String kernel · RBF kernel · Precision medicine

© Springer Nature Switzerland AG 2019
Y. Mu et al. (Eds.): CANS 2019, LNCS 11829, pp. 493–511, 2019.
https://doi.org/10.1007/978-3-030-31578-8_27

1   Introduction

By the recent development of next generation sequencing (NGS) technologies, DNA sequencing and RNA sequencing can be performed effectively and efficiently [23]. However; the generated sequence data contains private information about the host and one could infer many phenotypes of an individual, such as hair color, skin color and more importantly genetic diseases, from the relevant sequence data of that individual [5,11,16]. Such private information can be used against the owner of the sequence data potentially leading to increased health insurance premiums due to the genetic disease of that person. On the other hand, machine learning algorithms still require these sequence data to capture the underlying patterns of the diseases. In many of the real-world problems, machine learning algorithms need to have more data than one source can provide [17,20,21].

In this paper, we consider a scenario where we have two parties with sequence data, each of which we call *input-party* and one party, which we call *function-party* that wants to run machine learning algorithms on the data of these input-parties. To avoid the aforementioned privacy issues in our scenario due to the leakage of sequence data utilized in machine learning algorithms, we propose a framework utilizing randomized encoding [1,4] to enable machine learning algorithms involving a single type of basic arithmetic operation to use the data from two sources without sacrificing the privacy of participants. We provide the function-party with the private calculation of four basic arithmetic operations, that is addition/subtraction and division/multiplication. In order to demonstrate the performance of our framework, we chose support vector machine (SVM) with the oligo kernel and the radial basis function kernel on the prediction of coreceptor usage of HIV based on V3 loop sequences [13]. We computed the kernel matrices by employing our proposed framework and trained a prediction model on top of these kernel matrices.

In the rest of the paper, we will explain the similar studies in the literature in Sect. 2. We will give the background information about the oligo kernel, the radial basis kernel and randomized encoding in Sect. 3. Then, we will propose our framework in Sect. 4. Afterwards, we will discuss the security of the proposed framework in Sect. 5. Next, we explain the dataset that we utilized in the experiments in Sect. 6. We will then show and evaluate the results of these experiments in Sect. 7. Finally, we will conclude the paper in Sect. 8.

2   Related Work

In this context, we refer to a source with data, such as a clinic having sequence data, or an entity computing a function, such as a university conducting a study on the data of clinics, as party. In the literature, some approaches using an SVM are based on the distributed model, which assumes that each party of the computation has its own data and the function needs to be calculated by using the data of all parties. Vaidya et al. [25] proposed a privacy-preserving SVM classification algorithm. In the proposed approach, each party has its own data and they compute the gram matrix to train an SVM model by using a modified secure dot product calculation method. However; it focuses more on binary feature vectors and does not support advanced string kernels. Furthermore, there are also a number of studies which employ an outsourced model in which the data of the parties are stored on a cloud server as encrypted by the secret key of the owners of the data. Liu et al. [15] introduced an SVM algorithm which can be used to mine the encrypted outsourced data. Since the data is encrypted before outsourcing, this slows down the process of model training. Zhang et al. [28] also proposed a secure dot product calculation method to train an SVM model. The underlying idea is to transform the secret key of the dot product of two vectors encrypted by different keys into a known key by the server. In order to accomplish the transformation, the server collaborates with the owners of the keys of these vectors. The proposed approach fits into our scenario where we have two input-parties and one function-party. However; the approach utilizes integer vector encryption to preserve the privacy of the data. Therefore, it cannot handle the data having a large number of features within a reasonable time frame.

3 Preliminaries

In this section, we will explain the kernel functions that we utilized and the randomized encoding which is our base security scheme.

3.1 Oligo Kernel

Although the oligo kernel belongs to the family of string kernels, it is widely used to discover the patterns of biological sequences [10,18,19]. In the context of sequence analysis, the oligo kernel is designed to work with oligomers occurring in sequences such as DNA and protein. They can be varying lengths but general tendency is to keep the length of the oligomers short. The DNA oligomers with length 2, for instance, consist of all possible 2-length-monomer combinations of the DNA alphabet $\mathcal{A}$, where $\mathcal{A} = \{A, T, C, G\}$. To be more specific, the DNA oligomers with length 2 form a set $\mathcal{A}^2$ where $\mathcal{A}^2 = \{AA, AT, AC, AG, TA, TT, TC, TG, CA, CT, CC, CG, GA, GT, GC, GG\}$. In general, all possible combinations of K-length-monomers in an alphabet $\mathcal{A}$ of a sequence type are called K-mers and they form a set $\mathcal{A}^K$. These K-mers are utilized by the oligo kernel to determine the similarity between sequences. For each K-mer $\omega \in \mathcal{A}^K$ occurring in a sequence S, the corresponding oligo function μ is calculated as:

$$\mu_\omega(x) = \sum_{p \in S_\omega} \exp(-\frac{1}{2\sigma^2}(x - p)^2) \tag{1}$$

where S_ω is the set of occurrences of K-mer ω in sequence S, σ is the positional tolerance parameter for inexact matches. Based on the oligo function of each

K-mer, the mapping function Φ in the oligo kernel is defined as follows:

$$\Phi^K(s) = [\mu_{\omega_1}, \mu_{\omega_2}, \cdots, \mu_{\omega_n}]$$

where s is the sequence and n is the size of the set $\mathcal{A}^K$. Even though this representation is suitable for visualization and interpretation, we need to have the kernel function. As stated in [18], the kernel function is calculated as follows:

$$k(s_i, s_j) = \sqrt{\pi}\sigma \sum_{\omega \in \mathcal{A}^K} \sum_{p \in S_\omega^i} \sum_{q \in S_\omega^j} \exp(-\frac{1}{4\sigma^2}(p-q)^2) \qquad (2)$$

where s_i and s_j are the sequences, σ is the positional tolerance parameter, $\mathcal{A}^K$ is the set of all possible K-mers of monomers in alphabet $\mathcal{A}$, S_ω^i and S_ω^j are the set of occurrences of ω in sequence s_i and sequence s_j, respectively.

Unlike other similar approaches, the oligo kernel can be adjusted in a way that positional inexact matches of K-mers would also contribute to the similarity of sequences. The degree of this positional independence can be manipulated by the parameter σ. If σ is set close to 0, then only the exact matches of K-mers would contribute to the similarity. On the other hand, if σ is set to ∞, then there would be no importance of positions of K-mers.

Based on Eq. 2, the calculation of the oligo kernel requires the differences of the positions of K-mers in both sequences. In our framework, we address the required operation and enable the computation of the differences privately.

3.2   Radial Basis Function Kernel

The radial basis function (RBF) is one of the most popular kernel functions in kernel learning algorithms [24]. It is commonly used in many different areas [12, 22, 27]. For samples $x, y \in \mathbb{R}^n$, the RBF kernel can be formulated solely based on the dot product of samples as follows:

$$K(x, y) = \exp\left(-\frac{\|x \cdot x - 2x \cdot y + y \cdot y\|^2}{2\sigma^2}\right) \qquad (3)$$

where "$\cdot$" represents the dot product of vectors and σ is the parameter that adjusts the similarity level. As shown in Eq. 3, the calculation of the RBF kernel between the samples x and y can be done by the dot product, which consists of the element-wise multiplication and summing up the results of these multiplications. In our framework, we allow the private computation of the RBF kernel by enabling the element-wise multiplication of the vectors.

3.3   Randomized Encoding

In the cryptography literature, randomized encoding is proposed to compute a function $f(x)$ by a randomized function $\hat{f}(x; r)$, where r is a uniformly chosen random value, without revealing the input value x [2, 3]. The formal definition of the randomized encoding is as follows:

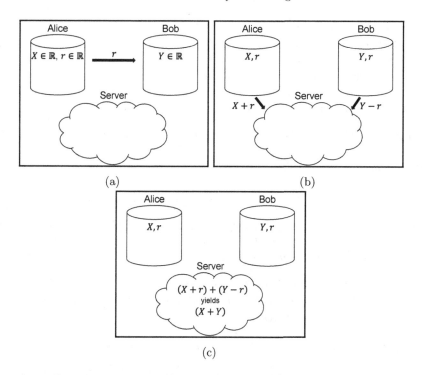

Fig. 1. In this setting, Alice and Bob are the input-parties and the server is the function-party. (a) Alice creates a uniformly chosen random value r. She shares it with Bob. (b) Once Bob receives the random value from Alice, he computes $Y - r$ and shares it with the server. Meanwhile, Alice computes $X + r$ and sends it to the server. (c) When the server receives the components of encoding, it calculates $(X+r)+(Y-r)$ to decode the result of the addition of the input values of Alice and Bob. At the end, the server obtains $X + Y$ without learning neither X nor Y. Similarly, none of the input-parties learns about the input value of the other input-party.

Definition 1 (Randomized Encoding [1]). *Let us define a function $f : X \to Y$. There exists a function $\hat{f} : X \times R \to Z$ which is a δ-correct, (t, ϵ)-private randomized encoding of f if randomized algorithms, decoder* Dec *and simulator* Sim, *can be defined and the followings hold for these algorithms:*

- *(δ-correctness) $\forall x \in X$:*

$$\Pr_{r \leftarrow R}[\mathsf{Dec}(\hat{f}(x;r)) \neq f(x)] \leq \delta.$$

- *$((t, \epsilon)$-privacy) $\forall x \in X$ and any circuit $\mathcal{A}$ of size t:*

$$\left| \Pr[\mathcal{A}(\mathsf{Sim}(f(x))) = 1] - \Pr_{r \leftarrow R}[\mathcal{A}(\hat{f}(x;r)) = 1] \right| \leq \epsilon.$$

where Dec *decodes the given encoding and* Sim *simulates the encoding such that simulation and real encoding are indistinguishable.*

Besides the formal definition of the randomized encoding, they also proposed two perfect decomposable and affine randomized encodings (DARE) for addition and multiplication of two values. In order for a randomized encoding to be affine and decomposable, each component of randomized encoding should be an affine function over the set that the function is defined and they should depend on only a single input value and a varying number of random values.

Definition 2 (Perfect RE for Addition [1]). *Let us define a function* $f(x_1, x_2) = x_1 + x_2$ *over some finite ring* R. *This addition function can be perfectly encoded by the following DARE:*

$$\hat{f}(x_1, x_2; r) = (x_1 + r, x_2 - r)$$

where r is a uniformly chosen random value. The encoding can be decoded by summing up the components of the encoding, and one can simulate the function by sampling two random values whose sum is y.

Definition 3 (Perfect RE for Multiplication [1]). *Let us define a function* $f(x_1, x_2) = x_1 \cdot x_2$ *over a ring* R. *This multiplication function can be perfectly encoded by the following DARE:*

$$\hat{f}(x_1, x_2; r_1, r_2, r_3) = (x_1 + r_1, x_2 + r_2, r_2 x_1 + r_3, r_1 x_2 + r_1 r_2 - r_3)$$

where r_1, r_2 *and* r_3 *are uniformly chosen random values. Given the encoding* (c_1, c_2, c_3, c_4), *we can recover* $f(x_1, x_2)$ *by computing* $c_1 \cdot c_2 - c_3 - c_4$. *The simulator* $\mathsf{Sim}(y; c_1, c_2, c_3) := (c_1, c_2, c_3, c_1 c_2 - y - c_3)$ *perfectly simulates* $\hat{f}$.

Randomized encoding preserves the privacy of the data by randomizing the input values and creating components by these values. At the end, it only allows the computation of the desired output and nothing else about the input values is revealed. Compared to the other methods in the literature such as homomorphic encryption and secure multi-party computation that result in a high overhead due to the use of computationally expensive cryptographic tools, the randomized encoding is faster and more efficient.

In the scenario where we have two input-parties and one function-party, the randomized encoding is also applicable. The randomized encoding of addition can be adapted to calculate the summation of two input values belonging to two different input-parties. The process is depicted in Fig. 1. The computation of the differences of two values can be done with the same encoding after multiplying the corresponding input value by -1. Similarly, one can employ the randomized encoding of multiplication in a function party in order to compute the multiplication of two input values owned by two distinct input-parties. The steps are demonstrated in Fig. 2. One can use the randomized encoding of multiplication to compute the division by simply using the reciprocal of the corresponding input value. It is worth to note that we assume that the value at the divisor should be non-zero.

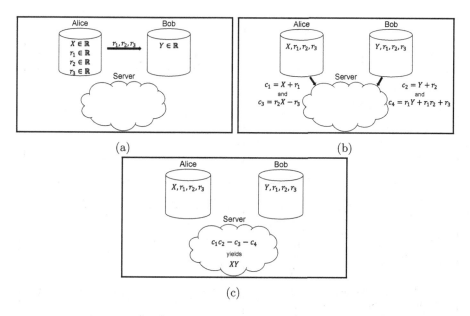

Fig. 2. In this setting, Alice and Bob are the input-parties and the server is the function-party. (a) Alice creates three uniformly chosen random values and shares them with Bob. (b) Once Bob receives the random values, he computes $c_2 = Y + r_2$ and $c_4 = r_1 Y + r_1 r_2 + r_3$. In the meantime, Alice computes $c_1 = X + r_1$ and $c_3 = r_2 X - r_3$. They send c_1, c_2, c_3 and c_4 to the server. (c) When the server receives the components of the encoding, it calculates $c_1 c_2 - c_3 - c_4$ to decode the result of the multiplication of the input values of Alice and Bob. At the end, the server obtains XY without learning neither X nor Y. Similarly, none of the input-parties learns about the input value of the other input-party.

4 Our Framework

In this paper, we propose a framework utilizing randomized encoding to support the computation of four basic arithmetic operations over the vectors of two input-parties in a function-party privately. In this application, we focus on the computation of differences and multiplications of these vectors. Our framework allows the server to learn only the intended outcome, either the element-wise differences or multiplications of the vectors, and nothing else about these vectors. Similarly, an input-party learns neither the input vector of the other input-party nor the result of the computation. Among the features of the framework, we use the element-wise differences of the vectors to compute the oligo kernel on the data owned by the input-parties. Additionally, we employ element-wise multiplication feature in order to compute the RBF kernel on the same setting. However; the framework is not limited to these methods. One can use the framework to compute a function which requires one of the features of the framework.

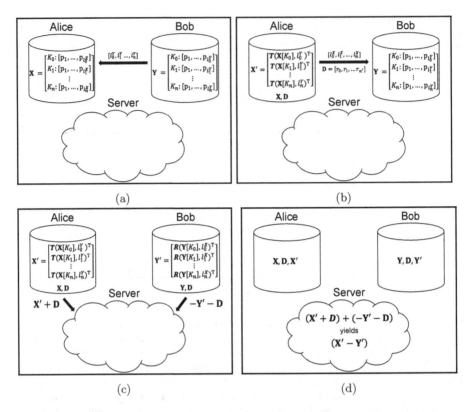

Fig. 3. In this figure, $\mathbf{K_i}$ represents the vector containing the positions of the occurrences of the i-th oligomer in the corresponding sequences. In those vectors, p_j of $\mathbf{K_i}$ represents the position of the j-th occurrence of the i-th oligomer. (a) Both input-parties find the positions of the oligomers in their sequences and insert a varying number of dummy positions into randomly chosen oligomers. Then, Bob shares the length of $\mathbf{K_j}$ $\forall j \in \{0, \cdots, n\}$ with Alice. (b) Afterwards, Alice creates the vector $X' \in \mathbb{R}^{n'}$ by applying the function $\mathbf{T}$ along with the vector $D \in \mathbb{R}^{n'}$ for encoding. Then, she sends D and the length of $\mathbf{K_i}$ $\forall i \in \{0, \cdots, n\}$ to Bob. (c) Bob creates the vector $Y' \in \mathbb{R}^{n'}$ by applying the function R. Next, they calculate $(-Y' - D) \in \mathbb{R}^{n'}$ and $(X' + D) \in \mathbb{R}^{n'}$, respectively, and share them with the server. (d) At the end, the server computes the element-wise summation of the given vectors to obtain all possible pairwise differences of the positions for each oligomer. Once the server prunes the entries involving dummy values, it can compute the oligo kernel function.

4.1   Computation of Oligo Kernel

We utilize the randomized encoding of addition in order to compute the oligo kernel. Since the calculation of difference is required to compute the oligo kernel, we adapt the encoding of addition by taking the negative of the input value of one of the input-parties. The process of the computation of the oligo kernel is depicted in Fig. 3. Let us assume that Alice and Bob have hashmap-like input vectors X and Y where $X, Y \in \mathbb{R}^n$ and n is the number of different oligomers. These vectors contain K_i representing the vector of positions of i-th oligomer in the sequence of the corresponding input-party for $i \in \{0, 1, \ldots, n\}$. The length of the vector K_i in X is denoted by l_i^X. It is worth to note that we have non-empty vectors for all oligomers in Fig. 3 due to the illustration purposes. In fact, K_i can be any length vector including empty vector, that is $K_i \in \{\emptyset, \mathbb{R}^1, \mathbb{R}^2, \cdots, \mathbb{R}^M\}$ $\forall i \in \{0, \cdots, n\}$ where M is the maximum possible number of oligomer that could occur in a specific length of sequences. First of all, the input-parties find the positions of oligomers in their sequences. Before Bob shares the length of vectors K_j $\forall j \in \{0, 1, \ldots, n\}$ with Alice, that is the vector $[l_0^Y, l_1^Y, \cdots, l_n^Y]$, Bob inserts varying number of dummy positions into randomly chosen oligomers where the position value is smaller than possible. In our case, it is -10^9. Similarly, Alice inserts some number of dummy positions into randomly chosen oligomers where the position is too large. For Alice, this value is 10^9. After receiving the number of occurrences of oligomers in the sequence of Bob, Alice applies the function T, which repeats the vector as a whole for the specified number of times, over her oligomers. As an example, $T([2, 5], 3)$ yields $A = [2, 5, 2, 5, 2, 5]$. Once Alice has the transpose of these repeated vectors, she concatenates them to create the vector $X' \in \mathbb{R}^{n'}$ where n' is the length of the vector after concatenation. Alice also creates the vector $D \in \mathbb{R}^{n'}$ having uniformly chosen random values for each entry of X'. Afterwards, Alice shares the vector D and the number of occurrences of her oligomers, that is the vector $[l_0^X, l_1^X, \cdots, l_n^X]$. It is important to note that all oligomers exist and they are common among input-parties due to the illustration purpose. However, in case of missing oligomers, Alice can work on and send only the common oligomers among the input-parties in order to reduce the communication cost. Once Bob gets the number of occurrences of oligomers in Alice and the vector D, he applies the function R, which repeats each entry of the given vector for the specified number of times, over his oligomer vectors. As a clarification, $R([2, 5], 3)$, for instances, yields $A = [2, 2, 2, 5, 5, 5]$. After Bob has the transpose of these repeated vectors, he concatenates them to create the vector $Y' \in \mathbb{R}^{n'}$ and calculates $(-Y' - D)$. In the meantime, Alice computes $(X' + D)$. Then, the input-parties share these vectors with the server. After the server receives the components of the encoding, it computes the summation of these components in order to obtain all possible pairwise differences of the positions for each oligomer in the sequences of Alice and Bob. At this point, the obtained vector contains the entries which are the results of the operation involving dummy values in addition to the actual entries. Since the server knows the artificial position values of Alice and Bob, it can detect whether a result is valid or not. If the absolute of a result is larger than or close to the artificial position value, then the server ignores it in the computation. At the end, the purified vector is utilized to compute the oligo kernel function over two sequences via Eq. 2. In order to compute the kernel matrix via the oligo kernel for multiple sequences in each input-party, we repeat the same process for all pairs of sequences from Alice and Bob. The differences among the sequences in the same party can be computed by that party and shared with the server to complete the kernel matrix. It is worth to note that sharing the number of occurrences of oligomers can be done once for all in order to reduce the communication cost.

4.2   Computation of RBF Kernel

We employ the randomized encoding for multiplication to compute the RBF kernel over the data of the input-parties. The first step of the computation is to calculate the element-wise multiplication of the vectors from different input-parties. We utilize the randomized encoding to overcome this problem. The process is demonstrated in Fig. 4. In the computation, let us assume that Alice and Bob have input matrices X and Y, respectively, whose columns represent samples and $A_{.i}$ shows the vector at the i-th column for any matrix A. In order to create the components of the encoding, Alice creates three vectors with uniformly chosen random values of the same length with a sample in the input vector X. Afterwards, Alice shares these vectors with Bob. When Bob receives the random values, he computes the random components M^2 and M^4. Meanwhile, Alice computes her shares of encoding, namely M^1 and M^3. Eventually, they send these components of randomized encoding to the server. Once the server receives the components, it computes the dot products between the i-th sample of Alice and the j-th sample of Bob by summing up the entries of the vector $(M^1_{.i} \odot M^2_{.j} - M^3_{.i} - M^4_{.j})$ where "$\odot$" represents the Hadamard product. The server repeats this process for all pairs of samples of Alice and Bob. In order to complete the *gram matrix* which indicates the inner product of the vectors, the input-parties compute the dot product among their own samples and send the resulting matrices to the server. Then, the server employs Eq. 3, in which the RBF kernel computation via the inner product is shown, to compute the RBF kernel matrix.

5   Security Analysis

In this section, we give the threat model that we used to assess the security of the framework. Based on the threat model, we evaluated the security of the oligo kernel and the RBF kernel calculations.

5.1   Threat Model

We use the semi-honest adversary model in which a computationally bounded adversary that is not allowed to deviate from the protocol description attempts to obtain a valuable information from the messages sent during the execution of the protocol. In other words, all parties follow the protocol specification, but the parties may try to obtain additional information about the private input values of other parties based on their views on the execution of the protocol. In our framework, the input-parties try to keep their data private so, they can be assumed to be trusted parties not to actively cheat. The function-party wants to perform arithmetic operations on the data of the input-parties. It is expected

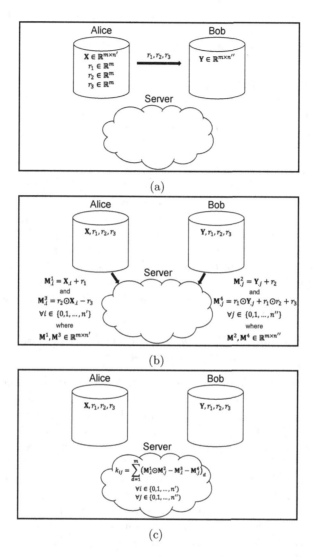

Fig. 4. (a) In order to create the components of the encoding, Alice creates three vectors with uniformly chosen random values of the same length with a sample in the input vector X. Then, she sends these vectors to Bob. (b) When Bob receives the random values, he computes the random components M^2 and M^4. Meanwhile, Alice computes her shares of encoding, namely M^1 and M^3. Eventually, they send these components of the randomized encoding to the server. (c) Once the server receives the components, it computes the dot products between the i-th sample of Alice and the j-th sample of Bob by summing up the entries of the vector $(M^1_{\cdot i} \odot M^2_{\cdot j} - M^3_{\cdot i} - M^4_{\cdot j})$. The server repeats this process for all pairs of samples of Alice and Bob.

that the function-party actively misbehaves in order to obtain the data of the input-parties. However, the function-party does not send messages to the input-parties in our framework. Therefore, it has to make a coalition with one of the input-parties in order to obtain the data of the other input-party.

5.2 Oligo Kernel Computation

Lemma 1. *Let $\mathcal{A}$ be a semi-honest adversary. The advantage of $\mathcal{A}$ of obtaining the positions of oligomers by analyzing the number of occurrences of oligomers is negligible.*

Proof. Alice and Bob share the number of occurrences of oligomers $[l_0^X, l_1^X, \cdots, l_n^X]$ and $[l_0^Y, l_1^Y, \cdots, l_n^Y]$ to each other. $\mathcal{A}$ can try to reconstruct Alice's sequence and Bob's sequence by using $[l_0^X, l_1^X, \cdots, l_n^X]$ and $[l_0^Y, l_1^Y, \cdots, l_n^Y]$, respectively. We assume that $\mathcal{A}$ can obtain some possible position information of the oligomers from all possible sequences that she can construct. The success rate of the attack increases with the increase in the size of oligomers, that is the value of K. Alice and Bob add a varying number of dummy positions for randomly chosen oligomers. This method increases the number of possible sequences that $\mathcal{A}$ can construct, and thus reduce the success rate of this attack to negligible levels.

Theorem 1. *The computation of the oligo kernel described in Sect. 4.1 is secure in the presence of a semi-honest adversary $\mathcal{A}$.*

Proof. In the computation of the oligo kernel, Alice and Bob compute $(X' + D)$ and $(-Y' - D)$, respectively, where X' and Y' are vectors of their secret values and D is the vector of random values. The calculation is directly based on the randomized encoding for addition. The only difference is that it is made over multiple values. Alice and Bob share the number of occurrences of oligomers $L^X = [l_0^X, l_1^X, \cdots, l_n^X]$ and $L^Y = [l_0^Y, l_1^Y, \cdots, l_n^Y]$ to each other during the calculation. We assume that there is a semi-honest adversary $\mathcal{A}$ that can obtain X' or Y' with non-negligible probability. $\mathcal{A}$ can use L^X or L^Y in order to obtain X' or Y', respectively. However, $\mathcal{A}$ cannot obtain X' or Y' from L^X or L^Y, respectively (Lemma 1). $\mathcal{A}$ has to decode X' and Y' from messages $(X' + D)$ and $(-Y' - D)$. However, this contradicts with the privacy property of the perfect randomized encoding of addition function (Definition 2).

5.3 RBF Kernel Computation

Theorem 2. *The computation of RBF kernel described in Sect. 4.2 is secure in the presence of semi-honest adversary $\mathcal{A}$.*

Proof. In the computation of the RBF kernel, Alice and Bob compute $\{(X_{\cdot i} + r_1), (r_2 \odot X_{\cdot i} - r_3)\}$ and $\{(Y_{\cdot j} + r_2), (r_1 \odot Y_{\cdot j} + r_1 \odot r_2 + r_3)\}$, respectively, where $X_{\cdot i}$ and $Y_{\cdot j}$ are the column vectors of their secret values, and r_1, r_2 and r_3 are the vectors of random values. The calculation is directly based on the randomized encoding of multiplication. The only difference is that it is made over multiple values. We assume that there is a polynomial-time adversary $\mathcal{A}$ that can decode $X_{.i}$ and $Y_{.j}$ from messages $(X_{.i} + r_1), (r_2 \odot X_{.i} + r_3), (Y_{.j} + r_2)$ and $(r_1 \odot Y_{.j} + r_1 \odot r_2 + r_3)$. This contradicts with the privacy property of perfect randomized encoding of the multiplication function (Definition 3).

6   Dataset

To show the benefit of privacy-preserving machine learning, we chose a genomic data set with clinical relevance in precision medicine. It comprises V3 loop sequences of HIV together with the coreceptor usage as the label. Coreceptor usage determines, how the viruses are entering the human cells. Since the most common variants can only use the human CCR5 coreceptor, which can be blocked by a drug, it is very important to determine the coreceptor usage of the viral population before prescribing those antiretroviral drugs [14]. Over recent years several successful applications based on predicting the phenotype from genetic data have been introduced [7]. We downloaded publically available data from the Los Alamos National Laboratory (LANL) HIV Sequence Database at http://www.hiv.lanl.gov/. We chose all amino acid sequences with coreceptor usage information and then classified the data into two classes (CCR5 only versus OTHER). We had 642 and 124 sequences in each class, respectively. The sequences were aligned with the HIVAlign tool from the LANL website with standard options. At the end, we obtained 766 sequences with 44 characters each.

7   Results and Discussion

We utilized the framework to calculate the oligo kernel and the RBF kernel. We applied these kernels on V3 loop sequences of HIV to predict the coreceptor usage by employing support vector machines [6]. We shared the labels of the samples with the server in plaintext domain since it does not reveal any additional information. We determined the best parameters according to F1-score by 5-fold cross-validation. In this process, we evaluated different values for the kernel parameters, which are the positional tolerance parameter $\sigma \in \{2^{-5}, 2^{-4}, \cdots, 2^{10}\}$ of the oligo kernel and the similarity adjustment parameter $\sigma \in \{2^{-5}, 2^{-4}, \cdots, 2^{10}\}$ of the RBF kernel. Similarly, we evaluated different values for the SVM parameters, which are the weight $w_1 \in \{2^0, 2^1, \cdots, 2^5\}$ of the minority class *OTHER* and the misclassification penalty parameter $C \in \{2^{-5}, 2^{-4}, \cdots, 2^{10}\}$. We repeated the parameter optimization step 10 times with different random folds and conducted the corresponding experiments with each set of optimal parameters separately. We evaluated the results of the experiments by employing F1-score and area under receiver operating characteristic (AUROC) curve. In the experiments, we selected the random values from the range $[1, 100]$.

In the experiments with the proposed framework, we employed three processes. Two of them are the input-parties and the last one is the function-party. We split the data into input-parties equally and used around 20% of the data of each input-party as test data. We conducted the experiments on a server having 512 GB memory, Intel Xeon E5-2650 processor and 64-bit operating system. We let the parties communicate with each other over TCP sockets and assumed that the communication is secure. We implemented the framework in Python. This holds also for the key switching approach [28] to which we compare our approach.

7.1   Oligo Kernel Experiments

In the oligo kernel experiments, we utilized the proposed framework to calculate the kernel function without sacrificing the privacy of the sequence data. Since we were not able to find a study which aimed for a similar scenario and utilized the oligo kernel, we compared our approach to the non-private oligo kernel computation in which we simply ran the oligo kernel in a single party where the whole data is available. The execution time of both private scheme (PP) and non-private scheme (NP) is shown in Fig. 5. Based on the figure, it can be stated that the privacy setup in the framework does not put too much extra burden on the execution time of the computation in PP. The private computation of the oligo kernel function is still done within a reasonable time. Moreover, the actual time required to obtain the results in PP is around one third of the given total execution time of PP since we utilized three processes, which are two processes for input-parties and one process for the function-party, to compute the function. Figure 5 displays the total execution time of these processes.

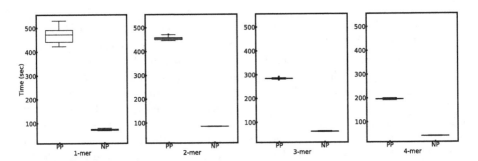

Fig. 5. We compare the execution time of our framework (PP) to the non-private scheme (NP) in the oligo kernel experiments. It shows that our framework is promising for real-life applications.

The execution time for the experiments of both schemes with larger K-mers tend to decrease with larger K values due to the decrease in the number of occurrences of K-mers in the sequences. The number of possible occurrences of

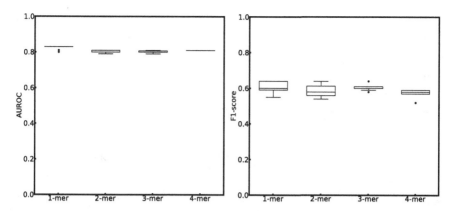

Fig. 6. In each K-mer, we have 10 different experiments. The experiments of both the private scheme and the non-private scheme with the oligo kernel yield the same results. Therefore, we give a single plot for each type of evaluation metric to display the results of both schemes. The results are better towards small K values due to the size of the alphabet of the protein sequences.

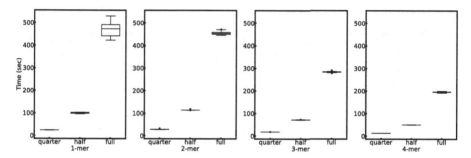

Fig. 7. The execution time of our approach in the oligo kernel experiment for a varying size of the dataset is depicted for 10 repetitions for each size and different K-mer lengths. It scales almost quadratically with the size of dataset for all K values.

K-mers in a sequence of length l is $l - K - 1$ and it gets lower by the increase of K. Furthermore, since the sequences that we employ have gap characters due to the alignment, the number of K-mers occurring in the sequences decreases parallel to the increase of K in our experiments considering that we do not allow gapped K-mers. Both approaches produced the same results in terms of F1-score and AUROC, and Fig. 6 demonstrates these results. Having the same results indicates that our framework is able to calculate the exact differences without sacrificing the privacy of the data.

We further conducted experiments on our proposed framework by using datasets of varying sizes to demonstrate the scalability of the framework. We used a quarter, a half and the full of the dataset. The execution time for these experiments are depicted in Fig. 7. The complexity of the framework grows almost quadratically with the size of dataset.

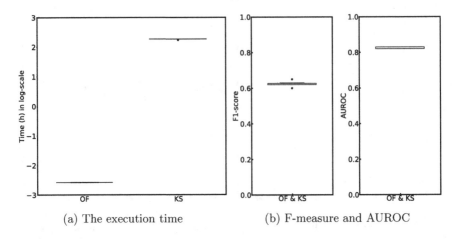

(a) The execution time   (b) F-measure and AUROC

Fig. 8. (a) In the RBF kernel experiment, we compare the execution time of our framework (OF) to the key switching approach (KS) in log-scale. In each approach, we have 10 repetitions of the experiments. (b) Both OF and KS experiments yielded the same results in terms of F1-score and AUROC. The RBF kernel yields comparable or even better results than the oligo kernel.

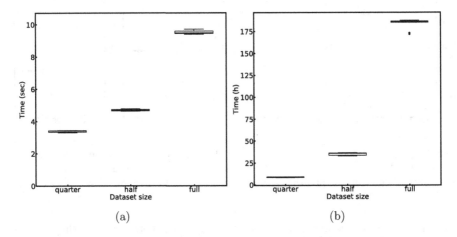

(a)   (b)

Fig. 9. (a) The execution time of our approach in the RBF kernel experiment for varying size of the dataset is shown. It scales almost quadratically with the size of dataset. (b) Similarly, the execution time of the key switching approach to compute the RBF kernel scales quadratically with the size of dataset. Note that the units are seconds and hours, respectively, in the figures.

7.2   RBF Kernel Experiments

As an alternative approach, we employed the RBF kernel to predict the coreceptor usage of HIV. In order to prepare the data for the RBF kernel, we encoded the sequences with one-hot-encoding such that each amino acid in a sequence was represented by 21 bits in which only one of those bits is 1 and the rest are 0. We utilized the formula given in Eq. 3 to calculate the RBF kernel on the samples of input-parties, which requires us to perform the element-wise multiplication of the feature vectors of these samples. In order to compare the performance of our proposed framework to existing approaches, we used key switching approach [28] which utilized the idea in [26,29]. We selected the length of the secret key in the key switching approach as 10 which was the length of the secret key in their experiments. In both experiments, the dot product among the samples in the same input-party is calculated directly in that input-party and the result is sent to the server. To compute the dot product of the samples from different input-parties, we utilized our proposed approach and the key switching approach separately. Both approaches gave the same results for the same set of parameters. F1-score and AUROC of these experiments are shown in Fig. 8b. Based on the results, the RBF kernel can be considered as a competitive alternative to the oligo kernel for this prediction scenario.

The execution time of both approaches are shown in Fig. 8a. It indicates that our approach is way faster than the key switching approach. The underlying reason for such a difference is that the number of features of the samples affects the execution time of the key switching approach drastically [28]. In our case, the length of the feature vectors is $21 * 44$, that is 924. Whereas, our approach can handle high dimensional vectors efficiently compared to the key switching approach and it does not involve any computationally expensive encryption. Moreover, we can conclude that our approach is more efficient than the computation of the dot product via HELib [8,9] since integer vector homomorphic encryption is faster than HELib based dot product calculation [26].

8   Conclusion

Due to the necessity of data sharing in biomedical studies, privacy preservation becomes very essential. Especially in genetics, the protection of the sequence data of patients in the studies involving two sources requires the development of new privacy methods. In order to address this, we propose a framework utilizing randomized encoding in order to enable the computation of machine learning algorithms with a single basic arithmetic operation on the data from two sources. During the computation, none of the input value of these sources is revealed to neither of the other parties in the computation. Moreover, the result of the computation is not revealed to the sources. We demonstrate the performance of the framework on the coreceptor usage prediction problem of HIV by utilizing V3 loop sequences. The results of experiments of the oligo kernel show that our framework can yield the same results as with non-private scheme. The execution time analysis of oligo kernel experiments shows that our framework is promising for real-life applications. In the experiments of the RBF kernel, we show that our framework can compute the kernel function without losing information. Moreover, it is significantly more efficient than the key switching approach and the dot product computation via HELib consequently. In addition to the utilized machine learning algorithms, our framework can be used on any problem requiring one type of four basic arithmetic operations on data from two sources. If two different basic arithmetic operations are required, then the function-party would infer the input values of the input-parties in the current setting by utilizing the information coming from different operations. As a future work, the proposed framework can be improved in a way that it could perform more than a single type of basic arithmetic operation in the same computation without sacrificing the privacy of the input values. Additionally, one can extend the idea to cover more than two input-parties in the computation.

Acknowledgement. This study is supported by the DFG Cluster of Excellence "Machine Learning – New Perspectives for Science", EXC 2064/1, project number 390727645. Furthermore, NP and MA acknowledge funding from the German Federal Ministry of Education and Research (BMBF) within the 'Medical Informatics Initiative' (DIFUTURE, reference number 01ZZ1804D).

References

1. Applebaum, B.: Garbled circuits as randomized encodings of functions: a primer. Tutorials on the Foundations of Cryptography. ISC, pp. 1–44. Springer, Cham (2017). https://doi.org/10.1007/978-3-319-57048-8_1
2. Applebaum, B., Ishai, Y., Kushilevitz, E.: Computationally private randomizing polynomials and their applications. Comput. Complex. **15**(2), 115–162 (2006)
3. Applebaum, B., Ishai, Y., Kushilevitz, E.: Cryptography in NC⁰. SIAM J. Comput. **36**(4), 845–888 (2006)
4. Applebaum, B., Ishai, Y., Kushilevitz, E.: How to garble arithmetic circuits. SIAM J. Comput. **43**(2), 905–929 (2014)
5. Ayday, E., De Cristofaro, E., Hubaux, J.P., Tsudik, G.: Whole genome sequencing: revolutionary medicine or privacy nightmare? Computer **48**(2), 58–66 (2015)
6. Cortes, C., Vapnik, V.: Support-vector networks. Mach. Learn. **20**(3), 273–297 (1995)
7. Döring, M., et al.: geno2pheno[ngs-freq]: a genotypic interpretation system for identifying viral drug resistance using next-generation sequencing data. Nucleic Acids Res. gky349 (2018). https://doi.org/10.1093/nar/gky349
8. Halevi, S., Shoup, V.: Algorithms in HElib. In: Garay, J.A., Gennaro, R. (eds.) CRYPTO 2014. LNCS, vol. 8616, pp. 554–571. Springer, Heidelberg (2014). https://doi.org/10.1007/978-3-662-44371-2_31
9. Halevi, S., Shoup, V.: HElib-an implementation of homomorphic encryption. Cryptology ePrint Archive, Report 2014/039 (2014)
10. Igel, C., Glasmachers, T., Mersch, B., Pfeifer, N., Meinicke, P.: Gradient-based optimization of kernel-target alignment for sequence kernels applied to bacterial gene start detection. IEEE/ACM Trans. Comput. Biol. Bioinform. **4**(2), 216–226 (2007)
11. Kale, G., Ayday, E., Tastan, O.: A utility maximizing and privacy preserving approach for protecting kinship in genomic databases. Bioinformatics **34**(2), 181–189 (2017)
12. Kauppi, J.P., et al.: Towards brain-activity-controlled information retrieval: decoding image relevance from MEG signals. NeuroImage **112**, 288–298 (2015)

13. Lengauer, T., Pfeifer, N., Kaiser, R.: Personalized HIV therapy to control drug resistance. Drug Discovery Today: Technol. **11**, 57–64 (2014)
14. Lengauer, T., Sander, O., Sierra, S., Thielen, A., Kaiser, R.: Bioinformatics prediction of HIV coreceptor usage. Nat. Biotechnol. **25**(12), 1407–1410 (2007). https://doi.org/10.1038/nbt1371
15. Liu, F., Ng, W.K., Zhang, W.: Encrypted SVM for outsourced data mining. In: 2015 IEEE 8th International Conference on Cloud Computing (CLOUD), pp. 1085–1092. IEEE (2015)
16. Lunshof, J.E., Chadwick, R., Vorhaus, D.B., Church, G.M.: From genetic privacy to open consent. Nat. Rev. Genet. **9**(5), 406 (2008)
17. Marouli, E., et al.: Rare and low-frequency coding variants alter human adult height. Nature **542**(7640), 186 (2017)
18. Meinicke, P., Tech, M., Morgenstern, B., Merkl, R.: Oligo kernels for datamining on biological sequences: a case study on prokaryotic translation initiation sites. BMC Bioinform. **5**(1), 169 (2004)
19. Mersch, B., Gepperth, A., Suhai, S., Hotz-Wagenblatt, A.: Automatic detection of exonic splicing enhancers (ESEs) using SVMs. BMC Bioinform. **9**(1), 369 (2008)
20. Michailidou, K., et al.: Genome-wide association analysis of more than 120,000 individuals identifies 15 new susceptibility loci for breast cancer. Nat. Genet. **47**(4), 373 (2015)
21. Ming, J., et al.: COINSTAC: decentralizing the future of brain imaging analysis. F1000Research **6** (2017)
22. Pfeifer, N., Kohlbacher, O.: Multiple instance learning allows MHC class II epitope predictions across alleles. In: Crandall, K.A., Lagergren, J. (eds.) WABI 2008. LNCS, vol. 5251, pp. 210–221. Springer, Heidelberg (2008). https://doi.org/10.1007/978-3-540-87361-7_18
23. Reis-Filho, J.S.: Next-generation sequencing. Breast Cancer Res. **11**(3), S12 (2009)
24. Schölkopf, B., Smola, A.J., et al.: Learning with Kernels: Support Vector Machines, Regularization, Optimization, and Beyond. MIT Press, Cambridge (2002)
25. Vaidya, J., Yu, H., Jiang, X.: Privacy-preserving SVM classification. Knowl. Inf. Syst. **14**(2), 161–178 (2008)
26. Yu, A., Lai, W.L., Payor, J.: Efficient integer vector homomorphic encryption (2015)
27. Zhang, J., Ma, K.K., Er, M.H., Chong, V.: Tumor segmentation from magnetic resonance imaging by learning via one-class support vector machine. In: International Workshop on Advanced Image Technology (IWAIT 2004), pp. 207–211 (2004)
28. Zhang, J., Wang, X., Yiu, S.M., Jiang, Z.L., Li, J.: Secure dot product of outsourced encrypted vectors and its application to SVM. In: Proceedings of the Fifth ACM International Workshop on Security in Cloud Computing, pp. 75–82. ACM (2017)
29. Zhou, H., Wornell, G.: Efficient homomorphic encryption on integer vectors and its applications. In: 2014 Information Theory and Applications Workshop (ITA), pp. 1–9. IEEE (2014)

Anonymous Asynchronous Payment Channel from k-Time Accountable Assertion

Yangguang Tian[1]([✉]), Yingjiu Li[1], Binanda Sengupta[1], Nan Li[2], and Yong Yu[3]

[1] School of Information System, Singapore Management University, Singapore, Singapore
{ygtian,yjli,binandas}@smu.edu.sg
[2] School of Electrical Engineering and Computing, University of Newcastle, Callaghan, NSW, Australia
nan.li@newcastle.edu.au
[3] School of Computer Science, Shaanxi Normal University, Xi'an, China
yuyong@snnu.edu.cn

Abstract. Accountable assertion enables a payer to make off-chain payments to a payee, and at the same time, the payer's secret credentials can be revealed if she equivocates (i.e., makes conflicting statements to others). In this paper, we introduce a new construction of accountable assertion that allows an assertion to be accountable for k times. We also present a new construction of anonymous payment channels for the cryptocurrency Bitcoin that allows a payer with k-time accountable assertions to anonymously make off-chain payments to the payee. In particular, we define formal security models for the new constructions, we also prove that the k-time assertion can achieve strong secrecy, and the asynchronous payment channel can achieve anonymity and untraceability. The proposed anonymous payment channel with k-time accountable assertions ensures that: (1) the payee can anonymously receive funds at asynchronous points of sale, and (2) the payee can trace the real identities of payers when they equivocate, and penalize them afterward.

Keywords: k-time accountability · Anonymity · Untraceability · Payment channel

1 Introduction

Bitcoin is a peer-to-peer payment system [13]. All payment transactions are appended in a public ledger (or blockchain), and each transaction is verified by network nodes in a peer-to-peer manner. Since the blockchain usually takes 10 min to confirm a set of valid payment transactions, the scalability becomes a bottleneck for its practicality. To address such scaling problem, the Lightning Network [19] (as well as Ethereum's Raiden Network [2]) was proposed, which aims to reduce the burden on the blockchain. That is, a massive amount of

© Springer Nature Switzerland AG 2019
Y. Mu et al. (Eds.): CANS 2019, LNCS 11829, pp. 512–524, 2019.
https://doi.org/10.1007/978-3-030-31578-8_28

transactions are performed *off-chain*, only two transactions (i.e., the initial and final ones) are appended in the blockchain.

Preventing the equivocation is important in the off-chain payments. One novel solution is to use the accountable assertion [21], which means that an off-chain payment transaction is confirmed as valid if the corresponding assertion is uniquely accountable for a statement (e.g., I buy a ticket with serial number r, where r is a unique value). More specifically, a payer is required to have a deposit in the initial transaction which is appended in the blockchain. Later, if a payer equivocates such as making any conflicting statements, then the payer's secret credential (or secret key) is revealed and her deposit is confiscated.

The accountable assertion can be further explored in a k-time setting (k-time accountable assertion kTAA), which is useful if a payer wishes to reuse her assertion multiple times. As a result, multiple payment transactions can be confirmed as valid using kTAA. A real-life example is, a user makes payment transactions with a set of service providers who belongs to the same organization. The user first derives a kTAA after a valid payment transaction with one of the service providers, then she continues to obtain services from remaining service providers using kTAA. That is, the kTAA ensures that an assertion can be accountable up to k number of service providers (or k-time accountability). In addition, the usage of kTAA can minimize user's computational and communication cost for off-chain transactions.

While kTAA offers a novel solution to the equivocation, it would trigger concerns on privacy when applying it to the asynchronous payment channel. Many privacy-preserving cryptocurrencies (as well as payment channels) have been studied in the literature [7,9,12,22]. However, to the best of our knowledge, there is no solution that can be directly applied to construct an anonymous asynchronous payment channel with kTAA. In particular, the anonymity guarantee for a payment channel is necessary [1,9] when privacy-sensitive payers are concerned about the privacy of off-chain payment transactions. Therefore, the complete goal of this work is to design an *anonymous asynchronous payment channel with kTAA*.

The anonymous asynchronous payment channel with kTAA is particularly useful in distributed applications such as public transport. Let us consider a number of passengers $\{A_i\}$ who wish to use a bus company B's services. Suppose a passenger A_i first appends a transaction into the Bitcoin network for *creating a payment channel* to B. The initial transaction includes a deposit ฿ d and a penalty ฿ p, which are anonymously and jointly signed by A_i and B. Then, a passenger A_i can use the payment channel to pay for multiple rides when she uses B's buses B_i up to the limit ฿ d (i.e., *paying through the channel*). Eventually, B's buses are synchronized every night when *closing the channel*. We let A_i keep the state (includes the current transaction amount, a revision number of the state and the deposit ฿ d) of the payment channel, which is anonymously signed by B. A_i generates a kTAA associates with the signed state, so this kTAA allows A_i to pay for a certain number (or set) of rides altogether. The anonymous payment channel ensures that any rides cannot disclose A_i's real identity. In addition, B

can trace and penalize any passenger who equivocates (e.g., A_i pays for a new set of rides using the old signed state) or misuses the kTAA (e.g., A_i uses the signed state for k+1 times).

Technical Challenges. The accountable assertion is essentially based on a chameleon hash function [10] in a (Merkle-style) tree setting, and all collisions in a tree path can be found using the secret key (or trapdoor) of a payer. Since we aim to construct a kTAA, we distribute the trapdoor into k shares by using Shamir's secret sharing scheme (SSS) [24]. We assume the height of the tree is k, and we find collisions in a tree path (from leaf to the root) using k secret shares, which is the key difference between accountable assertion and kTAA. The problem is, the beneficiary (e.g., payee) cannot reveal payer's trapdoor and penalize her if she equivocates or misuses kTAA in the asynchronous payment channel. Because the deposit ₿ d and penalty ₿ p are signed by the payer's trapdoor (*not* trapdoor shares) when *creating a payment channel*, the dishonest payer's trapdoor remains secure even when any trapdoor shares are extracted using the "extraction" algorithm of chameleon hash functions (see Sect. 2).

Overview of Techniques. First, we design a kTAA which is suitable for off-chain payments. The idea is to let the payer release a k-size set of trapdoor shares in a tree path, and the released k shares are not used to find collisions when generating a kTAA. Essentially, the kTAA has two trapdoor shares at each level: one is a secret value which is used to find a collision; the other one is a public value, which is used for tracing the dishonest payer. If equivocation or kTAA misuse occurs, then payer's trapdoor is revealed from the k publicly known shares and the *revealed* secret share (due to the "extraction" algorithm of chameleon hash function) by using the *reconstruction* of Shamir's SSS [24].

Second, we exploit cryptographic primitives to achieve anonymous payment channel with kTAA. Anonymous payment channel requires that the payee cannot identify the real identity of a payer when *paying through the channel*. We achieve it by using both ring signatures and blind signatures. That is, the deposit ₿ d and the penalty ₿ p are jointly signed by a payer and a payee using ring signatures [8,11] and Schnorr signatures [23], respectively. A payer derives a pseudonym/credential pair using blind signatures [4] when interacting with the payee, while the payee verifies the payment transactions between payer and payee using the pseudonym/credential pair. As a result, the payee cannot identify the payer from the pseudonym/credential pair (due to the untraceability of blind signatures), and the signed deposit and penalty (due to the anonymity of ring signatures).

Our Contributions. The main contributions of this work are summarized as follows.

- *New Accountable Assertion.* We propose a new k-time accountable assertion (kTAA) for non-equivocation contracts such that an assertion is accountable up to k times. In particular, the secret key used in the kTAA is revealed when equivocation or kTAA misuse occurs.

- *New Asynchronous Payment.* We propose a new asynchronous payment channel such that the real identity of a payer is hidden from the off-chain payee when paying through the channel.
- *Security Models.* We present the formal security definition for kTAA, and prove that it achieves strong secrecy. We also prove that the proposed asynchronous payment achieves anonymity and untraceability.

1.1 Related Work

Anonymous Cryptocurrency. Zerocoin (Miers et al.) is the first *anonymous* distributed e-cash system for Bitcoin [12]. Specifically, it includes some cryptographic techniques such as accumulators [6] and zero-knowledge proofs [23], in order to ensure unlinkability across multiple transactions. Meanwhile, the central "bank" [5] is not required, so Zerocoin is a *decentralized* e-cash scheme which is consistent with the Bitcoin network model. To ensure anonymity on sender/recipient and confidentiality on transaction amount, Sasson et al. [22] proposed a Zerocash system, which is an instantiation of decentralized anonymous payment (DAP) schemes. The DAP schemes allow users to pay each other directly and privately.

Saberhagen [26] introduced the Cryptonote architecture, which adds new privacy-related features to the Bitcoin networks. Specifically, the one-time ring signature [3] and non-interactive zero-knowledge (NIZK) proofs [23] are used to achieve untraceability (or sender anonymity) and unlinkability. Later, ring confidential transactions (RCT) [16] and RCT 2.0 [25] (a variant of RCT) were proposed to hide the transaction amount, which is not possible in Cryptonote. Both RCT and RCT 2.0 can be deployed in the cryptocurrency Monero.

Anonymous (Micro)Payments. Poon and Dryja [19] introduced a Lightning Network (or off-chain payments) to tackle the scalability issue of Bitcoin. Such payment channel allows payer transfers money to the payee without making all their signed transactions public on the blockchain (i.e., only the first and last transaction are appended to the blockchain). However, the Lightning network did not provide payment anonymity between payment channel participants (e.g., the payee can see the real identity of a payer who makes a payment). Later, an anonymous payment channel was introduced by Green and Miers [9]: Blind Off-chain Light-weight Transactions (Bolt). The Bolt can achieve unlinkability and confidentiality on payment amount by using cryptographic techniques, including NIZK and commitment scheme [18].

A separate line of research on probabilistic micropayment schemes [7,17] has evolved. Pass et al. [17] presented three probabilistic micropayments schemes for decentralized currencies. The probabilistic micropayment means that a payee receives a macro-value payment with a given probability so that, in expectation, he receives a micro-value payment [20]. However, these proposed probabilistic micropayments either have an on-line trusted party or do not support off-chain payments. Meanwhile, the probabilistic micropayments are linkable. To achieve the unlinkable probabilistic micropayments, Chiesa et al. [7] introduced the decentralized anonymous micropayment (DAM) scheme which allows users with access to a ledger to perform off-chain probabilistic payments with each other directly and privately.

Prevent the double-spending in the off-chain (micro)payments is important. In [7], the double-spend technique is based on 2-out-of-n Shamir's SSS [24], which is also used in [4]. By exploiting the chameleon hash function (HF) [10], Ruffing et al. [21] introduced the accountable assertion, which enables time-locked deposit in Bitcoin that are revoked when double spend occurs. In this work, we use chameleon HF to capture the double-spending payer.

To highlight our distinction, we show the feature (or function) difference between our proposed anonymous payments with kTAA and some closely-related works in Table 1: it shows that the proposed construction has k-time accountability and sender anonymity, it works at an unidirectional public channel (i.e., multiple payers and one payee) and avoids using expensive pairing operations. In particular, the proposed kTAA can be regarded as a generalization of accountable assertion [21], and we assume that the receiver (or payee) *behaves honestly* [7] in the asynchronous payment channel.

Table 1. The comparison between various cryptocurrencies and (micro)payments

Feature/Scheme	[12]	[26]	[22]	[21]	[16]	[19]	[9]	[7]	Ours
k-time accountability	×	×	×	×	×	×	×	×	✓
Deterministic pay[a]	✓	✓	✓	✓	✓	✓	✓	×	✓
Pairing free	✓	✓	×	✓	✓	✓	×	✓	✓
Unlinkability	✓	×[b]	✓	×	×	×	✓	✓	×
Sender anonymity	×	✓	✓	×	✓	×	✓	✓	✓
Recipient anonymity	×	×	✓	×	✓	×	✓	✓	×
Confidentiality	×	×	✓	×	✓	×	✓	✓	×

[a] Deterministic payments means non-probabilistic payments.
[b] Cryptonote transactions are linkable due to some attacks mentioned in [14].

2   Preliminaries

In this section, we present the chameleon hash function, which will be used in our proposed kTAA scheme. The chameleon hash function was proposed by Krawczyk and Rabin [10]. Together with an extraction algorithm **ExtractCsk**, the generic chameleon hash function is defined as follows [21].

- **KeyGen:** It takes a security parameter λ as input, outputs a secret/public key pair $(\mathsf{sk}, \mathsf{pk}) \leftarrow \mathsf{KeyGen}(1^\lambda)$.
- **CH:** It takes a public key pk, a message m and a randomness r as input, outputs an Chameleon hash $h \leftarrow \mathsf{CH}(\mathsf{pk}, m)$.

- **Col:** It takes a secret key sk, the old message/randomness pari (m, r), a hash h and a new message m' as input, outputs a new randomness r' such that $h = \mathsf{CH}(m, r) = \mathsf{CH}(m', r')$.
- **ExtractCsk:** It takes a public key pk, the old message/randomness pair (m, r) and new message/randomness pair (m', r'), outputs the secret key sk.

3 k-Time Accountable Assertions

In this section, we first define and construct a k-time accountable assertion (kTAA) based on accountable assertion [21] and Shamir's SSS [24]. Then, we present the strong secrecy model and provide the security of kTAA.

3.1 Definition

Definition 1 (k-time Accountable Assertions).

- **KeyGen:** *It takes the security parameter λ as input, outputs a public/secret key pair (pk, sk).*
- **Commit:** *It takes a secret key sk as input, outputs a set of commitments $C_1, \cdots, C_k$.*
- **Assert:** *It takes a secret key sk, a set of contexts $ct_1, \cdots, ct_k$ and a statement st, outputs an assertion τ.*
- **Verify:** *It takes a public key pk, an assertion τ, outputs $d \in \{0, 1\}$ to indicate whether τ is a valid assertion of a statement st in the set of contexts $ct_1, \cdots, ct_k$ and the set of commitments $C_1, \cdots, C_k$ under the public key pk.*
- **Extract:** *It takes a public key pk, a set of contexts $ct_1, \cdots, ct_k$, two statements st_0, st_1 and two assertions τ_0, τ_1, outputs the secret key sk.*

3.2 Security Models

A secure kTAA should have extractability and secrecy. The extractability means that a secret key can be extracted if two distinct statements have been asserted in the same set of contexts. We directly use the extractability model defined in [21]. Below we present the strong secrecy, which models an indistinguishability adversary in a decisional game.

Strong Secrecy. Informally, a secret key remains secure even if adversary asks the challenger to reveal a number of secret shares (which derived from the secret key) used in an assertion. We define a formal game between an adversary $\mathcal{A}$ and a simulator $\mathcal{S}$ as follows.

- Setup: $\mathcal{S}$ generates a secret/public key pair (sk, pk) for a user by running the corresponding **KeyGen** algorithm. $\mathcal{S}$ returns the public key pk to $\mathcal{A}$, and maintains a set $\mathcal{Q}$ which is initialized as empty. $\mathcal{S}$ also tosses a random coin b which will be used later in the game. Let $\mathcal{U}$ be another initially empty set for recording the corrupted secret shares. Let M be a deterministic algorithm which takes a secret key sk, a statement st and a set of contexts $\{ct_i\}$ as input, outputs a secret key share f_i.

- Training: $\mathcal{A}$ can make the following queries in arbitrary sequence to $\mathcal{S}$.
  - Assert: If $\mathcal{A}$ sends a statement st_i and a set of contexts $\{ct_i\}$ to $\mathcal{S}$, then $\mathcal{S}$ returns an assertion $\tau_i \leftarrow$ **Assert**($sk, \{ct_i\}, st_i$) associated with a set of commitments $\{C_i\}$ to $\mathcal{A}$. $\mathcal{S}$ includes the statement/context pair $(st_i, \{ct_i\})$ to $\mathcal{Q}$.
  - Corrupt: If $\mathcal{A}$ issues a corrupt query to an assertion τ_i with respect to a statement st_i, then $\mathcal{S}$ returns a secret share $f_i \leftarrow$ M($sk, st_i, \{ct_i\}$) to $\mathcal{A}$. $\mathcal{S}$ includes the corrupted secret share f_i to $\mathcal{U}$.
- Challenge: Upon receiving a challenge statement st^* and a challenge set of contexts $\{ct_i^*\}$ ($(st^*, \{ct_i^*\}) \notin \mathcal{Q}$) from $\mathcal{A}$, $\mathcal{S}$ simulates the challenge assertion using either a real secret share f^* to $\mathcal{A}$ if $b = 1$, or a random secret share $R \leftarrow \{0,1\}^\lambda$ if $b = 0$. $\mathcal{A}$ outputs a bit b', and wins if $b' = b$.
  We let the secret share $f^* \notin \mathcal{U}$ link to a context $ct^* \in \{ct_i^*\}$, and $\mathcal{A}$ is allowed to corrupt at most k secret shares $\{f_i\} \in \mathcal{U}$ with respect to the challenge statement and contexts set. We prohibit that $\mathcal{S}$ generates any assertions with distinct statements in the same set of contexts. We define the advantage of $\mathcal{A}$ in the above game as follows

$$\mathsf{Adv}_\mathcal{A}(\lambda) = |\Pr[\mathcal{S} \to 1] - 1/2|.$$

Definition 2. *We say a kTAA has strong secrecy if for any PPT $\mathcal{A}$, $\mathsf{Adv}_\mathcal{A}(\lambda)$ is a negligible function of the security parameter λ.*

3.3   Proposed Construction and Security Analysis

The kTAA construction is based on the discrete logarithms assumption, let $\mathbb{G}$ denote a cyclic group with prime order q and generator g. Let L be a hash function that maps bitstrings to leafs $\{1, \cdots, n^{\ell-1}\}$, where ℓ, n are the height and the branching factor of a tree. Let H, G be collision-resistant hash functions, and let F_κ be a pseudorandom function such that κ is a pseudo-random key.

- **KeyGen:** The key generation algorithm chooses a pseudo-random function key $\kappa \leftarrow \{0,1\}^\lambda$, and chooses an assertion secret/public key pair (ask, apk), where $ask = \alpha$ and $apk = (g^\alpha, z), z = $ H($y_1^1, \cdots, y_n^1$). The algorithm computes the entries for the chameleon hash function in the root node as $y_i^1 = $ CH(x_i^1, r_i^1), where $x_i^1 = F_\kappa(p, i, 0), r_i^1 = F_\kappa(p, i, 1), i \in [1, n]$ and p denotes the unique identifier for the position of the root node.
- **Commit:** The algorithm chooses a random polynomial f of degree at most k with coefficients $\{\alpha_l\} \in \mathbb{Z}_q$, for $1 \le l \le $ k, and publishes the k-size commitments $C_l = g^{\alpha_l}$, for $1 \le l \le $ k. Specifically, $f(x) \xleftarrow{k} \alpha + \sum_{l=1}^k \alpha_l \cdot x^l$, where α denotes the secret to be shared. *We assume the height of tree as $\ell = $ k.*
- **Assert:** The stateful assertion algorithm maintains an initially empty set L of used leaf positions. The algorithm verifies whether L($\{ct_i\}$) $\notin L$ ($i \in [1, \cdots, \ell]$) with respect to a statement st in a set of contexts $\{ct_i\}$, it outputs "0" when the verification fails. Otherwise, it adds L($\{ct_i\}$) to L, and the algorithm proceeds. Then, a user computes an assertion path

$(Y_\ell, a_\ell, Y_{\ell-1}, a_{\ell-1}, \cdots, Y_1, a_1)$ from a leaf Y_ℓ to the root Y_1. In the following, let $x_i^j = F_k(p_j, i, 0)$ and $r_i^j = F_k(p_j, i, 1)$, where p_j is a unique identifier of the position of the node Y_j. Note that $a_\ell \in \{1, \cdots, n\}$ is the position of the entry within Y_ℓ.

1. Compute Y_ℓ: Assert a statement st with respect to Y_ℓ by computing $r_{a_\ell}^{'\ell} = \mathsf{Col}(f(x_{(w,\ell)}), x_{a_\ell}^\ell, r_{a_\ell}^\ell, \mathsf{S}(st))$, where $f(x_{(w,\ell)}) \xleftarrow{k} \alpha + \sum_{l=1}^k \alpha_l \cdot x_{(w,\ell)}^l$, $x_{(w,\ell)} = \mathsf{H}(w, \mathsf{S}(st), ct_\ell)$ and $w \in [0,1]$ (its corresponding public key is $F(x_{(w,\ell)}) = g^{f(x_{(w,\ell)})}$).
2. Compute the entry $y_{a_\ell}^\ell = \mathsf{G}(\mathsf{CH}(\mathsf{S}(st)), r_{a_\ell}^{'\ell}) = \mathsf{G}(\mathsf{CH}(x_{a_\ell}^\ell, r_{a_\ell}^\ell), r_{a_\ell}^{'\ell})$.
3. Compute the remaining entries in node Y_ℓ as $y_i^\ell = \mathsf{CH}(x_i^\ell, r_i^\ell)$ for $i \in \{1, \cdots, n\}\backslash\{a_\ell\}$. The leaf Y_ℓ stores the entries $(y_1^\ell, \cdots, y_n^\ell)$.
4. Set $z_\ell = \mathsf{H}(y_1^\ell, \cdots, y_n^\ell)$ and $f_\ell = (y_1^\ell, \cdots, y_{a_\ell-1}^\ell, F(x_{(w,\ell)}), y_{a_\ell+1}^\ell, \cdots, y_n^\ell)$.

Next, it computes the remaining nodes up to the root for $h = \ell - 1, \cdots, 1$ as follows.

1. Assert the value z_{h+1} with respect to Y_h by computing $r_{a_\ell}^{'h} = \mathsf{Col}(f(x_{(w,h)}), x_{a_\ell}^h, r_{a_\ell}^h, z_{h+1})$, where $f(x_{(w,h)}) \xleftarrow{k} \alpha + \sum_{l=1}^k \alpha_l \cdot x_{(w,h)}^l$, $x_{(w,h)} = \mathsf{H}(w, r_{a_\ell}^{'h}, ct_h)$ and $w \in [0,1]$.
2. Compute the entry $y_{a_\ell}^h = \mathsf{CH}(z_{h+1}, r_{a_\ell}^{'h}) = \mathsf{CH}(x_{a_\ell}^h, r_{a_\ell}^h)$.
3. Compute the remaining entries in node Y_h as $y_i^h = \mathsf{CH}(x_i^p, r_i^p)$ for $i \in \{1, \cdots, n\}\backslash\{a_\ell\}$. The leaf Y_h stores the entries $(y_1^h, \cdots, y_n^h)$.
4. Set $z_h = \mathsf{H}(y_1^h, \cdots, y_n^h)$ and $f_h = (y_1^h, \cdots, y_{a_\ell-1}^h, F(x_{(w,h)}), y_{a_\ell+1}^h, \cdots, y_n^h)$

The assertion is $\tau = \{(r_{a_\ell}^{'\ell}, f_\ell, a_\ell, f(x_{(1-w,\ell)})), \cdots, (r_{a_1}^{'1}, f_1, a_1, f(x_{(1-w,1)}))\}$. At each level ℓ, the assertion outputs a secret share $f(x_{(1-w,\ell)})$ while the correct collision is found using a secret share $f(x_{(w,\ell)})$ (underline part).

Verify: It parses the assertion public key as (apk, z), and performs the following

1. apk is a valid chameleon hash function key.
2. Verify the released trapdoor shares, e.g., the trapdoor share at level ℓ: $f(x_{(1-w,\ell)})$

$$\underline{g^{f(x_{(1-w,\ell)})}} \stackrel{?}{=} \mathsf{apk} \cdot C_1^{x_{(1-w,\ell)}} \cdot C_2^{x_{(1-w,\ell)}^2} \cdots C_k^{x_{(1-w,\ell)}^k}$$

where $x_{(1-w,\ell)} = \mathsf{H}(w, \mathsf{S}(st), ct_\ell)$.

3. Verify $\mathsf{H}(y_1^1, \cdots, y_n^1) \stackrel{?}{=} z$ by reconstructing the nodes $(Y_\ell, Y_{\ell-1}, \cdots, Y_1)$ in a bottom-up manner (from the leaf Y_ℓ to the root Y_1), which includes the entries $y_1^1, \cdots, y_n^1$.

- **Extract:** It takes $(\{ct_i\}, st_0, st_1, \tau_0, \tau_1)$ as input, extracts a secret share first. For example, a context ct_ℓ at level ℓ: $f(x_{(w,\ell)}) \leftarrow \mathbf{ExtractCsk}(x_0, r_0, x_1, r_1)$, where $\mathbf{ExtractCsk}$ is an extraction algorithm of the chameleon hash function. That means a position (e.g, level ℓ) is found where two assertion paths form a collision in the chameleon hash function (i.e., a position where (x_0, r_0) is used in the assertion path of st_0 and (x_1, r_1) is used in the assertion path of st_1 such that $\mathsf{CH}(x_0, r_0) = \mathsf{CH}(x_1, r_1)$). Then, the secret $\underline{\alpha}$ is revealed using the *reconstruction* of Shamir's SSS [24] from the revealed secret share

$f(x_{(w,\ell)})$ and the publicly known k trapdoor shares (w.r.t. either st_0 or st_1). If no collision is found, then this algorithm outputs "0".

Theorem 1. *The k-time accountable assertion achieves strong secrecy in the random oracle model if the chameleon hash function is collision resistant and the DDH assumption is held in the underlying group* $\mathbb{G}$.

Due to page limitation, the detailed security proof is deferred to the full version of this work.

4   A New Asynchronous Payment Protocol

An asynchronous payment protocol consists of **Setup, Payment** and **Synchronization** algorithms. The overview of the proposed construction is described as follows. First, the deposit ฿ d and the penalty ฿ p with expiry time T is jointly signed by a user A and a recipient B using a ring signature scheme and Schnorr signature scheme in the **Setup** algorithm. In addition, user A derives a pseudonym/credential pair and a signed state from B (we assume the deposit specifies the beneficiary as B) using a blind signature scheme. In the **Payment** algorithm, user A first generates a kTAA when interacting with a point of sale B_i. Then, user A holds a signed state, a pseudonym/credential pair and a kTAA for receiving services from k points of sale. In the **Synchronization** algorithm, B synchronizes with $\{B_i\}$. If user A makes a conflicting statement on the old signed state, then B reveals A's secret key first, then penalizes her and revokes her's real identity accordingly. We denote ring signature, blind signature and digital signature as $\Sigma_1 = (\textbf{KeyGen}, \textbf{Sign}, \textbf{Verify})$, $\Sigma_2 = (\textbf{Setup}, \textbf{KeyGen}, \textbf{Sign}, \textbf{Verify})$ and $\Sigma_3 = (\textbf{Setup}, \textbf{KeyGen}, \textbf{Sign}, \textbf{Verify})$, respectively. The Σ_3 is used by B only for signing the deposit and the penalty in the **Setup** algorithm. We then show the detailed full protocol below.

- **Setup:** A and B wish to establish an asynchronous payment channel. Hence, A can make transactions with B under amount ฿ d, penalty ฿ p (we assume $d = p$ for convenience) and expiry time T. A and B perform the following
  1. A generates a public/secret key pair $(\text{pk}_A, \text{sk}_A)$ for asynchronous payments. B also generates a set of public/secret key pairs $\{(spk_{B_i}, ssk_{B_i})\}$ with respect to many points of sale.
  2. A creates a payment channel with B with total amount ฿ $(d + p)$ and expiry time T. The amount ฿ $(d+p)$ with expiry time T is jointly signed by $\textbf{pk}_A$ and ssk_B using $\Sigma_1.\textbf{Sign}$ and $\Sigma_3.\textbf{Sign}$, respectively. We denote $\textbf{pk}_A$ as a set of public keys involved in Σ_1, which includes a real public key pk_A and many dummy public keys.
  3. After confirming the payment channel by the Bitcoin network, A obtains a *blinded* signed statement $\sigma'_A \leftarrow \Sigma_2.\textbf{Sign}(ssk_B, state)$ from B, where $state = (T, d, \textbf{pk}_A, \{k\}, b = 0, spk_B)$. Then A derives a pseudonym/credential pair $(id_A, id_A.cred)$, where id_A is the pseudonym which includes user's public key pk_A, and $id_A.cred$ is the signature which

derives from σ'_A (we denote $\sigma_A \xleftarrow{\text{BS}} \sigma'_A$ for convenience). Note that $\{k\}$ denotes a set of contexts in **kTAA**.

4. A sets her assertion public key as $\text{apk}_A = id_A$ and its secret key ask is a blinded version of sk_A. In addition, A publishes a set of commitments $\{C\}^k$ by running the **kTAA.Commit** algorithm.

We remark that the *state* is a plain signed message in the algorithm $\Sigma_2.\textbf{Sign}$. The deposit ฿ d will specify an explicit beneficiary B who will receive the penalty ฿ p in case of equivocates. In the beginning, $b = 0$ means that all money in the payment channel belongs to A and none belongs to B. We stress that both A and B are honest entities in the **Setup** algorithm.

- **Payment:** If A wants to spend an amount ฿ x $(x \leq d)$ at a point of sale B_i, then A and B_i perform the interaction in Fig. 1, we provide further explanations afterwards.

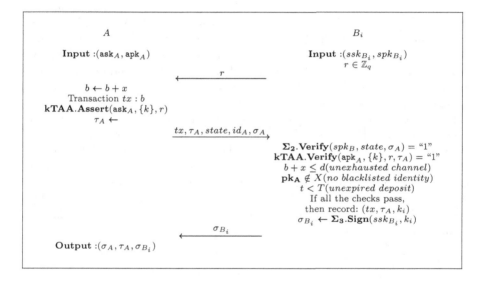

Fig. 1. Payment

1. To verify A's signed statement $(state, \sigma_A)$, B_i runs the $\Sigma_2.\textbf{Verify}$ algorithm, which takes pseudonym id_A as input. Meanwhile, to verify A's assertion, B_i runs the **kTAA.Verify** algorithm. Recall that the assertion public key $\text{apk}_A = id_A$.

2. The signed *state* includes a set of contexts $\{k\}$, each B_i needs to sign a used contexts k_i using $\sigma_{B_i} \leftarrow \Sigma_3.\textbf{Sign}(ssk_{B_i}, m_i)$, where $m_i = k_i$ denotes the used context. The remaining B_j $(j \neq i)$ perform the same checks as described in Fig. 1. For example, B_j verifies (k_i, σ_{B_i}), if the signature is valid and all the checks pass, then B_j records (tx, τ_A, m_j), returns $\sigma_{B_j} \leftarrow \Sigma_3.\textbf{Sign}(ssk_{B_j}, m_j)$ and provides services to A, where $m_j = (k_i, k_j)$.

3. The updated state/signature is $(state', \sigma'_A)$ if the old signed *state* is used up to k unsynchronous points of sale. Specifically, $b' = b + x$, $state' = (T, d, \mathbf{pk_A}, \{k\}, b', spk_{B_i})$ and $\sigma'_A \leftarrow \Sigma_2.\mathbf{Sign}(ssk_{B_i}, state')$.

– **Synchronization:** Before the expiry time T, B synchronizes with each B_i.
  1. B collects all transactions recorded by $\{B_i\}$.
  2. If B detects A has made conflicting statements, then B obtains A's secret key sk_A by running the **kTAA.Extract** algorithm.
  3. B uses the secret key sk_A to sign a transaction that spends the ฿ $(d+p)$ to an address under the control of B. Then B adds A's real identity $\mathbf{pk}_A$ to the blacklist X and broadcasts it to all B_i.
  4. Before expiry time T, B closes the channel. B also adds $\mathbf{pk}_A$ to the blacklist X and broadcasts it to all B_i.

Remark. The recipient B will eventually revoke A in the **Synchronization** algorithm (i.e., *closing the channel*). If A did not spend amount x for the committed k times, then B still receives the amount ฿ b when *closing the channel*. By contrast, if A reuses old signed state and its associated kTAA beyond the committed times (e.g., a k+1 TAA is generated without making conflict statement), then A's secret key is revealed using the *reconstruction* of Shamir's SSS. Due to page limitation, the security model (including anonymity and untraceability) and the detailed security proof are deferred to the full version of this work.

Instantiation. First, we use a linkable ring signature scheme in [11] or a traceable ring signature scheme in [8] to initiate the digital signature Σ_1 used in the deposit ฿ d and the penalty ฿ p with expiry time T. For example, we remove the public tracing property in [8] because the kTAA has such property. The simplified ring signature is essentially based on the non-interactive zero knowledge proof of knowledge (which is also used in CryptoNote protocol [26] and Monero [15]). Second, we use the digital signature in [4] to initiate the underlying blind signature Σ_2 as it does not involve expensive pairing operations. Last, Σ_3 can be instantiated to Schnorr signature scheme [23].

5   Conclusion

In this work, we have extended the accountable assertion [21] to a k-reusable setting (kTAA) using Shamir's secret sharing scheme [24]. We have also proposed a secure and anonymous asynchronous payment channel based on ring signatures and blind signatures. As for the future work, we would like to design: 1) an asynchronous payment channel with confidentiality on the transaction amount [7,22] while capturing double-spend based on accountable assertions; and 2) a bidirectional asynchronous payment channel that allows payer and payee to exchange payments in either direction [9].

Acknowledgements. This work is supported by the Singapore National Research Foundation under NCR Award Number NRF2014NCR-NCR001-012, National Natural Science Foundation of China (Grant No. 61872229 and 61872264), National Cryptography Development Fund during the 13th Five-year Plan Period (MMJJ20170216) and Fundamental Research Funds for the Central Universities (GK201702004).

References

1. Chainalysis. https://www.chainalysis.com
2. Raiden Network. https://raiden.network
3. Bender, A., Katz, J., Morselli, R.: Ring signatures: stronger definitions, and constructions without random oracles. In: Halevi, S., Rabin, T. (eds.) TCC 2006. LNCS, vol. 3876, pp. 60–79. Springer, Heidelberg (2006). https://doi.org/10.1007/11681878_4
4. Brands, S.: Untraceable off-line cash in wallet with observers. In: Stinson, D.R. (ed.) CRYPTO 1993. LNCS, vol. 773, pp. 302–318. Springer, Heidelberg (1994). https://doi.org/10.1007/3-540-48329-2_26
5. Camenisch, J., Hohenberger, S., Lysyanskaya, A.: Compact e-cash. In: Cramer, R. (ed.) EUROCRYPT 2005. LNCS, vol. 3494, pp. 302–321. Springer, Heidelberg (2005). https://doi.org/10.1007/11426639_18
6. Camenisch, J., Lysyanskaya, A.: Dynamic accumulators and application to efficient revocation of anonymous credentials. In: Yung, M. (ed.) CRYPTO 2002. LNCS, vol. 2442, pp. 61–76. Springer, Heidelberg (2002). https://doi.org/10.1007/3-540-45708-9_5
7. Chiesa, A., Green, M., Liu, J., Miao, P., Miers, I., Mishra, P.: Decentralized anonymous micropayments. In: Coron, J.-S., Nielsen, J.B. (eds.) EUROCRYPT 2017. LNCS, vol. 10211, pp. 609–642. Springer, Cham (2017). https://doi.org/10.1007/978-3-319-56614-6_21
8. Fujisaki, E., Suzuki, K.: Traceable ring signature. In: Okamoto, T., Wang, X. (eds.) PKC 2007. LNCS, vol. 4450, pp. 181–200. Springer, Heidelberg (2007). https://doi.org/10.1007/978-3-540-71677-8_13
9. Green, M., Miers, I.: Bolt: anonymous payment channels for decentralized currencies. In: CCS, pp. 473–489 (2017)
10. Krawczyk, H., Rabin, T.: Chameleon signatures. In: NDSS (2000)
11. Liu, J.K., Wei, V.K., Wong, D.S.: Linkable spontaneous anonymous group signature for ad hoc groups. In: Wang, H., Pieprzyk, J., Varadharajan, V. (eds.) ACISP 2004. LNCS, vol. 3108, pp. 325–335. Springer, Heidelberg (2004). https://doi.org/10.1007/978-3-540-27800-9_28
12. Miers, I., Garman, C., Green, M., Rubin, A.D.: Zerocoin: anonymous distributed e-cash from bitcoin. In: 2013 IEEE Symposium on Security and Privacy, pp. 397–411 (2013)
13. Nakamoto, S. : Bitcoin: a peer-to-peer electronic cash system (2008)
14. Noether, S.: Review of cryptonote white paper (2014)
15. Noether, S.: Ring signature confidential transactions for monero. IACR Cryptol. ePrint Arch. **2015**, 1098 (2015)
16. Noether, S., Mackenzie, A., et al.: Ring confidential transactions. Ledger **1**, 1–18 (2016)
17. Pass, R., Shelat, A.: Micropayments for decentralized currencies. In: CCS, pp. 207–218 (2015)
18. Pedersen, T.P.: Non-interactive and information-theoretic secure verifiable secret sharing. In: Feigenbaum, J. (ed.) CRYPTO 1991. LNCS, vol. 576, pp. 129–140. Springer, Heidelberg (1992). https://doi.org/10.1007/3-540-46766-1_9
19. Poon, J., Dryja, T.: The bitcoin lightning network: scalable off-chain instant payments (2016)
20. Rivest, R.L.: Electronic lottery tickets as micropayments. In: Hirschfeld, R. (ed.) FC 1997. LNCS, vol. 1318, pp. 307–314. Springer, Heidelberg (1997). https://doi.org/10.1007/3-540-63594-7_87

21. Ruffing, T., Kate, A., Schröder, D.: Liar, liar, coins on fire!: penalizing equivocation by loss of bitcoins. In: CCS, pp. 219–230 (2015)

22. E.B. Sasson, et al.: Decentralized anonymous payments from bitcoin. In: 2014 IEEE Symposium on Security and Privacy, pp. 459–474 (2014)

23. Schnorr, C.-P.: Efficient signature generation by smart cards. J. Cryptol. **4**(3), 161–174 (1991)

24. Shamir, A.: How to share a secret. Commun. ACM **22**(11), 612–613 (1979)

25. Sun, S.-F., Au, M.H., Liu, J.K., Yuen, T.H.: RingCT 2.0: a compact accumulator-based (linkable ring signature) protocol for blockchain cryptocurrency monero. In: Foley, S.N., Gollmann, D., Snekkenes, E. (eds.) ESORICS 2017. LNCS, vol. 10493, pp. 456–474. Springer, Cham (2017). https://doi.org/10.1007/978-3-319-66399-9_25

26. Van Saberhagen, N.: Cryptonote v 2.0 (2013)

Implicit Related-Key Factorization Problem on the RSA Cryptosystem

Mengce Zheng[(✉)] and Honggang Hu

Key Laboratory of Electromagnetic Space Information, CAS,
University of Science and Technology of China, Hefei, China
{mczheng,hghu2005}@ustc.edu.cn

Abstract. In this paper, we address the implicit related-key factorization problem on the RSA cryptosystem. Informally, we investigate under what condition it is possible to efficiently factor RSA moduli in polynomial time given the implicit information of related private keys. We propose lattice-based attacks using Coppersmith's techniques. We first analyze the special case given two RSA instances with known amounts of shared most significant bits (MSBs) and least significant bits (LSBs) of unknown related private keys. Subsequently a generic attack is proposed using a heuristic lattice construction when given more RSA instances. Furthermore, we conduct numerical experiments to verify the validity of the proposed attacks.

Keywords: RSA · Factorization · Implicit related-key · Lattice-based attack · Coppersmith's techniques

1 Introduction

The RSA public-key cryptosystem [18] plays an important role in the area of information security due to its simplicity and popularity. Its key equation is $ed \equiv 1 \mod \varphi(N)$, where N, e, d and $\varphi(N)$ are defined as follows. $N = pq$ is the product of two large primes of the same bit-size. e, d denote the public and private keys, which are also called the public/encryption and private/decryption exponents. $\varphi(N) = (p-1)(q-1)$ is Euler's totient function. One computes $c = m^e \mod N$ and $c^d \mod N$ for encryption and decryption operations, respectively.

In 1996, Coppersmith [4,5] made a significant breakthrough based on finding small roots of modular and integer polynomial equations. The fundamental works proposed novel and advanced lattice-based attacks on RSA. The main method is known as Coppersmith's techniques [6] and has been widely applied in the cryptanalytic field of RSA. Many researchers have proposed several effective attacks such as [1,7–10,15,21] etc. Among them, the *partial key exposure attack* has been extensively studied as an active attack scenario.

In 1998, Boneh et al. [2] proposed several attacks on RSA given a fraction of the private key bits with small public exponent e. The attacks employed some known most significant bits (MSBs) or least significant bits (LSBs) of d. In 2005,

© Springer Nature Switzerland AG 2019
Y. Mu et al. (Eds.): CANS 2019, LNCS 11829, pp. 525–537, 2019.
https://doi.org/10.1007/978-3-030-31578-8_29

Ernst et al. [10] presented improved lattice-based attacks that work up to full size exponents under a heuristic assumption. In our opinion, partial key exposure attack can be reduced to the problem of factoring RSA modulus with an oracle outputting some *explicit* information of d.

In 2009, May and Ritzenhofen [16] proposed the *implicit factorization problem*, which aims to factor RSA moduli with an oracle providing implicit information about the amount of shared LSBs of the primes. It is mainly considered for the malicious generation of RSA moduli like the construction of backdoor RSA moduli. Later, Sarkar and Maitra [19] proposed a better approach based on solving the approximate common divisor problem.

Inspired by the above attacks, we raise an interesting hybrid problem that aims to efficiently factor RSA moduli given some implicit information about the related private keys. We herein present the description of the *implicit related-key factorization problem* as follows. Let $(N_1, e_1, d_1), \ldots, (N_n, e_n, d_n)$ be n distinct key pairs, where $N_1, \ldots, N_n$ are of the same bit-size and the prime factors are also all of the same bit-size. Given the implicit information that certain portions of the bit pattern in private keys $d_1, \ldots, d_n$ are common, under what condition is it possible to efficiently factor $N_1, \ldots, N_n$. In this sense, the implicit factorization problem [16] can be refined into the implicit related-prime factorization problem accordingly.

There are several situations to use many RSA instances in practice like [20]. Once such RSA instances are generated with imperfect randomness or malicious backdoor keys, one may encounter the implicit related-key factorization problem. Our motivations come from two aspects. Mainly from the theoretical view, we study a new problem combing two existing attacks, which may further disclose the vulnerability of RSA with implicit information and enrich lattice-based cryptanalyses in the literature. Practically, side channel attacks may not give explicit information as expected. Instead, one may know the amounts of shared MSBs and LSBs of the private keys as some implicit information. The users' misuses with certain repeated bit patterns in the private keys may also lead to this problem.

We formulate the implicit related-key factorization problem with several RSA instances clearly. Given n key pairs of RSA parameters (N_i, e_i, d_i) for $1 \leq i \leq n$. We consider the full size case when $e_i \approx N$ for N denoting an integer of the same bit-size as N_i for simplicity. Besides, we assume $d_i \approx N^\delta$ share certain MSBs and LSBs like $d_j = d_i + d_{ji}D$ for $1 \leq i < j \leq n$, where D denotes the bit-length of shared LSBs-block and d_{ji} denotes the difference between every two unknown middle blocks with $|D| \approx N^\gamma$ and $|d_{ji}| \approx N^\beta$.

We follow Coppersmith's techniques [6] to handle the implicit related-key factorization problem. In addition, we adapt two subtle lattice techniques, namely the splitting technique and the linearization technique. Our attacks rely on a *heuristic assumption*, which works well in the literature. The assumption says that algebraically independent polynomials can be obtained by the lattice-based attacks and the common root can be efficiently extracted by the Gröbner basis computation [3].

Our main result is stated in Proposition 1, which will be proven in Sect. 3. We want to point out that the theoretical result is asymptotic since the corresponding lattice dimension is required preferably large.

Proposition 1. *Let* $N_1 = p_1 q_1$ *and* $N_2 = p_2 q_2$ *be given two RSA moduli of the same bit-size, where* p_1, q_1, p_2, q_2 *are large primes of the same bit-size. Let* e_1, d_1, e_2, d_2 *be some integers satisfying* $e_1 d_1 \equiv 1 \bmod (p_1 - 1)(q_1 - 1)$ *and* $e_2 d_2 \equiv 1 \bmod (p_2 - 1)(q_2 - 1)$ *such that* $e_1 \approx e_2 \approx N$ *and* $d_1 \approx d_2 \approx N^\delta$. *Given the implicit information that* $d_2 = d_1 + d_{21} D$ *for* $|d_{21}| \approx N^\beta$. *Then* N_1 *and* N_2 *can be factored in polynomial time if*

$$\delta < \frac{25 - 16\beta - \sqrt{177 - 96\beta}}{32}.$$

The rest of the paper is organized as follows. We provide basic knowledge of Coppersmith's techniques and Gaussian heuristic in Sect. 2. In Sect. 3, we propose a lattice-based attack for given two instances and further develop a notable lattice construction to analyze the case of n instances. We verify the validity of the proposed attacks by computer experiments in Sect. 4. Finally, concluding remarks are given in Sect. 5.

2   Preliminaries

In this section, we first briefly introduce lattice, the LLL reduction algorithm [14] and Coppersmith's techniques [6]. Then we give a rough condition for finding the small roots of modular polynomial equations. We also briefly describe the splitting technique [17] based on the Gaussian heuristic.

A lattice $\mathcal{L}$ spanned by linearly independent vectors $\boldsymbol{b}_1, \ldots, \boldsymbol{b}_m \in \mathbb{R}^n$ is the set of their integer linear combinations, which can be denoted by

$$\mathcal{L}(\boldsymbol{b}_1, \ldots, \boldsymbol{b}_m) = \left\{ \sum_{i=1}^{m} z_i \boldsymbol{b}_i : z_i \in \mathbb{Z} \right\}.$$

The basis vectors derive a basis matrix B by regarding each $\boldsymbol{b}_i$ as row (or column) vectors. The determinant of $\mathcal{L}$ is calculated as $\det(\mathcal{L}) = \sqrt{\det(BB^\mathsf{T})}$. The rank of $\mathcal{L}$ is m and we always consider a full-rank lattice for $m = n$. Thus, we have $\det(\mathcal{L}) = |\det(B)|$.

The LLL algorithm [14] is practically used for computing approximately short lattice vectors due to its efficient running outputs. We provide the following substratal lemma, whose proof refers to [15].

Lemma 1. *Let* $\mathcal{L}$ *be a lattice spanned by basis vectors* $(\boldsymbol{b}_1, \ldots, \boldsymbol{b}_m)$. *The LLL algorithm outputs a reduced basis* $(\boldsymbol{v}_1, \ldots, \boldsymbol{v}_m)$ *satisfying*

$$\|\boldsymbol{v}_1\|, \|\boldsymbol{v}_2\|, \ldots, \|\boldsymbol{v}_i\| \leq 2^{\frac{m(m-1)}{4(m+1-i)}} \det(\mathcal{L})^{\frac{1}{m+1-i}} \text{ for } 1 \leq i \leq m$$

in time polynomial in m *and in the bit-size of the entries of the basis matrix.*

Howgrave-Graham [13] refined on Coppersmith's techniques to propose a succinct lemma for judging whether the small roots of a modular equation are roots over $\mathbb{Z}$. For a given polynomial $g(x_1, \ldots, x_n) = \sum a_{i_1,\ldots,i_n} x_1^{i_1} \cdots x_n^{i_n}$, its norm is defined as $\|g(x_1, \ldots, x_n)\| := \sqrt{\sum |a_{i_1,\ldots,i_n}|^2}$.

Lemma 2. *Let $g(x_1, \ldots, x_n) \in \mathbb{Z}[x_1, \ldots, x_n]$ be an integer polynomial of at most m monomials. Suppose that*

1. $g(x_1', \ldots, x_n') \equiv 0 \bmod R$, *where* $|x_1'| \leq X_1, \ldots, |x_n'| \leq X_n$,
2. $\|g(x_1 X_1, \ldots, x_n X_n)\| < R/\sqrt{m}$.

Then $g(x_1', \ldots, x_n') = 0$ holds over the integers.

Combining Lemmas 1 and 2, one can solve modular/integer equations under a particular condition. One first constructs shift polynomials from a given equation and then generate a lattice basis matrix using the coefficient vectors. Once integer equations are derived from the first ℓ reduced vectors through the LLL algorithm, one can extract the root for $2^{\frac{m(m-1)}{4(m+1-\ell)}} \det(\mathcal{L})^{\frac{1}{m+1-\ell}} < R/\sqrt{m}$. It further leads to a rough condition $\det(\mathcal{L}) < R^m$ if ignoring the negligible lower terms. The first ℓ vectors are transformed into simultaneous equations sharing the common root over the integers. Hence, one can apply the Gröbner basis computation to extract the common root.

Recently, Peng et al. [17] proposed an improved lattice-attack on the Dual RSA scheme [20] using the splitting technique. It can split a variable of large norm into several variables of smaller norm by reducing a low-dimensional lattice. Concretely, it is based on the observation of Gaussian heuristic in random lattices, which says that the norm of the shortest non-zero vector s of a random m-dimensional lattice $\mathcal{L}$ satisfies $\|s\| \approx \sqrt{m/(2\pi e)} \det(\mathcal{L})^{\frac{1}{m}}$. Let the successive minimum $\lambda_i(\mathcal{L})$ denote the i-th minimum of $\mathcal{L}$, which is the radius of the smallest zero-centered ball containing at least i linearly independent lattice vectors. In this sense, $\|s\|$ can be written as $\lambda_1(\mathcal{L})$.

A further claim on this property can be found in [11]. The successive minima of a random m-dimensional lattice $\mathcal{L}$ are all asymptotically close to the Gaussian heuristic with an overwhelming probability. That is $\lambda_i(\mathcal{L}) \approx \sqrt{m/(2\pi e)} \det(\mathcal{L})^{\frac{1}{m}}$ for all $1 \leq i \leq m$. We adapt the splitting technique along with the linearization technique [12] to present convenient lattice construction in our lattice-based attacks. In this paper, we use the fact $|s_{i1}| \approx \det(\mathcal{L}_0)^{\frac{1}{m}}$, where s_i for $1 \leq i \leq m$ is a reduced basis vector after running the LLL algorithm on the constructed m-dimensional full-rank lattice $\mathcal{L}_0$.

3   Implicit Related-Key Factorization Attacks

We first propose a lattice-based attack for given two RSA instances, namely (N_1, e_1, d_1) and (N_2, e_2, d_2). Recall that we know $e_1 \approx e_2 \approx N$, where N denotes an integer with the same bit-size as N_1, N_2 and the private keys d_1, d_2 share some MSBs and LSBs leaving one different block in the middle. Moreover, we have $d_1 \approx d_2 \approx N^\delta$ and $d_2 = d_1 + d_{21}D$ for $|d_{21}| \approx N^\beta$ and $|D| \approx N^\gamma$ (assuming γ and β are given in advance).

We first perform the splitting technique to split one unknown private key into a linear combination of two smaller unknown variables. To do so, we construct a two-dimensional lattice $\mathcal{L}_0$ that is generated by the following basis matrix

$$B_0 = \begin{bmatrix} a_0 & e_1 \\ 0 & N_1 \end{bmatrix}$$

for a well-chosen integer a_0.

From the key equation $e_1 d_1 \equiv 1 \bmod \varphi(N_1)$ and $\varphi(N_1) = N_1 + 1 - p_1 - q_1$, we have $e_1 d_1 - k_1 N_1 = k_1(1 - p_1 - q_1) + 1$ for a positive integer k_1. Hence, we know $(d_1, -k_1)B_0 = (a_0 d_1, k_1(1 - p_1 - q_1) + 1)$ is a vector belonging to $\mathcal{L}_0$. We have $k_1 = (e_1 d_1 - 1)/\varphi(N_1) \approx N^\delta$. We set $a_0 = [N^{\frac{1}{2}}]$ to balance each coordinate of $(a_0 d_1, k_1(1 - p_1 - q_1) + 1)$, whose norm is $\|(a_0 d_1, k_1(1 - p_1 - q_1) + 1)\| \approx N^{\delta + \frac{1}{2}}$. The determinant of $\mathcal{L}_0$ is $\det(\mathcal{L}_0) = |\det(B_0)| = a_0 N_1 \approx N^{\frac{3}{2}}$ from our construction of the basis matrix B_0.

We can obtain two reduced basis vectors (s_{11}, s_{12}) and (s_{21}, s_{22}) through the lattice reduction algorithm. Further by applying the Gaussian heuristic, we have $\|(s_{11}, s_{12})\| = \|(s_{21}, s_{22})\| \approx \det(\mathcal{L}_0)^{\frac{1}{2}} \approx N^{\frac{3}{4}}$, which indicates the norms of s_{11}, s_{12}, s_{21} and s_{22} are roughly $N^{\frac{3}{4}}$. Actually, we have $s_{11} = a_0 a_1$ and $s_{21} = a_0 a_2$ as the reduced basis vectors are generated by

$$\begin{bmatrix} s_{11} & s_{12} \\ s_{21} & s_{22} \end{bmatrix} = \begin{bmatrix} a_1 & - \\ a_2 & - \end{bmatrix} \begin{bmatrix} a_0 & e_1 \\ 0 & N_1 \end{bmatrix} = \begin{bmatrix} a_0 a_1 & * \\ a_0 a_2 & * \end{bmatrix},$$

where known integers a_1 and a_2 are elements appearing in the first column vector of the unimodular transformation matrix. It can easily deduced that $|a_1| \approx |a_2| \approx |s_{21}/a_0| \approx N^{\frac{1}{4}}$.

On the other hand, we have $a_0 d_1 = s_{11} c_1 + s_{21} c_2$ since (s_{11}, s_{12}) and (s_{21}, s_{22}) are also basis vectors. Hence, we obtain $d_1 = a_1 c_1 + a_2 c_2$ for unknown c_1 and c_2. Combining it with $d_2 = d_1 + d_{21}D$, we finally have $d_2 = a_1 c_1 + a_2 c_2 + d_{21}D$. We want to figure out the norms of c_1 and c_2. As $|a_1| \approx |a_2| \approx N^{\frac{1}{4}}$, we have $|c_1| \approx |c_2| \approx |d_2/a_2| \approx N^{\delta - \frac{1}{4}}$. We substitute $d_2 = a_1 c_1 + a_2 c_2 + d_{21}D$ in another key equation $e_2 d_2 = k_2(N_2 + 1 - p_2 - q_2) + 1$ and have $e_2(a_1 c_1 + a_2 c_2 + d_{21}D) = k_2(N_2 + 1 - p_2 - q_2) + 1$. Therefore, we turn to solving $f(x, y, z, w) := x(y - N_2 - 1) + e_2 a_1 z + e_2 Dw - 1 \bmod e_2 a_2$ with the root $(k_2, p_2 + q_2, c_1, d_{21})$ for the implicit related-key factorization problem.

To provide an elegant lattice construction, we further apply the linearization technique introduced in [12]. Letting $u := xy - 1$, we have the linear polynomial $\bar{f}(x, z, w, u) := u - (N_2 + 1)x + e_2 a_1 z + e_2 Dw \bmod e_2 a_2$. The shift polynomials are defined as

$$g_{[i,j,k,l_1,l_2]}(x, y, z, w, u) := x^i y^j z^{l_1} w^{l_2} \bar{f}^k(x, z, w, u) E^{s-k}$$

for $E = e_2 a_2$, a positive integer s and $i, j, k, l_1, l_2 \in \mathbb{N}$. We denote the set of the shift polynomials by $\mathcal{G} := \mathcal{G}_1 \cup \mathcal{G}_2$, where

$$\mathcal{G}_1 := \{g_{[i,0,k,l_1,l_2]}(x, y, z, w, u) : k = 0, \ldots, s; i = 0, \ldots, s - k;$$
$$l_1 = 0, \ldots, s - k - i; l_2 = 0, \ldots, s - k - i - l_1.\},$$
$$\mathcal{G}_2 := \{g_{[0,j,k,l_1-l_2,l_2-k]}(x, y, z, w, u) : l_1 = 0, \ldots, s; j = 1, \ldots, \tau l_1;$$
$$l_2 = 0, \ldots, l_1; k = 0, \ldots, l_2.\}$$

for an optimizing parameter $0 \le \tau \le 1$ to be determined later. It is obvious that all the shift polynomials share the common root $(k_2, p_2 + q_2, c_1, d_0, k_2(p_2 + q_2) - 1)$ modulo E^s.

By defining auxiliary parameters $r = i + k + l_1 + l_2$ and $r' = i' + k' + l_1' + l_2'$, the polynomial and monomial orders $\prec$ are defined as $g_{[i,j,k,l_1,l_2]} \prec g_{[i',j',k',l_1',l_2']}$ and $x^i y^j u^k z^{l_1} w^{l_2} \prec x^{i'} y^{j'} u^{k'} z^{l_1'} w^{l_2'}$, respectively if $r < r'$ or $r = r'$, $i \ge i'$ or $r = r'$, $i = i'$, $l_1 \ge l_1'$ or $r = r'$, $i = i'$, $l_1 = l_1'$, $l_2 \ge l_2'$ or $r = r'$, $i = i'$, $l_1 = l_1'$, $l_2 = l_2'$, $j < j'$.

We can substitute each occurrence of xy by $u+1$. The lattice basis matrix B is constructed by taking the coefficient vectors of $g_{[i,j,k,l_1,l_2]}(xX, yY, zZ, wW, uU)$ in $\mathcal{G}$ as row vectors, where X, Y, Z, W and U denote the upper bounds on unknown variables. Additionally, the rows and columns of B are arranged according to the above polynomial and monomial orders. Two parameters s and τ can guarantee that B is square and triangular.

Table 1 shows a toy example of the lattice basis matrix B for $s = 1$ and $\tau = 1$, where each row can be viewed as the coefficient vector transformation from a shift polynomial. We are able to obtain the basis matrix B that generates the main lattice $\mathcal{L}$ directly from our construction.

Table 1. A toy example of the constructed lattice basis matrix B for $s = 1$ and $\tau = 1$ with $E = e_2 a_2$ and $C = -(N_2 + 1)$.

	1	x	z	yz	w	yw	u	yu
$g_{[0,0,0,0,0]}(xX, yY, zZ, wW, uU)$	E							
$g_{[1,0,0,0,0]}(xX, yY, zZ, wW, uU)$		EX						
$g_{[0,0,0,1,0]}(xX, yY, zZ, wW, uU)$			EZ					
$g_{[0,1,0,1,0]}(xX, yY, zZ, wW, uU)$				EYZ				
$g_{[0,0,0,0,1]}(xX, yY, zZ, wW, uU)$					EW			
$g_{[0,1,0,0,1]}(xX, yY, zZ, wW, uU)$						EYW		
$g_{[0,0,1,0,0]}(xX, yY, zZ, wW, uU)$		CX	$e_2 a_1 Z$		$e_2 DW$		U	
$g_{[0,1,1,0,0]}(xX, yY, zZ, wW, uU)$	C			$e_2 a_1 YZ$		$e_2 DYW$	CU	YU

Since we already know $X \approx N^\delta$, $Y \approx N^{\frac{1}{2}}$, $Z \approx N^{\delta - \frac{1}{4}}$, $W \approx N^\beta$, $U \approx N^{\delta + \frac{1}{2}}$ and $E \approx N^{\frac{5}{4}}$, we can calculate the determinant of $\mathcal{L}$ that is the product of the diagonal entries of the basis matrix B.

$$\det(\mathcal{L}) = \left(\prod_{k=0}^{s} \prod_{i=0}^{s-k} \prod_{l_1=0}^{s-k-i} \prod_{l_2=0}^{s-k-i-l_1} X^i Z^{l_1} W^{l_2} U^k E^{s-k} \right)$$

$$\times \left(\prod_{l_1=0}^{s} \prod_{j=1}^{\tau l_1} \prod_{l_2=0}^{l_1} \prod_{k=0}^{l_2} Y^j Z^{l_1-l_2} W^{l_2-k} U^k E^{s-k} \right)$$

$$= X^{s_x} Y^{s_y} Z^{s_z} W^{s_w} U^{s_u} E^{s_E},$$

where s_x, s_y, s_z, s_w, s_u and s_E are the respective exponent sums of the diagonal entries of the basis matrix B. The lattice dimension is

$$m = \sum_{k=0}^{s} \sum_{i=0}^{s-k} \sum_{l_1=0}^{s-k-i} \sum_{l_2=0}^{s-k-i-l_1} 1 + \sum_{l_1=0}^{s} \sum_{j=1}^{\tau l_1} \sum_{l_2=0}^{l_1} \sum_{k=0}^{l_2} 1 = \frac{1+3\tau}{24} s^4 + o(s^4).$$

Similarly, we calculate $s_x = \frac{1}{120} s^5$, $s_y = \frac{\tau^2}{20} s^5$, $s_z = s_w = s_u = \frac{1+4\tau}{120} s^5$ and $s_E = \frac{4+11\tau}{120} s^5$ when omitting $o(s^5)$ as it is negligible for sufficiently large s. From the rough condition $\det(\mathcal{L}) < R^m$ with $R = E^s$ for acquiring enough integer equations sharing the common root, we have

$$X^{s_x} Y^{s_y} Z^{s_z} W^{s_w} U^{s_u} E^{s_E} < E^{\frac{1+3\tau}{24} s^5}.$$

Moreover, we let s go to infinite and obtain the crucial condition

$$\frac{1}{120} \cdot \xi_x + \frac{\tau^2}{20} \cdot \xi_y + \frac{1+4\tau}{120} \cdot (\xi_z + \xi_w + \xi_u) + \frac{4+11\tau}{120} \cdot \xi_E < \frac{1+3\tau}{24} \cdot \xi_E,$$

where ξ_x, ξ_y, ξ_z, ξ_w, ξ_u and ξ_E denote the exponents of the respective upper bounds. We further reduce the crucial condition to a simplified one

$$\xi_x + 6\tau^2 \xi_y + (1+4\tau)(\xi_z + \xi_w + \xi_u - \xi_E) < 0.$$

We know $\xi_x = \delta$, $\xi_y = \frac{1}{2}$, $\xi_z = \delta - \frac{1}{4}$, $\xi_w = \beta$, $\xi_u = \delta + \frac{1}{2}$ and $\xi_E = \frac{5}{4}$. Thus, we obtain

$$\delta + 3\tau^2 + (1+4\tau)\left(\delta - \frac{1}{4} + \beta + \delta + \frac{1}{2} - \frac{5}{4} \right) < 0,$$

which leads to

$$\delta < \frac{(1-\beta)(1+4\tau) - 3\tau^2}{3 + 8\tau}.$$

The right side reaches its maximum by taking $\tau = (\sqrt{177 - 96\beta} - 9)/24$. We put it into the above inequality and hence derive the final condition

$$\delta < \frac{25 - 16\beta - \sqrt{177 - 96\beta}}{32}.$$

Once we extract the common root $(k_2, p_2 + q_2, c_1, d_{21}, k_2(p_2 + q_2) - 1)$, we can easily factorize N_2 using the value of $p_2 + q_2$. Then we have d_2 from $d_2 = e_2^{-1} \mod \varphi(N_2)$, which can be used to recover d_1 by $d_1 = d_2 - d_{21}D$. Thus, we

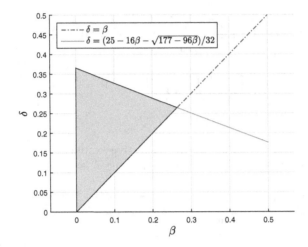

Fig. 1. The solid curve denotes the upper bound on δ and the dot-dash line denotes the lower bound on δ. The gray area indicates the vulnerable scenarios of the proposed implicit related-key factorization attack for given two RSA instances.

can factorize N_1 as knowing d_1 is equivalent to the factorization of N_1, which has been proved in [9].

The above result is illustrated in Fig. 1. We gain a significant improvement of the insecure bound on δ with the help of known implicit information about the related private keys. One may wonder whether our approach can handle the implicit related-key factorization problem for more than two RSA instances. We give an answer to this question below.

Recall the attack scenario for handling the implicit related-key factorization problem with n distinct RSA instances. Given n key pairs of RSA parameters (N_i, e_i, d_i) for $1 \leq i \leq n$. We assume $e_i \approx N$ and $d_i \approx N^\delta$ with $d_j = d_i + d_{ji}D$, where $|d_{ji}| \approx N^\beta$ and $|D| \approx N^\gamma$ for $1 \leq i < j \leq n$.

We first perform the splitting technique to split d_1 into a linear combination of several smaller unknown variables. We introduce a concise heuristic construction of a $(2n-1)$-dimensional lattice $\mathcal{L}_0$ that is generated by the basis matrix

$$
B_0 = \begin{bmatrix}
a_0 & 0 & \cdots & 0 & e_2 & \cdots & e_n \\
0 & b_0 & \cdots & 0 & e_2D & \cdots & 0 \\
\vdots & \vdots & \ddots & \vdots & \vdots & \ddots & \vdots \\
0 & 0 & \cdots & b_0 & 0 & \cdots & e_nD \\
0 & 0 & \cdots & 0 & N_2 & \cdots & 0 \\
\vdots & \vdots & \ddots & \vdots & \vdots & \ddots & \vdots \\
0 & 0 & \cdots & 0 & 0 & \cdots & N_n
\end{bmatrix}
$$

for two well-chosen integers a_0 and b_0. Hence, $(d_1, d_{21}, \ldots, d_{n1}, -k_2, \ldots, -k_n)B_0$ belongs to $\mathcal{L}_0$. That is $(a_0 d_1, b_0 d_{21}, \ldots, b_0 d_{n1}, k_2(1 - p_2 - q_2) + 1, \ldots, k_n(1 - p_n - q_n) + 1)$ as we know $e_i d_1 + e_i d_{i1} D - k_i N_i = e_i d_i - k_i N_i = k_i(1 - p_i - q_i) + 1$ for $2 \leq i \leq n$ from the related-key equations of d_i, d_j and the RSA key equations.

We know that $k_i = (e_i d_i - 1)/\varphi(N_i) \approx N^\delta$ for $1 \leq i \leq n$. To balance each coordinate of above vector, we set $a_0 = [N^{\frac{1}{2}}]$ and $b_0 = [N^{\frac{1}{2} + \delta - \beta}]$. The norm of the constructed vector is roughly estimated as $N^{\delta + \frac{1}{2}}$. The determinant of $\mathcal{L}_0$ is $\det(\mathcal{L}_0) = |\det(B_0)| = a_0 b_0^{n-1} \prod_{i=2}^{n} N_i \approx N^{\frac{3}{2} n - 1 + (n-1)(\delta - \beta)}$ from our construction of the basis matrix B_0.

When applying the Gaussian heuristic, the norm of the reduced basis vectors is roughly $\det(\mathcal{L}_0)^{\frac{1}{2n-1}} \approx N^{\frac{3n - 2 + 2(n-1)(\delta - \beta)}{2(2n-1)}}$. Similarly, we have $d_1 = a_1 c_1 + a_2 c_2 + \cdots + a_{2n-1} c_{2n-1}$ as an integer linear combination of $(2n - 1)$ unknown variables, where a_i's come from the first column vector of the unimodular transformation matrix. We have $|a_i| \approx \det(\mathcal{L}_0)^{\frac{1}{2n-1}}/a_0 \approx N^{\frac{3n - 2 + 2(n-1)(\delta - \beta)}{2(2n-1)} - \frac{1}{2}} = N^{\frac{(n-1)(2\delta - 2\beta + 1)}{2(2n-1)}}$ and hence $|c_i| \approx |d_1/a_1| \approx N^{\delta - \frac{(n-1)(2\delta - 2\beta + 1)}{2(2n-1)}} = N^{\frac{2n\delta + 2(n-1)\beta - n + 1}{2(2n-1)}}$.

Substituting the alternative expression of d_1 in $e_1 d_1 = k_1(N_1 + 1 - p_1 - q_1) + 1$, we try to solve $x(y - N_1 - 1) + e_1 a_1 z_1 + \cdots + e_1 a_{\hat{n}} z_{\hat{n}} - 1 \mod e_1 a_{\hat{n}+1}$ in $(\hat{n} + 2)$ variables with the root $(k_1, p_1 + q_1, c_1, \ldots, c_{\hat{n}})$ for $\hat{n} = 2n - 2$. Letting $u := xy - 1$, it can be rewritten in the linear form as $f_{\hat{n}}(x, z_1, \ldots, z_{\hat{n}}, u) := u - (N_1 + 1)x + e_1 a_1 z_1 + \cdots + e_1 a_{\hat{n}} z_{\hat{n}} \mod e_1 a_{\hat{n}+1}$. The shift polynomials are defined as $g_{[i,j,k,l_1,\ldots,l_{\hat{n}}]}(x, z_1, \ldots, z_{\hat{n}}, u) := x^i y^j z_1^{l_1} \cdots z_{\hat{n}}^{l_{\hat{n}}} f_{\hat{n}}^k E^{s-k}$ for $E = e_1 a_{\hat{n}+1}$, a positive integer s and $i, j, k, l_1, \ldots, l_{\hat{n}} \in \mathbb{N}$.

Analogous to the lattice-based solution applied to the case of two instances, we finally obtain the following proposition for the case of n instances.

Proposition 2. *Let $N_i = p_i q_i$ for $1 \leq i \leq n$ be given RSA moduli of the same bit-size, where p_i and q_i are large primes of the same bit-size. Let e_i and d_i be some integers satisfying $e_i d_i \equiv 1 \mod (p_i - 1)(q_i - 1)$ such that $e_i \approx N$ and $d_i \approx N^\delta$. Given the implicit information that $d_j = d_i + d_{ji} D$ for $1 \leq i < j \leq n$ with $|d_{ji}| \approx N^\beta$. Then given RSA moduli can be factored in polynomial time (but exponential in n) if*

$$\delta < \frac{1}{2} - \beta + \frac{2n^2 + n - 1 + 4n^2\beta - \sqrt{(2n-1)(6n^3 + 3n^2 - 1 - 8n^2(n-1)\beta)}}{4n^3}.$$

We illustrate the above result with respect to various β's in Fig. 2 and discuss more about it. On the one hand, we can achieve higher insecure bound as β decreases. On the other hand, exposing more RSA instances with implicit related-keys is more vulnerable. Let n go to infinity, the asymptotic upper bound converges to $\frac{1}{2} - \beta$. Consequently, it indicates that the proposed attack is effective for $\delta < \frac{1}{2}$ at best for $\beta = 0$, which is the same as the conjecture of the previous small exponent attack [1] unless there exist more effective attacks.

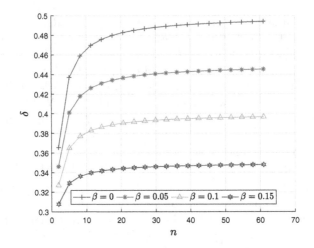

Fig. 2. The comparison of the upper bounds on δ of the proposed implicit related-key factorization attack for given n RSA instances with respect to $\beta = 0$, $\beta = 0.05$, $\beta = 0.1$ and $\beta = 0.15$.

4   Experimental Results

We verify the validity of the proposed attacks analyzed in Sect. 3 on the implicit related-key factorization problem for two instances. The experiments are carried out in `SageMath` under Windows 10 running on a laptop with Intel Core i7-8550U CPU 1.80 GHz. The numbers for generating the parameters of two RSA instances are chosen at random.

To be specific, we first generate two 1024-bit RSA moduli N_1 and N_2. Then we generate the implicit related-keys d_1 and d_2 with certain shared MSBs and LSBs according to the preset values of β and γ. Finally, we compute the corresponding public keys e_1 and e_2 from N_1, d_1 and N_2, d_2, respectively.

In each numerical experiment, we choose a suitable s with an optimal τ for constructing the lattice, which implies we shall first reduce a two-dimensional lattice and then another m-dimensional one. The comparison of the asymptotic and experimental results are given in Table 2. The γ and β-columns provide the concrete attack scenarios, by which we randomly generate two related private keys. The amounts (recorded in bits) of shared MSBs and LSBs are given in the MSBs and LSBs-columns. The δ_∞-column provides the asymptotic bounds on δ when s goes to infinity. The δ_e-column provides the experimental bounds for our lattice settings indicated by the s, τ and m-columns. The respective time consumption (recorded in seconds) of the LLL algorithm and the Gröbner basis computation are given in the TL and TG-columns.

During the experiments, we can collect sufficient polynomials satisfying our requirements. In other words, after running the LLL algorithm, we obtain enough short reduced basis vectors. The polynomial equations sharing the common root

Table 2. The comparison of asymptotic bounds and experimental results on δ of the proposed implicit related-key factorization attack for given two RSA instances.

γ	β	MSBs	LSBs	δ_∞	δ_e	s	τ	m	TL	TG
0.117	0.048	130	120	0.346	0.292	5	0.200	136	112.834	0.121
0.117	0.039	163	120	0.350	0.315	6	0.166	225	1933.569	0.151
0.043	0.058	199	44	0.342	0.295	5	0.200	136	140.853	0.122
0.034	0.063	204	35	0.340	0.296	6	0.166	225	1647.016	0.176
0.078	0.092	123	80	0.329	0.290	5	0.200	136	174.732	0.146
0.078	0.097	121	80	0.327	0.293	6	0.166	225	2336.019	0.162

over the integers are derived from the vector-to-equation transformation of the outputted lattice vectors. Based on the observation from Table 2, we briefly comment on the root-extraction procedure of the proposed attack. We put the derived polynomials into the Gröbner basis computation and obtain $p_2 + q_2$ that leads to the factorization of N_2. As mentioned before, we can also obtain the factorization of N_1. The time consumption of the Gröbner basis computation is much lower than that for running the LLL algorithm.

5   Concluding Remarks

In this paper, we propose the formulation of a new problem with respect to implicit related-key factorization, whose goal is to factor RSA moduli with the help of implicit information about related private keys. We then propose lattice-based attacks using Coppersmith's techniques, which are applied for solving modular polynomials as a powerful tool. Another technique we adapt is the splitting technique, which can split a variable of the large norm into some variables of the smaller norm.

We analyze the implicit related-key factorization problem for a special case when given two RSA instances. A lattice-based attack for such case is proposed and illustrated. We further verify the validity of the proposed attack by numerical experiments. For the case of more than two RSA instances, a similar attack is proposed based on a heuristic lattice construction. The concrete matrix construction with respect to the splitting technique may be improved (i.e. a_0 and b_0 can be further optimized).

Acknowledgments. The authors would like to thank the anonymous reviewers for their valuable comments and suggestions. This work was partially supported by the National Natural Science Foundation of China (Grant No. 61632013) and Anhui Initiative in Quantum Information Technologies under Grant AHY150400.

References

1. Boneh, D., Durfee, G.: Cryptanalysis of RSA with private key d less than $N^{0.292}$. In: Stern, J. (ed.) EUROCRYPT 1999. LNCS, vol. 1592, pp. 1–11. Springer, Heidelberg (1999). https://doi.org/10.1007/3-540-48910-X_1
2. Boneh, D., Durfee, G., Frankel, Y.: An attack on RSA given a small fraction of the private key bits. In: Ohta, K., Pei, D. (eds.) ASIACRYPT 1998. LNCS, vol. 1514, pp. 25–34. Springer, Heidelberg (1998). https://doi.org/10.1007/3-540-49649-1_3
3. Buchberger, B., Winkler, F.: Gröbner Bases and Applications. London Mathematical Society Lecture Note Series, vol. 251. Cambridge University Press, Cambridge (1998)
4. Coppersmith, D.: Finding a small root of a bivariate integer equation; factoring with high bits known. In: Maurer, U. (ed.) EUROCRYPT 1996. LNCS, vol. 1070, pp. 178–189. Springer, Heidelberg (1996). https://doi.org/10.1007/3-540-68339-9_16
5. Coppersmith, D.: Finding a small root of a univariate modular equation. In: Maurer, U. (ed.) EUROCRYPT 1996. LNCS, vol. 1070, pp. 155–165. Springer, Heidelberg (1996). https://doi.org/10.1007/3-540-68339-9_14
6. Coppersmith, D.: Small solutions to polynomial equations, and low exponent RSA vulnerabilities. J. Cryptol. **10**(4), 233–260 (1997)
7. Coron, J.-S.: Finding small roots of bivariate integer polynomial equations revisited. In: Cachin, C., Camenisch, J.L. (eds.) EUROCRYPT 2004. LNCS, vol. 3027, pp. 492–505. Springer, Heidelberg (2004). https://doi.org/10.1007/978-3-540-24676-3_29
8. Coron, J.-S.: Finding small roots of bivariate integer polynomial equations: a direct approach. In: Menezes, A. (ed.) CRYPTO 2007. LNCS, vol. 4622, pp. 379–394. Springer, Heidelberg (2007). https://doi.org/10.1007/978-3-540-74143-5_21
9. Coron, J.S., May, A.: Deterministic polynomial-time equivalence of computing the RSA secret key and factoring. J. Cryptol. **20**(1), 39–50 (2007)
10. Ernst, M., Jochemsz, E., May, A., de Weger, B.: Partial key exposure attacks on RSA up to full size exponents. In: Cramer, R. (ed.) EUROCRYPT 2005. LNCS, vol. 3494, pp. 371–386. Springer, Heidelberg (2005). https://doi.org/10.1007/11426639_22
11. Gama, N., Nguyen, P.Q.: Predicting lattice reduction. In: Smart, N. (ed.) EUROCRYPT 2008. LNCS, vol. 4965, pp. 31–51. Springer, Heidelberg (2008). https://doi.org/10.1007/978-3-540-78967-3_3
12. Herrmann, M., May, A.: Maximizing small root bounds by linearization and applications to small secret exponent RSA. In: Nguyen, P.Q., Pointcheval, D. (eds.) PKC 2010. LNCS, vol. 6056, pp. 53–69. Springer, Heidelberg (2010). https://doi.org/10.1007/978-3-642-13013-7_4
13. Howgrave-Graham, N.: Finding small roots of univariate modular equations revisited. In: Darnell, M. (ed.) Cryptography and Coding 1997. LNCS, vol. 1355, pp. 131–142. Springer, Heidelberg (1997). https://doi.org/10.1007/BFb0024458
14. Lenstra, A.K., Lenstra, H.W., Lovász, L.: Factoring polynomials with rational coefficients. Math. Ann. **261**(4), 515–534 (1982)
15. May, A.: New RSA vulnerabilities using lattice reduction methods. Ph.D. thesis, University of Paderborn (2003)
16. May, A., Ritzenhofen, M.: Implicit factoring: on polynomial time factoring given only an implicit hint. In: Jarecki, S., Tsudik, G. (eds.) PKC 2009. LNCS, vol. 5443, pp. 1–14. Springer, Heidelberg (2009). https://doi.org/10.1007/978-3-642-00468-1_1

17. Peng, L., Hu, L., Lu, Y., Xu, J., Huang, Z.: Cryptanalysis of dual RSA. Des. Codes Cryp. **83**(1), 1–21 (2017)
18. Rivest, R.L., Shamir, A., Adleman, L.: A method for obtaining digital signatures and public-key cryptosystems. Commun. ACM **21**(2), 120–126 (1978)
19. Sarkar, S., Maitra, S.: Approximate integer common divisor problem relates to implicit factorization. IEEE Trans. Inf. Theory **57**(6), 4002–4013 (2011)
20. Sun, H.M., Wu, M.E., Ting, W.C., Hinek, M.J.: Dual RSA and its security analysis. IEEE Trans. Inf. Theory **53**(8), 2922–2933 (2007)
21. Takayasu, A., Lu, Y., Peng, L.: Small CRT-exponent RSA revisited. In: Coron, J.-S., Nielsen, J.B. (eds.) EUROCRYPT 2017. LNCS, vol. 10211, pp. 130–159. Springer, Cham (2017). https://doi.org/10.1007/978-3-319-56614-6_5

Author Index

Printed in the United States
By Bookmasters